W9-ABY-642

Gmelin Handbook of Inorganic and Organometallic Chemistry

8th Edition

Gmelin Handbook of Inorganic and Organometallic Chemistry

8th Edition

Gmelin Handbuch der Anorganischen Chemie

Achte, völlig neu bearbeitete Auflage

PREPARED AND ISSUED BY

Gmelin-Institut für Anorganische Chemie
der Max-Planck-Gesellschaft
zur Förderung der Wissenschaften

Director: Ekkehard Fluck

FOUNDED BY

Leopold Gmelin

8TH EDITION

8th Edition begun under the auspices of the
Deutsche Chemische Gesellschaft by R. J. Meyer

CONTINUED BY

E. H. E. Pietsch and A. Kotowski, and by
Margot Becke-Goehring

Springer-Verlag
Berlin · Heidelberg · New York · London · Paris · Tokyo ·
Hong Kong · Barcelona · Budapest 1993

The following Gmelin Formula Index volumes have been published up to now:

Formula Index

Volume 1	Ac–Au
Volume 2	B–Br_2
Volume 3	Br_3–C_3
Volume 4	C_4–C_7
Volume 5	C_8–C_{12}
Volume 6	C_{13}–C_{23}
Volume 7	C_{24}–Ca
Volume 8	Cb–Cl
Volume 9	Cm–Fr
Volume 10	Ga–I
Volume 11	In–Ns
Volume 12	O–Zr
	Elements 104–132

Formula Index 1st Supplement

Volume 1	Ac–Au
Volume 2	B–$B_{1.9}$
Volume 3	B_2–B_{100}
Volume 4	Ba–C_7
Volume 5	C_8–C_{17}
Volume 6	C_{18}–C_x
Volume 7	Ca–I
Volume 8	In–Zr
	Elements 104–120

Formula Index 2nd Supplement

Volume 1	Ac–$B_{1.9}$
Volume 2	B_2–Br_x
Volume 3	C–$C_{6.9}$
Volume 4	C_7–$C_{11.4}$
Volume 5	C_{12}–$C_{16.5}$
Volume 6	C_{17}–$C5_{22.5}$
Volume 7	C_{23}–$C_{32.5}$
Volume 8	C_{33}–Cf
Volume 9	Cl–Ho
Volume 10	I–Zr

Formula Index 3rd Supplement

Volume 1	Ag–B_5 (present volume)

Gmelin Handbook of Inorganic and Organometallic Chemistry

8th Edition

INDEX

Formula Index

3rd Supplement Volume 1

$Ag-B_5$

AUTHORS Rainer Bohrer, Bernd Kalbskopf, Uwe Nohl, Hans-Jürgen Richter-Ditten, Paul Kämpf

CHIEF EDITORS Uwe Nohl, Gottfried Olbrich

Springer-Verlag
Berlin · Heidelberg · New York · London · Paris · Tokyo · Hong Kong · Barcelona · Budapest 1993

THE VOLUMES OF THE GMELIN HANDBOOK ARE EVALUATED FROM 1988 THROUGH 1992

Library of Congress Catalog Card Number: Agr 25-1383

ISBN 3-540-93668-8 Springer-Verlag, Berlin · Heidelberg · New York · London · Paris · Tokyo
ISBN 0-387-93668-8 Springer-Verlag, New York · Heidelberg · Berlin · London · Paris · Tokyo

Printed in Germany

Typesetting, printing, and bookbinding: Universitätsdruckerei H. Stürtz AG, Würzburg

Preface

The Gmelin Formula Index and the First and Second Supplement covered the volumes of the Eighth Edition of the Gmelin Handbook which appeared up to the end of 1987.

This Third Supplement extends the Gmelin Formula Index and includes the compounds from the volumes until 1992. The publication of the Third Supplement enables to locate all compounds described in the Gmelin Handbook of Inorganic and Organometallic Chemistry since 1924. The basic structure of the Formula Index remains the same as the previous editions.

Computer methods were employed during the preparation and the publication of the Third Supplement. Data acquisition, sorting, and data handling were performed using a suite of computer programs, developed originally by B. Roth, now at Chemplex GmbH. The SGML application for the final data processing for printing was developed in the computer department of the Gmelin Institute and at Universitätsdruckerei H. Stürtz AG, Würzburg.

Frankfurt am Main,
July 1993

U. Nohl, G. Olbrich

Instructions for Users of the Formula Index

First Column (Empirical Formula)

The empirical formulae are arranged in alphabetical order of the element symbols and by increasing values of the subscripts. Any indefinite subscripts are placed at the end of the respective sorting section. Ions always appear after the neutral species, positive ions preceding negative ones.

H_2O is included in empirical formulae only if it is an integral part of a complex, as indicated in the second column. For compounds which are described as solvates only both empirical formulae are given, with and without the solvent molecules. Multicomponent systems (solid solutions, melts, etc.) are listed under the empirical formulae of their respective components. However, solutions are found only under the solute, and polymers of the type $(AB)_n$ are sorted under AB.

Second Column (Linearized Formula)

The second column contains a linearized formula to indicate the constitution and configuration of a compound as close as possible. The formula given corresponds to that given in the handbook, except in cases where additional structural features can be described in more detail. For elements the names are included.

Entries with the same composition but with different structural formulae are arranged in the following order: elements or compounds, isotopic species, polymers, hydrates, and multicomponent systems.

For multicomponent systems the components are arranged in the sequence: inorganic components-organic components-water. The inorganic components are sorted alphabetically, the organic components according to the number of carbon atoms. If a component is a single element it is always represented by the unsubscripted atomic symbol. The term "system" is used in a restricted sense in this index; it represents mixtures described by phase diagrams or sometimes by, e.g., eutectic points.

Elements and compounds whose treatment in the handbook requires a larger amount of space are further characterized by topics like physical properties, preparation, electrochemical behaviour, etc.

Third Column (Volume and Page Numbers)

This column contains the volume descriptor and the page numbers, both separated by a hyphen. The volume descriptors consist of the atomic symbol of the element which is treated in a given volume, followed by an abbreviated form of the type of volume, including the part or section. The following abbreviations are used for the type of volume:

MVol.	Main Volume
SVol.	Supplement Volume
Org. Comp.	Organic Compounds
PFHOrg.	Perfluorohalogenoorganic Compounds of Main Group Elements
SVol.GD	Gmelin-Durrer, Metallurgy of Iron
Biol.Med.Ph.	Bor in Biologie, Medizin und Pharmazie

Volume descriptors like "3rd Suppl. Vol. 4" are abbreviated as "SVol. 3/4". For instance, the entry "B: B Comp.SVol. 3/4-345" indicates that the information can be found on page 345 of the boron volume "Boron Compounds 3rd Supplement Volume 4".

Ag–B_5

Ag . . . . Ag
Sorption on W . . . . W: SVol.A6b-237/54, 256/8
Surface diffusion on W . . . . W: SVol.A6b-254/6

– . . . . Ag alloys
Ag-As-Rh . . . . Rh: SVol.A1-261
Ag-Rh . . . . Rh: SVol.A1-259/61
Ag-Rh-Sn . . . . Rh: SVol.A1-261

– . . . . Ag systems
Ag-Dy-Rh . . . . Rh: SVol.A1-261
Ag-Rh . . . . Rh: SVol.A1-259
Ag-W . . . . W: SVol.A6b-235

$AgAsC_2F_{10}N_4S_2$ . . $[(F_2S{=}N{-}CN)_2Ag][AsF_6]$ . . . . S: S-N Comp.8-67
$AgAsC_{12}F_6H_{10}N_4O_4S_5$
$[Ag(C_6H_5{-}S(O)_2{-}N{=}S{=}NSN{=}S{=}N{-}S(O)_2{-}C_6H_5)][AsF_6]$
S: S-N Comp.7-298
$AgAsC_{12}H_{20}IMoO_2PS$
$(C_5H_5)Mo(CO)_2[P(CH_3)_3]{-}As(CH_3)_2{-}S{-}AgI$ . . . . Mo: Org.Comp.7-120, 136
$AgAs_3$ . . . . $AgAs_3$ solid solutions
$AgAs_3$-$RhAs_3$ . . . . Rh: SVol.A1-261
$AgAuCl_4$ . . . . $Ag[AuCl_4]$ . . . . Au: SVol.B1-213, 216
$AgAuCl_6Cs_2$ . . . . . $Cs_2[AgCl_2][AuCl_4]$ . . . . Au: SVol.B1-213
$AgAuF_4$ . . . . $Ag[AuF_4]$ . . . . Au: SVol.B1-125
$AgAu_2F_8$ . . . . $Ag[AuF_4]_2$ . . . . Au: SVol.B1-125
$AgBC_2H_2N_2$ . . . . . . $[H_2B(NC)_2]Ag$ . . . . B: B Comp.SVol.3/3-212
$AgBC_3HN_3$ . . . . . . $[HB(NC)_3]Ag$ . . . . B: B Comp.SVol.3/3-212
$AgBC_4F_{12}O_{12}S_4$ . . $Ag[B(O{-}S({=}O)_2{-}CF_3)_4]$ . . . . B: B Comp.SVol.4/4-165
$AgBC_8F_4H_{12}N_4$ . . . $Ag[BF_4] \cdot 4\ CH_3CN$ . . . . B: B Comp.SVol.4/3b-217
$AgB_{10}C_4H_{11}$ . . . . . 1,7-$C_2B_{10}H_{11}$-1-C≡CAg . . . . B: B Comp.SVol.3/4-222
$AgB_{11}CH_{12}$ . . . . . . $Ag[CB_{11}H_{12}]$ . . . . B: B Comp.SVol.4/4-307
$AgB_{11}C_{13}H_{24}$ . . . . $Ag[CB_{11}H_{12}] \cdot 2\ C_6H_6$ . . . . B: B Comp.SVol.4/4-307
$AgB_{11}C_{38}ClH_{42}IrOP_2$
$Ag(CB_{11}H_{12}) \cdot IrCl(CO)[P(C_6H_5)_3]_2 \cdot 0.5\ C_6H_5F$
B: B Comp.SVol.4/4-307
$AgB_{11}C_{41}ClF_{0.5}H_{44.5}IrOP_2$
$Ag(CB_{11}H_{12}) \cdot IrCl(CO)[P(C_6H_5)_3]_2 \cdot 0.5\ C_6H_5F$
B: B Comp.SVol.4/4-307

$AgBr_{0.8}F_{7.4}Pd$ .... $AgPdF_5 \cdot 0.8\ BrF_3$ .... Pd: SVol.B2-64
$AgBrF_4$ .... $Ag[BrF_4]$ .... Br: SVol.B3-110/5
$AgBrF_6$ .... $Ag[BrF_6]$ .... Br: SVol.B3-140/1
$AgBrO_3$ .... $Ag[BrO_3]$ .... Br: SVol.B2-121, 124, 157
$AgBr_{1.2}F_{7.6}Pd$ .... $AgPdF_4 \cdot 1.2\ BrF_3$ .... Pd: SVol.B2-64
$AgC_2F_4N_4S_2^+$ .... $[(F_2S{=}N{-}CN)_2Ag]^+$ .... S: S-N Comp.8-67
$AgC_4F_6H_6MoN_2$ .. $[Ag(NCCH_3)_2][MoF_6]$ .... Mo: SVol.B5-176
$AgC_4F_6NO_4S$ .... $OS(O)_2N(Ag)C(O)C(CF_3)_2$ .... F: PFHOrg.SVol.4-32, 57, 73
$AgC_6GeH_{13}O_2$ ... $Ag[Ge(CH_3)_3CH_2CH_2COO]$ .... Ge: Org.Comp.1-185
$AgC_8F_5H_{12}MoN_4O$ $AgMoOF_5 \cdot 4\ CH_3CN$ .... Mo: SVol.B5-176
$AgC_8F_{12}H_{12}Mo_2N_4$
$Ag[MoF_6]_2 \cdot 4\ CH_3CN$ .... Mo: SVol.B5-176
$AgC_{12}F_4N_4$ .... $Ag[2,5{-}C_6F_4{-}1,4{-}({=}C(CN)_2)_2]$, radical .... F: PFHOrg.SVol.6-106, 112/3
$AgC_{12}H_{10}N_4O_4S_5^+$
$[Ag(C_6H_5{-}S(O)_2{-}N{=}S{=}NSN{=}S{=}N{-}S(O)_2{-}C_6H_5)]^+$
S: S-N Comp.7-298
$AgC_{13}Cl_5H_8N_2O_4S_3$
$Ag[4{-}Cl{-}C_6H_4{-}S(O)_2{-}NS(CCl_3)N{-}S(O)_2{-}C_6H_4{-}4{-}Cl]$
S: S-N Comp.8-198/9
$AgC_{13}FeH_{16}N$ .... $(C_5H_5)Fe[C_5H_3(Ag){-}CH_2{-}N(CH_3)_2]$ .... Fe: Org.Comp.A10-175/6
$AgC_{14}Cl_3H_{12}N_2O_4S_3$
$Ag[C_6H_5{-}S(O)_2{-}NS(CCl_3)N{-}S(O)_2{-}C_6H_4CH_3{-}4]$
S: S-N Comp.8-198/9
$AgC_{14}Cl_4H_{11}N_2O_4S_3$
$Ag[4{-}CH_3{-}C_6H_4{-}S(O)_2{-}NS(CCl_3)N{-}S(O)_2{-}C_6H_4{-}4{-}Cl]$
S: S-N Comp.8-198/9
$AgC_{15}Fe_3H_{19}O_9P_2S$
$(CO)_9(H)Fe_3S[Ag(P(CH_3)_3)_2]$ .... Fe: Org.Comp.C6a-314, 316
$AgC_{20}F_6H_{20}MoN_4$ $[Ag(NC_5H_5)_4][MoF_6]$ .... Mo: SVol.B5-177
$AgC_{22}F_{18}H_{23}Mo_3N_5$
$[Ag(NC_5H_5)_4(NCCH_3)][MoF_6]_3$ .... Mo: SVol.B5-177
$AgC_{24}H_{20}NOPS^+$ $[Ag(P(C_6H_5)_3)(O{=}S{=}NC_6H_5)]^+$ .... S: S-N Comp.6-287
$AgC_{24}H_{20}N_2O_4PS$ $[Ag(P(C_6H_5)_3)(O{=}S{=}NC_6H_5)]NO_3$ .... S: S-N Comp.6-287
$AgC_{30}Fe_3H_{18}O_{11}P$ $(CO)_{10}Fe_3(C{-}O{-}CH_3)Ag[P(C_6H_5)_3]$ .... Fe: Org.Comp.C6b-79, 80, 82
$AgC_{45}Fe_3H_{31}O_9P_2S$
$(CO)_9(H)Fe_3S[Ag(P(C_6H_5)_3)_2]$ .... Fe: Org.Comp.C6a-314, 316
$AgC_{51}Fe_3H_{36}O_9P_3$ $(CO)_9(H)Fe_3(P{-}C_6H_5)[Ag(P(C_6H_5)_3)_2]$ .... Fe: Org.Comp.C6a-314, 316
AgCl .... AgCl systems
$AgCl{-}FeCl_3{-}PbCl_2{-}PdCl_2$ .... Pd: SVol.B2-170
$AgCl{-}NaCl{-}PdCl_2$ .... Pd: SVol.B2-169/70
$AgCl{-}PdCl_2$ .... Pd: SVol.B2-143, 144
$AgCl{-}PdCl_2{-}TeCl_4$ .... Pd: SVol.B2-168/9
$AgCl_7Pd_3$ .... $AgPd_3Cl_7$ .... Pd: SVol.B2-143
AgF_4Pd .... $AgPdF_4 \cdot 1.2\ BrF_3$ .... Pd: SVol.B2-64
AgF_5Pd .... $AgPdF_5 \cdot 0.8\ BrF_3$ .... Pd: SVol.B2-64
AgF_6Pd .... $AgPdF_6$ .... Pd: SVol.B2-64
$AgF_{11}PdZr_2$ .... $AgPdZr_2F_{11}$ .... Pd: SVol.B2-64
AgPo .... AgPo .... Po: SVol.1-329
$AgTh_2$ .... Th_2Ag solid solutions
$Th_2Ag{-}Th_2Al$ .... Th: SVol.B2-133/5

$Ag_2Au_3Cl_{17}H_{24}N_6$ $[NH_4]_6[Ag_2Cl_5][AuCl_4]_3$ .... Au: SVol.B1-213, 216
$Ag_2C_4H_{12}N_8Pd$ ... $[Ag(NH_3)_2]_2[Pd(CN)_4]$ .... Pd: SVol.B2-280
$Ag_2C_4N_4Pd$ ...... $Ag_2[Pd(CN)_4]$ .... Pd: SVol.B2-280
$Ag_2C_6N_6Pd$ ...... $Ag_2[Pd(CN)_6]$ .... Pd: SVol.B2-281, 284
$Ag_2C_{21}Fe_3H_{36}O_9P_4S$
$(CO)_9Fe_3S[Ag(P(CH_3)_3)_2]_2$ .... Fe: Org.Comp.C6a-314, 316
$Ag_2C_{81}Fe_3H_{60}O_9P_4S$
$(CO)_9Fe_3S[Ag(P(C_6H_5)_3)_2]_2$ .... Fe: Org.Comp.C6a-314, 316
$Ag_2C_{87}Fe_3H_{65}O_9P_5$
$(CO)_9Fe_3(P-C_6H_5)[Ag(P(C_6H_5)_3)_2]_2$ .... Fe: Org.Comp.C6a-314, 315
Ag_2Cl_4Pd .... $Ag_2[PdCl_4]$ .... Pd: SVol.B2-143
Ag_2Pd_2Te .... Ag_2Pd_2Te .... Pd: SVol.B2-254
Ag_2Pd_3S .... Ag_2Pd_3S .... Pd: SVol.B2-226
Ag_3Br_6Pd .... $Ag_3[PdBr_6]$ .... Pd: SVol.B2-195
$Ag_4As_4F_{24}N_{18}O_{18}S_{27}$
$[Ag_4((O{=}S{=}N)_2S)_9](AsF_6)_4 \cdot SO_2$ .... S: S-N Comp.6-266/8
$Ag_4As_4F_{24}N_{18}O_{20}S_{28}$
$[Ag_4((O{=}S{=}N)_2S)_9](AsF_6)_4 \cdot SO_2$ .... S: S-N Comp.6-266/8
$Ag_4C_{10}H_{17}I_6N_5ORe$
$[Re(CNCH_3)_5H_2O]I_2 \cdot 4\ AgI$ .... Re: Org.Comp.2-284, 287
$Ag_4N_{18}O_{18}S_{27}{}^{4+}$. $[Ag_4((O{=}S{=}N)_2S)_9]^{4+}$ .... S: S-N Comp.6-266/8
Ag_5Br_6Pd .... $Ag_5[PdBr_6]$ .... Pd: SVol.B2-179
$Ag_6B_{10}S_{18}$ ...... $Ag_6[B_{10}S_{18}]$ .... B: B Comp.SVol.3/4-108
$Ag_8C_{12}H_{18}I_{10}N_6Re$
$[Re(CNCH_3)_6]I_2 \cdot 8\ AgI$ .... Re: Org.Comp.2-291, 297

$Al_{0.5}Ni_{4.5}Th$ ..... $ThAl_{0.5}Ni_{4.5}$ .... Th: SVol.B2-117, 119/24
Al .... Al
Diffusion
in Th .... Th: SVol.B2-76/7
in U .... U: SVol.B2-85/8
of Th .... Th: SVol.B2-76/7
of U .... U: SVol.B2-85/8, 164/5
Sorption on W .... W: SVol.A6b-9/10
– .... Al alloys
Al-C-Th .... Th: SVol.C6-88
Al-Cu-Rh .... Rh: SVol.A1-257
Al-Hf-Rh .... Rh: SVol.A1-137
Al-Mg-Th .... Th: SVol.B2-58
Al-Mn-Rh .... Rh: SVol.A1-203
Al-Nb-Rh .... Rh: SVol.A1-166/7
Al-Rh .... Rh: SVol.A1-61/3
Al-Rh-Ru-U .... Rh: SVol.A1-273
Al-Rh-Th .... Rh: SVol.A1-139/40
Al-Rh-Ti .... Rh: SVol.A1-114
Al-Rh-U .... Rh: SVol.A1-188/90
Al-Rh-Zr .... Rh: SVol.A1-131
Al-Th .... Th: SVol.B2-77/103

Al . . . . Al solid solutions
Al-Rh. . . . . Rh: SVol.A1-61

– . . . . Al systems
Al-C-Th. . . . . Th: SVol.B2-103
Th: SVol.C6-88
Al-Ce-Th. . . . . Th: SVol.B2-105
Al-Ce-U . . . . U: SVol.B2-190
Al-Co-Th. . . . . Th: SVol.B2-124
Al-Co-U . . . . U: SVol.B2-224/5
Al-Cr-Th . . . . Th: SVol.B2-113
Al-Cr-U. . . . . U: SVol.B2-208
Al-Cu-U . . . . U: SVol.B2-238
Al-Dy-U . . . . U: SVol.B2-190/1
Al-$DyAl_2$-UAl_2 . . . . U: SVol.B2-190/1
Al-Fe-Th. . . . . Th: SVol.B2-126
Al-Fe-U. . . . . U: SVol.B2-230/2
Al-Gd-Th . . . . Th: SVol.B2-105
Al-Gd-U . . . . U: SVol.B2-190
Al-Ge-Th . . . . Th: SVol.B2-111
Al-Ge-U . . . . U: SVol.B2-207
Al-La-U. . . . . U: SVol.B2-190
Al-Li-U . . . . U: SVol.B2-180
Al-Lu-U. . . . . U: SVol.B2-190
Al-Mg-Th . . . . Th: SVol.B2-58
Al-Mg-U . . . . U: SVol.B2-73
Al-Mn-Th . . . . Th: SVol.B2-114
Al-Mn-U . . . . U: SVol.B2-211/2
Al-Mo-U . . . . U: SVol.B2-208/9
Al-Ni-Th . . . . Th: SVol.B2-117
Al-Ni-U. . . . . U: SVol.B2-215/7
Al-NiAl-UAl_2. . . . . U: SVol.B2-215/7
Al-Rh. . . . . Rh: SVol.A1-61
Al-Sc-U. . . . . U: SVol.B2-190
Al-Si-U . . . . U: SVol.B2-181/4
Al-Sm-U . . . . U: SVol.B2-190
Al-$SmAl_2$-UAl_2. . . . . U: SVol.B2-190
Al-Sn-U . . . . U: SVol.B2-207/8, 307/8
Al-Th. . . . . Th: SVol.B2-73/6
Al-Th-U. . . . . U: SVol.B2-203
Al-U. . . . . U: SVol.B2-82/5
Al-U-W . . . . U: SVol.B2-208
Al-U-Y . . . . U: SVol.B2-190
Al-U-Zr. . . . . U: SVol.B2-198/200
Al-UAl_2-$ZrAl_2$. . . . . U: SVol.B2-198/200
Al-W . . . . W: SVol.A6b-7/8

$AlAs_2C_{43}Cl_4H_{35}MoO_2$
$[(C_5H_5)Mo(CO)_2(As(C_6H_5)_3)_2][AlCl_4]$ . . . . Mo: Org.Comp.7-283, 292
$AlBBr_3C_5H_{13}N_2$ . . $CH_3B(-NCH_3-CH_2-CH_2-NCH_3-) \cdot AlBr_3$ . . . . . B: B Comp.SVol.3/3-113
$AlBBr_4C_4H_{10}N_2$ . . $(-NCH_3-CH_2-CH_2-NCH_3-)BBr \cdot AlBr_3$. . . . . . . B: B Comp.SVol.3/4-92

$AlBBr_4C_{10}H_{21}N$ .. $[(-C(CH_3)_2-(CH_2)_3-C(CH_3)_2-)N=BCH_3]AlBr_4$ .. B: B Comp.SVol.3/3-205
$AlBBr_4C_{11}H_{24}N_2$ .. $[(CH_3)_2N=B=NC_5H_6(CH_3)_4]AlBr_4$ ..... B: B Comp.SVol.3/3-204/6
$AlBBr_4C_{12}H_{28}N_2$ .. $[(i-C_3H_7)_2N=B=N(C_3H_7-i)_2][AlBr_4]$ ..... B: B Comp.SVol.4/3b-43
$AlBBr_4C_{13}H_{28}N_2$ .. $[(CH_3)_4C_5H_6N=B=N(C_2H_5)_2][AlBr_4]$ ..... B: B Comp.SVol.3/4-92
$AlBBr_4C_{15}H_{23}N$ .. $[(-C(CH_3)_2-(CH_2)_3-C(CH_3)_2-)N=BC_6H_5]AlBr_4$ .. B: B Comp.SVol.3/3-205
$AlBBr_4C_{19}H_{37}N_3$ .. $[2,6-(CH_3)_2-NC_5H_3-1-B(N(C_3H_7-i)_2)_2][AlBr_4]$ .. B: B Comp.SVol.4/3b-45
$AlBBr_4C_{27}H_{47}NO$ $[(CH_3)_4C_5H_6N=B=O-C_6H_2(C_4H_9-t)_3][AlBr_4]$ .... B: B Comp.SVol.3/3-205
B: B Comp.SVol.3/4-95
$AlBC_4Cl_4H_{12}N_2$ ... $ClB[N(CH_3)_2]_2 \cdot AlCl_3$ ..... B: B Comp.SVol.3/4-58
$AlBC_5Cl_3H_{13}N_2$ ... $CH_3B(-NCH_3-CH_2-CH_2-NCH_3-) \cdot AlCl_3$ ..... B: B Comp.SVol.3/3-113
$AlBC_5H_{13}I_3N_2$ .... $CH_3B(-NCH_3-CH_2-CH_2-NCH_3-) \cdot AlI_3$ ..... B: B Comp.SVol.3/3-113
$AlBC_8Cl_4H_{21}N_3$ ... $[(C_2H_5)_2N-B(-NHCH_3-CH_2-CH_2-NCH_3-)]AlCl_4$ B: B Comp.SVol.3/3-207
$AlBC_9Cl_2H_{19}NSi$ .. $[1-Cl_2Al-4,5-(C_2H_5)_2-2,2,3-(CH_3)_3-1,2,5-NSiBC_2]_2$
B: B Comp.SVol.4/3a-250/1
$AlBC_9Cl_3H_{17}S$ .... $[3.3.1]-9-BC_8H_{14}-9-SCH_3 \cdot AlCl_3$ ..... B: B Comp.SVol.4/4-132
$AlBC_9Cl_4H_{17}N_3$ ... $[NC_5H_5-1-B(N(CH_3)_2)_2][AlCl_4]$ ..... B: B Comp.SVol.4/3b-45
$AlBC_{11}Cl_4H_{19}N_3$ .. $[(-CCH_3(CH)_3CCH_3-)N-B(-NCH_3-CH_2CH_2-NCH_3-)]AlCl_4$
B: B Comp.SVol.3/3-207, 210
$AlBC_{12}Cl_4H_{28}N_2$ .. $[(i-C_3H_7)_2N=B=N(i-C_3H_7)_2]AlCl_4$ ..... B: B Comp.SVol.3/3-205
$AlBC_{13}Cl_4H_{28}N_2$ .. $[(2,2,6,6-(CH_3)_4NC_5H_6-1)=B=N(C_2H_5)_2][AlCl_4]$ B: B Comp.SVol.3/3-205
– ..... $[(2,2,6,6-(CH_3)_4NC_5H_6-1)=B=NH-C_4H_9-t][AlCl_4]$
B: B Comp.SVol.4/3b-45
$AlBC_{13}H_{29}NSi$ .... $[1-(C_2H_5)_2Al-4,5-(C_2H_5)_2-2,2,3-(CH_3)_3-1,2,5-NSiBC_2]_2$
B: B Comp.SVol.4/3a-250/1
$AlBC_{16}H_{36}N_2$ .... $[-C(CH_3)_2-CH_2CH_2CH_2-C(CH_3)_2-]N[$
$-Al(CH_3)_2-N(C_4H_9-t)-B(CH_3)-]$ ..... B: B Comp.SVol.4/3a-206/7
$AlBC_{17}Cl_4H_{33}N_3$ .. $[NC_5H_5-1-B(N(C_3H_7-i)_2)_2][AlCl_4]$ ..... B: B Comp.SVol.4/3b-45
$AlBC_{18}Cl_4H_{36}N_2$ .. $[(-C(CH_3)_2-(CH_2)_3-C(CH_3)_2-)N=B=N($
$-C(CH_3)_2-(CH_2)_3-C(CH_3)_2-)]AlCl_4$ ..... B: B Comp.SVol.3/3-205
$AlBC_{19}Cl_4H_{37}N_3$ .. $[2,6-(CH_3)_2-NC_5H_3-1-B(N(C_3H_7-i)_2)_2][AlCl_4]$ .. B: B Comp.SVol.4/3b-45
$AlBC_{19}H_{42}N_2$ .... $[-C(CH_3)_2-CH_2CH_2CH_2-C(CH_3)_2-]N[$
$-Al(C_2H_5)_2-N(C_4H_9-t)-B(C_2H_5)-]$ ..... B: B Comp.SVol.4/3a-206/7
$AlBC_{21}Cl_4H_{45}NOSi$
$[(-C(CH_3)_2-CH_2CH_2CH_2-C(CH_3)_2-)N=B=O$
$-Si(C_4H_9-t)_3][AlCl_4]$ ..... B: B Comp.SVol.4/3b-61
$AlBC_{22}Cl_4H_{32}N_2$ .. $[C_6H_5-CH_2-N(C_4H_9-t)=B=N(C_4H_9-t)-CH_2-C_6H_5][AlCl_4]$
B: B Comp.SVol.3/3-204/5
B: B Comp.SVol.4/3b-44
$AlBC_{25}Cl_4H_{17}N$ ... $[C_{13}H_9N-BC_{12}H_8][AlCl_4]$ ..... B: B Comp.SVol.3/3-207
B: B Comp.SVol.4/3b-45
$AlBC_{31}H_{42}N_2$ .... $[-C(CH_3)_2-CH_2CH_2CH_2-C(CH_3)_2-]N[$
$-Al(C_6H_5)_2-N(C_4H_9-t)-B(C_6H_5)-]$ ..... B: B Comp.SVol.4/3a-206/7
AlB_2 ..... AlB_2 ..... B: B Comp.SVol.4/3b-192
$AlB_2Br_3C_{16}H_{40}N_8$ $[(-NCH_3-CH_2-CH_2-NCH_3-)_2B]_2 \cdot AlBr_3$ ..... B: B Comp.SVol.3/3-118
$AlB_2Br_4C_{13}H_{28}N$ .. $[(C_2H_5)_2B=N=B(-C(CH_3)_2-(CH_2)_3-C(CH_3)_2-)]AlBr_4$
B: B Comp.SVol.3/3-205
$AlB_2Br_4C_{15}H_{33}N_2$ $[(-C(CH_3)_2-(CH_2)_3-C(CH_3)_2-)N(-BCH_3-N$
$(t-C_4H_9)-BCH_3-)]AlBr_4$ ..... B: B Comp.SVol.3/3-208

$AlB_2Br_5C_{14}H_{30}N_2$ $[CH_3B(-NC_5H_6(CH_3)_4-BBr-N(C_4H_9-t)-)][AlBr_4]$ B: B Comp.SVol.3/4-95

$AlB_2Br_6C_{13}H_{27}N_2$ $[BrB(-NC_5H_6(CH_3)_4-BBr-N(C_4H_9-t)-)][AlBr_4]$ B: B Comp.SVol.3/4-95

$AlB_2C_{13}H_{28}I_4N$ ... $[(C_2H_5)_2B{=}N{=}B(-C(CH_3)_2-(CH_2)_3-C(CH_3)_2-)]AlI_4$ B: B Comp.SVol.3/3-205

$AlB_2C_{16}Cl_3H_{40}N_8$ $[(-NCH_3-CH_2-CH_2-NCH_3-)_2B]_2 \cdot AlCl_3$ ...... B: B Comp.SVol.3/3-118

$AlB_2C_{16}H_{40}I_3N_8$ .. $[(-NCH_3-CH_2-CH_2-NCH_3-)_2B]_2 \cdot AlI_3$ .... B: B Comp.SVol.3/3-118

$AlB_2C_{17}Cl_4FeH_{22}MnO_3S$
$[(C_6H_6)Fe(1,2,5-SB_2C_2-3,4-(C_2H_5)_2-2,5-(CH_3)_2)Mn(CO)_3][AlCl_4]$ .... Fe: Org.Comp.B18-84

$AlB_2C_{19}Cl_5H_{32}N_2$ $[C_6H_5B(-NC_5H_6(CH_3)_4-BCl-N(C_4H_9-t)-)][AlCl_4]$ B: B Comp.SVol.3/4-62

$AlB_2C_{20}H_{43}N_2Si_2$ 4,5-$(C_2H_5)_2$-2,2,3-$(CH_3)_3$-1,2,5-$NSiBC_2$-1-$Al(C_2H_5)$-1-[1,2,5-$NSiBC_2$-2,2,3-$(CH_3)_3$-4,5-$(C_2H_5)_2$] B: B Comp.SVol.4/3a-253

$AlB_3Br_3C_6H_{18}N_3$ .. $(-BCH_3-NCH_3-)_3 \cdot AlBr_3$ .... B: B Comp.SVol.3/3-130

$AlB_3C_{27}H_{57}N_3Si_3$ [4,5-$(C_2H_5)_2$-2,2,3-$(CH_3)_3$-1,2,5-$NSiBC_2$-1$]_3$Al B: B Comp.SVol.4/3a-253

$AlB_4C_{12}H_{30}N$ .... 3,4-$(C_2H_5)_2$-6-$(C_2H_5)_3N$-3,4,6-$C_2AlB_4H_5$ ..... B: B Comp.SVol.4/4-194

$AlB_4C_{12}H_{32}N$ .... $(C_2H_5)_2$-2,3-$C_2B_4H_5-AlH_2-N(C_2H_5)_3$ .... B: B Comp.SVol.4/4-194

$AlB_8C_8H_{23}O$ ..... 6,9-$C_2B_8H_{10}-Al(C_2H_5)(OC_4H_8)$ .... B: B Comp.SVol.4/4-235

$AlB_8C_8H_{25}O$ ..... 6,9-$C_2B_8H_{10}-Al(C_2H_5)[O(C_2H_5)_2] \cdot 0.5\ C_6H_6$ .. B: B Comp.SVol.4/4-235

$AlB_8C_{11}H_{28}O$ .... 6,9-$C_2B_8H_{10}-Al(C_2H_5)[O(C_2H_5)_2] \cdot 0.5\ C_6H_6$ .. B: B Comp.SVol.4/4-235

$AlB_8C_{18}H_{44}N$ .... $[(C_2H_5)_2C_2B_4H_4]-Al[N(C_2H_5)_3]-[C_2B_4H_5(C_2H_5)_2]$ B: B Comp.SVol.4/4-194

$AlB_9C_6H_{20}$ ...... 1,2,3-$AlC_2B_9H_9$-1-C_2H_5-2,3-$(CH_3)_2$ .... B: B Comp.SVol.4/4-246

$AlB_{10}C_2H_{14}Li$ .... Li[1,2-$C_2B_{10}H_{11}$-1-AlH_3] .... B: B Comp.SVol.4/4-280

$AlB_{10}C_2H_{14}Na$ ... Na[1,2-$C_2B_{10}H_{11}$-1-AlH_3] .... B: B Comp.SVol.4/4-280

$AlB_{10}C_3H_{16}Li$ .... Li[1,2-$C_2B_{10}H_{10}$-1-CH_3-2-AlH_3] .... B: B Comp.SVol.4/4-280

$AlB_{10}C_3H_{16}Na$ ... Na[1,2-$C_2B_{10}H_{10}$-1-CH_3-2-AlH_3] .... B: B Comp.SVol.4/4-280

$AlB_{10}C_4Cl_{10}H_7$ ... 1,2-$C_2B_{10}Cl_{10}$-1-H-2-$Al(CH_3)_2$ .... B: B Comp.SVol.4/4-267

$AlB_{10}C_5H_{18}Li$ .... Li[1,2-$C_2B_{10}H_{10}$-1-$(C(CH_3){=}CH_2)$-2-AlH_3] .... B: B Comp.SVol.4/4-280

$AlB_{10}C_5H_{18}Na$ ... Na[1,2-$C_2B_{10}H_{10}$-1-$(C(CH_3){=}CH_2)$-2-AlH_3] ... B: B Comp.SVol.4/4-280

$AlB_{10}C_8H_{18}Li$ .... Li[1,2-$C_2B_{10}H_{10}$-1-C_6H_5-2-AlH_3] .... B: B Comp.SVol.4/4-280

$AlB_{10}C_8H_{18}Na$ ... Na[1,2-$C_2B_{10}H_{10}$-1-C_6H_5-2-AlH_3] .... B: B Comp.SVol.4/4-280

$AlB_{10}C_{10}Cl_{10}H_{19}$. 1,2-$C_2B_{10}Cl_{10}$-1-H-2-$Al(C_4H_9-i)_2$ .... B: B Comp.SVol.4/4-267

$AlB_{10}C_{16}H_{37}N_2$ ... 1,2-$C_2B_{10}H_{10}$-1-C_6H_5-2-$Al(CH_3)_2 \cdot (CH_3)_2NCH_2CH_2N(CH_3)_2$ .... B: B Comp.SVol.4/4-267

$AlB_{12}C_4H_{16}^-$ ..... $[Al(2,7-C_2B_6H_8)_2]^-$ .... B: B Comp.SVol.4/4-308

$AlB_{12}C_4H_{16}Na$ ... $Na[Al(2,7-C_2B_6H_8)_2]$ .... B: B Comp.SVol.4/4-308

$AlB_{16}C_4H_{20}^-$ ..... $[Al(6,9-C_2B_8H_{10})_2]^-$ .... B: B Comp.SVol.4/4-308/9

$AlB_{16}C_4H_{20}Na$ ... $Na[Al(6,9-C_2B_8H_{10})_2]$ .... B: B Comp.SVol.4/4-308/9

$AlB_{16}C_{40}H_{50}NP_2$ .. $[(C_6H_5)_3P{=}N{=}P(C_6H_5)_3][Al(6,9-C_2B_8H_{10})_2]$ .... B: B Comp.SVol.4/4-308

$AlB_{18}C_4H_{22}Tl$ .... $Tl[Al(1,2-C_2B_9H_{11})_2]$ .... B: B Comp.SVol.4/4-310

$AlBrC_{20}H_{24}Ti_2$ ... $[((C_5H_5)_2Ti)_2(AlH_4Br)]_n$ .... Ti: Org.Comp.5-330

$AlBr_3C_9H_8MoO_3$ .. $(C_5H_5)Mo(CO)_2[-C(CH_3){=}O-Al(Br)_2-Br-]$ ..... Mo: Org.Comp.8-156/7

$AlBr_3C_{25}H_{20}O_2PRe$
$(C_5H_5)Re(CO)_2[P(C_6H_5)_3] \cdot AlBr_3$ .... Re: Org.Comp.3-210/2, 214

$AlBr_3O_9$ .... $Al[BrO_3]_3 \cdot 9\ H_2O$ .... Br: SVol.B2-121

$AlBr_4C_8FeH_5O_3$ .. $[(C_5H_5)Fe(CO)_3][AlBr_4]$ .... Fe: Org.Comp.B15-36, 38, 40, 46

Formula	Compound	Reference
$AlBr_4C_9FeH_9O_2$	$[(C_5H_5)Fe(CO)_2(CH_2{=}CH_2)][AlBr_4]$	Fe: Org.Comp.B17-3
$AlBr_4C_{10}FeH_{11}O_2$	$[(C_5H_5)Fe(CO)_2(CH_2{=}CHCH_3)][AlBr_4]$	Fe: Org.Comp.B17-6
$AlBr_4C_{13}FeH_{15}O_2$	$[(C_5H_5)Fe(CO)_2(C_6H_{10}\text{-c})][AlBr_4]$	Fe: Org.Comp.B17-91
$AlBr_4C_{25}FeH_{41}O_2$	$[(C_5H_5)Fe(CO)_2(CH_2{=}CHC_{16}H_{33}\text{-n})][AlBr_4]$	Fe: Org.Comp.B17-15
AlC_2Th	$ThAlC_2$	Th: SVol.B2-103 Th: SVol.C6-88
$AlC_3H_{18}NSi_3$	$(SiH_3)_3N \cdot Al(CH_3)_3$	Si: SVol.B4-105
$AlC_4H_{18}NSi_2$	$(SiH_3)_2NCH_3 \cdot Al(CH_3)_3$	Si: SVol.B4-236
$AlC_5H_{18}NSi$	$SiH_3N(CH_3)_2 \cdot Al(CH_3)_3$	Si: SVol.B4-172
$AlC_6Cl_4O_6Re$	$[Re(CO)_6][AlCl_4]$	Re: Org.Comp.2-217, 222
$AlC_7Cl_3H_5MoNO_3$	$(C_5H_5)Mo(CO)_2NO \cdot AlCl_3$	Mo: Org.Comp.7-6, 9
$AlC_7H_{21}NOs^-$	$[N(C_4H_9\text{-n})_4][(CH_3)_4OsN\text{-}Al(CH_3)_3]$	Os: Org.Comp.A1-2, 28
$AlC_7H_{21}N_2OS$	$((CH_3)_2N)_2SO \cdot Al(CH_3)_3$	S: S-N Comp.8-341
$AlC_8Cl_4FeH_5O_3$	$[(C_5H_5)Fe(CO)_3][AlCl_4]$	Fe: Org.Comp.B15-36
$AlC_8Cl_4FeH_7O_3$	$[(CH_2CHCHCHCH_2)Fe(CO)_3][AlCl_4]$	Fe: Org.Comp.B15-18, 28
$AlC_9Cl_4FeH_7O_3$	$[(c\text{-}C_6H_7)Fe(CO)_3][AlCl_4]$	Fe: Org.Comp.B15-54/70
$AlC_{10}Cl_4GeH_{15}$	$[GeC_5(CH_3)_5][AlCl_4]$	Ge: Org.Comp.3-384
$AlC_{10}H_{12}Ti$	$[(C_5H_5)_2Ti(AlH_2) \cdot x\ O(C_2H_5)_2]_n$	Ti: Org.Comp.5-88
$AlC_{10}H_{14}Ti$	$[(C_5H_5)_2Ti(HAlH_3)]_n$	Ti: Org.Comp.5-330/1
$AlC_{10.8}H_{14}O_{0.2}Ti$	$[(C_5H_5)_2Ti(AlH_2) \cdot 0.2\ O(C_2H_5)_2]_n$	Ti: Org.Comp.5-321
$AlC_{11}Cl_4H_{11}MoN_2O_2$	$[(C_5H_5)Mo(CO)_2(NC\text{-}CH_3)_2][AlCl_4]$	Mo: Org.Comp.7-283, 284
$AlC_{12}H_{18}O_{12}Th$	$Al[Th(CH_3COO)_6]$	Th: SVol.C7-51/2
$AlC_{12}H_{42}Mo_6N_3O_{24}$	$(t\text{-}C_4H_9NH_3)_3[Al(OH)_6Mo_6O_{18}]$	Mo: SVol.B3b-253/4
$AlC_{14}H_{22}OTi$	$[(C_5H_5)_2Ti(AlH_2) \cdot O(C_2H_5)_2]_n$	Ti: Org.Comp.5-330
$AlC_{14}H_{42}InN_2P_2Si$	$[(CH_3)_2Si(N{=}P(CH_3)_3)_2Al(CH_3)_2][In(CH_3)_4]$	In: Org.Comp.1-337/8, 339, 342/3
$AlC_{15}Cl_4FeH_{15}O_6$	$[(1\text{-}CH_3OC(O)C(C(O)CH_3)(CH_3)\text{-}C_6H_6)Fe(CO)_3][AlCl_4]$	Fe: Org.Comp.B15-100
$AlC_{15}H_{44}InN_2P_2Si$	$[(CH_3)_2Si(N{=}P(CH_3)_3)Al(CH_3)_2(N{=}P(CH_3)_2C_2H_5)][In(CH_3)_4]$	In: Org.Comp.1-337/8, 339, 343
$AlC_{19}H_{53}NOsSi_4^-$	$[N(C_4H_9\text{-n})_4][((CH_3)_3Si\text{-}CH_2)_4OsN\text{-}Al(CH_3)_3]$	Os: Org.Comp.A1-32
$AlC_{20}ClH_{24}Ti_2$	$[(C_5H_5)_2Ti]_2(AlH_4Cl)$	Ti: Org.Comp.5-87/8
–	$[((C_5H_5)_2Ti)_2(AlH_4Cl)]_n$	Ti: Org.Comp.5-330
$AlC_{20}Cl_3H_{22}Ti_2$	$[(C_5H_5)_2Ti]_2(AlH_2Cl_3)$	Ti: Org.Comp.5-88
$AlC_{20}H_{23}Ti_2$	$(C_5H_5)_2TiAlH_4Ti(C_5H_5)(C_5H_4)$	Ti: Org.Comp.5-95/6
$AlC_{20}H_{24}Ti_2$	$[(C_5H_5)_2Ti]_2(AlH_4)$	Ti: Org.Comp.5-78
$AlC_{20}H_{25}Ti_2$	$[(C_5H_5)_2Ti]_2(AlH_5)$	Ti: Org.Comp.5-86/7
$AlC_{22}ClH_{29}O_{0.5}Ti_2$	$[(C_5H_5)_2Ti]_2(AlH_4Cl) \cdot 0.5\ O(C_2H_5)_2$	Ti: Org.Comp.5-88
$AlC_{22}H_{24}MoO_2P$	$(C_5H_5)Mo(CO)_2[CH_3P(C_6H_5)_2]Al(CH_3)_2$	Mo: Org.Comp.7-118, 122
$AlC_{23}H_{57}N_2Os$	$[N(C_4H_9\text{-n})_4][(CH_3)_4OsN\text{-}Al(CH_3)_3]$	Os: Org.Comp.A1-2, 28
$AlC_{24}Cl_3H_{20}MoNO_2P$	$C_5H_5Mo(P(C_6H_5)_3)(NO)(CO) \cdot AlCl_3$	Mo: Org.Comp.6-247
$AlC_{24}H_{28}MoO_2P$	$(C_5H_5)Mo(CO)_2[CH_3P(C_6H_5)_2]Al(C_2H_5)_2$	Mo: Org.Comp.7-118, 122
$AlC_{24}H_{31}Ti_2$	$[(C_5H_5)Ti]_2[(C_{10}H_8)(H_2Al(C_2H_5)_2)(H)]$	Ti: Org.Comp.5-261/2
$AlC_{25}Cl_3H_{20}O_2PRe$	$(C_5H_5)Re(CO)_2[P(C_6H_5)_3] \cdot AlCl_3$	Re: Org.Comp.3-210/2, 214
$AlC_{25}H_{55}O_4Sn$	$(C_8H_{17})_2Sn(OC_3H_7\text{-i})OAl(OC_3H_7\text{-i})_2$	Sn: Org.Comp.16-200

$AlC_{27}H_{26}MoO_2P$ .. $(C_5H_5)Mo(CO)_2[P(C_6H_5)_3]Al(CH_3)_2$ Mo:Org.Comp.7-118, 122
$AlC_{28}H_{36}MoO_2P$ .. $(C_5H_5)Mo(CO)_2[CH_3P(C_6H_5)_2]Al(C_4H_9\text{-}i)_2$ Mo:Org.Comp.7-118, 122
$AlC_{29}H_{28}MoO_3P$ .. $[(C_6H_5)_3P\text{-}C_5H_4]Mo(CO)_2\text{-}CO\text{-}Al(CH_3)_3$ Mo:Org.Comp.8-44
$AlC_{29}H_{30}MoO_2P$ .. $(C_5H_5)Mo(CO)_2[P(C_6H_5)_3]Al(C_2H_5)_2$ Mo:Org.Comp.7-118, 122
$AlC_{30}Cl_5H_{31}Ti_3$ ... $[(C_5H_5)_2Ti]_3(AlCl_5H) \cdot O(C_2H_5)_2$ .... Ti: Org.Comp.5-286
$AlC_{30}Cl_6H_{30}Ti_3$ ... $[(C_5H_5)_2Ti]_3(AlCl_6) \cdot O(C_2H_5)_2$ .... Ti: Org.Comp.5-286
$AlC_{33}Cl_4H_{44}OP_4Re$
$(CO)ReCl[P(CH_3)_2C_6H_5]_4 \cdot AlCl_3$ .... Re: Org.Comp.1-38
$AlC_{33}H_{38}MoO_2P$ .. $(C_5H_5)Mo(CO)_2[P(C_6H_5)_3]Al(C_4H_9\text{-}i)_2$ Mo:Org.Comp.7-118, 122
$AlC_{34}Cl_5H_{41}OTi_3$. $[(C_5H_5)_2Ti]_3(AlCl_5H) \cdot O(C_2H_5)_2$ .... Ti: Org.Comp.5-286
$AlC_{34}Cl_6H_{40}OTi_3$. $[(C_5H_5)_2Ti]_3(AlCl_6) \cdot O(C_2H_5)_2$ .... Ti: Org.Comp.5-286
$AlC_{35}H_{89}N_2OsSi_4$ $[N(C_4H_9\text{-}n)_4][((CH_3)_3Si\text{-}CH_2)_4OsN\text{-}Al(CH_3)_3]$ .. Os: Org.Comp.A1-32
$AlC_{36}ClH_{53}OP_4Re$ trans-$(CO)ReCl[P(CH_3)_2C_6H_5]_4 \cdot Al(CH_3)_3$ ... Re: Org.Comp.1-38
$AlC_{40}Cl_4H_{30}O_4P_2Re$
$[(CO)_4Re(P(C_6H_5)_3)_2][AlCl_4]$ .... Re: Org.Comp.1-481
$AlC_{40}H_{65}Ti_2$ ..... $[((CH_3)_5C_5)_2Ti]_2(AlH_5)$ .... Ti: Org.Comp.5-88/9
$AlC_{43}Cl_4H_{35}MoO_2P_2$
$[(C_5H_5)Mo(CO)_2(P(C_6H_5)_3)_2][AlCl_4]$ Mo:Org.Comp.7-283, 288
$AlC_{44}Cl_4H_{50}Ti_4$ ... $(C_{44}H_{50}AlCl_4)Ti_4$ .... Ti: Org.Comp.5-309
$AlC_{51}ClH_{59}OP_4Re$ $(CO)ReCl[P(CH_3)_2C_6H_5]_4 \cdot Al(C_6H_5)_3$ .... Re: Org.Comp.1-38
$AlC_{53}Cl_4H_{48}OP_4Re$
$(CO)ReCl[P(C_6H_5)_2CH_2CH_2P(C_6H_5)_2]_2 \cdot AlCl_3$ Re: Org.Comp.1-33
$AlC_{56}ClH_{57}OP_4Re$ $(CO)ReCl[P(C_6H_5)_2CH_2CH_2P(C_6H_5)_2]_2 \cdot Al(CH_3)_3$
Re: Org.Comp.1-33
$AlC_{71}ClH_{63}OP_4Re$ $(CO)ReCl[P(C_6H_5)_2CH_2CH_2P(C_6H_5)_2]_2 \cdot Al(C_6H_5)_3$
Re: Org.Comp.1-33
$AlCe_2$ .... Ce_2Al solid solutions
$Ce_2Al\text{-}Th_2Al$ .... Th: SVol.B2-105/6

$AlCl_3$ .... $AlCl_3$ melts
$AlCl_3\text{-}KCl\text{-}K_2MoO_4\text{-}LiCl$ Mo:SVol.B3b-267

$AlCoH_{1.2}U$ .... $UAlCoH_{1.2}$ .... U: SVol.B2-230
AlCoTh .... ThAlCo .... Th: SVol.B2-124
AlCoU .... UAlCo .... U: SVol.B2-224/30
$AlCo_2U_2$ .... U_2AlCo_2 .... U: SVol.B2-224/6
$AlCo_3U_2$ .... U_2AlCo_3 .... U: SVol.B2-224, 226
$AlCo_4Th$ .... $ThAlCo_4$ .... Th: SVol.B2-124/5
$AlCsF_6Pd$ .... $CsAlPdF_6$ .... Pd: SVol.B2-62
AlCuGd .... GdAlCu solid solutions
GdAlCu-ThAlCu .... Th: SVol.B2-130/2

AlCuTh .... ThAlCu solid solutions
ThAlCu-GdAlCu .... Th: SVol.B2-130/2
ThAlCu-YAlCu .... Th: SVol.B2-130/2

AlCuY .... YAlCu solid solutions
YAlCu-ThAlCu .... Th: SVol.B2-130/2

$AlCu_3U_2$ .... U_2AlCu_3 .... U: SVol.B2-238/42
$AlCu_9U_2$ .... U_2AlCu_9 .... U: SVol.B2-238/42
AlFeU .... UAlFe .... U: SVol.B2-231/3
$AlFe_3U_2$ .... U_2AlFe_3 .... U: SVol.B2-231, 233
AlGeU .... UAlGe .... U: SVol.B2-207

$AlH_{0.15}MnU$ ..... $UAlMnH_{0.15}$ ..... U: SVol.B2-214
$AlH_{1.35}NiU$ ..... $UAlNiH_{1.35}$ ..... U: SVol.B2-222
$AlH_{1.55}NiU$ ..... $UAlNiH_{1.55}$ ..... U: SVol.B2-222
$AlH_{1.9}NiU$ ..... $UAlNiH_{1.9}$ ..... U: SVol.B2-221/3
$AlH_{2.2}NiU$ ..... $UAlNiH_{2.2}$ ..... U: SVol.B2-222
$AlH_{2.3}NiU$ ..... $UAlNiH_{2.3}$ ..... U: SVol.B2-222
$AlH_{2.43}NiTh$ ..... $ThAlNiH_{2.43}$ ..... Th: SVol.B2-118
$AlH_{2.5}NiTh$ ..... $ThAlNiH_{2.5}$ ..... Th: SVol.B2-118/9
$AlH_{2.5}NiU$ ..... $UAlNiH_{2.5}$ ..... U: SVol.B2-222
$AlH_{2.5}Ni_4Th$ ..... $ThAlNi_4H_{2.5}$ ..... Th: SVol.B2-122/3
$AlH_{2.53}NiTh$ ..... $ThAlNiH_{2.53}$ ..... Th: SVol.B2-118
$AlH_{2.7}NiU$ ..... $UAlNiH_{2.7}$ ..... U: SVol.B2-221
$AlH_{2.74}NiU$ ..... $UAlNiH_{2.74}$ ..... U: SVol.B2-221/2
$AlH_{2.76}NiTh$ ..... $ThAlNiH_{2.76}$ ..... Th: SVol.B2-118
$AlH_{2.94}NiTh$ ..... $ThAlNiH_{2.94}$ ..... Th: SVol.B2-118
$AlH_{2.98}NiTh$ ..... $ThAlNiH_{2.98}$ ..... Th: SVol.B2-118
AlH_3Ni_4Th ..... $ThAlNi_4H_3$ ..... Th: SVol.B2-123
AlH_4Ni_4Th ..... $ThAlNi_4H_4$ ..... Th: SVol.B2-123
AlH_5Ni_4Th ..... $ThAlNi_4H_5$ ..... Th: SVol.B2-123
$AlH_6Mo_6O_{24}^{3-}$ ..... $Al(OH)_6Mo_6O_{18}^{3-}$ ..... Mo: SVol.B3b-60, 213, 253/4
AlH_6Ni_4Th ..... $ThAlNi_4H_6$ ..... Th: SVol.B2-123
AlHfRh ..... RhHfAl ..... Rh: SVol.A1-137
AlI_2Pd_5 ..... Pd_5AlI_2 ..... Pd: SVol.B2-214
$AlIn_2Th$ ..... $ThIn_2Al$ ..... Th: SVol.B2-151
AlIrTh ..... ThAlIr ..... Th: SVol.B2-135/6
AlIrU ..... UAlIr ..... U: SVol.B2-244/5
$AlMgN_3Si$ ..... $MgAlSiN_3$ ..... Si: SVol.B4-58
$AlMnRh_2$ ..... Rh_2MnAl ..... Rh: SVol.A1-203, 205/6
AlMnU ..... UAlMn ..... U: SVol.B2-211, 212/4
AlN ..... AlN solid solutions
AlN-BN ..... B: B Comp.SVol.3/3-90
AlNi ..... NiAl systems
$NiAl-Al-UAl_2$ ..... U: SVol.B2-215/7
$NiAl-UAl_2$ ..... U: SVol.B2-215/7
AlNiTh ..... ThAlNi ..... Th: SVol.A4-115
Th: SVol.B2-117/9
AlNiU ..... UAlNi ..... U: SVol.B2-217/21
$AlNi_4Th$ ..... $ThAlNi_4$ ..... Th: SVol.B2-117, 119/24
$AlNi_4U$ ..... $UAlNi_4$ solid solutions
UAl_xNi_{5-x} ..... U: SVol.B2-217/21
AlPdTh ..... ThAlPd ..... Th: SVol.B2-135/6
AlPo ..... AlPo ..... Po: SVol.1-318
AlPtTh ..... ThAlPt ..... Th: SVol.B2-135/6
AlPtU ..... UAlPt ..... U: SVol.B2-244/5
AlRh ..... RhAl ..... Rh: SVol.A1-61/3
AlRhTh ..... ThAlRh ..... Rh: SVol.A1-139/40
Th: SVol.B2-135/6
AlRhU ..... RhUAl ..... Rh: SVol.A1-188/90
U: SVol.B2-242/4

AlRhU . . . . RhUAl solid solutions
RhUAl-RuUAl . . . . Rh: SVol.A1-273

AlRuU . . . . RuUAl . . . . U: SVol.B2-242/4
– . . . . RuUAl solid solutions
RuUAl-RhUAl . . . . Rh: SVol.A1-273

AlSiU. . . . . UAlSi. . . . . U: SVol.B2-182, 186
$AlSi_2U_2$ . . . . U_2AlSi_2 . . . . U: SVol.B2-182, 186
AlTh . . . . ThAl . . . . Th: SVol.B2-73/4, 83/5, 86
– . . . . $ThAl_x$ . . . . Th: SVol.B2-73, 85/6
$AlTh_2$. . . . . Th_2Al. . . . . Th: SVol.B2-73/81, 86
– . . . . Th_2Al solid solutions
Th_2Al-Ce_2Al . . . . Th: SVol.B2-105/6
Th_2Al-Th_2Ag. . . . . Th: SVol.B2-133/5
Th_2Al-Th_2Ge. . . . . Th: SVol.B2-111/3

$Al_{1.25}Co_{3.75}Th$ . . . $ThAl_{1.25}Co_{3.75}$ . . . . Th: SVol.B2-124/5
$Al_{1.25}Fe_{3.75}Th$ . . . $ThAl_{1.25}Fe_{3.75}$ . . . . Th: SVol.B2-126

$Al_{1.5}Co_{3.5}Th$ . . . . . $ThAl_{1.5}Co_{3.5}$ . . . . Th: SVol.B2-124/5
$Al_{1.5}Fe_{3.5}Th$ . . . . . $ThAl_{1.5}Fe_{3.5}$ . . . . Th: SVol.B2-126
$Al_{1.5}H_3Ni_{3.5}Th$ . . . $ThAl_{1.5}Ni_{3.5}H_3$ . . . . Th: SVol.B2-122
$Al_{1.5}Mn_{3.5}Th$. . . . . $ThAl_{1.5}Mn_{3.5}$. . . . . Th: SVol.B2-114/5
$Al_{1.5}Ni_{3.5}Th$ . . . . . $ThAl_{1.5}Ni_{3.5}$ . . . . Th: SVol.B2-117, 119/24

$Al_{1.75}Co_{3.25}Th$ . . . $ThAl_{1.75}Co_{3.25}$ . . . . Th: SVol.B2-124/5
$Al_{1.75}Fe_{3.25}Th$ . . . $ThAl_{1.75}Fe_{3.25}$ . . . . Th: SVol.B2-126

$Al_2BF_3O_3$ . . . . $Al_2O_3 \cdot BF_3$ . . . . B: B Comp.SVol.4/3b-145
$Al_2B_2Br_6C_{16}H_{40}N_8$
$[(-NCH_3-CH_2-CH_2-NCH_3-)_2B]_2 \cdot 2\ AlBr_3$ . . . . B: B Comp.SVol.3/3-118
$Al_2B_2C_{16}Cl_6H_{40}N_8$ $[(-NCH_3-CH_2-CH_2-NCH_3-)_2B]_2 \cdot 2\ AlCl_3$ . . . . B: B Comp.SVol.3/3-118
$Al_2B_2C_{16}H_{40}I_6N_8$. $[(-NCH_3-CH_2-CH_2-NCH_3-)_2B]_2 \cdot 2\ AlI_3$. . . . . . B: B Comp.SVol.3/3-118
$Al_2B_2C_{18}Cl_4H_{38}N_2Si_2$
$[1-Cl_2Al-4,5-(C_2H_5)_2-2,2,3-(CH_3)_3-1,2,5-NSiBC_2]_2$
B: B Comp.SVol.4/3a-250/1
$Al_2B_2C_{26}H_{58}N_2Si_2$ $[1-(C_2H_5)_2Al-4,5-(C_2H_5)_2-2,2,3-(CH_3)_3-1,2,5-NSiBC_2]_2$
B: B Comp.SVol.4/3a-250/1
$Al_2B_{18}C_8H_{32}$ . . . . . $[1,2-C_2B_9H_{11}]Al[1,2-C_2B_9H_9(-H-)_2Al(C_2H_5)_2]$ B: B Comp.SVol.4/4-310
$Al_2Br_2C_{20}H_{26}Ti_2$ . . $[(C_5H_5)_2Ti(H_2Al(Br)H)]_2$ . . . . Ti: Org.Comp.5-90
$Al_2Br_4C_{10}H_{16}Ti_2$ . . $[(C_5H_5)(Br)Ti(H)_2]_2Al_2H_2Br_2$. . . . . Ti: Org.Comp.5-30
Al_2Br_8Pd. . . . . $PdAl_2Br_8$. . . . . Pd: SVol.B2-202
$Al_2C_5Th_4$. . . . . $Th_4Al_2C_5$. . . . . Th: SVol.B2-103
Th: SVol.C6-88
$Al_2C_{10}Cl_3H_{17}Ti_2$ . . $[(C_5H_5)(Cl)TiH_2]_2Al_2H_3Cl$ . . . . Ti: Org.Comp.5-30
$Al_2C_{10}Cl_4H_{16}Ti_2$ . . $[(C_5H_5)(Cl)TiH_2]_2Al_2H_2Cl_2$ . . . . Ti: Org.Comp.5-30
$Al_2C_{12}Cl_4H_{34}N_4Si$ $Si(N(CH_3)_2)_4 \cdot 2\ Al(C_2H_5)Cl_2$. . . . . Si: SVol.B4-212/3
$Al_2C_{12}Cl_8H_{18}MnN_6$
$[Mn(NCCH_3)_6][AlCl_4]_2$ . . . . Mn: MVol.D7-7/9
$Al_2C_{13}Cl_7FeH_{15}O_4$ $[1-HO-3,4,6,6-(CH_3)_4-C_6H_2Fe(CO)_3][Al_2Cl_7]$ . . Fe: Org.Comp.B15-133, 184
$Al_2C_{15}Cl_7FeH_{19}O_4$ $[1-HO-2,3,4,5,6,6-(CH_3)_6C_6Fe(CO)_3][Al_2Cl_7]$ . . Fe: Org.Comp.B15-134, 184/5
$Al_2C_{20}ClH_{27}Ti_2$ . . . $[(C_5H_5)_2Ti]_2(Al_2H_7Cl)$ . . . . Ti: Org.Comp.5-89
$Al_2C_{20}Cl_2H_{26}Ti_2$ . . $[(C_5H_5)_2Ti(H_2Al(Cl)H)]_2$ . . . . Ti: Org.Comp.5-89/90

$Al_2C_{22}H_{32}O_2Ti_2$ .. $[(C_5H_5)_2TiH_2Al(H)OCH_3]_2$ .... Ti: Org.Comp.5-90
$Al_2C_{22}H_{58}O_8Ti_2$ .. $[CH_3(C_2H_5O)_2Ti(OC_2H_5)_2Al(CH_3)_2]_2$ .... Ti: Org.Comp.5-6
$Al_2C_{24}Cl_2H_{30}Ti_2$ .. $[C_5H_4(C_5H_5)TiHAl(C_2H_5)Cl]_2$ .... Ti: Org.Comp.5-95
$Al_2C_{24}Cl_8H_{48}MnO_{12}S_6$
$[Mn((O=)_2SC_4H_8)_6][AlCl_4]_2$ .... Mn: MVol.D7-112/3
$Al_2C_{24}H_{30}Ti_2$ ..... $[(C_5H_4Al(CH_3)_2H)Ti]_2(C_{10}H_8)$ .... Ti: Org.Comp.5-265
$Al_2C_{24}H_{32}Ti_2$ ..... $[(C_5H_5)Ti(C_5H_5Al(CH_3)_2)]_2$ .... Ti: Org.Comp.5-94
$Al_2C_{24}H_{72}Mn_2P_8$ .. $[Mn(AlH_4)((CH_3)_2PCH_2CH_2P(CH_3)_2)_2]_2$ .... Mn: MVol.D8-76/7
$Al_2C_{26}H_{44}N_2Ti_2$ .. $[(C_5H_5)_2TiH_2AlH_2N(CH_3)_2CH_2]_2$ .... Ti: Org.Comp.5-90/1
$Al_2C_{28}H_{38}Ti_2$ ..... $[(C_5H_4Al(C_2H_5)_2H)Ti]_2(C_{10}H_8)$ .... Ti: Org.Comp.5-265
$Al_2C_{28}H_{40}Ti_2$ ..... $[(C_5H_5)Ti(C_5H_5Al(C_2H_5)_2)]_2$ .... Ti: Org.Comp.5-94/5
$Al_2C_{30}Cl_4H_{44}Ti_2$ .. $[(C_5H_5)_2Ti(C_2H_5)(Cl)Al(C_2H_5)Cl]_2(CH_2CH_2)$ .... Ti: Org.Comp.5-234
$Al_2C_{30}H_{36}Ti_3$ ..... $[(C_5H_5)_2Ti]_3(Al_2H_6) \cdot x\ O(C_2H_5)_2$ (x = 0.2 to 1.0)
Ti: Org.Comp.5-286
$Al_2C_{30}H_{48}N_2Ti_2$ .. $[(C_5H_4(CH_2)_2C_5H_4)TiH_2AlH_2]_2N(CH_3)_2CH_2CH_2N(CH_3)_2$
Ti: Org.Comp.5-273
$Al_2C_{30}H_{74}O_8Ti_2$ .. $[CH_3(C_3H_7O)_2Ti(OC_3H_7)_2Al(CH_3)_2]_2$ .... Ti: Org.Comp.5-6
$Al_2C_{30.8}H_{38}O_{0.2}Ti_3$
$[(C_5H_5)_2Ti]_3(Al_2H_6) \cdot 0.2\ O(C_2H_5)_2$ .... Ti: Org.Comp.5-286
$Al_2C_{32}H_{48}Ti_2$ ..... $[(C_5H_5)Ti(C_5H_5Al(C_3H_7)_2)]_2$ .... Ti: Org.Comp.5-94
$Al_2C_{32}H_{50}N_2Ti_2$ .. $[(C_5H_5)_2TiH_2AlH_2]_2N(CH_3)_2CH_2CH_2N(CH_3)_2 \cdot C_6H_6$
Ti: Org.Comp.5-91
$Al_2C_{32}H_{52}N_2Ti_2$ .. $[(C_5H_4(CH_2)_3C_5H_4)TiH_2AlH_2]_2N(CH_3)_2CH_2CH_2N(CH_3)_2$
Ti: Org.Comp.5-273
$Al_2C_{34}Cl_2H_{54}Ti_2$ .. $[(C_5H_5)_2Ti(C_2H_5)(Cl)Al(C_2H_5)_2]_2(CH_2CH_2)$ ..... Ti: Org.Comp.5-234
$Al_2C_{34}H_{46}OTi_3$ ... $[(C_5H_5)_2Ti]_3(Al_2H_6) \cdot O(C_2H_5)_2$ .... Ti: Org.Comp.5-286
$Al_2C_{35}H_{46}Ti_2$ ..... $[(C_5H_4Al(C_2H_5)_2H)Ti]_2(C_{10}H_8) \cdot C_6H_5CH_3$ .... Ti: Org.Comp.5-265
$Al_2C_{36}H_{56}Ti_2$ ..... $[(C_5H_5)Ti(C_5H_5Al(C_4H_9\text{-}i)_2)]_2$ .... Ti: Org.Comp.5-94
$Al_2C_{40}H_{46}Ti_4$ ..... $[(C_5H_5)_3Ti_2(H_4AlC_5H_4)]_2 \cdot C_6H_5CH_3$ .... Ti: Org.Comp.5-304/5
$Al_2C_{40}H_{66}OTi_2$ ... $[((CH_3)_5C_5)_2Ti]_2(Al_2H_6O)$ .... Ti: Org.Comp.5-93
$Al_2C_{40}H_{68}OTi_2$ ... $[((CH_3)_5C_5)_2Ti]_2Al_2H_7(OH)$ .... Ti: Org.Comp.5-93
$Al_2C_{40}H_{68}O_2Ti_2$ .. $[((CH_3)_5C_5)_2TiH_2Al(H)(OH)]_2$ .... Ti: Org.Comp.5-93
$Al_2C_{40}H_{68}Ti_2$ ..... $[((CH_3)_5C_5)_2TiH_2AlH_2]_2$ .... Ti: Org.Comp.5-92
$Al_2C_{47}H_{54}Ti_4$ ..... $[(C_5H_5)_3Ti_2(H_4AlC_5H_4)]_2 \cdot C_6H_5CH_3$ .... Ti: Org.Comp.5-304/5
$Al_2C_{48}H_{80}Ti_2$ ..... $[(C_5H_5)Ti(C_5H_5Al(C_7H_{15})_2)]_2$ .... Ti: Org.Comp.5-95
$Al_2C_{52}H_{72}Ti_2$ ..... $[(C_5H_5)Ti(C_5H_5Al(C_8H_{13})_2)]_2$ .... Ti: Org.Comp.5-95
$Al_2C_{52}H_{88}Ti_2$ ..... $[(C_5H_5)Ti(C_5H_5Al(C_8H_{17})_2)]_2$ .... Ti: Org.Comp.5-95
$Al_2C_{60}H_{88}Ti_2$ ..... $[(C_5H_5)Ti(C_5H_5Al(C_{10}H_{17})_2)]_2$ .... Ti: Org.Comp.5-95
Al_2Ce .... $CeAl_2$ solid solutions
$CeAl_2$-$ThAl_2$ .... Th: SVol.B2-106/9
$CeAl_2$-UAl_2 .... U: SVol.B2-191, 195
Al_2Cl_8Pd .... $Pd(AlCl_4)_2 = PdAl_2Cl_8$ .... Pd: SVol.B2-166/8
$Al_2Cl_{10}Pd$ .... $PdAl_2Cl_{10}$ .... Pd: SVol.B2-168
Al_2Co_3Th .... $ThAl_2Co_3$ .... Th: SVol.B2-124/5
Al_2CuU .... UAl_2Cu .... U: SVol.B2-238/42
Al_2Dy .... $DyAl_2$ solid solutions
$DyAl_2$-UAl_2 .... U: SVol.B2-191, 195
– .... $DyAl_2$ systems
$DyAl_2$-Al-UAl_2 .... U: SVol.B2-190/1

Al_2Fe_3Th .... $ThAl_2Fe_3$ .... Th: SVol.B2-126
Al_2Gd .... $GdAl_2$ solid solutions
$GdAl_2$-$ThAl_2$ .... Th: SVol.B2-106/9
$GdAl_2$-UAl_2 .... U: SVol.B2-191, 194/7

Al_2Ge_2U .... UAl_2Ge_2 .... U: SVol.B2-207
$Al_2H_{2.7}Ni_3Th$ ..... $ThAl_2Ni_3H_{2.7}$ .... Th: SVol.B2-122
Al_2I_8Pd .... $PdAl_2I_8$ .... Pd: SVol.B2-214
Al_2InTh .... $ThInAl_2$ .... Th: SVol.B2-151
Al_2La .... $LaAl_2$ solid solutions
$LaAl_2$-UAl_2 .... U: SVol.B2-191/6
$LaAl_2$-UAl_2-YAl_2 .... U: SVol.B2-191

Al_2Lu .... $LuAl_2$ solid solutions
$LuAl_2$-UAl_2 .... U: SVol.B2-191, 195

Al_2Mn_3Th .... $ThAl_2Mn_3$ .... Th: SVol.B2-114/5
Al_2Ni_3Th .... $ThAl_2Ni_3$ .... Th: SVol.B2-117, 119/24
Al_2Sc .... $ScAl_2$ solid solutions
$ScAl_2$-UAl_2 .... U: SVol.B2-191, 194/5

Al_2Si_2U .... UAl_2Si_2 .... U: SVol.B2-182, 186/8
$Al_2Si_3U_2$ .... $U_2Al_2Si_3$ .... U: SVol.B2-182, 186
Al_2Sm .... $SmAl_2$ solid solutions
$SmAl_2$-UAl_2 .... U: SVol.B2-191, 195

– .... $SmAl_2$ systems
$SmAl_2$-Al-UAl_2 .... U: SVol.B2-190

Al_2Th .... $ThAl_2$ .... Th: SVol.A4-114, 139/40
Th: SVol.B2-73/6, 86, 87/90
– .... $ThAl_2$ solid solutions
$ThAl_2$-$CeAl_2$ .... Th: SVol.B2-106/9
$ThAl_2$-$GdAl_2$ .... Th: SVol.B2-106/9
$ThAl_2$-$ThMn_2$ .... Th: SVol.B2-114
$ThAl_2$-UAl_2 .... U: SVol.B2-203/6

Al_2Th_3 .... Th_3Al_2 .... Th: SVol.B2-73/6, 79, 81/3, 86
Al_2U .... UAl_2
Chemical reactions .... U: SVol.B2-113/6
Crystallographic properties .... U: SVol.B2-92/5
Dispersions in Al. .... U: SVol.B2-116/8
Electrical properties .... U: SVol.B2-101/7
Formation and preparation. .... U: SVol.B2-82, 88/92
Magnetic properties .... U: SVol.B2-108/13
Mechanical properties .... U: SVol.B2-95/7
Nuclear fuels .... U: SVol.B2-116/8
Optical properties .... U: SVol.B2-107/8
Thermal properties .... U: SVol.B2-97/101

– .... UAl_2 solid solutions
UAl_2-$CeAl_2$ .... U: SVol.B2-191, 195
UAl_2-$DyAl_2$ .... U: SVol.B2-191, 195
UAl_2-$GdAl_2$ .... U: SVol.B2-191, 194/7
UAl_2-$LaAl_2$ .... U: SVol.B2-191/6
UAl_2-$LaAl_2$-YAl_2 .... U: SVol.B2-191

Al_2U .... UAl_2 solid solutions
UAl_2-$LuAl_2$ .... U: SVol.B2-191, 195
UAl_2-$ScAl_2$ .... U: SVol.B2-191, 194/5
UAl_2-$SmAl_2$ .... U: SVol.B2-191, 195
UAl_2-$ThAl_2$ .... U: SVol.B2-203/6
UAl_2-UCo_2 .... U: SVol.B2-224/30
UAl_2-UFe_2 .... U: SVol.B2-231/5
UAl_2-UFe_2-UMn_2 .... U: SVol.B2-238
UAl_2-UMn_2 .... U: SVol.B2-212/5
UAl_2-UMo_2 .... U: SVol.B2-208/9
UAl_2-USi_2 .... U: SVol.B2-181, 184/5
UAl_2-YAl_2 .... U: SVol.B2-191/5
UAl_2-$ZrAl_2$ .... U: SVol.B2-200/2

– .... UAl_2 systems
UAl_2-Al-$DyAl_2$ .... U: SVol.B2-190/1
UAl_2-Al-NiAl.......... U: SVol.B2-215/7
UAl_2-Al-$SmAl_2$ .... U: SVol.B2-190
UAl_2-Al-$ZrAl_2$ .... U: SVol.B2-198/200
UAl_2-Al_8Mo_3-Mo-U .... U: SVol.B2-208/9
UAl_2-NiAl .... U: SVol.B2-215/7
UAl_2-U-UCo_2 .... U: SVol.B2-224/5
UAl_2-U-UFe_2 .... U: SVol.B2-230/1
UAl_2-U-UFe_2-UMn_2 .... U: SVol.B2-237/8
UAl_2-U-UMn_2 .... U: SVol.B2-211
UAl_2-U-U_3Si_2 .... U: SVol.B2-182/4
UAl_2-UC .... U: SVol.B2-180
UAl_2-UCo_2 .... U: SVol.B2-224/5
UAl_2-UFe_2 .... U: SVol.B2-230/1
UAl_2-UFe_2-UMn_2 .... U: SVol.B2-237/8
UAl_2-UMn_2 .... U: SVol.B2-211/2
UAl_2-U_3Si_2 .... U: SVol.B2-182/4
UAl_2-$ZrAl_2$ .... U: SVol.B2-198, 200

Al_2Y .... YAl_2 solid solutions
YAl_2-$LaAl_2$-UAl_2 .... U: SVol.B2-191
YAl_2-UAl_2 .... U: SVol.B2-191/5

Al_2Zr .... $ZrAl_2$ solid solutions
UAl_2-$ZrAl_2$ .... U: SVol.B2-200/2

– .... $ZrAl_2$ systems
$ZrAl_2$-Al-UAl_2 .... U: SVol.B2-198/200
$ZrAl_2$-UAl_2 .... U: SVol.B2-198, 200

$Al_{2.25}Fe_{2.75}Th$ ... $ThAl_{2.25}Fe_{2.75}$ .... Th: SVol.B2-126

$Al_{2.5}Ni_{3.5}$ .... $NiAl_{2.5}Ni_{2.5}$ .... Th: SVol.B2-117, 119/24

Al_3Ce .... $CeAl_3$ solid solutions
$CeAl_3$-$ThAl_3$ .... Th: SVol.B2-110

$Al_3Cu_7U_2$ .... $U_2Al_3Cu_7$ .... U: SVol.B2-238/42
Al_3Dy .... $DyAl_3$ solid solutions
$DyAl_3$-UAl_3 .... U: SVol.B2-190, 197

$Al_3Fe_5U_4$ .... $U_4Al_3Fe_5$ .... U: SVol.B2-231/2
Al_3La .... $LaAl_3$ solid solutions
$LaAl_3$-$ThAl_3$ .... Th: SVol.B2-103/5
Al_3Ni_2Th .... $ThAl_3Ni_2$ .... Th: SVol.B2-117, 119/24
Al_3OsU_2 .... U_2Al_3Os .... U: SVol.B2-244/5
Al_3Rh .... $RhAl_3$ .... Rh: SVol.A1-61
Al_3Sm .... $SmAl_3$ solid solutions
$SmAl_3$-UAl_3 .... U: SVol.B2-190, 197
$Al_3Th_{0.5}U_{0.5}$ ..... $U_{0.5}Th_{0.5}Al_3$ .... U: SVol.B2-203/6
$Al_3Th_{0.8}U_{0.2}$ ..... $U_{0.2}Th_{0.8}Al_3$ .... U: SVol.B2-203
Al_3Th .... $ThAl_3$ .... Th: SVol.A4-115, 126, 185
Th: SVol.B2-73/6, 86, 91/5
– .... $ThAl_3$ solid solutions
$ThAl_3$-$CeAl_3$ .... Th: SVol.B2-110
$ThAl_3$-$LaAl_3$ .... Th: SVol.B2-103/5
$ThAl_3$-UAl_3 .... U: SVol.B2-203/6
$ThAl_3$-YAl_3 .... Th: SVol.B2-103/5
– .... $ThAl_3$ systems
$ThAl_3$-$ThIn_3$ .... Th: SVol.B2-151
Al_3Th_2 .... Th_2Al_3 .... Th: SVol.B2-73/4, 85/6
Al_3U .... UAl_3
Chemical reactions .... U: SVol.B2-130/4
Crystallographic properties .... U: SVol.B2-123/5, 240
Dispersions in Al .... U: SVol.B2-134/8
Electrical properties .... U: SVol.B2-128/30
Formation and preparation .... U: SVol.B2-82/3, 118/23
Magnetic properties .... U: SVol.B2-130
Mechanical properties .... U: SVol.B2-125/6
Nuclear fuels .... U: SVol.B2-134/8, 202
Thermal properties .... U: SVol.B2-126/8
– .... UAl_3 solid solutions
UAl_3-$DyAl_3$ .... U: SVol.B2-190, 197
UAl_3-$SmAl_3$ .... U: SVol.B2-190, 197
UAl_3-$ThAl_3$ .... U: SVol.B2-203/6
UAl_3-USi_3 .... U: SVol.B2-181/2, 185/6
UAl_3-USn_3 .... U: SVol.B2-207/8, 307/8
UAl_3-$ZrAl_3$ .... U: SVol.B2-200
Al_3Y .... YAl_3 solid solutions
YAl_3-$ThAl_3$ .... Th: SVol.B2-103/5
Al_3Zr .... $ZrAl_3$ solid solutions
$ZrAl_3$-UAl_3 .... U: SVol.B2-200
Al_4C_4Th .... $ThAl_4C_4$ .... Th: SVol.B2-103
Th: SVol.C6-88
$Al_4C_{48}H_{112}O_{16}Th$ $Th[Al(O\text{-}i\text{-}C_3H_7)_4]_4$ .... Th: SVol.C7-25/9
$Al_4C_{80}H_{124}O_2Ti_4$ .. $[(CH_3)_5C_5((CH_3)_4C_5CH_2)TiH_2Al]_4O_2 \cdot 2\ C_6H_6$ .. Ti: Org.Comp.5-305/6
$Al_4C_{92}H_{136}O_2Ti_4$ .. $[(CH_3)_5C_5((CH_3)_4C_5CH_2)TiH_2Al]_4O_2 \cdot 2\ C_6H_6$ .. Ti: Org.Comp.5-305/6
Al_4Dy .... $DyAl_4$ solid solutions
$DyAl_4$-UAl_4 .... U: SVol.B2-190, 197

Al_4Sm .... $SmAl_4$ solid solutions
  $SmAl_4$-UAl_4 .... U: SVol.B2-190, 197

Al_4U .... UAl_4
  Chemical reactions .... U: SVol.B2-151/3
  Crystallographic properties .... U: SVol.B2-143/6
  Dispersions in Al .... U: SVol.B2-153/6
  Formation and preparation .... U: SVol.B2-82/3, 139/43
  Nuclear fuels .... U: SVol.B2-153/6
  Physical properties .... U: SVol.B2-146/51

– .... UAl_4 solid solutions
  UAl_4-$DyAl_4$ .... U: SVol.B2-190, 197
  UAl_4-$SmAl_4$ .... U: SVol.B2-190, 197

Al_5Rh_2 .... Rh_2Al_5 .... Rh: SVol.A1-61
Al_5Th_3 .... Th_3Al_5 .... Th: SVol.B2-73, 85/6
Al_5U .... UAl_5 .... U: SVol.B2-83

$Al_6Ce_3Th_{10}$ .... $Th_{10}Ce_3Al_6$ .... Th: SVol.B2-106
Al_6Fe_6U .... UAl_6Fe_6 .... U: SVol.B2-232/6

Al_7Fe_5U .... UAl_7Fe_5 .... U: SVol.B2-232/6
Al_7Th .... $ThAl_7$ .... Th: SVol.B2-73
Al_7Th_2 .... Th_2Al_7 .... Th: SVol.A4-185
  Th: SVol.B2-73/4, 95/7
Al_7Th_4 .... Th_4Al_7 .... Th: SVol.B2-73, 85

$Al_{7.5}Cu_{4.5}U$ .... $UAl_{7.5}Cu_{4.5}$ .... U: SVol.B2-238/42

$Al_8Ce_2Th_{14}$ .... $Th_{14}Ce_2Al_8$ .... Th: SVol.B2-105/6
$Al_8Ce_3Th_{13}$ .... $Th_{13}Ce_3Al_8$ .... Th: SVol.B2-105/6
$Al_8Ce_4Th_{12}$ .... $Th_{12}Ce_4Al_8$ .... Th: SVol.B2-105/6
Al_8Cr_4Th .... $ThAl_8Cr_4$ .... Th: SVol.B2-113/4
Al_8Cr_4U .... UAl_8Cr_4 .... U: SVol.B2-210/1
Al_8Cu_4Th .... $ThAl_8Cu_4$ .... Th: SVol.B2-130
Al_8Cu_4U .... UAl_8Cu_4 .... U: SVol.B2-238/42
Al_8Fe_4Th .... $ThAl_8Fe_4$ .... Th: SVol.B2-126/9
Al_8Fe_4U .... UAl_8Fe_4 .... U: SVol.B2-232/6
Al_8Mn_4Th .... $ThAl_8Mn_4$ .... Th: SVol.B2-114/7
Al_8Mn_4U .... UAl_8Mn_4 .... U: SVol.B2-210, 215
Al_8Mo_3 .... Al_8Mo_3 systems
  Al_8Mo_3-Mo-U-UAl_2 .... U: SVol.B2-208/9

$Al_8Nb_{62}Rh_{30}$ .... $Rh_{30}Nb_{62}Al_8$ .... Rh: SVol.A1-166/7

Al_9Rh_2 .... Rh_2Al_9 .... Rh: SVol.A1-61

$Al_{10}Cu_7U_2$ .... $U_2Al_{10}Cu_7$ .... U: SVol.B2-238/42

$Al_{12}U$ .... UAl_{12} solid solutions
  UAl_{12}-UCu_{12} .... U: SVol.B2-238
  UAl_{12}-UFe_{12} .... U: SVol.B2-232/6

$Al_{13}Rh_4$ .... Rh_4Al_{13} .... Rh: SVol.A1-61
$Al_{13}U$ .... UAl_{13} solid solutions
  UAl_{13}-UBe_{13} .... U: SVol.B2-59

$Al_{15}Dy_3U_2$ .... $U_2Al_{15}Dy_3$ .... U: SVol.B2-190/1, 197

$Al_{19}Th_{12}$ .... $Th_{12}Al_{19}$ .... Th: SVol.B2-73, 85/6

$Al_{21}Cr_2U$ .... $UAl_{21}Cr_2$ .... U: SVol.B2-210
$Al_{21}Mo_2U$ .... $UAl_{21}Mo_2$ .... U: SVol.B2-210
$Al_{21}Ta_2U$ .... UTa_2Al_{21} .... U: SVol.B2-208
$Al_{21}UV_2$ .... UV_2Al_{21} .... U: SVol.B2-208
$Al_{21}UW_2$ .... $UAl_{21}W_2$ .... U: SVol.B2-210

$Al_{25}Nb_{65}Rh_{10}$ .... $Rh_{10}Nb_{65}Al_{25}$ .... Rh: SVol.A1-166/7

$Al_{75}Mo_{16}U_9$ ...... $U_9Al_{75}Mo_{16}$ .... U: SVol.B2-209

Am .... Am
Sorption on W .... W: SVol.A6b-336

Am^{3+} .... Am^{3+} .... Th: SVol.D1-22/36

Ar .... Ar
Permeation through Fe .... Fe: SVol.B1-10
Sorption
on Au .... Au: SVol.B1-6/8
on Fe .... Fe: SVol.B1-1/2

$ArBCl_3$ .... $Ar \cdot BCl_3$ .... B: B Comp.SVol.3/4-13
$ArBF_3$ .... $Ar \cdot BF_3$ .... B: B Comp.SVol.4/3b-107
ArBN .... ArBN .... B: B Comp.SVol.4/3a-43
ArBr .... ArBr .... Br: SVol.B1-4/8
$ArBr^+$ .... $ArBr^+$ .... Br: SVol.B1-26
ArBrH .... $Ar \cdot HBr$ .... Br: SVol.B1-511/2
– .... $Ar \cdot DBr$ .... Br: SVol.B1-65
$ArBrH^+$ .... $ArHBr^+$ .... Br: SVol.B1-512
ArBrKr .... ArKrBr .... Br: SVol.B1-23
ArBrXe .... ArXeBr .... Br: SVol.B1-23
$ArBr_2$ .... $Ar \cdot Br_2$ .... Br: SVol.B1-3

Ar_2Br^+ .... Ar_2Br^+ .... Br: SVol.B1-26

Ar_8Br_2 .... $Ar_8 \cdot Br_2$ .... Br: SVol.B1-3

$Ar_{13}Br_2$ .... $Ar_{13} \cdot Br_2$ .... Br: SVol.B1-3

$Ar_{20}Br_2$ .... $Ar_{20} \cdot Br_2$ .... Br: SVol.B1-3

$Ar_{70}Br_2$ .... $Ar_{70} \cdot Br_2$ .... Br: SVol.B1-3

$As_{0.5}In_{0.5}Ni_4U$ ... $U(In_{0.5}As_{0.5})Ni_4$ .... U: SVol.B2-292/3
As .... As alloys
As-Ag-Rh .... Rh: SVol.A1-261
As-Ce-Rh .... Rh: SVol.A1-73
As-Co-Rh .... Rh: SVol.A1-227
As-Fe-Rh .... Rh: SVol.A1-247
As-La-Rh .... Rh: SVol.A1-69/70
As-Mn-Rh .... Rh: SVol.A1-202
As-Mo-Ru .... Mo: SVol.A2b-27
As-Nd-Rh .... Rh: SVol.A1-79
As-Ni-Rh .... Rh: SVol.A1-221
As-Rh-Ru .... Rh: SVol.A1-269

As . . . . As alloys

As-Rh-Sb . . . . Rh: SVol.A1-44

As-Rh-Ti. . . . . Rh: SVol.A1-114

As-Rh-V . . . . Rh: SVol.A1-160/1

– . . . . As systems

As-Pd . . . . Pd: SVol.B2-343/4

As-Pd-Te . . . . Pd: SVol.B2-347

As-Rh . . . . Rh: SVol.A1-38

$AsAuC_{24}Cl_2H_{20}$ . . $[(C_6H_5)_4As][AuCl_2]$ . . . . Au: SVol.B1-185/6

$AsAuC_{24}Cl_4H_{20}$ . . $[(C_6H_5)_4As][AuCl_4]$ . . . . Au: SVol.B1-217, 269

$AsAuC_{24}H_{20}N_{12}$ . . $[(C_6H_5)_4As][Au(N_3)_4]$ . . . . Au: SVol.B1-105/7

$AsBBrCClH_5I$ . . . . $BBrClI \cdot CH_3AsH_2$ . . . . B: B Comp.SVol.4/4-115

$AsBBrCH_5I_2$ . . . . . $BBrI_2 \cdot CH_3AsH_2$ . . . . B: B Comp.SVol.4/4-115

$AsBBrC_2ClH_7I$. . . . $BBrClI \cdot (CH_3)_2AsH$. . . . . B: B Comp.SVol.4/4-115

$AsBBrC_2H_7I_2$ . . . . $BBrI_2 \cdot (CH_3)_2AsH$. . . . . B: B Comp.SVol.4/4-115

$AsBBrC_3ClH_9I$. . . . $BBrClI \cdot (CH_3)_3As$ . . . . B: B Comp.SVol.4/4-115

$AsBBrC_3H_9I_2$ . . . . $BBrI_2 \cdot (CH_3)_3As$ . . . . B: B Comp.SVol.4/4-115

$AsBBr_2CH_5I$ . . . . . $BBr_2I \cdot CH_3AsH_2$ . . . . B: B Comp.SVol.4/4-115

$AsBBr_2C_2H_7I$ . . . . $BBr_2I \cdot (CH_3)_2AsH$. . . . . B: B Comp.SVol.4/4-115

$AsBBr_2C_3H_9I$ . . . . $BBr_2I \cdot (CH_3)_3As$ . . . . B: B Comp.SVol.4/4-115

$AsBBr_2C_{11}H_{26}N_2$ $(i\text{-}C_3H_7)_2N\text{-}BBr\text{-}N(C_4H_9\text{-}t)\text{-}AsBr\text{-}CH_3$ . . . . B: B Comp.SVol.4/4-103

$AsBBr_2C_{16}H_{37}N_3$ $[(i\text{-}C_3H_7)_2N]_2B\text{-}N(C_4H_9\text{-}t)\text{-}AsBr_2$ . . . . B: B Comp.SVol.4/3a-155

$AsBBr_3CH_5$. . . . . . $CH_3AsH_2 \cdot BBr_3$ . . . . B: B Comp.SVol.4/4-81

$AsBBr_3C_2H_7$ . . . . . $(CH_3)_2AsH \cdot BBr_3$ . . . . B: B Comp.SVol.4/4-81

$AsBBr_3C_3H_9$ . . . . . $(CH_3)_3As \cdot BBr_3$ . . . . B: B Comp.SVol.3/4-78

B: B Comp.SVol.4/4-81

$AsBBr_3C_3H_9O_9S_3$ $(CH_3SO_3)_3As \cdot BBr_3$ . . . . B: B Comp.SVol.4/4-81

$AsBBr_3C_{13}H_{27}N_2$ $1\text{-}[Br_2As\text{-}N(C_4H_9\text{-}t)\text{-}BBr]\text{-}2,2,6,6\text{-}(CH_3)_4NC_5H_6$

B: B Comp.SVol.4/4-103

$AsBBr_3C_{18}H_{15}$ . . . $(C_6H_5)_3As\text{-}BBr_3$ . . . . B: B Comp.SVol.3/4-78

$AsBBr_3C_{21}H_{35}N_2$ $1\text{-}[2,6\text{-}(i\text{-}C_3H_7)_2C_6H_3\text{-}N(AsBr_2)BBr]\text{-}2,2,6,6\text{-}$

$(CH_3)_4NC_5H_6$ . . . . B: B Comp.SVol.4/4-103

$AsBCClH_5I_2$ . . . . . $BClI_2 \cdot CH_3AsH_2$ . . . . B: B Comp.SVol.4/4-115

$AsBCCl_2H_5I$ . . . . . $BCl_2I \cdot CH_3AsH_2$ . . . . B: B Comp.SVol.4/4-115

$AsBCCl_3H_5$ . . . . . . $CH_3AsH_2 \cdot BCl_3$. . . . . B: B Comp.SVol.4/4-34

$AsBCH_5I_3$ . . . . $CH_3AsH_2 \cdot BI_3$. . . . . B: B Comp.SVol.4/4-110/1, 115

$AsBC_2ClH_7I_2$. . . . . $BClI_2 \cdot (CH_3)_2AsH$ . . . . B: B Comp.SVol.4/4-115

$AsBC_2Cl_2H_7I$. . . . . $BCl_2I \cdot (CH_3)_2AsH$ . . . . B: B Comp.SVol.4/4-115

$AsBC_2Cl_3H_7$ . . . . . $(CH_3)_2AsH \cdot BCl_3$ . . . . B: B Comp.SVol.4/4-34

$AsBC_2H_7I_3$ . . . . . . $(CH_3)_2AsH \cdot BI_3$. . . . . B: B Comp.SVol.4/4-111, 115

$AsBC_3ClH_9I_2$. . . . . $BClI_2 \cdot (CH_3)_3As$ . . . . B: B Comp.SVol.4/4-115

$AsBC_3Cl_2H_9I$. . . . . $BCl_2I \cdot (CH_3)_3As$ . . . . B: B Comp.SVol.4/4-115

$AsBC_3Cl_3H_9$ . . . . . $(CH_3)_3As \cdot BCl_3$. . . . . B: B Comp.SVol.4/4-34

$AsBC_3H_9I_3$ . . . . . . $(CH_3)_3As \cdot BI_3$. . . . . B: B Comp.SVol.3/4-99

B: B Comp.SVol.4/4-111/2, 115

$AsBC_4H_{15}N$. . . . . . $H_3B\text{-}N(CH_3)_2\text{-}As(CH_3)_2$ . . . . B: B Comp.SVol.4/3b-15

Formula	Compound	Reference
$AsBC_5H_{18}Sn$	$(CH_3)_3Sn-As(CH_3)_2 \cdot BH_3$	Sn: Org.Comp.19-230, 235
$AsBC_6H_{19}N$	$H_3B-N(C_2H_5)_2-As(CH_3)_2$	B: B Comp.SVol.4/3b-15
$AsBC_8Cl_2H_{18}N_2$	$[-N(C_4H_9-t)-BCl-N(C_4H_9-t)-]AsCl$	B: B Comp.SVol.4/4-64/5
$AsBC_8H_{23}N$	$H_3B-N(C_3H_7-i)_2-As(CH_3)_2$	B: B Comp.SVol.4/3b-15
–	$H_3B-N(C_3H_7-n)_2-As(CH_3)_2$	B: B Comp.SVol.4/3b-15
$AsBC_9ClH_{21}N_2$	$[-N(C_4H_9-t)-BCl-N(C_4H_9-t)-]AsCH_3$	B: B Comp.SVol.4/4-64/5
$AsBC_{10}H_{24}N_2$	$CH_3-B[-N(C_4H_9-t)-As(CH_3)-N(C_4H_9-t)-]$	B: B Comp.SVol.4/3a-208
$AsBC_{11}ClH_{25}N_2$	$[-N(C_4H_9-t)-BCl-N(C_4H_9-t)-]As-C_3H_7-i$	B: B Comp.SVol.4/4-64/5
$AsBC_{11}H_{27}N_3$	$(CH_3)_2N-B[-N(C_4H_9-t)-As(CH_3)-N(C_4H_9-t)-]$	B: B Comp.SVol.4/3a-208
$AsBC_{12}ClH_{27}N_2$	$[-N(C_4H_9-t)-BCl-N(C_4H_9-t)-]As-C_4H_9-n$	B: B Comp.SVol.4/4-64/5
–	$[-N(C_4H_9-t)-BCl-N(C_4H_9-t)-]As-C_4H_9-t$	B: B Comp.SVol.4/4-64/5
$AsBC_{12}H_{23}MoO_2P$	$(C_5H_5)Mo(CO)_2[P(CH_3)_3]As(CH_3)_2 \cdot BH_3$	Mo: Org.Comp.7-133, 150
$AsBC_{13}Cl_3H_{27}N_2$	$[-C(CH_3)_2-CH_2CH_2CH_2-C(CH_3)_2-]N[-BCl-N(C_4H_9-t)-AsCl_2-]$	B: B Comp.SVol.4/4-65
$AsBC_{13}F_3H_{27}N_2$	$2,2,6,6-(CH_3)_4-NC_5H_6-1-BF-N(C_4H_9-t)-AsF_2$	B: B Comp.SVol.4/3b-246
$AsBC_{13}F_5H_{13}$	$[(C_6H_5)_2AsF-CH_3][BF_4]$	B: B Comp.SVol.4/3b-108
$AsBC_{13}H_{30}N_2O$	$t-C_4H_9-O-B[-N(C_4H_9-t)-As(CH_3)-N(C_4H_9-t)-]$	B: B Comp.SVol.4/3a-208
$AsBC_{14}ClH_{32}N_3$	$[-N(C_4H_9-t)-BCl-N(C_4H_9-t)-]As-N(C_3H_7-i)_2$	B: B Comp.SVol.4/4-64/5
$AsBC_{14}H_{19}N$	$(C_2H_5)_2N-B(-CH=CH-AsC_6H_5-CH=CH-)$	B: B Comp.SVol.3/3-151
$AsBC_{14}H_{21}N$	$(C_2H_5)_2N-B(-CH=CH-AsC_6H_5-CH_2-CH_2-)$	B: B Comp.SVol.3/3-151
$AsBC_{16}Cl_2H_{37}N_3$	$(i-C_3H_7)_2N-BCl-N(C_4H_9-t)-AsCl-N(C_3H_7-i)_2$	B: B Comp.SVol.4/4-63
$AsBC_{16}H_{34}N_2O_3$	$1,3,2-O_2AsC_2H_4-2-[N(C_4H_9-t)-B(OCH_3)-1-NC_5H_6-2,2,6,6-(CH_3)_4]$	B: B Comp.SVol.4/3b-52
$AsBC_{16}H_{36}N_2$	$n-C_4H_9-B[-N(C_4H_9-t)-As(C_4H_9-n)-N(C_4H_9-t)-]$	B: B Comp.SVol.4/3a-208
$AsBC_{16}H_{36}N_2O_3$	$(CH_3O)_2As-N(C_4H_9-t)-B(OCH_3)-1-NC_5H_6-2,2,6,6-(CH_3)_4$	B: B Comp.SVol.4/3b-52
$AsBC_{17}H_{36}N_2$	$1-[(t-C_4H_9)_2As-N=B]-2,2,6,6-(CH_3)_4-NC_5H_6$	B: B Comp.SVol.4/3a-160/5
$AsBC_{17}H_{37}N_3O_2$	$1-[1,3,2-O_2AsC_2H_4-2-N(C_4H_9-t)-B(N(CH_3)_2)]-2,2,6,6-(CH_3)_4-NC_5H_6$	B: B Comp.SVol.4/3a-155
$AsBC_{18}H_{15}I_3$	$(C_6H_5)_3As-BI_3$	B: B Comp.SVol.3/4-99
$AsBC_{18}H_{43}N_3$	$[(i-C_3H_7)_2N]_2B-N(C_4H_9-t)-As(CH_3)_2$	B: B Comp.SVol.4/3a-155
$AsBC_{19}FH_{31}N_2O_2$	$1-[1,3,2-O_2AsC_6H_4-2-N(C_4H_9-t)-BF]-NC_5H_6-2,2,6,6-(CH_3)_4$	B: B Comp.SVol.4/3b-246
$AsBC_{19}H_{45}N_5$	$1-[((CH_3)_2N)_2As-N(C_4H_9-t)-B(N(CH_3)_2)]-2,2,6,6-(CH_3)_4-NC_5H_6$	B: B Comp.SVol.4/3a-155
$AsBC_{20}F_4H_{30}$	$[((CH_3)_5C_5)_2As][BF_4]$	B: B Comp.SVol.3/3-252
$AsBC_{21}Cl_3H_{35}N_2$	$1-[2,6-(i-C_3H_7)_2C_6H_3-N(AsCl_2)-BCl]-2,2,6,6-(CH_3)_4-NC_5H_6$	B: B Comp.SVol.4/4-64
$AsBC_{21}F_3H_{35}N_2$	$2,2,6,6-(CH_3)_4NC_5H_6-1-BF-N(AsF_2)-C_6H_3(C_3H_7-i)_2-2,6$	B: B Comp.SVol.4/3b-246
$AsBC_{21}H_{29}NSi$	$1-(C_6H_5)_2As-4,5-(C_2H_5)_2-2,2,3-(CH_3)_3-1,2,5-NSiBC_2$	B: B Comp.SVol.4/3a-250/1
$AsBC_{24}H_{44}N_2O_3$	$2,6-(i-C_3H_7)_2-C_6H_3-N[As(OCH_3)_2]-B(OCH_3)-1-NC_5H_6-2,2,6,6-(CH_3)_4$	B: B Comp.SVol.4/3b-52
$AsBC_{27}FH_{39}N_2O_2$	$1-[1,3,2-O_2AsC_6H_4-2-N(C_6H_3-2,6-(C_3H_7-i)_2)-BF]-NC_5H_6-2,2,6,6-(CH_3)_4$	B: B Comp.SVol.4/3b-246
$AsBC_{27}H_{53}N_5$	$1-[((CH_3)_2N)_2As-N(C_6H_3-2,6-(C_3H_7-i)_2)-B(N(CH_3)_2)]-2,2,6,6-(CH_3)_4-NC_5H_6$	B: B Comp.SVol.4/3a-155

$AsBC_{33}F_4H_{28}MoN_2O_2$
$[(CH_2CHCH_2)Mo(CO)_2(NC_5H_4\text{-}2\text{-}(2\text{-}C_5H_4N))(As(C_6H_5)_3)][BF_4]$ .... Mo: Org.Comp.5-200, 202

$AsBC_{33}Fe_3H_{24}O_9$ $[As(C_6H_5)_4][(CO)_9(H)Fe_3(BH_3)]$ .... Fe: Org.Comp.C6a-299

$AsBC_{34}F_4H_{30}MoN_2O_2$
$[(CH_2C(CH_3)CH_2)Mo(CO)_2(NC_5H_4\text{-}2\text{-}(2\text{-}C_5H_4N))(As(C_6H_5)_3)][BF_4]$ .... Mo: Org.Comp.5-200, 204

$AsBC_{34}Fe_3H_{26}O_9$ $[As(C_6H_5)_4][(CO)_9(H)Fe_3(BH_2\text{-}CH_3)]$ .... Fe: Org.Comp.C6a-299

$AsBC_{50}FeH_{42}O_2$ .. $[(C_5H_5)Fe(CO)_2CH_2As(C_6H_5)_3][B(C_6H_5)_4]$ .... Fe: Org.Comp.B14-141

$AsB_{10}C_2Cl_2H_{11}$ ... $1,2\text{-}C_2B_{10}H_{11}\text{-}9\text{-}AsCl_2$ .... B: B Comp.SVol.3/4-202

– .... $1,7\text{-}C_2B_{10}H_{11}\text{-}9\text{-}AsCl_2$ .... B: B Comp.SVol.3/4-219

$AsB_{10}C_2H_{12}O_2$ ... $1,7\text{-}C_2B_{10}H_{11}\text{-}9\text{-}AsO_2H$ .... B: B Comp.SVol.3/4-219

$AsB_{10}C_3Cl_2H_{13}$ ... $1,7\text{-}C_2B_{10}H_{10}\text{-}1\text{-}CH_3\text{-}7\text{-}AsCl_2$ .... B: B Comp.SVol.3/4-219

$AsB_{10}C_3H_{13}O$ .... $1,7\text{-}C_2B_{10}H_{10}\text{-}1\text{-}CH_3\text{-}7\text{-}AsO$ .... B: B Comp.SVol.3/4-219

$AsB_{10}C_5Cl_2H_{15}$ ... $1,2\text{-}C_2B_{10}H_{10}\text{-}1\text{-}C(CH_3){=}CH_2\text{-}2\text{-}AsCl_2$ .... B: B Comp.SVol.3/4-202

$AsB_{10}C_5Cl_2H_{17}$ ... $1,2\text{-}C_2B_{10}H_{10}\text{-}1\text{-}CH(CH_3)_2\text{-}2\text{-}AsCl_2$ .... B: B Comp.SVol.3/4-202

$AsB_{10}C_5H_{17}$ .... $1,2\text{-}C_2B_{10}H_{10}\text{-}1,2\text{-}[\text{-}CH_2As(CH_3)CH_2\text{-}]$ .... B: B Comp.SVol.3/4-194

$AsB_{10}C_5H_{17}O$ .... $1,2\text{-}C_2B_{10}H_{10}\text{-}1\text{-}CH(CH_3)_2\text{-}2\text{-}AsO$ .... B: B Comp.SVol.3/4-202

$AsB_{10}C_5H_{18}Li$ .... $1,2\text{-}C_2B_{10}H_{10}\text{-}1\text{-}CH_2As(CH_3)_2\text{-}2\text{-}Li$ .... B: B Comp.SVol.3/4-194

$AsB_{10}C_5H_{19}O_3$ ... $1,2\text{-}C_2B_{10}H_{10}\text{-}1\text{-}CH(CH_3)_2\text{-}2\text{-}AsO_3H_2$ .... B: B Comp.SVol.3/4-202

$AsB_{10}C_{10}H_{19}$ .... $1,2\text{-}C_2B_{10}H_{10}\text{-}1,2\text{-}[\text{-}CH_2As(C_6H_5)CH_2\text{-}]$ .... B: B Comp.SVol.3/4-194

$AsB_{18}C_{28}CoH_{42}$ .. $[(C_6H_5)_4As][Co(1,2\text{-}C_2B_9H_{11})_2]$ .... B: B Comp.SVol.4/4-312

$AsB_{20}C_6ClH_{26}$ ... $(1,7\text{-}C_2B_{10}H_{10}\text{-}1\text{-}CH_3\text{-}7\text{-})_2AsCl$ .... B: B Comp.SVol.3/4-248

$AsB_{20}C_6H_{27}O_2$ ... $(1,7\text{-}C_2B_{10}H_{10}\text{-}1\text{-}CH_3\text{-}7\text{-})_2AsO_2H$ .... B: B Comp.SVol.3/4-248/9

$AsB_{20}C_{10}ClH_{34}$ ... $(1,2\text{-}C_2B_{10}H_{10}\text{-}1\text{-}CH(CH_3)_2\text{-}2\text{-})_2AsCl$ .... B: B Comp.SVol.3/4-249

$AsB_{20}C_{10}H_{35}O_2$ .. $(1,2\text{-}C_2B_{10}H_{10}\text{-}1\text{-}CH(CH_3)_2\text{-}2\text{-})_2AsO_2H$ .... B: B Comp.SVol.3/4-249

$AsB_{30}C_{12}H_{45}$ .... $(1,7\text{-}C_2B_{10}H_{10}\text{-}1\text{-}CH_3\text{-}7\text{-}CH_2)_3As$ .... B: B Comp.SVol.3/4-253

$AsBiC_{32}H_{38}O_2Sn$ $(C_6H_5)_2Bi\text{-}Sn(C_3H_7\text{-}i)_2\text{-}As(C_6H_3\text{-}OH\text{-}CH_3)_2$ ... Sn: Org.Comp.19-246/7

$AsBrC_5F_6NO_5ReS$ $[Re(CO)_5(NSBr)][AsF_6]$ .... Re: Org.Comp.2-154
S: S-N Comp.5-252

$AsBrC_5H_6O_3ReS$. $(CO)_3ReBr[As(S)(CH_3)_2]$ .... Re: Org.Comp.1-190

$AsBrC_6H_4MnO_3$ .. $Mn(O_3AsC_6H_4Br\text{-}4)$ .... Mn: MVol.D8-214/6

$AsBrC_6H_{11}MnO_7$. $[Mn(O_3AsC_6H_4\text{-}Br\text{-}4)(OH)(H_2O)_3]$ .... Mn: MVol.D8-217/8

$AsBrC_7H_{21}In$ .... $[As(CH_3)_4][(CH_3)_3InBr]$ .... In: Org.Comp.1-349/50, 353

$AsBrC_{10}H_{15}O_4Re$ $(CO)_4Re(Br)As(C_2H_5)_3$ .... Re: Org.Comp.1-457

$AsBrC_{12}H_{11}O_4Re$ cis-$(CO)_4Re(Br)As(CH_3)_2C_6H_5$ .... Re: Org.Comp.1-457

$AsBrC_{12}H_{20}MoO_2P^+$
$[(C_5H_5)Mo(CO)_2(P(CH_3)_3)(AsBr(CH_3)_2)]^+$ .... Mo: Org.Comp.7-283, 293

$AsBrC_{14}H_{22}O_3Sn$ $(C_4H_9)_2SnOAs(O)(C_6H_4Br\text{-}4)O$ .... Sn: Org.Comp.15-363, 364/5

$AsBrC_{16}H_{17}IMoO_2$ $(C_5H_5)Mo(CO)_2(I)[(CH_3)_2As\text{-}CH_2C_6H_4\text{-}2\text{-}Br]$ .. Mo: Org.Comp.7-56/7, 99

– .... $(C_5H_5)Mo(CO)_2(I)[(CH_3)_2As\text{-}C_6H_4\text{-}2\text{-}CH_2Br]$ .. Mo: Org.Comp.7-57, 98/9

$AsBrC_{17}H_{20}MoO_3$ $(C_5H_5)Mo(CO)_2(Br)[(CH_3)_2As\text{-}C_6H_4\text{-}2\text{-}CH_2\text{-}O\text{-}CH_3]$
Mo: Org.Comp.7-56/7, 98

$AsBrC_{19}H_{38}MoO_2Si_3$
$(C_5H_5)Mo(CO)_2(Br)[As(CH_2Si(CH_3)_3)_3]$ .... Mo: Org.Comp.7-57, 96

$AsBrC_{20}H_{18}O_3Sn$ $(C_6H_5CH_2)_2Sn[\text{-}OAs(O)(C_6H_4Br\text{-}4)O\text{-}]$ .... Sn: Org.Comp.16-81/2

$AsBrC_{20}H_{27}MoO_3PS$
$[(C_5H_5)Mo(CO)_2(P(CH_3)_3)((CH_3)_2As\text{-}S\text{-}CH_2C(O)C_6H_5)]Br$
Mo: Org.Comp.7-283, 294

$AsBrC_{22}H_{15}O_4Re$ $(CO)_4Re(Br)As(C_6H_5)_3$ .... Re: Org.Comp.1-457

Formula	Compound	Reference
$AsBrC_{22}H_{38}O_3Sn$	$(C_8H_{17})_2Sn[-OAs(O)(C_6H_4Br-4)O-]$	Sn: Org.Comp.16-59
$AsBrC_{24}H_{20}O_4$	$[(C_6H_5)_4As][BrO_4]$	Br: SVol.B2-162
$AsBrC_{25}H_{20}MoO_2$	$(C_5H_5)Mo(CO)_2(Br)[As(C_6H_5)_3]$	Mo: Org.Comp.7-56/7, 101
$AsBrC_{28}H_{20}NO_4Re$	$(CO)_3ReBr[-O=C(CH=As(C_6H_5)_3)-(2,1-NC_5H_4)-]$	Re: Org.Comp.1-202
$AsBrC_{28}H_{26}MoO_2$	$(C_5H_5)Mo(CO)_2(Br)[As(CH_2C_6H_5)_3]$	Mo: Org.Comp.7-56/7, 96/7
$AsBrC_{36}Cl_3H_{30}Sb$	$[As(C_6H_5)_4][(C_6H_5)_2Sb(Cl_3)Br]$	Sb: Org.Comp.5-157
$AsBrF_6O_2$	$[BrO_2][AsF_6]$	Br: SVol.B2-87/8
$AsBrF_8$	$[BrF_2][AsF_6]$	Br: SVol.B3-41/5
$AsBrF_8O$	$[BrOF_2][AsF_6]$	Br: SVol.B3-155/6
$AsBrF_{10}OXe$	$[F-Xe(-F-)BrOF_2][AsF_6]$	Br: SVol.B3-165
$AsBrF_{12}$	$[BrF_6][AsF_6]$	Br: SVol.B3-138/40
$AsBrH_4$	$HBr \cdot AsH_3$	Br: SVol.B1-372/7
$AsBrH_4MnO_4^-$	$[MnO_3(AsH_3)(OH)Br]^-$	Mn: MVol.D8-202
$AsBr_2C_6H_{18}In$	$[As(CH_3)_4][(CH_3)_2In(Br)_2]$	In: Org.Comp.1-349/50, 353, 359/60
$AsBr_2C_8H_{12}MnN$	$Mn[(NH_2-C_6H_4-As(CH_3)_2-2)Br_2]$	Mn: MVol.D8-204
$AsBr_2C_8H_{21}N_2PtS$	$[(As(C_2H_5)_3)PtBr_2(CH_3N=S=NCH_3)]$	S: S-N Comp.7-328/9
$AsBr_2C_{12}H_{20}MoO_2P$	$[(C_5H_5)Mo(CO)_2(P(CH_3)_3)(AsBr(CH_3)_2)]Br$	Mo: Org.Comp.7-283, 293
$AsBr_2C_{14}H_{15}MnN_2$	$[Mn(NC_5H_4-2-CH=N-C_6H_4-2-As(CH_3)_2)Br_2]$	Mn: MVol.D6-79/80
$AsBr_2C_{15}H_{17}MnN_2$	$[Mn(6-CH_3-NC_5H_3-2-CH=N-C_6H_4-2-As(CH_3)_2)Br_2]$	Mn: MVol.D6-79/80
$AsBr_2C_{16}H_{17}MoO_2$	$(C_5H_5)Mo(CO)_2(Br)[(CH_3)_2As-C_6H_4-2-CH_2Br]$	Mo: Org.Comp.7-57, 98
$AsBr_2C_{16}H_{19}MnN_2$	$[Mn(NC_5H_4-2-CH=N-C_6H_4-2-As(C_2H_5)_2)Br_2]$	Mn: MVol.D6-79/80
$AsBr_2C_{17}H_{21}MnN_2$	$[Mn(6-CH_3-NC_5H_3-2-CH=N-C_6H_4-2-As(C_2H_5)_2)Br_2] \cdot 0.5\ C_6H_6$	Mn: MVol.D6-79/80
$AsBr_2C_{20}H_{24}MnN_2$	$[Mn(6-CH_3-NC_5H_3-2-CH=N-C_6H_4-2-As(C_2H_5)_2)Br_2] \cdot 0.5\ C_6H_6$	Mn: MVol.D6-79/80
$AsBr_2C_{23}H_{20}MoNO$	$C_5H_5Mo(NO)(As(C_6H_5)_3)Br_2$	Mo: Org.Comp.6-33
$AsBr_3C_5H_{15}In$	$[As(CH_3)_4][CH_3In(Br)_3]$	In: Org.Comp.1-349/50, 354
$AsBr_4C_{27}H_{15}O_5Re$	$(C_6H_5)_3As-Re(CO)_3[-O-1,2-(C_6Br_4-3,4,5,6)-O-]$, radical	Re: Org.Comp.1-148
$AsBr_4C_{36}H_{30}Sb$	$[As(C_6H_5)_4][(C_6H_5)_2SbBr_4]$	Sb: Org.Comp.5-175/6
$AsBr_5C_{30}H_{25}Sb$	$[As(C_6H_5)_4][(C_6H_5)SbBr_5]$	Sb: Org.Comp.5-262
$AsBr_6C_{12}Ga_3H_{24}O_3$	$[(OC_4H_8)(Br)_2Ga]_3As$	Ga: SVol.D1-112/3
$AsCCl_2F_6NOS$	$[Cl_2SN=C=O][AsF_6]$	S: S-N Comp.8-128
$AsCF_7N_2S$	$F_2S=N-CN \cdot AsF_5$	S: S-N Comp.8-66/7
$AsCF_8NOS$	$[F_2SN=C=O][AsF_6]$	S: S-N Comp.8-78
$AsCF_{10}H_2N$	$[CF_3-NFH_2][AsF_6]$	F: PFHOrg.SVol.6-1, 14
$AsCH_3Mo_6O_{21}^{2-}$	$(CH_3AsO_3)Mo_6O_{18}(H_2O)_6^{2-}$	Mo: SVol.B3b-126/7

$AsCH_4Mo_7O_{27}^{7-}$ $(CH_3AsO_3)Mo_7O_{24}H^{7-}$ Mo:SVol.B3b-125, 130/2
$AsCH_{15}Mo_6O_{27}^{2-}$ $(CH_3AsO_3)Mo_6O_{18}(H_2O)_6^{2-}$ Mo:SVol.B3b-126/7
$AsC_2F_6H_6NOS$ ... $[OSN(CH_3)_2][AsF_6]$ .... S: S-N Comp.6-289, 292
$AsC_2F_8H_6NS$..... $[F_2SN(CH_3)_2][AsF_6]$ .... S: S-N Comp.8-74/5
$AsC_2F_9H_3NOS$ ... $[OSN(CH_3)CF_3][AsF_6]$..................... S: S-N Comp.6-289, 297
$AsC_2F_{11}H_3NS$.... $[F_2SN(CH_3)CF_3][AsF_6]$ .... S: S-N Comp.8-76
$AsC_2F_{12}H_2N$ ..... $[(CF_3)_2NH_2][AsF_6]$ .... F: PFHOrg.SVol.5-12/3, 14, 43
$AsC_2F_{12}H_2NO$.... $[CF_3-NH_2-O-CF_3][AsF_6]$ .... F: PFHOrg.SVol.5-17, 49
$AsC_2H_7Mo_4O_{15}^{2-}$ $((CH_3)_2AsO_2)Mo_4O_{12}(OH)^{2-}$................ Mo:SVol.B3b-122/3, 132, 247
AsC_3ClF_9NO..... $(CF_3)_2N-O-AsCl-CF_3$..................... F: PFHOrg.SVol.5-118/9, 130
$AsC_3ClF_{12}N_2S_2$... $[(CF_3)_2C=NSNSCl][AsF_6]$ .... F: PFHOrg.SVol.6-48
$AsC_3F_6H_9O_2SSn$.. $(CH_3)_3SnAsF_6 \cdot SO_2$ .... Sn: Org.Comp.19-229, 232
$AsC_3F_6H_9Sn$..... $(CH_3)_3SnAsF_6$........................... Sn: Org.Comp.19-229, 232
$AsC_3F_9H_6N_2S$.... $[(CH_3)_2NS=NCF_3][AsF_6]$.................. S: S-N Comp.7-333
$AsC_3F_{10}NO$...... $(CF_3)_2N-O-AsF-CF_3$ .... F: PFHOrg.SVol.5-118/9, 130
$AsC_3H_{11}Sn$...... $(CH_3)_3Sn-AsH_2$.......................... Sn: Org.Comp.19-229, 233
$AsC_3H_{27}Mo_6N_2O_{27}$
$(CH_6N)_2[(CH_3AsO_3)Mo_6O_{18}(H_2O)_6] \cdot 6\ H_2O$... Mo:SVol.B3b-127
$AsC_4F_3H_{10}Sn$.... $(CH_3)_3Sn-AsH-CF_3$ .... Sn: Org.Comp.19-229, 233
$AsC_4F_7H_{12}N_2S$... $[(CH_3)_2NS(F)N(CH_3)_2][AsF_6]$ .... S: S-N Comp.8-182
$AsC_4F_{12}NO$...... $(CF_3)_2N-O-As(CF_3)_2$ .... F: PFHOrg.SVol.5-118/9, 130, 133
$AsC_4F_{16}H_2NO$.... $[CF_3-NH_2-O-C_3F_7-i][AsF_6]$................. F: PFHOrg.SVol.5-17, 49
$AsC_4H_9N_2O_2S_2$... $(O=S=N)_2AsC_4H_9-t$....................... S: S-N Comp.6-80
$AsC_4H_{11}Mo_4O_{15}^{2-}$
$((C_2H_5)_2AsO_2)Mo_4O_{12}(OH)^{2-}$............... Mo:SVol.B3b-122/3
$AsC_4H_{12}In$....... $(CH_3)_2In-As(CH_3)_2$....................... In: Org.Comp.1-312, 316, 319/21
$AsC_4H_{12}InO_2$ .... $(CH_3)_2In-OAs(O)(CH_3)_2$ .... In: Org.Comp.1-211, 215/6
$AsC_4H_{14}In$....... $(CH_3)_3In \cdot CH_3-AsH_2$..................... In: Org.Comp.1-27, 40
$AsC_4H_{19}Mo_4N_6O_{15}$
$(CH_6N_3)_2[((CH_3)_2AsO_2)Mo_4O_{12}(OH)] \cdot H_2O$... Mo:SVol.B3b-123
$AsC_5ClF_6NO_5ReS$ $[Re(CO)_5(NSCl)][AsF_6]$.................... Re: Org.Comp.2-153/4
S: S-N Comp.5-252
$AsC_5ClH_6O_3ReS$.. $(CO)_3ReCl[As(=S)(CH_3)_2]$.................. Re: Org.Comp.1-189
$AsC_5Cl_3H_{15}In$.... $[As(CH_3)_4][CH_3-In(Cl)_3]$................... In: Org.Comp.1-349/50, 352, 357/8
– .... $[CD_3-As(CH_3)_3][CD_3-In(Cl)_3]$............... In: Org.Comp.1-349/50, 352
$AsC_5F_6HMnNO_6S$ $[Mn(CO)_5(O=S=NH)][AsF_6]$.................. S: S-N Comp.6-251
$AsC_5F_6HNO_6ReS$ $[(CO)_5ReN(H)SO][AsF_6]$ .... Re: Org.Comp.2-153
S: S-N Comp.6-252
$AsC_5F_6H_2O_6Re$... $[(CO)_5ReOH_2][AsF_6]$...................... Re: Org.Comp.2-156/7, 159/60
$AsC_5F_6H_9Sn$..... $(CH_3)_3Sn-As(CF_3)_2$....................... Sn: Org.Comp.19-231, 235/6
$AsC_5F_6INO_5ReS$.. $[Re(CO)_5(NSI)][AsF_6]$ .... S: S-N Comp.5-252
$AsC_5F_6O_5Re$..... $(CO)_5ReFAsF_5$ .... Re: Org.Comp.2-34
$AsC_5F_6O_7ReS$.... $[(CO)_5ReSO_2][AsF_6]$...................... Re: Org.Comp.2-158, 160
$AsC_5F_7NO_5ReS$ .. $[Re(CO)_5(NSF)][AsF_6]$ .... Re: Org.Comp.2-153, 159
S: S-N Comp.5-245/6
$AsC_5F_8HNO_6ReS$ $[(CO)_5ReN(H)SOF_2][AsF_6]$ .... Re: Org.Comp.2-153

Formula	Compound	Reference
$AsC_5F_9NO_5ReS$	$[(CO)_5ReNSF_3][AsF_6]$	Re: Org.Comp.2-153
$AsC_5F_{10}N$	$NCFCFCFCFCF \cdot AsF_5$	F: PFHOrg.SVol.4-85, 89
$AsC_5F_{11}HN$	$[NHCFCFCFCFCF][AsF_6]$	F: PFHOrg.SVol.4-85, 90
$AsC_5F_{11}N$	$(C_5F_5N)(AsF_6)$	F: PFHOrg.SVol.4-85, 90
$AsC_5F_{12}N$	$NCF_2CF_2CFCFCF \cdot AsF_5$	F: PFHOrg.SVol.4-85
$AsC_5F_{15}N_2O_2$	$[(CF_3)_2N-O]_2As-CF_3$	F: PFHOrg.SVol.5-118/9
$AsC_5H_6IO_3ReS$	$(CO)_3ReI[As(S)(CH_3)_2]$	Re: Org.Comp.1-191
$AsC_5H_{15}Sn$	$(CH_3)_3Sn-As(CH_3)_2$	Sn: Org.Comp.19-228/30, 233/5
$AsC_6ClF_{18}N_2O_2$	$[(CF_3)_2N-O]_2AsCl(CF_3)_2$	F: PFHOrg.SVol.5-118/9, 131/2
$AsC_6ClH_4MnO_3$	$Mn(O_3AsC_6H_4Cl-4) \cdot 0.5\ H_2O$	Mn: MVol.D8-214/6
$AsC_6ClH_{11}MnO_7$	$[Mn(O_3AsC_6H_4-Cl-4)(OH)(H_2O)_3] \cdot H_2O$	Mn: MVol.D8-217/8
$AsC_6Cl_2H_{18}In$	$[As(CH_3)_4][(CH_3)_2In(Cl)_2]$	In: Org.Comp.1-349/50, 351
–	$[As(CH_3)_4][(CD_3)_2In(Cl)_2]$	In: Org.Comp.1-349/50, 352
$AsC_6Cl_5H_6O_3Sb^-$	$[(4-(HO)_2(O)AsC_6H_4)SbCl_5]^-$	Sb: Org.Comp.5-255
$AsC_6F_6H_4O_6Re$	$[(CO)_5ReO(H)CH_3][AsF_6]$	Re: Org.Comp.2-157
$AsC_6F_6O_6Re$	$[Re(CO)_6][AsF_6]$	Re: Org.Comp.2-217, 224
$AsC_6F_8MnN_2O_5S$	$[(F_2S=N-CN)Mn(CO)_5][AsF_6]$	S: S-N Comp.8-67
$AsC_6F_8N_2O_5ReS$	$[(CO)_5ReNCNSF_2][AsF_6]$	Re: Org.Comp.2-154 S: S-N Comp.8-67
$AsC_6F_8N_2O_6ReS$	$[(CO)_5ReNCNSOF_2][AsF_6]$	Re: Org.Comp.2-154
$AsC_6F_{18}N_3O_3$	$[(CF_3)_2N-O]_3As$	F: PFHOrg.SVol.5-118/9, 131, 133, 145/6
$AsC_6F_{19}N_2O_2$	$[(CF_3)_2N-O]_2AsF(CF_3)_2$	F: PFHOrg.SVol.5-118/9, 130, 132
$AsC_6GeH_{15}N_2$	$Ge(CH_3)_3C[As(CH_3)_2]N_2$	Ge: Org.Comp.1-146
$AsC_6H_4MnNO_5$	$Mn(2-NO_2-C_6H_4-AsO_3) \cdot H_2O$	Mn: MVol.D8-214/6
$AsC_6H_4MnNO_6$	$Mn(4-HO-3-NO_2-C_6H_3-AsO_3) \cdot H_2O$	Mn: MVol.D8-214/6
$AsC_6H_5MnO_3$	$Mn(C_6H_5-AsO_3)$	Mn: MVol.D8-214/6
–	$Mn(C_6H_5-AsO_3) \cdot H_2O$	Mn: MVol.D8-214/6
$AsC_6H_5Mo_7O_{25}{}^{4-}$	$[(C_6H_5-AsO_3)(MoO_4)Mo_6O_{18}]^{4-}$	Mo: SVol.B3b-126, 130
$AsC_6H_5Mo_{11}O_{39}P^{5-}$	$[PMo_{11}As(C_6H_5)O_{39}]^{5-}$	Mo: SVol.B3b-128/9
$AsC_6H_5Mo_{11}O_{39}Si^{6-}$	$[SiMo_{11}As(C_6H_5)O_{39}]^{6-}$	Mo: SVol.B3b-128/9
$AsC_6H_5N_2O_2S_2$	$(O=S=N)_2AsC_6H_5$	S: S-N Comp.6-80
$AsC_6H_6Mo_7O_{27}{}^{7-}$	$[(C_6H_5-AsO_3)Mo_7O_{24}H]^{7-}$	Mo: SVol.B3b-125, 130/2
$AsC_6H_6O_4ReS_2$	$(CO)_4Re[-S-As(CH_3)_2-S-]$	Re: Org.Comp.1-364
$AsC_6H_7Mo_7NO_{27}{}^{7-}$	$[(NH_2-C_6H_4-AsO_3)Mo_7O_{24}H]^{7-}$	Mo: SVol.B3b-125, 130/2
$AsC_6H_8O_6Sb$	$4-[(HO)_2As(O)]-C_6H_4-Sb(O)(OH)_2$	Sb: Org.Comp.5-293
$AsC_6H_8O_7Sb$	$2-HO-5-[(HO)_2As(O)]-C_6H_3-Sb(O)(OH)_2$	Sb: Org.Comp.5-296
$AsC_6H_{11}MnNO_9$	$[Mn(O_3As-C_6H_4-NO_2-2)(OH)(H_2O)_3] \cdot 2\ H_2O$	Mn: MVol.D8-217/8
–	$[Mn(O_3As-C_6H_4-NO_2-3)(OH)(H_2O)_3]$	Mn: MVol.D8-217/8
–	$[Mn(O_3As-C_6H_4-NO_2-4)(OH)(H_2O)_3]$	Mn: MVol.D8-217/8
$AsC_6H_{12}MnO_7$	$[Mn(O_3As-C_6H_5)(OH)(H_2O)_3]$	Mn: MVol.D8-217/8
$AsC_6H_{17}Mo_6O_{27}{}^{2-}$	$[(C_6H_5-AsO_3)Mo_6O_{18}(H_2O)_6]^{2-}$	Mo: SVol.B3b-126/7
$AsC_6H_{18}I_2In$	$[As(CH_3)_4][(CH_3)_2In(I)_2]$	In: Org.Comp.1-349/50, 355

Formula	Compound	Reference
$AsC_6H_{18}In$	$(CH_3)_3In \cdot As(CH_3)_3$	In: Org.Comp.1-27, 40, 46
–	$(C_2H_5)_3In \cdot AsH_3$	In: Org.Comp.1-71
$AsC_7ClH_{21}In$	$[As(CH_3)_4][(CH_3)_3InCl]$	In: Org.Comp.1-349/51, 356
–	$[As(CH_3)_4][(CD_3)_3InCl]$	In: Org.Comp.1-349/50, 351
$AsC_7F_6H_5MnNO_2S$	$[(C_5H_5)Mn(NS)(CO)_2][AsF_6]$	S: S-N Comp.5-50, 60
$AsC_7F_6H_6N_2O_5ReS$	$[(CO)_5ReNSN(CH_3)_2][AsF_6]$	Re: Org.Comp.2-154, 159 S: S-N Comp.5-255/6
$AsC_7F_6O_5Re$	$(CO)_5ReAs(CF_3)_2$	Re: Org.Comp.2-18, 20
$AsC_7F_{12}MnN_2O_6S$	$[Mn(CO)_5(NS\text{-}ON(CF_3)_2)][AsF_6]$	S: S-N Comp.5-254
$AsC_7F_{12}N_2O_6ReS$	$[(CO)_5ReNSON(CF_3)_2][AsF_6]$	Re: Org.Comp.2-154 S: S-N Comp.5-254
$AsC_7GeH_{21}Sn$	$(CH_3)_3Sn\text{-}As(CH_3)\text{-}Ge(CH_3)_3$	Sn: Org.Comp.19-229, 232, 236/7
$AsC_7H_5MnO_5$	$Mn(O_3AsC_6H_4COOH\text{-}2) \cdot H_2O$	Mn: MVol.D8-214/6
$AsC_7H_6O_5Re$	$(CO)_5ReAs(CH_3)_2$	Re: Org.Comp.2-18, 20
$AsC_7H_7MnO_3$	$Mn(O_3AsC_6H_4CH_3\text{-}4)$	Mn: MVol.D8-214/6
$AsC_7H_7MnO_4$	$Mn(O_3As\text{-}C_6H_4\text{-}OCH_3\text{-}2) \cdot H_2O$	Mn: MVol.D8-214/6
–	$Mn(O_3As\text{-}C_6H_4\text{-}OCH_3\text{-}4)$	Mn: MVol.D8-214/6
$AsC_7H_{14}MnO_8$	$[Mn(O_3As\text{-}C_6H_4\text{-}OCH_3\text{-}2)(OH)(H_2O)_3]$	Mn: MVol.D8-217/8
–	$[Mn(O_3As\text{-}C_6H_4\text{-}OCH_3\text{-}4)(OH)(H_2O)_3]$	Mn: MVol.D8-217/8
$AsC_7H_{21}IIn$	$[As(CH_3)_4][(CH_3)_3InI]$	In: Org.Comp.1-349/50, 354
$AsC_7H_{23}O_5Sn$	$(n\text{-}_4H_9)_2Sn[\text{-}O\text{-}As(O)(C_6H_4COOH\text{-}2)\text{-}O\text{-}]$	Sn: Org.Comp.15-364, 366
$AsC_8Cl_2H_{21}N_2PtS$	$[((C_2H_5)_3As)PtCl_2(CH_3N{=}S{=}NCH_3)]$	S: S-N Comp.7-316/7, 321/2
$AsC_8F_6H_6O_6Re$	$[(CO)_5ReOC(CH_3)_2][AsF_6]$	Re: Org.Comp.2-157
$AsC_8F_{24}N_3O_3$	$[(CF_3)_2N\text{-}O]_3As(CF_3)_2$	F: PFHOrg.SVol.5-118/9, 131, 133
AsC_8GeH_{13}	$Ge(CH_3)_3C_5H_4As\text{-}4$	Ge: Org.Comp.2-104, 107
$AsC_8H_{15}NiO_3Sn$	$(CH_3)_3Sn\text{-}As(CH_3)_2Ni(CO)_3$	Sn: Org.Comp.19-229/30
$AsC_8H_{18}KN_2S$	$K[(t\text{-}C_4H_9)_2AsN{=}S{=}N]$	S: S-N Comp.7-115
$AsC_8H_{18}N_2S^-$	$[(t\text{-}C_4H_9)_2AsN{=}S{=}N]^-$	S: S-N Comp.7-115
$AsC_8H_{21}I_2N_2PtS$	$[(As(C_2H_5)_3)PtI_2(CH_3N{=}S{=}NCH_3)]$	S: S-N Comp.7-328/9
$AsC_8H_{24}In$	$[As(CH_3)_4][In(CH_3)_4]$	In: Org.Comp.1-337, 339, 342/3
$AsC_9ClH_{10}MnO_2$	$Mn[OC(O)\text{-}C_6H_4\text{-}As(CH_3)_2\text{-}2]Cl \cdot 0.5\ H_2O$	Mn: MVol.D8-203/4
$AsC_9ClH_{12}MoO_3$	$(C_5H_5)Mo(CO)_2(Cl)[HO\text{-}As(CH_3)_2]$	Mo: Org.Comp.7-57, 99
$AsC_9ClH_{14}N_2SSi$	$(CH_3)_3SiN{=}S{=}NAsCl\text{-}C_6H_5$	S: S-N Comp.7-152
$AsC_9ClH_{14}N_4S_3Si$	$(CH_3)_3SiN{=}S{=}NSN{=}S{=}NAsCl\text{-}C_6H_5$	S: S-N Comp.7-149
$AsC_9Cl_2H_{11}MoO_2$	$(C_5H_5)Mo(CO)_2(Cl)[ClAs(CH_3)_2]$	Mo: Org.Comp.7-56/7, 99
$AsC_9CrH_9N_2O_7S_2$	$(CO)_5Cr(t\text{-}C_4H_9As(N{=}S{=}O)_2)$	S: S-N Comp.6-90/1
$AsC_9F_6H_9O_4PRe$	$cis\text{-}(CO)_4Re[P(CH_3)_3]As(CF_3)_2$	Re: Org.Comp.1-439
$AsC_9F_6H_{12}NS$	$[SN][AsF_6] \cdot 1,3,5\text{-}(CH_3)_3C_6H_3$	S: S-N Comp.5-47/9
$AsC_9F_6H_{12}N_2O_5ReSSi$	$[(CO)_5ReNSN(CH_3)Si(CH_3)_3][AsF_6]$	Re: Org.Comp.2-154
$AsC_9F_6H_{18}N_2O_4Re$	$[(CO)_3Re(H_2O)(N(CH_3)_2C_2H_4\text{-}N(CH_3)_2)][AsF_6]$	Re: Org.Comp.1-300, 304, 305
$AsC_9Ge_2H_{27}Sn$	$(CH_3)_3Sn\text{-}As[Ge(CH_3)_3]_2$	Sn: Org.Comp.19-229, 232, 236/7
$AsC_9H_9MoO_2S_2$	$(C_5H_5)Mo(CO)_2[\text{-}1,2\text{-}(1,3,2\text{-}S_2AsC_2H_4)\text{-}]$	Mo: Org.Comp.7-213, 243, 244
$AsC_9H_{11}MoN_2O_2$	$(C_5H_5)Mo(CO)_2(N_2)As(CH_3)_2$	Mo: Org.Comp.7-133

Formula	Compound	Reference
$AsC_9H_{11}MoO_2$	$(C_5H_5)Mo(CO)_2As(CH_3)_2$	Mo:Org.Comp.7-2, 36
$AsC_9H_{15}Sn$	$(CH_3)_3Sn-AsH-C_6H_5$	Sn: Org.Comp.19-229, 233
$AsC_9H_{16}In$	$(CH_3)_3In \cdot C_6H_5-AsH_2$	In: Org.Comp.1-27, 40
$AsC_9H_{21}O_3Sn$	$(C_4H_9)_2SnOAs(O)(CH_3)O$	Sn: Org.Comp.15-363, 364/5
$AsC_9H_{24}In$	$(C_2H_5)_3In \cdot As(CH_3)_3$	In: Org.Comp.1-71
$AsC_{10}ClH_{15}O_4Re$	cis-$(CO)_4Re(Cl)As(C_2H_5)_3$	Re: Org.Comp.1-456
$AsC_{10}Cl_2H_{17}N_2PdS$	$[(C_6H_5-As(CH_3)_2)PdCl_2(CH_3N=S=NCH_3)]$	S: S-N Comp.7-309/11
$AsC_{10}Cl_2H_{25}N_2PtS$	$[((C_2H_5)_3As)PtCl_2(C_2H_5-N=S=N-C_2H_5)]$	S: S-N Comp.7-316/7, 319, 321/2
$AsC_{10}CrH_{15}O_5Sn$	$(CH_3)_3Sn-As(CH_3)_2Cr(CO)_5$	Sn: Org.Comp.19-229/30, 235
$AsC_{10}F_6H_8NS$	$[SN][AsF_6] \cdot C_{10}H_8$	S: S-N Comp.5-47/9
$AsC_{10}FeH_6O_8Re$	$(CO)_4Re[-As(CH_3)_2-Fe(CO)_4-]$	Re: Org.Comp.1-500
$AsC_{10}Fe_3HO_9$	$(CO)_9Fe_3AsC-H$	Fe: Org.Comp.C6a-266/7, 268/9
$AsC_{10}Fe_3H_3O_9S$	$(CO)_9Fe_3SAs-CH_3$	Fe: Org.Comp.C6a-202, 205
$AsC_{10}H_{11}NO_3ReS_2$	$C_5H_5N-1-Re(CO)_3[-S-As(CH_3)_2=S-]$	Re: Org.Comp.1-129
$AsC_{10}H_{15}MoO_5Sn$	$(CH_3)_3Sn-As(CH_3)_2Mo(CO)_5$	Sn: Org.Comp.19-230, 235
$AsC_{10}H_{15}O_4Re$	$(CO)_4Re[As(C_2H_5)_3]$, radical	Re: Org.Comp.1-476/8
$AsC_{10}H_{15}O_5SnW$	$(CH_3)_3Sn-As(CH_3)_2W(CO)_5$	Sn: Org.Comp.19-229/30
$AsC_{10}H_{24}In$	$(CH_3)_2In-As(C_4H_9-t)_2$	In: Org.Comp.1-312, 316, 322
$AsC_{11}Cl_5H_{12}NO_3Sb$	$[C_5H_6N][H_2AsO_3-C_6H_4-SbCl_5] \cdot 1.33\ H_2O$	Sb: Org.Comp.5-255
$AsC_{11}CrH_5N_2O_7S_2$	$(CO)_5Cr(C_6H_5As(N=S=O)_2)$	S: S-N Comp.6-90/1
$AsC_{11}F_6FeH_{11}$	$[(C_6H_6)Fe(C_5H_5)][AsF_6]$	Fe: Org.Comp.B18-142/6, 150/1, 154, 157, 174/81
$AsC_{11}F_6H_6O_5Re$	$(CO)_4Re[-As(CH_3)_2-C(CF_3)=C(CF_3)-C(=O)-]$	Re: Org.Comp.1-417, 422
$AsC_{11}F_6H_{15}N_2Sn$	$(CH_3)_3SnN=C=C(CN)C(CF_3)_2As(CH_3)_2$	Sn: Org.Comp.18-107, 109
$AsC_{11}F_6H_{18}O_3P_2Re$	fac-$(CO)_3Re[P(CH_3)_3]_2As(CF_3)_2$	Re: Org.Comp.1-255
$AsC_{11}FeH_6O_9Re$	$(CO)_5ReAs(CH_3)_2-Fe(CO)_4$	Re: Org.Comp.2-176
$AsC_{11}FeH_{13}$	$C_5H_5FeC_4H_2As((CH_3)_2-2,5)$	Fe: Org.Comp.B17-212
$AsC_{11}H_{13}N_2PdS_3$	$Pd(SCN)_2[CH_3S-2-C_6H_4-As(CH_3)_2]$	Pd: SVol.B2-350
$AsC_{11}H_{13}N_3O_3Re$	$(CO)_3Re(N_2)[As(CH_3)_2C_6H_5]NH_2$	Re: Org.Comp.1-291
$AsC_{11}H_{24}O_3P_2Re$	$(CO)_3Re[P(CH_3)_3]_2As(CH_3)_2$	Re: Org.Comp.1-254
$AsC_{11}H_{24}O_3P_2ReS$	fac-$(CO)_3Re[P(CH_3)_3]_2-S-As(CH_3)_2$	Re: Org.Comp.1-258
–	mer-trans-$(CO)_3Re[P(CH_3)_3]_2-S-As(CH_3)_2$	Re: Org.Comp.1-258
$AsC_{11}H_{24}O_3P_2ReSSe$	fac-$(CO)_3Re[P(CH_3)_3]_2-S-As(Se)(CH_3)_2$	Re: Org.Comp.1-259
$AsC_{11}H_{24}O_3P_2ReS_2$	fac-$(CO)_3Re[P(CH_3)_3]_2-S-As(S)(CH_3)_2$	Re: Org.Comp.1-259
–	mer-trans-$(CO)_3Re[P(CH_3)_3]_2-S-As(S)(CH_3)_2$	Re: Org.Comp.1-258
$AsC_{11}H_{27}Sn$	$(CH_3)_3Sn-As(C_4H_9-t)_2$	Sn: Org.Comp.19-229/30
$AsC_{11}H_{28}NSn$	$(CH_3)_3SnNHAs(C_4H_9-t)_2$	Sn: Org.Comp.18-20, 22
$AsC_{12}ClH_{11}O_4Re$	cis-$(CO)_4Re(Cl)As(CH_3)_2C_6H_5$	Re: Org.Comp.1-456
$AsC_{12}ClH_{17}IMoO_2$	$(C_5H_5)Mo(CO)_2(I)[(CH_3)_2As-CH_2CH_2CH_2-Cl]$	Mo:Org.Comp.7-56/7, 97
$AsC_{12}ClH_{20}N_2SSi$	$(CH_3)_3SiN=S=NAsCl-C_6H_2(CH_3)_3-2,4,6$	S: S-N Comp.7-152

$AsC_{12}ClH_{20}N_4S_3Si$
$(CH_3)_3SiN{=}S{=}NSN{=}S{=}NAsCl{-}C_6H_2(CH_3)_3{-}2,4,6$ S: S-N Comp.7-149
$AsC_{12}Cl_2H_{21}N_2PdS$
$[(C_6H_5{-}As(CH_3)_2)PdCl_2(C_2H_5{-}N{=}S{=}N{-}C_2H_5)]$ . . S: S-N Comp.7-309/11
$AsC_{12}Cl_2H_{29}N_2PtS$
$[((C_2H_5)_3As)PtCl_2(i{-}C_3H_7{-}N{=}S{=}N{-}C_3H_7{-}i)]$. . . . S: S-N Comp.7-316/7, 319, 321/2
$AsC_{12}CrH_6O_{10}Re$ $(CO)_5ReAs(CH_3)_2{-}Cr(CO)_5$ . . . . Re: Org.Comp.2-174
$AsC_{12}F_6H_{16}MoO_2P$
$[(C_5H_5)Mo(CO)_2(CH_2{=}CHCH_2As(CH_3)_2)][PF_6]$ . . Mo: Org.Comp.8-185, 188
$AsC_{12}F_6H_{18}NS$ . . . $[SN][AsF_6] \cdot (CH_3)_6C_6$ . . . . S: S-N Comp.5-47/9
$AsC_{12}F_{12}MnO_8PRe$
$(CO)_4Re[{-}P(CF_3)_2{-}Mn(CO)_4{-}As(CF_3)_2{-}]$ . . . . . . Re: Org.Comp.1-487
$AsC_{12}FeH_{24}N_2O_6P_2Re$
$(CO)_3Re[P(CH_3)_3]_2{-}As(CH_3)_2{-}Fe(NO)_2(CO)$ . . . Re: Org.Comp.1-316
$AsC_{12}FeH_{27}N_2O_3Sn$
$(CH_3)_3Sn{-}As(C_4H_9{-}t)_2Fe(CO)(NO)_2$. . . . . Sn: Org.Comp.19-229, 231
$AsC_{12}GeH_{23}Sn$ . . . $(CH_3)_3Sn{-}As(C_6H_5){-}Ge(CH_3)_3$. . . . . Sn: Org.Comp.19-229, 232, 236/7
$AsC_{12}H_6MoO_{10}Re$ $(CO)_5ReAs(CH_3)_2{-}Mo(CO)_5$. . . . . Re: Org.Comp.2-175
$AsC_{12}H_6O_{10}ReW$ $(CO)_5ReAs(CH_3)_2{-}W(CO)_5$. . . . . Re: Org.Comp.2-175
$AsC_{12}H_{10}NOS$. . . . $O{=}S{=}NAs(C_6H_5)_2$ . . . . S: S-N Comp.6-79/80
$AsC_{12}H_{11}Mo_4O_{15}{}^{2-}$
$((C_6H_5)_2AsO_2)Mo_4O_{12}(OH)^{2-}$ . . . . Mo: SVol.B3b-122/3
$AsC_{12}H_{15}N_3O_4Re$ mer-$(CO)_3Re(NH_2NH_2)[As(CH_3)_2C_6H_5]NCO$ . . . Re: Org.Comp.1-292
$AsC_{12}H_{16}IMoO_2$ . . $(C_5H_5)Mo(CO)_2(I)[(CH_3)_2AsCH_2CH{=}CH_2]$ . . . . . Mo: Org.Comp.7-56/7, 97
$AsC_{12}H_{16}MoO_2{}^+$ $[(C_5H_5)Mo(CO)_2(CH_2{=}CHCH_2As(CH_3)_2)][PF_6]$ . . Mo: Org.Comp.8-185, 188
$AsC_{12}H_{17}MoO_2$. . . $(C_5H_5)Mo(CO)_2[{-}CH_2CH_2CH_2{-}As(CH_3)_2{-}]$ . . . . Mo: Org.Comp.8-113, 155, 172/3
$AsC_{12}H_{20}MoO_2P$. $(C_5H_5)Mo(CO)_2[P(CH_3)_3]As(CH_3)_2$. . . . . Mo: Org.Comp.7-119, 133, 148/50
$AsC_{12}H_{20}MoO_2PS$ $(C_5H_5)Mo(CO)_2[P(CH_3)_3]{-}As({=}S)(CH_3)_2$ . . . . . . Mo: Org.Comp.7-120/1, 136, 151
$AsC_{12}H_{20}MoO_5P$. $(C_5H_5)Mo(CO)_2[As(CH_3)_3]{-}P({=}O)(OCH_3)_2$ . . . . . Mo: Org.Comp.7-131
– . . . . $(C_5H_5)Mo(CO)_2[P(OCH_3)_3]{-}As(CH_3)_2$ . . . . Mo: Org.Comp.7-119, 137
$AsC_{12}H_{21}N_2O_4Sn$ $(CH_3)_3SnN_2C_3(COOCH_3)_2As(CH_3)_2$ . . . . Sn: Org.Comp.18-84, 90, 101/2
$AsC_{12}H_{21}Sn$ . . . . . $(CH_3)_3Sn{-}As(C_6H_5){-}C_3H_7{-}i$ . . . . Sn: Org.Comp.19-229, 232, 236
$AsC_{12}H_{30}NSn$ . . . . $(CH_3)_3SnN{=}As(C_3H_7{-}i)_3$ . . . . Sn: Org.Comp.18-107, 114
$AsC_{13}CoH_{24}NO_6P_2Re$
$(CO)_3Re[P(CH_3)_3]_2{-}As(CH_3)_2{-}Co(CO)_2(NO)$ . . . Re: Org.Comp.1-316
$AsC_{13}CoH_{27}NO_3Sn$
$(CH_3)_3Sn{-}As(C_4H_9{-}t)_2Co(CO)_2(NO)$. . . . . Sn: Org.Comp.19-229, 231
$AsC_{13}Co_2H_{11}MoO_6S$
$C_5H_5Mo(CO)(S)(As(CH_3)_2)Co(CO)_2Co(CO)_3$ . . . Mo: Org.Comp.6-392/3
$AsC_{13}F_6FeH_{15}$ . . . $[(1,4{-}(CH_3)_2{-}C_6H_4)Fe(C_5H_5)][AsF_6]$. . . . . Fe: Org.Comp.B19-1, 5, 11, 81/7

$AsC_{13}F_6H_8N_2O_3Re$
$(CO)_3Re[NC_5H_4\text{-}2\text{-}(2\text{-}C_5H_4N)]FAsF_5 \cdot 0.5\ H_2O$
Re: Org.Comp.1-157

$AsC_{13}F_6H_{10}N_2O_4Re$
$[(CO)_3Re(NC_5H_4\text{-}C_5H_4N)(H_2O)][AsF_6]$ .... Re: Org.Comp.1-301/2

$AsC_{13}FeH_{15}O_8PRe$
$(CO)_4Re[P(CH_3)_3]\text{-}Fe(CO)_4\text{-}As(CH_3)_2$ .... Re: Org.Comp.1-495

$AsC_{13}FeH_{15}O_{11}PRe$
$(CO)_4Re[P(OCH_3)_3]\text{-}As(CH_3)_2\text{-}Fe(CO)_4$ ...... Re: Org.Comp.1-484

$AsC_{13}Fe_3H_5N_2O_9$ $(CO)_7(NO)_2Fe_3As\text{-}C_6H_5$ .... Fe: Org.Comp.C6a-38

$AsC_{13}Fe_3H_9O_9S$ .. $(CO)_9Fe_3SAs\text{-}C_4H_9\text{-}t$ .... Fe: Org.Comp.C6a-202/3, 205, 209

$AsC_{13}H_{11}MnO_6Re$ $(CO)_4Re[\text{-}As(CH_3)_2\text{-}Mn(CO)_2(C_5H_5)\text{-}]$ .... Re: Org.Comp.1-498/9

$AsC_{13}H_{14}N_2O_6SSb$
$H_2O_3As\text{-}C_6H_4\text{-}NHC(S)NH\text{-}C_6H_4\text{-}Sb(O)(OH)_2$ .. Sb: Org.Comp.5-290

$AsC_{13}H_{16}I_2MoNO$ $C_5H_5Mo(NO)(As(CH_3)_2C_6H_5)I_2$ .... Mo: Org.Comp.6-36

$AsC_{13}H_{16}MoNO_2$ $(C_5H_5)Mo(CO)_2(CN)[(CH_3)_2AsCH_2CH{=}CH_2]$.... Mo: Org.Comp.7-57, 97

$AsC_{13}H_{19}MoO_6Si$ $(CH_3)_3Si\text{-}O\text{-}C[CH{=}As(CH_3)_3]{=}Mo(CO)_5$....... Mo: Org.Comp.5-104, 111

$AsC_{13}H_{23}IMoO_2PS$
$[(C_5H_5)Mo(CO)_2(P(CH_3)_3)(CH_3\text{-}S\text{-}As(CH_3)_2)]I$.. Mo: Org.Comp.7-283, 294

$AsC_{13}H_{23}MoNO_2P$ $(C_5H_5)Mo(CO)_2[P(CH_3)_2N(CH_3)_2]\text{-}As(CH_3)_2$ ... Mo: Org.Comp.7-119, 137

$AsC_{13}H_{23}MoO_2PS^+$
$[(C_5H_5)Mo(CO)_2(P(CH_3)_3)(CH_3\text{-}S\text{-}As(CH_3)_2)]^+$ Mo: Org.Comp.7-283, 294

$AsC_{14}ClH_{22}O_3Sn$ $(C_4H_9)_2SnOAs(O)(C_6H_4Cl\text{-}4)O$.............. Sn: Org.Comp.15-363, 364/5

$AsC_{14}Cl_2H_{15}MnN_2$ $[Mn(NC_5H_4\text{-}2\text{-}CH{=}N\text{-}C_6H_4\text{-}2\text{-}As(CH_3)_2)Cl_2]$... Mn: MVol.D6-79/80

$AsC_{14}Cl_2H_{25}N_2PdS$
$[(C_6H_5\text{-}As(CH_3)_2)PdCl_2(i\text{-}C_3H_7\text{-}N{=}S{=}N\text{-}C_3H_7\text{-}i)]$
S: S-N Comp.7-309/11

$AsC_{14}Cl_2H_{33}N_2PtS$
$[((C_2H_5)_3As)PtCl_2(t\text{-}C_4H_9\text{-}N{=}S{=}N\text{-}C_4H_9\text{-}t)]$.... S: S-N Comp.7-316/8, 321/2

$AsC_{14}F_6FeH_{17}$ ... $[(1,3,5\text{-}(CH_3)_3\text{-}C_6H_3)Fe(C_5H_5)][AsF_6]$ .... Fe: Org.Comp.B19-103, 133/6

– .... $[(i\text{-}C_3H_7\text{-}C_6H_5)Fe(C_5H_5)][AsF_6]$............. Fe: Org.Comp.B18-142/6, 197, 216, 273/4

$AsC_{14}FeH_{18}NO_8PRe$
$(CO)_4Re[P(CH_3)_2N(CH_3)_2]\text{-}Fe(CO)_4\text{-}As(CH_3)_2$ Re: Org.Comp.1-495

$AsC_{14}Fe_2H_{12}O_{10}ReS_2$
$(CO)_5ReAs(CH_3)_2\text{-}Fe(CO)_2(SCH_3)_2Fe(CO)_3$ ... Re: Org.Comp.2-176

$AsC_{14}H_{11}MnO_7Re$ $(CO)_5ReAs(CH_3)_2\text{-}Mn(CO)_2C_5H_5$............ Re: Org.Comp.2-175

$AsC_{14}H_{15}I_2MnN_2$ $[Mn(NC_5H_4\text{-}2\text{-}CH{=}N\text{-}C_6H_4\text{-}2\text{-}As(CH_3)_2)I_2] \cdot 0.5\ C_2H_5OH$
Mn: MVol.D6-79/80

$AsC_{14}H_{16}In$...... $(CH_3)_2In\text{-}As(C_6H_5)_2$ .... In: Org.Comp.1-312, 316

$AsC_{14}H_{22}NO_5Sn$.. $(n\text{-}C_4H_9)_2Sn[\text{-}O\text{-}As(O)(C_6H_4NO_2\text{-}2)\text{-}O\text{-}]$ ..... Sn: Org.Comp.15-363, 364/5

– .... $(n\text{-}C_4H_9)_2Sn[\text{-}O\text{-}As(O)(C_6H_4NO_2\text{-}4)\text{-}O\text{-}]$ ..... Sn: Org.Comp.15-363

$AsC_{14}H_{22}NO_6Sn$.. $(n\text{-}C_4H_9)_2Sn[\text{-}O\text{-}As(O)(C_6H_3(NO_2\text{-}3)OH\text{-}4)\text{-}O\text{-}]$
Sn: Org.Comp.15-364

$AsC_{14}H_{23}N_2O_5Sn$ $(n\text{-}C_4H_9)_2Sn[\text{-}O\text{-}As(O)(C_6H_3(NO_2\text{-}3)NH_2\text{-}4)\text{-}O\text{-}]$
Sn: Org.Comp.15-364

$AsC_{14}H_{23}OSn$.... $2\text{-}(t\text{-}C_4H_9)\text{-}3\text{-}(CH_3)_3Sn\text{-}1,3\text{-}OAsC_7H_5$ .... Sn: Org.Comp.19-232, 236

$AsC_{14}H_{23}O_3Sn$ ... $(n\text{-}C_4H_9)_2Sn[\text{-}O\text{-}As(O)(C_6H_5)\text{-}O\text{-}]$ .... Sn: Org.Comp.15-363, 364/5

$AsC_{14}H_{23}O_4Sn$ ... $(n\text{-}C_4H_9)_2Sn[\text{-}O\text{-}As(O)(C_6H_4OH\text{-}4)\text{-}O\text{-}]$...... Sn: Org.Comp.15-363, 364/5

$AsC_{14}H_{24}NO_3Sn$ .. $(n\text{-}C_4H_9)_2Sn[\text{-}O\text{-}As(O)(C_6H_4NH_2\text{-}2)\text{-}O\text{-}]$ ..... Sn: Org.Comp.15-363, 364/5
– .... $(n\text{-}C_4H_9)_2Sn[\text{-}O\text{-}As(O)(C_6H_4NH_2\text{-}4)\text{-}O\text{-}]$ ..... Sn: Org.Comp.15-363
$AsC_{14}H_{24}NSn$ .... $1,2,2\text{-}(C_2H_5)_3\text{-}3\text{-}C_6H_5\text{-}1,3,2\text{-}NAsSnC_2H_4$ ..... Sn: Org.Comp.19-242
$AsC_{14}H_{40}InSi_4$ ... $[(CH_3)_3Si\text{-}CH_2]_2In\text{-}As[Si(CH_3)_3]_2$ .... In: Org.Comp.1-312, 316, 322
$AsC_{15}Cl_2H_{17}MnN_2$ $[Mn(6\text{-}CH_3\text{-}NC_5H_3\text{-}2\text{-}CH{=}N\text{-}C_6H_4\text{-}2\text{-}As(CH_3)_2)Cl_2]$
Mn: MVol.D6-79/80
$AsC_{15}Co_2FeH_{11}MoO_8S$
$C_5H_5Mo(CO)(CO)(S)FeCo_2(CO)_6(As(CH_3)_2)$ ... Mo: Org.Comp.6-395/6
$AsC_{15}CrFe_3HO_{14}$ $(CO)_9Fe_3As(CH)[Cr(CO)_5]$ .... Fe: Org.Comp.C6a-266/7, 270
$AsC_{15}F_6FeH_{13}$ ... $[(C_{10}H_8)Fe(C_5H_5)][AsF_6]$ .... Fe: Org.Comp.B19-220, 233, 302/5
$AsC_{15}FeH_{20}MoO_5P$
$C_5H_5Mo(CO)(P(CH_3)_3)(CO)(As(CH_3)_2)Fe(CO)_3$ Mo: Org.Comp.6-377
$AsC_{15}FeH_{24}O_7P_2ReS$
$fac\text{-}(CO)_3Re[P(CH_3)_3]_2\text{-}S\text{-}As(CH_3)_2\text{-}Fe(CO)_4$ .. Re: Org.Comp.1-319
$AsC_{15}Fe_3HMoO_{14}$ $(CO)_9Fe_3As(CH)[Mo(CO)_5]$ .... Fe: Org.Comp.C6a-266, 268
$AsC_{15}Fe_3HO_{14}W$ .. $(CO)_9Fe_3As(CH)[W(CO)_5]$ .... Fe: Org.Comp.C6a-266, 268
$AsC_{15}Fe_3H_5O_9S$ .. $(CO)_9Fe_3SAs\text{-}C_6H_5$ .... Fe: Org.Comp.C6a-202/3, 206, 209/10
$AsC_{15}Fe_3H_{11}O_9S$ $(CO)_9Fe_3SAs\text{-}C_6H_{11}\text{-}c$ .... Fe: Org.Comp.C6a-202/3, 206
$AsC_{15}Fe_3H_{15}O_9S$ $(CO)_9Fe_3[As(CH_3)_2](S\text{-}C_4H_9\text{-}t)$ .... Fe: Org.Comp.C6a-213/5
$AsC_{15}H_{16}HgMoNO_5$
$[(C_5H_5)Mo(CO)_2(C_6H_5\text{-}As(CH_3)_2)HgNO_3]$ ..... Mo: Org.Comp.7-121/2, 144, 154
$AsC_{15}H_{16}IMoO_2$ .. $(C_5H_5)Mo(CO)_2(I)[(CH_3)_2As\text{-}C_6H_5]$ .... Mo: Org.Comp.7-56/7, 97
$AsC_{15}H_{17}I_2MnN_2$ $[Mn(6\text{-}CH_3\text{-}NC_5H_3\text{-}2\text{-}CH{=}N\text{-}C_6H_4\text{-}2\text{-}As(CH_3)_2)I_2] \cdot 0.5\ C_2H_5OH$ .... Mn: MVol.D6-79/80
$AsC_{15}H_{17}O_3Sn$ ... $(C_6H_5CH_2)_2Sn[\text{-}OAs(O)(CH_3)O\text{-}]$ .... Sn: Org.Comp.16-81/2
$AsC_{15}H_{18}I_2MnN_2O_{0.5}$
$[Mn(NC_5H_4\text{-}2\text{-}CH{=}N\text{-}C_6H_4\text{-}2\text{-}As(CH_3)_2)I_2] \cdot 0.5\ C_2H_5OH$
Mn: MVol.D6-79/80
$AsC_{15}H_{19}Sn$ ..... $(CH_3)_3Sn\text{-}As(C_6H_5)_2$ .... Sn: Org.Comp.19-228/9, 231, 236
$AsC_{15}H_{20}In$ ...... $(CH_3)_3In \cdot HAs(C_6H_5)_2$ .... In: Org.Comp.1-27, 40
$AsC_{15}H_{23}MoO_2$ ... $(C_5H_5)Mo(CO)_2{=}As(C_4H_9\text{-}t)_2$ .... Mo: Org.Comp.7-36/7
$AsC_{15}H_{23}O_5Sn$ ... $(n\text{-}C_4H_9)_2Sn[\text{-}O\text{-}As(O)(C_6H_4COOH\text{-}4)\text{-}O\text{-}]$ ... Sn: Org.Comp.15-364, 366
$AsC_{15}H_{25}O_3Sn$ ... $(n\text{-}C_4H_9)_2Sn[\text{-}O\text{-}As(O)(CH_2C_6H_5)\text{-}O\text{-}]$ .... Sn: Org.Comp.15-363, 364/5
– .... $(n\text{-}C_4H_9)_2Sn[\text{-}O\text{-}As(O)(C_6H_4CH_3\text{-}4)\text{-}O\text{-}]$ ..... Sn: Org.Comp.15-364/5
$AsC_{15}H_{25}O_4Sn$ ... $(n\text{-}C_4H_9)_2Sn[\text{-}O\text{-}As(O)(C_6H_4OCH_3\text{-}2)\text{-}O\text{-}]$ .... Sn: Org.Comp.15-363, 364/5
– .... $(n\text{-}C_4H_9)_2Sn[\text{-}O\text{-}As(O)(C_6H_4OCH_3\text{-}4)\text{-}O\text{-}]$ .... Sn: Org.Comp.15-363, 364/5
$AsC_{15}H_{29}N_4S_2Si_2$ $((CH_3)_3SiN{=}S{=}N)_2As\text{-}C_6H_2(CH_3)_3\text{-}2,4,6$ .... S: S-N Comp.7-152
$AsC_{15}H_{36}NSn$ .... $(CH_3)_3SnN{=}As(C_4H_9\text{-}t)_3$ .... Sn: Org.Comp.18-107, 114
$AsC_{16}ClH_{17}IMoO_2$ $(C_5H_5)Mo(CO)_2(I)[(CH_3)_2As\text{-}CH_2C_6H_4\text{-}2\text{-}Cl]$ ... Mo: Org.Comp.7-56/7, 98
$AsC_{16}Cl_2H_{19}MnN_2$ $[Mn(NC_5H_4\text{-}2\text{-}CH{=}N\text{-}C_6H_4\text{-}2\text{-}As(C_2H_5)_2)Cl_2]$ .. Mn: MVol.D6-79/80
$AsC_{16}Cl_2H_{29}N_2PdS$
$[(C_6H_5\text{-}As(CH_3)_2)PdCl_2(t\text{-}C_4H_9\text{-}N{=}S{=}N\text{-}C_4H_9\text{-}t)]$
S: S-N Comp.7-309/13

$AsC_{16}Cl_2H_{37}N_2PtS$
$[((C_2H_5)_3As)PtCl_2((CH_3)_3CCH_2\text{-}N{=}S{=}N\text{-}CH_2C(CH_3)_3)]$ S: S-N Comp.7-316/7, 319, 321/2

$AsC_{16}CrH_{24}O_8P_2ReS$
fac-$(CO)_3Re[P(CH_3)_3]_2\text{-}S\text{-}As(CH_3)_2\text{-}Cr(CO)_5$ . . Re: Org.Comp.1-319/20

$AsC_{16}F_6H_{14}NS$ . . . $[SN][AsF_6]\cdot 9,10\text{-}(CH_3)_2C_{14}H_8$ . . . . S: S-N Comp.5-47/9

$AsC_{16}FeH_{20}MoO_6P$
$(C_5H_5)Mo(CO)_2[P(CH_3)_3]As(CH_3)_2Fe(CO)_4$ . . . . Mo: Org.Comp.7-120/1, 133, 150

$AsC_{16}FeH_{24}N$ . . . . $(C_5H_5)Fe[C_5H_3(As(CH_3)_2)\text{-}CH(CH_3)\text{-}N(CH_3)_2]$ . . Fe: Org.Comp.A10-94

$AsC_{16}Fe_3H_8O_9P$ . . $(CO)_9Fe_3(P\text{-}C_6H_5)(As\text{-}CH_3)$ . . . . Fe: Org.Comp.C6a-172

$AsC_{16}H_{11}N_2O_{11}S_2Th^{2+}$
$Th[(HO)_2(O_3S)_2C_{10}H_3NNC_6H_4(AsO)(OH)_2]^{2+}$ . . Th: SVol.D1-106/7, 116

$AsC_{16}H_{15}MnN_4S_2$ $[Mn(NC_5H_4\text{-}2\text{-}CH{=}N\text{-}C_6H_4\text{-}2\text{-}As(CH_3)_2)(NCS)_2]_n\cdot 2\ C_2H_5OH$ . . . . Mn: MVol.D6-79/80

$AsC_{16}H_{16}HgMoNO_2$
$[(C_5H_5)Mo(CO)_2(C_6H_5\text{-}As(CH_3)_2)HgCN]$ . . . . . . Mo: Org.Comp.7-121, 143, 152, 154

$AsC_{16}H_{16}HgMoNO_2S$
$[(C_5H_5)Mo(CO)_2(C_6H_5\text{-}As(CH_3)_2)HgSCN]$ . . . . . Mo: Org.Comp.7-121/2, 144

$AsC_{16}H_{16}HgMoNO_3$
$[(C_5H_5)Mo(CO)_2(C_6H_5\text{-}As(CH_3)_2)HgOCN]$ . . . . . Mo: Org.Comp.7-121/2, 144

$AsC_{16}H_{17}MoO_2$ . . . $(C_5H_5)Mo(CO)_2[\text{-}CH_2\text{-}C_6H_4\text{-}2\text{-}As(CH_3)_2\text{-}]$ . . . . Mo: Org.Comp.8-113, 155/6

– . . . . $(C_5H_5)Mo(CO)_2[\text{-}C_6H_4\text{-}2\text{-}CH_2\text{-}As(CH_3)_2\text{-}]$ . . . . Mo: Org.Comp.8-113, 156

$AsC_{16}H_{18}HgIMoO_2$
$[(CH_3\text{-}C_5H_4)Mo(CO)_2(C_6H_5\text{-}As(CH_3)_2)HgI]$ . . . . Mo: Org.Comp.7-147, 153, 154

$AsC_{16}H_{18}IMoO_2$ . . $(C_5H_5)Mo(CO)_2(I)[(CH_3)_2As\text{-}C_6H_4\text{-}2\text{-}CH_3]$ . . . . Mo: Org.Comp.7-56/7, 97/8

$AsC_{16}H_{19}I_2MnN_2$ $[Mn(NC_5H_4\text{-}2\text{-}CH{=}N\text{-}C_6H_4\text{-}2\text{-}As(C_2H_5)_2)I_2]\cdot C_2H_5OH$ Mn: MVol.D6-79/80

$AsC_{16}H_{19}O_3Sn$ . . . $(C_6H_5)_2SnOAs(O)(C_4H_9)O$ . . . . Sn: Org.Comp.16-151/2

$AsC_{16}H_{20}I_2MnN_2O_{0.5}$
$[Mn(6\text{-}CH_3\text{-}NC_5H_3\text{-}2\text{-}CH{=}N\text{-}C_6H_4\text{-}2\text{-}As(CH_3)_2)I_2]\cdot 0.5\ C_2H_5OH$ . . . . Mn: MVol.D6-79/80

$AsC_{16}H_{24}MoO_8P_2ReS$
fac-$(CO)_3Re[P(CH_3)_3]_2\text{-}S\text{-}As(CH_3)_2\text{-}Mo(CO)_5$ Re: Org.Comp.1-320

$AsC_{16}H_{24}O_8P_2ReSW$
fac-$(CO)_3Re[P(CH_3)_3]_2\text{-}S\text{-}As(CH_3)_2\text{-}W(CO)_5$ . . Re: Org.Comp.1-320

$AsC_{16}H_{25}O_4Sn$ . . . $(C_4H_9)_2SnOAs(O)(C_6H_5)CH_2COO$ . . . . Sn: Org.Comp.15-367

$AsC_{16}H_{26}MoO_7P$. $(C_5H_5)Mo(CO)_2[P(OCH_3)_3]\text{-}2\text{-}[1,3,2\text{-}O_2AsC_2(CH_3)_4\text{-}4,4,5,5]$ . . . . Mo: Org.Comp.7-119, 138

$AsC_{16}H_{36}N_2PS$ . . . $(t\text{-}C_4H_9)_2AsN{=}S{=}NP(C_4H_9\text{-}t)_2$ . . . . S: S-N Comp.7-115/6

$AsC_{16}H_{36}N_2PSSe$ $(t\text{-}C_4H_9)_2AsN{=}S{=}NP({=}Se)(C_4H_9\text{-}t)_2$ . . . . S: S-N Comp.7-116/7

$AsC_{16}H_{36}N_2PSTe$ $(t\text{-}C_4H_9)_2AsN{=}S{=}NP({=}Te)(C_4H_9\text{-}t)_2$ . . . . S: S-N Comp.7-116/7

$AsC_{16}H_{36}N_2PS_2$ . . $(t\text{-}C_4H_9)_2AsN{=}S{=}NP({=}S)(C_4H_9\text{-}t)_2$ . . . . S: S-N Comp.7-116/7

$AsC_{17}Cl_2H_{21}MnN_2$ $[Mn(6\text{-}CH_3\text{-}NC_5H_3\text{-}2\text{-}CH{=}N\text{-}C_6H_4\text{-}2\text{-}As(C_2H_5)_2)Cl_2]\cdot H_2O$ . . . . Mn: MVol.D6-79/80

$AsC_{17}Co_2H_{20}MoO_6P$
$C_5H_5Mo(CO)(PC_4H_9\text{-}t)Co(CO)_2(As(CH_3)_2)Co(CO)_3$ Mo: Org.Comp.6-399

$AsC_{17}CrH_{10}NO_6S$ $(CO)_5Cr((C_6H_5)_2AsN{=}S{=}O)$ . . . . S: S-N Comp.6-89/91

$AsC_{17}CrH_{20}MoO_7PS$
$(C_5H_5)Mo(CO)_2[P(CH_3)_3]-As(CH_3)_2-S-Cr(CO)_5$ Mo:Org.Comp.7-120, 136

$AsC_{17}FeH_{23}MoNO_6P$
$(C_5H_5)Mo(CO)_2[P(CH_3)_2N(CH_3)_2]-As(CH_3)_2-Fe(CO)_4$
Mo:Org.Comp.7-121, 137

$AsC_{17}Fe_3H_{10}O_9P$ $(CO)_9Fe_3(P-C_6H_4-4-CH_3)(As-CH_3)$ . . . . Fe: Org.Comp.C6a-172

$AsC_{17}H_{17}MnN_4S_2$ $[Mn(6-CH_3-NC_5H_3-2-CH{=}N-C_6H_4-2-As(CH_3)_2)(NCS)_2]_n$
Mn:MVol.D6-79/80

$AsC_{17}H_{19}HgMoO_4$ $[(C_5H_5)Mo(CO)_2(C_6H_5-As(CH_3)_2)HgOC(O)CH_3]$ Mo:Org.Comp.7-121/2, 144

$AsC_{17}H_{20}IMoO_3$ . . $(C_5H_5)MoI(CO)_2[(CH_3)_2As-C_6H_4-2-CH_2OCH_3]$ Mo:Org.Comp.7-56/7, 98

$AsC_{17}H_{20}MoO_7PW$
$(C_5H_5)Mo(CO)_2[P(CH_3)_3]As(CH_3)_2W(CO)_5$. Mo:Org.Comp.7-120, 133

$AsC_{17}H_{21}I_2MnN_2$ $[Mn(6-CH_3-NC_5H_3-2-CH{=}N-C_6H_4-2-As(C_2H_5)_2)I_2]$
Mn:MVol.D6-79/80

$AsC_{17}H_{22}MoO_2P$. $(C_5H_5)Mo(CO)_2[C_6H_5-P(CH_3)_2]-As(CH_3)_2$. Mo:Org.Comp.7-119, 137

$AsC_{17}H_{37}O_3Sn$ . . . $(C_8H_{17})_2Sn[-OAs(O)(CH_3)O-]$ . . . . Sn: Org.Comp.16-59

$AsC_{18}Cl_2H_{23}I_2N_2PtS$
$[(As(C_2H_5)_3)PtCl_2(4-I-C_6H_4-N{=}S{=}N-C_6H_4-I-4)]$
S: S-N Comp.7-323/6

$AsC_{18}Cl_4H_{23}N_2PtS$
$[(As(C_2H_5)_3)PtCl_2(4-Cl-C_6H_4-N{=}S{=}N-C_6H_4-Cl-4)]$
S: S-N Comp.7-323/6

$AsC_{18}Cl_6H_{21}N_2PtS$
$[(As(C_2H_5)_3)PtCl_2(3,5-Cl_2-C_6H_3-N{=}S{=}N-C_6H_3-Cl_2-3,5)]$
S: S-N Comp.7-323/6

$AsC_{18}FeH_{29}MoO_5P_2$
$(C_5H_5)Mo(CO)_2[P(CH_3)_3]-As(CH_3)_2-Fe(CO)_3-P(CH_3)_3$
Mo:Org.Comp.7-120/1, 134, 151

$AsC_{18}FeH_{29}MoO_{11}P_2$
$(C_5H_5)Mo(CO)_2[P(OCH_3)_3]-As(CH_3)_2-Fe(CO)_3-P(OCH_3)_3$
Mo:Org.Comp.7-121, 138

$AsC_{18}Fe_3H_8MnO_{11}$
$(CO)_9Fe_3As(CH)[Mn(CO)_2(C_5H_4-CH_3)]$ . . . . Fe: Org.Comp.C6a-266/7, 268

$AsC_{18}Fe_3H_9O_{14}SW$
$(CO)_9Fe_3SAs-(C_4H_9-t)[W(CO)_5]$ . . . . Fe: Org.Comp.C6a-211/3

$AsC_{18}H_{15}MnN_3O_3$ $[Mn(NO)_3((C_6H_5)_3As)]$ Mn:MVol.D8-202/3

$AsC_{18}H_{18}O_9Sb_3$ . . $As(C_6H_4)_3[Sb(O)(OH)_2-3]_3$ . . . . Sb: Org.Comp.5-317

$AsC_{18}H_{19}MnN_4S_2$ $[Mn(NC_5H_4-2-CH{=}N-C_6H_4-2-As(C_2H_5)_2)$
$(NCS)_2]_n \cdot 0.5\ C_6H_6$. Mn:MVol.D6-79/80

$AsC_{18}H_{21}OS_3Ti_2$. . $[(CH_3C_5H_4)_3Ti_2](O)(AsS_3)$ . . . . Ti: Org.Comp.5-65/6

$AsC_{18}H_{25}I_2MnN_2O$ $[Mn(NC_5H_4-2-CH{=}N-C_6H_4-2-As(C_2H_5)_2)I_2] \cdot C_2H_5OH$
Mn:MVol.D6-79/80

$AsC_{18}H_{25}Sn$ . . . . . $(C_2H_5)_3Sn-As(C_6H_5)_2$. Sn: Org.Comp.19-238

$AsC_{18}H_{29}MnO_5P_2ReS$
$(CO)_3Re[P(CH_3)_3]_2-S-As(CH_3)_2-Mn(CO)_2C_5H_5$ Re: Org.Comp.1-319

$AsC_{18}H_{42}NSn$ . . . . $(n-C_4H_9)_3Sn-As(C_2H_5)-CH_2CH_2-NH-C_2H_5$. . . . Sn: Org.Comp.19-238

$AsC_{19}CoFeH_{16}MoO_7SW$
$C_5H_5Mo(CO)(CO)(S)W(C_5H_5)FeCo(CO)_5(As(CH_3)_2)$ Mo: Org.Comp.6-397

$AsC_{19}Co_2H_{16}MoO_6P$
$C_5H_5Mo(CO)(PC_6H_5)(As(CH_3)_2)Co(CO)_2Co(CO)_3$ Mo: Org.Comp.6-395

$AsC_{19}Fe_3H_{11}O_{11}$ . . $(CO)_{11}Fe_3[As(CH_3)_2-C_6H_5]$ . . . . Fe: Org.Comp.C6b-128, 133, 135

$AsC_{19}H_{21}MnN_4S_2$ $[Mn(6-CH_3-NC_5H_3-2-CH=N-C_6H_4-2-As(C_2H_5)_2)(NCS)_2]_n \cdot C_2H_5OH$ . . . . Mn: MVol.D6-79/80

$AsC_{20}ClH_{18}O_3Sn$ $(C_6H_5CH_2)_2Sn[-OAs(O)(C_6H_4Cl-4)O-]$ . . . . Sn: Org.Comp.16-81/2

$AsC_{20}ClH_{21}InO$ . . . $(CH_3)_2InCl \cdot OAs(C_6H_5)_3$ . . . . In: Org.Comp.1-118, 120

$AsC_{20}Cl_2H_{21}N_2PtS$
$[((C_6H_5)_3As)PtCl_2(CH_3N=S=NCH_3)] \cdot C_6H_6$ . . . . S: S-N Comp.7-316/7, 322

$AsC_{20}Cl_2H_{29}N_2O_2PtS$
$[(As(C_2H_5)_3)PtCl_2(4-CH_3O-C_6H_4-N=S=N-C_6H_4-OCH_3-4)]$ S: S-N Comp.7-323/6

$AsC_{20}Cl_2H_{29}N_2PtS$
$[(As(C_2H_5)_3)PtCl_2(4-CH_3-C_6H_4-N=S=N-C_6H_4-CH_3-4)]$ S: S-N Comp.7-323/6

$AsC_{20}CoH_{16}MoO_8RuSW$
$C_5H_5Mo(CO)(S)W(CO)_2(C_5H_5)Ru(CO)_2(As(CH_3)_2)Co(CO)_3$ Mo: Org.Comp.6-397/8

$AsC_{20}Co_2H_{22}MoO_5PS$
$(C_5H_5)(CO)MoS[As(CH_3)_2]Co(CO)_2Co[P(CH_3)_2C_6H_5](CO)_2$ . . . . Mo: Org.Comp.6-394

– . . . . $(C_5H_5)(CO)MoS[As(CH_3)_2]Co[P(CH_3)_2C_6H_5](CO)Co(CO)_3$ Mo: Org.Comp.6-393/4

$AsC_{20}CrFe_3H_5O_{14}S$
$(CO)_9Fe_3SAs-(C_6H_5)[Cr(CO)_5]$ . . . . Fe: Org.Comp.C6a-211/3

$AsC_{20}FeH_{35}MoN_2O_5P_2$
$(C_5H_5)Mo(CO)_2[P(CH_3)_2N(CH_3)_2]-As(CH_3)_2-Fe(CO)_3-P(CH_3)_2N(CH_3)_2$ . . . . Mo: Org.Comp.7-121, 138

$AsC_{20}Fe_3H_5O_{14}SW$
$(CO)_9Fe_3SAs-(C_6H_5)[W(CO)_5]$ . . . . Fe: Org.Comp.C6a-211/3

$AsC_{20}Fe_3H_{11}O_{14}SW$
$(CO)_9Fe_3SAs-(C_6H_{11}-c)[W(CO)_5]$ . . . . Fe: Org.Comp.C6a-211/3

$AsC_{20}GeH_{25}$ . . . . . $Ge(C_2H_5)_3C{\equiv}CAs(C_6H_5)_2$ . . . . Ge: Org.Comp.2-253

$AsC_{20}H_{18}IMoO_2$ . . $(C_5H_5)Mo(CO)_2(I)[CH_3As(C_6H_5)_2]$ . . . . Mo: Org.Comp.7-56/7, 99/100

$AsC_{20}H_{18}NO_5Sn$ . . $(C_6H_5CH_2)_2Sn[-OAs(O)(C_6H_4NO_2-2)O-]$ . . . . . . Sn: Org.Comp.16-81/2

$AsC_{20}H_{19}O_3Sn$ . . . $(C_6H_5CH_2)_2Sn[-OAs(O)(C_6H_5)O-]$ . . . . Sn: Org.Comp.16-81/2

$AsC_{20}H_{19}O_4Sn$ . . . $(C_6H_5CH_2)_2Sn[-OAs(O)(C_6H_4OH-4)O-]$ . . . . Sn: Org.Comp.16-81/2

$AsC_{20}H_{20}NO_3Sn$ . . $(C_6H_5CH_2)_2Sn[-OAs(O)(C_6H_4NH_2-2)O-]$ . . . . . . Sn: Org.Comp.16-81/2

$AsC_{20}H_{21}Sn$ . . . . . $(C_6H_5)_3Sn-As(CH_3)_2$ . . . . Sn: Org.Comp.19-239/40

$AsC_{20}H_{27}MnN_4O_2S_2$
$[Mn(NC_5H_4-2-CH=N-C_6H_4-2-As(CH_3)_2)(NCS)_2]_n \cdot 2\ C_2H_5OH$ . . . . Mn: MVol.D6-79/80

$AsC_{20}H_{27}MoO_3PS^+$
$[(C_5H_5)Mo(CO)_2(P(CH_3)_3)((CH_3)_2As-S-CH_2C(O)C_6H_5)]^+$ Mo: Org.Comp.7-283, 294

$AsC_{21}ClFeH_{27}NPd$ $(C_5H_5)Fe[C_5H_3(PdCl(As(CH_3)_2-C_6H_5))-CH_2-N(CH_3)_2]$
Fe: Org.Comp.A10-178/80

$AsC_{21}ClH_{27}NO_3Sn$ $(C_4H_9)_2SnOAs(O)(C_6H_4(NCHC_6H_4Cl-2)-2)O$ ... Sn: Org.Comp.15-364, 366

$AsC_{21}FFe_3H_{14}O_{11}$ $(CO)_{11}Fe_3[As(C_2H_5)_2-C_6H_4-4-F]$ ... Fe: Org.Comp.C6b-128, 133

$AsC_{21}FeH_{27}INPd$. $(C_5H_5)Fe[C_5H_3(PdI(As(CH_3)_2-C_6H_5))-CH_2-N(CH_3)_2]$
Fe: Org.Comp.A10-179, 182

$AsC_{21}H_{19}O_5Sn$ ... $(C_6H_5CH_2)_2Sn[-OAs(O)(C_6H_4COOH-2)O-]$ ... Sn: Org.Comp.16-81, 83

$AsC_{21}H_{21}MnN_3O_6$ $[Mn(NO)_3((CH_3O-4-C_6H_4)_3As)]$ ... Mn:MVol.D8-202/3

$AsC_{21}H_{21}O_3Sn$ ... $(C_6H_5CH_2)_2Sn[-OAs(O)(CH_2C_6H_5)O-]$ ... Sn: Org.Comp.16-81/2

– ... $(C_6H_5CH_2)_2Sn[-OAs(O)(C_6H_4CH_3-4)O-]$ ... Sn: Org.Comp.16-81, 83

$AsC_{21}H_{21}O_4Sn$ ... $(C_6H_5CH_2)_2Sn[-OAs(O)(C_6H_4-OCH_3-4)O-]$ ... Sn: Org.Comp.16-81/2

$AsC_{21}H_{22}MnN_4S_2$ $[Mn(NC_5H_4-2-CH{=}N-C_6H_4-2-As(C_2H_5)_2)$
$(NCS)_2]_n \cdot 0.5\ C_6H_6$ ... Mn:MVol.D6-79/80

$AsC_{21}H_{27}MnN_4OS_2$
$[Mn(6-CH_3-NC_5H_3-2-CH{=}N-C_6H_4-2-$
$As(C_2H_5)_2)(NCS)_2]_n \cdot C_2H_5OH$ ... Mn:MVol.D6-79/80

$AsC_{21}H_{27}N_2O_5Sn$ $(n-C_4H_9)_2Sn[-O-As(O)(C_6H_4(NCHC_6H_4NO_2-2)-2)-O-]$
Sn: Org.Comp.15-364, 366

– ... $(n-C_4H_9)_2Sn[-O-As(O)(C_6H_4(NCHC_6H_4NO_2-3)-2)-O-]$
Sn: Org.Comp.15-364, 366

$AsC_{21}H_{31}Sn$ ... $(n-C_3H_7)_3Sn-As(C_6H_5)_2$ ... Sn: Org.Comp.19-238

$AsC_{22}ClH_{15}O_4Re$ cis-$(CO)_4Re(Cl)As(C_6H_5)_3$ ... Re: Org.Comp.1-456

$AsC_{22}ClH_{21}IMoO_2$ $(C_5H_5)Mo(CO)_2(I)[Cl-CH_2CH_2CH_2-As(C_6H_5)_2]$ Mo:Org.Comp.7-56/7, 100

$AsC_{22}ClH_{38}O_3Sn$ $(C_8H_{17})_2Sn[-OAs(O)(C_6H_4Cl-4)O-]$ ... Sn: Org.Comp.16-59

$AsC_{22}ClH_{62}In_2Si_6$ $[-In(CH_2Si(CH_3)_3)_2-As(Si(CH_3)_3)_2-In(CH_2Si(CH_3)_3)_2-Cl-]$
In: Org.Comp.1-325/6

$AsC_{22}Cl_2H_{25}N_2PtS$
$[((C_6H_5)_3As)PtCl_2(C_2H_5-N{=}S{=}N-C_2H_5)]$ ... S: S-N Comp.7-316/7, 319

$AsC_{22}Cl_2H_{33}N_2PtS$
$[(As(C_2H_5)_3)PtCl_2(3,5-(CH_3)_2C_6H_3-N{=}S{=}N$
$-C_6H_3(CH_3)_2-3,5)]$ ... S: S-N Comp.7-323/6

$AsC_{22}CrH_{22}MoO_7P$
$(C_5H_5)Mo(CO)_2[C_6H_5-P(CH_3)_2]-As(CH_3)_2-Cr(CO)_5$
Mo:Org.Comp.7-137

$AsC_{22}F_6H_{20}MoO_2P$
$[(C_5H_5)Mo(CO)_2(CH_2{=}CHCH_2As(C_6H_5)_2)][PF_6]$ Mo:Org.Comp.8-185, 188

$AsC_{22}FeH_{18}Li$ ... $(Li-C_5H_4)Fe[C_5H_4-As(C_6H_5)_2]$ ... Fe: Org.Comp.A10-113/7

$AsC_{22}Fe_3H_{17}O_{11}$. $(CO)_{11}Fe_3[As(C_2H_5)_2-CH_2C_6H_5]$ ... Fe: Org.Comp.C6b-128, 133, 135, 139

$AsC_{22}GeH_{29}$ ... $(C_2H_5)_3Ge-C{\equiv}C-As(C_6H_4CH_3-2)_2$ ... Ge:Org.Comp.2-253

– ... $(C_2H_5)_3Ge-C{\equiv}C-As(C_6H_4CH_3-4)_2$ ... Ge:Org.Comp.2-253

$AsC_{22}H_{20}MoO_2^+$ $[(C_5H_5)Mo(CO)_2(CH_2{=}CHCH_2As(C_6H_5)_2)][PF_6]$ Mo:Org.Comp.8-185, 188

$AsC_{22}H_{21}MoO_2$ ... $(C_5H_5)Mo(CO)_2[-CH_2CH_2CH_2-As(C_6H_5)_2-]$ ... Mo:Org.Comp.8-113, 156

$AsC_{22}H_{38}NO_5Sn$ .. $(C_8H_{17})_2Sn[-OAs(O)(C_6H_4NO_2-2)O-]$ ... Sn: Org.Comp.16-60

$AsC_{22}H_{39}O_3Sn$ ... $(C_8H_{17})_2Sn[-OAs(O)(C_6H_5)O-]$ ... Sn: Org.Comp.16-59

$AsC_{22}H_{39}O_4Sn$ ... $(C_8H_{17})_2Sn[-OAs(O)(C_6H_4OH-4)O-]$ ... Sn: Org.Comp.16-59

$AsC_{22}H_{40}NO_3Sn$ .. $(C_8H_{17})_2Sn[-OAs(O)(C_6H_4NH_2-2)O-]$ ... Sn: Org.Comp.16-59

$AsC_{23}ClFeH_{32}NO_4Rh$
$[((C_5H_5)Fe(C_5H_3(As(CH_3)_2)-CH(CH_3)$
$-N(CH_3)_2))Rh(C_7H_8)][ClO_4]$ ... Fe: Org.Comp.A10-94

$AsC_{23}ClH_{26}MoO_2Si$
$(C_5H_5)Mo(CO)_2(Cl)[(C_6H_5)_2AsCH_2Si(CH_3)_3]$ ... Mo:Org.Comp.7-56/7, 100
$AsC_{23}Cl_2H_{20}MoNO$
$C_5H_5Mo(NO)(As(C_6H_5)_3)Cl_2$ ... Mo:Org.Comp.6-32
$AsC_{23}F_6FeH_{35}$ ... $[((C_2H_5)_6C_6)Fe(C_5H_5)][AsF_6]$ ... Fe: Org.Comp.B19-173, 205/9
$AsC_{23}F_6H_{20}MoN_2O_2P$
$[C_5H_5Mo(NO)_2As(C_6H_5)_3][PF_6]$ ... Mo:Org.Comp.6-61
$AsC_{23}FeH_{32}NRh^+$ $[((C_5H_5)Fe(C_5H_3(As(CH_3)_2)-CHCH_3-N(CH_3)_2))Rh(C_7H_8)]^+$ ... Fe: Org.Comp.A10-94
$AsC_{23}H_{20}I_2MoNO$ $C_5H_5Mo(NO)(As(C_6H_5)_3)I_2$ ... Mo:Org.Comp.6-36
$AsC_{23}H_{20}MoN_2O_2^+$
$[C_5H_5Mo(NO)_2As(C_6H_5)_3]^+$ ... Mo:Org.Comp.6-61
$AsC_{23}H_{27}Sn$ ... $(C_6H_5)_2As-Sn(CH_3)_2-CH_2-CH(CH_3)-C_6H_5$ ... Sn: Org.Comp.19-240/1
$AsC_{23}H_{39}O_5Sn$ ... $(n-C_8H_{17})_2Sn[-OAs(O)(C_6H_4-COOH-2)O-]$ ... Sn: Org.Comp.16-60
– ... $(n-C_8H_{17})_2Sn[-OAs(O)(C_6H_4-COOH-4)O-]$ ... Sn: Org.Comp.16-60
$AsC_{23}H_{41}O_3Sn$ ... $(n-C_8H_{17})_2Sn[-OAs(O)(CH_2C_6H_5)O-]$ ... Sn: Org.Comp.16-59
– ... $(n-C_8H_{17})_2Sn[-OAs(O)(C_6H_4CH_3-4)O-]$ ... Sn: Org.Comp.16-60
$AsC_{24}ClH_{25}IMoO_2$ $(C_5H_5)Mo(CO)_2(I)[Cl-CH_2CH_2CH_2-As(C_6H_4CH_3-2)_2]$
Mo:Org.Comp.7-56/7, 100
$AsC_{24}Cl_2H_{29}N_2PtS$
$[((C_6H_5)_3As)PtCl_2(i-C_3H_7-N=S=N-C_3H_7-i)]$ ... S: S-N Comp.7-316/7, 319
$AsC_{24}Cl_2H_{37}N_2PtS$
$[(As(C_2H_5)_3)PtCl_2(2,4,6-(CH_3)_3C_6H_2-N=S=N-C_6H_2(CH_3)_3-2,4,6)]$ ... S: S-N Comp.7-323/6
$AsC_{24}Cl_5H_{20}NOsS$ $[(C_6H_5)_4As][Os(NS)Cl_5]$ ... S: S-N Comp.5-84
$AsC_{24}Cl_5H_{20}NReS$ $[(C_6H_5)_4As][Re(NS)Cl_5]$ ... S: S-N Comp.5-51, 65/6
$AsC_{24}Cl_5H_{20}N_2OsS_2$
$[(C_6H_5)_4As][Os((NS)_2Cl)Cl_4]$ ... S: S-N Comp.5-51, 84/5
$AsC_{24}Cl_6H_{20}MoNS$
$[(C_6H_5)_4As][MoCl_5(NSCl)]$ ... S: S-N Comp.5-260/1
$AsC_{24}Cl_6H_{20}NSW$ $[(C_6H_5)_4As][WCl_5(NSCl)]$ ... S: S-N Comp.5-263/4
$AsC_{24}Cl_6H_{20}N_2ReS_2$
$[(C_6H_5)_4As][ReCl_4(NSCl)_2] \cdot CH_2Cl_2$ ... S: S-N Comp.5-267/9
$AsC_{24}F_5H_{20}Si$ ... $[(C_6H_5)_4As][SiF_5]$ ... Si: SVol.B7-272
$AsC_{24}F_6H_{24}MoO_2P$
$[(C_5H_5)Mo(CO)_2(CH_2=CH-CH_2-As(C_6H_4-4-CH_3)_2)][PF_6]$
Mo:Org.Comp.8-185, 188/9
$AsC_{24}FeH_{20}IS$ ... $C_5H_5Fe(CS)(As(C_6H_5)_3)I$ ... Fe: Org.Comp.B15-262
$AsC_{24}H_{20}NS_3$ ... $[(C_6H_5)_4As][S_3N]$ ... S: S-N Comp.6-299/303
– ... $[(C_6H_5)_4As][S_3{}^{15}N]$ ... S: S-N Comp.6-300
$AsC_{24}H_{20}NS_4$ ... $[(C_6H_5)_4As][S_4N]$ ... S: S-N Comp.6-304, 306/8, 310/2
$AsC_{24}H_{24}MoO_2^+$ $[(C_5H_5)Mo(CO)_2(CH_2=CHCH_2-As(C_6H_4CH_3-4)_2)][PF_6]$
Mo:Org.Comp.8-185, 188/9
$AsC_{24}H_{25}MoO_2$ ... $(C_5H_5)Mo(CO)_2[-CH_2CH_2CH_2As(C_6H_4CH_3-4)_2-]$
Mo:Org.Comp.8-113, 156
$AsC_{24}H_{37}Sn$ ... $(n-C_4H_9)_3Sn-As(C_6H_5)_2$ ... Sn: Org.Comp.19-238
$AsC_{24}H_{41}O_4Sn$ ... $(C_8H_{17})_2Sn[-OAs(O)(C_6H_5)CH_2COO-]$ ... Sn: Org.Comp.16-61
$AsC_{25}ClH_{20}MoO_2$ $(C_5H_5)Mo(CO)_2(Cl)[As(C_6H_5)_3]$ ... Mo:Org.Comp.7-56/7, 100/1

$AsC_{25}ClH_{21}N_2O_7Re$
$[(CO)_3Re(As(C_6H_5)_3)(NCCH_3)_2][ClO_4]$ .... Re: Org.Comp.1-311
$AsC_{25}Cl_8H_{22}N_2ReS_2$
$[(C_6H_5)_4As][ReCl_4(NSCl)_2] \cdot CH_2Cl_2$ .... S: S-N Comp.5-267/9
$AsC_{25}F_6FeH_{20}OPS$
$[(C_5H_5)Fe(CS)(CO)As(C_6H_5)_3][PF_6]$ .... Fe: Org.Comp.B15-272
$AsC_{25}FeH_{20}OS^+$ .. $[C_5H_5Fe(CS)(CO)As(C_6H_5)_3]^+$ .... Fe: Org.Comp.B15-272
$AsC_{25}FeH_{30}MoO_6P$
$(C_5H_5)Mo(CO)_2[P(CH_3)_3]$-$As(CH_3)_2$-$Fe(CO)_3$ $[C_6H_5CH{=}CHC({=}O)CH_3]$ .... Mo: Org.Comp.7-134, 150
$AsC_{25}Fe_3H_{18}NO_9S$
$(CO)_9Fe_3[10\text{-}(10,5\text{-}AsNC_{12}H_9)](S\text{-}C_4H_9\text{-}t)$ .... Fe: Org.Comp.C6a-213, 215
$AsC_{25}Fe_3H_{19}O_9S$ $(CO)_9Fe_3[As(C_6H_5)_2](S\text{-}C_4H_9\text{-}t)$ .... Fe: Org.Comp.C6a-213/4
$AsC_{25}H_{18}NO_7Th$ .. $(O_2)Th[2,6\text{-}(OOC)_2\text{-}NC_5H_3] \cdot (C_6H_5)_3AsO$ .... Th: SVol.D4-158
$AsC_{25}H_{20}IMoO_2$ .. $(C_5H_5)Mo(CO)_2(I)[As(C_6H_5)_3]$ .... Mo: Org.Comp.7-56/7, 101
$AsC_{25}H_{20}MoNaO_2$ $Na[(C_5H_5)Mo(CO)_2(As(C_6H_5)_3)]$ .... Mo: Org.Comp.7-50
$AsC_{25}H_{20}MoO_2^-$ $[(C_5H_5)Mo(CO)_2(As(C_6H_5)_3)]^-$ .... Mo: Org.Comp.7-50
$AsC_{25}H_{20}O_2Re$ ... $(C_5H_5)Re(CO)_2[As(C_6H_5)_3]$ .... Re: Org.Comp.3-210, 213
$AsC_{25}H_{21}N_2O_3Re^+$
$[(CO)_3Re(As(C_6H_5)_3)(NCCH_3)_2]^+$ .... Re: Org.Comp.1-311
$AsC_{25}H_{25}O_3PReS_2$
$(C_6H_5)_3As$-$Re(CO)_3[\text{-}S\text{-}P(C_2H_5)_2{=}S\text{-}]$ .... Re: Org.Comp.1-134
$AsC_{25}H_{44}NO_3Sn$ .. $(C_4H_9)_3SnN(C_6H_5)COAs(OC(CH_3)_2)_2$ .... Sn: Org.Comp.18-181, 188
$AsC_{26}Cl_2H_{26}In$ ... $[As(C_6H_5)_4][(CH_3)_2In(Cl)_2]$ .... In: Org.Comp.1-349/50, 352
$AsC_{26}Cl_2H_{27}N_2PtS$
$[((C_6H_5)_3As)PtCl_2(CH_3N{=}S{=}NCH_3)] \cdot C_6H_6$.... S: S-N Comp.7-316/7, 322
$AsC_{26}Cl_2H_{33}N_2PtS$
$[((C_6H_5)_3As)PtCl_2(t\text{-}C_4H_9\text{-}N{=}S{=}N\text{-}C_4H_9\text{-}t)]$.... S: S-N Comp.7-316/8
$AsC_{26}Cl_4H_{26}Sb$... $[As(C_6H_5)_4][(CH_3)_2SbCl_4]$.................. Sb: Org.Comp.5-141
$AsC_{26}Cl_7H_{26}Ti_2$ .. $[(C_6H_5)_4As][(CH_3)_2Ti_2Cl_7]$ .... Ti: Org.Comp.5-4
$AsC_{26}F_6FeH_{23}NOP$
$[(C_5H_5)Fe(CNCH_3)(CO)As(C_6H_5)_3][PF_6]$....... Fe: Org.Comp.B15-301
$AsC_{26}FeH_{22}O_2^+$ .. $[C_5H_5Fe(CO)_2CH_2As(C_6H_5)_3]^+$.............. Fe: Org.Comp.B14-141
$AsC_{26}FeH_{23}NO^+$. $[C_5H_5Fe(CNCH_3)(CO)As(C_6H_5)_3]^+$........... Fe: Org.Comp.B15-301
$AsC_{26}FeH_{28}N$ .... $(C_5H_5)Fe[C_5H_3(As(C_6H_5)_2)\text{-}CH(CH_3)\text{-}N(CH_3)_2]$ Fe: Org.Comp.A10-94
$AsC_{26}Fe_3H_{15}O_8SSe$
$(CO)_8Fe_3SSe[As(C_6H_5)_3]$ .... Fe: Org.Comp.C6a-42, 45
$AsC_{26}Fe_3H_{15}O_8STe$
$(CO)_8Fe_3STe[As(C_6H_5)_3]$ .... Fe: Org.Comp.C6a-42, 45
$AsC_{26}Fe_3H_{15}O_8S_2$ $(CO)_8Fe_3S_2[As(C_6H_5)_3]$ .... Fe: Org.Comp.C6a-42, 43, 47
$AsC_{26}Fe_3H_{15}O_8SeTe$
$(CO)_8Fe_3SeTe[As(C_6H_5)_3]$ .... Fe: Org.Comp.C6a-42, 46
$AsC_{26}Fe_3H_{15}O_8Se_2$
$(CO)_8Fe_3Se_2[As(C_6H_5)_3]$ .... Fe: Org.Comp.C6a-42, 44
$AsC_{26}Fe_3H_{15}O_8Te_2$
$(CO)_8Fe_3Te_2[As(C_6H_5)_3]$ .... Fe: Org.Comp.C6a-42, 45
$AsC_{26}H_{21}MnN_4$... $Mn[\text{-}As(CH_3)\text{-}2\text{-}C_6H_4\text{-}N{=}C(C_6H_5)C(C_6H_5){=}N$ $\text{-}N(C_5H_4N\text{-}2)\text{-}]$ .... Mn: MVol.D6-268/9
$AsC_{26}H_{22}O_5Re$ ... $(C_6H_5)_3As$-$Re(CO)_3[\text{-}O\text{-}CCH_3{=}CH\text{-}CCH_3{=}O\text{-}]$ Re: Org.Comp.1-134
$AsC_{26}H_{23}MoO_2$... $(C_5H_5)Mo(CO)_2[As(C_6H_5)_3]CH_3$ .... Mo: Org.Comp.8-77, 84

$AsC_{26}H_{38}N_2O_{10}Os_3PS$
$[H_2Os_3(CO)_{10}P(C_4H_9\text{-t})_2N{=}SNAs(C_4H_9\text{-t})_2]$ .... S: S-N Comp.7-123/6

$AsC_{27}ClH_{23}NO_3Sn$ $(C_6H_5CH_2)_2Sn[\text{-}OAs(O)(C_6H_4(N{=}CHC_6H_4Cl\text{-}2)\text{-}2)O\text{-}]$
Sn: Org.Comp.16-81, 83/4

$AsC_{27}Cl_2H_{27}MnN_4O_9$
$[Mn(2\text{-}(CH_3)_2As\text{-}C_6H_4\text{-}N{=}C(C_6H_5)C(C_6H_5){=}N$
$\text{-}NH\text{-}2\text{-}C_5H_4N)(H_2O)][ClO_4]_2$ Mn:MVol.D6-268/9

$AsC_{27}Co_2H_{33}MoO_4P_2S$
$C_5H_5Mo(CO)(S)(As(CH_3)_2)Co(P(CH_3)_2C_6H_5)$
$(CO)Co(P(CH_3)_2C_6H_5)(CO)_2$ Mo:Org.Comp.6-394

$AsC_{27}F_6FeH_{26}N_2P$ $[(C_5H_5)Fe(CNCH_3)_2As(C_6H_5)_3][PF_6]$ .... Fe: Org.Comp.B15-327

$AsC_{27}F_6FeH_{26}OPS_2$
$[C_5H_5(CO)(As(C_6H_5)_3)Fe{=}C(SCH_3)_2][PF_6]$ ..... Fe: Org.Comp.B16a-67, 72

$AsC_{27}F_6H_{24}MoO_2P$
$[(C_5H_5)Mo(CO)_2(CH_2{=}CH_2)(P(C_6H_5)_3)][AsF_6]$ .. Mo:Org.Comp.8-188

$AsC_{27}F_6H_{24}MoO_3P$
$[(C_5H_5)Mo(CO)_2(P(C_6H_5)_3){=}CH\text{-}OCH_3][AsF_6]$ .. Mo:Org.Comp.8-106

$AsC_{27}FeH_{26}N_2^+$ .. $[(C_5H_5)Fe(CNCH_3)_2As(C_6H_5)_3]^+$ .... Fe: Org.Comp.B15-327

$AsC_{27}FeH_{26}OS_2^+$ $[C_5H_5(CO)(As(C_6H_5)_3)Fe{=}C(SCH_3)_2]^+$ .... Fe: Org.Comp.B16a-67, 72

$AsC_{27}Fe_3H_{15}O_9Te_2$
$(CO)_9Fe_3Te_2[As(C_6H_5)_3]$ .... Fe: Org.Comp.C6a-121/3

$AsC_{27}Fe_3H_{21}O_9S$ $(CO)_9Fe_3[As(C_6H_5)_2](S\text{-}C_6H_{11}\text{-}c)$ .... Fe: Org.Comp.C6a-213/4

$AsC_{27}H_{23}MoO_3$... $(C_5H_5)Mo(CO)_2[As(C_6H_5)_3]\text{-}C({=}O)CH_3$ Mo:Org.Comp.8-45, 57

$AsC_{27}H_{23}N_2O_5Sn$ $(C_6H_5CH_2)_2Sn[\text{-}OAs(O)(C_6H_4(N{=}CHC_6H_4NO_2\text{-}2)\text{-}2)O\text{-}]$
Sn: Org.Comp.16-81, 83/4

– .... $(C_6H_5CH_2)_2Sn[\text{-}OAs(O)(C_6H_4(N{=}CHC_6H_4NO_2\text{-}3)\text{-}2)O\text{-}]$
Sn: Org.Comp.16-81, 83/4

$AsC_{28}F_4Fe_3H_{16}O_{10}P$
$(CO)_{10}Fe_3[1\text{-}(CH_3)_2As\text{-}2\text{-}(C_6H_5)_2P\text{-}C_4F_4\text{-}3,3,4,4]$
Fe: Org.Comp.C6b-6, 7, 10

$AsC_{28}FeH_{21}O_8PRe$
$(CO)_4Re[P(C_6H_5)_3]\text{-}As(CH_3)_2\text{-}Fe(CO)_4$ .... Re: Org.Comp.1-484

$AsC_{28}H_{20}N_2O_{10}Re$ $[As(C_6H_5)_4][cis\text{-}(CO)_4Re(ONO_2)_2]$ .... Re: Org.Comp.1-342

$AsC_{28}H_{28}MnN_2S_4$ $[As(C_6H_5)_4][Mn(SCH{=}CHS)_2N_2H_4]$ Mn:MVol.D7-45

$AsC_{28}H_{29}MoO_2Sn$ $(C_5H_5)Mo(CO)_2[As(C_6H_5)_3]Sn(CH_3)_3$ Mo:Org.Comp.7-119, 122, 125

$AsC_{29}ClH_{43}NO_3Sn$ $(C_8H_{17})_2Sn[\text{-}OAs(O)(C_6H_4(N{=}CHC_6H_4Cl\text{-}2)\text{-}2)O\text{-}]$
Sn: Org.Comp.16-60

$AsC_{29}Fe_2H_{23}O_2$ .. $[(C_5H_5)(CO)_2Fe\text{-}C_5H_4]Fe[C_5H_4\text{-}As(C_6H_5)_2]$ ... Fe: Org.Comp.A10-173/4

$AsC_{29}Fe_3H_{15}O_{11}$. $(CO)_{11}Fe_3[As(C_6H_5)_3]$ .... Fe: Org.Comp.C6b-128, 133, 134

$AsC_{29}H_{43}N_2O_5Sn$ $(C_8H_{17})_2Sn[\text{-}OAs(O)(C_6H_4(N{=}CHC_6H_4NO_2\text{-}2)\text{-}2)O\text{-}]$
Sn: Org.Comp.16-60

– .... $(C_8H_{17})_2Sn[\text{-}OAs(O)(C_6H_4(N{=}CHC_6H_4NO_2\text{-}3)\text{-}2)O\text{-}]$
Sn: Org.Comp.16-60

$AsC_{29}H_{44}NO_3Sn$ .. $(C_8H_{17})_2Sn[\text{-}OAs(O)(C_6H_4N{=}CHC_6H_5\text{-}2)O\text{-}]$ ... Sn: Org.Comp.16-60

$AsC_{30}Cl_5H_{25}Sb$... $[As(C_6H_5)_4][C_6H_5SbCl_5]$ .... Sb: Org.Comp.5-245

$AsC_{30}F_5H_{25}Sb$ ... $[As(C_6H_5)_4][C_6H_5SbF_5]$ .... Sb: Org.Comp.5-238

$AsC_{30}Fe_3H_{25}O_8S$ $(CO)_8(H)Fe_3S\text{-}(C_4H_9\text{-}t)[As(C_6H_5)_3]$ .... Fe: Org.Comp.C6a-49/51

$AsC_{30}H_{20}N_2O_4Re$ $[As(C_6H_5)_4][(CO)_4Re(CN)_2]$ .... Re: Org.Comp.1-343

$AsC_{30}H_{25}Sn$ ..... $(C_6H_5)_3Sn\text{-}As(C_6H_5)_2$ .... Sn: Org.Comp.19-239/40

$AsC_{30}H_{46}Mn_2N_2O_4PS$
$[(C_5H_5)Mn(CO)_2(P(C_4H_9\text{-}t)_2N{=}S{=}N\text{-}$
$As(C_4H_9\text{-}t)_2)Mn(CO)_2(C_5H_5)]$ . . . . S: S-N Comp.7-127, 130
$AsC_{31}ClFeH_{31}NPd$ $(C_5H_5)Fe[C_5H_3(PdCl(As(C_6H_5)_3))CH_2N(CH_3)_2]$ Fe: Org.Comp.A10-178/80, 184
$AsC_{31}ClFeH_{31}NPd^+$
$[(C_5H_5)Fe(C_5H_3(Pd(AsCl(C_6H_5)_3))\text{-}CH_2N(CH_3)_2)]^+$
Fe: Org.Comp.A10-184
$AsC_{31}ClH_{27}NO_7PRe$
$[(CO)_3Re(P(C_6H_5)_2CH_2CH_2As(C_6H_5)_2)(NCCH_3)][ClO_4]$
Re: Org.Comp.1-303
$AsC_{31}ClH_{29}MoO_2P$
$(CH_2CHCH_2)Mo(Cl)(CO)_2[(C_6H_5)_2P\text{-}CH_2CH_2\text{-}As(C_6H_5)_2]$
Mo: Org.Comp.5-271, 277
$AsC_{31}F_6FeH_{26}OP$ $[C_5H_5(CO)(P(C_6H_5)_3)Fe{=}CH(C_6H_5)][AsF_6]$ . . . . . Fe: Org.Comp.B16a-21, 28
$AsC_{31}F_6H_{30}MoO_4P$
$[(C_5H_5)Mo(CO)_2(P(C_6H_5)_3){=}CH\text{-}OC(O)\text{-}C_4H_9\text{-}t][AsF_6]$
Mo: Org.Comp.8-107
$AsC_{31}FeH_{25}O_2$ . . . $[As(C_6H_5)_4][C_5H_5Fe(CO)_2]$ . . . . Fe: Org.Comp.B14-111
$AsC_{31}H_{20}O_3Re$ . . . $(C_6H_5)_4C_4AsRe(CO)_3$ . . . . Re: Org.Comp.2-417
$AsC_{31}H_{21}O_5Re$ . . . $(C_6H_5)_3As\text{-}Re(CO)_3[\text{-}O\text{-}(1,2\text{-}C_{10}H_6)\text{-}O\text{-}]$, radical
Re: Org.Comp.1-147
$AsC_{31}H_{25}MoN_2O_2$ $[As(C_6H_5)_4][C_5H_5Mo(NO)(CO)CN]$ . . . . Mo: Org.Comp.6-246
$AsC_{31}H_{27}NO_3PRe^+$
$[(CO)_3Re(P(C_6H_5)_2CH_2CH_2As(C_6H_5)_2)(NCCH_3)]^+$
Re: Org.Comp.1-303
$AsC_{31}H_{29}IMoO_2P$ $(CH_2CHCH_2)Mo(I)(CO)_2[(C_6H_5)_2P\text{-}CH_2CH_2\text{-}As(C_6H_5)_2]$
Mo: Org.Comp.5-271, 272, 277
$AsC_{32}ClFeH_{32}N_2Pd$
$(C_5H_5)Fe[C_5H_3(PdCl(As(C_6H_5)_3))\text{-}C(CH_3){=}N\text{-}N(CH_3)_2]$
Fe: Org.Comp.A10-178, 181
$AsC_{32}F_6FeH_{27}NOP$
$[(C_5H_5)Fe(CN\text{-}CH_2C_6H_5)(CO)As(C_6H_5)_3][PF_6]$. . Fe: Org.Comp.B15-303
$AsC_{32}FeH_{27}NO^+$. $[(C_5H_5)Fe(CN\text{-}CH_2C_6H_5)(CO)As(C_6H_5)_3]^+$ . . . . Fe: Org.Comp.B15-303
$AsC_{32}Fe_3H_{27}O_8S$ $(CO)_8(H)Fe_3S\text{-}(C_6H_{11}\text{-}c)[As(C_6H_5)_3]$ . . . . Fe: Org.Comp.C6a-49/51
$AsC_{32}GeH_{25}O$. . . . $Ge(C_6H_5)_3C(As(C_6H_5)_2){=}C{=}O$. . . . . Ge: Org.Comp.3-110
$AsC_{32}H_{20}N_2O_4ReS_2$
$[As(C_6H_5)_4][(CO)_4Re(\text{-}S{=}C(CN)\text{-}C(CN){=}S\text{-})]$ . . . Re: Org.Comp.1-369, 370
$AsC_{32}H_{24}NO_6Th$ . . $(O_2)Th[2\text{-}(2\text{-}O\text{-}C_6H_4\text{-}CH{=}N)\text{-}C_6H_4\text{-}COO]\cdot(C_6H_5)_3AsO$
Th: SVol.D4-158
$AsC_{32}H_{24}N_2PPdS_3$ $Pd(SCN)(NCS)[(C_6H_5)_2As\text{-}2\text{-}C_6H_4\text{-}P(S)(C_6H_5)_2]$
Pd: SVol.B2-350/1
– . . . . $Pd(SCN)_2[(C_6H_5)_2As\text{-}2\text{-}C_6H_4\text{-}P(S)(C_6H_5)_2]$. . . . Pd: SVol.B2-350
$AsC_{32}H_{24}N_2PdS_2Sb$
$Pd(SCN)_2[(C_6H_5)_2As\text{-}2\text{-}C_6H_4\text{-}Sb(C_6H_5)_2]$ . . . . . Pd: SVol.B2-352
$AsC_{33}ClFeH_{36}NO_4Rh$
$[((C_5H_5)Fe(C_5H_3(As(C_6H_5)_2)\text{-}CH(CH_3)$
$\text{-}N(CH_3)_2))Rh(C_7H_8)][ClO_4]$ . . . . Fe: Org.Comp.A10-94/5

$AsC_{33}F_6H_{28}MoO_3P$
$[(C_5H_5)Mo(CO)_2(P(C_6H_5)_3)=CH-O-CH_2C_6H_5][AsF_6]$
Mo:Org.Comp.8-107
$AsC_{33}FeH_{25}$ ..... $C_5H_5FeC_4As((C_6H_5)_4-2,3,4,5)$ .... Fe: Org.Comp.B17-212
$AsC_{33}FeH_{36}NRh^+$ $[((C_5H_5)Fe(C_5H_3(As(C_6H_5)_2)-CHCH_3-N(CH_3)_2))Rh(C_7H_8)]^+$ .... Fe: Org.Comp.A10-94/5
$AsC_{33}H_{20}O_5Re$ ... $(CO)_5ReAsC_4(C_6H_5)_4$-c .... Re: Org.Comp.2-19, 20
$AsC_{33}H_{25}O_3PReS_2$
$(C_6H_5)_3As-Re(CO)_3[-S-P(C_6H_5)_2=S-]$........ Re: Org.Comp.1-134
$AsC_{33}H_{28}MoN_2O_2^+$
$[(CH_2CHCH_2)Mo(CO)_2(NC_5H_4-2-(2-C_5H_4N))(As(C_6H_5)_3)][BF_4]$...................... Mo:Org.Comp.5-200, 202
$AsC_{33}H_{29}MnNS_4$.. $[As(C_6H_5)_4][Mn(SCH=CHS)_2NC_5H_5]$.......... Mn:MVol.D7-45/6
$AsC_{34}ClF_6H_{26}MoO_4$
$[As(C_6H_5)_4][(CH_2CHCH_2)Mo(Cl)(CO)_2(CF_3-C(O)CHC(O)-CF_3)]$............. Mo:Org.Comp.5-269/71
$AsC_{34}Cl_2H_{33}N_2PtS$
$[(As(C_6H_5)_3)PtCl_2(3,5-(CH_3)_2C_6H_3-N=S=N-C_6H_3(CH_3)_2-3,5)]$ .... S: S-N Comp.7-324/6
$AsC_{34}FeH_{28}P$ .... $[(C_6H_5)_2As-C_5H_4]Fe[C_5H_4-P(C_6H_5)_2]$ .... Fe: Org.Comp.A10-88/9
$AsC_{34}Ge_2H_{38}O_4Re$
$[As(C_6H_5)_4][(CO)_4Re(Ge(CH_3)_3)_2]$ .... Re: Org.Comp.1-343/4
$AsC_{34}H_{20}O_{10}ReW$ $[(C_6H_5)_4As][(CO)_5ReW(CO)_5]$............... Re: Org.Comp.2-197, 204
$AsC_{34}H_{28}MoNO_3$ $[As(C_6H_5)_4][(C_5H_5)Mo(CO)_2(CN)-C(=O)CH_3]$... Mo:Org.Comp.8-1, 3/4
$AsC_{34}H_{30}MoN_2O_2^+$
$[(CH_2C(CH_3)CH_2)Mo(CO)_2(NC_5H_4-2-(2-C_5H_4N))(As(C_6H_5)_3)][BF_4]$.............. Mo:Org.Comp.5-200, 204
$AsC_{34}H_{38}O_4ReSn_2$
$[As(C_6H_5)_4][(CO)_4Re(Sn(CH_3)_3)_2]$ .... Re: Org.Comp.1-344
$AsC_{35}ClF_6H_{28}MoO_4$
$[As(C_6H_5)_4][(CH_2C(CH_3)CH_2)Mo(Cl)(CO)_2(CF_3-C(O)CHC(O)-CF_3)]$............. Mo:Org.Comp.5-269/71
– .... $[As(C_6H_5)_4][(CH_3CHCHCH_2)Mo(Cl)(CO)_2(CF_3-C(O)CHC(O)-CF_3)]$............. Mo:Org.Comp.5-269/71
$AsC_{35}Fe_3H_{21}O_{10}$. $[As(C_6H_5)_4][(CO)_{10}Fe_3(CH)]$................ Fe: Org.Comp.C6b-60, 61, 64
$AsC_{35}H_{30}MoNO_3$ $[As(C_6H_5)_4][(C_5H_5)Mo(CO)_2(CN)-C(=O)-C_2H_5]$ Mo:Org.Comp.8-1, 4
$AsC_{35}H_{32}MoO_3P$. $(C_5H_5)Mo(CO)_2[P(C_6H_5)_2-CH_2CH_2-As(C_6H_5)_2]-C(=O)CH_3$
Mo:Org.Comp.8-45, 56
$AsC_{35}H_{35}O_5Re$ ... $(C_6H_5)_3As-Re(CO)_3[-O-1,2-(C_6H_2(C_4H_9-t)_2-3,5)-O-]$, radical............... Re: Org.Comp.1-146
$AsC_{36}Cl_2H_{30}NPRhS$
$Rh(NS)Cl_2(P(C_6H_5)_3)(As(C_6H_5)_3)$ .... S: S-N Comp.5-76/9
$AsC_{36}Cl_3H_{30}NPRuS$
$Ru(NS)Cl_3(P(C_6H_5)_3)(As(C_6H_5)_3)$ .... S: S-N Comp.5-51/2, 75
$AsC_{36}Cl_3H_{30}NRuSSb$
$Ru(NS)Cl_3(As(C_6H_5)_3)(Sb(C_6H_5)_3)$ .... S: S-N Comp.5-51/2, 75
$AsC_{36}Cl_3H_{30}N_3Sb$ $[As(C_6H_5)_4][(C_6H_5)_2Sb(Cl_3)N_3]$.............. Sb: Org.Comp.5-158
$AsC_{36}Cl_4H_{30}Sb$... $[As(C_6H_5)_4][(C_6H_5)_2SbCl_4]$................. Sb: Org.Comp.5-154/5
$AsC_{36}F_4H_{30}Sb$ ... $[As(C_6H_5)_4][(C_6H_5)_2SbF_4]$ .... Sb: Org.Comp.5-136
$AsC_{36}F_{15}H_{15}In$ ... $(C_6F_5)_3In \cdot As(C_6H_5)_3$ .... In: Org.Comp.1-89, 93

$AsC_{36}F_{15}H_{15}InO$ . . $(C_6F_5)_3In \cdot OAs(C_6H_5)_3$ . . . . In: Org.Comp.1-89, 92/3

$AsC_{36}H_{30}N_{12}Sb$ . . $[As(C_6H_5)_4][(C_6H_5)_2Sb(N_3)_4]$ . . . . Sb: Org.Comp.5-177

$AsC_{37}Cl_3H_{30}NSSb$ $[As(C_6H_5)_4][(C_6H_5)_2Sb(Cl_3)NCS]$ . . . . Sb: Org.Comp.5-158/9

$AsC_{37}H_{31}MoN_2O_3$ $[As(C_6H_5)_4][(C_5H_5)Mo(CO)_2(CN)-C(O)CH_2CH_2CH_2-CN]$ Mo:Org.Comp.8-1, 5

$AsC_{37}H_{32}MoOP$ . . $C_5H_5Mo(P(C_6H_5)_2C_6H_4As(C_6H_5)_2-2)(CO)CH_3$ . . Mo:Org.Comp.6-256, 258/9

$AsC_{37}H_{32}N_2O_3Re$ $(C_6H_5)_3As-Re(CO)_3[-N(C_6H_4CH_3-4)-C(CH_3)=N(C_6H_4CH_3-4)-]$ . . . . Re: Org.Comp.1-134

$AsC_{38}FeH_{37}NP$ . . . $[(C_6H_5)_2As-C_5H_4]Fe[C_5H_3(P(C_6H_5)_2)-CH(CH_3)-N(CH_3)_2]$ Fe: Org.Comp.A10-325

– . . . . $[(C_6H_5)_2P-C_5H_4]Fe[C_5H_3(As(C_6H_5)_2)-CH(CH_3)-N(CH_3)_2]$ Fe: Org.Comp.A10-325

$AsC_{38}Fe_3H_{25}O_8P_2$ $(CO)_8Fe_3(P-C_6H_5)_2[As(C_6H_5)_3]$ . . . . Fe: Org.Comp.C6a-62/7

$AsC_{38}Fe_3H_{25}O_8P_2^{-}$ $[(CO)_8Fe_3(P-C_6H_5)_2(As(C_6H_5)_3)]^{-}$, radical anion Fe: Org.Comp.C6a-65

$AsC_{38}Fe_3H_{30}NO_9$ $[As(C_6H_5)_4][(CO)_9Fe_3(N(C_4H_9-t)=CH)]$. . . . . Fe: Org.Comp.C6a-287, 288

$AsC_{38}H_{32}O_6ReS$ . . $[(C_6H_5)_4As][H_8C_4O-1-Re(CO)_3(-S-(1,2-C_6H_4)-COO-)]$ Re: Org.Comp.1-137

$AsC_{39}ClH_{30}O_3PRe$ cis-$(CO)_3ReCl[P(C_6H_5)_3][As(C_6H_5)_3]$. . . . . Re: Org.Comp.1-293

$AsC_{39}H_{35}MoN_2O$ $C_5H_5Mo(As(C_6H_5)_3)(CO)(NC_6H_4CH_3-4)_2CH$ . . . Mo:Org.Comp.6-254

$AsC_{40}GeH_{30}O_4Re$ $(CO)_4Re[As(C_6H_5)_3]Ge(C_6H_5)_3$ . . . . Re: Org.Comp.1-455

$AsC_{40}H_{30}N_4S_4Sb$ $[As(C_6H_5)_4][(C_6H_5)_2Sb(NCS)_4]$. . . . . Sb: Org.Comp.5-177

$AsC_{40}H_{30}O_4ReSn$ $(CO)_4Re[As(C_6H_5)_3]Sn(C_6H_5)_3$. . . . . Re: Org.Comp.1-456

$AsC_{40}H_{32}MoNO_3$ $[As(C_6H_5)_4][(C_5H_5)Mo(CO)_2(CN)-C(=O)-CH_2-C_6H_5]$ Mo:Org.Comp.8-1, 4

$AsC_{42}H_{35}N_2Sn$ . . . $(C_6H_5)_3Sn-N(C_6H_5)-As(=N-C_6H_5)(C_6H_5)_2$ . . . . . Sn: Org.Comp.19-30, 32/3

$AsC_{43}H_{44}O_5PRe$ . . $[(C_6H_5)_2PCH_2CH_2As(C_6H_5)_2]Re(CO)_3[-O-1,2-(C_6H_2(C_4H_9-t)_2-3,5)-O-]$, radical . . . . Re: Org.Comp.1-146

$AsC_{45}Fe_3H_{31}O_9P_2$ $[As(C_6H_5)_4][(CO)_9Fe_3(PH-C_6H_5)(P-C_6H_5)]$ . . . . Fe: Org.Comp.C6a-228

$AsC_{45}H_{29}N_3O_{12}Th$ $[As(C_6H_5)_4][Th(NC_5H_3(COO)_2-2,6)_3]$ . . . . Th: SVol.C7-154, 156

– . . . . $[As(C_6H_5)_4][Th(NC_5H_3(COO)_2-2,6)_3] \cdot 3\,H_2O$ . . Th: SVol.C7-154, 156

$AsC_{45}H_{41}MnNS_4$. . $[As(C_6H_5)_4][Mn(1,2-S_2C_6H_2(CH_3)_2-4,5)_2NC_5H_5]$ Mn:MVol.D7-45/6

$AsC_{46}Cl_2H_{44}PTi_2$ $[(C_5H_5)_2TiCl]_2((C_6H_5)_2PCH_2CH_2As(C_6H_5)_2)$. . . . Ti: Org.Comp.5-179

$AsC_{55}H_{59}N_4OOs$. . $(CO)Os[N_4C_{20}H_4(C_2H_5)_8][As(C_6H_5)_3]$ . . . . Os: Org.Comp.A1-204

$AsC_{80}H_{60}In$. . . . . . $[As(C_6H_5)_4][In(C_4(C_6H_5)_4)_2]$. . . . . In: Org.Comp.1-340, 345

$AsClH_4MnO_4^{-}$ . . . . $[MnO_3(AsH_3)(OH)Cl]^{-}$. Mn:MVol.D8-202

$AsCoH_2Mo_{11}O_{40}^{5-}$ $AsMo_{11}Co(H_2O)O_{39}^{5-}$. Mo:SVol.B3b-117

AsCoRh. . . . . CoRhAs. . . . . Rh: SVol.A1-227

AsF_6NS . . . . $[SN][AsF_6]$. . . . . S: S-N Comp.5-44/5

$AsF_{10}NOSTe$. . . . . $[SN][F_5TeOAsF_5]$ . . . . S: S-N Comp.5-45

$AsF_{10}NS_2$ . . . . $[F_2SNSF_2][AsF_6]$. . . . . S: S-N Comp.8-71/3

$AsFeHMo_{11}O_{40}^{5-}$ $AsMo_{11}Fe(OH)O_{39}^{5-}$. Mo:SVol.B3b-117

AsFeRh . . . . RhFeAs. . . . . Rh: SVol.A1-247

AsGa. . . . . GaAs

Sorption of NH_3 . . . . Ga: SVol.D1-214

$AsHMoO_6$ . . . . $HMoO_2(AsO_4)$. Mo:SVol.B3b-188

$AsHMo_{11}O_{39}^{6-}$ . . . $HAsMo_{11}O_{39}^{6-}$ Mo:SVol.B3b-94/6

$AsH_2Mo_{18}O_{60}^{7-}$ .. $(AsO_3)(H_2O_3)Mo_{18}O_{54}^{7-}$ Mo:SVol.B3b-50, 52
$AsH_3Mo_9O_{34}^{6-}$ ... $H_3AsMo_9O_{34}^{6-}$ Mo:SVol.B3b-94/6
$AsH_4IMnO_4^-$ ..... $[MnO_3(AsH_3)(OH)I]^-$ Mn:MVol.D8-202
$AsH_4Mo_9O_{34}^{5-}$ ... $H_4AsMo_9O_{34}^{5-}$ Mo:SVol.B3b-94/6
$AsH_5Mo_9O_{34}^{4-}$ ... $H_5AsMo_9O_{34}^{4-}$ Mo:SVol.B3b-94/6
$AsH_6Mo_9Na_3O_{34}$.. $Na_3[AsMo_9O_{31}(H_2O)_3] \cdot x\ H_2O$ (x = 12 to 13)
Mo:SVol.B3b-50
$AsH_6Mo_9O_{34}^{3-}$ ... $H_6AsMo_9O_{34}^{3-}$ Mo:SVol.B3b-94/6
AsMnRh .... RhMnAs .... Rh: SVol.A1-202
$AsMo_9O_{34}^{9-}$ ..... $AsO_4Mo_9O_{30}^{9-}$ Mo:SVol.B3b-48/50, 117
$AsMo_{11}O_{39}^{7-}$ .... $AsO_4Mo_{11}O_{35}^{7-}$ Mo:SVol.B3b-47/8, 116/7
$AsMo_{12}O_{40}^{3-}$ .... $AsO_4Mo_{12}O_{36}^{3-}$ Mo:SVol.B3b-44/7, 115/6, 120
$AsNdRh_2$ .... Rh_2NdAs .... Rh: SVol.A1-79
AsNiRh .... RhNiAs .... Rh: SVol.A1-221
AsPd .... PdAs .... Pd: SVol.B2-346
AsPdS .... PdAsS .... Pd: SVol.B2-347
AsPdSe .... PdAsSe .... Pd: SVol.B2-244/5, 347
$AsPd_2$ .... Pd_2As .... Pd: SVol.B2-343, 345
$AsPd_{2.5}$ .... $Pd_{2.5}As$ = Pd_5As_2 .... Pd: SVol.B2-343/5
$AsPd_{2.65}$ .... $Pd_{2.65}As$ .... Pd: SVol.B2-343/4
$AsPd_3$ .... Pd_3As .... Pd: SVol.B2-343/4
$AsPd_5$ .... Pd_5As .... Pd: SVol.B2-343/4
$AsPd_7$ .... Pd_7As .... Pd: SVol.B2-343/4
AsPo .... AsPo .... Po: SVol.1-312
AsRh .... RhAs .... Rh: SVol.A1-38/41
AsRhSb .... RhSbAs .... Rh: SVol.A1-44
AsRhTi .... RhTiAs .... Rh: SVol.A1-114
AsRhV .... RhVAs .... Rh: SVol.A1-160/1
$AsRh_{1.5}$ .... $Rh_{1.5}As$ .... Rh: SVol.A1-38/41
$AsRh_{1.7}$ .... $Rh_{1.7}As$ .... Rh: SVol.A1-38
$AsRh_2$ .... Rh_2As .... Rh: SVol.A1-38/41
AsTh .... ThAs
Thermodynamic data .... Th: SVol.A4-183
$As_2BC_{43}F_4H_{35}MoO_2$
$[(C_5H_5)Mo(CO)_2(As(C_6H_5)_3)_2][BF_4]$ Mo:Org.Comp.7-283, 292
$As_2B_2C_{78}H_{74}MnN_4$
$[Mn(6\text{-}CH_3\text{-}NC_5H_3\text{-}2\text{-}CH{=}N\text{-}C_6H_4\text{-}2\text{-}$
$As(CH_3)_2)_2][B(C_6H_5)_4]_2 \cdot 4\ H_2O$ Mn:MVol.D6-79/80
$As_2B_2C_{82}H_{82}MnN_4$
$[Mn(6\text{-}CH_3\text{-}NC_5H_3\text{-}2\text{-}CH{=}N\text{-}C_6H_4\text{-}2\text{-}$
$As(C_2H_5)_2)_2][B(C_6H_5)_4]_2 \cdot 4\ H_2O$ Mn:MVol.D6-79/80
$As_2B_{20}C_{10}H_{36}Pd$. $[1,2\text{-}C_2B_{10}H_{10}\text{-}1\text{-}CH_2As(CH_3)_2\text{-}2\text{-}]_2Pd$ .... B: B Comp.SVol.3/4-194
$As_2B_{20}C_{10}H_{36}Pt$ .. $[1,2\text{-}C_2B_{10}H_{10}\text{-}1\text{-}CH_2As(CH_3)_2\text{-}2\text{-}]_2Pt$ .... B: B Comp.SVol.3/4-194
As_2BaPd_2 .... $BaPd_2As_2$ .... Pd: SVol.B2-347
$As_2BrC_9H_{18}O_3ReS_2$
fac-$(CO)_3ReBr[As(CH_3)_2SCH_3]_2$ .... Re: Org.Comp.1-283
$As_2BrC_{13}H_{16}O_3Re$ $(CO)_3ReBr[C_6H_4\text{-}1,2\text{-}(As(CH_3)_2)_2]$ .... Re: Org.Comp.1-190
$As_2BrC_{19}H_{22}O_3Re$ fac-$(CO)_3ReBr[As(CH_3)_2C_6H_5]_2$ .... Re: Org.Comp.1-283
$As_2BrC_{28}H_{22}O_3Re$ $(CO)_3ReBr[(C_6H_5)_2AsCH_2As(C_6H_5)_2]$ .... Re: Org.Comp.1-190
$As_2BrC_{29}H_{22}O_4Re$ $(CO)_4Re(Br)As(C_6H_5)_2CH_2As(C_6H_5)_2$ .... Re: Org.Comp.1-457

$As_2BrC_{31}H_{29}MoO_2$
$(CH_2CHCH_2)Mo(Br)(CO)_2[(C_6H_5)_2As-CH_2CH_2-As(C_6H_5)_2]$ Mo:Org.Comp.5-271, 278, 280/1
As_2BrF_{13} . . . . $[BrF_2][As_2F_{11}]$ . . . . Br: SVol.B3-41/5
$As_2Br_2C_{10}ClH_{18}MnO_5$
$[Mn(C_6H_4(As(CH_3)_2)_2)(H_2O)Br_2][ClO_4]$ Mn:MVol.D8-205/6
$As_2Br_2C_{14}FeH_{20}Pd$
$Fe[C_5H_4-As(CH_3)_2]_2PdBr_2$ . . . . Fe: Org.Comp.A10-89, 90, 93
$As_2Br_2C_{14}FeH_{20}Pt$ $Fe[C_5H_4-As(CH_3)_2]_2PtBr_2$ . . . . Fe: Org.Comp.A10-89, 90, 93
$As_2Br_2C_{20}H_{32}MnSb_2$
$[Mn((CH_3)_2As-C_6H_4-Sb(CH_3)_2-2)_2Br_2]$ Mn:MVol.D8-219
$As_2Br_2C_{26}H_{26}MnO_2$
$[Mn(O=As(C_6H_5)_2CH_3)_2Br_2]$ Mn:MVol.D8-212
$As_2Br_2C_{34}FeH_{28}Pd$
$Fe[C_5H_4-As(C_6H_5)_2]_2PdBr_2$ . . . . Fe: Org.Comp.A10-89, 91/2, 93
$As_2Br_2C_{36}ClH_{30}NOsS$
$Os(NS)Br_2Cl(As(C_6H_5)_3)_2$ . . . . S: S-N Comp.5-79, 82/3
$As_2Br_2C_{36}ClH_{30}NRuS$
$Ru(NS)Br_2Cl(As(C_6H_5)_3)_2$ . . . . S: S-N Comp.5-51/2, 75
$As_2Br_2C_{36}H_{30}Mn$ $[Mn((C_6H_5)_3As)_2Br_2]$ Mn:MVol.D8-202
$As_2Br_2C_{36}H_{30}MnO_2$
$[Mn(O=As(C_6H_5)_3)_2Br_2]$ Mn:MVol.D8-208/9
$As_2Br_2C_{42}H_{42}MnO_2$
$[Mn(O=As(CH_2C_6H_5)_3)_2]Br_2$ Mn:MVol.D8-213
$As_2Br_4C_{11}H_{16}OOs^-$
$[N(C_4H_9-n)_4][(CO)OsBr_4(((CH_3)_2As)_2-1,2-C_6H_4)]$
Os: Org.Comp.A1-200, 207
$As_2Br_4C_{27}H_{52}NOOs$
$[N(C_4H_9-n)_4][(CO)OsBr_4(((CH_3)_2As)_2-1,2-C_6H_4)]$
Os: Org.Comp.A1-200, 207
$As_2C_2H_6Mo_6O_{24}{}^{4-}$
$(CH_3AsO_3)_2Mo_6O_{18}{}^{4-}$ Mo:SVol.B3b-125/8, 130
$As_2C_2H_7Mo_6O_{25}{}^{5-}$
$(CH_3As)_2Mo_6O_{24}(OH)^{5-}$ Mo:SVol.B3b-127/8
$As_2C_2H_8Mo_6O_{25}{}^{4-}$
$(CH_3AsO_3)_2Mo_6O_{18}(OH_2)^{4-}$ Mo:SVol.B3b-125/8, 130
$As_2C_4H_{12}MnS_4$ . . . $[Mn(S_2As(CH_3)_2)_2]$ Mn:MVol.D8-218
$As_2C_4H_{12}N_2S$ . . . . $(CH_3)_2AsN=S=NAs(CH_3)_2$ . . . . S: S-N Comp.7-117/21
$As_2C_5F_{12}NO_5ReS$ $[Re(NS)(CO)_5][AsF_6]_2$ . . . . Re: Org.Comp.2-154/5
S: S-N Comp.5-50/1, 70
$As_2C_6F_{12}H_6N_4O_8S_4Sn$
$3,3-(CH_3)_2-2,4,6,7-(CF_3SO_2)_4-2,4,6,7,1,5,3-[3.1.1]-N_4As_2Sn$ . . . . Sn: Org.Comp.19-79, 80
$As_2C_6H_{18}Sn$ . . . . . $(CH_3)_2As-Sn(CH_3)_2-As(CH_3)_2$ . . . . Sn: Org.Comp.19-241
$As_2C_6H_{36}Mo_{18}N_{18}O_{62}$
$(CH_6N_3)_6[As_2Mo_{18}O_{62}] \cdot 9\ H_2O$ Mo:SVol.B3b-51/2
$As_2C_7F_{12}FeH_5NO_2S$
$[(C_5H_5)Fe(NS)(CO)_2][AsF_6]_2$ . . . . S: S-N Comp.5-50, 70

$As_2C_8CuF_{36}N_8O_4S_4$
$[Cu(NS\text{-}ON(CF_3)_2)_4][AsF_6]_2$ . . . . S: S-N Comp.5-254
$As_2C_8F_6H_{24}Si$ . . . . $[(CH_3)_4As]_2[SiF_6]$ . . . . Si: SVol.B7-298
$As_2C_8F_{12}H_{18}Sn$. . . $(n\text{-}C_4H_9)_2Sn(AsF_6)_2$ . . . . Sn: Org.Comp.19-242
$As_2C_8H_{18}NiO_2Sn$ $(CH_3)_2Sn[-As(CH_3)_2\text{-}Ni(CO)_2\text{-}As(CH_3)_2\text{-}]$. . . . . Sn: Org.Comp.19-241
$As_2C_8H_{24}In_2O_4$ . . . $[(CH_3)_2InOAs(O)(CH_3)_2]_2$ . . . . In: Org.Comp.1-215/6
$As_2C_8H_{25}Mo_6N_2Na_3O_{26}$
$Na_3[(CH_3)_4N]_2[HAs_2Mo_6O_{26}] \cdot 7\ H_2O$. . . . . Mo: SVol.B3b-62/3
$As_2C_9ClH_{18}O_3ReS_2$
fac-$(CO)_3ReCl[CH_3SAs(CH_3)_2]_2$ . . . . Re: Org.Comp.1-282
$As_2C_9Fe_3O_9$ . . . . . $(CO)_9Fe_3As_2$ . . . . Fe: Org.Comp.C6a-140/2
$As_2C_9H_{18}IO_3ReS_2$ fac-$(CO)_3ReI[As(CH_3)_2SCH_3]_2$. . . . . Re: Org.Comp.1-283
$As_2C_{10}Cl_2H_{16}MnO_2$
$[Mn((O{=}As(CH_3)_2)_2\text{-}1,2\text{-}C_6H_4)Cl_2]$. . . . . Mn: MVol.D8-214
$As_2C_{10}Cl_3H_{18}MnO_5$
$[Mn(C_6H_4(As(CH_3)_2)_2)(H_2O)Cl_2][ClO_4]$. . . . . Mn: MVol.D8-205/6
$As_2C_{10}CrH_{18}O_4Sn$ $(CH_3)_2Sn[-As(CH_3)_2\text{-}Cr(CO)_4\text{-}As(CH_3)_2\text{-}]$ . . . . Sn: Org.Comp.19-241
$As_2C_{10}H_{30}Mo_6N_2Na_2O_{24}$
$[(CH_3)_4N]_2Na_2[(CH_3AsO_3)_2Mo_6O_{18}] \cdot 6\ H_2O$ . . Mo: SVol.B3b-126
$As_2C_{11}Cl_4H_{16}OOs^-$
$[N(C_4H_9\text{-}n)_4][(CO)OsCl_4(((CH_3)_2As)_2\text{-}1,2\text{-}C_6H_4)]$
Os: Org.Comp.A1-200, 207
$As_2C_{11}F_{12}H_3N_3O_{10}Re_2S_2$
$[(CO)_5Re\text{-}NS\text{-}N(CH_3)\text{-}SN\text{-}Re(CO)_5][AsF_6]_2$ . . . S: S-N Comp.5-256
$As_2C_{11}Fe_3O_{11}$. . . . $(CO)_{11}Fe_3As_2$ . . . . Fe: Org.Comp.C6b-156
$As_2C_{11}H_{16}I_4OOs^-$ $[N(C_4H_9\text{-}n)_4][(CO)OsI_4(((CH_3)_2As)_2\text{-}1,2\text{-}C_6H_4)]$ Os: Org.Comp.A1-200, 208
$As_2C_{12}CoF_{48}N_{12}O_6S_6$
$[Co(NS\text{-}ON(CF_3)_2)_6][AsF_6]_2$ . . . . S: S-N Comp.5-253/4
$As_2C_{12}F_{12}MnO_8Re$
$(CO)_4Re[-As(CF_3)_2\text{-}Mn(CO)_4\text{-}As(CF_3)_2\text{-}]$. . . . . Re: Org.Comp.1-487
$As_2C_{12}F_{48}FeN_{12}O_6S_6$
$[Fe(NS\text{-}ON(CF_3)_2)_6][AsF_6]_2$ . . . . S: S-N Comp.5-253/4
$As_2C_{12}F_{48}N_{12}NiO_6S_6$
$[Ni(NS\text{-}ON(CF_3)_2)_6][AsF_6]_2$ . . . . S: S-N Comp.5-253/4
$As_2C_{12}Fe_3H_8O_9$ . . $(CO)_9Fe_3(As\text{-}CH_3)_2(CH_2)$ . . . . Fe: Org.Comp.C6a-281
– . . . . $(CO)_9Fe_3As[As(CH_3)_2](CH_2)$. . . . . Fe: Org.Comp.C6a-281
$As_2C_{12}H_{10}Mo_6O_{24}{}^{4-}$
$(C_6H_5AsO_3)_2Mo_6O_{18}{}^{4-}$ . . . . Mo: SVol.B3b-125/6, 130
$As_2C_{12}H_{10}O_6Th$ . . $Th(C_6H_5AsO_3)_2$ . . . . Th: SVol.D1-133
$As_2C_{12}H_{12}Mo_6N_2O_{24}{}^{4-}$
$(NH_2C_6H_4AsO_3)_2Mo_6O_{18}{}^{4-}$. . . . . Mo: SVol.B3b-125/6, 130
$As_2C_{12}H_{12}Mo_6O_{25}{}^{4-}$
$(C_6H_5AsO_3)_2Mo_6O_{18}(OH_2)^{4-}$ . . . . Mo: SVol.B3b-125/6, 130
$As_2C_{12}H_{14}Mo_6N_2O_{25}{}^{4-}$
$(NH_2C_6H_4AsO_3)_2Mo_6O_{18}(OH_2)^{4-}$ . . . . Mo: SVol.B3b-125/6,130
$As_2C_{13}ClCrH_{12}O_9Re$
$(CO)_5ReAs(CH_3)_2\text{-}Cr(As(CH_3)_2Cl)(CO)_4$-cis . . . Re: Org.Comp.2-174
$As_2C_{13}ClH_{12}MoO_9Re$
$(CO)_5ReAs(CH_3)_2\text{-}Mo(As(CH_3)_2Cl)(CO)_4$-cis . . Re: Org.Comp.2-175

$As_2C_{13}ClH_{12}O_9ReW$
$(CO)_5ReAs(CH_3)_2\text{-}W(As(CH_3)_2Cl)(CO)_4\text{-}cis$ ... Re: Org.Comp.2-175

$As_2C_{13}ClH_{16}O_3Re$ $(CO)_3ReCl[1,2\text{-}((CH_3)_2As)_2C_6H_4]$ ... Re: Org.Comp.1-188

$As_2C_{13}Fe_3H_{10}O_9$. $(CO)_9Fe_3(As\text{-}C_2H_5)_2$ ... Fe: Org.Comp.C6a-170

$As_2C_{13}H_{16}IO_3Re$ .. $(CO)_3ReI[C_6H_4\text{-}1,2\text{-}(As(CH_3)_2)_2]$ ... Re: Org.Comp.1-191

$As_2C_{14}ClH_{16}O_4Re$ $(CO)_4Re(Cl)As(CH_3)_2C_6H_4\text{-}2\text{-}As(CH_3)_2$ ... Re: Org.Comp.1-456

$As_2C_{14}Cl_2FeH_{20}Pd$
$Fe[C_5H_4\text{-}As(CH_3)_2]_2PdCl_2$ ... Fe: Org.Comp.A10-89, 90, 93

$As_2C_{14}Cl_2FeH_{20}Pt$ $Fe[C_5H_4\text{-}As(CH_3)_2]_2PtCl_2$ ... Fe: Org.Comp.A10-89, 90, 93

$As_2C_{14}FeH_{20}$ .... $Fe[C_5H_4\text{-}As(CH_3)_2]_2$ ... Fe: Org.Comp.A10-86

$As_2C_{14}FeH_{20}I_2Ni$ $Fe[C_5H_4\text{-}As(CH_3)_2]_2NiI_2$ ... Fe: Org.Comp.A10-91, 93

$As_2C_{14}FeH_{20}I_2Pd$ $Fe[C_5H_4\text{-}As(CH_3)_2]_2PdI_2$ ... Fe: Org.Comp.A10-89, 90, 93

$As_2C_{14}FeH_{20}I_2Pt$. $Fe[C_5H_4\text{-}As(CH_3)_2]_2PtI_2$ ... Fe: Org.Comp.A10-89, 91, 93

$As_2C_{14}FeH_{20}S_2$ .. $Fe[C_5H_4\text{-}AsS(CH_3)_2]_2$ ... Fe: Org.Comp.A10-86

$As_2C_{14}Ge_4H_{42}Sn$ $[(CH_3)_3Ge]_2As\text{-}Sn(CH_3)_2\text{-}As[Ge(CH_3)_3]_2$ ... Sn: Org.Comp.19-241

$As_2C_{15}FeH_{20}I_2NiO$ $Fe[C_5H_4\text{-}As(CH_3)_2]_2Ni(CO)I_2$ ... Fe: Org.Comp.A10-90, 92

$As_2C_{15}FeH_{23}{}^+$ ... $[((CH_3)_2As\text{-}C_5H_4)Fe(C_5H_4\text{-}As(CH_3)_3)]^+$ ... Fe: Org.Comp.A10-86

$As_2C_{15}FeH_{23}I$ .... $[((CH_3)_2As\text{-}C_5H_4)Fe(C_5H_4\text{-}As(CH_3)_3)]I$ ... Fe: Org.Comp.A10-86

$As_2C_{16}Cl_2H_{24}MnN_2O_8$
$Mn[(NH_2\text{-}C_6H_4\text{-}As(CH_3)_2\text{-}2)_2(ClO_4)_2]$ ... Mn: MVol.D8-204

$As_2C_{16}CoFeH_{12}O_{12}Re$
$(CO)_5ReAs(CH_3)_2\text{-}Co(CO)_3As(CH_3)_2\text{-}Fe(CO)_4$ Re: Org.Comp.2-176

$As_2C_{16}F_4H_{12}I_2MnO_8Re$
$1\text{-}[(CO)_4Re(I)\text{-}As(CH_3)_2]\text{-}2\text{-}[As(CH_3)_2\text{-}Mn(I)$
$(CO)_4]\text{-}C_4F_4\text{-}3,3,4,4$ ... Re: Org.Comp.1-499/500

$As_2C_{16}F_4H_{12}MnO_8Re$
$(CO)_4Re[\text{-}As(CH_3)_2\text{-}(1,2\text{-}(C_4F_4\text{-}3,3,4,4))$
$\text{-}As(CH_3)_2\text{-}Mn(CO)_4\text{-}]$ ... Re: Org.Comp.1-499/500

$As_2C_{16}FeH_{26}{}^{2+}$ .. $[Fe(C_5H_4\text{-}As(CH_3)_3)_2]^{2+}$ ... Fe: Org.Comp.A10-86/7

$As_2C_{16}FeH_{26}I_2$ ... $[Fe(C_5H_4\text{-}As(CH_3)_3)_2]I_2$ ... Fe: Org.Comp.A10-86/7

$As_2C_{16}H_{20}MnO_6$ .. $[MnH_2(O_3AsCH_2C_6H_4CH_3\text{-}4)_2]$ ... Mn: MVol.D8-214/6

$As_2C_{16}H_{21}IMoO$ .. $C_5H_5Mo(CO)(As(CH_3)_2C_6H_4As(CH_3)_2\text{-}2)I$ ... Mo: Org.Comp.6-231

$As_2C_{16}H_{36}Mo_6N_{12}O_{25}$
$(CH_6N_3)_4[(C_6H_5AsO_3)_2Mo_6O_{18}(OH_2)] \cdot 4\ H_2O$ Mo: SVol.B3b-126

$As_2C_{16}H_{36}N_2S$ ... $(t\text{-}C_4H_9)_2AsN{=}S{=}NAs(C_4H_9\text{-}t)_2$ ... S: S-N Comp.7-117/21

$As_2C_{16}H_{36}N_2SSe$ $(t\text{-}C_4H_9)_2As({=}Se)N{=}S{=}NAs(C_4H_9\text{-}t)_2$ ... S: S-N Comp.7-122

$As_2C_{16}H_{36}N_2SSe_2$ $(t\text{-}C_4H_9)_2As({=}Se)N{=}S{=}NAs({=}Se)(C_4H_9\text{-}t)_2$ ... S: S-N Comp.7-122

$As_2C_{16}H_{36}N_2S_3$ .. $(t\text{-}C_4H_9)_2As({=}S)N{=}S{=}NAs({=}S)(C_4H_9\text{-}t)_2$ ... S: S-N Comp.7-122

$As_2C_{17}Cl_2FeH_{26}OPt$
$Fe(C_5H_4\text{-}As(CH_3)_2)_2PtCl_2 \cdot O{=}C(CH_3)_2$ ... Fe: Org.Comp.A10-93

$As_2C_{17}H_{21}IMoO_2$ $[(C_5H_5)Mo(CO)_2(1,2\text{-}((CH_3)_2As)_2\text{-}C_6H_4)]I$ ... Mo: Org.Comp.7-277

$As_2C_{17}H_{21}MoO_2{}^+$ $[(C_5H_5)Mo(CO)_2(1,2\text{-}((CH_3)_2As)_2\text{-}C_6H_4)]^+$ ... Mo: Org.Comp.7-277

$As_2C_{18}ClH_{28}N_2PtS^+$
$[PtCl(CH_3N{=}S{=}NCH_3)(As(CH_3)_2C_6H_5)_2]^+$ ... S: S-N Comp.7-329/30

$As_2C_{18}Cl_2H_{28}N_2O_4PtS$
$[PtCl(CH_3N{=}S{=}NCH_3)(As(CH_3)_2C_6H_5)_2][ClO_4]$ .. S: S-N Comp.7-329/30

$As_2C_{18}CrFeH_{20}O_4$ $Fe[C_5H_4\text{-}As(CH_3)_2]_2Cr(CO)_4$ ... Fe: Org.Comp.A10-89, 92

$As_2C_{18}CrH_{12}MnO_{14}Re$
$(CO)_5ReAs(CH_3)_2\text{-}Cr(CO)_4As(CH_3)_2Mn(CO)_5$ .. Re: Org.Comp.2-175

$As_2C_{18}F_4Fe_3H_{12}O_{10}$
$(CO)_{10}Fe_3[1,2-((CH_3)_2As)_2-C_4F_4]$ .... Fe: Org.Comp.C6b-6/7, 10/1
$As_2C_{18}F_{12}FeH_{24}$ .. $[(1,3,5-(CH_3)_3-C_6H_3)_2Fe][AsF_6]_2$ .... Fe: Org.Comp.B19-364, 378, 385
$As_2C_{18}FeH_{20}MoO_4$
$Fe[C_5H_4-As(CH_3)_2]_2Mo(CO)_4$ .... Fe: Org.Comp.A10-89/90, 92
$As_2C_{18}FeH_{20}O_4V^-$
$[Fe(C_5H_4-As(CH_3)_2)_2V(CO)_4]^-$ .... Fe: Org.Comp.A10-89, 91, 93
$As_2C_{18}FeH_{20}O_4W$ $Fe[C_5H_4-As(CH_3)_2]_2W(CO)_4$ .... Fe: Org.Comp.A10-89/90, 92
$As_2C_{19}ClH_{22}O_3Re$ fac-$(CO)_3ReCl[C_6H_5-As(CH_3)_2]_2$ .... Re: Org.Comp.1-282
$As_2C_{19}Cr_2Fe_3O_{19}$ $(CO)_9Fe_3As_2[Cr(CO)_5]_2$ .... Fe: Org.Comp.C6a-144/6
$As_2C_{19}Fe_2H_{30}NO_{10}P_2ReS$
$(CO)_3Re[P(CH_3)_3]_2-S-As(CH_3)_2-Fe(NO)(CO)_2Fe(CO)_4As(CH_3)_2$ .... Re: Org.Comp.1-320
$As_2C_{19}H_{22}O_3Re$ .. cis-$(CO)_3Re[C_6H_5As(CH_3)_2]_2$ .... Re: Org.Comp.1-284
$As_2C_{20}Cl_2H_{32}MnSb_2$
$[Mn((CH_3)_2As-C_6H_4-Sb(CH_3)_2-2)_2Cl_2]$ .... Mn: MVol.D8-219
$As_2C_{20}Cl_2H_{42}N_2Pd_2S_2$
$Pd_2(SCN)_2Cl_2[As(C_3H_7-n)_3]_2$ .... Pd: SVol.B2-348
$As_2C_{20}CoFeH_{30}O_{10}P_2Re$
$(CO)_3Re[P(CH_3)_3]_2-As(CH_3)_2-Co(CO)_3-As(CH_3)_2-Fe(CO)_4$ .... Re: Org.Comp.1-316
$As_2C_{20}CoFeH_{30}O_{10}P_2ReS$
$(CO)_3Re[P(CH_3)_3]_2-S-As(CH_3)_2-Co(CO)_3Fe(CO)_4As(CH_3)_2$ .... Re: Org.Comp.1-320
$As_2C_{20}FeH_{20}Ni_2O_6$
$Fe[C_5H_4-As(CH_3)_2]_2(Ni(CO)_3)_2$ .... Fe: Org.Comp.A10-90
$As_2C_{20}Fe_2H_{26}MoNO_9P$
$(C_5H_5)Mo(CO)_2[P(CH_3)_3]-As(CH_3)_2-Fe(CO)_2(NO)-As(CH_3)_2-Fe(CO)_4$ .... Mo: Org.Comp.7-120/1, 135
$As_2C_{20}Fe_3H_{16}O_{10}$ $(CO)_{10}Fe_3[1,2-((CH_3)_2As)_2-C_6H_4]$ .... Fe: Org.Comp.C6b-7, 11/2
$As_2C_{20}H_{28}N_2O_{10}Sn$
$(n-C_4H_9)_2Sn[O-As(O)(OH)-C_6H_4-NO_2-4]_2$ .... Sn: Org.Comp.15-362
$As_2C_{20}H_{28}N_2O_{12}Sn$
$(n-C_4H_9)_2Sn[O-As(O)(OH)-C_6H_3(NO_2-3)OH-4]_2$ Sn: Org.Comp.15-362
$As_2C_{20}H_{30}N_4O_{10}Sn$
$(n-C_4H_9)_2Sn[O-As(O)(OH)-C_6H_3(NO_2-3)NH_2-4]_2$ Sn: Org.Comp.15-362
$As_2C_{20}H_{30}O_6Sn$ .. $(n-C_4H_9)_2Sn[O-As(O)(OH)-C_6H_5]_2$ .... Sn: Org.Comp.15-361
– .... $(C_6H_5)_2Sn[O-As(O)(OH)-C_4H_9-n]_2$ .... Sn: Org.Comp.16-151/2
$As_2C_{20}H_{32}I_2MnSb_2$
$[Mn((CH_3)_2As-C_6H_4-Sb(CH_3)_2-2)_2I_2]$ .... Mn: MVol.D8-219
$As_2C_{20}H_{42}N_2PdS_2$ $Pd(SCN)_2[As(C_3H_7-n)_3]_2$ .... Pd: SVol.B2-348
$As_2C_{21}CoFeH_{26}MoO_9P$
$(C_5H_5)Mo(CO)_2[P(CH_3)_3]-As(CH_3)_2-Co(CO)_3-As(CH_3)_2-Fe(CO)_4$ .... Mo: Org.Comp.7-120/1, 135
$As_2C_{21}Fe_3H_{10}O_9$. $(CO)_9Fe_3(As-C_6H_5)_2$ .... Fe: Org.Comp.C6a-170/2

$As_2C_{22}FeH_{26}MnMoO_{10}P$
$(C_5H_5)Mo(CO)_2[P(CH_3)_3]-As(CH_3)_2-Mn$
$(CO)_4-As(CH_3)_2-Fe(CO)_4$ Mo:Org.Comp.7-120/1, 134
$As_2C_{22}Fe_2H_{22}MoO_8$
$(C_5H_5)Fe(CO)_2-As(CH_3)_2-(C_5H_5)Mo(CO)_2-As$
$(CH_3)_2Fe(CO)_4$ Mo:Org.Comp.7-121, 139
$As_2C_{22}H_{10}N_4O_{14}S_2Th^{4-}$
$Th[(O)_2(O_3S)_2C_{10}H_2(NNC_6H_4AsO_3)_2]^{4-}$ . . . . Th: SVol.D1-116
$As_2C_{22}H_{32}MnO_6$ . . $[Mn(OOC-C_6H_4-As(C_2H_5)_2-2)_2(H_2O)_2] \cdot H_2O$ Mn:MVol.D8-203/4
$As_2C_{22}H_{32}O_4Sn$ . . $(C_4H_9)_2Sn[-OAs(O)(C_6H_5)CH_2CH_2As(O)(C_6H_5)O-]$
Sn: Org.Comp.15-367
$As_2C_{22}H_{42}N_4Pd_2S_4$
$Pd_2(SCN)_4[As(C_3H_7-n)_3]_2$ . . . . Pd: SVol.B2-348
$As_2C_{23}CoFeH_{35}MoO_8P_2$
$(C_5H_5)Mo(CO)_2[P(CH_3)_3]-As(CH_3)_2-Co$
$(CO)_3-As(CH_3)_2-Fe(CO)_3-P(CH_3)_3$ Mo:Org.Comp.7-120/1, 136
$As_2C_{23}CrFeH_{22}MoO_9$
$(C_5H_5)Cr(CO)_3-As(CH_3)_2-(C_5H_5)Mo(CO)_2-As$
$(CH_3)_2Fe(CO)_4$ Mo:Org.Comp.7-121, 138
$As_2C_{23}FeH_{22}MoO_9W$
$(C_5H_5)Mo(CO)_2[As(CH_3)_2W(CO)_5]$
$-As(CH_3)_2-Fe(CO)_2(C_5H_5)$ Mo:Org.Comp.7-140
– . . . . $(C_5H_5)W(CO)_3-As(CH_3)_2-(C_5H_5)Mo(CO)_2-As$
$(CH_3)_2Fe(CO)_4$ Mo:Org.Comp.7-121, 139
$As_2C_{24}CoH_{31}MnMoO_7P$
$(C_5H_5)Mo(CO)_2[P(CH_3)_3]-As(CH_3)_2-Co$
$(CO)_3-As(CH_3)_2-Mn(CO)_2(C_5H_5)$ Mo:Org.Comp.7-120/1, 135
$As_2C_{24}F_{12}H_{60}MnO_{18}P_8$
$[Mn(P(P(=O)(OC_2H_5)_2)_3)_2][AsF_6]_2$ Mn:MVol.D8-72/3
$As_2C_{24}FeH_{31}MoO_7PW$
$(C_5H_5)Mo(CO)_2[P(CH_3)_3]-As(CH_3)_2-(OC)_3Fe[$
$-As(CH_3)_2-W(CO)_2(C_5H_5)-]$ Mo:Org.Comp.7-120, 121, 135
$As_2C_{24}Fe_2H_{31}MoO_7P$
$(CH_3)_3P-Fe(C_5H_5)(CO)-As(CH_3)_2-(C_5H_5)Mo$
$(CO)_2-As(CH_3)_2Fe(CO)_4$ Mo:Org.Comp.7-121, 139
$As_2C_{24}H_{20}N_2S$ . . . $(C_6H_5)_2AsN=S=NAs(C_6H_5)_2$ . . . . S: S-N Comp.7-117/21
$As_2C_{24}H_{42}MoN_2O_3S$
$(C_5H_5)Mo(CO)_2[-C(=O)-N(S-NH$
$-As(C_4H_9-t)_2)-As(C_4H_9-t)_2-]$ Mo:Org.Comp.8-112, 131/2
$As_2C_{24}H_{44}N_2S$ . . . $(c-C_6H_{11})_2AsN=S=NAs(C_6H_{11}-c)_2$ . . . . S: S-N Comp.7-117/21
$As_2C_{25}CrFeH_{31}MoO_8P$
$(CH_3)_3P-Cr(C_5H_5)(CO)_2-As(CH_3)_2-(C_5H_5)Mo$
$(CO)_2-As(CH_3)_2Fe(CO)_4$ Mo:Org.Comp.7-121, 139
$As_2C_{25}FeH_{31}MoO_8PW$
$(CH_3)_3P-W(C_5H_5)(CO)_2-As(CH_3)_2-(C_5H_5)Mo$
$(CO)_2-As(CH_3)_2Fe(CO)_4$ Mo:Org.Comp.7-121, 139
– . . . . $(C_5H_5)Mo(CO)_2[P(CH_3)_3]-As(CH_3)_2-Fe$
$(CO)_3-As(CH_3)_2-W(CO)_3(C_5H_5)$ Mo:Org.Comp.7-120, 121,
134/5

$As_2C_{26}Cl_2H_{24}MnO_2$
$[Mn(O{=}As(C_6H_5)_2CH_2CH_2As(C_6H_5)_2{=}O)Cl_2]$ . . . Mn:MVol.D8-214
$As_2C_{26}Cl_2H_{26}MnO_2$
$[Mn(O{=}As(C_6H_5)_2CH_3)_2Cl_2]$ Mn:MVol.D8-212
$As_2C_{26}Cl_2H_{54}N_2Pd_2S_2$
$Pd_2(SCN)_2Cl_2[As(C_4H_9\text{-}n)_3]_2$ . . . . Pd: SVol.B2-348
$As_2C_{26}FeH_{40}NO_4V$
$[N(C_2H_5)_4][Fe(C_5H_4\text{-}As(CH_3)_2)_2V(CO)_4]$ . . . . . . Fe: Org.Comp.A10-89, 91, 93
$As_2C_{26}Fe_3H_{22}O_{10}$ $(CO)_{10}Fe_3[As(CH_3)_2\text{-}C_6H_5]_2$ . . . . Fe: Org.Comp.C6b-5, 7
$As_2C_{26}H_{38}N_2O_{10}Os_3S$
$[H_2Os_3(CO)_{10}As(C_4H_9\text{-}t)_2N{=}SNAs(C_4H_9\text{-}t)_2]$ . . S: S-N Comp.7-123/5
$As_2C_{26}H_{54}N_2PdS_2$ $Pd(NCS)_2[As(C_4H_9\text{-}n)_3]_2$ . . . . Pd: SVol.B2-348
– . . . . $Pd(SCN)_2[As(C_4H_9\text{-}n)_3]_2$ . . . . Pd: SVol.B2-348
$As_2C_{27}Cl_4H_{52}NOOs$
$[N(C_4H_9\text{-}n)_4][(CO)OsCl_4(((CH_3)_2As)_2\text{-}1,2\text{-}C_6H_4)]$
Os: Org.Comp.A1-200, 207
$As_2C_{27}H_{22}N_2PdS_2$ $Pd(SCN)_2[(C_6H_5)_2AsCH_2As(C_6H_5)_2]$ . . . . Pd: SVol.B2-350
$As_2C_{27}H_{36}N_2O_{11}Os_3S$
$[Os_3(CO)_{11}As(C_4H_9\text{-}t)_2N{=}SNAs(C_4H_9\text{-}t)_2]$. . . . . S: S-N Comp.7-123/6
$As_2C_{27}H_{52}I_4NOOs$ $[N(C_4H_9\text{-}n)_4][(CO)OsI_4(((CH_3)_2As)_2\text{-}1,2\text{-}C_6H_4)]$ Os: Org.Comp.A1-200, 208
$As_2C_{28}ClH_{22}O_3Re$ $(CO)_3ReCl[(C_6H_5)_2AsCH_2As(C_6H_5)_2]$. . . . . Re: Org.Comp.1-188
$As_2C_{28}Fe_3H_{42}O_{10}$ $(CO)_{10}Fe_3[As(C_3H_7)_3]_2$. . . . . Fe: Org.Comp.C6b-5, 7
$As_2C_{28}H_{20}IO_4Re$. . $(CO)_4Re(I)As(C_6H_5)_2As(C_6H_5)_2$ . . . . Re: Org.Comp.1-457
$As_2C_{28}H_{22}IO_3Re$. . $(CO)_3ReI[(C_6H_5)_2AsCH_2As(C_6H_5)_2]$ . . . . Re: Org.Comp.1-191
$As_2C_{28}H_{24}N_2PdS_2$ $Pd(SCN)_2[(C_6H_5)_2AsCH_2CH_2As(C_6H_5)_2]$ . . . . . . Pd: SVol.B2-350
$As_2C_{28}H_{28}O_4Sn$ . . $(C_6H_5CH_2)_2Sn[\text{-}OAs(O)(C_6H_5)CH_2CH_2As(O)(C_6H_5)O\text{-}]$
Sn: Org.Comp.16-84
$As_2C_{28}H_{32}In_2$ . . . . $[(CH_3)_2In\text{-}As(C_6H_5)_2]_2$ . . . . In: Org.Comp.1-316
$As_2C_{28}H_{54}N_4Pd_2S_4$
$Pd_2(SCN)_4[As(C_4H_9\text{-}n)_3]_2$ . . . . Pd: SVol.B2-348
$As_2C_{28}H_{80}In_2Si_8$. . $[((CH_3)_3Si\text{-}CH_2)_2In\text{-}As(Si(CH_3)_3)_2]_2$ . . . . In: Org.Comp.1-322
$As_2C_{29}ClH_{22}O_4Re$ $(CO)_4Re(Cl)As(C_6H_5)_2CH_2As(C_6H_5)_2$. . . . . Re: Org.Comp.1-456
$As_2C_{29}H_{22}IO_4Re$. . $(CO)_4Re(I)As(C_6H_5)_2CH_2As(C_6H_5)_2$ . . . . Re: Org.Comp.1-457
$As_2C_{29}H_{26}N_2PdS_2$ $Pd(SCN)_2[(C_6H_5)_2AsCH_2CH_2CH_2As(C_6H_5)_2]$ . . . Pd: SVol.B2-350
$As_2C_{30}Cl_2H_{24}MnN_6O_8$
$[Mn((2\text{-}NC_5H_4)_3As)_2][ClO_4]_2$ Mn:MVol.D8-205
$As_2C_{30}Cl_2H_{34}MnN_4O_8$
$[Mn(6\text{-}CH_3\text{-}NC_5H_3\text{-}2\text{-}CH{=}N\text{-}C_6H_4\text{-}2\text{-}As(CH_3)_2)_2][ClO_4]_2$
Mn:MVol.D6-79/80
$As_2C_{30}F_2Fe_3H_{28}O_{10}$
$(CO)_{10}Fe_3[As(C_2H_5)_2\text{-}C_6H_4\text{-}4\text{-}F]_2$. . . . . Fe: Org.Comp.C6b-5, 7
$As_2C_{30}H_{46}Mn_2N_2O_4S$
$[(C_5H_5)Mn(CO)_2(As(C_4H_9\text{-}t)_2N{=}S{=}NAs$
$(C_4H_9\text{-}t)_2)Mn(CO)_2(C_5H_5)]$ . . . . S: S-N Comp.7-127, 131, 133/4
$As_2C_{30}H_{48}O_4Sn$ . . $(C_8H_{17})_2Sn[\text{-}OAs(O)(C_6H_5)CH_2CH_2As(O)(C_6H_5)O\text{-}]$
Sn: Org.Comp.16-61
$As_2C_{31}ClH_{27}NO_7Re$
$[(CO)_3Re(As(C_6H_5)_2CH_2CH_2As(C_6H_5)_2)(NCCH_3)][ClO_4]$
Re: Org.Comp.1-303

$As_2C_{31}ClH_{29}MoO_2$ $(CH_2CHCH_2)MoCl(CO)_2[(C_6H_5)_2AsCH_2CH_2As(C_6H_5)_2]$ Mo:Org.Comp.5-271, 277/8, 280

$As_2C_{31}ClH_{29}MoO_2^+$ $[(CH_2CHCH_2)MoCl(CO)_2((C_6H_5)_2AsCH_2CH_2As(C_6H_5)_2)]^+$ Mo:Org.Comp.5-280

$As_2C_{31}H_{27}NO_3Re^+$ $[(CO)_3Re(As(C_6H_5)_2CH_2CH_2As(C_6H_5)_2)(NCCH_3)]^+$ Re: Org.Comp.1-303

$As_2C_{31}H_{29}IMoO_2$ $(CH_2CHCH_2)MoI(CO)_2[(C_6H_5)_2AsCH_2CH_2As(C_6H_5)_2]$ Mo:Org.Comp.5-271, 278

$As_2C_{32}Fe_3H_{34}O_{10}$ $(CO)_{10}Fe_3[As(C_2H_5)_2-CH_2C_6H_5]_2$ . . . . Fe: Org.Comp.C6b-5, 7

$As_2C_{32}H_{16}N_4O_{20}S_4Th^{6-}$ $Th[O_3AsC_6H_4NNC_{10}H_4(O)(SO_3)_2]_2^{6-}$ . . . . Th: SVol.D1-123

$As_2C_{32}H_{24}N_2PdS_2$ $Pd(SCN)(NCS)[(C_6H_5)_2As-2-C_6H_4-As(C_6H_5)_2]$ Pd: SVol.B2-351

– . . . . $Pd(SCN)_2[(C_6H_5)_2As-2-C_6H_4-As(C_6H_5)_2]$ . . . . . Pd: SVol.B2-351

$As_2C_{32}H_{31}IMoO$ . . $C_5H_5Mo(CO)(As(C_6H_5)_2CH_3)_2I$ Mo:Org.Comp.6-227

$As_2C_{33}H_{30}MoO_3$ . . $Mo(CO)_3[As(C_6H_5)_2CH_2CH_2CH=CHCH_2CH_2As(C_6H_5)_2]$ Mo:Org.Comp.5-156, 160

$As_2C_{34}Cl_2FeH_{28}Pd$ $Fe[C_5H_4-As(C_6H_5)_2]_2PdCl_2$ . . . . Fe: Org.Comp.A10-89, 91, 93

$As_2C_{34}Cl_2FeH_{28}Pt$ $Fe[C_5H_4-As(C_6H_5)_2]_2PtCl_2$ . . . . Fe: Org.Comp.A10-89, 92, 93

$As_2C_{34}Cl_2H_{42}MnN_4O_8$ $[Mn(6-CH_3-NC_5H_3-2-CH=N-C_6H_4-2-As(C_2H_5)_2)_2][ClO_4]_2 \cdot H_2O$ Mn:MVol.D6-79/80

$As_2C_{34}FeH_{28}$ . . . . $Fe[C_5H_4-As(C_6H_5)_2]_2$ . . . . Fe: Org.Comp.A10-87/8

$As_2C_{34}FeH_{28}I_2Pd$ $Fe[C_5H_4-As(C_6H_5)_2]_2PdI_2$ . . . . Fe: Org.Comp.A10-89, 92, 93

$As_2C_{34}FeH_{28}S_2$ . . $Fe[C_5H_4-AsS(C_6H_5)_2]_2$ . . . . Fe: Org.Comp.A10-88

$As_2C_{36}Cl_2H_{30}Mn$ $[Mn((C_6H_5)_3As)_2Cl_2]$ Mn:MVol.D8-202

$As_2C_{36}Cl_2H_{30}MnO_2$ $[Mn(O=As(C_6H_5)_3)_2Cl_2]$ Mn:MVol.D8-208/9

$As_2C_{36}Cl_2H_{30}NRhS$ $Rh(NS)Cl_2(As(C_6H_5)_3)_2$ . . . . S: S-N Comp.5-76/9

$As_2C_{36}Cl_3H_{30}MnO_2$ $[Mn(O=As(C_6H_5)_3)_2Cl_3]$ Mn:MVol.D8-211

$As_2C_{36}Cl_3H_{30}NOsS$ $Os(NS)Cl_3(As(C_6H_5)_3)_2$ . . . . S: S-N Comp.5-79, 82/3

$As_2C_{36}Cl_3H_{30}NRuS$ $Ru(NS)Cl_3(As(C_6H_5)_3)_2$ . . . . S: S-N Comp.5-51/2, 74

$As_2C_{36}Cl_4H_{30}N_2Rh_2S_2$ $(Rh(NS)Cl_2(As(C_6H_5)_3))_2$ . . . . S: S-N Comp.5-76/9

$As_2C_{36}FeH_{28}NiO_2$ $Fe[C_5H_4-As(C_6H_5)_2]_2Ni(CO)_2$ . . . . Fe: Org.Comp.A10-91, 93

$As_2C_{36}FeH_{34}^{2+}$ . . $[Fe(C_5H_4-As(C_6H_5)_2CH_3)_2]^{2+}$ . . . . Fe: Org.Comp.A10-88

$As_2C_{36}FeH_{34}I_2$ . . . $[Fe(C_5H_4-As(C_6H_5)_2CH_3)_2]I_2$ . . . . Fe: Org.Comp.A10-88

$As_2C_{36}H_{30}IMnN_2O_2$ $[Mn(NO)_2((C_6H_5)_3As)_2I]$ Mn:MVol.D8-202

$As_2C_{36}H_{30}I_2Mn$. . . $[Mn((C_6H_5)_3As)_2I_2]$ Mn:MVol.D8-202

$As_2C_{36}H_{30}I_2MnO_2$ $[Mn(O=As(C_6H_5)_3)_2I_2]$ Mn:MVol.D8-208/9

$As_2C_{36}H_{30}O_4Sn$ . . $(C_6H_5)_2Sn(OAs(O)(C_6H_5)_2)_2$ . . . . Sn: Org.Comp.16-151

$As_2C_{36}H_{30}Sn$ . . . . $(C_6H_5)_2As-Sn(C_6H_5)_2-As(C_6H_5)_2$ . . . . Sn: Org.Comp.19-242

$As_2C_{37}ClH_{35}N_2O_5Os$
$[(CO)OsH(NC_5H_4\text{-}2\text{-}(2\text{-}C_5H_4N))(As(C_6H_5)_2CH_3)_2][ClO_4]$
Os: Org.Comp.A1-200, 202

$As_2C_{37}Cl_2H_{30}NORhS$
$Rh(NS)(CO)Cl_2(As(C_6H_5)_3)_2$ . . . . S: S-N Comp.5-76/9

$As_2C_{37}Cl_2H_{30}NO_2Re$
$(CO)ReCl_2[As(C_6H_5)_3]_2(NO)$ . . . . Re: Org.Comp.1-46

$As_2C_{37}H_{35}N_2OOs^+$
$[(CO)OsH(NC_5H_4\text{-}2\text{-}(2\text{-}C_5H_4N))(As(C_6H_5)_2CH_3)_2][ClO_4]$
Os: Org.Comp.A1-200, 202

$As_2C_{38}Cl_4H_{30}N_2O_2Rh_2S_2$
$(Rh(NS)(CO)Cl_2(As(C_6H_5)_3))_2$ . . . . S: S-N Comp.5-76/9

$As_2C_{38}Cl_4H_{36}Pd$ . . $[CH_3As(C_6H_5)_3]_2[PdCl_4]$ . . . . Pd: SVol.B2-124

$As_2C_{38}Cl_6H_{36}Pd_2$ $[(C_6H_5)_3AsCH_3]_2[Pd_2Cl_6]$ . . . . Pd: SVol.B2-164

$As_2C_{38}CrFeH_{28}O_4$ $Fe[C_5H_4\text{-}As(C_6H_5)_2]_2Cr(CO)_4$ . . . . Fe: Org.Comp.A10-89, 91, 92

$As_2C_{38}FeH_{28}MoO_4$
$Fe[C_5H_4\text{-}As(C_6H_5)_2]_2Mo(CO)_4$ . . . . Fe: Org.Comp.A10-89, 91, 92

$As_2C_{38}FeH_{28}O_4W$ $Fe[C_5H_4\text{-}As(C_6H_5)_2]_2W(CO)_4$ . . . . Fe: Org.Comp.A10-89, 91, 92

$As_2C_{38}FeH_{36}$ . . . . $Fe[C_5H_4\text{-}CH_2CH_2\text{-}As(C_6H_5)_2]_2$ . . . . Fe: Org.Comp.A10-88

$As_2C_{38}H_{30}MnN_2O_2S_2$
$[Mn(O{=}As(C_6H_5)_3)_2(NCS)_2]$ . . . . Mn: MVol.D8-208/9

$As_2C_{38}H_{30}N_2O_2PdS_2$
$Pd(NCS)_2[OAs(C_6H_5)_3]_2$ . . . . Pd: SVol.B2-301

$As_2C_{38}H_{30}N_2PdS_2$ $Pd(NCS)_2[As(C_6H_5)_3]_2$ . . . . Pd: SVol.B2-349

– . . . . $Pd(SCN)_2[As(C_6H_5)_3]_2$ . . . . Pd: SVol.B2-348/9

$As_2C_{38}H_{32}MnO_6$ . . $[Mn(OC(O)\text{-}C_6H_4\text{-}As(C_6H_5)_2\text{-}2)_2(H_2O)_2]$ . . . . . . Mn: MVol.D8-203/4

$As_2C_{38}H_{36}N_2O_{22}Os_6S$
$[Os_3(CO)_{11}(As(C_4H_9\text{-}t)_2N{=}S{=}NAs(C_4H_9\text{-}t)_2)Os_3(CO)_{11}]$
S: S-N Comp.7-127, 131, 133

$As_2C_{39}ClH_{30}O_3Re$ $(CO)_3ReCl[As(C_6H_5)_3]_2$ . . . . Re: Org.Comp.1-283

$As_2C_{39}ClH_{33}O_3Os$ $(CO)(Cl)Os[As(C_6H_5)_3]_2[\text{-}OC(CH_3)O\text{-}]$ . . . . Os: Org.Comp.A1-205

$As_2C_{39}ClH_{35}N_2O_5Os$
$[(CO)OsH(1,10\text{-}N_2C_{12}H_8)(As(C_6H_5)_2\text{-}CH_3)_2][ClO_4]$
Os: Org.Comp.A1-200, 202

$As_2C_{39}ClH_{39}N_2O_5Os$
$[(CO)OsH(NC_5H_4\text{-}2\text{-}(2\text{-}C_5H_4N))$
$(As(C_6H_5)_2\text{-}C_2H_5)_2][ClO_4]$ . . . . Os: Org.Comp.A1-200, 203

$As_2C_{39}Cl_3H_{37}N_2O_{22}Os_6S$
$[Os_3(CO)_{11}(As(C_4H_9\text{-}t)_2N{=}S{=}NAs(C_4H_9\text{-}t)_2)$
$Os_3(CO)_{11}] \cdot CDCl_3$ . . . . S: S-N Comp.7-133/5

$As_2C_{39}F_6FeH_{40}NPPd$
$[(C_5H_5)Fe(C_5H_3(Pd(As(C_6H_5)_2\text{-}CH_2CH_2\text{-}As$
$(C_6H_5)_2))\text{-}CH_2\text{-}N(CH_3)_2)][PF_6]$ . . . . Fe: Org.Comp.A10-187, 188

$As_2C_{39}FeH_{40}NPd^+$
$[(C_5H_5)Fe(C_5H_3(Pd(As(C_6H_5)_2\text{-}CH_2CH_2\text{-}As$
$(C_6H_5)_2))\text{-}CH_2\text{-}N(CH_3)_2)]^+$ . . . . Fe: Org.Comp.A10-187, 188

$As_2C_{39}H_{30}IO_3Re$ . . $(CO)_3ReI[As(C_6H_5)_3]_2$ . . . . Re: Org.Comp.1-284

$As_2C_{39}H_{30}NO_6Re$ fac-$(CO)_3Re[As(C_6H_5)_3]_2ONO_2$ . . . . Re: Org.Comp.1-282

$As_2C_{39}H_{35}N_2OOs^+$
$[(CO)OsH(1,10-N_2C_{12}H_8)(As(C_6H_5)_2-CH_3)_2][ClO_4]$ Os: Org.Comp.A1-200, 202
$As_2C_{39}H_{39}N_2OOs^+$
$[(CO)OsH(NC_5H_4-2-(2-C_5H_4N))(As(C_6H_5)_2-C_2H_5)_2][ClO_4]$ . . . . . Os: Org.Comp.A1-200, 203
$As_2C_{41}ClH_{38}MoO_2P$
$[(C_5H_5)Mo(CO)_2((C_6H_5)_2As-CH_2CH_2-P(C_6H_5)-CH_2CH_2-As(C_6H_5)_2)]Cl$ Mo:Org.Comp.7-277
$As_2C_{41}ClH_{39}N_2O_5Os$
$[(CO)OsH(1,10-N_2C_{12}H_8)(As(C_6H_5)_2-C_2H_5)_2][ClO_4]$ Os: Org.Comp.A1-200, 203
$As_2C_{41}F_6H_{38}MoO_2P_2$
$[(C_5H_5)Mo(CO)_2((C_6H_5)_2As-CH_2CH_2-P(C_6H_5)-CH_2CH_2-As(C_6H_5)_2)][PF_6]$ Mo:Org.Comp.7-277
$As_2C_{41}H_{38}MoO_2P^+$
$[(C_5H_5)Mo(CO)_2((C_6H_5)_2As-CH_2CH_2-P(C_6H_5)-CH_2CH_2-As(C_6H_5)_2)]^+$ Mo:Org.Comp.7-277
$As_2C_{41}H_{39}N_2OOs^+$
$[(H)(CO)Os(1,10-N_2C_{12}H_8)(As(C_6H_5)_2-C_2H_5)_2][ClO_4]$ Os: Org.Comp.A1-200, 203
$As_2C_{42}ClH_{35}MoO$ $C_5H_5Mo(CO)(As(C_6H_5)_3)_2Cl$ Mo:Org.Comp.6-227/8, 233
$As_2C_{42}Cl_2H_{48}O_{11}Sb_2$
$[((CH_3)_3SbOAs(C_6H_5)_3)_2O][ClO_4]_2$ . . . . . Sb: Org.Comp.5-101
$As_2C_{42}H_{40}MnO_6$ . . $[Mn(OC(O)-C_6H_4-As(C_6H_4CH_3-4)_2-2)_2(H_2O)_2]$ Mn:MVol.D8-203/4
$As_2C_{42}H_{48}O_3Sb_2{}^{2+}$
$[((CH_3)_3SbOAs(C_6H_5)_3)_2O]^{2+}$ . . . . . Sb: Org.Comp.5-101
$As_2C_{43}ClH_{35}MoO_2$ $[(C_5H_5)Mo(CO)_2(As(C_6H_5)_3)_2]Cl$ Mo:Org.Comp.7-283, 292
$As_2C_{43}ClH_{36}O_4OsP$
$(H)(CO)(Cl)Os[As(C_6H_5)_2]_2[P(O-C_6H_5)_3]$ . . . . . Os: Org.Comp.A1-205
$As_2C_{43}F_6FeH_{38}NP$ $[(C_5H_5)Fe(CNCH_3)(As(C_6H_5)_3)_2][PF_6]$ . . . . . Fe: Org.Comp.B15-281
$As_2C_{43}F_6H_{35}MoO_2P$
$[(C_5H_5)Mo(CO)_2(As(C_6H_5)_3)_2][PF_6]$ Mo:Org.Comp.7-283, 292
$As_2C_{43}FeH_{38}N^+$ . . $[C_5H_5Fe(CNCH_3)(As(C_6H_5)_3)_2]^+$ . . . . . Fe: Org.Comp.B15-281
$As_2C_{43}Fe_3H_{30}O_7STe$
$(CO)_7Fe_3STe[As(C_6H_5)_3]_2$ . . . . . Fe: Org.Comp.C6a-16/7, 20
$As_2C_{43}Fe_3H_{30}O_7S_2$
$(CO)_7Fe_3S_2[As(C_6H_5)_3]_2$ . . . . . Fe: Org.Comp.C6a-18, 21
$As_2C_{43}Fe_3H_{30}O_7Te_2$
$(CO)_7Fe_3Te_2[As(C_6H_5)_3]_2$ . . . . . Fe: Org.Comp.C6a-16/7, 20
$As_2C_{43}H_{35}MoO_2{}^+$ $[(C_5H_5)Mo(CO)_2(As(C_6H_5)_3)_2]^+$ Mo:Org.Comp.7-283, 292
$As_2C_{43}H_{41}MoO_3P$ $(C_5H_5)Mo(CO)_2[C(=O)CH_3][As(C_6H_5)_2-CH_2CH_2-P(C_6H_5)-CH_2CH_2-As(C_6H_5)_2]$ Mo:Org.Comp.8-45, 57
$As_2C_{45}ClH_{35}MoN_2$ $C_5H_5Mo(As(C_6H_5)_3)_2(Cl)=C=C(CN)_2$ Mo:Org.Comp.6-68
$As_2C_{45}ClH_{40}OOsP$ $(H)(CO)(Cl)Os[As(C_6H_5)_2-CH_2CH_2-P(C_6H_5)_2][As(C_6H_5)_3]$ Os: Org.Comp.A1-204
$As_2C_{46}H_{46}O_4Sn$ . . $(C_4H_9)_2Sn(OOCC_6H_4As(C_6H_5)_2-2)_2$ . . . . . Sn: Org.Comp.15-291
$As_2C_{47}Fe_3H_{40}O_7S$ $(CO)_7(H)Fe_3S-(C_4H_9-t)[As(C_6H_5)_3]_2$ . . . . . Fe: Org.Comp.C6a-24/6
$As_2C_{48}Cl_2H_{40}N_2O_2RuS_2$
$RuCl_2(As(C_6H_5)_3)_2(O=S=NC_6H_5)_2$ . . . . . S: S-N Comp.6-274

$As_2C_{48}Cl_5H_{40}N_2ReS_2$
$[(C_6H_5)_4As]_2[Re(NS)(NSCl)Cl_4] \cdot CH_2Cl_2$ ..... S: S-N Comp.5-51, 67/8

$As_2C_{48}Cl_6H_{40}Pd_2$ $[(C_6H_5)_4As]_2[Pd_2Cl_6]$ ..... Pd: SVol.B2-163

$As_2C_{48}Cl_{10}H_{40}N_2SW_2$
$[(C_6H_5)_4As]_2[Cl_5W{=}N{=}S{=}N{=}WCl_5]$ ..... S: S-N Comp.7-292/3

$As_2C_{48}Cl_{14}H_{40}Mo_6$
$[(C_6H_5)_4As]_2Mo_6Cl_{14}$ Mo:SVol.B5-280

$As_2C_{48}H_{40}MoS_4$ .. $[(C_6H_5)_4As]_2[MoS_4]$ Mo:SVol.B7-278

$As_2C_{48}H_{40}Mo_2S_{10}$ $[(C_6H_5)_4As]_2[(S_2)SMo(S)_2MoS(S_4)] \cdot 0.5\ NC\text{-}CH_3$
Mo:SVol.B7-316

$As_2C_{48}H_{40}Mo_2S_{11}$ $[(C_6H_5)_4As]_2[(S_2)_2SMo(S)MoS(S_2)_2] \cdot NC\text{-}CH_3$ Mo:SVol.B7-318/9

$As_2C_{49}Cl_7H_{42}N_2ReS_2$
$[(C_6H_5)_4As]_2[Re(NS)(NSCl)Cl_4] \cdot CH_2Cl_2$ ..... S: S-N Comp.5-51, 67/8

$As_2C_{49}Fe_3H_{42}O_7S$ $(CO)_7(H)Fe_3S\text{-}(C_6H_{11}\text{-}c)[As(C_6H_5)_3]_2$ ..... Fe: Org.Comp.C6a-24/6

$As_2C_{49}H_{41.5}Mo_2N_{0.5}S_{10}$
$[(C_6H_5)_4As]_2[(S_2)SMo(S)_2MoS(S_4)] \cdot 0.5\ NC\text{-}CH_3$
Mo:SVol.B7-316

$As_2C_{50}H_{40}Mo_2O_5$ $[(C_5H_5)Mo(CO_2)(As(C_6H_5)_3)_2][(C_5H_5)Mo(CO)_3]$ Mo:Org.Comp.7-283

$As_2C_{50}H_{43}Mo_2NS_{11}$
$[(C_6H_5)_4As]_2[(S_2)_2SMo(S)MoS(S_2)_2] \cdot NC\text{-}CH_3$ Mo:SVol.B7-318/9

$As_2C_{50}H_{54}O_4Sn$ .. $(C_4H_9)_2Sn(OOCC_6H_4As(C_6H_4CH_3\text{-}4)_2\text{-}2)_2$ ..... Sn: Org.Comp.15-291

$As_2C_{52}H_{40}N_4PdS_4$ $[(C_6H_5)_4As]_2[Pd(SCN)_4]$ ..... Pd: SVol.B2-295

$As_2C_{56}H_{66}Mn_2O_{10}$ $[Mn(O{=}As(C_6H_5)_3)(t\text{-}C_4H_9COO)_2]_2$ Mn:MVol.D8-210

$As_2C_{59}Fe_3H_{40}O_{11}$ $[As(C_6H_5)_4]_2[Fe_3(CO)_{11}]$ ..... Fe: Org.Comp.C6b-104

$As_2C_{60}H_{40}MnN_6S_6$
$[As(C_6H_5)_4]_2[Mn(SC(CN){=}C(CN)S)_3]$ Mn:MVol.D7-48

$As_2C_{60}H_{40}Mn_2S_{12}$ $[As(C_6H_5)_4]_2[Mn_2(C_4({=}S)_4)_3]$ Mn:MVol.D7-57

$As_2C_{62}Fe_3H_{40}O_{13}W$
$[As(C_6H_5)_4]_2[CFe_3W(CO)_{13}]$ ..... Fe: Org.Comp.C6b-75

$As_2C_{64}H_{48}N_4P_2Pd_2S_4$
$[Pd((C_6H_5)_2PC_6H_4As(C_6H_5)_2)_2][Pd(SCN)_4]$ ..... Pd: SVol.B2-296

$As_2C_{68}H_{54}MnN_2O_6S_4$
$[As(C_6H_5)_4]_2[Mn(NC_9H_4(CH_3\text{-}2)(S({=}O)_2O\text{-}5)(S\text{-}8))_2]$
Mn:MVol.D7-68

$As_2C_{72}Cl_2H_{60}O_{11}Sb_2$
$[((C_6H_5)_3SbOAs(C_6H_5)_3)_2O][ClO_4]_2$ ..... Sb: Org.Comp.5-102

$As_2C_{72}H_{60}O_3Sb_2{}^{2+}$
$[((C_6H_5)_3SbOAs(C_6H_5)_3)_2O]^{2+}$ ..... Sb: Org.Comp.5-102

As_2CaPd_2 ..... $CaPd_2As_2$ ..... Pd: SVol.B2-347

As_2CeRh_2 ..... $CeRh_2As_2$ ..... Rh: SVol.A1-73

$As_2CoF_{18}N_6S_6$ ... $[Co(NSF)_6][AsF_6]_2$ ..... S: S-N Comp.5-243/5

$As_2F_{12}H_4N_4NiO_4S_4$
$[Ni(O{=}S{=}NH)_4](AsF_6)_2$ ..... S: S-N Comp.6-252

$As_2F_{12}MnO_4S_2$ ... $[Mn(SO_2)_2(AsF_6)_2]$ Mn:MVol.D7-112

$As_2F_{12}MnO_{2x}S_x$ ... $Mn(AsF_6)_2 \cdot n\ SO_2$ Mn:MVol.D7-111/2

$As_2F_{12}N_4O_4S_6Zn$ $[Zn((O{=}S{=}N)_2S)_2](AsF_6)_2$ ..... S: S-N Comp.6-264

$As_2F_{12}N_4O_8S_8Zn$ $[Zn((O{=}S{=}N)_2S)_2](AsF_6)_2 \cdot 2\ SO_2$ ..... S: S-N Comp.6-264/5

$As_2F_{18}N_6NiS_6$ .... $[Ni(NSF)_6][AsF_6]_2$ ..... S: S-N Comp.5-243/5

$As_2F_{24}MnN_4S_4$ ... $[Mn(NSF_3)_4(AsF_6)_2]$ Mn:MVol.D7-220/1

$As_2F_{24}MnO_4P_4$ ... $[Mn(O{=}PF_3)_4][AsF_6]_2$ Mn:MVol.D8-184

$As_2HMo_6O_{26}{}^{5-}$ ... $[HAs_2Mo_6O_{26}]^{5-}$ Mo:SVol.B3b-62/3, 94/6
$As_2H_2Mo_6O_{26}{}^{4-}$ .. $[H_2As_2Mo_6O_{26}]^{4-}$ Mo:SVol.B3b-62/3, 94/6
$As_2H_2Mo_6O_{27}{}^{6-}$ .. $[(OAsO_3)_2Mo_6O_{18}(H_2O)]^{6-}$ Mo:SVol.B3b-62/3
$As_2H_4Mo_6O_{27}{}^{4-}$ .. $[H_2As_2Mo_6O_{26}(H_2O)]^{4-}$ Mo:SVol.B3b-62/3
– .... $[H_2As_2Mo_6O_{27}H_2]^{4-}$ Mo:SVol.B3b-94/6
$As_2H_6Mo_{18}O_{62}$ ... $H_6As_2Mo_{18}O_{62} \cdot 25\ H_2O$ Mo:SVol.B3b-50/1
– .... $H_6As_2Mo_{18}O_{62} \cdot 35\ H_2O$ Mo:SVol.B3b-50/1
As_2K_2Pd .... K_2PdAs_2 .... Pd: SVol.B2-347
As_2LaRh_2 .... Rh_2LaAs_2 .... Rh: SVol.A1-69/70
$As_2Mo_5O_{23}{}^{6-}$ .... $[As_2Mo_5O_{23}]^{6-}$ Mo:SVol.B3b-94/6
$As_2Mo_6O_{26}{}^{6-}$ .... $[(AsO_4)_2Mo_6O_{18}]^{6-}$ Mo:SVol.B3b-62/3, 94/6, 118/9
$As_2Mo_{17}O_{61}{}^{10-}$ ... $[As_2Mo_{17}O_{61}]^{10-}$ Mo:SVol.B3b-52, 94/6
$As_2Mo_{18}O_{62}{}^{6-}$ ... $[(AsO_4)_2Mo_{18}O_{54}]^{6-}$ Mo:SVol.B3b-50/2, 94/6
As_2Pd .... $PdAs_2$ .... Pd: SVol.B2-343, 346
As_2Pd_2Sr .... $SrPd_2As_2$ .... Pd: SVol.B2-347
As_2Pd_3 .... Pd_3As_2 .... Pd: SVol.B2-345
As_2Pd_4S .... Pd_4As_2S .... Pd: SVol.B2-347
As_2Pd_5 .... $Pd_5As_2 = Pd_{2.5}As$ .... Pd: SVol.B2-343/5
As_2Rh .... $RhAs_2$ .... Rh: SVol.A1-38/41
As_2Th .... $ThAs_2$
Thermodynamic data .... Th: SVol.A4-183
$As_3BC_{38}H_{47}O_3Re$ $[(CO)_3Re(CH_3C(CH_2As(CH_3)_2)_3)][B(C_6H_5)_4]$ ... Re: Org.Comp.1-218
$As_3BrC_{40}H_{40}OOs$ $(H)(CO)(Br)Os[As(C_6H_5)_2-CH_3]_3$ .... Os: Org.Comp.A1-200, 201/2
$As_3BrC_{43}H_{46}OOs$ $(H)(CO)(Br)Os[As(C_6H_5)_2-C_2H_5]_3$ .... Os: Org.Comp.A1-200, 203
$As_3BrC_{55}H_{46}OOs$ $(H)(CO)(Br)Os[As(C_6H_5)_3]_3$ .... Os: Org.Comp.A1-200, 206
$As_3BrC_{64}H_{64}OOs$ $(H)(CO)(Br)Os[As(CH_2-C_6H_5)_3]_3$ .... Os: Org.Comp.A1-200, 207
– .... $(H)(CO)(Br)Os[As(C_6H_4-4-CH_3)_3]_3$ .... Os: Org.Comp.A1-200, 207
$As_3Br_2C_{40}H_{39}OOs$ $(CO)(Br)_2Os[As(C_6H_5)_2-CH_3]_3$ .... Os: Org.Comp.A1-200, 201
$As_3Br_2C_{43}H_{45}OOs$ $(CO)(Br)_2Os[As(C_6H_5)_2-C_2H_5]_3$ .... Os: Org.Comp.A1-200, 202
$As_3Br_2C_{46}H_{51}OOs$ $(CO)(Br)_2Os[As(C_6H_5)_2-C_3H_7-n]_3$ .... Os: Org.Comp.A1-200, 204
$As_3C_{12}H_{15}MoO_2$ .. $[(CH_3)_5C_5]Mo(CO)_2(As_3)$ Mo:Org.Comp.7-39/40
$As_3C_{12}H_{36}In_3$ .... $[(CH_3)_2In-As(CH_3)_2]_3$ .... In: Org.Comp.1-316, 319/21
$As_3C_{14}ClH_{27}O_3Re$ $(CO)_3ReCl[CH_3C(CH_2As(CH_3)_2)_3]$ .... Re: Org.Comp.1-188/9
$As_3C_{14}ClH_{27}O_7Re$ $[(CO)_3Re(CH_3C(CH_2As(CH_3)_2)_3)][ClO_4]$ .... Re: Org.Comp.1-218
$As_3C_{14}F_6FeH_{20}OPS$
$[(C_5H_5)Fe(CS)(OC(CH_3)_2)(As_3C_5H_9)][PF_6]$ .... Fe: Org.Comp.B15-265
$As_3C_{14}FeH_{20}OS^+$ $[(C_5H_5)Fe(CS)(OC(CH_3)_2)-As_3C_5H_9]^+$ .... Fe: Org.Comp.B15-265
$As_3C_{14}H_{27}O_3Re^+$ $[(CO)_3Re(CH_3C(CH_2As(CH_3)_2)_3)]^+$ .... Re: Org.Comp.1-218
$As_3C_{40}ClH_{40}OOs$ $(H)(CO)(Cl)Os[As(C_6H_5)_2-CH_3]_3$ .... Os: Org.Comp.A1-200, 201
$As_3C_{40}Cl_2H_{39}OOs$ $(CO)(Cl)_2Os[As(C_6H_5)_2-CH_3]_3$ .... Os: Org.Comp.A1-200, 201
$As_3C_{42}H_{35}O_6Sn$ .. $C_6H_5Sn(OAs(O)(C_6H_5)_2)_3$ .... Sn: Org.Comp.17-66, 68
$As_3C_{42}H_{35}Sn$ .... $[(C_6H_5)_2As]_3Sn-C_6H_5$ .... Sn: Org.Comp.19-242/3
$As_3C_{43}ClH_{46}OOs$ $(H)(CO)(Cl)Os[As(C_6H_5)_2-C_2H_5]_3$ .... Os: Org.Comp.A1-200, 203
$As_3C_{43}Cl_2H_{45}OOs$ $(CO)(Cl)_2Os[As(C_6H_5)_2-C_2H_5]_3$ .... Os: Org.Comp.A1-200, 202
$As_3C_{43}H_{47}OOs$ ... $(H)_2(CO)Os[As(C_6H_5)_2-C_2H_5]_3$ .... Os: Org.Comp.A1-203
$As_3C_{46}Cl_2H_{51}OOs$ $(CO)(Cl)_2Os[As(C_6H_5)_2-C_3H_7-n]_3$ .... Os: Org.Comp.A1-200, 204
$As_3C_{54}Cl_3H_{45}MnO_3$
$[Mn(O=As(C_6H_5)_3)_3Cl_3]$ Mn:MVol.D8-211

$As_3C_{54}H_{45}I_2MnO_7S_2$
$[Mn(O{=}As(C_6H_5)_3)_3(SO_2)_2]I_2$ Mn:MVol.D8-210
$As_3C_{55}ClH_{46}OOs$ $(H)(CO)(Cl)Os[As(C_6H_5)_3]_3$ .... Os: Org.Comp.A1-200, 205/6, 211
$As_3C_{55}Cl_2H_{45}OOs$ $(CO)(Cl)_2Os[As(C_6H_5)_3]_3$ .... Os: Org.Comp.A1-200, 205
$As_3C_{55}H_{46}IOOs$ .. $(H)(CO)(I)Os[As(C_6H_5)_3]_3$ .... Os: Org.Comp.A1-206
$As_3C_{55}H_{47}OOs$ ... $(H)_2(CO)Os[As(C_6H_5)_3]_3$ .... Os: Org.Comp.A1-206
$As_3C_{60}ClFe_3H_{45}N_3O_9Sn$
$Cl-Sn[Fe(CO)_2(NO)As(C_6H_5)_3]_3$ .... Fe: Org.Comp.C6a-15
$As_3C_{63}H_{63}MnN_2O_9$
$[Mn(O{=}As(CH_2C_6H_5)_3)_3][NO_3]_2$ Mn:MVol.D8-213
$As_3C_{64}ClH_{64}OOs$ $(H)(CO)(Cl)Os[As(CH_2-C_6H_5)_3]_3$ .... Os: Org.Comp.A1-200, 207
– .... $(H)(CO)(Cl)Os[As(C_6H_4-4-CH_3)_3]_3$ .... Os: Org.Comp.A1-200, 207
As_3Pd_8 .... Pd_8As_3 .... Pd: SVol.B2-343/4
As_3Rh .... $RhAs_3$ .... Rh: SVol.A1-38/41
– .... $RhAs_3$ solid solutions
$RhAs_3-AgAs_3$ .... Rh: SVol.A1-261
$RhAs_3-RuAs_3$ .... Rh: SVol.A1-269
As_3Ru .... $RuAs_3$ solid solutions
$RuAs_3-RhAs_3$ .... Rh: SVol.A1-269
$As_4BBrC_{45}H_{52}OOs$
$[(CO)OsBr(((CH_3)_2As)_2-1,2-C_6H_4)_2][B(C_6H_5)_4]$ Os: Org.Comp.A1-200, 209/10
$As_4BC_{45}ClH_{52}OOs$
$[(CO)OsCl(((CH_3)_2As)_2-1,2-C_6H_4)_2][B(C_6H_5)_4]$ Os: Org.Comp.A1-200, 209
$As_4BC_{45}H_{52}IOOs$ $[(CO)OsI(((CH_3)_2As)_2-1,2-C_6H_4)_2][B(C_6H_5)_4]$ .. Os: Org.Comp.A1-200, 210
$As_4BrC_{21}H_{32}OOs^+$
$[(CO)OsBr(((CH_3)_2As)_2-1,2-C_6H_4)_2][B(C_6H_5)_4]$ Os: Org.Comp.A1-200, 209/10
$As_4BrC_{21}H_{32}ORe$ $(CO)ReBr[As(CH_3)_2-C_6H_4-2-As(CH_3)_2]_2$ ..... Re: Org.Comp.1-34
$As_4BrC_{53}H_{44}O_3Re$ $(CO)_3ReBr[(C_6H_5)_2AsCH_2As(C_6H_5)_2]_2$ .... Re: Org.Comp.1-283
$As_4Br_2C_{20}H_{32}Mn$ $[Mn(C_6H_4(As(CH_3)_2)_2-1,2)_2Br_2]$ Mn:MVol.D8-205/6
$As_4Br_2C_{21}H_{32}OOs$ $(CO)(Br)_2Os[((CH_3)_2As)_2-1,2-C_6H_4]_2$ .... Os: Org.Comp.A1-200/1, 208, 211
$As_4Br_2C_{21}H_{32}ORe^+$
$[(CO)ReBr_2(As(CH_3)_2-C_6H_4-2-As(CH_3)_2)_2]^+$ .. Re: Org.Comp.1-35
$As_4Br_5C_{21}H_{32}ORe$ $[(CO)ReBr_2(As(CH_3)_2-C_6H_4-2-As(CH_3)_2)_2][Br_3]$
Re: Org.Comp.1-35
$As_4C_4H_{12}Mo_{12}O_{46}{}^{4-}$
$(CH_3AsO_3)_4Mo_{12}O_{34}{}^{4-}$ Mo:SVol.B3b-127, 128
$As_4C_8H_{20}Mo_{12}O_{50}{}^{4-}$
$(HOC_2H_4AsO_3)_4Mo_{12}O_{34}{}^{4-}$ Mo:SVol.B3b-127
$As_4C_{12}Cl_2H_{36}MnO_{12}$
$[Mn(O{=}As(CH_3)_3)_4(ClO_4)_2]$ Mn:MVol.D8-207
$As_4C_{20}Cl_2H_{32}Mn$ $[Mn(C_6H_4(As(CH_3)_2)_2-1,2)_2Cl_2]$ Mn:MVol.D8-205/6
$As_4C_{20}Cl_{12}H_{32}Mo_6$
$[(Mo_6Cl_8)Cl_2(C_6H_4(As(CH_3)_2)_2)_2]Cl_2$ Mo:SVol.B5-267
$As_4C_{20}H_{32}I_2Mn$... $[Mn(C_6H_4(As(CH_3)_2)_2-1,2)_2I_2]$ Mn:MVol.D8-205/6
$As_4C_{21}ClH_{32}I_2O_5Re$
$[(CO)ReI_2(As(CH_3)_2-C_6H_4-2-As(CH_3)_2)_2][ClO_4]$
Re: Org.Comp.1-35

$As_4C_{21}ClH_{32}OOs^+$
$[(CO)OsCl(((CH_3)_2As)_2\text{-}1,2\text{-}C_6H_4)_2][B(C_6H_5)_4]$ Os: Org.Comp.A1-200, 209
$As_4C_{21}ClH_{32}ORe$ $(CO)ReCl[As(CH_3)_2\text{-}C_6H_4\text{-}2\text{-}As(CH_3)_2]_2$ ...... Re: Org.Comp.1-34
$As_4C_{21}H_{32}IOOs^+$ $[(CO)(I)Os(((CH_3)_2As)_2\text{-}1,2\text{-}C_6H_4)_2][B(C_6H_5)_4]$ Os: Org.Comp.A1-200, 210
$As_4C_{21}H_{32}IORe$ .. $(CO)ReI[As(CH_3)_2\text{-}C_6H_4\text{-}2\text{-}As(CH_3)_2]_2$ .... Re: Org.Comp.1-34
$As_4C_{21}H_{32}I_2OOs$ .. $(CO)(I)_2Os[((CH_3)_2As)_2\text{-}1,2\text{-}C_6H_4]_2$ .... Os: Org.Comp.A1-200/1, 208
$As_4C_{21}H_{32}I_2ORe^+$ $[(CO)ReI_2(As(CH_3)_2\text{-}C_6H_4\text{-}2\text{-}As(CH_3)_2)_2]^+$ .... Re: Org.Comp.1-35
$As_4C_{21}H_{32}I_5ORe$ .. $[(CO)ReI_2(As(CH_3)_2\text{-}C_6H_4\text{-}2\text{-}As(CH_3)_2)_2][I_3]$ .. Re: Org.Comp.1-35
$As_4C_{24}H_{20}Mo_{12}O_{46}{}^{4-}$
$(C_6H_5AsO_3)_4Mo_{12}O_{34}{}^{4-}$ Mo:SVol.B3b-127
$As_4C_{24}H_{28}Mo_{12}N_4O_{46}$
$[(NH_3C_6H_4As)_4Mo_{12}O_{46}] \cdot 10\ CH_3CN \cdot 6\ H_2O$ Mo:SVol.B3b-127
$As_4C_{44}H_{20}N_8O_{28}S_4Th^{12-}$
$Th[(O)_2(O_3S)_2C_{10}H_2(NNC_6H_4AsO_3)_2]_2{}^{12-}$ ..... Th: SVol.D1-116
$As_4C_{44}H_{58}Mo_{12}N_{14}O_{46}$
$[(NH_3C_6H_4As)_4Mo_{12}O_{46}] \cdot 10\ CH_3CN \cdot 6\ H_2O$ Mo:SVol.B3b-127
$As_4C_{52}Cl_2H_{52}MnO_{12}$
$[Mn(O{=}As(C_6H_5)_2CH_3)_4(ClO_4)][ClO_4]$ Mn:MVol.D8-212
$As_4C_{53}ClH_{44}O_3Re$ $(CO)_3ReCl[(C_6H_5)_2AsCH_2As(C_6H_5)_2]_2$ .... Re: Org.Comp.1-282
$As_4C_{53}ClH_{48}ORe$ $(CO)ReCl[As(C_6H_5)_2CH_2CH_2As(C_6H_5)_2]_2$ ..... Re: Org.Comp.1-34
$As_4C_{53}H_{44}IO_3Re$ .. $(CO)_3ReI[(C_6H_5)_2AsCH_2As(C_6H_5)_2]_2$ .... Re: Org.Comp.1-283
$As_4C_{72}CdH_{60}I_4Mn$ $[Mn((C_6H_5)_3As)_4][CdI_4]$ Mn:MVol.D8-202
$As_4C_{72}Cl_2H_{60}MnO_{12}$
$[Mn(O{=}As(C_6H_5)_3)_4(ClO_4)][ClO_4]$ Mn:MVol.D8-207/8
$As_4C_{72}Cl_4H_{60}MoO_4$
$MoCl_4 \cdot 4\ (C_6H_5)_3AsO$ Mo:SVol.B5-325
$As_4C_{72}H_{60}I_2MnO_4$ $[Mn(O{=}As(C_6H_5)_3)_4I]I$ Mn:MVol.D8-207/8
$As_4C_{72}H_{60}I_6Mn$ ... $[Mn((C_6H_5)_3As)_4][I_3]_2$ Mn:MVol.D8-202
$As_4C_{74}H_{60}MnN_2O_4S_2$
$[Mn(O{=}As(C_6H_5)_3)_4(NCS)][NCS]$ Mn:MVol.D8-207/8
$As_4C_{76}Cl_2H_{68}In_2{}^{4+}$
$[((C_6H_5)_3AsCH_2)_2InCl_2In(CH_2As(C_6H_5)_3)_2]^{4+}$ .. In: Org.Comp.1-368
$As_4C_{76}Cl_6H_{68}In_2$ .. $[((C_6H_5)_3AsCH_2)_2InCl_2In(CH_2As(C_6H_5)_3)_2]Cl_4$ .. In: Org.Comp.1-368
$As_4C_{76}H_{60}I_3Mn_2O_8$
$[Mn(O{=}As(C_6H_5)_3)_4I][cis\text{-}Mn(CO)_4I_2]$ Mn:MVol.D8-208
$As_4C_{84}Cl_2H_{84}MnO_{12}$
$[Mn(O{=}As(CH_2C_6H_5)_3)_4][ClO_4]_2$ Mn:MVol.D8-213
$As_4HMo_{12}O_{50}{}^{7-}$ .. $HAs_4Mo_{12}O_{50}{}^{7-}$ Mo:SVol.B3b-73
$As_4H_2Mo_{12}O_{50}{}^{6-}$ $H_2As_4Mo_{12}O_{50}{}^{6-}$ Mo:SVol.B3b-73
$As_4H_3Mo_{12}O_{50}{}^{5-}$ $H_3As_4Mo_{12}O_{50}{}^{5-}$ Mo:SVol.B3b-73
$As_4H_4Mo_4Na_4O_{26}$ $Na_4[H_4As_4Mo_4O_{26}] \cdot 6\ H_2O$ Mo:SVol.B3b-70/1
$As_4H_4Mo_4O_{26}{}^{4-}$ .. $H_4As_4Mo_4O_{26}{}^{4-}$ Mo:SVol.B3b-70/1
$As_4H_4Mo_{12}O_{50}{}^{4-}$ $H_4As_4Mo_{12}O_{50}{}^{4-}$ Mo:SVol.B3b-73, 94/6
$As_4H_{20}Mo_{12}N_4O_{50}$ $(NH_4)_4[H_4As_4Mo_{12}O_{50}] \cdot 4\ H_2O$ Mo:SVol.B3b-73
$As_4Mo_{12}O_{50}{}^{8-}$ ... $As_4Mo_{12}O_{50}{}^{8-}$ Mo:SVol.B3b-73
$As_4O_{16}Th_3$ ...... $Th_3(AsO_4)_4$ .... Th: SVol.D1-2
As_4Th_3 .... Th_3As_4
Thermodynamic data .... Th: SVol.A4-183

$As_5C_{15}Cl_2H_{45}MnO_{13}$
$[Mn(O{=}As(CH_3)_3)_5][ClO_4]_2$ Mn:MVol.D8-206

$As_5C_{90}Cl_4H_{75}O_{21}Th$
$Th(ClO_4)_4 \cdot 5\ (C_6H_5)_3AsO \cdot 3\ H_2O$ .... Th: SVol.D4-164
– .... $Th(ClO_4)_4 \cdot 5\ (C_6H_5)_3AsO \cdot 4\ H_2O$ .... Th: SVol.D4-164

As_7Rh_9 .... Rh_9As_7 .... Rh: SVol.A1-38/41
As_7Rh_{12} .... $Rh_{12}As_7$ .... Rh: SVol.A1-38/41

$As_8Pd_{70}Te_{22}$ ..... $Pd_{70}As_8Te_{22}$ .... Pd: SVol.B2-347

Au .... Au
Chemical reactions
with F .... Au: SVol.B1-160/4
with H, D, and T .... Au: SVol.B1-28/42
with N .... Au: SVol.B1-101/2
with noble gases .... Au: SVol.B1-5/27
with O .... Au: SVol.B1-59/66

Compounds
Survey .... Au: SVol.B1-1/4
Uses .... Au: SVol.B1-1/4

Diffusion
in Rh .... Rh: SVol.A1-262, 266
in W .... W: SVol.A6b-286
of D .... Au: SVol.B1-33/4
of H .... Au: SVol.B1-32/3
of He .... Au: SVol.B1-18/9
of N .... Au: SVol.B1-101
of O .... Au: SVol.B1-62
of T .... Au: SVol.B1-34

Sorption
of Cl .... Au: SVol.B1-160/2
of D .... Au: SVol.B1-31/2
of H .... Au: SVol.B1-28/31
of N .... Au: SVol.B1-100/1
of noble gases .... Au: SVol.B1-5/15
of O .... Au: SVol.B1-60/2
on W .... W: SVol.A6b-260/86

Surface diffusion on W .... W: SVol.A6b-286/7

– .... Au alloys
Au-Nb-Rh .... Rh: SVol.A1-266
Au-Rh .... Rh: SVol.A1-262/6
Au-Rh-Ti .... Rh: SVol.A1-266

– .... Au systems
Au-Cl .... Au: SVol.B1-162/4
Au-Po .... Po: SVol.1-329
Au-Rh .... Rh: SVol.A1-262/3
Au-W .... W: SVol.A6b-259

$AuB_2C_{30}CoH_{33}P$ .. $(C_5H_5)Co[C_3B_2H(CH_3)_4]AuP(C_6H_5)_3$ .... B: B Comp.SVol.4/4-180
$AuB_2C_{35}H_{43}PRh$ .. $(C_5H_5)Rh[C_3B_2(CH_3)(C_2H_5)_4AuP(C_6H_5)_3]$ ..... B: B Comp.SVol.4/4-180
$AuB_8C_{25}H_{30}NiP$ .. $(C_5H_5)[(C_6H_5)_3PAu]$-7,8,9-$C_2NiB_8H_{10}$ .... B: B Comp.SVol.3/4-176

$AuB_9C_{32}Fe_2H_{31}O_8PW$
$AuFe_2W[C(CH_3)][C_2B_9H_7(CH_3)_2](CO)_8[P(C_6H_5)_3]$
B: B Comp.SVol.4/4-248

$AuB_9C_{39}H_{47}O_2PPtW$
$AuPtW[C(C_6H_5)](CO)_2[P(C_6H_5)_3](C_8H_{12})[C_2B_9H_9\text{-}(CH_3)_2]$
B: B Comp.SVol.4/4-248

$AuB_9C_{40}H_{49}O_2PPtW$
$AuPtW[C(C_6H_4\text{-}4\text{-}CH_3)](CO)_2[P(C_6H_5)_3]$
$(C_8H_{12})[C_2B_9H_9\text{-}(CH_3)_2]$ ... B: B Comp.SVol.4/4-248

$AuB_9C_{60}H_{59}NO_4P_2W_2$
$[(C_6H_5)_3P{=}N{=}P(C_6H_5)_3][AuW_2(C$
$\text{-}C_6H_4\text{-}4\text{-}CH_3)_2(CO)_4(C_2B_9H_9\text{-}(CH_3)_2)]$ ... B: B Comp.SVol.4/4-249

$AuB_{10}BrC_{10}H_{34}N_4P_2$
$[1\text{-}P(N(CH_3)_2)_2\text{-}1,2\text{-}C_2B_{10}H_{10}\text{-}2\text{-}P(N(CH_3)_2)_2]AuBr$
B: B Comp.SVol.4/4-264

$AuB_{10}BrC_{18}H_{32}N_2P_2$
$[1,2\text{-}C_2B_{10}H_{10}\text{-}1\text{-}P(C_6H_5)_2\text{-}2\text{-}P(N(CH_3)_2)_2]AuBr$
B: B Comp.SVol.4/4-264

$AuB_{10}BrC_{26}H_{30}P_2$ $[1,2\text{-}C_2B_{10}H_{10}\text{-}1,2\text{-}(P(C_6H_5)_2)_2]AuBr$ ... B: B Comp.SVol.4/4-264

$AuB_{20}C_{20}H_{68}N_8P_4^+$
$[(1,2\text{-}C_2B_{10}H_{10}\text{-}1,2\text{-}((CH_3)_2N\text{-}P\text{-}N(CH_3)_2)_2)_2Au]^+$
B: B Comp.SVol.4/4-264

$AuB_{20}C_{36}H_{64}N_4P_4^+$
$[(1,2\text{-}C_2B_{10}H_{10}\text{-}1,2\text{-}(P(C_6H_5)_2)_2)Au(1,2\text{-}$
$((CH_3)_2N\text{-}P\text{-}N(CH_3)_2)_2\text{-}1,2\text{-}C_2B_{10}H_{10})]^+$ ... B: B Comp.SVol.4/4-264

– ... $[(1,2\text{-}C_2B_{10}H_{10}\text{-}1\text{-}P(C_6H_5)_2\text{-}2\text{-}P(N(CH_3)_2)_2)_2Au]^+$
B: B Comp.SVol.4/4-264

$AuB_{20}C_{52}H_{60}P_4^+$ $[(1,2\text{-}C_2B_{10}H_{10}\text{-}1,2\text{-}(P(C_6H_5)_2)_2)_2Au]^+$ ... B: B Comp.SVol.4/4-264

$AuBrC_{11}Cl_4H_{12}N_2O$
$[1,2\text{-}N_2C_3H\text{-}1,5\text{-}(CH_3)_2\text{-}2\text{-}(C_6H_5)\text{-}4\text{-}(Br)\text{-}3\text{-}(=O)][AuCl_4]$
Au: SVol.B1-294

$AuBrF_6$ ... $[BrF_2][AuF_4]$ ... Au: SVol.B1-125, 128/9
Br: SVol.B3-41/5

$AuBrF_{10}$ ... $[BrF_4][AuF_6]$ ... Au: SVol.B1-139/40, 147/8

$AuBrF_{12}$ ... $[BrF_6][AuF_6]$ ... Au: SVol.B1-139/40, 143/4, 147/8
Br: SVol.B3-138/40

$AuCCl_4H_6N_3$ ... $[C(NH_2)_3][AuCl_4]$ ... Au: SVol.B1-216, 289

$AuC_2Cl_4H_4N_3$ ... $[C_2H_4N_3][AuCl_4]$ ... Au: SVol.B1-295

$AuC_3Cl_4H_8NO_2$ ... $[CH_3\text{-}CHNH_3\text{-}COOH][AuCl_4]$ ... Au: SVol.B1-298

$AuC_3Cl_4H_8NO_3$ ... $[HO\text{-}CH_2CHNH_3\text{-}COOH][AuCl_4]$ ... Au: SVol.B1-298

$AuC_3Cl_4H_{10}N$ ... $[(CH_3)_3NH][AuCl_4]$ ... Au: SVol.B1-213

$AuC_4Cl_4H_7N_2$ ... $[1,2\text{-}N_2C_3H_4\text{-}3\text{-}CH_3][AuCl_4]$ ... Au: SVol.B1-294

$AuC_4Cl_4H_8NO_2S$ .. $[1,3\text{-}SNC_3H_7\text{-}4\text{-}COOH][AuCl_4]$ ... Au: SVol.B1-300

$AuC_4Cl_4H_8NO_4$ ... $[HOOC\text{-}CH_2CHNH_3\text{-}COOH][AuCl_4]$ ... Au: SVol.B1-298

$AuC_4Cl_4H_{10}NO_2$ .. $[CH_3\text{-}OC(=O)\text{-}CH(CH_3)\text{-}NH_3][AuCl_4]$ ... Au: SVol.B1-216

$AuC_4Cl_4H_{12}N$ ... $[(CH_3)_4N][AuCl_4]$ ... Au: SVol.B1-226

$AuC_4Cl_7H_9P$ ... $[t\text{-}C_4H_9\text{-}PCl_3][AuCl_4]$ ... Au: SVol.B1-339

$AuC_5Cl_4H_5N_4O$ ... $[1,3,7,9\text{-}N_4C_5H_5\text{-}6\text{-}(=O)][AuCl_4] \cdot 2\,H_2O$ ... Au: SVol.B1-217, 295

$AuC_5Cl_4H_6N$	$[C_5H_6N][AuCl_4]$	Au: SVol.B1-213, 216, 229, 291
$AuC_5Cl_4H_9N_2$	$[1,2\text{-}N_2C_3H_3\text{-}3,5\text{-}(CH_3)_2][AuCl_4]$	Au: SVol.B1-294
$AuC_5Cl_4H_{12}N$	$[C_5H_{12}N][AuCl_4]$	Au: SVol.B1-291
$AuC_5Cl_4H_{12}NO_2$	$[i\text{-}C_3H_7\text{-}CHNH_3\text{-}COOH][AuCl_4]$	Au: SVol.B1-298
$AuC_5Cl_4H_{14}N_2$	$[1,4\text{-}N_2C_4H_{11}\text{-}1\text{-}CH_3][AuCl_4]$	Au: SVol.B1-295
$AuC_5Cl_5H_5N$	$[2\text{-}Cl\text{-}C_5H_5N][AuCl_4]$	Au: SVol.B1-291
–	$[4\text{-}Cl\text{-}C_5H_5N][AuCl_4]$	Au: SVol.B1-291
$AuC_5Cl_7H_{11}P$	$[C_2H_5\text{-}C(CH_3)_2\text{-}PCl_3][AuCl_4]$	Au: SVol.B1-339
$AuC_6Cl_4H_4S_4$	$[S_2C_3H_2(C_3H_2S_2)][AuCl_4]$	Au: SVol.B1-336
$AuC_6Cl_4H_6N_3$	$[C_6H_6N_3][AuCl_4]$	Au: SVol.B1-295
$AuC_6Cl_4H_8N$	$[2\text{-}CH_3\text{-}C_5H_5N][AuCl_4]$	Au: SVol.B1-291
–	$[3\text{-}CH_3\text{-}C_5H_5N][AuCl_4]$	Au: SVol.B1-291
–	$[4\text{-}CH_3\text{-}C_5H_5N][AuCl_4]$	Au: SVol.B1-291
$AuC_6Cl_4H_{14}N$	$[2\text{-}CH_3\text{-}C_5H_{11}N][AuCl_4]$	Au: SVol.B1-291
–	$[3\text{-}CH_3\text{-}C_5H_{11}N][AuCl_4]$	Au: SVol.B1-291
–	$[4\text{-}CH_3\text{-}C_5H_{11}N][AuCl_4]$	Au: SVol.B1-291
$AuC_6Cl_4H_{16}N_2O$	$[1,4\text{-}N_2C_4H_{11}\text{-}1\text{-}CH_2CH_2OH][AuCl_4]$	Au: SVol.B1-295
$AuC_6Cl_4H_{17}N_3$	$[1,4\text{-}N_2C_4H_{11}\text{-}1\text{-}CH_2CH_2\text{-}NH_2][AuCl_4]$	Au: SVol.B1-295
$AuC_6Cl_7H_{11}P$	$[C_6H_{11}\text{-}PCl_3][AuCl_4]$	Au: SVol.B1-339
$AuC_7Cl_4H_9N_4O_2$	$[1,3,7,9\text{-}N_4C_5H_3\text{-}2,6\text{-}(=O)_2\text{-}1,3\text{-}(CH_3)_2][AuCl_4]$	Au: SVol.B1-295
–	$[1,3,7,9\text{-}N_4C_5H_3\text{-}2,6\text{-}(=O)_2\text{-}3,7\text{-}(CH_3)_2][AuCl_4]$	Au: SVol.B1-295
–	$[1,3,7,9\text{-}N_4C_5H_3\text{-}2,6\text{-}(=O)_2\text{-}3,8\text{-}(CH_3)_2][AuCl_4]$	Au: SVol.B1-295
$AuC_7Cl_4H_{10}N$	$[2,4\text{-}(CH_3)_2\text{-}C_5H_4N][AuCl_4]$	Au: SVol.B1-292
–	$[2,5\text{-}(CH_3)_2\text{-}C_5H_4N][AuCl_4]$	Au: SVol.B1-292
–	$[2,6\text{-}(CH_3)_2\text{-}C_5H_4N][AuCl_4]$	Au: SVol.B1-292
–	$[3,5\text{-}(CH_3)_2\text{-}C_5H_4N][AuCl_4]$	Au: SVol.B1-292
–	$[3,6\text{-}(CH_3)_2\text{-}C_5H_4N][AuCl_4]$	Au: SVol.B1-292
$AuC_7Cl_4H_{19}N_3$	$[1,4\text{-}N_2C_4H_{11}\text{-}1\text{-}CHCH_3CH_2\text{-}NH_2][AuCl_4]$	Au: SVol.B1-295
$AuC_7N_2O_5Re$	$(CO)_5ReNCAuCN$	Re: Org.Comp.2-174
$AuC_8Cl_2H_{20}N$	$[(C_2H_5)_4N][AuCl_2]$	Au: SVol.B1-185/6
$AuC_8Cl_4H_{11}N_2O_2$	$[H(2\text{-}(O=CH\text{-}NCH_3\text{-}CH_2)\text{-}C_5H_4N(O)\text{-}1)][AuCl_4]$	Au: SVol.B1-293
$AuC_8Cl_4H_{11}N_2S$	$[C_6H_5CH_2\text{-}S\text{-}C(NH_2)_2][AuCl_4]$	Au: SVol.B1-216, 269
$AuC_8Cl_4H_{11}N_4O_2$	$[1,3,7,9\text{-}N_4C_5H_2\text{-}2,6\text{-}(=O)_2\text{-}1,3,7\text{-}(CH_3)_3][AuCl_4]$	Au: SVol.B1-295
–	$[1,3,7,9\text{-}N_4C_5H_2\text{-}2,6\text{-}(=O)_2\text{-}1,3,8\text{-}(CH_3)_3][AuCl_4]$	Au: SVol.B1-295
–	$[1,3,7,9\text{-}N_4C_5H_3\text{-}2,6\text{-}(=O)_2\text{-}8\text{-}(C_2H_5)\text{-}3\text{-}(CH_3)][AuCl_4]$	Au: SVol.B1-295
$AuC_8Cl_4H_{12}N$	$[2,4,6\text{-}(CH_3)_3\text{-}C_5H_3N][AuCl_4]$	Au: SVol.B1-292
$AuC_8Cl_4H_{16}NO$	$[1,3,3\text{-}(CH_3)_3\text{-}2\text{-}(O=)\text{-}C_5H_7N][AuCl_4]$	Au: SVol.B1-291
–	$[1,5,5\text{-}(CH_3)_3\text{-}2\text{-}(O=)\text{-}C_5H_7N][AuCl_4]$	Au: SVol.B1-291
$AuC_8Cl_4H_{19}N_2O_2$	$[(CH_3\text{-}C(=O)\text{-}N(CH_3)_2)_2H][AuCl_4]$	Au: SVol.B1-217, 290
$AuC_8Cl_4H_{20}N$	$[(C_2H_5)_4N][AuCl_4]$	Au: SVol.B1-213, 222/9, 268
$AuC_8Cl_4H_{20}N_2O_2$	$[1,4\text{-}N_2C_4H_{10}\text{-}1,4\text{-}(CH_2CH_2OH)_2][AuCl_4]$	Au: SVol.B1-295
$AuC_8H_{20}N_{13}$	$[(C_2H_5)_4N][Au(N_3)_4]$	Au: SVol.B1-105/7
$AuC_9Cl_4H_8N$	$[1\text{-}NC_9H_8][AuCl_4]$	Au: SVol.B1-293
$AuC_9Cl_4H_{13}N_2O_3$	$[H_3O][AuCl_4] \cdot [2,2\text{-}(CH_3)_2\text{-}1,3\text{-}(O)_2\text{-}1,3\text{-}N_2C_7H_4]$	Au: SVol.B1-339

$AuC_9Cl_4H_{13}N_4O_2$ $[1,3,7,9\text{-}N_4C_5H_2\text{-}2,6\text{-}(=O)_2\text{-}8\text{-}(C_2H_5)\text{-}1,3\text{-}(CH_3)_2][AuCl_4]$ Au: SVol.B1-295
$AuC_9Cl_4H_{23}N_3O$ .. $[1,4\text{-}N_2C_4H_{10}\text{-}1\text{-}CH_2CH_2OH\text{-}4\text{-}CH_2CH_2CH_2NH_2][AuCl_4]$ Au: SVol.B1-295
$AuC_{10}Cl_4H_9N_2$ ... $[H(NC_5H_4\text{-}2\text{-}(2\text{-}C_5H_4N))][AuCl_4]$ ... Au: SVol.B1-293
$AuC_{10}Cl_4H_9N_2O_2$ $[H(1\text{-}(O)NC_5H_4\text{-}2\text{-}(2\text{-}C_5H_4N(O)\text{-}1))][AuCl_4]$ ... Au: SVol.B1-293
$AuC_{10}Cl_4H_9N_4O_6$ $[H(4\text{-}O_2N\text{-}C_5H_4N(O)\text{-}1)_2][AuCl_4]$ ... Au: SVol.B1-292
$AuC_{10}Cl_4H_{10}N$ ... $[1\text{-}NC_9H_7\text{-}2\text{-}CH_3][AuCl_4]$ ... Au: SVol.B1-294
– ... $[1\text{-}NC_9H_7\text{-}6\text{-}CH_3][AuCl_4]$ ... Au: SVol.B1-294
– ... $[1\text{-}NC_9H_7\text{-}8\text{-}CH_3][AuCl_4]$ ... Au: SVol.B1-294
$AuC_{10}Cl_4H_{11}N_2O_2$ $[(C_5H_5N(O)\text{-}1)_2H][AuCl_4]$ ... Au: SVol.B1-217, 335
$AuC_{10}Cl_4H_{15}N_2O_3$ $[H_3O][AuCl_4] \cdot [2\text{-}(C_2H_5)\text{-}2\text{-}(CH_3)\text{-}1,3\text{-}(O)_2\text{-}1,3\text{-}N_2C_7H_4]$ ... Au: SVol.B1-339
$AuC_{10}Cl_4H_{19}N_2O_2$ $[H(NC_4H_6\text{-}1\text{-}CH_3\text{-}2\text{-}(=O))_2][AuCl_4]$ ... Au: SVol.B1-291
$AuC_{10}Cl_4H_{24}N_2O_2$ $[1,4\text{-}N_2C_4H_{10}\text{-}1,4\text{-}(CH_2CHCH_3\text{-}OH)_2][AuCl_4]$ .. Au: SVol.B1-295
$AuC_{10}Cl_6H_{14}P$ ... $[t\text{-}C_4H_9\text{-}PCl_2\text{-}C_6H_5][AuCl_4]$ ... Au: SVol.B1-339
$AuC_{11}Cl_4H_{12}N$ ... $[1\text{-}NC_9H_6\text{-}2,4\text{-}(CH_3)_2][AuCl_4]$ ... Au: SVol.B1-294
– ... $[1\text{-}NC_9H_6\text{-}2,5\text{-}(CH_3)_2][AuCl_4]$ ... Au: SVol.B1-294
– ... $[1\text{-}NC_9H_6\text{-}2,6\text{-}(CH_3)_2][AuCl_4]$ ... Au: SVol.B1-294
$AuC_{11}Cl_4H_{13}N_2O$ $[1,2\text{-}N_2C_3H_2\text{-}1,5\text{-}(CH_3)_2\text{-}2\text{-}(C_6H_5)\text{-}3\text{-}(=O)][AuCl_4]$ Au: SVol.B1-294
$AuC_{11}Cl_4H_{14}N_3O$ $[1,2\text{-}N_2C_3H\text{-}1,5\text{-}(CH_3)_2\text{-}2\text{-}(C_6H_5)\text{-}4\text{-}(NH_2)\text{-}3\text{-}(=O)][AuCl_4]$ ... Au: SVol.B1-294
$AuC_{11}Cl_6H_{16}P$ ... $[C_2H_5\text{-}C(CH_3)_2\text{-}PCl_2\text{-}C_6H_5][AuCl_4]$ ... Au: SVol.B1-339
$AuC_{12}Cl_2H_{14}N_3O$ $[(NH_3)(O=)C_4N_2(CH_3)_2(C_6H_5)][AuCl_2]$ ... Au: SVol.B1-188
$AuC_{12}Cl_4H_9N_2$ ... $[1,10\text{-}N_2C_{12}H_9][AuCl_4]$ ... Au: SVol.B1-293
$AuC_{12}Cl_4H_{10}N_3S$. $[5,10\text{-}SNC_{12}H_6\text{-}7\text{-}(NH_3)\text{-}2\text{-}(=NH)][AuCl_4]$ ... Au: SVol.B1-314
$AuC_{12}Cl_4H_{11}N_2$ ... $[C_6H_5\text{-}NHN\text{-}C_6H_5][AuCl_4]$ ... Au: SVol.B1-336
$AuC_{12}Cl_4H_{11}N_2O_6$ $[H(3\text{-}HOOC\text{-}C_5H_4N(O)\text{-}1)_2][AuCl_4]$ ... Au: SVol.B1-292
$AuC_{12}Cl_4H_{15}N_2O_2$ $[H(2\text{-}CH_3\text{-}C_5H_4N(O)\text{-}1)_2][AuCl_4]$ ... Au: SVol.B1-217, 292, 335
– ... $[H(3\text{-}CH_3\text{-}C_5H_4N(O)\text{-}1)_2][AuCl_4]$ ... Au: SVol.B1-217, 292
– ... $[H(4\text{-}CH_3\text{-}C_5H_4N(O)\text{-}1)_2][AuCl_4]$ ... Au: SVol.B1-217, 292
$AuC_{12}Cl_4H_{16}N_3O$ $[1,2\text{-}N_2C_3H\text{-}1,5\text{-}(CH_3)_2\text{-}2\text{-}(C_6H_5)\text{-}4\text{-}(NHCH_3)\text{-}3\text{-}(=O)][AuCl_4]$ ... Au: SVol.B1-294
$AuC_{12}Cl_4H_{19}N_2O_3$ $[H_3O][AuCl_4] \cdot [2\text{-}(t\text{-}C_4H_9)\text{-}2\text{-}(CH_3)\text{-}1,3\text{-}(O)_2\text{-}1,3\text{-}N_2C_7H_4]$ ... Au: SVol.B1-339
$AuC_{12}Cl_4H_{19}N_4O_2$ $[1,3,7,9\text{-}N_4C_5H_2\text{-}2,6\text{-}(=O)_2\text{-}8\text{-}(C_5H_{11})\text{-}1,3\text{-}(CH_3)_2][AuCl_4]$ ... Au: SVol.B1-295
$AuC_{12}Cl_4H_{28}N$ ... $[(C_3H_7)_4N][AuCl_4]$ ... Au: SVol.B1-225, 226
$AuC_{12}Cl_6H_{16}P$ ... $[C_6H_{11}\text{-}PCl_2\text{-}C_6H_5][AuCl_4]$ ... Au: SVol.B1-339
$AuC_{13}Cl_4H_{18}N_3O$ $[1,2\text{-}N_2C_3H\text{-}1,5\text{-}(CH_3)_2\text{-}2\text{-}(C_6H_5)\text{-}4\text{-}(N(CH_3)_2)\text{-}3\text{-}(=O)][AuCl_4]$ ... Au: SVol.B1-294
$AuC_{14}Cl_4H_{19}N_2O_2$ $[H(2,6\text{-}(CH_3)_2\text{-}C_5H_3N(O)\text{-}1)_2][AuCl_4]$ ... Au: SVol.B1-292
$AuC_{14}Cl_4H_{21}NO_5$ $[C_{14}H_{21}NO_5][AuCl_4]$ ... Au: SVol.B1-336
$AuC_{14}Cl_4H_{30}O_4S_2$ $[(CH_3)_2C_{12}H_{24}O_4S_2][AuCl_4]$ ... Au: SVol.B1-336
$AuC_{16}Cl_2H_{36}N$ ... $[(C_4H_9)_4N][AuCl_2]$ ... Au: SVol.B1-185/6
$AuC_{16}Cl_4H_{36}N$ ... $[(C_4H_9)_4N][AuCl_4]$ ... Au: SVol.B1-213, 222, 226, 230, 268/9
$AuC_{18}Cl_4H_{13}N_2O_2$ $[H(1\text{-}(O)NC_9H_6\text{-}2\text{-}(2\text{-}C_9H_6N(O)\text{-}1))][AuCl_4]$ ... Au: SVol.B1-335
$AuC_{18}Cl_4H_{16}O_2P_2$ $[(C_6H_5)_3P_2O_2H][AuCl_4]$ ... Au: SVol.B1-217

$AuC_{19}Cl_7H_{15}P$ .... $[(C_6H_5)_3C-PCl_3][AuCl_4]$ .... Au: SVol.B1-339

$AuC_{20}Cl_4H_{44}N$ ... $[(C_5H_{11})_4N][AuCl_4]$ .... Au: SVol.B1-225, 226

$AuC_{21}Cl_4H_{38}N$ ... $[1-C_{16}H_{33}-C_5H_5N][AuCl_4]$ .... Au: SVol.B1-296

$AuC_{22}Cl_4H_{25}N_4O_2$ $[H(1,2-N_2C_3H-1,5-(CH_3)_2-2-(C_6H_5)-3-(O))_2][AuCl_4]$ Au: SVol.B1-294

$AuC_{23}H_{15}O_5PRe$ .. $(CO)_5ReAuP(C_6H_5)_3$ .... Re: Org.Comp.2-186

$AuC_{24}Cl_4H_{20}P$ .... $[(C_6H_5)_4P][AuCl_4]$ .... Au: SVol.B1-229

$AuC_{24}Cl_4H_{29}N_2O_4$ $[H(NC_4H_6-1-(CH_2-C(OH)C_6H_5)-2-(O))_2][AuCl_4]$ Au: SVol.B1-291

$AuC_{24}Cl_4H_{52}N$ ... $[(i-C_8H_{17})_3NH][AuCl_4]$ .... Au: SVol.B1-222

$AuC_{25}Cl_6H_{20}P$ .... $[(C_6H_5)_3C-PCl_2-C_6H_5][AuCl_4]$ .... Au: SVol.B1-339

$AuC_{28}Cl_2FeH_{22}P$. $(Cl-C_5H_4)Fe[C_5H_3(Cl)-AuP(C_6H_5)_3]$ .... Fe: Org.Comp.A10-332/3

$AuC_{28}Cl_4H_{47}N_2O_2$ $[(2-C_9H_{19}-C_5H_4N(O)-1)_2H][AuCl_4]$ .... Au: SVol.B1-217

$AuC_{28}FeH_{23}NO_2P$ $(C_5H_5)Fe[C_5H_3(AuP(C_6H_5)_3)-NO_2]$ .... Fe: Org.Comp.A10-177

$AuC_{29}FeH_{26}P$ .... $FeC_{10}H_8(AuP(C_6H_5)_3)CH_3$ .... Fe: Org.Comp.A10-177

$AuC_{30}FeH_{26}OP$ ... $(C_5H_5)Fe[C_5H_3(AuP(C_6H_5)_3)-C(=O)CH_3]$ .... Fe: Org.Comp.A10-177

$AuC_{30}Fe_3H_{18}O_{10}P$ $(CH_3-C)Fe_3(CO)_{10}Au[P(C_6H_5)_3]$ .... Fe: Org.Comp.C6b-79, 81

$AuC_{30}Fe_3H_{18}O_{11}P$ $(CO)_{10}Fe_3(C-O-CH_3)Au[P(C_6H_5)_3]$ .... Fe: Org.Comp.C6b-79, 80, 82

$AuC_{31}H_{27}N_2O_2PRe$ $(C_5H_5)Re(CO)(N=N-C_6H_4-OCH_3-4)-AuP(C_6H_5)_3$ Re: Org.Comp.3-133/5, 138, 140

$AuC_{32}Fe_3H_{25}NO_9P$ $(CO)_9Fe_3[N(C_4H_9-t)=CH][Au-P(C_6H_5)_3]$ .... Fe: Org.Comp.C6a-313/4

$AuC_{33}Fe_3H_{21}O_9P_2$ $(CO)_9(H)Fe_3(P-C_6H_5)[Au-P(C_6H_5)_3]$ .... Fe: Org.Comp.C6a-315, 317

$AuC_{33}H_{32}MnMoO_4P_2$ $C_5H_5Mo(CO)(P(C_6H_5)_2)(CHCHCHCH_3)$ $(AuP(CH_3)_2C_6H_5)Mn(CO)_3$ .... Mo: Org.Comp.6-373/4

$AuC_{34}FeH_{27}NO_2P$ $FeC_{10}H_8(AuP(C_6H_5)_3)-C_6H_4-NO_2-4$ .... Fe: Org.Comp.A10-177

$AuC_{34}FeH_{28}P$ .... $FeC_{10}H_8(AuP(C_6H_5)_3)-C_6H_5$ .... Fe: Org.Comp.A10-177

$AuC_{36}Cl_4H_{31}O_2P_2$ $HAuCl_4 \cdot 2\ OP(C_6H_5)_3$ .... Au: SVol.B1-338

$AuC_{39}H_{35}IO_3OsP_3$ $(CO)Os(PH_2-AuI)[-OC(CH_3)O-][P(C_6H_5)_3]_2$ ... Os: Org.Comp.A1-139/40, 155/6

$AuC_{40}H_{38}INOOsP_3S_2$ $(AuI-PH_2)(CO)Os[S_2C-N(CH_3)_2][P(C_6H_5)_3]_2$ ... Os: Org.Comp.A1-153

$AuC_{45}H_{35}MoO_4P_2$ $C_5H_4C(O)HMo(CO)(CO)_2Au(P(C_6H_5)_3)_2$ .... Mo: Org.Comp.6-383

$AuC_{72}Cl_4H_{60}LiO_4P_4$ $Li[AuCl_4] \cdot 4\ OP(C_6H_5)_3$ .... Au: SVol.B1-316

$AuC_{73}F_6H_{62}OOsP_5$ $[((C_6H_5)_3P)_3(CO)Os(-H-Au(P(C_6H_5)_3)-H-)][PF_6]$ Os: Org.Comp.A1-218/9

$AuC_{73}H_{62}OOsP_4^+$ $[((C_6H_5)_3P)_3(CO)Os(-H-Au(P(C_6H_5)_3)-H-)][PF_6]$ Os: Org.Comp.A1-218/9

$AuCl$ .... $AuCl$ .... Au: SVol.B1-164/79

– .... Au_2Cl_2 .... Au: SVol.B1-164/8

$AuClO$ .... $AuOCl$ .... Au: SVol.B1-344/6

$AuCl_2$ .... $AuCl_2$ .... Au: SVol.B1-189/90

$AuCl_2^-$ .... $[AuCl_2]^-$ .... Au: SVol.B1-179/88

$AuCl_2H$ .... $HAuCl_2$ .... Au: SVol.B1-189

$AuCl_3$ .... $AuCl_3$ .... Au: SVol.B1-190/212

$AuCl_3$ .... Au_2Cl_6 .... Au: SVol.B1-193, 197/9, 203
$AuCl_3O_{12}$ .... $Au[ClO_4]_3$ .... Au: SVol.B1-346
$AuCl_4^-$ .... $[AuCl_4]^-$
Catalytic properties .... Au: SVol.B1-271/2
Chemical reactions
Exchange reactions .... Au: SVol.B1-248/57
Hydrolysis .... Au: SVol.B1-238/48
Irradiation effects .... Au: SVol.B1-235/7
Precipitation reactions .... Au: SVol.B1-267/70
Redox reactions .... Au: SVol.B1-257/67
with inorganic compounds .... Au: SVol.B1-238/71
with organic compounds .... Au: SVol.B1-272/330
Diffusion .... Au: SVol.B1-219/20
Electrochemistry .... Au: SVol.B1-232/4
Formation .... Au: SVol.B1-213/5
Ion .... Au: SVol.B1-215/9
Spectra .... Au: SVol.B1-220/31
Thermodynamic properties .... Au: SVol.B1-219/20
Uses .... Au: SVol.B1-271/2

$AuCl_4Cs$ .... $Cs[AuCl_4]$ .... Au: SVol.B1-223/4, 226, 231, 268
$AuCl_4H$ .... $HAuCl_4$ .... Au: SVol.B1-330/44
– .... $HAuCl_4 \cdot 3\,H_2O$ .... Au: SVol.B1-330/44
– .... $HAuCl_4 \cdot 4\,H_2O$ .... Au: SVol.B1-330/44
– .... $DAuCl_4 \cdot 4\,D_2O$ .... Au: SVol.B1-330/4
– .... $HAuCl_4 \cdot n\,H_2O$ .... Au: SVol.B1-330/44
– .... $HAuCl_4$ solutions
Chemical reactions
Exchange reactions .... Au: SVol.B1-249/57
Hydrolysis .... Au: SVol.B1-238/48
Irradiation effects .... Au: SVol.B1-235/7
Precipitation reactions .... Au: SVol.B1-267/70
Redox reactions .... Au: SVol.B1-257/67
with inorganic compounds .... Au: SVol.B1-238/71, 334
with organic compounds .... Au: SVol.B1-272/330, 334/41

$AuCl_4H_3O$ .... $[H_3O][AuCl_4]$ .... Au: SVol.B1-216
$AuCl_4H_4N$ .... $[NH_4][AuCl_4]$ .... Au: SVol.B1-213, 229/30, 268, 282
– .... $[NH_4][AuCl_4] \cdot 0.67\,H_2O$ .... Au: SVol.B1-216
– .... $[NH_4][AuCl_4] \cdot x\,H_2O$ .... Au: SVol.B1-231
$AuCl_4K$ .... $K[AuCl_4]$ .... Au: SVol.B1-216, 226, 229/31, 236, 246/55, 261, 268/70, 297, 300/1, 310/7
– .... $K[AuCl_4] \cdot 2\,H_2O$ .... Au: SVol.B1-216, 230
$AuCl_4NO$ .... $[NO][AuCl_4]$ .... Au: SVol.B1-344
$AuCl_4Na$ .... $Na[AuCl_4]$ .... Au: SVol.B1-216, 222/3, 229/30, 235, 255/69, 297/320

$AuCl_4Na$ . . . . $Na[AuCl_4] \cdot 2\ H_2O$ . . . . Au: SVol.B1-216, 222/3, 230, 268, 298/312
– . . . . $Na[AuCl_4] \cdot 3\ H_2O$ . . . . Au: SVol.B1-308
– . . . . $Na[AuCl_4] \cdot x\ H_2O$ . . . . Au: SVol.B1-230
$AuCl_4Rb$ . . . . $Rb[AuCl_4]$ . . . . Au: SVol.B1-216, 226, 262, 268/9
$AuCl_4Tl$ . . . . $Tl[AuCl_4]$ . . . . Au: SVol.B1-213, 216, 268
$AuCl_6CoH_{18}N_6$ . . . $[Co(NH_3)_6][AuCl_4]Cl_2$ . . . . Au: SVol.B1-262
$AuCl_7Cs_2Pd$ . . . . . $Cs_2PdAuCl_7$ . . . . Pd: SVol.B2-171
$AuCl_7PdRb_2$ . . . . . $Rb_2PdAuCl_7$ . . . . Pd: SVol.B2-171
$AuCl_7S$ . . . . $[SCl_3][AuCl_4]$ . . . . Au: SVol.B1-217
$AuCl_7Se$ . . . . $[SeCl_3][AuCl_4]$ . . . . Au: SVol.B1-217
$AuCl_7Te$ . . . . $[TeCl_3][AuCl_4]$ . . . . Au: SVol.B1-217
$AuCl_8P$ . . . . $[PCl_4][AuCl_4]$ . . . . Au: SVol.B1-217, 226
$AuCsF_4$ . . . . $Cs[AuF_4]$ . . . . Au: SVol.B1-125/9
$AuCsF_6$ . . . . $Cs[AuF_6]$ . . . . Au: SVol.B1-139/48
$AuCs_2F_6$ . . . . $Cs_2[AuF_6]$ . . . . Au: SVol.B1-133
$AuCuF_5$ . . . . $[CuF][AuF_4]$ . . . . Au: SVol.B1-125, 127
AuF . . . . AuF . . . . Au: SVol.B1-113/4
AuF^- . . . . $[AuF]^-$ . . . . Au: SVol.B1-115
AuF_2 . . . . AuF_2 . . . . Au: SVol.B1-115/6
AuF_2^- . . . . $[AuF_2]^-$ . . . . Au: SVol.B1-115
AuF_3 . . . . AuF_3 . . . . Au: SVol.B1-116/25
AuF_3^- . . . . $[AuF_3]^-$ . . . . Au: SVol.B1-116
AuF_4 . . . . AuF_4 . . . . Au: SVol.B1-132
AuF_4^- . . . . $[AuF_4]^-$ . . . . Au: SVol.B1-125/30
AuF_4K . . . . $K[AuF_4]$ . . . . Au: SVol.B1-125/9
AuF_4Li . . . . $Li[AuF_4]$ . . . . Au: SVol.B1-125
AuF_4NO . . . . $[NO][AuF_4]$ . . . . Au: SVol.B1-125, 129, 131
AuF_4NO_2 . . . . $[NO_2][AuF_4]$ . . . . Au: SVol.B1-132
AuF_4Na . . . . $Na[AuF_4]$ . . . . Au: SVol.B1-125/9
AuF_4Rb . . . . $Rb[AuF_4]$ . . . . Au: SVol.B1-125/9
AuF_5 . . . . AuF_5 . . . . Au: SVol.B1-133/9
– . . . . $(AuF_5)_n$ . . . . Au: SVol.B1-132, 138
AuF_5^- . . . . $[AuF_5]^-$ . . . . Au: SVol.B1-132
AuF_5Xe . . . . $[XeF][AuF_4]$ . . . . Au: SVol.B1-125, 129, 130
AuF_6 . . . . AuF_6 . . . . Au: SVol.B1-158/9
AuF_6^- . . . . $[AuF_6]^-$ . . . . Au: SVol.B1-139/47
AuF_6^{3-} . . . . $[AuF_6]^{3-}$ . . . . Au: SVol.B1-130
AuF_6H . . . . $HAuF_6$ . . . . Au: SVol.B1-139, 147, 155
AuF_6K . . . . $K[AuF_6]$ . . . . Au: SVol.B1-139/40, 143/4, 147/8
AuF_6K_2 . . . . $K_2[AuF_6]$ . . . . Au: SVol.B1-133
AuF_6Li . . . . $Li[AuF_6]$ . . . . Au: SVol.B1-143/4, 147/8
AuF_6NO . . . . $[NO][AuF_6]$ . . . . Au: SVol.B1-139/40, 143/5, 147/8, 157/8
$AuF_6N_2O_2$ . . . . $[NO]_2[AuF_6]$ . . . . Au: SVol.B1-132/3
AuF_6Na . . . . $Na[AuF_6]$ . . . . Au: SVol.B1-139/40, 143/4, 147/8

Formula	Compound	Reference
AuF_6O_2	$[O_2][AuF_6]$	Au: SVol.B1-139/40, 143/5, 147/8, 156/7
AuF_6Rb	$Rb[AuF_6]$	Au: SVol.B1-139/40, 143/4, 147/8
AuF_7	AuF_7	Au: SVol.B1-159
AuF_7Kr	$[KrF][AuF_6]$	Au: SVol.B1-139/49
AuF_7Xe	$[XeF][AuF_6]$	Au: SVol.B1-139/40, 143, 147/8, 150
AuF_9Kr_2	$[Kr_2F_3][AuF_6]$	Au: SVol.B1-139/40, 143, 147/50
AuF_9Xe	$[XeF_5][AuF_4]$	Au: SVol.B1-129, 131
AuF_9Xe_2	$[Xe_2F_3][AuF_6]$	Au: SVol.B1-139/40, 147/8, 150/1
$AuF_{11}Xe$	$[XeF_5][AuF_6]$	Au: SVol.B1-139/43, 146/8, 151/2
$AuF_{12}I$	$[IF_6][AuF_6]$	Au: SVol.B1-139/44, 147/8
$AuF_{12}Re$	$[ReF_6][AuF_6]$	Au: SVol.B1-143/4
$AuF_{17}Xe_2$	$[Xe_2F_{11}][AuF_6]$	Au: SVol.B1-139/43, 147/8, 153/5
$AuH_{0.43}$	$AuH_{0.43}$	Au: SVol.B1-43
$AuH_{0.5}$	$AuH_{0.5}$	Au: SVol.B1-43
AuH	AuH	Au: SVol.B1-43/55
–	AuD	Au: SVol.B1-43/55
AuH^+	$[AuH]^+$	Au: SVol.B1-56
AuH^-	$[AuH]^-$	Au: SVol.B1-57
$AuHN_4O_{12}$	$HAu(NO_3)_4 \cdot 3\ H_2O$	Au: SVol.B1-107/8
$AuHO$	$AuOH$	Au: SVol.B1-59, 84/5
–	$AuOH \cdot H_2O$	Au: SVol.B1-84/5
$AuHO_2$	$AuO(OH)$	Au: SVol.B1-88/96
$AuHO_3^{2-}$	$[HAuO_3]^{2-}$	Au: SVol.B1-59, 97
AuH_2	AuH_2	Au: SVol.B1-57
AuH_2^+	$[AuH_2]^+$	Au: SVol.B1-56
AuH_2^-	$[AuH_2]^-$	Au: SVol.B1-58
AuH_2O_2	$Au(OH)_2$	Au: SVol.B1-59, 87/8
$AuH_2O_2^-$	$[Au(OH)_2]^-$	Au: SVol.B1-86
$AuH_2O_3^-$	$[H_2AuO_3]^-$	Au: SVol.B1-97
AuH_3	AuH_3	Au: SVol.B1-43
AuH_3^+	$[AuH_3]^+$	Au: SVol.B1-56
AuH_3O_3	$Au(OH)_3$	Au: SVol.B1-59, 88/96
AuH_4^+	$[AuH_4]^+$	Au: SVol.B1-56
$AuH_4O_4^-$	$[Au(OH)_4]^- = [AuO_2 \cdot 2\ H_2O]^-$	Au: SVol.B1-97
$AuKN_4O_{12}$	$K[Au(NO_3)_4]$	Au: SVol.B1-108/9
AuN_3	AuN_3	Au: SVol.B1-103
AuN_3O_9	$Au(NO_3)_3 \cdot n\ H_2O$	Au: SVol.B1-107
$AuN_4O_{12}^-$	$[Au(NO_3)_4]^-$	Au: SVol.B1-108/9
AuN_5O_{13}	$[NO][Au(NO_3)_4]$	Au: SVol.B1-109/11
AuN_5O_{14}	$[NO_2][Au(NO_3)_4]$	Au: SVol.B1-110/2
AuN_6^-	$[Au(N_3)_2]^-$	Au: SVol.B1-103/4
AuN_9	$Au(N_3)_3$	Au: SVol.B1-104
AuN_{12}^-	$[Au(N_3)_4]^-$	Au: SVol.B1-105/7

$AuNb_3$ .... Nb_3Au solid solutions
Nb_3Au-Nb_3Rh .... Rh: SVol.A1-266
$AuNe^+$ .... $[AuNe]^+$ .... Au: SVol.B1-27
AuO .... AuO .... Au: SVol.B1-59, 68/9
– .... AuO_x .... Au: SVol.B1-81
AuO^+ .... $[AuO]^+$ .... Au: SVol.B1-81/2
AuO^- .... $[AuO]^-$ .... Au: SVol.B1-82
AuO_2 .... AuO_2 .... Au: SVol.B1-59, 80/1
– .... $Au(O_2)$ .... Au: SVol.B1-59, 82
AuO_2^+ .... $[AuO_2]^+$ .... Au: SVol.B1-81/2
AuO_2^- .... $[AuO_2]^-$ .... Au: SVol.B1-82
– .... $[AuO_2 \cdot 2 H_2O]^- = [Au(OH)_4]^-$ .... Au: SVol.B1-97
– .... $[AuO_2 \cdot n H_2O]^-$ .... Au: SVol.B1-96/7
AuO_3^+ .... $[AuO_3]^+$ .... Au: SVol.B1-81/2
AuO_3^{3-} .... $[AuO_3]^{3-}$ .... Au: SVol.B1-97
AuO_6 .... $Au(O_2)_3$ .... Au: SVol.B1-83
AuPo .... AuPo .... Po: SVol.1-329
AuRh .... RhAu .... Rh: SVol.A1-264
AuSnU .... USnAu .... U: SVol.B2-313/4

$Au_2BC_{46}F_4FeH_{38}NO_2P_2$
$[(C_5H_5)Fe(C_5H_3((AuP(C_6H_5)_3))_2-NO_2)][BF_4]$ ... Fe: Org.Comp.A10-177
$Au_2BC_{47}F_4FeH_{41}P_2$
$[FeC_{10}H_8(AuP(C_6H_5)_3)_2CH_3][BF_4]$ .... Fe: Org.Comp.A10-177
$Au_2BC_{48}F_4FeH_{41}OP_2$
$[(C_5H_5)Fe(C_5H_3((AuP(C_6H_5)_3)_2)-C(O)CH_3)][BF_4]$
Fe: Org.Comp.A10-177
$Au_2BC_{52}F_4FeH_{42}NO_2P_2$
$[FeC_{10}H_8(AuP(C_6H_5)_3)_2-C_6H_4-NO_2-4][BF_4]$ ... Fe: Org.Comp.A10-177
$Au_2BC_{52}F_4FeH_{43}P_2$
$[FeC_{10}H_8(AuP(C_6H_5)_3)_2-C_6H_5][BF_4]$ .... Fe: Org.Comp.A10-177
$Au_2B_2C_{56}H_{76}N_2P_2Si_2$
$2,2-[(C_6H_5)_3PAu]_2-1,3-[(i-C_3H_7)_2N]_2-1,3-B_2$
$C_2-4,4-[Si(CH_3)_3]_2$ .... B: B Comp.SVol.4/3a-245
$Au_2B_{10}Br_2C_{10}H_{34}N_4P_2$
$1,2-C_2B_{10}H_{10}-1,2-[P(N(CH_3)_2)_2(AuBr)]_2$ .... B: B Comp.SVol.4/4-264
$Au_2B_{10}Br_2C_{18}H_{32}N_2P_2$
$1,2-C_2B_{10}H_{10}-1-P(C_6H_5)_2(AuBr)-2-P[N(CH_3)_2]_2(AuBr)$
B: B Comp.SVol.4/4-264
$Au_2B_{10}Br_2C_{26}H_{30}P_2$
$1,2-C_2B_{10}H_{10}-1,2-[P(C_6H_5)_2(AuBr)]_2$ .... B: B Comp.SVol.4/4-264
$Au_2B_{20}Br_2C_{20}H_{68}N_8P_4$
$[1-P(N(CH_3)_2)_2-1,2-C_2B_{10}H_{10}-2-P(N(CH_3)_2)_2]_2(AuBr)_2$
B: B Comp.SVol.4/4-264
$Au_2B_{20}Br_2C_{36}H_{64}N_4P_4$
$[1,2-C_2B_{10}H_{10}-1-P(C_6H_5)_2-2-P(N(CH_3)_2)_2]_2(AuBr)_2$
B: B Comp.SVol.4/4-264
$Au_2B_{20}Br_2C_{52}H_{60}P_4$
$[1,2-C_2B_{10}H_{10}-1,2-(P(C_6H_5)_2)_2]_2(AuBr)_2$ .... B: B Comp.SVol.4/4-264
Au_2BaF_8 .... $Ba[AuF_4]_2$ .... Au: SVol.B1-125, 127

Au_2BaF_{12} .... $Ba[AuF_6]_2$ .... Au: SVol.B1-143/5, 147/8
$Au_2C_5Cl_8H_{14}N_2O_2$ $[NH_3-(CH_2)_3-CHNH_3-COOH][AuCl_4]_2$ .... Au: SVol.B1-298
$Au_2C_6Cl_8H_{16}N_4O_2$ $[NH_3C(=NH)NH-(CH_2)_3-CHNH_3-COOH][AuCl_4]_2$
Au: SVol.B1-298
$Au_2C_8Cl_8CoH_{23}N_6O_2$
$[(NH_2-(C_2H_4-NH)_3-C_2H_4-NH_2)Co(NO_2)][AuCl_4]_2$
Au: SVol.B1-297
$Au_2C_{45}Fe_3H_{30}O_9P_2S$
$(CO)_9Fe_3S[Au-P(C_6H_5)_3]_2$ .... Fe: Org.Comp.C6a-315, 316, 317
$Au_2C_{46}FeH_{38}NO_2P_2^+$
$[(C_5H_5)Fe(C_5H_3((AuP(C_6H_5)_3))_2-(NO_2))]^+$ .... Fe: Org.Comp.A10-177
$Au_2C_{47}FeH_{41}P_2^+$ $[FeC_{10}H_8(AuP(C_6H_5)_3)_2CH_3]^+$ .... Fe: Org.Comp.A10-177
$Au_2C_{48}FeH_{41}OP_2^+$
$[(C_5H_5)Fe(C_5H_3((AuP(C_6H_5)_3)_2)-C(=O)CH_3)]^+$. Fe: Org.Comp.A10-177
$Au_2C_{51}Fe_3H_{35}O_9P_3$
$(CO)_9Fe_3(P-C_6H_5)[Au-P(C_6H_5)_3]_2$ .... Fe: Org.Comp.C6a-315, 316
$Au_2C_{52}FeH_{42}NO_2P_2^+$
$[FeC_{10}H_8(AuP(C_6H_5)_3)_2-C_6H_4-NO_2-4]^+$ .... Fe: Org.Comp.A10-177
$Au_2C_{52}FeH_{43}P_2^+$ $[FeC_{10}H_8(AuP(C_6H_5)_3)_2-C_6H_5]^+$ .... Fe: Org.Comp.A10-177
$Au_2C_{52}Fe_3H_{37}O_9P_3$
$(CO)_9Fe_3(P-C_6H_4-CH_3)[Au-P(C_6H_5)_3]_2$ .... Fe: Org.Comp.C6a-315, 316
Au_2CaF_8 .... $Ca[AuF_4]_2$ .... Au: SVol.B1-125
Au_2CaF_{12} .... $Ca[AuF_6]_2$ .... Au: SVol.B1-139/40, 143/5, 147/8
Au_2CdF_8 .... $Cd[AuF_4]_2$ .... Au: SVol.B1-125
Au_2Cl_2 .... Au_2Cl_2 .... Au: SVol.B1-164/8
Au_2Cl_6 .... Au_2Cl_6 .... Au: SVol.B1-193, 197/9, 203
$Au_2Cl_6Cs_2$ .... $Cs_2[AuCl_2][AuCl_4]$ .... Au: SVol.B1-213, 216
Au_2Cl_8Zn .... $Zn[AuCl_4]_2$ .... Au: SVol.B1-213
– .... $Zn[AuCl_4]_2 \cdot 2\ H_2O$ .... Au: SVol.B1-216
$Au_2Cl_{12}Cs_4Pd$ .... $Cs_4PdAu_2Cl_{12}$ .... Pd: SVol.B2-171
$Au_2Cl_{12}PdRb_4$ .... $Rb_4PdAu_2Cl_{12}$ .... Pd: SVol.B2-171
Au_2CoF_8 .... $Co[AuF_4]_2$ .... Au: SVol.B1-125
Au_2F_8Hg .... $Hg[AuF_4]_2$ .... Au: SVol.B1-125
Au_2F_8Mg .... $Mg[AuF_4]_2$ .... Au: SVol.B1-125, 127
Au_2F_8Mn .... $Mn[AuF_4]_2$ .... Au: SVol.B1-125
Au_2F_8Ni .... $Ni[AuF_4]_2$ .... Au: SVol.B1-125
Au_2F_8Pb .... $Pb[AuF_4]_2$ .... Au: SVol.B1-125
Au_2F_8Pd .... $Pd[AuF_4]_2$ .... Au: SVol.B1-125
Au_2F_8Sr .... $Sr[AuF_4]_2$ .... Au: SVol.B1-125
Au_2F_8Zn .... $Zn[AuF_4]_2$ .... Au: SVol.B1-125, 127
$Au_2F_{12}Mg$ .... $Mg[AuF_6]_2$ .... Au: SVol.B1-143/5, 147/8
$Au_2F_{12}Sr$ .... $Sr[AuF_6]_2$ .... Au: SVol.B1-139/40, 143/5, 147/8
Au_2H^+ .... $[Au_2H]^+$ .... Au: SVol.B1-58
$Au_2H_9O_9^{3-}$ .... $[Au_2(OH)_9]^{3-}$ .... Au: SVol.B1-97/9
– .... $[Au_2O_3(OH)_3(H_2O)_3]_n^{3n-}$ .... Au: SVol.B1-98/9
Au_2InTh .... $ThInAu_2$ .... Th: SVol.B2-155
Au_2O .... Au_2O .... Au: SVol.B1-59, 66/7

Au_2O_3 .... Au_2O_3 .... Au: SVol.B1-59, 69/80
– .... $Au_2O_3 \cdot n\ H_2O$ .... Au: SVol.B1-88/96

$Au_3C_4Cl_{12}H_{16}N_3$ .. $[NH_3CH_2CH_2NH_2CH_2CH_2NH_3][AuCl_4]_3 \cdot H_2O$. Au: SVol.B1-288
$Au_3C_{30}Cl_{12}H_{25}N_6$ $[H(NC_5H_3\text{-}2,6\text{-}(2\text{-}C_5H_4N)_2)][H_2(NC_5H_3\text{-}2,6\text{-}(2\text{-}C_5H_4N)_2)][AuCl_4]_3$ .... Au: SVol.B1-293
$Au_3C_{58}H_{45}O_4P_3Re$ $(CO)_4Re[AuP(C_6H_5)_3]_3$ .... Re: Org.Comp.1-500/1
$Au_3Cl_8Rb_3$ .... $Rb_3[AuCl_2]_2[AuCl_4]$ .... Au: SVol.B1-213, 216
$Au_3H_3N_2$ .... $Au_3(NH_3)N$ .... Au: SVol.B1-103
$Au_3H_6N_2O_3$ .... $(AuOH)_3(NH_3)N$ .... Au: SVol.B1-103
Au_3N .... Au_3N .... Au: SVol.B1-103
Au_3O_2 .... Au_3O_2 .... Au: SVol.B1-83

$Au_4B_2C_{82}F_8FeH_{68}P_4$
$[Fe(C_5H_4(AuP(C_6H_5)_3)_2)_2][BF_4]_2$ .... Fe: Org.Comp.A10-176/7
$Au_4C_{82}FeH_{68}P_4{}^{2+}$ $[Fe(C_5H_4(AuP(C_6H_5)_3)_2)_2]^{2+}$ .... Fe: Org.Comp.A10-176/7
Au_4Cl_8 .... Au_4Cl_8 .... Au: SVol.B1-189/90

$Au_5Bi_2F_{21}$ .... $[Bi_2F][AuF_4]_5$ .... Au: SVol.B1-125
$Au_5Ce_2F_{21}$ .... $[Ce_2F][AuF_4]_5$ .... Au: SVol.B1-125
$Au_5Dy_2F_{21}$ .... $[Dy_2F][AuF_4]_5$ .... Au: SVol.B1-125
$Au_5Er_2F_{21}$ .... $[Er_2F][AuF_4]_5$ .... Au: SVol.B1-125/7
$Au_5Eu_2F_{21}$ .... $[Eu_2F][AuF_4]_5$ .... Au: SVol.B1-125
$Au_5F_{21}Gd_2$ .... $[Gd_2F][AuF_4]_5$ .... Au: SVol.B1-125
$Au_5F_{21}In_2$ .... $[In_2F][AuF_4]_5$ .... Au: SVol.B1-125
$Au_5F_{21}La_2$ .... $[La_2F][AuF_4]_5$ .... Au: SVol.B1-125
$Au_5F_{21}Lu_2$ .... $[Lu_2F][AuF_4]_5$ .... Au: SVol.B1-125
$Au_5F_{21}Nd_2$ .... $[Nd_2F][AuF_4]_5$ .... Au: SVol.B1-125
$Au_5F_{21}Pr_2$ .... $[Pr_2F][AuF_4]_5$ .... Au: SVol.B1-125, 127
$Au_5F_{21}Sm_2$ .... $[Sm_2F][AuF_4]_5$ .... Au: SVol.B1-125
$Au_5F_{21}Tb_2$ .... $[Tb_2F][AuF_4]_5$ .... Au: SVol.B1-125
$Au_5F_{21}Tl_2$ .... $[Tl_2F][AuF_4]_5$ .... Au: SVol.B1-125
$Au_5F_{21}Tm_2$ .... $[Tm_2F][AuF_4]_5$ .... Au: SVol.B1-125
$Au_5F_{21}Yb_2$ .... $[Yb_2F][AuF_4]_5$ .... Au: SVol.B1-125
$Au_5F_{31}La_2$ .... $[La_2F][AuF_6]_5$ .... Au: SVol.B1-139/40, 147/8
$Au_5F_{31}Pr_2$ .... $[Pr_2F][AuF_6]_5$ .... Au: SVol.B1-139/40, 147/8

B .... B alloys
B-Mo Mo:SVol.A2b-31
B-Mo-Nb-Si Mo:SVol.A2b-27
B-Mo-P Mo:SVol.A2b-27
B-Mo-P-Ru Mo:SVol.A2b-27
B-Mo-Ru Mo:SVol.A2b-27
B-Mo-Ru-Si Mo:SVol.A2b-27
B-Mo-Si Mo:SVol.A2b-27

– .... B solid solutions
B-Eu .... Sc: MVol.C11b-215
B-Pd .... Pd: SVol.B2-257/8
B-Sc .... Sc: MVol.C11a-119, 128/9

– .... B systems
B-C-Ce .... Sc: MVol.C11b-3
B-C-Th .... Th: SVol.C6-132/4

B ... B systems
B-Ce ... Sc: MVol.C11b-1/2
B-Dy ... Sc: MVol.C11b-316
B-Er ... Sc: MVol.C11b-330/1
B-Eu ... Sc: MVol.C11b-215
B-Gd ... Sc: MVol.C11b-264/5
B-Ho ... Sc: MVol.C11b-323/4
B-La ... Sc: MVol.C11a-162/4
B-Lu ... Sc: MVol.C11b-377
B-Nd ... Sc: MVol.C11b-115
B-Pd ... Pd: SVol.B2-256/7
B-Pm ... Sc: MVol.C11b-136
B-Pr ... Sc: MVol.C11b-92
B-Sc ... Sc: MVol.C11a-119/20
B-Sm ... Sc: MVol.C11b-136/8
B-Tb ... Sc: MVol.C11b-307/8
B-Tm ... Sc: MVol.C11b-340/1
B-Y ... Sc: MVol.C11a-129/31
B-Yb ... Sc: MVol.C11b-346/7

$BBaS_2$ ... $Ba[BS_2]$ ... B: B Comp.SVol.3/4-108
$BBeH_6N$ ... $H_3B-NH_2(BeH)$ ... B: B Comp.SVol.4/3b-4
BBe_2H_6N ... $H_3B-NH(BeH)_2$ ... B: B Comp.SVol.4/3b-4
BBe_3H_6N ... $H_3B-N(BeH)_3$ ... B: B Comp.SVol.4/3b-4
$BBiBr_3C_3H_9O_9S_3$ $(CH_3SO_3)_3Bi \cdot BBr_3$ ... B: B Comp.SVol.4/4-81
$BBi_2C_{43}F_4H_{35}MoO_2$
$[(C_5H_5)Mo(CO)_2(Bi(C_6H_5)_3)_2][BF_4]$ ... Mo: Org.Comp.7-283, 293
BBr ... BBr ... B: B Comp.SVol.3/4-71
B: B Comp.SVol.4/4-76
$BBrCClH_3$ ... $(CH_3)BBrCl$ ... B: B Comp.SVol.3/4-95
$BBrC_2F_3H_6N$ ... $BF_3-NBr(CH_3)_2$ ... B: B Comp.SVol.4/4-107
$BBrC_2H_4S_2$ ... $2-Br-1,3,2-S_2BC_2H_4$ ... B: B Comp.SVol.3/4-138
B: B Comp.SVol.4/4-162/3
$BBrC_2H_6$ ... $(CH_3)_2BBr$ ... B: B Comp.SVol.3/4-87
B: B Comp.SVol.4/4-93
$BBrC_2H_6O_2$ ... $(CH_3O)_2BBr$ ... B: B Comp.SVol.4/4-97
$BBrC_2H_8S$ ... $(CH_3)_2S \cdot BH_2Br$ ... B: B Comp.SVol.3/4-137
B: B Comp.SVol.4/4-88, 162
$BBrC_3ClFH_9N$ ... $N(CH_3)_3 \cdot BBrClF$ ... B: B Comp.SVol.3/4-65/7
$BBrC_3F_2H_9N$ ... $N(CH_3)_3 \cdot BBrF_2$ ... B: B Comp.SVol.3/4-65/7
$BBrC_3H_6S_2$ ... $2-Br-1,3,2-S_2BC_3H_6$ ... B: B Comp.SVol.4/4-163
$BBrC_3H_9N$ ... $[(CH_3)_2N][CH_3]BBr$ ... B: B Comp.SVol.3/4-93
$BBrC_3H_{11}P$ ... $(CH_3)_3P \cdot BH_2Br$ ... B: B Comp.SVol.4/4-88
$BBrC_4ClFH_{11}N$ ... $N(CH_3)_2(C_2H_5) \cdot BBrClF$ ... B: B Comp.SVol.3/4-65/7
$BBrC_4F_2H_{11}N$ ... $N(CH_3)_2(C_2H_5) \cdot BBrF_2$ ... B: B Comp.SVol.3/4-65/7
$BBrC_4F_6H_7N$ ... $(CF_3)_2BBr \cdot NH(CH_3)_2$ ... B: B Comp.SVol.4/4-93, 106
$BBrC_4F_{12}N_2S_4$ ... $[(CF_3S)_2N]_2BBr$ ... B: B Comp.SVol.4/4-102, 165/6
$BBrC_4H_{10}$ ... $(C_2H_5)_2BBr$ ... B: B Comp.SVol.3/4-87
B: B Comp.SVol.4/4-93

Formula	Compound	Reference
$BBrC_4H_{10}N_2$	1,3-$(CH_3)_2$-2-Br-1,3,2-$N_2BC_2H_4$	B: B Comp.SVol.3/4-92
		B: B Comp.SVol.4/4-102
$BBrC_4H_{12}N_2$	$[(CH_3)_2N]_2BBr$	B: B Comp.SVol.3/4-92
		B: B Comp.SVol.4/4-102
$BBrC_4H_{12}S$	$C_2H_5BHBr \cdot (CH_3)_2S$	B: B Comp.SVol.3/4-86
$BBrC_5ClFH_{13}N$	$N(CH_3)(C_2H_5)_2 \cdot BBrClF$	B: B Comp.SVol.3/4-65/7
$BBrC_5F_2H_6O_2$	$[-B(F)_2OC(CH_3)C(Br)C(CH_3)O-]$	B: B Comp.SVol.4/3b-233
$BBrC_5F_2H_{13}N$	$N(CH_3)(C_2H_5)_2 \cdot BBrF_2$	B: B Comp.SVol.3/4-65/7
$BBrC_5F_3H_4N$	3-BrC_5H_4N-BF_3	B: B Comp.SVol.3/3-294
$BBrC_5H_6$	[-BBr-CH=CH-CH_2-CH=CH-]	B: B Comp.SVol.4/4-94
$BBrC_5H_7N$	3-Br-C_5H_4N-BH_3	B: B Comp.SVol.3/3-177
–	4-Br-C_5H_4N-BH_3	B: B Comp.SVol.3/3-177
$BBrC_5H_8S_2$	BrB[-CH-C(C_3H_7-n)-S-S-]	B: B Comp.SVol.3/4-140
$BBrC_5H_9N$	(-BCH_3-NCH_3-CH_2-CBr=CH-)	B: B Comp.SVol.3/3-162
–	HC≡C-CH_2-NCH_3-BCH_3-Br	B: B Comp.SVol.3/4-93
$BBrC_5H_9N_3S_2$	BrB(-NCH_3-CS-NCH_3-CS-NCH_3-)	B: B Comp.SVol.3/4-93
$BBrC_5H_{13}N$	$(C_2H_5)_2N$-BBr-CH_3	B: B Comp.SVol.4/4-104
–	t-C_4H_9-NH-BBr-CH_3	B: B Comp.SVol.4/4-105
$BBrC_5H_{14}S$	n-$C_3H_7BHBr \cdot (CH_3)_2S$	B: B Comp.SVol.3/4-86
$BBrC_6ClFH_{15}N$	$N(C_2H_5)_3 \cdot BBrClF$	B: B Comp.SVol.3/4-65/7
$BBrC_6Cl_2H_4$	2-Br-C_6H_4-BCl_2	B: B Comp.SVol.4/4-43
–	3-Br-C_6H_4-BCl_2	B: B Comp.SVol.4/4-43
–	4-Br-C_6H_4-BCl_2	B: B Comp.SVol.4/4-43
$BBrC_6F_2H_7N$	4-CH_3-$C_5H_4N \cdot BBrF_2$	B: B Comp.SVol.3/4-65/7
$BBrC_6F_2H_{15}N$	$N(C_2H_5)_3 \cdot BBrF_2$	B: B Comp.SVol.3/4-65/7
$BBrC_6F_2H_{18}N_2$	$[((CH_3)_3N)_2BF_2]Br$	B: B Comp.SVol.3/3-379
		B: B Comp.SVol.3/4-96
$BBrC_6Fe_2O_6S_2$	$BFe_2S_2(CO)_6(Br)$	B: B Comp.SVol.4/4-163
$BBrC_6H_4O_2$	2-Br-1,3,2-$O_2BC_6H_4$	B: B Comp.SVol.4/4-97
$BBrC_6H_{10}S_2$	BrB[-C(C_2H_5)=C(C_2H_5)-S-S-]	B: B Comp.SVol.3/4-140
–	BrB[-CH=C(C_4H_9-n)-S-S-]	B: B Comp.SVol.3/4-140
–	BrB[-CCH_3=C(C_3H_7-n)-S-S-]	B: B Comp.SVol.3/4-140
–	BrB[-S-C(C_2H_5)=C(C_2H_5)-S-]	B: B Comp.SVol.3/4-139
$BBrC_6H_{11}N$	HC≡C-$CH_2N(C_2H_5)$-B(CH_3)-Br	B: B Comp.SVol.3/4-93
–	[-B(CH_3)-N(C_2H_5)-CH_2-CBr=CH-]	B: B Comp.SVol.3/3-162
$BBrC_6H_{14}$	(n-$C_3H_7)_2$BBr	B: B Comp.SVol.3/4-88
–	(i-$C_3H_7)_2$BBr	B: B Comp.SVol.4/4-93
$BBrC_6H_{15}N$	t-C_4H_9-NH-BBr-C_2H_5	B: B Comp.SVol.4/4-105
$BBrC_6H_{15}NO_2$	$(CH_3)_3N$-BH_2-COO-CH_2-CH_2Br	B: B Comp.SVol.4/3b-25
$BBrC_6H_{16}S$	$(C_2H_5)_2BBr \cdot (CH_3)_2S$	B: B Comp.SVol.3/4-87
$BBrC_7ClFH_{13}N$	$HC(C_2H_4)_3N \cdot BBrClF$	B: B Comp.SVol.3/4-65/7
$BBrC_7Cl_2H_6$	2-$BrCH_2$-C_6H_4-BCl_2	B: B Comp.SVol.4/4-43
$BBrC_7FH_{18}NSn$	$(CH_3)_3SnN(C_4H_9$-t)BFBr	Sn: Org.Comp.18-64
$BBrC_7F_2H_{13}N$	$HC(C_2H_4)_3N \cdot BBrF_2$	B: B Comp.SVol.3/4-65/7
$BBrC_7H_6S_2$	BrB[-S-(1,2-C_6H_3-5-CH_3)-S-]	B: B Comp.SVol.3/4-140
$BBrC_7H_{11}N$	$(CH_3)_2BBr \cdot NC_5H_5$	B: B Comp.SVol.4/4-93
$BBrC_7H_{13}N$	[-BCH_3-N(n-C_3H_7)-CH_2-CBr=CH-]	B: B Comp.SVol.3/3-162
$BBrC_7H_{15}NO_4$	$(CH_3)_3N$-BH_2-C(O)-O-C(O)-O-CH_2-CH_2Br	B: B Comp.SVol.4/3b-25
$BBrC_7H_{16}S$	c-$C_5H_9BHBr \cdot (CH_3)_2S$	B: B Comp.SVol.3/4-86
$BBrC_7H_{18}S$	n-$C_5H_{11}BHBr \cdot (CH_3)_2S$	B: B Comp.SVol.3/4-86

Formula	Compound	Reference
$BBrC_8ClFH_{11}N$	$N(CH_3)_2(C_6H_5) \cdot BBrClF$	B: B Comp.SVol.3/4-65/7
$BBrC_8ClH_9$	CH_3-BCl-C_6H_4-2-CH_2Br	B: B Comp.SVol.4/4-49
$BBrC_8F_2H_{11}N$	$N(CH_3)_2(C_6H_5) \cdot BBrF_2$	B: B Comp.SVol.3/4-65/7, 95
$BBrC_8F_2H_{19}N$	$N(C_2H_5)(i\text{-}C_3H_7)_2 \cdot BBrF_2$	B: B Comp.SVol.3/4-65/7
$BBrC_8F_2H_{22}N_2$	$[((CH_3)_2(C_2H_5)N)_2BF_2][Br]$	B: B Comp.SVol.3/3-379
$BBrC_8H_6S_2$	BrB[-CH=C(C_6H_5)-S-S-]	B: B Comp.SVol.3/4-140
–	BrB[-S-CH=C(C_6H_5)-S-]	B: B Comp.SVol.3/4-139
$BBrC_8H_{11}N$	$[(CH_3)_2N][C_6H_5]BBr$	B: B Comp.SVol.3/4-93
$BBrC_8H_{14}$	[3.3.1]-9-BC_8H_{14}-9-Br	B: B Comp.SVol.3/4-88
		B: B Comp.SVol.4/4-94
$BBrC_8H_{14}S_2$	BrB[-C(C_3H_7-n)=C(C_3H_7-n)-S-S-]	B: B Comp.SVol.3/4-140
$BBrC_8H_{18}$	$(n\text{-}C_4H_9)_2BBr$	B: B Comp.SVol.3/4-88
		B: B Comp.SVol.4/4-93
–	$(t\text{-}C_4H_9)_2BBr$	B: B Comp.SVol.3/4-88
		B: B Comp.SVol.4/4-93/4
$BBrC_8H_{19}N$	t-C_4H_9-NH-BBr-C_4H_9-t	B: B Comp.SVol.4/4-105
$BBrC_8H_{20}N_2$	$[(C_2H_5)_2N]_2BBr$	B: B Comp.SVol.3/4-92
$BBrC_8H_{20}S$	i-C_3H_7-C$(CH_3)_2$-BHBr $\cdot$ $(CH_3)_2S$	B: B Comp.SVol.3/4-86
		B: B Comp.SVol.4/4-93
–	n-C_3H_7-CH(CH_3)-CH_2-BHBr $\cdot$ $(CH_3)_2S$	B: B Comp.SVol.3/4-86
		B: B Comp.SVol.4/4-90
–	n-C_3H_7-CH(C_2H_5)-BHBr $\cdot$ $(CH_3)_2S$	B: B Comp.SVol.3/4-86
–	n-C_6H_{13}-BHBr $\cdot$ $(CH_3)_2S$	B: B Comp.SVol.3/4-86, 137
		B: B Comp.SVol.4/4-90
$BBrC_8H_{21}NSn$	$(CH_3)_3SnN(C_4H_9\text{-}t)BBrCH_3$	Sn: Org.Comp.18-65
$BBrC_8H_{24}N_2Sn_2$	$(CH_3)_2$N-BBr-N$[Sn(CH_3)_3]_2$	B: B Comp.SVol.4/4-103
$BBrC_9ClFH_{13}N$	4-CH_3-C_6H_4-N$(CH_3)_2$ $\cdot$ BBrClF	B: B Comp.SVol.3/4-65/7
–	C_6H_5-NCH_3-C_2H_5 $\cdot$ BBrClF	B: B Comp.SVol.3/4-65/7
$BBrC_9ClFH_{21}N$	$N(n\text{-}C_3H_7)_3 \cdot BBrClF$	B: B Comp.SVol.3/4-65/7
$BBrC_9F_2H_{13}N$	4-$CH_3C_6H_4N(CH_3)_2$ $\cdot$ $BBrF_2$	B: B Comp.SVol.3/4-65/7
$BBrC_9F_2H_{21}N$	$N(n\text{-}C_3H_7)_3 \cdot BBrF_2$	B: B Comp.SVol.3/4-65/7
$BBrC_9H_8$	2-Br-2-BC_9H_8	B: B Comp.SVol.4/4-94
$BBrC_9H_9N$	H_3B $\cdot$ 1-NC_9H_6-3-Br	B: B Comp.SVol.4/3b-14
$BBrC_9H_9O_6Re$	fac-$(CO)_3$Re[-C(CH_3)O-$]_3$BBr	Re: Org.Comp.2-379, 381
$BBrC_9H_{16}$	$(Br)BC_8H_{13}(CH_3)$	B: B Comp.SVol.3/4-89
$BBrC_9H_{21}N_2PS$	[BBr-N(C_4H_9-t)-P(=S)(CH_3)-N(C_4H_9-t)-]	B: B Comp.SVol.4/4-106
$BBrC_9H_{23}N$	$(n\text{-}C_3H_7)_3N \cdot BH_2Br$	B: B Comp.SVol.4/4-88
–	$(i\text{-}C_3H_7)_3N \cdot BH_2Br$	B: B Comp.SVol.4/4-88
$BBrC_9H_{23}NSn$	$(CH_3)_3SnN(C_4H_9\text{-}t)BBrC_2H_5$	Sn: Org.Comp.18-65
$BBrC_{9.6}EuH_{23.2}O_{2.4}$		
	$EuBr[BH_4] \cdot 2.4\ C_4H_8O$	Sc: MVol.C11b-491
$BBrC_{10}ClFH_{15}N$	C_6H_5-CH_2-NCH_3-C_2H_5 $\cdot$ BBrClF	B: B Comp.SVol.3/4-65/7
–	C_6H_5-N$(C_2H_5)_2$ $\cdot$ BBrClF	B: B Comp.SVol.3/4-65/7
$BBrC_{10}F_2H_{10}N_2$	$[(C_5H_5N)_2BF_2][Br]$	B: B Comp.SVol.3/3-379
$BBrC_{10}F_2H_{15}N$	$N(C_2H_5)_2(C_6H_5) \cdot BBrF_2$	B: B Comp.SVol.3/4-65/7
$BBrC_{10}F_2H_{22}N_2$	$[((CH_3)_3N)(HC(\text{-}CH_2\text{-}CH_2\text{-})_3N)BF_2][Br]$	B: B Comp.SVol.3/3-379
$BBrC_{10}F_4FeH_{10}O_2$	$[(C_5H_5)Fe(CO)_2(CH_2\text{=}CHCH_2Br)][BF_4]$	Fe: Org.Comp.B17-23
$BBrC_{10}FeH_{10}O_2$	$(HO)_2B$-$C_5H_4FeC_5H_4Br$	Fe: Org.Comp.A9-279, 283
$BBrC_{10}H_{11}N$	(-BCH_3-NC_6H_5-CH_2-CBr=CH-)	B: B Comp.SVol.3/3-162
$BBrC_{10}H_{18}$	$(c\text{-}C_5H_9)_2BBr$	B: B Comp.SVol.3/4-88

$BBrC_{10}H_{18}$ ...... $(c\text{-}C_5H_9)_2BBr$ .... B: B Comp.SVol.4/4-94

$BBrC_{10}H_{20}$ ...... $t\text{-}C_4H_9CH{=}C(C_2H_5)B(Br)C_2H_5$ .... B: B Comp.SVol.3/4-88

$BBrC_{10}H_{21}N$ ..... $(CH_3)_4C_5H_6N\text{-}BBrCH_3$ .... B: B Comp.SVol.3/4-93

$BBrC_{10}H_{21}NO$.... $2,2,6,6\text{-}(CH_3)_4\text{-}NC_5H_6\text{-}1\text{-}O\text{-}BBr\text{-}CH_3$ .... B: B Comp.SVol.4/4-97

$BBrC_{10}H_{22}$ ...... $(n\text{-}C_5H_{11})_2BBr$ .... B: B Comp.SVol.3/4-88

$BBrC_{10}H_{24}S$ ..... $n\text{-}C_8H_{17}BHBr \cdot (CH_3)_2S$ .... B: B Comp.SVol.3/4-86

$BBrC_{11}ClH_{15}$..... $t\text{-}C_4H_9\text{-}BCl\text{-}C_6H_4\text{-}2\text{-}CH_2Br$ .... B: B Comp.SVol.4/4-49

$BBrC_{11}F_2H_{24}N_2$ .. $[((CH_3)_2(C_2H_5)N)(HC(\text{-}CH_2\text{-}CH_2\text{-})_3N)BF_2][Br]$ B: B Comp.SVol.3/3-379

$BBrC_{11}F_4FeH_{10}$ .. $[(Br\text{-}C_6H_5)Fe(C_5H_5)][BF_4]$ .... Fe: Org.Comp.B18-142/6, 197, 200, 247

$BBrC_{11}F_4FeH_{12}O_2$ $[(C_5H_5)Fe(CO)_2(CH_2{=}C(CH_3)CH_2Br)][BF_4]$ .... Fe: Org.Comp.B17-62

$BBrC_{11}H_{13}N$ ..... $HC{\equiv}C\text{-}CH_2N(CH_2C_6H_5)B(CH_3)Br$ .... B: B Comp.SVol.3/4-93

$BBrC_{11}H_{24}$ ...... $(n\text{-}C_5H_{11})(n\text{-}C_6H_{13})BBr$ .... B: B Comp.SVol.3/4-88

$BBrC_{11}H_{24}N_2$ .... $(CH_3)_4C_5H_6N\text{-}BBr\text{-}N(CH_3)_2$ .... B: B Comp.SVol.3/4-92

$BBrC_{11}H_{24}N_2O$ ... $2,2,6,6\text{-}(CH_3)_4\text{-}NC_5H_6\text{-}1\text{-}O\text{-}BBr\text{-}N(CH_3)_2$..... B: B Comp.SVol.4/4-107

$BBrC_{12}F_2H_{18}N_2$ .. $[(C_5H_5N)(HC(\text{-}CH_2\text{-}CH_2\text{-})_3N)BF_2][Br]$ .... B: B Comp.SVol.3/3-379

$BBrC_{12}F_2H_{26}N_2$ .. $[((C_2H_5)_2(CH_3)N)(HC(\text{-}CH_2\text{-}CH_2\text{-})_3N)BF_2][Br]$ B: B Comp.SVol.3/3-379

$BBrC_{12}F_2H_{27}N$ ... $N(n\text{-}C_4H_9)_3 \cdot BBrF_2$ .... B: B Comp.SVol.3/4-65/7

$BBrC_{12}F_4FeH_{12}$ .. $[(2\text{-}Br\text{-}C_6H_4\text{-}CH_3)Fe(C_5H_5)][BF_4]$ .... Fe: Org.Comp.B19-1, 4/5, 45

– .... $[(3\text{-}Br\text{-}C_6H_4\text{-}CH_3)Fe(C_5H_5)][BF_4]$ .... Fe: Org.Comp.B19-1, 4/5, 45/6

– .... $[(4\text{-}Br\text{-}C_6H_4\text{-}CH_3)Fe(C_5H_5)][BF_4]$ .... Fe: Org.Comp.B19-1, 4/5, 46

$BBrC_{12}F_4H_{10}$ .... $[(C_6H_5)_2Br][BF_4]$........ B: B Comp.SVol.3/3-354
B: B Comp.SVol.4/4-107

$BBrC_{12}H_8$ .... $5\text{-}Br\text{-}5\text{-}BC_{12}H_8$ .... B: B Comp.SVol.4/4-94

$BBrC_{12}H_8O$...... $6\text{-}Br\text{-}5,6\text{-}OBC_{12}H_8$........ B: B Comp.SVol.4/4-97

$BBrC_{12}H_{10}$ ...... $(C_6H_5)_2BBr$ .... B: B Comp.SVol.3/4-87
B: B Comp.SVol.4/4-93

$BBrC_{12}H_{22}$ ...... $(Br)BC_9H_{13}(CH_3)_3$........ B: B Comp.SVol.3/4-89

– .... $(c\text{-}C_6H_{11})_2BBr$ .... B: B Comp.SVol.3/4-88

$BBrC_{12}H_{26}$ ...... $(n\text{-}C_6H_{13})_2BBr$ .... B: B Comp.SVol.3/4-88

– .... $[n\text{-}C_3H_7CH(CH_3)CH_2]_2BBr$........ B: B Comp.SVol.3/4-88

$BBrC_{12}H_{28}N_2$ .... $[(i\text{-}C_3H_7)_2N]_2BBr$ .... B: B Comp.SVol.4/4-102

$BBrC_{12}H_{29}N$ ..... $(n\text{-}C_4H_9)_3N \cdot BH_2Br$ .... B: B Comp.SVol.4/4-88

– .... $(i\text{-}C_4H_9)_3N \cdot BH_2Br$........ B: B Comp.SVol.4/4-88

$BBrC_{13}ClH_{20}N$ ... $2\text{-}BrCH_2\text{-}C_6H_4\text{-}BCl\text{-}N(C_3H_7\text{-}i)_2$ .... B: B Comp.SVol.4/4-60

$BBrC_{13}F_2H_{28}N_2$ .. $[((C_2H_5)_3N)(HC(\text{-}CH_2\text{-}CH_2\text{-})_3N)BF_2][Br]$ ...... B: B Comp.SVol.3/3-379

$BBrC_{13}F_4FeH_{12}O_3$ $[(BrC_{10}H_{12})Fe(CO)_3][BF_4]$ .... Fe: Org.Comp.B15-230

$BBrC_{13}F_4FeH_{14}O_4$ $[(C_5H_5)Fe(CO)_2(BrCH_2CH{=}CH\text{-}COO\text{-}C_2H_5)][BF_4]$
Fe: Org.Comp.B17-59

– .... $[(C_5H_5)Fe(CO)_2(CH_2{=}CHCH(Br)\text{-}COO\text{-}C_2H_5)][BF_4]$
Fe: Org.Comp.B17-29

$BBrC_{13}H_{23}N$ ..... $(t\text{-}C_4H_9)_2BBr \cdot NC_5H_5$........ B: B Comp.SVol.4/4-94

$BBrC_{13}H_{28}N_2$ .... $2,2,6,6\text{-}(CH_3)_4\text{-}NC_5H_6\text{-}1\text{-}BBr\text{-}N(C_2H_5)_2$ ...... B: B Comp.SVol.3/4-92

– .... $2,2,6,6\text{-}(CH_3)_4\text{-}NC_5H_6\text{-}1\text{-}BBr\text{-}NHC_4H_9\text{-}t$ ..... B: B Comp.SVol.4/4-103

$BBrC_{13}H_{30}S$ ..... $(CH_3)_2S\text{-}B(C_5H_{11}\text{-}n)(C_6H_{13}\text{-}n)Br$ .... B: B Comp.SVol.3/4-137

$BBrC_{14}F_2H_{26}N_2$ .. $[(HC(\text{-}CH_2\text{-}CH_2\text{-})_3N)_2BF_2][Br]$........ B: B Comp.SVol.3/3-379

$BBrC_{14}H_{10}S_2$ .... $BrB(\text{-}S\text{-}CC_6H_5{=}CC_6H_5\text{-}S\text{-})$ .... B: B Comp.SVol.3/4-139

$BBrC_{14}H_{19}N$ ..... $(C_2H_5\text{-}CH{=}CH)_2B\text{-}NH\text{-}C_6H_4\text{-}4\text{-}Br$........ B: B Comp.SVol.3/3-153

– .... $[CH_3CH{=}C(CH_3)]_2B\text{-}NH\text{-}C_6H_4\text{-}2\text{-}Br$ .... B: B Comp.SVol.3/3-153

$BBrC_{14}H_{32}N_2Si$ ... $[(CH_3)_3Si][CH_3(CH_2)_2CH{=}]C{-}N(n{-}C_4H_9){-}B[N(CH_3)_2]Br$ B: B Comp.SVol.3/4-94

$BBrC_{15}ClH_{15}$ ..... $3{-}Cl{-}C_6H_4{-}BBr{-}C_6H_2{-}2,4,6{-}(CH_3)_3$ .... B: B Comp.SVol.4/4-94

$BBrC_{15}H_{20}MoN_4O_2$ $(CH_2CBrCH_2)Mo(CO)_2[{-}N_2C_3H(CH_3)_2{-}]_2BH_2$ .. Mo: Org.Comp.5-213, 215

$BBrC_{15}H_{23}N$ ..... $(CH_3)_4C_5H_6N{-}B(Br)C_6H_5$ .... B: B Comp.SVol.3/4-93

$BBrC_{15}H_{31}N_2O_2P$ $2,2,6,6{-}(CH_3)_4{-}NC_5H_6{-}1{-}B[{-}N(C_4H_9{-}t){-}P(Br){-}O{-}CH_2CH_2{-}O{-}]$ .... B: B Comp.SVol.4/3b-54

$BBrC_{16}FeH_{14}N_2$ .. $HNC_6H_4HNB{-}C_5H_4FeC_5H_4Br$ .... Fe: Org.Comp.A9-280, 284

$BBrC_{16}H_{12}S$ ..... $BrB({-}CH{=}CC_6H_5{-}S{-}CC_6H_5{=}CH{-})$ .... B: B Comp.SVol.3/4-89, 140

$BBrC_{16}H_{23}N$ ..... $2{-}(BrCH_2CH_2){-}[3.3.1.1^{3,7}]{-}1{-}BC_9H_{14} \cdot NC_5H_5$ B: B Comp.SVol.4/3b-30

$BBrC_{16}H_{39}N$ ..... $[N(n{-}C_4H_9)_4][H_3BBr]$ .... B: B Comp.SVol.3/4-94

$BBrC_{17}F_4FeH_{16}O_4$ $[(C_5H_5)Fe(CO)_2(CH_2{=}CHCH_2C_6H_2(Br{-}5)(OH{-}4)OCH_3{-}3)][BF_4]$ .... Fe: Org.Comp.B17-32

$BBrC_{17}F_4H_{32}MoO_6P_2$ $[C_5H_5Mo(P(OCH_3)_3)_2Br{-}CC{-}C_4H_9{-}t][BF_4]$ .... Mo: Org.Comp.6-117, 128

$BBrC_{17}H_{16}MoN_8O_2$ $(CH_2CBrCH_2)Mo(CO)_2(N_2C_3H_3)_4B$ .... Mo: Org.Comp.5-225/6, 228

$BBrC_{17}H_{33}N_3$ .... $[NC_5H_5{-}1{-}B(N(C_3H_7{-}i)_2)_2]Br$ .... B: B Comp.SVol.3/3-207; B: B Comp.SVol.4/3b-45

$BBrC_{18}H_{12}O$ ..... $6{-}Br{-}4{-}C_6H_5{-}5,6{-}OBC_{12}H_7$ .... B: B Comp.SVol.4/4-97

$BBrC_{18}H_{20}$ ...... $t{-}C_4H_9{-}C(C_6H_5){=}CH{-}BBr{-}C_6H_5$ .... B: B Comp.SVol.3/4-88

– .... $n{-}C_4H_9{-}CH{=}C(C_6H_5){-}BBr{-}C_6H_5$ .... B: B Comp.SVol.3/4-88

– .... $t{-}C_4H_9{-}CH{=}C(C_6H_5){-}BBr{-}C_6H_5$ .... B: B Comp.SVol.3/4-88

$BBrC_{18}H_{22}MoN_6O_2$ $BrCMo(CO)_2[{-}N_2C_3H(CH_3)_2{-}]_3BH$ .... Mo: Org.Comp.5-94

$BBrC_{18}H_{31}N$ ..... $2,6{-}(i{-}C_3H_7)_2{-}C_6H_3{-}NH{-}BBr{-}C(CH_3)_2{-}C_3H_7{-}i$ B: B Comp.SVol.4/4-105

$BBrC_{18}H_{36}N_2O_2$ .. $[2,2,6,6{-}(CH_3)_4{-}NC_5H_6{-}1{-}O]_2BBr$ .... B: B Comp.SVol.4/4-97

$BBrC_{18}H_{38}$ ...... $(n{-}C_6H_{13})(n{-}C_{12}H_{25})BBr$ .... B: B Comp.SVol.3/4-88

$BBrC_{19}F_4H_{28}MoO_6P_2$ $[C_5H_5Mo(P(OCH_3)_3)_2Br{-}CC{-}C_6H_5][BF_4]$ ...... Mo: Org.Comp.6-117/8, 128

$BBrC_{19}F_4H_{30}MoO_6P_2$ $[(P(OCH_3)_3)_2(Br)(C_5H_5)MoC{-}CH_2C_6H_5{-}t][BF_4]$ Mo: Org.Comp.6-75

$BBrC_{19}H_{15}N_3$ .... $3{-}(4{-}Br{-}C_6H_4){-}2,5{-}(C_6H_5)_2{-}1,2,4,3{-}N_3BCH$ ... B: B Comp.SVol.4/3a-174/5

$BBrC_{19}H_{31}N_2O_2P$ $2,2,6,6{-}(CH_3)_4{-}NC_5H_6{-}1{-}B[{-}N(C_4H_9{-}t){-}P(Br){-}O{-}(1,2{-}C_6H_4){-}O{-}]$ .... B: B Comp.SVol.4/3b-54

$BBrC_{19}H_{37}N_3$ .... $[2,6{-}(CH_3)_2{-}NC_5H_3{-}1{-}B(N(C_3H_7{-}i)_2)_2]Br$ ..... B: B Comp.SVol.4/3b-45

$BBrC_{20}H_{16}N_4O_3$ .. $2{-}(4{-}Br{-}C_6H_4){-}3{-}(4{-}CH_3O{-}C_6H_4){-}5{-}(4{-}NO_2{-}C_6H_4){-}1,2,4,3{-}N_3BCH$ .... B: B Comp.SVol.4/3a-174/5

$BBrC_{20}H_{34}$ ...... $[2,7,7{-}(CH_3)_3{-}(3.1.1){-}C_7H_8{-}3{-}]_2BBr$ .... B: B Comp.SVol.4/4-94

$BBrC_{20}H_{44}S$ ..... $(CH_3)_2S{-}B(C_6H_{13}{-}n)(C_{12}H_{25}{-}n)Br$ .... B: B Comp.SVol.3/4-137

$BBrC_{21}H_{28}Si$ ..... $(CH_3)_3SiC(C_4H_9{-}n){=}C(C_6H_5)B(Br)C_6H_5$ .... B: B Comp.SVol.3/4-88

$BBrC_{21}H_{36}N_2O$ ... $1{-}[2,6{-}(i{-}C_3H_7)_2C_6H_3{-}NHBBr{-}O]{-}2,2,6,6{-}(CH_3)_4NC_5H_6$ B: B Comp.SVol.4/4-107

$BBrC_{22}F_4H_{22}NSb$ $[({-}C_6H_4CH_2N(CH_3)CH_2C_6H_4{-})SbBr{-}C_6H_4CH_3][BF_4]$ Sb: Org.Comp.5-87

$BBrC_{22}H_{27}N$ ..... $1{-}(C_{13}H_9{-}9{-}BBr){-}2,2,6,6{-}(CH_3)_4{-}NC_5H_6$ ..... B: B Comp.SVol.4/4-105

$BBrC_{22}H_{29}N$ ..... $2,2,6,6{-}(CH_3)_4{-}NC_5H_6{-}1{-}BBr{-}CH(C_6H_5)_2$ ..... B: B Comp.SVol.4/4-105

$BBrC_{22}H_{32}N_2$ .... $C_6H_5CH_2{-}N(C_4H_9{-}t){-}BBr{-}N(C_4H_9{-}t){-}CH_2C_6H_5$ B: B Comp.SVol.4/4-102

$BBrC_{25}F_4H_{25}NOPRe$
$[(C_5H_5)Re(NO)(P(C_6H_5)_3)(Br-C_2H_5)][BF_4]$ ..... Re: Org.Comp.3-29, 31
$BBrC_{26}F_4FeH_{22}LiO_2P$
$[C_5H_5Fe(CO)_2CH_2P(C_6H_5)_3]BF_4 \cdot LiBr$ .... Fe: Org.Comp.B14-137, 143
$BBrC_{35}FeH_{30}$ .... $[(Br-C_6H_5)Fe(C_5H_5)][B(C_6H_5)_4]$ .... Fe: Org.Comp.B18-142/6, 197, 248
$BBrC_{40}FeH_{36}$ .... $[(Br-C_5H_6-C_6H_5)Fe(C_5H_5)][B(C_6H_5)_4]$ .... Fe: Org.Comp.B18-142/6, 219
$BBrC_{41}FeH_{34}O$ ... $[(4-Br-C_6H_4-O-C_6H_5)Fe(C_5H_5)][B(C_6H_5)_4]$ .... Fe: Org.Comp.B18-142/6, 198, 252
$BBrC_{42}FeH_{36}O$ ... $[(2-CH_3C_6H_4-O-C_6H_4Br-4)Fe(C_5H_5)][B(C_6H_5)_4]$ Fe: Org.Comp.B19-2, 49
– .... $[(3-CH_3C_6H_4-O-C_6H_4Br-4)Fe(C_5H_5)][B(C_6H_5)_4]$ Fe: Org.Comp.B19-2, 49
$BBrC_{77}H_{116}OOsP_4$
$[(CO)OsBr((c-C_6H_{11})_2PCH_2CH_2P(C_6H_{11}-c)_2)_2][B(C_6H_5)_4]$ Os: Org.Comp.A1-123/4
BBrClF .... BBrClF
Adducts with tertiary amines .... B: B Comp.SVol.3/4-65/7
BBrClI .... BBrClI .... B: B Comp.SVol.4/4-115
$BBrCl_2$ .... $BBrCl_2$ .... B: B Comp.SVol.3/4-95/6
B: B Comp.SVol.4/4-107
$BBrEuH_4$ .... $EuBr[BH_4] \cdot n\ C_4H_8O$ .... Sc: MVol.C11b-491
$BBrEu_2O_3$ .... Eu_2BO_3Br .... Sc: MVol.C11b-450
$BBrF_2$ .... $BBrF_2$ .... B: B Comp.SVol.3/4-95/6
B: B Comp.SVol.4/4-107
Adducts with tertiary amines .... B: B Comp.SVol.3/4-65/7, 96
$BBrF_4O_2$ .... $[BrO_2][BF_4]$ .... Br: SVol.B2-87/8
$BBrF_6$ .... $[BrF_2][BF_4]$ .... Br: SVol.B3-41/5
$BBrF_6O$ .... $[BrOF_2][BF_4]$ .... Br: SVol.B3-155/6
$BBrH_2$ .... BH_2Br .... B: B Comp.SVol.3/4-83/4
B: B Comp.SVol.4/4-88
$BBrH_2S_2$ .... $BrB(SH)_2$ .... B: B Comp.SVol.4/4-161
$BBrI_2$ .... $BBrI_2$ .... B: B Comp.SVol.3/4-101
B: B Comp.SVol.4/4-115
BBrO .... BrBO .... B: B Comp.SVol.3/4-91
BBrS .... BrB=S .... B: B Comp.SVol.4/4-156/7
BBr_2 .... BBr_2 .... B: B Comp.SVol.3/4-71
B: B Comp.SVol.4/4-76
BBr_2CH_3 .... CH_3-BBr_2 .... B: B Comp.SVol.3/4-84
B: B Comp.SVol.4/4-88/90
– .... $CH_3-B^{79}Br_2$ .... B: B Comp.SVol.4/4-88/9
– .... $CH_3-B^{79}Br^{81}Br$ .... B: B Comp.SVol.4/4-88/9
– .... $CH_3-B^{81}Br_2$ .... B: B Comp.SVol.4/4-88/9
– .... $CH_3-{}^{10}BBr_2$ .... B: B Comp.SVol.3/4-84
– .... $CH_3-{}^{10}B^{79}Br^{81}Br$ .... B: B Comp.SVol.4/4-88/9
– .... $CD_3-{}^{10}B^{79}Br^{81}Br$ .... B: B Comp.SVol.4/4-88
BBr_2CH_3O .... $CH_3-O-BBr_2$ .... B: B Comp.SVol.4/4-97
$BBr_2C_2F_6NS_2$ .... $(CF_3-S)_2N-BBr_2$ .... B: B Comp.SVol.4/4-99, 166
$BBr_2C_2H_5$ .... $C_2H_5-BBr_2$ .... B: B Comp.SVol.3/4-84

$BBr_2C_2H_5$ .... C_2H_5-BBr_2 .... B: B Comp.SVol.4/4-90

$BBr_2C_2H_6N$ ...... $(CH_3)_2N$-BBr_2 .... B: B Comp.SVol.3/4-91

B: B Comp.SVol.4/4-98

$BBr_2C_2H_7S$ ...... $(CH_3)_2S \cdot BHBr_2$ .... B: B Comp.SVol.3/4-83, 137

B: B Comp.SVol.4/4-87, 161/2

$BBr_2C_3FH_9N$ ..... $N(CH_3)_3 \cdot BBr_2F$ .... B: B Comp.SVol.3/4-65/7

$BBr_2C_3H_3S_2$ ..... $BrB(-S-CH=CCH_2Br-S-)$ .... B: B Comp.SVol.3/4-139

$BBr_2C_3H_7$ .... n-$C_3H_7BBr_2$ .... B: B Comp.SVol.3/4-85

$BBr_2C_4FH_{11}N$ .... $N(CH_3)_2(C_2H_5) \cdot BBr_2F$ .... B: B Comp.SVol.3/4-65/7

$BBr_2C_4F_6HN_2O$ ... $OBBr_2NHC(CF_3)NC(CF_3)$ .... F: PFHOrg.SVol.4-197, 211

$BBr_2C_4H_9$ .... n-C_4H_9-BBr_2 .... B: B Comp.SVol.3/4-85

– .... t-C_4H_9-BBr_2 .... B: B Comp.SVol.4/4-91

$BBr_2C_4H_{10}N$ ..... $(C_2H_5)_2N$-BBr_2 .... B: B Comp.SVol.4/4-98

– .... $(C_2H_5)_2N$-$^{10}BBr_2$ .... B: B Comp.SVol.3/4-91

– .... t-C_4H_9-NH-BBr_2 .... B: B Comp.SVol.4/4-99

$BBr_2C_5FH_{13}N$ .... $N(CH_3)(C_2H_5)_2 \cdot BBr_2F$ .... B: B Comp.SVol.3/4-65/7

$BBr_2C_5F_3H_{14}N_2$ .. $[(CH_3)_2NH$-$BBr(CF_3)$-$NH(CH_3)_2]Br$ .... B: B Comp.SVol.4/3a-168

B: B Comp.SVol.4/4-106

$BBr_2C_5H_7O_2$ ..... $Br_2B(-O=CCH_3-CH=CCH_3-O-)$ .... B: B Comp.SVol.3/4-91

$BBr_2C_5H_9$ .... c-$C_5H_9BBr_2$ .... B: B Comp.SVol.3/4-85

$BBr_2C_5H_9OSi$ .... $[(CH_3)_3Si][Br_2B]C=C=O$ .... B: B Comp.SVol.3/4-84/5

$BBr_2C_5H_{11}$ ...... i-C_3H_7-$CH(CH_3)$-$BBr_2 \cdot S(CH_3)_2$ .... B: B Comp.SVol.4/4-91

– .... n-C_5H_{11}-BBr_2 .... B: B Comp.SVol.3/4-85

$BBr_2C_5H_{12}N$ ..... t-C_4H_9-NBr-BBr-CH_3 .... B: B Comp.SVol.4/4-104

$BBr_2C_6ClH_4$ ..... 3-Cl-C_6H_4-BBr_2 .... B: B Comp.SVol.4/4-91

– .... 4-Cl-C_6H_4-BBr_2 .... B: B Comp.SVol.4/4-91

$BBr_2C_6FH_7N$ ..... 4-$CH_3C_5H_4N \cdot BBr_2F$ .... B: B Comp.SVol.3/4-65/7

$BBr_2C_6FH_{15}N$ .... $N(C_2H_5)_3 \cdot BBr_2F$ .... B: B Comp.SVol.3/4-65/7

$BBr_2C_6H_5$ .... C_6H_5-BBr_2 .... B: B Comp.SVol.3/4-85

B: B Comp.SVol.4/4-90

$BBr_2C_6H_{11}$ ...... 2-CH_3-c-C_5H_8-$BBr_2 \cdot S(CH_3)_2$ .... B: B Comp.SVol.4/4-91

– .... n-C_4H_9-CH=CHBBr_2 .... B: B Comp.SVol.3/4-85

– .... t-C_4H_9-CH=CHBBr_2 .... B: B Comp.SVol.3/4-85

– .... c-C_6H_{11}-BBr_2 .... B: B Comp.SVol.3/4-85

$BBr_2C_6H_{13}$ ...... i-C_3H_7-$C(CH_3)_2$-BBr_2 .... B: B Comp.SVol.4/4-91

– .... n-C_3H_7-$CH(CH_3)CH_2$-BBr_2 .... B: B Comp.SVol.3/4-85

– .... n-C_3H_7-$CH(C_2H_5)$-BBr_2 .... B: B Comp.SVol.3/4-86

B: B Comp.SVol.4/4-91

– .... n-C_6H_{13}-BBr_2 .... B: B Comp.SVol.3/4-85

$BBr_2C_6H_{14}N$ ..... $(i-C_3H_7)_2N$-BBr_2 .... B: B Comp.SVol.3/4-91

B: B Comp.SVol.4/4-98

– .... t-C_4H_9-NBr-BBr-C_2H_5 .... B: B Comp.SVol.4/4-104

$BBr_2C_6H_{18}NSi_2$ ... $[(CH_3)_3Si]_2N$-BBr_2 .... B: B Comp.SVol.4/4-99

$BBr_2C_6H_{18}NSn_2$ .. $[(CH_3)_3Sn]_2N$-BBr_2 .... B: B Comp.SVol.4/4-99

$BBr_2C_7FH_{13}N$ .... $HC(C_2H_4)_3N \cdot BBr_2F$ .... B: B Comp.SVol.3/4-65/7

$BBr_2C_7H_7$ .... 3-CH_3-C_6H_4-BBr_2 .... B: B Comp.SVol.4/4-91

– .... 4-CH_3-C_6H_4-BBr_2 .... B: B Comp.SVol.4/4-91

– .... $C_6H_5CH_2$-BBr_2 .... B: B Comp.SVol.3/4-86

B: B Comp.SVol.4/4-91

Formula	Compound	Reference
$BBr_2C_7H_{13}$	$n\text{-}C_4H_9\text{-}C(CH_3)\text{=}CH\text{-}BBr_2$	B: B Comp.SVol.4/4-92
$BBr_2C_7H_{14}N$	$CH_3(Br)B[\text{-}NH(C_3H_7\text{-}i)\text{-}CH_2\text{-}CBr\text{=}CH\text{-}]$	B: B Comp.SVol.3/4-93
–	$CH_3(Br)B[\text{-}NH(C_3H_7\text{-}n)\text{-}CH_2\text{-}CBr\text{=}CH\text{-}]$	B: B Comp.SVol.3/4-93
$BBr_2C_7H_{15}S$	$(CH_3)_2S\text{-}B(C_5H_9)Br_2$	B: B Comp.SVol.3/4-137
$BBr_2C_7H_{16}N$	$t\text{-}C_4H_9\text{-}NBr\text{-}BBr\text{-}C_3H_7\text{-}n$	B: B Comp.SVol.4/4-104
–	$t\text{-}C_4H_9\text{-}NH\text{-}BBr\text{-}CH_2CH_2\text{-}CH_2Br$	B: B Comp.SVol.4/4-105
$BBr_2C_7H_{17}S$	$i\text{-}C_3H_7\text{-}CH(CH_3)\text{-}BBr_2 \cdot S(CH_3)_2$	B: B Comp.SVol.4/4-91
–	$n\text{-}C_5H_{11}\text{-}BBr_2 \cdot S(CH_3)_2$	B: B Comp.SVol.4/4-91
$BBr_2C_7H_{18}NSn$	$(CH_3)_2Sn(Br)\text{-}N(C_4H_9\text{-}t)\text{-}BBr\text{-}CH_3$	B: B Comp.SVol.4/4-105
		Sn: Org.Comp.19-124/6
–	$(CH_3)_3Sn\text{-}N(C_4H_9\text{-}t)\text{-}BBr_2$	B: B Comp.SVol.4/4-99
		Sn: Org.Comp.18-58, 65
$BBr_2C_8FH_{11}N$	$N(CH_3)_2(C_6H_5) \cdot BBr_2F$	B: B Comp.SVol.3/4-65/7, 95
$BBr_2C_8H_6O_6Re$	$cis\text{-}(CO)_4Re[(CH_3CO)_2]BBr_2$	Re: Org.Comp.2-344/5, 347
$BBr_2C_8H_7NO_5Re$	$cis\text{-}(CO)_4Re[(CH_3CO)(CH_3CNH)]BBr_2$	Re: Org.Comp.2-343
$BBr_2C_8H_{12}N$	$(Br)_2NBC_6H_6(CH_3)_2$	B: B Comp.SVol.3/3-166
$BBr_2C_8H_{15}$	$n\text{-}C_6H_{13}CH\text{=}CHBBr_2$	B: B Comp.SVol.3/4-86
$BBr_2C_8H_{17}$	$n\text{-}C_8H_{17}BBr_2$	B: B Comp.SVol.3/4-86
$BBr_2C_8H_{17}S$	$2\text{-}CH_3\text{-}c\text{-}C_5H_8\text{-}BBr_2 \cdot S(CH_3)_2$	B: B Comp.SVol.3/4-137
		B: B Comp.SVol.4/4-91
–	$c\text{-}C_6H_{11}\text{-}BBr_2 \cdot S(CH_3)_2$	B: B Comp.SVol.4/4-91
$BBr_2C_8H_{18}N$	$t\text{-}C_4H_9\text{-}NBr\text{-}BBr\text{-}C_4H_9\text{-}n$	B: B Comp.SVol.4/4-104
–	$t\text{-}C_4H_9\text{-}NBr\text{-}BBr\text{-}C_4H_9\text{-}t$	B: B Comp.SVol.4/4-104
–	$t\text{-}C_4H_9\text{-}NH\text{-}BBr\text{-}C(CH_3)_2\text{-}CH_2Br$	B: B Comp.SVol.4/4-105
–	$t\text{-}C_4H_9\text{-}NH\text{-}BBr\text{-}CH_2CH_2\text{-}CHBr\text{-}CH_3$	B: B Comp.SVol.4/4-105
$BBr_2C_8H_{19}S$	$i\text{-}C_3H_7\text{-}C(CH_3)_2\text{-}BBr_2 \cdot S(CH_3)_2$	B: B Comp.SVol.4/4-162
–	$n\text{-}C_3H_7\text{-}CH(CH_3)\text{-}CH_2\text{-}BBr_2 \cdot S(CH_3)_2$	B: B Comp.SVol.3/4-137
		B: B Comp.SVol.4/4-91
–	$n\text{-}C_3H_7\text{-}CH(C_2H_5)\text{-}BBr_2 \cdot S(CH_3)_2$	B: B Comp.SVol.3/4-137
		B: B Comp.SVol.4/4-91
–	$n\text{-}C_6H_{13}\text{-}BBr_2 \cdot S(CH_3)_2$	B: B Comp.SVol.3/4-137
		B: B Comp.SVol.4/4-91, 162
$BBr_2C_8H_{20}NSn$	$(CH_3)_2Sn(Br)\text{-}N(C_4H_9\text{-}t)\text{-}B(Br)\text{-}C_2H_5$	Sn: Org.Comp.19-124/6
$BBr_2C_9FH_{13}N$	$4\text{-}CH_3\text{-}C_6H_4\text{-}N(CH_3)_2 \cdot BBr_2F$	B: B Comp.SVol.3/4-65/7
–	$C_6H_5\text{-}N(CH_3)C_2H_5 \cdot BBr_2F$	B: B Comp.SVol.3/4-65/7
$BBr_2C_9FH_{21}N$	$N(n\text{-}C_3H_7)_3 \cdot BBr_2F$	B: B Comp.SVol.3/4-65/7
$BBr_2C_9H_{11}$	$2,4,6\text{-}(CH_3)_3\text{-}C_6H_2\text{-}BBr_2$	B: B Comp.SVol.3/4-86
		B: B Comp.SVol.4/4-92
$BBr_2C_9H_{13}Si$	$(CH_3)_3Si\text{-}2\text{-}C_6H_4\text{-}BBr_2$	B: B Comp.SVol.4/4-92
–	$(CH_3)_3Si\text{-}3\text{-}C_6H_4\text{-}BBr_2$	B: B Comp.SVol.4/4-92
–	$(CH_3)_3Si\text{-}4\text{-}C_6H_4\text{-}BBr_2$	B: B Comp.SVol.4/4-92
$BBr_2C_9H_{14}N$	$(Br)_2NBC_6H_6(CH_3)(C_2H_5)$	B: B Comp.SVol.3/3-166
$BBr_2C_9H_{15}$	$(Br)BC_8H_{13}(CH_2Br)$	B: B Comp.SVol.3/4-89
$BBr_2C_9H_{17}$	$4\text{-}(n\text{-}C_3H_7)\text{-}c\text{-}C_6H_{10}\text{-}BBr_2$	B: B Comp.SVol.4/4-92
$BBr_2C_9H_{18}N$	$2,2,6,6\text{-}(CH_3)_4\text{-}NC_5H_6\text{-}1\text{-}BBr_2$	B: B Comp.SVol.4/4-99
$BBr_2C_9H_{18}NO$	$2,2,6,6\text{-}(CH_3)_4\text{-}NC_5H_6\text{-}1\text{-}O\text{-}BBr_2$	B: B Comp.SVol.4/4-97
$BBr_2C_9H_{20}N$	$t\text{-}C_4H_9\text{-}NBr\text{-}BBr\text{-}C_5H_{11}\text{-}n$	B: B Comp.SVol.4/4-104
–	$t\text{-}C_4H_9\text{-}NH\text{-}BBr\text{-}CH_2CH_2\text{-}CHBr\text{-}C_2H_5$	B: B Comp.SVol.4/4-105
$BBr_2C_9H_{21}S$	$(CH_3)_2S\text{-}B(C_7H_{15}\text{-}n)Br_2$	B: B Comp.SVol.3/4-137
$BBr_2C_{10}FH_{15}N$	$C_6H_5\text{-}CH_2\text{-}N(CH_3)C_2H_5 \cdot BBr_2F$	B: B Comp.SVol.3/4-65/7

$BBr_2C_{10}FH_{15}N$ ... $C_6H_5\text{-}N(C_2H_5)_2 \cdot BBr_2F$ ... B: B Comp.SVol.3/4-65/7
$BBr_2C_{10}F_5H_9N$ ... $t\text{-}C_4H_9\text{-}NBr\text{-}BBr\text{-}C_6F_5$ ... B: B Comp.SVol.4/4-104
$BBr_2C_{10}H_{10}O_6Re$ $cis\text{-}(CO)_4Re[(CH_3CO)(CH(CH_3)_2CO)]BBr_2$ ... Re: Org.Comp.2-346
$BBr_2C_{10}H_{15}$ ... $(CH_3)_5C_5\text{-}BBr_2$ ... B: B Comp.SVol.4/4-92
$BBr_2C_{10}H_{16}N$ ... $(Br)_2\text{-}BNC_6H_6\text{-}(CH_3)(C_3H_7\text{-}i)$ ... B: B Comp.SVol.3/3-166
– ... $(Br)_2\text{-}BNC_6H_6\text{-}(CH_3)(C_3H_7\text{-}n)$ ... B: B Comp.SVol.3/3-166
$BBr_2C_{10}H_{17}$ ... $2,7,7\text{-}(CH_3)_3\text{-}[3.1.1]\text{-}C_7H_8\text{-}3\text{-}BBr_2 \cdot S(CH_3)_2$ ... B: B Comp.SVol.4/4-92
$BBr_2C_{10}H_{19}$ ... $(n\text{-}C_4H_9)_2C{=}CH\text{-}BBr_2$ ... B: B Comp.SVol.4/4-92
$BBr_2C_{10}H_{21}Si$ ... $(CH_3)_3Si\text{-}CH_2\text{-}C(C_4H_9\text{-}n){=}CH\text{-}BBr_2$ ... B: B Comp.SVol.4/4-92
$BBr_2C_{11}H_{16}N$ ... $C_6H_5\text{-}CH_2\text{-}N(C_4H_9\text{-}t)\text{-}BBr_2$ ... B: B Comp.SVol.4/4-99
$BBr_2C_{11}H_{18}N$ ... $(Br)_2NBC_6H_6(CH_3)(C_4H_9\text{-}n)$ ... B: B Comp.SVol.3/3-166
$BBr_2C_{11}H_{27}N_2Si$ ... $BrCH_2\text{-}C(CH_3)_2\text{-}N[Si(CH_3)_3]\text{-}BBr\text{-}NH\text{-}C_4H_9\text{-}t$ B: B Comp.SVol.4/4-105
– ... $t\text{-}C_4H_9\text{-}N[Si(CH_3)_3]\text{-}BBr\text{-}NBr\text{-}C_4H_9\text{-}t$ ... B: B Comp.SVol.4/4-103
$BBr_2C_{12}FH_{27}N$ ... $N(n\text{-}C_4H_9)_3 \cdot BBr_2F$ ... B: B Comp.SVol.3/4-65/7
$BBr_2C_{12}H_9$ ... $3\text{-}C_6H_5\text{-}C_6H_4\text{-}BBr_2$ ... B: B Comp.SVol.4/4-92
– ... $4\text{-}C_6H_5\text{-}C_6H_4\text{-}BBr_2$ ... B: B Comp.SVol.4/4-92
$BBr_2C_{12}H_{11}N_2$ ... $C_{12}H_{11}N_2\text{-}BBr_2$ ... B: B Comp.SVol.3/4-92
$BBr_2C_{12}H_{15}$ ... $n\text{-}C_4H_9\text{-}C(C_6H_5){=}CH\text{-}BBr_2$ ... B: B Comp.SVol.4/4-92
$BBr_2C_{12}H_{18}N$ ... $2,6\text{-}(i\text{-}C_3H_7)_2\text{-}C_6H_3\text{-}NH\text{-}BBr_2$ ... B: B Comp.SVol.4/4-99
$BBr_2C_{12}H_{19}$ ... $n\text{-}C_4H_9\text{-}CC\text{-}C(C_4H_9\text{-}n){=}CH\text{-}BBr_2$ ... B: B Comp.SVol.4/4-92
$BBr_2C_{12}H_{20}NSn$ ... $(CH_3)_2Sn(Br)\text{-}N(C_4H_9\text{-}t)\text{-}B(Br)\text{-}C_6H_5$ ... Sn: Org.Comp.19-124/6
$BBr_2C_{12}H_{21}$ ... $(Br)BC_9H_{13}(CH_3)_2CH_2Br$ ... B: B Comp.SVol.3/4-89
$BBr_2C_{12}H_{23}$ ... $n\text{-}C_5H_{11}\text{-}CH{=}C(C_5H_{11}\text{-}n)\text{-}BBr_2$ ... B: B Comp.SVol.4/4-92
$BBr_2C_{12}H_{23}S$ ... $2,7,7\text{-}(CH_3)_3\text{-}[3.1.1]\text{-}C_7H_8\text{-}3\text{-}BBr_2 \cdot S(CH_3)_2$ ... B: B Comp.SVol.4/4-92
$BBr_2C_{12}H_{27}S$ ... $(CH_3)_2S\text{-}B(C_{10}H_{21})Br_2$ ... B: B Comp.SVol.3/4-137
$BBr_2C_{13}H_{17}$ ... $t\text{-}C_4H_9\text{-}CBr{=}CH\text{-}BBr\text{-}CH_2\text{-}C_6H_5$ ... B: B Comp.SVol.3/4-88
– ... $t\text{-}C_4H_9\text{-}CH{=}C(BBr_2)\text{-}CH_2\text{-}C_6H_5$ ... B: B Comp.SVol.3/4-86
$BBr_2C_{13}H_{23}N_2Si_2$ $[\text{-}BBr_2\text{-}N(Si(CH_3)_3)C(C_6H_5)N(Si(CH_3)_3)\text{-}]$ ... B: B Comp.SVol.4/4-99
$BBr_2C_{13}H_{27}$ ... $n\text{-}C_{13}H_{27}BBr_2$ ... B: B Comp.SVol.3/4-86
$BBr_2C_{13}H_{29}S$ ... $(CH_3)_2S\text{-}B(C_{11}H_{23})Br_2$ ... B: B Comp.SVol.3/4-137
$BBr_2C_{14}F_4FeH_{13}O_2$
$[(C_5H_5)Fe(CO)_2(C_7H_8Br_2)][BF_4]$ ... Fe: Org.Comp.B17-99
$BBr_2C_{14}H_{10}O_6Re$ $cis\text{-}(CO)_4Re[(CH_3CO)(CH_2(C_6H_5)CO)]BBr_2$ ... Re: Org.Comp.2-345
$BBr_2C_{14}H_{25}$ ... $C_2H_5\text{-}CH{=}C(C_2H_5)\text{-}BBr\text{-}CH{=}CBr\text{-}C_6H_{13}\text{-}n$ ... B: B Comp.SVol.4/4-94
– ... $n\text{-}C_4H_9\text{-}CH{=}CH\text{-}BBr\text{-}CH{=}CBr\text{-}C_6H_{13}\text{-}n$ ... B: B Comp.SVol.4/4-94
$BBr_2C_{16}H_{37}N_3P$ ... $(i\text{-}C_3H_7)_2N\text{-}BBr\text{-}N(C_4H_9\text{-}t)\text{-}PBr\text{-}N(C_3H_7\text{-}i)_2$ ... B: B Comp.SVol.4/4-103
$BBr_2C_{16}H_{38}N$ ... $[N(n\text{-}C_4H_9)_4][H_2BBr_2]$ ... B: B Comp.SVol.3/4-94
$BBr_2C_{18}F_4H_{23}MoO$
$[C_5H_5Mo(CO)(t\text{-}C_4H_9\text{-}CC\text{-}Br)_2][BF_4]$ ... Mo: Org.Comp.6-320
$BBr_2C_{18}H_{16}P$ ... $(C_6H_5)_3P \cdot BHBr_2$ ... B: B Comp.SVol.4/4-88
$BBr_2C_{22}F_4H_{15}MoO$
$[C_5H_5Mo(CO)(C_6H_5\text{-}CC\text{-}Br)_2][BF_4]$ ... Mo: Org.Comp.6-321
BBr_2Cl ... BBr_2Cl ... B: B Comp.SVol.3/4-95/6
B: B Comp.SVol.4/4-107
BBr_2F ... BBr_2F ... B: B Comp.SVol.3/4-95/6
B: B Comp.SVol.4/4-107
Adducts with tertiary amines ... B: B Comp.SVol.3/4-65/7, 96
BBr_2H ... $HBBr_2$ ... B: B Comp.SVol.3/4-83
B: B Comp.SVol.4/4-87
– ... $x\ HBBr_2 \cdot [\text{-}CH_2\text{-}CHCH_3\text{-}]_nS[\text{-}CH_2\text{-}CHCH_3\text{-}]_n$ B: B Comp.SVol.4/4-88

BBr_2HS . . . . $HS-BBr_2$ . . . . B: B Comp.SVol.4/4-161
BBr_2H_2N . . . . H_2N-BBr_2 . . . . B: B Comp.SVol.4/4-98
BBr_2I . . . . BBr_2I . . . . B: B Comp.SVol.3/4-101
B: B Comp.SVol.4/4-115
BBr_3 . . . . BBr_3 . . . . B: B Comp.SVol.3/4-9, 66, 71/8, 84
B: B Comp.SVol.4/4-76/80
– . . . . $^{10}BBr_3$ . . . . B: B Comp.SVol.3/4-84
B: B Comp.SVol.4/4-78
– . . . . $^{11}BBr_3$ . . . . B: B Comp.SVol.4/4-78
– . . . . $BBr_3 \cdot H_2O$ . . . . B: B Comp.SVol.4/3b-148
– . . . . BBr_3 systems
$BBr_3-O-SiCl_4$ . . . . B: B Comp.SVol.4/4-77
$BBr_3CH_3NO_2$ . . . . . $CH_3NO_2-BBr_3$ . . . . B: B Comp.SVol.3/4-77
$BBr_3C_2ClH_6N$ . . . . $(CH_3)_2NCl \cdot BBr_3$ . . . . B: B Comp.SVol.4/4-80
$BBr_3C_2Cl_2H_2O_2$ . . $Cl_2CHCOOH \cdot BBr_3$ . . . . B: B Comp.SVol.3/4-76
$BBr_3C_2FH_6N$ . . . . . $(CH_3)_2NF \cdot BBr_3$ . . . . B: B Comp.SVol.4/4-80
$BBr_3C_2H_2$ . . . . $BrCH=CH-BBr_2$ . . . . B: B Comp.SVol.4/4-91
$BBr_3C_2H_6O$ . . . . . . $(CH_3)_2O-BBr_3$ . . . . B: B Comp.SVol.3/4-76/7
– . . . . $(CD_3)_2O-BBr_3$ . . . . B: B Comp.SVol.3/4-76/7
$BBr_3C_2H_6S$ . . . . . . $(CH_3)_2S-BBr_3$ . . . . B: B Comp.SVol.4/4-81, 161
$BBr_3C_2H_7P$ . . . . . . $(CH_3)_2PH-BBr_3$ . . . . B: B Comp.SVol.3/4-78
– . . . . $(CH_3)_2PD-BBr_3$ . . . . B: B Comp.SVol.3/4-78
– . . . . $(CD_3)_2PH-BBr_3$ . . . . B: B Comp.SVol.3/4-78
$BBr_3C_3H_4$ . . . . $CH_3CBr=CHBBr_2$ . . . . B: B Comp.SVol.3/4-85
$BBr_3C_3H_5N$ . . . . . . $C_2H_5CN-BBr_3$ . . . . B: B Comp.SVol.3/4-77
$BBr_3C_3H_9N$ . . . . . . $(CH_3)_3N-BBr_3$ . . . . B: B Comp.SVol.4/4-80
$BBr_3C_3H_9O_9S_3Sb$ $(CH_3SO_3)_3Sb \cdot BBr_3$ . . . . B: B Comp.SVol.4/4-81
$BBr_3C_3H_9P$ . . . . . . $(CH_3)_3P-BBr_3$ . . . . B: B Comp.SVol.3/4-78
$BBr_3C_3N_3OPS_3$ . . . $(S=C=N)_3PO \cdot BBr_3$ . . . . B: B Comp.SVol.4/4-81
$BBr_3C_4Cl_4H_4O_4$ . . $[Cl_2CHCOOH_2][Cl_2CHCOOBBr_3]$ . . . . B: B Comp.SVol.3/4-76
$BBr_3C_4F_3H_9O_3SSi$ $(CH_3)_3SiOSO_2CF_3 \cdot BBr_3$ . . . . B: B Comp.SVol.3/4-77
$BBr_3C_4H_4S$ . . . . . . $SC_4H_4 \cdot BBr_3$ . . . . B: B Comp.SVol.4/4-81, 161
$BBr_3C_4H_8S$ . . . . . . $SC_4H_8 \cdot BBr_3$ . . . . B: B Comp.SVol.4/4-81, 161
$BBr_3C_4H_9NO$ . . . . . $CH_3CON(CH_3)_2 \cdot BBr_3$ . . . . B: B Comp.SVol.3/4-77
$BBr_3C_4H_{10}O$ . . . . . $(C_2H_5)_2O \cdot BBr_3$ . . . . B: B Comp.SVol.4/4-80
$BBr_3C_5ClH_4N$ . . . . $3-Cl-C_5H_4N-BBr_3$ . . . . B: B Comp.SVol.3/4-77
$BBr_3C_5FH_4N$ . . . . . $3-F-C_5H_4N-BBr_3$ . . . . B: B Comp.SVol.3/4-77
$BBr_3C_5GaH_{13}N_2$ . . $CH_3B(-NCH_3-CH_2-CH_2-NCH_3-) \cdot GaBr_3$ . . . . . B: B Comp.SVol.3/3-113
$BBr_3C_5H_5In$ . . . . . . $In(C_5H_5-c) \cdot BBr_3$ . . . . In: Org.Comp.1-377/8
$BBr_3C_5H_5N$ . . . . . . $C_5H_5N \cdot BBr_3$ . . . . B: B Comp.SVol.3/4-77
B: B Comp.SVol.4/3b-149/50
B: B Comp.SVol.4/4-81
$BBr_3C_5H_{12}Si$ . . . . . $(CH_3)_2SiBr(CH_2)_3BBr_2$ . . . . B: B Comp.SVol.3/4-86
$BBr_3C_6F_6H_{10}O_6S_2Si$
$(C_2H_5)_2Si(OSO_2CF_3)_2 \cdot BBr_3$ . . . . B: B Comp.SVol.3/4-77
$BBr_3C_6H_4N_2$ . . . . . $3-NC-C_5H_4N-BBr_3$ . . . . B: B Comp.SVol.3/4-77
– . . . . $4-NC-C_5H_4N-BBr_3$ . . . . B: B Comp.SVol.3/4-77
$BBr_3C_6H_5NO_2$ . . . . $C_6H_5NO_2-BBr_3$ . . . . B: B Comp.SVol.3/4-77
$BBr_3C_6H_{10}$ . . . . . . $C_2H_5-CBr=C(C_2H_5)-BBr_2$ . . . . B: B Comp.SVol.4/4-91

$BBr_3C_6H_{10}$ ...... $n\text{-}C_4H_9\text{-}CBr{=}CH\text{-}BBr_2$ ...... B: B Comp.SVol.3/4-85
B: B Comp.SVol.4/4-92
– ...... $t\text{-}C_4H_9\text{-}CBr{=}CH\text{-}BBr_2$ ...... B: B Comp.SVol.3/4-85
$BBr_3C_6H_{11}N$ ...... $Br_2B[\text{-}NH(C_3H_7\text{-}i)\text{-}CH_2\text{-}CBr{=}CH\text{-}]$ ...... B: B Comp.SVol.3/4-93
$BBr_3C_6H_{14}S$ ...... $(n\text{-}C_3H_7)_2S \cdot BBr_3$ ...... B: B Comp.SVol.4/4-81
$BBr_3C_6H_{16}Si_2$ ...... $(CH_3)_2SiBr\text{-}CH_2Si(CH_3)(Br)\text{-}CH_2\text{-}B(Br)CH_3$ ...... B: B Comp.SVol.3/4-88
– ...... $(CH_3)_2SiBr\text{-}CH_2Si(CH_3)_2\text{-}CH_2\text{-}BBr_2$ ...... B: B Comp.SVol.3/4-86
$BBr_3C_7F_3H_{15}O_3SSi$
$(C_2H_5)_3SiOSO_2CF_3 \cdot BBr_3$ ...... B: B Comp.SVol.3/4-77
$BBr_3C_7H_8S$ ...... $C_6H_5\text{-}S\text{-}CH_3 \cdot BBr_3$ ...... B: B Comp.SVol.4/4-81, 161
$BBr_3C_7H_9N$ ...... $(Br)_3NBC_6H_6(CH_3)$ ...... B: B Comp.SVol.3/4-93
$BBr_3C_7H_{12}Si$ ...... $(C_2H_3)_2SiBr(CH_2)_3BBr_2$ ...... B: B Comp.SVol.3/4-86
$BBr_3C_7H_{13}N$ ...... $HC(C_2H_4)_3N \cdot BBr_3$ ...... B: B Comp.SVol.3/4-66
$BBr_3C_8H_6$ ...... $C_6H_5CBr{=}CHBBr_2$ ...... B: B Comp.SVol.3/4-86
$BBr_3C_8H_{11}N$ ...... $(Br)_3NBC_6H_6(C_2H_5)$ ...... B: B Comp.SVol.3/4-93
$BBr_3C_8H_{12}N_4$ ...... $(1,2\text{-}N_2C_3Br_3)\text{-}1\text{-}B(\text{-}NCH_3\text{-}CH_2\text{-}CH_2\text{-}CH_2\text{-}NCH_3\text{-})$
B: B Comp.SVol.3/3-95
$BBr_3C_8H_{14}$ ...... $n\text{-}C_3H_7\text{-}CBr{=}C(C_3H_7\text{-}n)\text{-}BBr_2$ ...... B: B Comp.SVol.4/4-92
– ...... $n\text{-}C_6H_{13}\text{-}CBr{=}CH\text{-}BBr_2$ ...... B: B Comp.SVol.3/4-86
B: B Comp.SVol.4/4-92
$BBr_3C_9H_9N$ ...... $B(CHCBrCH_2)_3N$ ...... B: B Comp.SVol.3/3-166
$BBr_3C_9H_{11}NO$ ...... $C_6H_5CON(CH_3)_2 \cdot BBr_3$ ...... B: B Comp.SVol.3/4-77
$BBr_3C_9H_{12}Si$ ...... $2\text{-}[(CH_3)_2SiBr\text{-}CH_2]\text{-}C_6H_4\text{-}BBr_2$ ...... B: B Comp.SVol.4/4-92
$BBr_3C_9H_{13}N$ ...... $(Br)_3\text{-}BNC_6H_6\text{-}C_3H_7\text{-}i$ ...... B: B Comp.SVol.3/4-93
– ...... $(Br)_3\text{-}BNC_6H_6\text{-}C_3H_7\text{-}n$ ...... B: B Comp.SVol.3/4-93
$BBr_3C_9H_{15}P$ ...... $[(CH_3)_3PC_6H_5][HBBr_3]$ ...... B: B Comp.SVol.3/4-84
$BBr_3C_{10}H_{11}N$ ...... $Br_2B[\text{-}NHCH_2C_6H_5\text{-}CH_2\text{-}CBr{=}CH\text{-}]$ ...... B: B Comp.SVol.3/4-93
$BBr_3C_{10}H_{14}Si$ ...... $CH_3(C_6H_5)SiBr(CH_2)_3BBr_2$ ...... B: B Comp.SVol.3/4-86
$BBr_3C_{10}H_{15}N$ ...... $(Br)_3NBC_6H_6(C_4H_9\text{-}n)$ ...... B: B Comp.SVol.3/4-93
$BBr_3C_{10}H_{18}$ ...... $n\text{-}C_4H_9\text{-}CBr{=}C(C_4H_9\text{-}n)\text{-}BBr_2$ ...... B: B Comp.SVol.4/4-92
– ...... $n\text{-}C_8H_{17}\text{-}CBr{=}CH\text{-}BBr_2$ ...... B: B Comp.SVol.4/4-92
$BBr_3C_{10}H_{23}N_2P$ ...... $(i\text{-}C_3H_7)_2N\text{-}BBr\text{-}N(C_4H_9\text{-}t)\text{-}PBr_2$ ...... B: B Comp.SVol.4/4-103
$BBr_3C_{12}H_8OS$ ...... $5,10\text{-}OSC_{12}H_8 \cdot BBr_3$ ...... B: B Comp.SVol.4/4-161
$BBr_3C_{12}H_8S$ ...... $5\text{-}SC_{12}H_8 \cdot BBr_3$ ...... B: B Comp.SVol.4/4-81
$BBr_3C_{12}H_{10}S$ ...... $(C_6H_5)_2S \cdot BBr_3$ ...... B: B Comp.SVol.4/4-81, 161
$BBr_3C_{13}H_{11}N$ ...... $C_6H_5CH{=}N(C_6H_5) \cdot BBr_3$ ...... B: B Comp.SVol.3/4-77
$BBr_3C_{13}H_{11}NO$ ...... $C_6H_5\text{-}C(O)\text{-}C_6H_4\text{-}NH_2\text{-}4 \cdot BBr_3$ ...... B: B Comp.SVol.3/4-77
– ...... $HC(O)\text{-}N(C_6H_5)_2 \cdot BBr_3$ ...... B: B Comp.SVol.3/4-77
$BBr_3C_{13}H_{27}N_2P$ ...... $1\text{-}[Br_2P\text{-}N(C_4H_9\text{-}t)\text{-}BBr]\text{-}2,2,6,6\text{-}(CH_3)_4NC_5H_6$ B: B Comp.SVol.4/4-103
$BBr_3C_{14}H_{12}MoN_6O_2$
$(CH_2CHCH_2)Mo(CO)_2[\text{-}N_2C_3H_2(Br)\text{-}]_3BH$ Mo:Org.Comp.5-226, 227
$BBr_3C_{14}H_{14}S$ ...... $(C_6H_5\text{-}CH_2)_2S \cdot BBr_3$ ...... B: B Comp.SVol.4/4-81
$BBr_3C_{15}H_{14}MoN_6O_2$
$[CH_2C(CH_3)CH_2]Mo(CO)_2[\text{-}N_2C_3H_2(Br)\text{-}]_3BH$ Mo:Org.Comp.5-226, 228
$BBr_3C_{15}H_{16}Si$ ...... $(C_6H_5)_2SiBr(CH_2)_3BBr_2$ ...... B: B Comp.SVol.3/4-86
$BBr_3C_{16}H_{28}$ ...... $(n\text{-}C_6H_{13}\text{-}CBr{=}CH)_2BBr$ ...... B: B Comp.SVol.4/4-94
$BBr_3C_{16}H_{37}N$ ...... $[N(n\text{-}C_4H_9)_4][HBBr_3]$ ...... B: B Comp.SVol.3/4-84, 94
$BBr_3C_{18}H_{15}N$ ...... $H_3N\text{-}B(C_6H_4\text{-}4\text{-}Br)_3$ ...... B: B Comp.SVol.3/3-191
$BBr_3C_{18}H_{15}P$ ...... $(C_6H_5)_3P\text{-}BBr_3$ ...... B: B Comp.SVol.3/4-78

$BBr_3C_{18}H_{19}N_6O_3Re$
$(CO)_3Re[1,2-N_2C_3(Br-4)(CH_3)_2-3,5]_3BH$ ...... Re: Org.Comp.1-115, 118
$BBr_3C_{19}H_{15}NO$ ... $C_6H_5CON(C_6H_5)_2 \cdot BBr_3$ ... B: B Comp.SVol.3/4-77
$BBr_3C_{21}H_{35}N_2P$ .. 1-[2,6-(i-$C_3H_7)_2C_6H_3$-N(PBr_2)BBr]-2,2,6,6-$(CH_3)_4NC_5H_6$
B: B Comp.SVol.4/4-103
$BBr_3C_{21}H_{35}N_2Sb$ 1-[2,6-(i-$C_3H_7)_2C_6H_3$-N($SbBr_2$)BBr]-2,2,6,6-$(CH_3)_4NC_5H_6$ ... B: B Comp.SVol.4/4-103
$BBr_3C_{22}H_{19}N_3$ ... $(-N=CCH_3-CH=CNH_2-)NH-B(C_6H_4-4-Br)_3$ ... B: B Comp.SVol.3/3-192
BBr_3Cl_3P ... $BBr_3 \cdot PCl_3$ ... B: B Comp.SVol.3/4-77
BBr_3F_3P ... $BBr_3 \cdot PF_3$ ... B: B Comp.SVol.3/4-77
BBr_3H_2S ... $BBr_3 \cdot H_2S$ ... B: B Comp.SVol.4/3b-148
B: B Comp.SVol.4/4-161
BBr_3H_3N ... $BBr_3 \cdot NH_3$ ... B: B Comp.SVol.3/4-77
B: B Comp.SVol.4/3b-148
– ... $BBr_3 \cdot {}^{15}NH_3$ ... B: B Comp.SVol.3/4-77
– ... $BBr_3 \cdot ND_3$ ... B: B Comp.SVol.3/4-77
BBr_3H_3P ... $BBr_3 \cdot PH_3$ ... B: B Comp.SVol.3/4-77
B: B Comp.SVol.4/3b-148
– ... $BBr_3 \cdot PD_3$ ... B: B Comp.SVol.3/4-77
BBr_4^- ... $[BBr_4]^-$ ... B: B Comp.SVol.3/4-78/9
B: B Comp.SVol.4/4-81/3
$BBr_4C_2H_5S$ ... $CH_3(CH_2Br)S-BBr_3$ ... B: B Comp.SVol.3/4-136
$BBr_4C_5H_4N$ ... $3-Br-C_5H_4N-BBr_3$ ... B: B Comp.SVol.3/4-77
$BBr_4C_7H_7$ ... $[C_7H_7][BBr_4]$ ... B: B Comp.SVol.4/4-81
$BBr_4C_8H_{20}N$ ... $[N(C_2H_5)_4][BBr_4]$ ... B: B Comp.SVol.3/4-79
B: B Comp.SVol.4/4-81
$BBr_4C_{12}GaH_{28}N_2$ $[(i-C_3H_7)_2N=B=N(C_3H_7-i)_2][GaBr_4]$ ... B: B Comp.SVol.4/3b-43
$BBr_4C_{13}H_{27}N_2Si$ .. 2,2,6,6-$(CH_3)_4NC_5H_6$-1-BBr-N(C_4H_9-t)-$SiBr_3$ B: B Comp.SVol.4/4-103
$BBr_4C_{14}H_{10}MnO_2$ $[(C_5H_5)(CO)_2Mn\equiv CC_6H_5][BBr_4]$ ... B: B Comp.SVol.3/4-79
$BBr_4C_{14}H_{10}O_2Re$ $[(C_5H_5)(CO)_2Re\equiv C-C_6H_5][BBr_4]$ ... B: B Comp.SVol.3/4-79
B: B Comp.SVol.4/4-81
$BBr_4C_{15}F_3H_9MnO_2$
$[(C_5H_5)(CO)_2Mn\equiv CC_6H_4-4-CF_3][BBr_4]$ ... B: B Comp.SVol.3/4-79
$BBr_4C_{15}F_3H_9O_2Re$ $[(C_5H_5)(CO)_2Re\equiv CC_6H_4-4-CF_3][BBr_4]$ ... B: B Comp.SVol.3/4-79
$BBr_4C_{16}H_{36}N$ ... $[(n-C_4H_9)_4N][BBr_4]$ ... B: B Comp.SVol.3/4-94
B: B Comp.SVol.4/4-81
$BBr_4C_{21}H_{35}N_2Si$ .. 1-[2,6-(i-$C_3H_7)_2C_6H_3$-N($SiBr_3$)BBr]-2,2,6,6-$(CH_3)_4NC_5H_6$ ... B: B Comp.SVol.4/4-103
$BBr_6C_2H_6P$ ... $[(CH_3)_2PBr_2][BBr_4]$ ... B: B Comp.SVol.4/4-83
BBr_6P ... Br_3P-BBr_3 ... B: B Comp.SVol.3/4-77/8
BBr_7CH_3P ... $[CH_3-PBr_3][BBr_4]$ ... B: B Comp.SVol.4/4-83
BBr_8P ... $[PBr_4][BBr_4]$ ... B: B Comp.SVol.4/4-81
$BCCl_2F_3HNS$ ... $Cl_2B-NH-SCF_3$ ... B: B Comp.SVol.4/4-59, 166
$BCCl_2H_3$ ... CH_3-BCl_2 ... B: B Comp.SVol.3/4-39/40
B: B Comp.SVol.4/4-41
– ... $CH_3-B^{35}Cl^{37}Cl$ ... B: B Comp.SVol.3/4-40
– ... $CH_3-{}^{10}BCl_2$ ... B: B Comp.SVol.3/4-40
$BCCl_2H_3O$ ... Cl_2B-OCH_3 ... B: B Comp.SVol.3/4-50
$BCCl_2H_6NSi$ ... $SiH_3(CH_3)NBCl_2$ ... Si: SVol.B4-318
$BCCl_3HN$ ... $Cl_3B-N\equiv CH$ ... B: B Comp.SVol.4/3b-148

$BCCl_3HN$ .... $Cl_3B-N{\equiv}CH$ .... B: B Comp.SVol.4/4-32
$BCCl_3H_4S$ .... $(CH_3)HS-BCl_3$ .... B: B Comp.SVol.3/4-136
– .... $(CH_3)HS-{}^{10}BCl_3$ .... B: B Comp.SVol.3/4-136
– .... $(CH_3)HS-{}^{11}BCl_3$ .... B: B Comp.SVol.3/4-136
$BCCl_5H_5P$ .... $[CH_3-PH_2Cl][BCl_4]$ .... B: B Comp.SVol.4/4-35
$BCF_2H_2^-$ .... $[F_2B{=}CH_2]^-$ .... B: B Comp.SVol.3/3-362
BCF_2H_3 .... CH_3-BF_2 .... B: B Comp.SVol.3/3-362
B: B Comp.SVol.4/3b-226
$BCF_2H_3N_2Si$ ..... $SiH_3(CN)NBF_2$ .... Si: SVol.B4-318
BCF_2H_3O .... F_2B-OCH_3 .... B: B Comp.SVol.3/3-366
B: B Comp.SVol.4/3b-232
BCF_2H_4N .... $F_2B-NH-CH_3$ .... B: B Comp.SVol.4/3b-243
– .... $F_2{}^{10}B-{}^{14}NH-{}^{12}CH_3$ .... B: B Comp.SVol.4/3b-243
– .... $F_2{}^{11}B-{}^{14}NH-{}^{12}CH_3$ .... B: B Comp.SVol.4/3b-243
BCF_2H_6NSi ...... $SiH_3(CH_3)NBF_2$ .... Si: SVol.B4-318
BCF_3HN .... $F_3B-N{\equiv}CH$ .... B: B Comp.SVol.4/3b-148
BCF_3H_2O .... F_3B-OCH_2 .... B: B Comp.SVol.4/3b-144
BCF_3H_3KO ...... $K[F_3B-OCH_3]$ .... B: B Comp.SVol.3/3-372
B: B Comp.SVol.4/3b-237
$BCF_3H_3O^-$ .... $[F_3B-OCH_3]^-$ .... B: B Comp.SVol.3/3-372
B: B Comp.SVol.4/3b-237
BCF_3H_4O .... $F_3B-O(H)-CH_3$ .... B: B Comp.SVol.3/3-286
B: B Comp.SVol.4/3b-143
BCF_3H_5N .... $F_3B-NH_2-CH_3$ .... B: B Comp.SVol.3/3-293
B: B Comp.SVol.4/3b-148, 243
BCF_3H_7NO ...... $NH_4[BF_3(OCH_3)]$ .... B: B Comp.SVol.3/3-372
$BCF_3H_9NSi_2$ ..... $(SiH_3)_2NCH_3 \cdot BF_3$ .... Si: SVol.B4-236
BCF_3N^- .... $[CN-BF_3]^-$ .... B: B Comp.SVol.3/3-378
– .... $[NC-BF_3]^-$ .... B: B Comp.SVol.3/3-378
BCF_3O .... F_3B-CO .... B: B Comp.SVol.3/3-295
B: B Comp.SVol.4/3b-107, 145
– .... F_3B-OC .... B: B Comp.SVol.4/3b-145
$BCF_4H_5N_2O$ ..... $H[BF_4] \cdot O{=}C(NH_2)_2$ .... B: B Comp.SVol.4/3b-165
$BCF_4H_5N_2S$ ...... $H[BF_4] \cdot S{=}C(NH_2)_2$ .... B: B Comp.SVol.4/3b-165
BCF_4H_6N .... $[CH_3-NH_3][BF_4]$ .... B: B Comp.SVol.3/3-343
B: B Comp.SVol.4/3b-206, 208, 209
$BCF_4H_6N_3$ .... $[C(NH_2)_3][BF_4]$ .... B: B Comp.SVol.4/3b-206/7
$BCF_4H_6N_3S$ ...... $H[BF_4] \cdot NH_2-NH-CS-NH_2$ .... B: B Comp.SVol.4/3b-165
BCF_4N_3 .... $[N{\equiv}C-N{\equiv}N][BF_4]$ .... B: B Comp.SVol.4/3b-218
BCH_2N .... $(BH_2CN)_n$ .... B: B Comp.SVol.3/3-184
BCH_3I_2O .... I_2B-OCH_3 .... B: B Comp.SVol.4/4-114
BCH_3N^- .... $[H_3B-C{\equiv}N]^-$ .... B: B Comp.SVol.3/3-212
– .... $[H_3B-N{\equiv}C]^-$ .... B: B Comp.SVol.3/3-212
BCH_3NNaS ...... $Na[H_3B-SCN]$ .... B: B Comp.SVol.4/4-151
BCH_3S .... CH_3BS .... B: B Comp.SVol.3/4-110
BCH_3S^+ .... $[CH_3BS]^+$ .... B: B Comp.SVol.4/4-129
BCH_4N .... $(-BH-NH-CH_2-)$ .... B: B Comp.SVol.3/3-162

Formula	Compound	Reference
BCH_4N	$HC{\equiv}N-BH_3$	B: B Comp.SVol.3/3-172
–	$H_2B-N{=}CH_2$	B: B Comp.SVol.3/3-172
		B: B Comp.SVol.4/3a-223/4
BCH_5N^+	$[H_2N{=}BCH_3]^+$	B: B Comp.SVol.3/3-206
$BCH_5NO_2^-$	$[H_3N-{}^{10}BH_2-COO]^-$	B: B Comp.SVol.4/3b-50
BCH_5N_2	H_3N-BH_2CN	B: B Comp.SVol.3/3-184
BCH_6N	$[-BH_2-CH_2-NH_2-]$	B: B Comp.SVol.4/3b-25
BCH_6NO_2	NH_3-BH_2-COOH	B: B Comp.SVol.3/3-185/6
		B: B Comp.SVol.4/3b-23
BCH_7N	CH_3NH-BH_3, radical	B: B Comp.SVol.4/3b-4
BCH_7N_2O	$NH_3-{}^{10}BH_2-C(O)-NH_2$	B: B Comp.SVol.4/3b-23
BCH_8N	$H_3B-NH_2-CH_3$	B: B Comp.SVol.3/3-172
		B: B Comp.SVol.4/3b-4
BCH_8NSi	$SiH_3(CH_3)NBH_2$	Si: SVol.B4-317
$BCH_{12}NSi_2$	$(SiH_3)_2NCH_3 \cdot BH_3$	Si: SVol.B4-236
$BCTh$	$ThBC$	Th: SVol.C6-132/4
$BC_2ClF_3H_2N$	$ClCH_2CN-BF_3$	B: B Comp.SVol.3/3-295
$BC_2ClF_3H_6N$	$(CH_3)_2NCl \cdot BF_3$	B: B Comp.SVol.4/4-72
$BC_2ClF_6H_2N_2S_2$	$ClB(NH-S-CF_3)_2$	B: B Comp.SVol.4/4-63, 166
$BC_2ClH_4O_2$	$ClB[-O-(CH_2)_2-O-]$	B: B Comp.SVol.3/4-50
$BC_2ClH_4S_2$	$ClB(-S-CH_2-CH_2-S-)$	B: B Comp.SVol.3/4-138/9
BC_2ClH_6	$(CH_3)_2BCl$	B: B Comp.SVol.3/4-42
		B: B Comp.SVol.4/4-45
$BC_2ClH_6O_2$	$(CH_3O)_2BCl$	B: B Comp.SVol.3/4-50
		B: B Comp.SVol.4/4-54/5
BC_2ClH_8S	$(CH_3)_2S-BH_2Cl$	B: B Comp.SVol.3/4-38, 137
		B: B Comp.SVol.4/4-39, 162
$BC_2Cl_2F_3HN$	$Cl_2CHCN-BF_3$	B: B Comp.SVol.3/3-295
$BC_2Cl_2F_3HNO$	$[CF_3-C({=}O)NH-BCl_2]_n$	F: PFHOrg.SVol.5-25, 59
$BC_2Cl_2F_6NS_2$	$Cl_2B-N(SCF_3)_2$	B: B Comp.SVol.4/4-59, 166
$BC_2Cl_2H_3$	$H_2C{=}CHBCl_2$	B: B Comp.SVol.3/4-40
$BC_2Cl_2H_5$	$C_2H_5-BCl_2$	B: B Comp.SVol.3/4-40
		B: B Comp.SVol.4/4-41
$BC_2Cl_2H_5O$	$Cl_2B-O-C_2H_5$	B: B Comp.SVol.4/4-54
$BC_2Cl_2H_6N$	$Cl_2B-N(CH_3)_2$	B: B Comp.SVol.3/4-53
		B: B Comp.SVol.4/4-57
$BC_2Cl_2H_7S$	$(CH_3)_2S-BHCl_2$	B: B Comp.SVol.3/4-37, 137
		B: B Comp.SVol.4/4-39, 161
$BC_2Cl_3FH_6N$	$(CH_3)_2NF \cdot BCl_3$	B: B Comp.SVol.4/4-71/2
$BC_2Cl_3H_2$	$CHCl{=}CH-BCl_2$	B: B Comp.SVol.3/4-41
		B: B Comp.SVol.4/4-41
$BC_2Cl_3H_6O$	$(CH_3)_2O-BCl_3$	B: B Comp.SVol.3/4-32
		B: B Comp.SVol.4/4-34
–	$(CD_3)_2O-BCl_3$	B: B Comp.SVol.3/4-32
$BC_2Cl_3H_6S$	$(CH_3)_2S-BCl_3$	B: B Comp.SVol.4/4-34, 161
–	$C_2H_5-SH-{}^{10}BCl_3$	B: B Comp.SVol.3/4-136
–	$C_2H_5-SH-{}^{11}BCl_3$	B: B Comp.SVol.3/4-136
$BC_2Cl_3H_6Se$	$(CH_3)_2Se-BCl_3$	B: B Comp.SVol.4/4-34
$BC_2Cl_3H_6Te$	$(CH_3)_2Te-BCl_3$	B: B Comp.SVol.4/4-34
$BC_2Cl_4H_5O$	$ClCH_2-O-CH_3 \cdot BCl_3$	B: B Comp.SVol.4/4-34

$BC_2Cl_4H_5S$ ...... $CH_3(CH_2Cl)S-BCl_3$ ...... B: B Comp.SVol.3/4-136
$BC_2Cl_4H_6N$ ...... $(CH_3)_2NCl \cdot BCl_3$ ...... B: B Comp.SVol.4/4-32/3
$BC_2Cl_5H_7P$ ...... $[(CH_3)_2PHCl][BCl_4]$ ...... B: B Comp.SVol.4/4-35
$BC_2CuF_3H_6Li$ .... $(CH_3)_2CuLi \cdot BF_3$ ...... B: B Comp.SVol.4/3b-151
BC_2FH_6 .... $(CH_3)_2BF$ ...... B: B Comp.SVol.3/3-363
$BC_2FH_6O_2$ .... $(CH_3O)_2BF$ ...... B: B Comp.SVol.3/3-367
B: B Comp.SVol.4/3b-232
BC_2F_2H .... $HC{\equiv}C-BF_2$ ...... B: B Comp.SVol.4/3b-226
$BC_2F_2H_3$ .... $H_2C{=}CH-BF_2$ ...... B: B Comp.SVol.3/3-362
B: B Comp.SVol.4/3b-226
$BC_2F_2H_5$ .... $C_2H_5-BF_2$ ...... B: B Comp.SVol.3/3-363
$BC_2F_2H_5O$ .... $C_2H_5-O-BF_2$ ...... B: B Comp.SVol.3/3-366
$BC_2F_2H_6KO_2$ ..... $K[BF_2(OCH_3)_2]$ ...... B: B Comp.SVol.3/3-372
$BC_2F_2H_6N$ .... $F_2B-N(CH_3)_2$ ...... B: B Comp.SVol.4/3b-243
$BC_2F_2H_6O_2^-$ ..... $[F_2B(OCH_3)_2]^-$ ...... B: B Comp.SVol.3/3-372
$BC_2F_2H_{10}NO_2$ .... $NH_4[BF_2(OCH_3)_2]$ ...... B: B Comp.SVol.3/3-372
$BC_2F_3H_3N$ .... $F_3B-N{\equiv}C-CH_3$ ...... B: B Comp.SVol.3/3-295
B: B Comp.SVol.4/3b-148
$BC_2F_3H_3O_2^-$ ..... $[F_3B-OC(O)-CH_3]^-$ ...... B: B Comp.SVol.4/3b-237
$BC_2F_3H_3O_3^-$ ..... $[F_3B-OO-C(O)-CH_3]^-$ ...... B: B Comp.SVol.4/3b-237
$BC_2F_3H_4O$ .... $(-CH_2CH_2-)O-BF_3$ ...... B: B Comp.SVol.3/3-287
– .... $CH_3-CHO-BF_3$ ...... B: B Comp.SVol.4/3b-144
$BC_2F_3H_4O_2$ ...... $CH_3-COOH \cdot BF_3$ ...... B: B Comp.SVol.3/3-288
B: B Comp.SVol.4/3b-145
– .... $HCOO-CH_3 \cdot BF_3$ ...... B: B Comp.SVol.3/3-288
$BC_2F_3H_5N$ .... $(-CH_2-CH_2-NH-)BF_3$ ...... B: B Comp.SVol.3/3-294
$BC_2F_3H_6O$ .... $F_3B-O(CH_3)_2$ ...... B: B Comp.SVol.3/3-286
B: B Comp.SVol.4/3b-144
– .... $F_3B-O(H)-C_2H_5$ ...... B: B Comp.SVol.4/3b-143
$BC_2F_3H_6OS$ ...... $(CH_3)_2S(O)-BF_3$ ...... B: B Comp.SVol.3/3-285
B: B Comp.SVol.3/4-142
– .... $(CD_3)_2S(O)-BF_3$ ...... B: B Comp.SVol.3/4-142
$BC_2F_3H_6S$ .... $(CH_3)_2S-BF_3$ ...... B: B Comp.SVol.3/4-136
$BC_2F_3H_7N$ .... $(CH_3)_2NH \cdot BF_3$ ...... B: B Comp.SVol.3/3-294
B: B Comp.SVol.4/3b-148
– .... $C_2H_5-NH_2 \cdot BF_3$ ...... B: B Comp.SVol.3/3-293/4
B: B Comp.SVol.4/3b-149
$BC_2F_3H_7NO$ ..... $F_3B-NH_2-CH_2CH_2-OH$ ...... B: B Comp.SVol.3/3-293
B: B Comp.SVol.4/3b-149
$BC_2F_3H_8N_2$ ...... $(CH_2NH_2)_2 \cdot BF_3$ ...... B: B Comp.SVol.3/3-294
$BC_2F_3H_9NO$ ..... $[C_2H_5-NH_3][F_3B-OH]$ ...... B: B Comp.SVol.4/3b-149
$BC_2F_3H_9NSi$ ..... $SiH_3N(CH_3)_2 \cdot BF_3$ ...... Si: SVol.B4-171
$BC_2F_3N_2$ .... $NCCN-BF_3$ ...... B: B Comp.SVol.3/3-293
$BC_2F_4H_3O$ .... $[CH_3CO][BF_4]$ ...... B: B Comp.SVol.3/3-336
B: B Comp.SVol.4/3b-203
$BC_2F_4H_6N$ .... $F_3B-NF(CH_3)_2$ ...... B: B Comp.SVol.4/3b-149
$BC_2F_4H_6NO$ ..... $H[BF_4] \cdot CH_3-CO-NH_2$ ...... B: B Comp.SVol.4/3b-165
$BC_2F_4H_6NOS$ .... $[OSN(CH_3)_2][BF_4]$ ...... S: S-N Comp.6-289/93
$BC_2F_4H_6NOS_2$ .... $O{=}S{=}NS(CH_3)_2(BF_4)$ ...... S: S-N Comp.6-39
$BC_2F_4H_6N_3O_2$ .... $[F_2B-NH_2-C(O)-NH-C(O)-NH_2][HF_2]$ ...... B: B Comp.SVol.4/3b-250

Formula	Compound	Reference
$BC_2F_4H_8N$	$[(CH_3)_2NH_2][BF_4]$	B: B Comp.SVol.3/3-343
		B: B Comp.SVol.4/3b-208, 210
–	$[C_2H_5-NH_3][BF_4]$	B: B Comp.SVol.3/3-343
		B: B Comp.SVol.4/3b-149, 207/8, 209
$BC_2F_4H_9O_2$	$[(CH_3OH)_2H][BF_4]$	B: B Comp.SVol.3/3-307
$BC_2F_4H_{10}NOS$	$[NH_4][BF_4] \cdot (CH_3)_2SO \cdot 0.6\ H_2O$	B: B Comp.SVol.3/3-342
$BC_2F_6H_2NO$	$CF_3-C(=O)NH_2 \cdot BF_3$	F: PFHOrg.SVol.5-17/8, 52
$BC_2F_6H_6NS$	$[F_2SN(CH_3)_2][BF_4]$	S: S-N Comp.8-74
$BC_2F_6N_7S_2$	$(N_3)_2B-N(S-CF_3)_2$	B: B Comp.SVol.4/3a-156
		B: B Comp.SVol.4/4-165/6
$BC_2F_8NS_2$	$F_2B-N(S-CF_3)_2$	B: B Comp.SVol.4/3b-243
		B: B Comp.SVol.4/4-166
BC_2F_9KN	$K[F_3B-N(CF_3)_2]$	B: B Comp.SVol.4/3b-250
$BC_2F_{11}Te$	$[(CF_3)_2TeF][BF_4]$	B: B Comp.SVol.4/3b-108
$BC_2F_{15}H_3NO_3Te_3$	$(F_5TeO)_3B \cdot CH_3CN$	B: B Comp.SVol.3/4-150
$BC_2H_5N_2$	$CH_2=N-BH-N=CH_2$	B: B Comp.SVol.4/3a-166, 167
$BC_2H_6I_2N$	$I_2BN(CH_3)_2$	B: B Comp.SVol.3/4-100
$BC_2H_6I_3S$	$(CH_3)_2S \cdot BI_3$	B: B Comp.SVol.4/4-161
BC_2H_6N	$CH_3-B=N-CH_3$	B: B Comp.SVol.3/3-141
		B: B Comp.SVol.4/3a-211/2
–	$CH_3-C{\equiv}N-BH_3$	B: B Comp.SVol.3/3-172
–	$CH_3-N{\equiv}C-BH_3$	B: B Comp.SVol.3/3-172
$BC_2H_6N_3$	$(CH_3)_2B-N_3$	B: B Comp.SVol.4/3a-218
$BC_2H_6NaS_2$	$Na[H_2B(-S-CH_2-CH_2-S-)]$	B: B Comp.SVol.3/4-117
$BC_2H_7I_2S$	$(CH_3)_2S-BHI_2$	B: B Comp.SVol.3/4-137
		B: B Comp.SVol.4/4-112
$BC_2H_7I_3P$	$(CH_3)_2PH-BI_3$	B: B Comp.SVol.3/4-98/9
–	$(CH_3)_2PD-BI_3$	B: B Comp.SVol.3/4-98/9
–	$(CD_3)_2PH-BI_3$	B: B Comp.SVol.3/4-98/9
$BC_2H_7N^-$	$[H_3BCH_2CHNH]^-$	B: B Comp.SVol.4/3b-6
BC_2H_7OS	$O(CH_2)_2S-BH_3$	B: B Comp.SVol.3/4-116
BC_2H_8LiS	$LiCH_2(CH_3)S-BH_3$	B: B Comp.SVol.3/4-116
BC_2H_8N	$(CH_3)_2B-NH_2$	B: B Comp.SVol.3/3-147
		B: B Comp.SVol.4/3a-225
–	$(CH_3)_2N-BH_2$	B: B Comp.SVol.3/3-147
		B: B Comp.SVol.4/3a-224
$BC_2H_8NO_2$	$CH_3-NH_2-BH_2-COOH$	B: B Comp.SVol.3/3-186
		B: B Comp.SVol.4/3b-24
–	$NH_3-BH_2-COO-CH_3$	B: B Comp.SVol.3/3-186/7
$BC_2H_8NSi_2$	$(SiH_3)_2N-B(-CH=CH-)$	Si: SVol.B4-149
$BC_2H_8N_2^+$	$[H_2N=B=N(CH_3)_2]^+$	B: B Comp.SVol.3/3-206
BC_2H_8S	$(CH_3)_2S-BH_2$, radical	B: B Comp.SVol.4/4-134
BC_2H_9N	$H_2B-NH(CH_3)_2$, radical	B: B Comp.SVol.4/3b-5, 8
–	$H_3B-N(CH_3)_2$, radical	B: B Comp.SVol.4/3b-5
BC_2H_9S	$(CH_3)_2S-BH_3$	B: B Comp.SVol.3/4-116
		B: B Comp.SVol.4/4-133/4
$BC_2H_{10}N$	$(CH_3)_2HN-BH_3$	B: B Comp.SVol.3/3-173/4

Formula	Compound	Reference
$BC_2H_{10}N$	$(CH_3)_2HN-BH_3$	B: B Comp.SVol.4/3b-5/6
–	$C_2H_5-NH_2-BH_3$	B: B Comp.SVol.3/3-173
$BC_2H_{10}NNa$	$Na[(CH_3)_2HNBH_3]$	B: B Comp.SVol.3/3-211
$BC_2H_{12}NSi$	$SiH_3N(CH_3)_2 \cdot BH_3$	Si: SVol.B4-171
BC_2Th	$ThBC_2$	Th: SVol.C6-132/4
$BC_3ClF_2H_6O_2$	$CH_3C(O)OCH_3-BClF_2$	B: B Comp.SVol.3/4-64
$BC_3ClF_2H_9N$	$N(CH_3)_3 \cdot BClF_2$	B: B Comp.SVol.3/4-65/7
$BC_3ClH_6O_2$	$2-Cl-1,3,2-O_2BC_3H_6$	B: B Comp.SVol.3/4-52
–	$2-Cl-4-CH_3-1,3,2-O_2BC_2H_3$	B: B Comp.SVol.3/4-52
$BC_3ClH_6S_2$	$2-Cl-1,3,2-S_2BC_3H_6$	B: B Comp.SVol.4/4-163
BC_3ClH_9N	$(CH_3)ClBN(CH_3)_2$	B: B Comp.SVol.3/4-55
$BC_3ClH_{11}N$	$(CH_3)_3N-BH_2Cl$	B: B Comp.SVol.3/4-37
		B: B Comp.SVol.4/4-39
$BC_3ClH_{11}P$	$(CH_3)_3P-BH_2Cl$	B: B Comp.SVol.4/4-39
$BC_3Cl_2FH_6O_2$	$CH_3C(O)OCH_3-BCl_2F$	B: B Comp.SVol.3/4-64
$BC_3Cl_2FH_9N$	$(CH_3)_3N-BCl_2F$	B: B Comp.SVol.3/4-65/7
$BC_3Cl_2H_5$	$c-C_3H_5-BCl_2$	B: B Comp.SVol.3/4-41
$BC_3Cl_2H_7$	$n-C_3H_7-BCl_2$	B: B Comp.SVol.3/4-41
		B: B Comp.SVol.4/4-42
–	$i-C_3H_7-BCl_2$	B: B Comp.SVol.3/4-41
		B: B Comp.SVol.4/4-42
$BC_3Cl_2H_9NOP$	$(CH_3)_2B-N(CH_3)POCl_2$	B: B Comp.SVol.3/3-154
$BC_3Cl_2H_{10}N$	$(CH_3)_3N-BHCl_2$	B: B Comp.SVol.3/4-37
$BC_3Cl_3H_5N$	$C_2H_5CN-BCl_3$	B: B Comp.SVol.3/4-32
$BC_3Cl_3H_6O_2$	$CH_3C(O)OCH_3 \cdot BCl_3$	B: B Comp.SVol.3/4-52
$BC_3Cl_3H_7NO$	$(CH_3)_2N-CHO-BCl_3$	B: B Comp.SVol.4/4-34
$BC_3Cl_3H_8S$	$n-C_3H_7-SH-{}^{10}BCl_3$	B: B Comp.SVol.3/4-136
–	$n-C_3H_7-SH-{}^{11}BCl_3$	B: B Comp.SVol.3/4-136
$BC_3Cl_3H_9N$	$(CH_3)_3N-BCl_3$	B: B Comp.SVol.3/4-32
		B: B Comp.SVol.4/4-32
$BC_3Cl_3H_{11}PSn$	$(CH_3)_3Sn-PH_2-BCl_3$	Sn: Org.Comp.19-162
$BC_3Cl_5H_9P$	$[(CH_3)_3PCl][BCl_4]$	B: B Comp.SVol.4/4-35
$BC_3FH_9KO_3$	$K[BF(OCH_3)_3]$	B: B Comp.SVol.3/3-372
$BC_3FH_9O_3^-$	$[BF(OCH_3)_3]^-$	B: B Comp.SVol.3/3-372
		B: B Comp.SVol.4/3b-237
$BC_3FH_{13}NO_3$	$NH_4[BF(OCH_3)_3]$	B: B Comp.SVol.3/3-372
$BC_3F_2H_9IN$	$N(CH_3)_3 \cdot BF_2I$	B: B Comp.SVol.3/4-65/7
$BC_3F_3GaH_{12}N$	$GaH_3[N(CH_3)_3] \cdot BF_3$	Ga: SVol.D1-227
$BC_3F_3H_3N$	$H_2C{=}CHCN-BF_3$	B: B Comp.SVol.3/3-295
$BC_3F_3H_4O$	$F_3B-OCH-CH{=}CH_2$	B: B Comp.SVol.4/3b-144
$BC_3F_3H_5N$	$C_2H_5CN-BF_3$	B: B Comp.SVol.3/3-295
$BC_3F_3H_6O$	$F_3B-OC(CH_3)_2$	B: B Comp.SVol.3/3-285
–	$F_3B-OCH-C_2H_5$	B: B Comp.SVol.4/3b-144
$BC_3F_3H_6O_2$	$F_3B-OC(OCH_3)-CH_3$	B: B Comp.SVol.3/3-285, 288
		B: B Comp.SVol.4/3b-145
–	$F_3B-OC(O-C_2H_5)-H$	B: B Comp.SVol.3/3-288
$BC_3F_3H_8O$	$i-C_3H_7-OH-BF_3$	B: B Comp.SVol.3/3-287
–	$(CD_3)_2CH-OH-BF_3$	B: B Comp.SVol.3/3-287
$BC_3F_3H_9N$	$(CH_3)_3N-BF_3$	B: B Comp.SVol.3/4-65/7

Formula	Compound		Reference
$BC_3F_3H_9N$	$(CH_3)_3N-BF_3$	B:	B Comp.SVol.4/3b-148, 149
–	$C_2H_5-NH(CH_3)-BF_3$	B:	B Comp.SVol.3/3-294
$BC_3F_3H_9O_4P$	$(CH_3)_3PO_4-BF_3$	B:	B Comp.SVol.3/3-285
$BC_3F_4H_3Te_2$	$[1,2-Te_2C_3H_3][BF_4]$	B:	B Comp.SVol.3/4-150/1
$BC_3F_4H_5N_4OS$	$[(4-O=S=N)(1-CH_3)C_2H_2N_3-1,2,4][BF_4]$	S:	S-N Comp.6-74
$BC_3F_4H_8NOS$	$[OSN(CH_3)C_2H_5][BF_4]$	S:	S-N Comp.6-289, 296
$BC_3F_4H_9N_2OS$	$[O=S=NN(CH_3)_3][BF_4]$	S:	S-N Comp.6-68/9
$BC_3F_4H_9O$	$[(CH_3)_3O][BF_4]$	B:	B Comp.SVol.3/3-334/5
		B:	B Comp.SVol.4/3b-202/3
$BC_3F_4H_9S$	$[(CH_3)_3S][BF_4]$	B:	B Comp.SVol.3/4-140
$BC_3F_4H_9S_2$	$[(CH_3)_2S-S-CH_3][BF_4]$	B:	B Comp.SVol.4/4-163
$BC_3F_4H_{10}N$	$[(CH_3)_3NH][BF_4]$	B:	B Comp.SVol.3/3-343
		B:	B Comp.SVol.4/3b-208, 210
–	$[n-C_3H_7-NH_3][BF_4]$	B:	B Comp.SVol.4/3b-207, 209
–	$[i-C_3H_7-NH_3][BF_4]$	B:	B Comp.SVol.4/3b-207
$BC_3F_5H_7N$	$(CH_3)_2NH-BF_2-CF_3$	B:	B Comp.SVol.4/3b-226
BC_3F_7	$F_2C=C(BF_2)CF_3$	B:	B Comp.SVol.3/3-363
$BC_3F_7H_5N$	$[CH_3N(CHF_2)=CHF][BF_4]$	B:	B Comp.SVol.3/3-252
$BC_3F_9H_3N_3S_3$	$B(NH-SCF_3)_3$	B:	B Comp.SVol.4/3a-153
		B:	B Comp.SVol.4/4-166
$BC_3F_9O_9S_3$	$B[O-S(=O)_2-CF_3]_3$	B:	B Comp.SVol.3/4-142
		B:	B Comp.SVol.4/4-165
$BC_3F_9Se_3$	$B(SeCF_3)_3$	B:	B Comp.SVol.4/4-170
$BC_3F_{12}H_9O_6Te_3$	$B(OTeF_4OCH_3)_3$	B:	B Comp.SVol.3/4-150
$BC_3H_5N_2$	$1-H_2B-1,2-N_2C_3H_3$	B:	B Comp.SVol.4/3a-224
$BC_3H_5S_2$	$CH_3B(-S-CH=CH-S-)$	B:	B Comp.SVol.3/4-114
$BC_3H_6NS_3$	$CH_3B(-S-CS-NCH_3-S-)$	B:	B Comp.SVol.3/4-128
$BC_3H_6N_3$	$B(N=CH_2)_3$	B:	B Comp.SVol.4/3a-152/3
$BC_3H_6N_3O_2$	$CH_3B(-NH-CO-NH-CO-NH-)$	B:	B Comp.SVol.3/3-114
$BC_3H_7S_2$	$CH_3B(-S-CH_2-CH_2-S-)$	B:	B Comp.SVol.3/4-112
BC_3H_8N	$H_2C=B-N(CH_3)_2$	B:	B Comp.SVol.4/3a-225
BC_3H_8NS	$3-CH_3-1,3,2-SNBC_2H_5$	B:	B Comp.SVol.4/4-145/6
$BC_3H_9I_3N$	$(CH_3)_3N-BI_3$	B:	B Comp.SVol.3/4-98
		B:	B Comp.SVol.4/4-110
BC_3H_9N	$(-CHCH_3-CH_2-)HN-BH_2$, radical	B:	B Comp.SVol.3/3-174
–	$H_2B-NH-CH(CH_3)-CH_2$, radical	B:	B Comp.SVol.3/3-147
$BC_3H_9N_2$	$(CH_3)_2HN-BH_2CN$	B:	B Comp.SVol.3/3-184
$BC_3H_9O_9S_3$	$B[O-S(=O)_2-CH_3]_3$	B:	B Comp.SVol.4/4-138
$BC_3H_9S_2$	$1,3-S_2C_3H_6 \cdot BH_3$	B:	B Comp.SVol.4/4-134
–	$CH_3-B(SCH_3)_2$	B:	B Comp.SVol.3/4-110
$BC_3H_9S_3$	$B(SCH_3)_3$	B:	B Comp.SVol.3/4-110
$BC_3H_9Se_3$	$B(SeCH_3)_3$	B:	B Comp.SVol.3/4-147
		B:	B Comp.SVol.4/4-170
$BC_3H_{10}N$	$(CH_3)_2B-NH-CH_3$	B:	B Comp.SVol.3/3-91/2, 148
		B:	B Comp.SVol.4/3a-229
–	$H_3B-N(CH_3)[-CH_2CH_2-]$	B:	B Comp.SVol.4/3b-11

Formula	Compound	Reference
$BC_3H_{10}N$	$H_3B\text{-}NH[\text{-}CHCH_3\text{-}CH_2\text{-}]$	B: B Comp.SVol.3/3-174
$BC_3H_{10}NO$	$(CH_3)_2N(CHO)\text{-}BH_3$	B: B Comp.SVol.3/3-176
$BC_3H_{10}NO_2$	$(CH_3)_2NH\text{-}BH_2\text{-}COOH$	B: B Comp.SVol.3/3-186
		B: B Comp.SVol.4/3b-24
–	$CH_3\text{-}NH_2\text{-}BH_2\text{-}COO\text{-}CH_3$	B: B Comp.SVol.3/3-187
		B: B Comp.SVol.4/3b-24
$BC_3H_{10}NS$	$1,3\text{-}SNC_3H_7 \cdot BH_3$	B: B Comp.SVol.4/3b-7
		B: B Comp.SVol.4/4-147/8
–	$(CH_3)_2N\text{-}CH{=}S \cdot BH_3$	B: B Comp.SVol.3/3-176
$BC_3H_{11}IN$	$(CH_3)_3N\text{-}BH_2I$	B: B Comp.SVol.3/4-100
		B: B Comp.SVol.4/4-112
$BC_3H_{11}IP$	$(CH_3)_3P\text{-}BH_2I$	B: B Comp.SVol.4/4-112
$BC_3H_{11}N$	$(CH_3)_3N\text{-}BH_2$, radical	B: B Comp.SVol.4/3b-26
$BC_3H_{11}N_2$	$CH_3B(NHCH_3)_2$	B: B Comp.SVol.3/3-91/2, 103
–	$H_3B\text{-}NH[\text{-}CH_2CH_2\text{-}NH\text{-}CH_2\text{-}]$	B: B Comp.SVol.4/3b-7
$BC_3H_{11}N_2O$	$H_3N\text{-}BH_2\text{-}C(O)NH\text{-}C_2H_5$	B: B Comp.SVol.3/3-186/7
$BC_3H_{12}N$	$(CH_3)_3N\text{-}BH_3$	B: B Comp.SVol.3/3-174/5
		B: B Comp.SVol.4/3b-7/8
–	$n\text{-}C_3H_7\text{-}NH_2\text{-}BH_3$	B: B Comp.SVol.3/3-173
–	$i\text{-}C_3H_7\text{-}NH_2\text{-}BH_3$	B: B Comp.SVol.3/3-173
		B: B Comp.SVol.4/3b-4
–	$H_3N\text{-}B(CH_3)_3$	B: B Comp.SVol.4/3b-27
$BC_3H_{12}N_3$	$B(NH\text{-}CH_3)_3$	B: B Comp.SVol.3/3-91/2
		B: B Comp.SVol.4/3a-153
$BC_4ClF_2H_8O_2$	$CH_3COOC_2H_5 \cdot BClF_2$	B: B Comp.SVol.3/4-64
$BC_4ClF_2H_{11}N$	$N(CH_3)_2C_2H_5 \cdot BClF_2$	B: B Comp.SVol.3/4-65/7
$BC_4ClF_6H_2N_2O_2$	$[CF_3\text{-}C({=}O)NH]_2BCl$	F: PFHOrg.SVol.5-25, 59
$BC_4ClF_6H_7N$	$(CF_3)_2BCl \cdot NH(CH_3)_2$	B: B Comp.SVol.4/4-70
$BC_4ClF_{12}N_2S_4$	$[(CF_3S)_2N]_2BCl$	B: B Comp.SVol.4/4-63, 165/6
BC_4ClH_6	$(H_2C{=}CH)_2BCl$	B: B Comp.SVol.3/4-43
–	$[\text{-}CH_2CH_2CH{=}CH\text{-}]BCl$	B: B Comp.SVol.4/4-47
BC_4ClH_{10}	$(C_2H_5)_2BCl$	B: B Comp.SVol.3/4-42
		B: B Comp.SVol.4/4-45
$BC_4ClH_{10}N_2$	$1,3\text{-}(CH_3)_2\text{-}2\text{-}Cl\text{-}1,3,2\text{-}N_2BC_2H_4$	B: B Comp.SVol.3/4-58
		B: B Comp.SVol.4/4-62
$BC_4ClH_{10}O$	$C_4H_8O\text{-}BH_2Cl$	B: B Comp.SVol.3/4-38
$BC_4ClH_{10}O_2$	$ClB(O\text{-}C_2H_5)_2$	B: B Comp.SVol.3/4-51
		B: B Comp.SVol.4/4-55
$BC_4ClH_{12}N_2$	$ClB[N(CH_3)_2]_2$	B: B Comp.SVol.3/4-58
		B: B Comp.SVol.4/4-62
$BC_4ClH_{12}O$	$(C_2H_5)_2O\text{-}BH_2Cl$	B: B Comp.SVol.3/4-38
		B: B Comp.SVol.4/4-39
$BC_4ClH_{14}N_2O_2$	$(CH_3)_3N\text{-}BH_2\text{-}C(O)\text{-}NHOH \cdot HCl$	B: B Comp.SVol.4/3b-24, 25
$BC_4Cl_2FH_8O_2$	$CH_3COOC_2H_5 \cdot BCl_2F$	B: B Comp.SVol.3/4-64
$BC_4Cl_2FH_{11}N$	$N(CH_3)_2C_2H_5 \cdot BCl_2F$	B: B Comp.SVol.3/4-65/7
$BC_4Cl_2F_6HN_2O$	$OBCl_2NHC(CF_3)NC(CF_3)$	F: PFHOrg.SVol.4-197, 211
$BC_4Cl_2H_9$	$n\text{-}C_4H_9\text{-}BCl_2$	B: B Comp.SVol.3/4-41
		B: B Comp.SVol.4/4-42

Formula	Compound		Reference
$BC_4Cl_2H_9$	$s\text{-}C_4H_9\text{-}BCl_2$	B:	B Comp.SVol.4/4-42
–	$t\text{-}C_4H_9\text{-}BCl_2$	B:	B Comp.SVol.3/4-41
		B:	B Comp.SVol.4/4-42
$BC_4Cl_2H_{10}N$	$Cl_2B\text{-}N(C_2H_5)_2$	B:	B Comp.SVol.3/4-53
		B:	B Comp.SVol.4/4-57
$BC_4Cl_2H_{11}O$	$(C_2H_5)_2O \cdot BHCl_2$	B:	B Comp.SVol.4/4-39
$BC_4Cl_2H_{12}NSi$	$Cl_2B\text{-}N(CH_3)\text{-}Si(CH_3)_3$	B:	B Comp.SVol.4/4-59
$BC_4Cl_3F_3H_9O_3SSi$	$(CH_3)_3SiO(BCl_3)SO_2CF_3$	B:	B Comp.SVol.3/4-32
$BC_4Cl_3H_8S$	$SC_4H_8 \cdot BCl_3$	B:	B Comp.SVol.4/4-161
$BC_4Cl_3H_{10}S$	$(C_2H_5)_2S\text{-}^{10}BCl_3$	B:	B Comp.SVol.3/4-136
–	$(C_2H_5)_2S\text{-}^{11}BCl_3$	B:	B Comp.SVol.3/4-136
$BC_4Cl_5H_{10}S_2$	$[H_3C\text{-}S\text{-}CH_2\text{-}S(CH_3)CH_2Cl][BCl_4]$	B:	B Comp.SVol.3/4-34
$BC_4Cl_6H_{10}P$	$[(C_2H_5)_2PCl_2][BCl_4]$	B:	B Comp.SVol.3/4-34
BC_4FH_6	$(CH_2{=}CH)_2BF$	B:	B Comp.SVol.3/3-363
$BC_4FH_{10}N_2$	$(\text{-}NCH_3\text{-}CH_2\text{-}CH_2\text{-}NCH_3\text{-})BF$	B:	B Comp.SVol.3/3-377
$BC_4F_2H_9$	$n\text{-}C_4H_9\text{-}BF_2$	B:	B Comp.SVol.3/3-363
–	$t\text{-}C_4H_9\text{-}BF_2$	B:	B Comp.SVol.3/3-363
$BC_4F_2H_{10}N$	$F_2B\text{-}N(C_2H_5)_2$	B:	B Comp.SVol.3/3-375
$BC_4F_2H_{11}IN$	$N(CH_3)_2(C_2H_5) \cdot BF_2I$	B:	B Comp.SVol.3/4-65/7
$BC_4F_2H_{12}NSi$	$F_2B\text{-}N(CH_3)\text{-}Si(CH_3)_3$	B:	B Comp.SVol.4/3b-244
$BC_4F_3H_4N_2$	$1,3\text{-}N_2C_4H_4 \cdot BF_3$	B:	B Comp.SVol.4/3b-149
–	$1,4\text{-}N_2C_4H_4 \cdot BF_3$	B:	B Comp.SVol.4/3b-149
$BC_4F_3H_6O_2$	$[\text{-}CH_2CH_2CH_2\text{-}O\text{-}]CO\text{-}BF_3$	B:	B Comp.SVol.3/3-285
$BC_4F_3H_6O_3$	$[\text{-}O\text{-}CH_2CH_2CH_2\text{-}O\text{-}]CO\text{-}BF_3$	B:	B Comp.SVol.3/3-285
–	$[CH_3C(O)]_2O\text{-}BF_3$	B:	B Comp.SVol.3/3-285
$BC_4F_3H_8O$	$C_2H_5\text{-}C(CH_3)O\text{-}BF_3$	B:	B Comp.SVol.3/3-285
–	$C_4H_8O\text{-}BF_3$	B:	B Comp.SVol.3/3-285, 287
		B:	B Comp.SVol.4/3b-144
$BC_4F_3H_8O_2$	$C_4H_8O_2\text{-}BF_3$	B:	B Comp.SVol.3/3-285
–	$F_3B\text{-}OC(O\text{-}C_2H_5)\text{-}CH_3$	B:	B Comp.SVol.3/3-285, 288
		B:	B Comp.SVol.4/3b-145
$BC_4F_3H_8O_4$	$F_3B \cdot 2\ HOOC\text{-}CH_3$	B:	B Comp.SVol.3/3-288
		B:	B Comp.SVol.4/3b-145
$BC_4F_3H_{10}O$	$(C_2H_5)_2O\text{-}BF_3$	B:	B Comp.SVol.3/3-267/9, 285
		B:	B Comp.SVol.4/3b-123/7
–	$n\text{-}C_4H_9\text{-}OH\text{-}BF_3$	B:	B Comp.SVol.3/3-287
		B:	B Comp.SVol.4/3b-143
$BC_4F_3H_{11}N$	$F_3B\text{-}N(CH_3)_2\text{-}C_2H_5$	B:	B Comp.SVol.3/4-65/7
–	$F_3B\text{-}NH(C_2H_5)_2$	B:	B Comp.SVol.3/3-294
		B:	B Comp.SVol.4/3b-149
–	$F_3B\text{-}NH_2\text{-}C_4H_9\text{-}n$	B:	B Comp.SVol.4/3b-149
–	$F_3B\text{-}NH_2\text{-}C_4H_9\text{-}i$	B:	B Comp.SVol.4/3b-149
–	$F_3B\text{-}NH_2\text{-}C_4H_9\text{-}t$	B:	B Comp.SVol.3/3-94
		B:	B Comp.SVol.4/3b-149
$BC_4F_3H_{11}NS$	$(CH_3)_3N\text{-}BH_2\text{-}SCF_3$	B:	B Comp.SVol.4/3b-23
$BC_4F_3H_{11}NSe$	$(CH_3)_3N\text{-}BH_2\text{-}SeCF_3$	B:	B Comp.SVol.4/3b-23
$BC_4F_4H_5Te_2$	$[1,2\text{-}Te_2C_3H_2\text{-}3\text{-}CH_3][BF_4]$	B:	B Comp.SVol.3/4-150/1

$BC_4F_4H_7S_2$ ...... $[-CH_2CH_2CH_2-SCHS-][BF_4]$ ...... B: B Comp.SVol.4/4-163
$BC_4F_4H_8N$ .... $[CH_3-C{\equiv}N-C_2H_5][BF_4]$ ...... B: B Comp.SVol.4/3b-210
$BC_4F_4H_{10}NOS$.... $[OSN(C_2H_5)_2][BF_4]$ ...... S: S-N Comp.6-289, 294
$BC_4F_4H_{11}O$ ...... $[(CH_3)_2O-C_2H_5][BF_4]$ ...... B: B Comp.SVol.3/3-334
– .... $[HO(C_2H_5)_2][BF_4]$ ...... B: B Comp.SVol.4/3b-203
$BC_4F_4H_{12}N$ ...... $[(CH_3)_4N][BF_4]$ ...... B: B Comp.SVol.3/3-343
B: B Comp.SVol.4/3b-208, 210
– .... $[(C_2H_5)_2NH_2][BF_4]$ ...... B: B Comp.SVol.3/3-343
B: B Comp.SVol.4/3b-208, 210
– .... $[n-C_4H_9-NH_3][BF_4]$ ...... B: B Comp.SVol.4/3b-207, 209
– .... $[t-C_4H_9-NH_3][BF_4]$ ...... B: B Comp.SVol.4/3b-207
$BC_4F_5H_{12}N_2S$ .... $[(CH_3)_2NS(F)N(CH_3)_2][BF_4]$ ...... S: S-N Comp.8-182
$BC_4F_6H_6N$ .... $(CF_3)_2B-N(CH_3)_2$ ...... B: B Comp.SVol.4/3a-225/6
B: B Comp.SVol.4/3b-227
$BC_4F_6H_8NO$ ..... $HO-B(CF_3)_2-NH(CH_3)_2$ ...... B: B Comp.SVol.4/3b-69
$BC_4F_6H_8NOS$ .... $[F_2SN(CH_2)_4O][BF_4]$ ...... S: S-N Comp.8-76
$BC_4F_6H_{10}NS$ ..... $[F_2SN(C_2H_5)_2][BF_4]$ ...... S: S-N Comp.8-75
$BC_4F_7H_7N$ .... $(CH_3)_2NH-BF(CF_3)_2$ ...... B: B Comp.SVol.4/3b-227
$BC_4F_{12}H_2N_3S_4$ ... $(CF_3-S-NH)_2B-N(S-CF_3)_2$ ...... B: B Comp.SVol.4/3a-156
B: B Comp.SVol.4/4-166
$BC_4F_{12}N_5S_4$ ..... $(CF_3-S)_2N-B(N_3)-N(S-CF_3)_2$ ...... B: B Comp.SVol.4/3a-156
B: B Comp.SVol.4/4-165/6
$BC_4H_4NO_8$ ...... $NH_4[B(-O-CO-CO-O-)_2]$ ...... B: B Comp.SVol.3/3-227
$BC_4H_5O_2S$ .... $SC_4H_3-3-B(OH)_2$ ...... B: B Comp.SVol.3/4-121
B: B Comp.SVol.4/4-137
$BC_4H_5O_2Se$ ...... $SeC_4H_3-B(OH)_2$ ...... B: B Comp.SVol.4/4-174
$BC_4H_6N_3$ .... $1,2-N_2C_3H_4-1-BH_2CN$ ...... B: B Comp.SVol.3/3-184
$BC_4H_7N^-$ .... $[H_3B(NC_4H_4)]^-$ ...... B: B Comp.SVol.3/3-213
$BC_4H_7N_2$ .... $H_3B \cdot 1,2-N_2C_4H_4$ ...... B: B Comp.SVol.4/3b-14
$BC_4H_8I_3S$ .... $SC_4H_8 \cdot BI_3$ ...... B: B Comp.SVol.4/4-161
BC_4H_8N .... $(-BCH_3-NH-CH_2-CH{=}CH-)$ ...... B: B Comp.SVol.3/3-162
$BC_4H_8NS_2$ .... $(CH_3)_2NB(-S-CH{=}CH-S-)$ ...... B: B Comp.SVol.3/4-126
$BC_4H_8NS_3$ .... $CH_3B(-S-CS-NC_2H_5-S-)$ ...... B: B Comp.SVol.3/4-128
$BC_4H_8N_3$ .... $H_3B \cdot 1,3-N_2C_4H_3-2-NH_2$ ...... B: B Comp.SVol.4/3b-14
$BC_4H_8N_3O_2$ ...... $CH_3B(-NCH_3-CO-NH-CO-NH-)$ ...... B: B Comp.SVol.3/3-115
$BC_4H_9I_2$ .... $t-C_4H_9-BI_2$ ...... B: B Comp.SVol.4/4-112
$BC_4H_9N^-$ .... $[(CH_3)_3B-N{\equiv}C]^-$ ...... B: B Comp.SVol.3/3-212
$BC_4H_9N_2OS$ ..... $[-N(CH_3)-C(=S)-O-B(CH_3)-N(CH_3)-]$ ...... B: B Comp.SVol.4/3b-64
B: B Comp.SVol.4/4-152
$BC_4H_9N_2O_2$ ...... $[-N(CH_3)-C(=O)-O-B(CH_3)-N(CH_3)-]$ ...... B: B Comp.SVol.4/3b-64
$BC_4H_9N_2S_2$ ...... $[-N(CH_3)-C(=S)-S-B(CH_3)-N(CH_3)-]$ ...... B: B Comp.SVol.4/3a-251
B: B Comp.SVol.4/4-145
$BC_4H_9O_{39}SnW_{11}{}^{6-}$
$[C_4H_9SnO_5W_{11}BO_{34}]^{6-}$ ...... Sn: Org.Comp.17-50
$BC_4H_{10}I$ .... $(C_2H_5)_2BI$ ...... B: B Comp.SVol.4/4-113
$BC_4H_{10}N$ .... $C_2H_5-B{=}N-C_2H_5$ ...... B: B Comp.SVol.3/3-141
$BC_4H_{10}NO_2$ ...... $[-O-CH_2CH_2-O-]B-N(CH_3)_2$ ...... B: B Comp.SVol.3/3-150

Formula	Compound	Reference
$BC_4H_{10}NO_2$	$[-O-CH_2CH_2-O-]B-N(CH_3)_2$	B: B Comp.SVol.4/3b-63
$BC_4H_{10}NS_2$	$[-S-CH_2CH_2-S-]B-N(CH_3)_2$	B: B Comp.SVol.3/4-124/5
		B: B Comp.SVol.4/4-147
$BC_4H_{10}N_3OS$	$O=S=N-B(-NCH_3-CH_2-CH_2-NCH_3-)$	B: B Comp.SVol.3/3-97
		B: B Comp.SVol.3/4-134
		S: S-N Comp.6-80
$BC_4H_{10}N_5$	$N_3-B(-NCH_3-CH_2-CH_2-NCH_3-)$	B: B Comp.SVol.3/3-97
$BC_4H_{11}NO_2^-$	$[(CH_3)_3N-^{10}BH_2-COO]^-$	B: B Comp.SVol.4/3b-50
$BC_4H_{11}N_2$	$(CH_3)_3N-BH_2-CN$	B: B Comp.SVol.3/3-184
		B: B Comp.SVol.4/3b-22
–	$(CH_3)_3N-^{10}BH_2-CN$	B: B Comp.SVol.3/3-184
		B: B Comp.SVol.4/3b-23
–	$n-C_3H_7-NH_2-BH_2-CN$	B: B Comp.SVol.4/3b-22
–	$i-C_3H_7-NH_2-BH_2-CN$	B: B Comp.SVol.4/3b-22
–	$HB(-NCH_3-CH_2CH_2-NCH_3-)$	B: B Comp.SVol.3/3-113
		B: B Comp.SVol.4/3a-174
$BC_4H_{11}N_2S$	$(CH_3)_3N-BH_2-N=C=S$	B: B Comp.SVol.4/3b-23
$BC_4H_{11}N_8$	$HN_3-B(N_3)[-NCH_3-CH_2-CH_2-NCH_3-]$	B: B Comp.SVol.3/3-190
$BC_4H_{11}O$	$H_3B \cdot OC_4H_8$	B: B Comp.SVol.3/3-267
		B: B Comp.SVol.4/3b-123
$BC_4H_{11}S_2$	$2-CH_3-1,3-S_2C_3H_5 \cdot BH_3$	B: B Comp.SVol.4/4-134
$BC_4H_{12}IN_2$	$IB[N(CH_3)_2]_2$	B: B Comp.SVol.3/4-101
$BC_4H_{12}Li$	$Li[(n-C_4H_9)BH_3]$	B: B Comp.SVol.3/3-175
$BC_4H_{12}LiN_2$	$(CH_3)_2B-NLi-N(CH_3)_2$	B: B Comp.SVol.3/3-153
$BC_4H_{12}N$	$(CH_3)_2B-N(CH_3)_2$	B: B Comp.SVol.3/3-148
		B: B Comp.SVol.4/3a-225
–	$(C_2H_5)_2B-NH_2$	B: B Comp.SVol.3/3-148
		B: B Comp.SVol.4/3a-225
–	$H_2B-NH-C_4H_9-t$	B: B Comp.SVol.4/3a-224
–	$H_3B \cdot NC_4H_9$	B: B Comp.SVol.4/3b-7
$BC_4H_{12}NO$	$1,4-ONC_4H_9 \cdot BH_3$	B: B Comp.SVol.3/3-174
–	$CH_3O-B(CH_3)-N(CH_3)_2$	B: B Comp.SVol.4/3b-61
$BC_4H_{12}NOS$	$(CH_3)_2B-N(CH_3)-S(=O)CH_3$	B: B Comp.SVol.3/3-154
		B: B Comp.SVol.3/4-134
$BC_4H_{12}NO_2$	$(CH_3)_2HN-BH_2-COO-CH_3$	B: B Comp.SVol.3/3-187
–	$(CH_3)_2NH-BH_2-COO-CH_3$	B: B Comp.SVol.4/3b-24
–	$(CH_3)_3N-BH_2-COOH$	B: B Comp.SVol.3/3-186
		B: B Comp.SVol.4/3b-24
–	$(CH_3)_3N-^{10}BH_2-COOH$	B: B Comp.SVol.4/3b-24
–	$(CH_3O)_2B-N(CH_3)_2$	B: B Comp.SVol.3/3-149
–	$n-C_3H_7-NH_2-BH_2-COOH$	B: B Comp.SVol.4/3b-24
–	$i-C_3H_7-NH_2-BH_2-COOH$	B: B Comp.SVol.4/3b-24
$BC_4H_{12}NS$	$2-CH_3-1,3-SNC_3H_6 \cdot BH_3$	B: B Comp.SVol.4/3b-7
		B: B Comp.SVol.4/4-147/8
–	$(CH_3)_2N-B(SCH_3)-CH_3$	B: B Comp.SVol.4/4-139
$BC_4H_{12}NS_2$	$(CH_3S)_2BN(CH_3)_2$	B: B Comp.SVol.3/4-122
$BC_4H_{12}N_2^+$	$[(CH_3)_2N=B=N(CH_3)_2]^+$	B: B Comp.SVol.3/3-206
$BC_4H_{12}N_2PS$	$CH_3-B[-N(CH_3)-P(CH_3)(=S)-N(CH_3)-]$	B: B Comp.SVol.4/3a-207/8
$BC_4H_{12}N_3$	$(t-C_4H_9)N_3-BH_3$, radical	B: B Comp.SVol.3/3-172
$BC_4H_{12}N_3OS$	$[(CH_3)_2N]_2B-N=S=O$	B: B Comp.SVol.3/3-97

Formula	Compound	Reference
$BC_4H_{12}N_3OS$	$[(CH_3)_2N]_2B\text{-}N{=}S{=}O$	B: B Comp.SVol.3/4-134
		S: S-N Comp.6-80
$BC_4H_{12}O_4^-$	$[B(OCH_3)_4]^-$	B: B Comp.SVol.4/3b-237
$BC_4H_{12}S$	$(C_2H_5)_2S \cdot BH_2$, radical	B: B Comp.SVol.4/4-134
$BC_4H_{13}N_2$	$HB[N(CH_3)_2]_2$	B: B Comp.SVol.3/3-102/3
–	$DB[N(CH_3)_2]_2$	B: B Comp.SVol.3/3-102/3
$BC_4H_{13}N_2O$	$CH_3\text{-}NH_2\text{-}BH_2\text{-}C(O)\text{-}NH\text{-}C_2H_5$	B: B Comp.SVol.3/3-187
		B: B Comp.SVol.4/3b-24
$BC_4H_{13}N_2O_2$	$(CH_3)_3N\text{-}BH_2\text{-}C(O)\text{-}NHOH \cdot HCl$	B: B Comp.SVol.4/3b-24, 25
$BC_4H_{13}S$	$(C_2H_5)_2S\text{-}BH_3$	B: B Comp.SVol.4/4-134
$BC_4H_{14}N$	$(C_2H_5)_2NH\text{-}BH_3$	B: B Comp.SVol.3/3-174
		B: B Comp.SVol.4/3b-6
–	$n\text{-}C_4H_9\text{-}NH_2\text{-}BH_3$	B: B Comp.SVol.3/3-173
–	$i\text{-}C_4H_9\text{-}NH_2\text{-}BH_3$	B: B Comp.SVol.4/3b-4
–	$t\text{-}C_4H_9\text{-}NH_2\text{-}BH_3$	B: B Comp.SVol.3/3-173
		B: B Comp.SVol.4/3b-4
$BC_4H_{14}NNa$	$Na[(t\text{-}C_4H_9)H_2NBH_3]$	B: B Comp.SVol.3/3-211
$BC_4H_{14}NO$	$H_3B\text{-}N(CH_3)_2\text{-}CH_2CH_2\text{-}OH$	B: B Comp.SVol.3/3-176
		B: B Comp.SVol.4/3b-15
$BC_4H_{15}NP$	$H_3B\text{-}N(CH_3)_2\text{-}P(CH_3)_2$	B: B Comp.SVol.4/3b-15
$BC_4H_{15}N_2$	$H_3B\text{-}N(CH_3)_2\text{-}N(CH_3)_2$	B: B Comp.SVol.4/3b-9
$BC_4H_{17}N_2Si$	$SiH_2(N(CH_3)_2)_2 \cdot BH_3$	Si: SVol.B4-186
$BC_5ClF_2H_6O_2$	$[\text{-}B(F)_2OC(CH_3)C(Cl)C(CH_3)O\text{-}]$	B: B Comp.SVol.4/3b-233
$BC_5ClF_2H_{13}N$	$N(C_2H_5)_2CH_3 \cdot BClF_2$	B: B Comp.SVol.3/4-65/7
$BC_5ClF_3H_4N$	$3\text{-}ClC_5H_4N\text{-}BF_3$	B: B Comp.SVol.3/3-294
$BC_5ClF_3H_{14}N_2^+$	$[CF_3\text{-}BCl(NH(CH_3)_2)_2]^+$	B: B Comp.SVol.4/4-70
BC_5ClH_6	$[\text{-}CH{=}CHCH_2CH{=}CH\text{-}]BCl$	B: B Comp.SVol.4/4-47
BC_5ClH_7N	$3\text{-}Cl\text{-}C_5H_4N\text{-}BH_3$	B: B Comp.SVol.3/3-177
–	$4\text{-}Cl\text{-}C_5H_4N\text{-}BH_3$	B: B Comp.SVol.3/3-177
$BC_5ClH_9N_3O_2$	$ClB(\text{-}NCH_3\text{-}CO\text{-}NCH_3\text{-}CO\text{-}NCH_3\text{-})$	B: B Comp.SVol.3/4-59
$BC_5ClH_9N_3S_2$	$ClB(\text{-}NCH_3\text{-}CS\text{-}NCH_3\text{-}CS\text{-}NCH_3\text{-})$	B: B Comp.SVol.3/4-59
BC_5ClH_{10}	$1\text{-}Cl\text{-}3\text{-}CH_3\text{-}BC_4H_7$	B: B Comp.SVol.4/4-47
–	$1\text{-}Cl\text{-}BC_5H_{10}$	B: B Comp.SVol.3/4-44
		B: B Comp.SVol.4/4-47
$BC_5ClH_{12}N_2$	$ClB[\text{-}NCH_3\text{-}(CH_2)_3\text{-}NCH_3\text{-}]$	B: B Comp.SVol.3/4-58
$BC_5ClH_{13}NO$	$CH_3N(C_2H_4)_2O\text{-}BH_2Cl$	B: B Comp.SVol.3/4-38
$BC_5ClH_{15}NSi$	$[(CH_3)_3Si]ClBN(CH_3)_2$	B: B Comp.SVol.3/4-56
$BC_5ClH_{15}NSn$	$[(CH_3)_3Sn]ClBN(CH_3)_2$	B: B Comp.SVol.3/4-56
$BC_5Cl_2FH_{13}N$	$N(C_2H_5)_2CH_3 \cdot BCl_2F$	B: B Comp.SVol.3/4-65/7
$BC_5Cl_2F_3H_{14}N_2$	$[(CH_3)_2NH\text{-}BCl(CF_3)\text{-}NH(CH_3)_2]Cl$	B: B Comp.SVol.4/3a-168
		B: B Comp.SVol.4/4-70
$BC_5Cl_2F_6H_9N_2S_2$	$(CF_3S)_2N\text{-}BCl_2 \cdot N(CH_3)_3$	B: B Comp.SVol.4/4-59, 166
$BC_5Cl_2H_5$	$c\text{-}C_5H_5BCl_2$	B: B Comp.SVol.3/4-41
$BC_5Cl_2H_5S$	$2\text{-}Cl_2B\text{-}5\text{-}CH_3\text{-}C_4H_2S$	B: B Comp.SVol.3/4-140
$BC_5Cl_2H_7O_2$	$Cl_2B(\text{-}O{=}CCH_3\text{-}CH{=}CCH_3\text{-}O\text{-})$	B: B Comp.SVol.3/4-52
$BC_5Cl_2H_9$	$c\text{-}C_5H_9\text{-}BCl_2$	B: B Comp.SVol.4/4-43
$BC_5Cl_2H_9OSi$	$Cl_2B[(CH_3)_3Si]C{=}C{=}O$	B: B Comp.SVol.3/4-41
$BC_5Cl_2H_{12}N$	$Cl_2B\text{-}N(CH_3)\text{-}C_4H_9\text{-}n$	B: B Comp.SVol.3/4-54
–	$Cl_2B\text{-}N(CH_3)\text{-}C_4H_9\text{-}t$	B: B Comp.SVol.3/4-54
$BC_5Cl_2H_{13}NOP$	$(C_2H_5)_2B\text{-}N(CH_3)POCl_2$	B: B Comp.SVol.3/3-154

Formula	Compound	Reference
$BC_5Cl_2H_{13}N_3OP$	$Cl_2P(O)N(CH_3)B(-NCH_3-CH_2-CH_2-NCH_3-)$	B: B Comp.SVol.3/3-96
$BC_5Cl_2H_{15}N_3OP$	$[(CH_3)_2N]_2BN(CH_3)POCl_2$	B: B Comp.SVol.3/3-96
$BC_5Cl_3GaH_{13}N_2$	$CH_3B(-NCH_3-CH_2-CH_2-NCH_3-) \cdot GaCl_3$	B: B Comp.SVol.3/3-113
$BC_5Cl_3H_5In$	$In(C_5H_5-c) \cdot BCl_3$	In: Org.Comp.1-377/8
$BC_5Cl_3H_5N$	$NC_5H_5 \cdot BCl_3$	B: B Comp.SVol.3/4-32
		B: B Comp.SVol.4/3b-149/50
		B: B Comp.SVol.4/4-32
$BC_5Cl_3H_{12}Si$	$(CH_3)_2SiCl(CH_2)_3BCl_2$	B: B Comp.SVol.3/4-42
$BC_5Cl_3H_{17}NSn$	$[(CH_3)_2NH_2][(CH_3)_3SnBCl_3]$	B: B Comp.SVol.3/4-64
$BC_5Cl_5F_2$	$C_5Cl_5BF_2$	B: B Comp.SVol.3/3-363
BC_5FH_6	$H_2C(-CH=CH-)_2BF$	B: B Comp.SVol.4/3b-228
BC_5FH_7N	$3-F-C_5H_4N-BH_3$	B: B Comp.SVol.3/3-177
$BC_5FH_{13}I_2N$	$N(CH_3)(C_2H_5)_2 \cdot BFI_2$	B: B Comp.SVol.3/4-65/7
$BC_5F_2H_5$	$C_5H_5BF_2$	B: B Comp.SVol.3/3-363
$BC_5F_2H_7O_2$	$F_2B-O-C(CH_3)=CH-C(O)-CH_3$	B: B Comp.SVol.4/3b-232
–	$[-B(F)_2OC(CH_3)CHC(CH_3)O-]$	B: B Comp.SVol.4/3b-233/4
$BC_5F_2H_7O_3$	$F_2B-O-C(CH_3)=CH-C(O)-OCH_3$	B: B Comp.SVol.4/3b-232
$BC_5F_2H_{13}IN$	$N(CH_3)(C_2H_5)_2 \cdot BF_2I$	B: B Comp.SVol.3/4-65/7
$BC_5F_2H_{14}NSi$	$F_2B-N(C_2H_5)-Si(CH_3)_3$	B: B Comp.SVol.4/3b-244
$BC_5F_3HO_6Re$	$(CO)_5ReO(H)BF_3$	Re: Org.Comp.2-21/2, 29
–	$(CO)_5ReO(D)BF_3$	Re: Org.Comp.2-22
$BC_5F_3H_5In$	$In(C_5H_5-c) \cdot BF_3$	In: Org.Comp.1-377/8
$BC_5F_3H_5N$	$NC_5H_5 \cdot BF_3$	B: B Comp.SVol.3/3-294/5
		B: B Comp.SVol.4/3b-149/50
$BC_5F_3H_9N$	$NC_5H_9 \cdot BF_3$	B: B Comp.SVol.4/3b-150
$BC_5F_3H_{10}N_2$	$CF_3-B[-N(CH_3)-CH_2CH_2-N(CH_3)-]$	B: B Comp.SVol.4/3a-174
$BC_5F_3H_{10}N_2O_3S$	$CF_3SO_3B(-NCH_3-CH_2-CH_2-NCH_3-)$	B: B Comp.SVol.3/3-113
$BC_5F_3H_{10}O$	$(C_2H_5)_2CO-BF_3$	B: B Comp.SVol.3/3-285
–	$C_3H_7-C(CH_3)O-BF_3$	B: B Comp.SVol.3/3-285
–	$i-C_3H_7-C(CH_3)O-BF_3$	B: B Comp.SVol.3/3-285
$BC_5F_3H_{10}O_2$	$n-C_3H_7-OC(CH_3)O-BF_3$	B: B Comp.SVol.4/3b-145
$BC_5F_3H_{11}N$	$[-(CH_2)_5-NH-] \cdot BF_3$	B: B Comp.SVol.3/3-294
$BC_5F_3H_{12}N_2$	$(CH_3)_2N-B(CF_3)-N(CH_3)_2$	B: B Comp.SVol.4/3a-167/8
$BC_5F_3H_{12}N_2S$	$[(CH_3)_2N]_2CS-BF_3$	B: B Comp.SVol.3/3-288
$BC_5F_3H_{13}N$	$N(C_2H_5)_2CH_3 \cdot BF_3$	B: B Comp.SVol.3/4-65/7
$BC_5F_4H_2O_5ReS$	$[(CO)_5ReSH_2][BF_4]$	Re: Org.Comp.2-158
$BC_5F_4H_2O_6Re$	$[(CO)_5ReOH_2][BF_4]$	Re: Org.Comp.2-156
$BC_5F_4H_4N$	$3-FC_5H_4N-BF_3$	B: B Comp.SVol.3/3-294
$BC_5F_4H_5Se$	$[c-C_5H_5Se][BF_4]$	B: B Comp.SVol.3/4-148
$BC_5F_4H_5Te$	$[c-C_5H_5Te][BF_4]$	B: B Comp.SVol.3/4-151
$BC_5F_4H_9O$	$[i-C_4H_9CO][BF_4]$	B: B Comp.SVol.4/3b-203
$BC_5F_4H_{15}N_2O_2S_2$	$[(CH_3)_2N-S(O)O-S(CH_3)N(CH_3)_2][BF_4]$	S: S-N Comp.8-313
$BC_5F_4O_5Re$	$(CO)_5ReFBF_3$	Re: Org.Comp.2-32/3
$BC_5F_6H_{10}NS$	$[F_2SN(CH_2)_5][BF_4]$	S: S-N Comp.8-75/6
BC_5F_7	$C_5F_5BF_2$	B: B Comp.SVol.3/3-363
$BC_5F_{15}HN_3S_5$	$[(CF_3S)_2N]_2B-NH-SCF_3$	B: B Comp.SVol.4/3a-156
		B: B Comp.SVol.4/4-166
$BC_5F_{15}H_2O_{15}S_5$	$[CF_3SO_2OH_2][B(OSO_2CF_3)_4]$	B: B Comp.SVol.3/4-142
$BC_5H_3NaO_5Re$	$Na[(CO)_5ReBH_3]$	Re: Org.Comp.2-168
$BC_5H_3O_5Re^-$	$[(CO)_5ReBH_3]^-$	Re: Org.Comp.2-168/9

Formula	Compound	Reference
$BC_5H_5I_3N$	$I_3B \cdot NC_5H_5$	B: B Comp.SVol.4/3b-149
$BC_5H_5N_2$	$[-B(N_2)=CH-CH=CH-CH=CH-]$	B: B Comp.SVol.4/3b-26
$BC_5H_5O_3S$	$2-(O=CH)-SC_4H_2-3-B(OH)_2$	B: B Comp.SVol.4/4-137
–	$4-(O=CH)-SC_4H_2-3-B(OH)_2$	B: B Comp.SVol.4/4-137
$BC_5H_7N_2O_2$	$H_3B \cdot NC_5H_4-4-NO_2$	B: B Comp.SVol.4/3b-13
BC_5H_8N	$NC_5H_5 \cdot BH_3$	B: B Comp.SVol.3/3-177
		B: B Comp.SVol.4/3b-9
$BC_5H_9N_2$	$2-NH_2-NC_5H_4 \cdot BH_3$	B: B Comp.SVol.3/3-177
		B: B Comp.SVol.4/3b-13
–	$(C_5H_5)B(NH_2)_2$	B: B Comp.SVol.3/3-102
$BC_5H_{10}N$	$(-CH=CH-CH=CH-)B(NH_3)-CH_3$	B: B Comp.SVol.4/3b-29
$BC_5H_{10}NOS_2$	$CH_3B[-S-CO-N(C_3H_7-n)-S-]$	B: B Comp.SVol.3/4-133
$BC_5H_{10}NS_3$	$CH_3B[-S-CS-N(C_3H_7-i)-S-]$	B: B Comp.SVol.3/4-128
$BC_5H_{10}N_3O_2$	$1,3,5-(CH_3)_3-1,3,5,2-N_3BC_2H(=O)_2-4,6$	B: B Comp.SVol.4/3a-176
$BC_5H_{11}N$	$[-BH_2-(CH)_3-N(CH_3)_2-]$	B: B Comp.SVol.3/3-165
$BC_5H_{11}N_2$	$H_3B-N[=C(CH_3)-CH=C(CH_3)-NH-]$	B: B Comp.SVol.3/3-176
–	$H_3B-N[=C(CH_3)-N(CH_3)-CH=CH-]$	B: B Comp.SVol.4/3b-12
$BC_5H_{11}N_2O$	$O(-CH_2-CH_2-)_2NH-BH_2CN$	B: B Comp.SVol.3/3-184
$BC_5H_{11}N_4^-$	$[(t-C_4H_9)NNN-BH_2CN]^-$, radical anion	B: B Comp.SVol.3/3-184
$BC_5H_{11}OS_2$	$2-(i-C_3H_7-O)-1,3,2-S_2BC_2H_4$	B: B Comp.SVol.4/4-137
$BC_5H_{11}S_2$	$i-C_3H_7B(-S-CH_2-CH_2-S-)$	B: B Comp.SVol.3/4-121
$BC_5H_{12}IO$	$t-C_4H_9-BI-OCH_3$	B: B Comp.SVol.4/4-114
$BC_5H_{12}N$	$CH_3-B=N-C_4H_9-t$	B: B Comp.SVol.3/3-143
		B: B Comp.SVol.4/3a-212
–	$t-C_4H_9-B=N-CH_3$	B: B Comp.SVol.4/3a-213
$BC_5H_{12}NO_3S$	$[-O-CHCH_3-CH_2-O-]B-N=S(O)(CH_3)_2$	B: B Comp.SVol.3/3-156
		B: B Comp.SVol.3/4-134
$BC_5H_{13}N_2$	$CH_3-B(-NCH_3-CH_2CH_2-NCH_3-)$	B: B Comp.SVol.3/3-113
		B: B Comp.SVol.4/3a-174
–	$(C_2H_5)_2NH-BH_2-CN$	B: B Comp.SVol.4/3b-22
–	$n-C_4H_9-NH_2-BH_2-CN$	B: B Comp.SVol.4/3b-22
–	$HB(-NCH_3-CH_2CH_2CH_2-NCH_3-)$	B: B Comp.SVol.3/3-113
$BC_5H_{13}N_2Se$	$1,3-(CH_3)_2-1,3,2-N_2BC_2H_4-2-SeCH_3$	B: B Comp.SVol.4/4-174
$BC_5H_{13}O_3S$	$(C_2H_5)_2BOSO_2CH_3$	B: B Comp.SVol.3/4-121
$BC_5H_{13}S_2$	$2,2-(CH_3)_2-1,3-S_2C_3H_4 \cdot BH_3$	B: B Comp.SVol.4/4-134
$BC_5H_{13}S_3$	$(CH_3)_2S-BH[-S-CH_2CH_2CH_2-S-]$	B: B Comp.SVol.4/4-134
$BC_5H_{14}N$	$CH_2=CH-CH_2-N(CH_3)_2-BH_3$	B: B Comp.SVol.3/3-178
–	$(CH_3)_2B-NH-C_3H_7-i$	B: B Comp.SVol.4/3a-229
–	$(C_2H_5)_2B-NH-CH_3$	B: B Comp.SVol.3/3-148
		B: B Comp.SVol.4/3a-230
–	$H_3B-NC_4H_8-1-CH_3$	B: B Comp.SVol.4/3b-11
–	$H_3B-NC_5H_{11}$	B: B Comp.SVol.3/3-174
$BC_5H_{14}NO$	$O(-CH_2-CH_2-)_2NCH_3-BH_3$	B: B Comp.SVol.3/3-177
$BC_5H_{14}NO_2$	$(CH_3)_3N-BH_2-COO-CH_3$	B: B Comp.SVol.3/3-186/7
		B: B Comp.SVol.4/3b-25
–	$(C_2H_5)_2NH-BH_2-COOH$	B: B Comp.SVol.4/3b-24
–	$n-C_4H_9-NH_2-BH_2-COOH$	B: B Comp.SVol.4/3b-24
–	$s-C_4H_9-NH_2-BH_2-COOH$	B: B Comp.SVol.4/3b-24
$BC_5H_{14}NS$	$2,2-(CH_3)_2-1,3-SNC_3H_5 \cdot BH_3$	B: B Comp.SVol.4/3b-7
		B: B Comp.SVol.4/4-147/8

Formula	Compound	Reference
$BC_5H_{14}N_3$	2-CH_3NH-1,3-$(CH_3)_2$-1,3,2-$N_2BC_2H_4$	B: B Comp.SVol.4/3a-154
$BC_5H_{15}LiNSi$	Li[$(CH_3)_2$N-B(Si$(CH_3)_3$)]	B: B Comp.SVol.3/3-211
$BC_5H_{15}NOP$	$(CH_3)_2$B-N(CH_3)PO$(CH_3)_2$	B: B Comp.SVol.3/3-154
$BC_5H_{15}N_2$	1,3-$(CH_3)_2$-1,3-$N_2C_3H_6$ · BH_3	B: B Comp.SVol.4/3b-12
–	$CH_3B[N(CH_3)_2]_2$	B: B Comp.SVol.3/3-103/4
–	$CD_3B[N(CH_3)_2]_2$	B: B Comp.SVol.3/3-103/4
$BC_5H_{15}N_2O$	$(CH_3)_2$HN-BH_2CONHC_2H_5	B: B Comp.SVol.3/3-186/7
$BC_5H_{15}N_2Si$	[-B(CH_3)-N(CH_3)-Si$(CH_3)_2$-N(CH_3)-]	B: B Comp.SVol.4/3a-205/6
$BC_5H_{16}N$	$(CH_3)_2$NH-B$(CH_3)_3$	B: B Comp.SVol.4/3b-27
–	H_3B-NH(CH_3)-C_4H_9-t	B: B Comp.SVol.4/3b-6
$BC_5H_{16}NO$	$(CH_3)_2$HN-B$(CH_3)_2$OCH_3	B: B Comp.SVol.3/3-216
$BC_5H_{16}NS$	$(CH_3)_3$N-$BH_2CH_2SCH_3$	B: B Comp.SVol.3/3-188
		B: B Comp.SVol.3/4-130
$BC_5H_{16}NSi$	$(CH_3)_2$B-NH-Si$(CH_3)_3$	B: B Comp.SVol.3/3-148
		B: B Comp.SVol.4/3a-230
$BC_5H_{18}NSi$	$SiH_3N(CH_3)_2$ · B$(CH_3)_3$	Si: SVol.B4-171/2
$BC_5H_{20}NSi_2$	$(SiH_3)_2NC_2H_5$ · B$(CH_3)_3$	Si: SVol.B4-241
$BC_6ClF_2H_7N$	4-CH_3-C_5H_4N · $BClF_2$	B: B Comp.SVol.3/4-65/7
$BC_6ClF_2H_{15}N$	$(C_2H_5)_3$N · $BClF_2$	B: B Comp.SVol.3/4-65/7
		B: B Comp.SVol.4/4-71
$BC_6ClFe_2O_6S_2$	$BFe_2S_2(CO)_6(Cl)$	B: B Comp.SVol.4/4-163
$BC_6ClH_4O_2$	2-Cl-1,3,2-$O_2BC_6H_4$	B: B Comp.SVol.3/4-52
		B: B Comp.SVol.4/4-55
$BC_6ClH_6N_2$	4-Cl-C_5H_4N-BH_2CN	B: B Comp.SVol.3/3-184
BC_6ClH_{12}	1-Cl-BC_6H_{12}	B: B Comp.SVol.4/4-47
$BC_6ClH_{12}S$	5-Cl-BC_6H_{12}S	B: B Comp.SVol.4/4-163
BC_6ClH_{14}	(n-C_3H_7)$_2$BCl	B: B Comp.SVol.3/4-43
		B: B Comp.SVol.4/4-46
–	(i-C_3H_7)$_2$BCl	B: B Comp.SVol.3/4-43
		B: B Comp.SVol.4/4-46
–	i-C_3H_7-C$(CH_3)_2$-BHCl	B: B Comp.SVol.3/4-38
		B: B Comp.SVol.4/4-40
$BC_6ClH_{15}N$	i-C_3H_7-BCl-NH-C_3H_7-i	B: B Comp.SVol.3/4-56
$BC_6ClH_{15}NO_2$	$(CH_3)_3$N-BH_2-COO-CH_2CH_2Cl	B: B Comp.SVol.3/3-187
		B: B Comp.SVol.4/3b-25
$BC_6ClH_{17}N$	$(C_2H_5)_3$N-BH_2Cl	B: B Comp.SVol.3/3-175
		B: B Comp.SVol.3/4-37
–	(i-C_3H_7)$_2$NH-BH_2Cl	B: B Comp.SVol.4/4-39
$BC_6Cl_2FH_7N$	4-CH_3-C_5H_4N-BCl_2F	B: B Comp.SVol.3/4-65/7
$BC_6Cl_2FH_{15}N$	$(C_2H_5)_3$N-BCl_2F	B: B Comp.SVol.3/4-65/7
		B: B Comp.SVol.4/4-71
$BC_6Cl_2F_5$	C_6F_5-BCl_2	B: B Comp.SVol.3/4-41
$BC_6Cl_2H_5$	C_6H_5-BCl_2	B: B Comp.SVol.3/4-40/1
		B: B Comp.SVol.4/4-41/2
$BC_6Cl_2H_{11}$	2-CH_3-c-C_5H_8-BCl_2	B: B Comp.SVol.4/4-43
–	C_2H_5-CH=C(C_2H_5)-BCl_2	B: B Comp.SVol.4/4-43
–	n-C_4H_9-CH=CH-BCl_2	B: B Comp.SVol.4/4-43
–	c-C_6H_{11}-BCl_2	B: B Comp.SVol.4/4-43
$BC_6Cl_2H_{13}$	i-C_3H_7-C$(CH_3)_2$-BCl_2	B: B Comp.SVol.3/4-42
		B: B Comp.SVol.4/4-43

Formula	Compound	Reference
$BC_6Cl_2H_{13}$	$n-C_3H_7-CH(C_2H_5)-BCl_2$	B: B Comp.SVol.4/4-43
$BC_6Cl_2H_{14}N$	$Cl_2B-N(C_3H_7-i)_2$	B: B Comp.SVol.3/4-54
		B: B Comp.SVol.4/4-58
$BC_6Cl_2H_{16}N$	$(C_2H_5)_3N-BHCl_2$	B: B Comp.SVol.3/3-175
$BC_6Cl_2H_{18}NSi_2$	$Cl_2B-N[Si(CH_3)_3]_2$	B: B Comp.SVol.3/4-54
		B: B Comp.SVol.4/4-58
$BC_6Cl_2H_{18}NSn_2$	$Cl_2B-N[Sn(CH_3)_3]_2$	B: B Comp.SVol.4/4-58
$BC_6Cl_3F_3H_{13}O_3SSi$	$(i-C_3H_7)(CH_3)_2SiO(BCl_3)SO_2CF_3$	B: B Comp.SVol.3/4-32
$BC_6Cl_3H_5NO_2$	$C_6H_5-NO_2 \cdot BCl_3$	B: B Comp.SVol.4/4-34
$BC_6Cl_3H_5NO_3$	$2-NO_2-C_6H_4-OH \cdot BCl_3$	B: B Comp.SVol.4/4-34
$BC_6Cl_3H_6N_2O_2$	$2-NO_2-C_6H_4-NH_2 \cdot BCl_3$	B: B Comp.SVol.4/4-34
$BC_6Cl_3H_{10}$	$n-C_4H_9-CCl=CH-BCl_2$	B: B Comp.SVol.4/4-43
$BC_6Cl_3H_{11}N$	$Cl_2B[-NH(C_3H_7-i)-CH_2-CCl=CH-]$	B: B Comp.SVol.3/4-55
$BC_6Cl_3H_{14}O$	$t-C_4H_9-CH_2-O-CH_3 \cdot BCl_3$	B: B Comp.SVol.4/4-34
$BC_6Cl_3H_{15}N$	$(C_2H_5)_3N-BCl_3$	B: B Comp.SVol.3/4-32
$BC_6Cl_3H_{15}NSn$	$(CH_3)_2Sn(Cl)-N(C_4H_9-t)-BCl_2$	Sn: Org.Comp.19-118, 123
$BC_6Cl_3H_{16}Si_2$	$(CH_3)_2SiCl-CH_2Si(CH_3)_2-CH_2-BCl_2$	B: B Comp.SVol.3/4-42
–	$(CH_3)_2SiCl-CH_2SiCl(CH_3)-CH_2-BCl-CH_3$	B: B Comp.SVol.3/4-45
$BC_6Cl_4H_{11}N_2$	$[1-CH_3-3-C_2H_5-1,3-N_2C_3H_3][BCl_4]$	B: B Comp.SVol.4/4-35
$BC_6Cl_6Ga_2H_{18}N_3$	$[((CH_3)_2N)_3BGaCl_2]GaCl_4$	B: B Comp.SVol.3/3-94
BC_6FH_6	$(-CH=CH-CH=CH-CH=CH-)BF$	B: B Comp.SVol.3/3-363
BC_6FH_{14}	$(i-C_3H_7)_2BF$	B: B Comp.SVol.4/3b-227
$BC_6FH_{15}I_2N$	$N(C_2H_5)_3 \cdot BFI_2$	B: B Comp.SVol.3/4-65/7
$BC_6F_2H_5$	$C_6H_5-BF_2$	B: B Comp.SVol.4/3b-226
$BC_6F_2H_9O_2$	$[-B(F)_2OC(CH_3)C(CH_3)C(CH_3)O-]$	B: B Comp.SVol.4/3b-232/3
$BC_6F_2H_{14}N$	$F_2B-N(C_3H_7-i)_2$	B: B Comp.SVol.4/3b-243, 244
$BC_6F_2H_{15}IN$	$N(C_2H_5)_3 \cdot BF_2I$	B: B Comp.SVol.3/4-65/7
$BC_6F_2H_{16}NSi$	$F_2B-N(C_3H_7-i)-Si(CH_3)_3$	B: B Comp.SVol.4/3b-244
$BC_6F_2H_{18}NSi_2$	$F_2B-N[Si(CH_3)_3]_2$	B: B Comp.SVol.3/3-375
		B: B Comp.SVol.4/3b-243, 244
$BC_6F_2H_{20}N_3Si_3$	$FB[-NH-Si(CH_3)_2-NH-Si(CH_3)_2-N(SiF(CH_3)_2)-]$	B: B Comp.SVol.3/3-377
–	$FB[-NH-Si(CH_3)_2-N(SiF(CH_3)_2)-Si(CH_3)_2-NH-]$	B: B Comp.SVol.3/3-377
$BC_6F_3H_4N_2$	$3-NC-C_5H_4N-BF_3$	B: B Comp.SVol.3/3-294
–	$4-NC-C_5H_4N-BF_3$	B: B Comp.SVol.3/3-294
$BC_6F_3H_5IO$	$F_3B-OI-C_6H_5$	B: B Comp.SVol.4/3b-145
$BC_6F_3H_6O$	$F_3B-O(H)-C_6H_5$	B: B Comp.SVol.3/3-286
		B: B Comp.SVol.4/3b-144
$BC_6F_3H_7N$	$F_3B-NC_5H_4-4-CH_3$	B: B Comp.SVol.3/4-65/7
		B: B Comp.SVol.4/3b-150
–	$F_3B-NH_2-C_6H_5$	B: B Comp.SVol.3/3-294
		B: B Comp.SVol.4/3b-149
$BC_6F_3H_{10}O$	$(c-C_5H_{10}CO)-BF_3$	B: B Comp.SVol.3/3-285
$BC_6F_3H_{11}NOS$	$O=S=NC_6H_{11}-c \cdot BF_3$	S: S-N Comp.6-103
$BC_6F_3H_{12}N_2O_2$	$[(CH_3)_2N]_2BOC(O)CF_3$	B: B Comp.SVol.3/3-106
$BC_6F_3H_{12}O$	$(t-C_4H_9)(CH_3)CO-BF_3$	B: B Comp.SVol.3/3-285

Formula	Compound	Reference
$BC_6F_3H_{15}N$	$N(C_2H_5)_3 \cdot BF_3$	B: B Comp.SVol.3/3-294
		B: B Comp.SVol.3/4-65/7
$BC_6F_4H_5INO_3$	$[C_6H_5\text{-}I\text{-}O\text{-}NO_2][BF_4]$	B: B Comp.SVol.4/3b-219
$BC_6F_4H_5OS$	$C_6H_5S(F)\text{=}O\text{-}BF_3$	B: B Comp.SVol.3/4-142
$BC_6F_4H_6IO$	$C_6H_5\text{-}I\text{-}O\text{-}H[BF_4]$	B: B Comp.SVol.4/3b-167
$BC_6F_4H_8N$	$[C_6H_5\text{-}NH_3][BF_4]$	B: B Comp.SVol.4/3b-208
$BC_6F_4H_9Te_2$	$[1,2\text{-}Te_2C_3H_2\text{-}3\text{-}(C_3H_7\text{-}i)][BF_4]$	B: B Comp.SVol.3/4-150/1
$BC_6F_4H_{11}S_2$	$[\text{-}CH_2CH_2CH_2\text{-}SC(C_2H_5)S\text{-}][BF_4]$	B: B Comp.SVol.4/4-163
$BC_6F_4H_{14}In$	$[(i\text{-}C_3H_7)_2In][BF_4]$	In: Org.Comp.1-336/7
$BC_6F_4H_{15}N_2S_2$	$[(CH_3)_2SN\text{=}S\text{=}N\text{-}C_4H_9\text{-}t][BF_4]$	S: S-N Comp.7-22
$BC_6F_4H_{15}O$	$[(C_2H_5)_3O][BF_4]$	B: B Comp.SVol.3/3-335/6
		B: B Comp.SVol.4/3b-203, 219
$BC_6F_4H_{15}S$	$[(C_2H_5)_3S][BF_4]$	B: B Comp.SVol.3/4-140
$BC_6F_4H_{16}N$	$[(C_2H_5)_3NH][BF_4]$	B: B Comp.SVol.3/3-343
		B: B Comp.SVol.4/3b-208, 210
–	$[(n\text{-}C_3H_7)_2NH_2][BF_4]$	B: B Comp.SVol.4/3b-210
$BC_6F_4H_{18}N_3S$	$[((CH_3)_2N)_3S][BF_4]$	S: S-N Comp.8-222/5
$BC_6F_4H_{18}N_3S_2$	$[((CH_3)_2N)_2S\text{-}N\text{=}S(CH_3)_2][BF_4]$	S: S-N Comp.8-230, 239
$BC_6F_4O_6Re$	$[Re(CO)_6][BF_4]$	Re: Org.Comp.2-217, 222
$BC_6F_5H_4OS$	$4\text{-}F\text{-}C_6H_4S(F)\text{=}O\text{-}BF_3$	B: B Comp.SVol.3/4-142
$BC_6F_8H_2N$	$C_6F_5\text{-}NH_2 \cdot BF_3$	F: PFHOrg.SVol.5-8/9, 40
$BC_6F_{18}N_3S_6$	$B[N(SCF_3)_2]_3$	B: B Comp.SVol.4/3a-153
		B: B Comp.SVol.4/4-148, 165/6
$BC_6H_5I_2$	$C_6H_5\text{-}BI_2$	B: B Comp.SVol.4/4-113
$BC_6H_5N_3Na$	$Na[HB(NC_4H_4)(CN)_2]$	B: B Comp.SVol.3/3-213
BC_6H_6NO	$HB[\text{-}NH\text{-}(1,2\text{-}C_6H_4)\text{-}O\text{-}]$	B: B Comp.SVol.3/3-164
$BC_6H_6NO_6$	$B[\text{-}O\text{-}CO\text{-}CH_2\text{-}]_3N$	B: B Comp.SVol.3/3-219
$BC_6H_7N^+$	$[H_2N\text{=}BC_6H_5]^+$	B: B Comp.SVol.3/3-206
$BC_6H_7N_2$	$3\text{-}NC\text{-}C_5H_4N\text{-}BH_3$	B: B Comp.SVol.3/3-177
–	$4\text{-}NC\text{-}C_5H_4N\text{-}BH_3$	B: B Comp.SVol.3/3-177
–	$C_5H_5N\text{-}BH_2CN$	B: B Comp.SVol.3/3-184
$BC_6H_8KN_4$	$K[H_2B(1,3\text{-}N_2C_3H_3)_2]$	B: B Comp.SVol.3/3-213
BC_6H_8N	$[\text{-}CH\text{=}CH\text{-}CH\text{=}CH\text{-}CH\text{=}CH\text{-}]B\text{-}NH_2$	B: B Comp.SVol.3/3-150
$BC_6H_8NO_2$	$C_5H_5N\text{-}BH_2COOH$	B: B Comp.SVol.3/3-187
$BC_6H_8NO_5$	$B[\text{-}O\text{-}CH_2\text{-}CH_2\text{-}][\text{-}O\text{-}CO\text{-}CH_2\text{-}]_2N$	B: B Comp.SVol.3/3-219
$BC_6H_8NO_8$	$NH_4[B(\text{-}O\text{-}CO\text{-}CH_2\text{-}CO\text{-}O\text{-})_2]$	B: B Comp.SVol.3/3-227
$BC_6H_8N_3$	$4\text{-}NH_2\text{-}C_5H_4N\text{-}BH_2CN$	B: B Comp.SVol.3/3-184
BC_6H_8NaSe	$Na[C_6H_5\text{-}Se\text{-}BH_3]$	B: B Comp.SVol.4/4-174
$BC_6H_{10}ISe_2$	$4,5\text{-}(C_2H_5)_2\text{-}3\text{-}I\text{-}1,2,3\text{-}Se_2BC_2$	B: B Comp.SVol.4/4-171/2
$BC_6H_{10}N$	$1\text{-}(CH_3)_2B\text{-}NC_4H_4$	B: B Comp.SVol.4/3a-233
–	$2\text{-}CH_3\text{-}C_5H_4N\text{-}BH_3$	B: B Comp.SVol.3/3-177
–	$4\text{-}CH_3\text{-}C_5H_4N\text{-}BH_3$	B: B Comp.SVol.3/3-177
		B: B Comp.SVol.4/3b-13
–	$C_5H_5N\text{-}BH_2\text{-}CH_3$	B: B Comp.SVol.4/3b-25
–	$C_6H_5\text{-}NH_2\text{-}BH_3$	B: B Comp.SVol.3/3-173
		B: B Comp.SVol.4/3b-5, 9
$BC_6H_{10}NO$	$H_3B \cdot NC_5H_4\text{-}2\text{-}CH_2OH$	B: B Comp.SVol.4/3b-13

Formula	Compound		Reference
$BC_6H_{10}NO$	$H_3B \cdot NC_5H_4$-4-OCH_3	B:	B Comp.SVol.4/3b-13
$BC_6H_{10}NO_4$	B[-O-CH_2-CH_2-]$_2$[-O-C(O)-CH_2-]N	B:	B Comp.SVol.3/3-219
–	NH_4[$(HO)_2$B(-O-C_6H_4-O-)] $\cdot$ H_2O	B:	B Comp.SVol.3/3-226/7
$BC_6H_{10}N_5$	6-($H_3B \cdot NH_2$)-9-CH_3-1,3,7,9-$N_4C_5H_2$	B:	B Comp.SVol.4/3b-15
$BC_6H_{11}N_2$	2-NH_2-3-CH_3-C_5H_3N-BH_3	B:	B Comp.SVol.3/3-177
		B:	B Comp.SVol.4/3b-14
–	2-NH_2-4-CH_3-C_5H_3N-BH_3	B:	B Comp.SVol.3/3-177
		B:	B Comp.SVol.4/3b-14
–	2-NH_2-5-CH_3-C_5H_3N-BH_3	B:	B Comp.SVol.3/3-177
		B:	B Comp.SVol.4/3b-14
–	2-NH_2-6-CH_3-C_5H_3N-BH_3	B:	B Comp.SVol.3/3-177
		B:	B Comp.SVol.4/3b-14
–	C_6H_5-NH-NH_2-BH_3	B:	B Comp.SVol.3/3-173
–	H_3B-NH_2-C_6H_4-2-NH_2	B:	B Comp.SVol.4/3b-11
–	H_3B-NH_2-C_6H_4-3-NH_2	B:	B Comp.SVol.4/3b-11
–	H_3B-NH_2-C_6H_4-4-NH_2	B:	B Comp.SVol.4/3b-5, 11
$BC_6H_{11}O_2$	[-O-$(CH_2)_3$-]$BOCH_2CH$=CH_2	B:	B Comp.SVol.3/3-176
$BC_6H_{12}N$	$(CH_3)_2N$-B[-CH=CH-CH_2-CH_2-]	B:	B Comp.SVol.4/3a-228
–	$(CH_3)_2N$-B[-CH_2-CH=CH-CH_2-]	B:	B Comp.SVol.4/3a-228
$BC_6H_{12}N_3$	NC-CH_2CH_2-$N(CH_3)_2$-BH_2-CN	B:	B Comp.SVol.3/3-184
		B:	B Comp.SVol.4/3b-23
$BC_6H_{12}N_3O_2$	CH_3B(-NCH_3-CO-NCH_3-CO-NCH_3-)	B:	B Comp.SVol.3/3-114
$BC_6H_{13}N_2$	$C_5H_{10}NH$-BH_2CN	B:	B Comp.SVol.3/3-184
$BC_6H_{13}N_2O$	[-CH_2CH_2-O-CH_2CH_2-]$N(CH_3)$-BH_2-CN	B:	B Comp.SVol.4/3b-23
$BC_6H_{13}N_2S$	CH_3B[-NH-C(S)-N(C_4H_9-t)-]	B:	B Comp.SVol.3/3-112
		B:	B Comp.SVol.3/4-131
$BC_6H_{14}I$	(i-C_3H_7)$_2$BI	B:	B Comp.SVol.4/4-113
$BC_6H_{14}N$	1-$(CH_3)_2N$-BC_4H_8	B:	B Comp.SVol.3/3-150
–	C_2H_5-B=N-C_4H_9-t	B:	B Comp.SVol.3/3-143
		B:	B Comp.SVol.4/3a-213
–	i-C_3H_7-B=N-C_3H_7-i	B:	B Comp.SVol.4/3a-213
–	n-C_3H_7-B=N-C_3H_7-n	B:	B Comp.SVol.3/3-141
–	[-B(n-C_3H_7)-N(n-C_3H_7)-]$_n$	B:	B Comp.SVol.3/3-134
$BC_6H_{14}NO_2$	[-B$(C_2H_5)_2$-O-C(=O)-CH_2-NH_2-]	B:	B Comp.SVol.4/3b-70
–	[-O-CH_2CH_2-O-]B-$N(C_2H_5)_2$	B:	B Comp.SVol.4/3b-63
$BC_6H_{14}NO_4$	$(CH_3)_3N$-BH_2-C(O)-O-C(O)-O-CH_3	B:	B Comp.SVol.4/3b-25
$BC_6H_{14}NS_2$	$(C_2H_5)_2NB$(-S-CH_2-CH_2-S-)	B:	B Comp.SVol.3/4-124/5, 126
$BC_6H_{14}N_3$	$(C_3H_7)_2B$-N_3	B:	B Comp.SVol.3/3-144
–	(n-C_3H_7)$_2$B-N_3	B:	B Comp.SVol.4/3a-218
–	(i-C_3H_7)$_2$B-N_3	B:	B Comp.SVol.3/3-144
$BC_6H_{15}N_2O$	$(CH_3)_2B$-NCH_3-C(O)$N(CH_3)_2$	B:	B Comp.SVol.3/3-155
$BC_6H_{15}N_2O_2$	[$(CH_3)_2N$]$_2$BOC(O)CH_3	B:	B Comp.SVol.3/3-106
$BC_6H_{15}N_4$	H_3N-$BC_6H_{12}N_3$	B:	B Comp.SVol.3/3-178, 191
$BC_6H_{15}S_3$	$B(SC_2H_5)_3$	B:	B Comp.SVol.3/4-110
$BC_6H_{16}Li$	Li[$(C_2H_5)_3BH$]	B:	B Comp.SVol.4/3b-123
$BC_6H_{16}N$	$(CH_3)_2B$-$N(C_2H_5)_2$	B:	B Comp.SVol.3/3-148
–	$(CH_3)_2B$-NH-C_4H_9-t	B:	B Comp.SVol.3/3-148
		B:	B Comp.SVol.4/3a-229

Formula	Compound	Reference
$BC_6H_{16}N$	$(C_2H_5)_2B-N(CH_3)_2$	B: B Comp.SVol.3/3-148
		B: B Comp.SVol.4/3a-226
–	$(n-C_3H_7)_2B-NH_2$	B: B Comp.SVol.3/3-149
–	$(i-C_3H_7)_2B-NH_2$	B: B Comp.SVol.4/3a-226
–	$H_3B-NC_5H_{10}-1-CH_3$	B: B Comp.SVol.3/3-177
$BC_6H_{16}NO$	$O(-CH_2-CH_2-)_2NC_2H_5-BH_3$	B: B Comp.SVol.3/3-177
$BC_6H_{16}NOS$	$(CH_3)_2B-N(i-C_3H_7)SOCH_3$	B: B Comp.SVol.3/3-154
$BC_6H_{16}NO_2$	$(CH_3)_3N-BH_2-COO-C_2H_5$	B: B Comp.SVol.3/3-187
		B: B Comp.SVol.4/3b-25
–	$H_3B-N(CH_3)_2-CH_2CH_2-OOC-CH_3$	B: B Comp.SVol.4/3b-15
$BC_6H_{16}N_3$	$(CH_3)_2NB(-NCH_3-CH_2-CH_2-NCH_3-)$	B: B Comp.SVol.3/3-95
$BC_6H_{16}N_3OS$	$O=S(CH_3)_2=N-B(-NCH_3-CH_2-CH_2-NCH_3-)$	B: B Comp.SVol.3/3-97
		B: B Comp.SVol.3/4-134
$BC_6H_{17}N$	$(C_2H_5)_3N-BH_2$, radical	B: B Comp.SVol.3/3-175
		B: B Comp.SVol.4/3b-8/9
–	$(i-C_3H_7)_2NH-BH_2$, radical	B: B Comp.SVol.4/3b-26
$BC_6H_{17}N_2$	$C_2H_5-B[N(CH_3)_2]_2$	B: B Comp.SVol.3/3-104
–	$n-C_4H_9-B(NH-CH_3)_2$	B: B Comp.SVol.4/3a-169
–	$[-N(CH_3)-CH(CH_3)-N(CH_3)-CH_2CH_2-] \cdot BH_3$	B: B Comp.SVol.4/3b-12
$BC_6H_{17}N_2O$	$(CH_3)_3N-BH_2CONHC_2H_5$	B: B Comp.SVol.3/3-186/7
$BC_6H_{17}N_3^+$	$[(CH_3)_2N-B(-NHCH_3-CH_2-CH_2-NCH_3-)]^+$	B: B Comp.SVol.3/3-207
$BC_6H_{18}N$	$(CH_3)_3N-B(CH_3)_3$	B: B Comp.SVol.4/3b-28
–	$(C_2H_5)_3N-BH_3$	B: B Comp.SVol.3/3-175/6
		B: B Comp.SVol.4/3b-8/9
–	$(C_2H_5)_3N-^{10}BH_3$	B: B Comp.SVol.4/3b-8
–	$(C_2H_5)_3N-BH_3$ systems	
	$(C_2H_5)_3N-BH_3$ - $1,2-C_2B_{10}H_{11}-1-CH(CH_3)_2$	B: B Comp.SVol.3/4-192
–	$(i-C_3H_7)_2NH-BH_3$	B: B Comp.SVol.4/3b-6
–	$t-C_4H_9-N(CH_3)_2-BH_3$	B: B Comp.SVol.4/3b-10
$BC_6H_{18}NO$	$(HOCH_2-CH_2)(C_2H_5)_2N-BH_3$	B: B Comp.SVol.3/3-176
$BC_6H_{18}NOSi$	$[(CH_3)_3Si][CH_3O]B-N(CH_3)_2$	B: B Comp.SVol.3/3-155
$BC_6H_{18}NSi$	$(CH_3)_2B-N(CH_3)Si(CH_3)_3$	B: B Comp.SVol.3/3-151
$BC_6H_{18}N_3$	$B[N(CH_3)_2]_3$	B: B Comp.SVol.3/3-92/4
		B: B Comp.SVol.4/3a-153
$BC_6H_{18}N_3OS$	$[(CH_3)_2N]_2B-NCH_3-S(O)CH_3$	B: B Comp.SVol.3/3-96
		B: B Comp.SVol.3/4-134
$BC_6H_{19}INS$	$[(CH_3)_3NBH_2CH_2S(CH_3)_2]I$	B: B Comp.SVol.3/4-145
$BC_6H_{19}NP$	$(CH_3)_3P-BH_2CH_2N(CH_3)_2$	B: B Comp.SVol.3/3-187
$BC_6H_{19}N_2$	$(CH_3)_2N-CH_2CH_2-N(CH_3)_2-BH_3$	B: B Comp.SVol.3/3-176
–	$(CH_3)_3N-BH_2CH_2-N(CH_3)_2$	B: B Comp.SVol.3/3-187
$BC_6H_{20}NSi_2$	$H_2B-N[Si(CH_3)_3]_2$	B: B Comp.SVol.3/3-147
$BC_6H_{22}N_3Si$	$SiH(N(CH_3)_2)_3 \cdot BH_3$	Si: SVol.B4-193
$BC_7ClF_2H_{13}N$	$HC(C_2H_4)_3N \cdot BClF_2$	B: B Comp.SVol.3/4-65/7
$BC_7ClH_6O_2S$	$ClB[-O-(1,2-C_6H_3(OCH_3-4))-S-]$	B: B Comp.SVol.3/4-142
$BC_7ClH_8N_4$	$Cl-B[-N(C_6H_5)-N=N-N(CH_3)-]$	B: B Comp.SVol.4/4-67
BC_7ClH_8O	$ClB(C_6H_5)(OCH_3)$	B: B Comp.SVol.3/4-50
$BC_7ClH_{11}InN_4$	$CH_3-In(Cl)[-N_2C_3H_3-BH_2-N_2C_3H_3-]$	In: Org.Comp.1-296/7
$BC_7ClH_{15}N$	$HC(-CH_2-CH_2-)_3N-BH_2Cl$	B: B Comp.SVol.3/3-177
		B: B Comp.SVol.3/4-38
$BC_7ClH_{15}NO_4$	$(CH_3)_3N-BH_2-C(O)-O-C(O)-O-CH_2-CH_2Cl$	B: B Comp.SVol.4/3b-25

$BC_7ClH_{19}NSn$ .... $[(CH_3)_3Sn]ClBN(C_2H_5)_2$ .... B: B Comp.SVol.3/4-56
$BC_7Cl_2FH_{13}N$ .... $HC(C_2H_4)_3N \cdot BCl_2F$ .... B: B Comp.SVol.3/4-65/7
$BC_7Cl_2H_7$ .... 2-CH_3-C_6H_4-BCl_2 .... B: B Comp.SVol.4/4-43
– .... C_6H_5-CH_2-BCl_2 .... B: B Comp.SVol.3/4-42
B: B Comp.SVol.4/4-43
$BC_7Cl_2H_8N$ .... $Cl_2BN(CH_3)(C_6H_5)$ .... B: B Comp.SVol.3/4-54
$BC_7Cl_2H_{11}$ .... [2.2.1]-C_7H_{11}-BCl_2 .... B: B Comp.SVol.4/4-43
$BC_7Cl_2H_{13}N_2O_2$ .. $Cl_2B[-O=C(N(CH_3)_2)-CH=C(N(CH_3)_2)-O-]$ .... B: B Comp.SVol.3/4-52
$BC_7Cl_2H_{14}N$ .... Cl_2B-NCH_3-C_6H_{11}-c .... B: B Comp.SVol.3/4-54
– .... $HC(-CH_2-CH_2-)_3N$-$BHCl_2$ .... B: B Comp.SVol.3/3-177
$BC_7Cl_2H_{17}S$ .... n-$C_5H_{11}BCl_2 \cdot (CH_3)_2S$ .... B: B Comp.SVol.3/4-42
$BC_7Cl_2H_{18}NSi$ .... Cl_2B-$N(C_4H_9$-t)-$Si(CH_3)_3$ .... B: B Comp.SVol.4/4-59
$BC_7Cl_2H_{18}NSn$ ... Cl_2B-$N(C_4H_9$-t)-$Sn(CH_3)_3$ .... B: B Comp.SVol.4/4-59
Sn: Org.Comp.18-57, 64
$BC_7Cl_3F_3H_{15}O_3SSi$
$(C_2H_5)_3SiO(BCl_3)SO_2CF_3$ .... B: B Comp.SVol.3/4-32
$BC_7Cl_3H_{12}Si$ .... $(CH_2=CH)_2SiCl(CH_2)_3BCl_2$ .... B: B Comp.SVol.3/4-42
$BC_7Cl_3H_{13}N$ .... $HC(C_2H_4)_3N \cdot BCl_3$ .... B: B Comp.SVol.3/4-66
$BC_7F_2H_7$ .... 2-CH_3-C_6H_4-BF_2 .... B: B Comp.SVol.4/3b-226
$BC_7F_2H_{13}N_2O_2$ ... $F_2B[-NH=C(O-C_2H_5)-CH=C(O-C_2H_5)-NH-]$ ... B: B Comp.SVol.3/3-377
– .... $F_2B[-O-C(N(CH_3)_2)=CH-C(N(CH_3)_2)=O-]$ .... B: B Comp.SVol.3/3-368
$BC_7F_2H_{18}NSi$ .... F_2B-$N(C_4H_9$-t)-$Si(CH_3)_3$ .... B: B Comp.SVol.4/3b-244
$BC_7F_3H_6O$ .... F_3B-OCH-C_6H_5 .... B: B Comp.SVol.3/3-288
B: B Comp.SVol.4/3b-144
$BC_7F_3H_9N$ .... F_3B-NC_5H_3-2,6-$(CH_3)_2$ .... B: B Comp.SVol.4/3b-150
– .... F_3B-$NH(CH_3)$-C_6H_5 .... B: B Comp.SVol.3/3-294
– .... F_3B-NH_2-CH_2-C_6H_5 .... B: B Comp.SVol.3/3-294
B: B Comp.SVol.4/3b-149
$BC_7F_3H_{13}N$ .... $HC(C_2H_4)_3N \cdot BF_3$ .... B: B Comp.SVol.3/4-65/7
$BC_7F_3H_{14}O$ .... (i-$C_3H_7)_2CO$-BF_3 .... B: B Comp.SVol.3/3-285
$BC_7F_3H_{14}O_3S$ .... (n-$C_3H_7)_2B(OSO_2CF_3)$ .... B: B Comp.SVol.3/4-142
$BC_7F_3H_{17}N_3O_3S$ .. $[(CH_3)_2N-B(-NHCH_3-CH_2CH_2-NCH_3-)]O_3SCF_3$ B: B Comp.SVol.3/3-207
BC_7F_4 .... $C_7[BF_4]$ .... B: B Comp.SVol.3/3-302
B: B Comp.SVol.4/3b-107
$BC_7F_4H_3NO_5Re$ .. $[(CO)_5ReNCCH_3][BF_4]$ .... Re: Org.Comp.2-152
$BC_7F_4H_4O_5Re$ .... $[(CO)_5Re(C_2H_4)][BF_4]$ .... Re: Org.Comp.2-350, 352
$BC_7F_4H_4O_6Re$ .... $[(CO)_5ReOC(CH_3)H][BF_4]$ .... Re: Org.Comp.2-157
$BC_7F_4H_5NO_3Re$ .. $[(C_5H_5)Re(CO)_2NO][BF_4]$ .... Re: Org.Comp.3-189/93
$BC_7F_4H_8NO$ .... $H[BF_4] \cdot C_6H_5$-CO-NH_2 .... B: B Comp.SVol.4/3b-165
$BC_7F_4H_9N_2O$ .... $H[BF_4] \cdot O=C(NH_2)$-NH-C_6H_5 .... B: B Comp.SVol.4/3b-165
$BC_7F_4H_{13}IMoN_3O$ $[C_5H_5Mo(NO)(NH_2N(CH_3)_2)I][BF_4]$ .... Mo: Org.Comp.6-39
$BC_7FeH_5O_3$ .... $C_4BH_5Fe(CO)_3$ .... B: B Comp.SVol.3/4-155
$BC_7GeH_{17}O$ .... $(CH_3)_2Ge(-CH_2CH_2B(OCH_3)CH_2CH_2-)$ .... Ge: Org.Comp.3-295, 298
$BC_7H_6N_3$ .... 4-CN-C_5H_4N-BH_2CN .... B: B Comp.SVol.3/3-184
$BC_7H_7I_4$ .... $[C_7H_7][BI_4]$ .... B: B Comp.SVol.4/4-110
BC_7H_7NS .... 3-CH_3-1,3,2-$SNBC_6H_4$ .... B: B Comp.SVol.4/4-146
$BC_7H_7N_2O$ .... $C_6H_5B(-O-N=CH-NH-)$ .... B: B Comp.SVol.3/3-163
$BC_7H_7N_2O_2$ .... 4-HOOC-C_5H_4N-BH_2CN .... B: B Comp.SVol.3/3-184
$BC_7H_7O_3S$ .... 2-HO-5-CH_3O-1,3,2-$OSBC_6H_3$ .... B: B Comp.SVol.3/4-120
$BC_7H_7S_2$ .... 2-CH_3-1,3,2-$S_2BC_6H_4$ .... B: B Comp.SVol.3/4-114

BC_7H_8NO .... 2-CH_3-1,3,2-$ONBC_6H_5$ .... B: B Comp.SVol.3/3-164
– .... 3-CH_3-1,3,2-$ONBC_6H_5$ .... B: B Comp.SVol.3/3-164
BC_7H_8NS........ 1,3-SNC_7H_5 · BH_3 .... B: B Comp.SVol.4/4-147/8
– .... 2-CH_3-1,3,2-$SNBC_6H_5$ .... B: B Comp.SVol.3/4-127/8
$BC_7H_9N_2$ .... 2-CH_3-1,3,2-$N_2BC_6H_6$ .... B: B Comp.SVol.3/3-117
– .... $C_6H_5NH_2$-BH_2-CN .... B: B Comp.SVol.3/3-184
$BC_7H_9O_2S$....... C_6H_5-S-B(OCH_3)-OH .... B: B Comp.SVol.3/4-119
– .... C_6H_5-S-CH_2-B$(OH)_2$ .... B: B Comp.SVol.3/4-121
$BC_7H_{10}NO$ ...... H_3B · NC_5H_4-2-C(O)CH_3 .... B: B Comp.SVol.4/3b-13
– .... [-O-BH_2-1-NC_5H_4-2-CH_2CH_2-] .... B: B Comp.SVol.4/3b-72
$BC_7H_{10}NS$....... 1,3-SNC_7H_7 · BH_3 .... B: B Comp.SVol.4/3b-7
B: B Comp.SVol.4/4-147/8
$BC_7H_{12}N$........ BH_3 · NC_5H_3-2,6-$(CH_3)_2$ .... B: B Comp.SVol.3/3-177
B: B Comp.SVol.4/3b-13
– .... BH_3 · NC_5H_4-2-C_2H_5 .... B: B Comp.SVol.4/3b-13
– .... BH_3 · NH(CH_3)-C_6H_5 .... B: B Comp.SVol.4/3b-5, 10
– .... BH_3 · NH_2-CH_2-C_6H_5 .... B: B Comp.SVol.4/3b-9
– .... $(CH_3)_2BH$ · NC_5H_5 .... B: B Comp.SVol.4/3b-26
– .... $(CH_3)_2N$-B[-CH=CH-CH=CH-CH_2-] .... B: B Comp.SVol.4/3a-228
– .... $(CH_3)_2N$-B[-CH=CH-CH_2-CH=CH-] .... B: B Comp.SVol.4/3a-228
$BC_7H_{12}NO$ ...... BH_3 · NC_5H_4-2-CH(OH)-CH_3 .... B: B Comp.SVol.4/3b-13
– .... BH_3 · NC_5H_4-2-CH_2CH_2-OH .... B: B Comp.SVol.4/3b-13
$BC_7H_{13}N_2$ .... BH_3 · NC_5H_4-2-N$(CH_3)_2$ .... B: B Comp.SVol.4/3b-13
$BC_7H_{13}N_2O$...... 1-HO-2,3,1-$N_2BC_7H_{12}$ .... B: B Comp.SVol.4/3b-67
$BC_7H_{13}N_4$ .... (1,2-$N_2C_3H_3$)-1-B(-NCH_3-CH_2-CH_2-NCH_3-) .. B: B Comp.SVol.3/3-95
– .... (1,3-$N_2C_3H_3$)-1-B(-NCH_3-CH_2-CH_2-NCH_3-) .. B: B Comp.SVol.3/3-96
$BC_7H_{13}SSe$...... 5-CH_3-3,4-$(C_2H_5)_2$-1,2,5-$SSeBC_2$ .... B: B Comp.SVol.4/4-172/3
$BC_7H_{13}Se_2$ ...... 4,5-$(C_2H_5)_2$-3-CH_3-1,2,3-Se_2BC_2 .... B: B Comp.SVol.4/4-171/2
$BC_7H_{14}N$........ $(CH_3)_2B$-N(C_2H_5)CH_2-C≡CH .... B: B Comp.SVol.3/3-151
$BC_7H_{14}N_3O$...... $(CH_3)_3N$-$^{10}BH_2$-CO-1-(1,3-$N_2C_3H_3$) .... B: B Comp.SVol.4/3b-25
$BC_7H_{14}N_3O_2$..... CH_3B(-NCH_3-CO-NC_2H_5-CO-NCH_3-) .... B: B Comp.SVol.3/3-115
$BC_7H_{14}N_5$ .... (1,2,4-$N_3C_2H_2$)-1-B(-NCH_3-CH_2-CH_2-CH_2-NCH_3-)
B: B Comp.SVol.3/3-95
$BC_7H_{15}LiNSi$..... Li[1-$(CH_3)_3Si$-2-CH_3-1,2-NBC_3H_3] .... B: B Comp.SVol.3/3-213
B: B Comp.SVol.4/3b-48
$BC_7H_{15}NSi^-$ ..... [1-$(CH_3)_3Si$-2-CH_3-1,2-NBC_3H_3]$^-$ .... B: B Comp.SVol.4/3a-249
$BC_7H_{15}N_2$ .... 3-CH_3-1,3-$N_2C_3H_3$ · B$(CH_3)_3$ .... B: B Comp.SVol.4/3b-28
– .... $(CH_3)_3N$ · BH_2-1-NC_4H_4 .... B: B Comp.SVol.4/3b-23/4
$BC_7H_{15}N_2O_2S_2$ ... $(CH_3O)_2B$[-NH-C(SCH_3)=CH-C(SCH_3)=NH-] .. B: B Comp.SVol.3/3-122/3
B: B Comp.SVol.4/4-152
$BC_7H_{15}N_4$ .... [$(CH_3)_2N]_2B$[1-(1,2-$N_2C_3H_3$)] .... B: B Comp.SVol.3/3-94
$BC_7H_{15}OS$....... 5-CH_3-2-(i-C_3H_7)-1,3,2-$OSBC_3H_5$ .... B: B Comp.SVol.4/4-137
$BC_7H_{15}O_2S$...... 2-($CH_3SCH_2CH_2$)-4,5-$(CH_3)_2$-1,3,2-$O_2BC_2H_2$ B: B Comp.SVol.4/4-138
$BC_7H_{15}O_3S$...... 1-[CH_3-S(=O)$_2$O]-2,5-$(CH_3)_2$-BC_4H_6 .... B: B Comp.SVol.4/4-137/8
$BC_7H_{15}S_2$ .... $C_5H_{11}B$(-S-CH_2-CH_2-S-) .... B: B Comp.SVol.3/4-112
$BC_7H_{16}N$........ n-C_3H_7-B=N-C_4H_9-t .... B: B Comp.SVol.3/3-143
B: B Comp.SVol.4/3a-213
– .... i-C_3H_7-B=N-C_4H_9-t .... B: B Comp.SVol.3/3-142/3
B: B Comp.SVol.4/3a-213
– .... HC(-CH_2-CH_2-$)_3$N-BH_3 .... B: B Comp.SVol.3/3-177

Formula	Compound	Reference
$BC_7H_{16}NO$	2-(i-C_4H_9)-1,3,2-$ONBC_3H_7$	B: B Comp.SVol.4/3b-66
–	3-HO-[2.2.2]-1-NC_7H_{12} · BH_3	B: B Comp.SVol.4/3b-13
–	$(CH_3)_2NH$ · $B(OCH_3)[-CH=CH-CH_2CH_2-]$	B: B Comp.SVol.4/3b-70
–	$(CH_3)_2NH$ · $B(OCH_3)[-CH_2-CH=CH-CH_2-]$	B: B Comp.SVol.4/3b-70
$BC_7H_{16}NO_4$	$(CH_3)_3N-BH_2-C(O)-O-C(O)-O-C_2H_5$	B: B Comp.SVol.4/3b-25
$BC_7H_{16}NSi$	1-$(CH_3)_3Si$-2-CH_3-1,2-NBC_3H_4	B: B Comp.SVol.3/3-162
		B: B Comp.SVol.4/3a-249
$BC_7H_{17}N_2$	$(CH_3)_3N-BH_2-C(CH_3)_2-CN$	B: B Comp.SVol.4/3b-25
–	$(C_2H_5)_3N-BH_2-CN$	B: B Comp.SVol.4/3b-23
–	(n-$C_3H_7)_2NH-BH_2-CN$	B: B Comp.SVol.4/3b-22
–	(i-$C_3H_7)_2NH-BH_2-CN$	B: B Comp.SVol.4/3b-22
$BC_7H_{17}N_2O$	$H_3B-N[=CH-C(CH_3)_2-N(OH)-C(CH_3)_2-]$	B: B Comp.SVol.4/3b-12
$BC_7H_{17}N_2S$	$(C_2H_5)_3N-BH_2-N=C=S$	B: B Comp.SVol.4/3b-23
–	(i-$C_3H_7)_2NH-BH_2-N=C=S$	B: B Comp.SVol.4/3b-23
$BC_7H_{18}N$	$(-CH_2CH_2-CH_2CH_2-)NH-B(CH_3)_3$	B: B Comp.SVol.4/3b-28
–	(i-$C_3H_7)_2B-NH-CH_3$	B: B Comp.SVol.4/3a-230
$BC_7H_{18}NO_2$	(i-$C_3H_7)_2NH-BH_2-COOH$	B: B Comp.SVol.4/3b-24
$BC_7H_{18}NSi$	t-$C_4H_9-B=N-Si(CH_3)_3$	B: B Comp.SVol.4/3a-214
$BC_7H_{18}N_3$	1,3-$(CH_3)_2$-2-$(CH_3)_2N$-1,3,2-$N_2BC_3H_6$	B: B Comp.SVol.3/3-95
–	1-C_2H_5-2-$(CH_3)_2N$-1,3,2-$N_2BC_3H_7$	B: B Comp.SVol.4/3a-154
$BC_7H_{19}NOSi$	(t-$C_4H_9O)BH-NSi(CH_3)_3$	B: B Comp.SVol.3/3-154
$BC_7H_{19}N_2$	$[-N(CH_3)-C(CH_3)_2-N(CH_3)-CH_2CH_2-]$ · BH_3	B: B Comp.SVol.4/3b-12
$BC_7H_{19}N_2Si$	$(CH_3)_3Si-B(-NCH_3-CH_2CH_2-NCH_3-)$	B: B Comp.SVol.3/3-113
		B: B Comp.SVol.4/3a-174
$BC_7H_{19}N_2Sn$	$(CH_3)_3SnB(-NCH_3-CH_2CH_2-NCH_3-)$	B: B Comp.SVol.3/3-113
$BC_7H_{19}N_4SSi$	$(CH_3)_3Si-N=S=N-B(-NCH_3-CH_2CH_2-NCH_3-)$	B: B Comp.SVol.3/3-97
		S: S-N Comp.7-178
$BC_7H_{20}NO_2Si$	$(CH_3)_3N-BH_2COOSi(CH_3)_3$	B: B Comp.SVol.3/3-186/7
$BC_7H_{20}NSi$	1-SiH_3-NC_4H_8 · $B(CH_3)_3$	Si: SVol.B4-178
–	$(C_2H_5)_2B-NH-Si(CH_3)_3$	B: B Comp.SVol.3/3-148
		B: B Comp.SVol.4/3a-230
$BC_7H_{20}NSn$	$(CH_3)_3SnN(C_2H_5)B(CH_3)_2$	Sn: Org.Comp.18-62, 73
$BC_7H_{21}N_2Si$	$(CH_3)_2B-N[Si(CH_3)_3]-N(CH_3)_2$	B: B Comp.SVol.3/3-153
–	$(CH_3)_3Si-B[N(CH_3)_2]_2$	B: B Comp.SVol.3/3-105
$BC_7H_{21}N_2Sn$	$(CH_3)_3Sn-B[N(CH_3)_2]_2$	B: B Comp.SVol.3/3-106
$BC_7H_{21}N_3OP$	$(CH_3)_2B-N(CH_3)PO[N(CH_3)_2]_2$	B: B Comp.SVol.3/3-154
$BC_7H_{22}NSi$	$SiH_3N(C_2H_5)_2$ · $B(CH_3)_3$	Si: SVol.B4-175
$BC_7H_{23}N_2Si$	$SiH_2(N(CH_3)_2)_2$ · $B(CH_3)_3$	Si: SVol.B4-186
$BC_8ClF_2H_7NO_4$	4-NO_2-$C_6H_4COOCH_3$ · $BClF_2$	B: B Comp.SVol.3/4-64
$BC_8ClF_2H_8O_2$	$C_6H_5-C(O)OCH_3$ · $BClF_2$	B: B Comp.SVol.3/4-64
–	$C_6H_5-OC(O)CH_3$ · $BClF_2$	B: B Comp.SVol.3/4-64
$BC_8ClF_2H_{11}N$	$N(CH_3)_2C_6H_5$ · $BClF_2$	B: B Comp.SVol.3/4-65/7
$BC_8ClF_2H_{19}N$	$N(C_2H_5)$(i-$C_3H_7)_2$ · $BClF_2$	B: B Comp.SVol.3/4-65/7
$BC_8ClF_3H_7O_2$	4-F-$C_6H_4-C(O)OCH_3$ · $BClF_2$	B: B Comp.SVol.3/4-64
–	4-F-$C_6H_4-OC(O)CH_3$ · $BClF_2$	B: B Comp.SVol.3/4-64
BC_8ClH_8	1-Cl-1-BC_8H_8	B: B Comp.SVol.4/4-48
–	2-Cl-2-BC_8H_8	B: B Comp.SVol.4/4-48
BC_8ClH_9N	$(H_2C=CH)ClBNHC_6H_5$	B: B Comp.SVol.3/4-56
$BC_8ClH_{10}MnN_2O_4$	Mn[(2,5-$(O=)_2$-NC_4H_4-1$)_2BH_2$]Cl	Mn: MVol.D8-7/8
$BC_8ClH_{11}N$	$C_6H_5-BCl-N(CH_3)_2$	B: B Comp.SVol.3/4-55

Formula	Compound	Reference
$BC_8ClH_{11}N$	C_6H_5-BCl-$N(CH_3)_2$	B: B Comp.SVol.4/4-60
$BC_8ClH_{12}MnN_4$	Mn[(2-CH_3-1,3-$N_2C_3H_2$-1)$_2BH_2$]Cl	Mn: MVol.D8-18/20
BC_8ClH_{14}	9-Cl-[3.3.1]-9-BC_8H_{14}	B: B Comp.SVol.4/4-47
–	$(CH_3CH{=}CCH_3)_2BCl$	B: B Comp.SVol.3/4-43
–	$(C_2H_5$-$CH{=}CH)_2BCl$	B: B Comp.SVol.3/4-43
$BC_8ClH_{14}O_2$	(n-C_3H_7)ClB(-O=CCH_3-CH=CCH_3-O-)	B: B Comp.SVol.3/4-52
$BC_8ClH_{16}O_2$	i-$C_3H_7C(CH_3)_2$-BCl-OC(O)CH_3	B: B Comp.SVol.4/4-55
$BC_8ClH_{16}O_2S$	4,5-$(CH_3)_2$-1,3,2-$O_2BC_2H_2$-2-(CHCl-CH_2CH_2-SCH_3)	B: B Comp.SVol.4/4-165
BC_8ClH_{18}	(n-$C_4H_9)_2$BCl	B: B Comp.SVol.3/4-43
		B: B Comp.SVol.4/4-46
–	(i-$C_4H_9)_2$BCl	B: B Comp.SVol.3/4-43
		B: B Comp.SVol.4/4-46
–	(s-$C_4H_9)_2$BCl	B: B Comp.SVol.3/4-43
		B: B Comp.SVol.4/4-46
–	(t-$C_4H_9)_2$BCl	B: B Comp.SVol.3/4-43
		B: B Comp.SVol.4/4-46
$BC_8ClH_{19}N$	n-C_4H_9-BCl-NH-C_4H_9-t	B: B Comp.SVol.3/4-56
–	HClB-N(s-$C_4H_9)_2$	B: B Comp.SVol.3/4-56
$BC_8ClH_{20}N_2$	ClB[$N(C_2H_5)_2]_2$	B: B Comp.SVol.3/4-58
		B: B Comp.SVol.4/4-62
$BC_8ClH_{20}S$	i-C_3H_7-$C(CH_3)_2$-BHCl · $S(CH_3)_2$	B: B Comp.SVol.3/4-137
		B: B Comp.SVol.4/4-162
$BC_8ClH_{21}N$	(n-$C_4H_9)_2$NH · BH_2Cl	B: B Comp.SVol.4/4-39
–	(i-$C_4H_9)_2$NH · BH_2Cl	B: B Comp.SVol.4/4-39
$BC_8ClH_{21}NSi$	t-C_4H_9-BCl-N(CH_3)-Si$(CH_3)_3$	B: B Comp.SVol.4/4-60
$BC_8ClH_{22}N_2Si_2$	ClB[-N(C_2H_5)-Si$(CH_3)_2$-Si$(CH_3)_2$-N(C_2H_5)-]	B: B Comp.SVol.4/4-65
$BC_8ClH_{23}N$	[$(C_2H_5)_4$N][H_3BCl]	B: B Comp.SVol.4/4-39
$BC_8ClH_{24}N_2Sn_2$	$N(CH_3)_2$-BCl-N[Sn$(CH_3)_3]_2$	B: B Comp.SVol.4/4-63
$BC_8Cl_2FH_7NO_4$	4-NO_2-$C_6H_4COOCH_3$ · BCl_2F	B: B Comp.SVol.3/4-64
$BC_8Cl_2FH_8O_2$	C_6H_5-C(O)OCH_3 · BCl_2F	B: B Comp.SVol.3/4-64
–	C_6H_5-OC(O)CH_3 · BCl_2F	B: B Comp.SVol.3/4-64
$BC_8Cl_2FH_{11}N$	$N(CH_3)_2C_6H_5$ · BCl_2F	B: B Comp.SVol.3/4-65/7
$BC_8Cl_2FH_{19}N$	N(C_2H_5)(i-$C_3H_7)_2$ · BCl_2F	B: B Comp.SVol.3/4-65/7
$BC_8Cl_2F_2H_7O_2$	2-Cl-C_6H_4-C(O)OCH_3 · $BClF_2$	B: B Comp.SVol.3/4-64
–	4-F-C_6H_4-C(O)OCH_3 · BCl_2F	B: B Comp.SVol.3/4-64
–	4-F-C_6H_4-OC(O)CH_3 · BCl_2F	B: B Comp.SVol.3/4-64
$BC_8Cl_2H_6O_6Re$	cis-$(CO)_4$Re[$(CH_3CO)_2$]BCl_2	Re: Org.Comp.2-344
$BC_8Cl_2H_7$	4-(CH_2=CH)-C_6H_4-BCl_2	B: B Comp.SVol.4/4-44
–	2-(CHD=CH)-C_6H_4-BCl_2	B: B Comp.SVol.4/4-44
$BC_8Cl_2H_7NO_5Re$	cis-$(CO)_4$Re[(CH_3CO)(CH_3CNH)]BCl_2	Re: Org.Comp.2-343
$BC_8Cl_2H_9$	2-CH_3-C_6H_4-CH_2-BCl_2	B: B Comp.SVol.4/4-43
–	2-C_2H_5-C_6H_4-BCl_2	B: B Comp.SVol.4/4-43
–	C_6H_5-CH_2CH_2-BCl_2	B: B Comp.SVol.4/4-43
$BC_8Cl_2H_{17}$	n-C_8H_{17}-BCl_2	B: B Comp.SVol.4/4-44
$BC_8Cl_2H_{17}S$	$(CH_3)_2$S-B(c-C_5H_8-2-CH_3)Cl_2	B: B Comp.SVol.3/4-137
$BC_8Cl_2H_{18}N$	Cl_2B-N(C_4H_9-n)$_2$	B: B Comp.SVol.3/4-54
–	Cl_2B-N(C_4H_9-s)$_2$	B: B Comp.SVol.3/4-54
–	Cl_2B-N(C_4H_9-t)$_2$	B: B Comp.SVol.4/4-58
$BC_8Cl_2H_{18}N_2P$	[-N(C_4H_9-t)-BCl-N(C_4H_9-t)-]PCl	B: B Comp.SVol.4/4-64/5

Formula	Compound	Reference
$BC_8Cl_2H_{19}S$	$n\text{-}C_3H_7\text{-}CH(C_2H_5)\text{-}BCl_2 \cdot (CH_3)_2S$	B: B Comp.SVol.3/4-42
–	$n\text{-}C_6H_{13}\text{-}BCl_2 \cdot (CH_3)_2S$	B: B Comp.SVol.3/4-137
$BC_8Cl_2H_{20}NSn$	$(CH_3)_2Sn(Cl)\text{-}N(C_4H_9\text{-}t)\text{-}B(Cl)\text{-}C_2H_5$	Sn: Org.Comp.19-118, 123
$BC_8Cl_3FH_7O_2$	$2\text{-}Cl\text{-}C_6H_4\text{-}C(O)OCH_3 \cdot BCl_2F$	B: B Comp.SVol.3/4-64
–	$4\text{-}F\text{-}C_6H_4\text{-}C(O)OCH_3 \cdot BCl_3$	B: B Comp.SVol.3/4-52
–	$4\text{-}F\text{-}C_6H_4\text{-}OC(O)CH_3 \cdot BCl_3$	B: B Comp.SVol.3/4-52
$BC_8Cl_3H_8O_2$	$C_6H_5\text{-}C(O)OCH_3 \cdot BCl_3$	B: B Comp.SVol.3/4-52
–	$C_6H_5\text{-}OC(O)CH_3 \cdot BCl_3$	B: B Comp.SVol.3/4-52
$BC_8Cl_3H_{11}N$	$C_6H_5\text{-}N(CH_3)_2 \cdot BCl_3$	B: B Comp.SVol.3/4-66
–	$(Cl)_3NBC_6H_6(C_2H_5)$	B: B Comp.SVol.3/4-57
$BC_8Cl_3H_{14}$	$n\text{-}C_6H_{13}\text{-}CCl{=}CH\text{-}BCl_2$	B: B Comp.SVol.4/4-43
$BC_8Cl_3H_{18}N_2Si$	$[\text{-}N(C_4H_9\text{-}t)\text{-}BCl\text{-}N(C_4H_9\text{-}t)\text{-}]SiCl_2$	B: B Comp.SVol.4/4-64/5
$BC_8Cl_4FeH_5O_3$	$[(C_5H_5)Fe(CO)_3][BCl_4]$	Fe: Org.Comp.B15-36, 38, 46
$BC_8Cl_6FeH_5O_2$	$[C_5H_5(CO)_2Fe{=}CCl_2][BCl_4]$	Fe: Org.Comp.B16a-135, 148/9
$BC_8CoH_8O_2$	$(CO)_2Co(C_5H_5BCH_3)$	B: B Comp.SVol.4/4-180
$BC_8FH_7N_3O_2$	$FB[\text{-}O\text{-}(1,2\text{-}C_6H_4)\text{-}CH{=}N\text{-}][\text{-}O\text{-}CNH_2{=}N\text{-}]$	B: B Comp.SVol.3/3-378
$BC_8FH_{11}I_2N$	$N(CH_3)_2(C_6H_5) \cdot BFI_2$	B: B Comp.SVol.3/4-65/7
$BC_8FH_{11}N$	$F(C_6H_5)B\text{-}N(CH_3)_2$	B: B Comp.SVol.3/3-376
BC_8FH_{18}	$(n\text{-}C_4H_9)_2BF$	B: B Comp.SVol.3/3-364
–	$(t\text{-}C_4H_9)_2BF$	B: B Comp.SVol.4/3b-227
$BC_8FH_{21}NSi$	$F(CH_3)B\text{-}N[Si(CH_3)_3][t\text{-}C_4H_9]$	B: B Comp.SVol.3/3-376
$BC_8F_2H_5O_6Re^-$	$[cis\text{-}(CO)_4ReCH_2COCO(CH_3)BF_2]^-$	Re: Org.Comp.2-388
$BC_8F_2H_6O_6Re$	$cis\text{-}(CO)_4Re[(CH_3CO)_2]BF_2$	Re: Org.Comp.2-344, 347
$BC_8F_2H_8NO_2$	$F_2B[\text{-}O\text{-}(1,2\text{-}C_6H_4)\text{-}CH{=}NOCH_3\text{-}]$	B: B Comp.SVol.3/3-378
$BC_8F_2H_{11}IN$	$N(CH_3)_2(C_6H_5) \cdot BF_2I$	B: B Comp.SVol.3/4-65/7
$BC_8F_2H_{11}O_2$	$OC_5H_9\text{-}2\text{-}O\text{-}CH_2\text{-}C{\equiv}C\text{-}BF_2$	B: B Comp.SVol.4/3b-226
$BC_8F_2H_{17}$	$n\text{-}C_8H_{17}\text{-}BF_2$	B: B Comp.SVol.4/3b-226
$BC_8F_2H_{18}N$	$F_2B\text{-}N(C_4H_9\text{-}t)_2$	B: B Comp.SVol.4/3b-243
$BC_8F_3H_{10}N_2O_2$	$4\text{-}O_2NC_6H_4N(CH_3)_2 \cdot BF_3$	B: B Comp.SVol.3/4-65
$BC_8F_3H_{11}N$	$N(CH_3)_2C_6H_5 \cdot BF_3$	B: B Comp.SVol.3/4-65/7
$BC_8F_3H_{18}O$	$F_3B\text{-}O(C_4H_9\text{-}n)_2$	B: B Comp.SVol.3/3-287 B: B Comp.SVol.4/3b-144
$BC_8F_3H_{19}N$	$N(C_2H_5)(i\text{-}C_3H_7)_2 \cdot BF_3$	B: B Comp.SVol.3/4-65/7
$BC_8F_3H_{25}N_3Si_4$	$FB[\text{-}N(SiF(CH_3)_2)\text{-}Si(CH_3)_2\text{-}N(SiF(CH_3)_2)\text{-}Si(CH_3)_2\text{-}NH\text{-}]$	B: B Comp.SVol.3/3-377
–	$F_2B\text{-}N[\text{-}Si(CH_3)_2\text{-}NH\text{-}Si(CH_3)_2\text{-}N(Si(CH_3)_2F)\text{-}Si(CH_3)_2\text{-}]$	B: B Comp.SVol.3/3-375
$BC_8F_4FeH_5O_3$	$[(C_5H_5)Fe(CO)_3][BF_4]$	Fe: Org.Comp.B15-36/8, 40/1, 45, 46
$BC_8F_4FeH_5O_4$	$[(HO\text{-}C_5H_4)Fe(CO)_3][BF_4]$	Fe: Org.Comp.B15-50/1
$BC_8F_4FeH_7O_2$	$[C_5H_5(CO)_2Fe{=}CH_2][BF_4]$	Fe: Org.Comp.B16a-87, 93/4
$BC_8F_4FeH_7O_3$	$[(CH_2CHCHCHCH_2)Fe(CO)_3][BF_4]$	Fe: Org.Comp.B15-15/6, 18, 28
$BC_8F_4H_4N_2O_5Re$	$[(CO)_5ReN(\text{-}CHCHNHCH\text{-})][BF_4]$	Re: Org.Comp.2-151
–	$[(CO)_5ReN(\text{-}NHCHCHCH\text{-})][BF_4]$	Re: Org.Comp.2-151
$BC_8F_4H_4O_7Re$	$[(CO)_5Re{=}C(OCH_2)_2][BF_4]$	Re: Org.Comp.2-162
$BC_8F_4H_6O_5Re$	$[(CO)_5Re(CH_2{=}CHCH_3)][BF_4]$	Re: Org.Comp.2-350, 352/3
$BC_8F_4H_6O_6Re$	$[(CO)_5ReOC(CH_3)_2][BF_4]$	Re: Org.Comp.2-157
$BC_8F_4H_8N_2O_2Re$	$[(C_5H_5)Re(CO)(NO)(NC\text{-}CH_3)][BF_4]$	Re: Org.Comp.3-154, 155

Formula	Compound	Reference
$BC_8F_4H_{11}N_2S_2$	$[(CH_3)_2SN{=}S{=}N{-}C_6H_5][BF_4]$	S: S-N Comp.7-23
$BC_8F_4H_{11}SSe$	$[(C_6H_5Se)(CH_3)_2S][BF_4]$	B: B Comp.SVol.3/4-147
$BC_8F_4H_{15}IMoN_3O$	$[C_5H_5Mo(NO)(N(CH_3)HN(CH_3)_2)I][BF_4]$	Mo: Org.Comp.6-39
$BC_8F_4H_{17}IN_4O_2Re$	$[(CO)Re(1,4,7{-}c{-}N_3C_6H_{15})(NO)CH_2I][BF_4]$	Re: Org.Comp.1-52/3
$BC_8F_4H_{18}N_4O_2Re$	$[(CO)Re(1,4,7{-}c{-}N_3C_6H_{15})(NO)CH_3][BF_4]$	Re: Org.Comp.1-52
$BC_8F_4H_{20}N$	$[(C_2H_5)_4N][BF_4]$	B: B Comp.SVol.3/3-343/4
		B: B Comp.SVol.4/3b-208, 210
–	$[(n{-}C_4H_9)_2NH_2][BF_4]$	B: B Comp.SVol.4/3b-208, 210
$BC_8F_6FeH_5O_2$	$[C_5H_5(CO)_2Fe{=}CF_2][BF_4]$	Fe: Org.Comp.B16a-135, 148
$BC_8F_6H_{12}N_3S_2$	$1,9{-}(CF_3S)_2{-}1,9,5,10{-}N_3BC_6H_{12}$	B: B Comp.SVol.4/4-167
$BC_8F_6H_{14}NO_6S_2$	$(i{-}C_3H_7)_2N{-}B[O{-}S({=}O)_2{-}CF_3]_2$	B: B Comp.SVol.4/3b-61
		B: B Comp.SVol.4/4-167
$BC_8F_7H_8Se$	$[C_6H_5Se(CH_3)CF_3][BF_4]$	B: B Comp.SVol.3/4-148
$BC_8F_9H_6N_2S_2$	$[(CH_3)_2SN{=}S{=}N{-}C_6F_5][BF_4]$	S: S-N Comp.7-23
$BC_8F_{10}FeH_{11}P_2$	$[(C_5H_5)Fe(CH_2{=}CHCH_3)(PF_3)_2][BF_4]$	Fe: Org.Comp.B16b-5
$BC_8H_6I_2O_6Re$	$cis{-}(CO)_4Re[(CH_3CO)_2]BI_2$	Re: Org.Comp.2-345, 347/8
$BC_8H_6MnO_3$	$(OC)_3Mn(C_5BH_6)$	B: B Comp.SVol.3/4-155
$BC_8H_7I_2NO_5Re$	$I_2B[{-}NH{=}CCH_3{-}Re(CO)_4{=}CCH_3{-}O{-}]$	B: B Comp.SVol.3/4-99
		Re: Org.Comp.2-343/4
$BC_8H_7O_3Ru$	$(CO)_3Ru(C_4H_4BCH_3)$	B: B Comp.SVol.4/4-180
$BC_8H_7O_4Ru$	$(CO)_3Ru(C_4H_4BOCH_3)$	B: B Comp.SVol.4/4-180
$BC_8H_7S_2$	$C_6H_5B({-}S{-}CH{=}CH{-}S{-})$	B: B Comp.SVol.3/4-114
$BC_8H_8KN_4$	$K[HB(1,2{-}N_2C_3H_3)(NC_4H_4)CN]$	B: B Comp.SVol.3/3-213
–	$K[HB(1,3{-}N_2C_3H_3)(NC_4H_4)CN]$	B: B Comp.SVol.3/3-213
$BC_8H_8N_3O_2$	$C_6H_5B({-}NH{-}CO{-}NH{-}CO{-}NH{-})$	B: B Comp.SVol.3/3-114
BC_8H_9NS	$3{-}C_2H_5{-}1,3,2{-}SNBC_6H_4$	B: B Comp.SVol.4/4-146
$BC_8H_9N_2O$	$4{-}CH_3C(O){-}C_5H_4N{-}BH_2CN$	B: B Comp.SVol.3/3-184
–	$C_6H_5{-}B({-}O{-}N{=}CH{-}NCH_3{-})$	B: B Comp.SVol.3/3-163
$BC_8H_9N_4$	$({-}CH{=}N{-}CH{=}CH{-})NH{-}BH(CN)(C_4H_4N)$	B: B Comp.SVol.3/3-190
–	$({-}N{=}CH{-}CH{=}CH{-})NH{-}BH(CN)(C_4H_4N)$	B: B Comp.SVol.3/3-190
$BC_8H_9O_2$	$2{-}C_6H_5{-}1,3,2{-}O_2BC_2H_4$	B: B Comp.SVol.4/3b-69
$BC_8H_9S_2$	$2,5{-}(CH_3)_2{-}1,3,2{-}S_2BC_6H_3$	B: B Comp.SVol.3/4-114
–	$2{-}C_6H_5{-}1,3,2{-}S_2BC_2H_4$	B: B Comp.SVol.3/4-112/3
$BC_8H_{10}NO$	$1{-}HO{-}2{-}CH_3{-}2,1{-}NBC_7H_6$	B: B Comp.SVol.3/3-163
$BC_8H_{10}NO_2$	$2{-}(CH_3)_2N{-}1,3,2{-}O_2BC_6H_4$	B: B Comp.SVol.3/3-151
$BC_8H_{10}NO_8$	$C_2H_5NH[B({-}O{-}CO{-}CH_2{-}CO{-}O{-})_2]$	B: B Comp.SVol.3/3-227
$BC_8H_{10}NS$	$2{-}CH_3{-}1,3{-}SNC_7H_4 \cdot BH_3$	B: B Comp.SVol.4/4-147/8
$BC_8H_{10}NS_2$	$2{-}(CH_3)_2N{-}1,3,2{-}S_2BC_6H_4$	B: B Comp.SVol.3/4-127
		B: B Comp.SVol.4/4-147
$BC_8H_{10}N_2^-$	$[H_2B(NC_4H_4)_2]^-$	B: B Comp.SVol.3/3-213
$BC_8H_{10}N_3O$	$3{-}CH_3NHCO{-}C_5H_4N{-}BH_2CN$	B: B Comp.SVol.3/3-184
$BC_8H_{11}S_2$	$C_6H_5B(SCH_3)_2$	B: B Comp.SVol.3/4-110
$BC_8H_{12}N$	$(CH_3)_2B{-}NH{-}C_6H_5$	B: B Comp.SVol.4/3a-229
–	$(HC{\equiv}C)_2B{-}N(C_2H_5)_2$	B: B Comp.SVol.3/3-149
		B: B Comp.SVol.4/3a-226
$BC_8H_{12}NO$	$H_2B[{-}N(CH_3)_2{-}(1,2{-}C_6H_4){-}O{-}]$	B: B Comp.SVol.3/3-219
$BC_8H_{12}NS$	$2{-}CH_3{-}1,3{-}SNC_7H_6 \cdot BH_3$	B: B Comp.SVol.4/3b-7
		B: B Comp.SVol.4/4-147/8

$BC_8H_{12}N_3$ . . . $1\text{-}C_4H_4N\text{-}B[\text{-}NCH_3\text{-}CH{=}CH\text{-}NCH_3\text{-}]$ . . . B: B Comp.SVol.3/3-112/3
– . . . $4\text{-}(CH_3)_2N\text{-}C_5H_4N\text{-}BH_2CN$ . . . B: B Comp.SVol.3/3-184
$BC_8H_{12}N_5$ . . . $(CH_3)_2N\text{-}B[1\text{-}(1,2\text{-}N_2C_3H_3)]_2$ . . . B: B Comp.SVol.3/3-94
$BC_8H_{14}I$ . . . $9\text{-}I\text{-}[3.3.1]\text{-}9\text{-}BC_8H_{14}$ . . . B: B Comp.SVol.3/4-100
B: B Comp.SVol.4/4-113
$BC_8H_{14}In$ . . . $In(C_5H_5\text{-}c) \cdot B(CH_3)_3$ . . . In: Org.Comp.1-377/8
$BC_8H_{14}InN_4$ . . . $(CH_3)_2In[\text{-}(1,2\text{-}N_2C_3H_3)\text{-}BH_2\text{-}(1,2\text{-}N_2C_3H_3)\text{-}]$ . . . In: Org.Comp.1-276, 279, 281/2
$BC_8H_{14}Li_2N$ . . . $[3.3.1]\text{-}9\text{-}BC_8H_{14}\text{-}9\text{-}NLi_2$ . . . B: B Comp.SVol.4/3a-228
$BC_8H_{14}N$ . . . $1\text{-}(C_2H_5)_2B\text{-}NC_4H_4$ . . . B: B Comp.SVol.4/3a-233
– . . . $(CH_3)_3B\text{-}NC_5H_5$ . . . B: B Comp.SVol.4/3b-28
– . . . $H_3B\text{-}N(CH_3)_2\text{-}C_6H_5$ . . . B: B Comp.SVol.3/3-178
B: B Comp.SVol.4/3b-5, 10
– . . . $H_3B\text{-}NH(C_2H_5)\text{-}C_6H_5$ . . . B: B Comp.SVol.4/3b-5, 10
– . . . $H_3B\text{-}NH_2\text{-}CH(CH_3)\text{-}C_6H_5$ . . . B: B Comp.SVol.3/3-173
B: B Comp.SVol.4/3b-9
– . . . $H_3B\text{-}NH_2\text{-}CH_2CH_2\text{-}C_6H_5$ . . . B: B Comp.SVol.4/3b-9
– . . . $H_3B\text{-}NH_2\text{-}C_6H_3(CH_3)_2\text{-}2,6$ . . . B: B Comp.SVol.3/3-173
$BC_8H_{14}NOS_2$ . . . $CH_3B[\text{-}S\text{-}CO\text{-}N(C_6H_{11}\text{-}c)\text{-}S\text{-}]$ . . . B: B Comp.SVol.3/4-133
$BC_8H_{14}N_3$ . . . $(1\text{-}H_4C_4N)B(\text{-}NCH_3\text{-}CH_2\text{-}CH_2\text{-}NCH_3\text{-})$ . . . B: B Comp.SVol.3/3-95
$BC_8H_{15}LiN$ . . . $Li[1\text{-}(t\text{-}C_4H_9)\text{-}2\text{-}CH_3\text{-}1,2\text{-}NBC_3H_3]$ . . . B: B Comp.SVol.3/3-213
B: B Comp.SVol.4/3b-48
– . . . $[3.3.1]\text{-}9\text{-}BC_8H_{14}\text{-}9\text{-}NHLi$ . . . B: B Comp.SVol.4/3a-228
$BC_8H_{15}N_2$ . . . $HC(\text{-}CH_2\text{-}CH_2\text{-})_3N\text{-}BH_2CN$ . . . B: B Comp.SVol.3/3-184
– . . . $H_3B\text{-}N(CH_3)_2\text{-}C_6H_4\text{-}2\text{-}NH_2$ . . . B: B Comp.SVol.4/3b-11
– . . . $H_3B\text{-}N(CH_3)_2\text{-}C_6H_4\text{-}4\text{-}NH_2$ . . . B: B Comp.SVol.4/3b-11
– . . . $H_3B\text{-}NH_2\text{-}C_6H_4\text{-}2\text{-}N(CH_3)_2$ . . . B: B Comp.SVol.4/3b-11
– . . . $H_3B\text{-}NH_2\text{-}C_6H_4\text{-}4\text{-}N(CH_3)_2$ . . . B: B Comp.SVol.4/3b-11
$BC_8H_{15}N_4$ . . . $(1,2\text{-}N_2C_3H_3)\text{-}1\text{-}B(\text{-}NCH_3\text{-}CH_2CH_2CH_2\text{-}NCH_3\text{-})$ B: B Comp.SVol.3/3-95
– . . . $(1,3\text{-}N_2C_3H_3)\text{-}1\text{-}B(\text{-}NCH_3\text{-}CH_2CH_2CH_2\text{-}NCH_3\text{-})$ B: B Comp.SVol.3/3-96
– . . . $1\text{-}[B(\text{-}NCH_3\text{-}CH_2CH_2\text{-}NCH_3\text{-})]\text{-}3\text{-}CH_3\text{-}1,2\text{-}N_2C_3H_2$ B: B Comp.SVol.3/3-96
$BC_8H_{15}N_4O_3$ . . . $CH_3B[\text{-}NCH_3\text{-}(CO\text{-}NCH_3\text{-})_3]$ . . . B: B Comp.SVol.3/3-116
$BC_8H_{15}SSe$ . . . $3,4,5\text{-}(C_2H_5)_3\text{-}1,2,5\text{-}SSeBC_2$ . . . B: B Comp.SVol.4/4-172, 174
$BC_8H_{15}Se$ . . . $[3.3.1]\text{-}9\text{-}BC_8H_{14}\text{-}9\text{-}SeH$ . . . B: B Comp.SVol.4/4-170
$BC_8H_{15}Se_2$ . . . $3,4,5\text{-}(C_2H_5)_3\text{-}1,2,3\text{-}Se_2BC_2$ . . . B: B Comp.SVol.4/4-171/2
$BC_8H_{16}N$ . . . $1\text{-}(t\text{-}C_4H_9)\text{-}2\text{-}CH_3\text{-}1,2\text{-}NBC_3H_4$ . . . B: B Comp.SVol.4/3a-249
– . . . $(CH_3)_2B\text{-}N(C_3H_7\text{-}n)\text{-}CH_2\text{-}C{\equiv}CH$ . . . B: B Comp.SVol.3/3-151
– . . . $(C_2H_5)_2N\text{-}B[\text{-}CH{=}CH\text{-}CH_2\text{-}CH_2\text{-}]$ . . . B: B Comp.SVol.4/3a-228
– . . . $(C_2H_5)_2N\text{-}B[\text{-}CH_2\text{-}CH{=}CH\text{-}CH_2\text{-}]$ . . . B: B Comp.SVol.4/3a-228
– . . . $[(3.3.1)\text{-}9\text{-}BC_8H_{14}\text{-}9\text{-}NH_2]_2$ . . . B: B Comp.SVol.4/3a-228
$BC_8H_{16}NO_2$ . . . $HC(\text{-}CH_2\text{-}CH_2\text{-})_3N\text{-}BH_2COOH$ . . . B: B Comp.SVol.3/3-186
$BC_8H_{16}NO_3S$ . . . $[\text{-}O\text{-}CH_2CH_2\text{-}O\text{-}]B[\text{-}O{=}CCH_3\text{-}NH\text{-}CH(CH_2CH_2SCH_3)\text{-}]$ B: B Comp.SVol.4/4-153/4
$BC_8H_{16}NO_4$ . . . $[\text{-}B(C_2H_5)_2\text{-}O\text{-}C({=}O)\text{-}CH_2\text{-}NH(CH_2\text{-}COOH)\text{-}]$ . . . B: B Comp.SVol.4/3b-71
$BC_8H_{16}N_3$ . . . $CH_2{=}CHNHB(NC_3H_6)_2$ . . . B: B Comp.SVol.3/3-95
$BC_8H_{16}N_3O_2$ . . . $CH_3B(\text{-}NC_2H_5\text{-}CO\text{-}NCH_3\text{-}CO\text{-}NC_2H_5\text{-})$ . . . B: B Comp.SVol.3/3-115

$BC_8H_{16}N_5$ .... $N_3B(NC_4H_8)_2$ .... B: B Comp.SVol.3/3-97
$BC_8H_{17}N_2O$ ...... $[(CH_3)_2N]_2B-C≡COC_2H_5$ .... B: B Comp.SVol.3/3-106
$BC_8H_{17}N_4O_3$ ..... $[-N(CH_3)-C(=O)-N(CH_3)-C(=O)-N(CH_3)-]B[$
$-O-CH_2CH_2-NH(CH_3)-]$ .... B: B Comp.SVol.4/3b-56
$BC_8H_{17}OS$ .... $5,5-(CH_3)_2-2-(i-C_3H_7)-1,3,2-OSBC_3H_4$ .... B: B Comp.SVol.4/4-137
$BC_8H_{17}S_2$ .... $C_6H_{13}B(-S-CH_2-CH_2-S-)$ .... B: B Comp.SVol.3/4-112
$BC_8H_{18}I$ .... $(t-C_4H_9)_2BI$ .... B: B Comp.SVol.4/4-113
$BC_8H_{18}I_2S$ .... $(n-C_3H_7)(C_2H_5)CBI_2 \cdot S(CH_3)_2$ .... B: B Comp.SVol.3/4-100
$BC_8H_{18}N$ .... $(-CH=CH-CH_2-CH_2-)B(CH_3)-N(CH_3)_3$ .... B: B Comp.SVol.4/3b-29
– .... $(-CH_2-CH=CH-CH_2-)B(CH_3)-N(CH_3)_3$ .... B: B Comp.SVol.4/3b-29
– .... $n-C_4H_9-B=N-C_4H_9-t$ .... B: B Comp.SVol.3/3-143
B: B Comp.SVol.4/3a-213
– .... $i-C_4H_9-B=N-C_4H_9-i$ .... B: B Comp.SVol.4/3a-213
– .... $s-C_4H_9-B=N-C_4H_9-s$ .... B: B Comp.SVol.3/3-142
– .... $s-C_4H_9-B=N-C_4H_9-t$ .... B: B Comp.SVol.4/3a-214
– .... $t-C_4H_9-B=N-C_4H_9-t$ .... B: B Comp.SVol.3/3-142/4
B: B Comp.SVol.4/3a-214
– .... $[-B(n-C_4H_9)-N(n-C_4H_9)-]_n$ .... B: B Comp.SVol.3/3-134
– .... $[-B(i-C_4H_9)-N(i-C_4H_9)-]_n$ .... B: B Comp.SVol.3/3-134
$BC_8H_{18}NS_2$ ...... $(i-C_3H_7)_2NB(-S-CH_2-CH_2-S-)$ .... B: B Comp.SVol.3/4-124/5, 126
$BC_8H_{18}N_3$ .... $(n-C_4H_9)_2B-N_3$ .... B: B Comp.SVol.3/3-144
B: B Comp.SVol.4/3a-218
– .... $(s-C_4H_9)_2B-N_3$ .... B: B Comp.SVol.3/3-144
$BC_8H_{19}N_2$ .... $CH_3CH=CHCH_2B[N(CH_3)_2]_2$ .... B: B Comp.SVol.3/3-108
$BC_8H_{19}N_2O$ ...... $(CH_3)_2B-NCH_3-C(O)N(C_2H_5)_2$ .... B: B Comp.SVol.3/3-155
– .... $(CH_3)_2B-N(i-C_3H_7)-C(O)N(CH_3)_2$ .... B: B Comp.SVol.3/3-156
– .... $CH_3-O-CH=CH-CH_2-B[N(CH_3)_2]_2$ .... B: B Comp.SVol.3/3-108
B: B Comp.SVol.4/3a-168
– .... $H_3B-N[=C(CH_3)-C(CH_3)_2-N(OH)-C(CH_3)_2-]$ ... B: B Comp.SVol.4/3b-12
$BC_8H_{19}N_2S$ ...... $(CH_3)_2B-NCH_3-C(S)-N(C_2H_5)_2$ .... B: B Comp.SVol.3/3-156
B: B Comp.SVol.3/4-130
$BC_8H_{19}N_4S$ ...... $2-(t-C_4H_9-N=S=N)-1,3,2-N_2BC_2H_4(CH_3)_2-1,3$ S: S-N Comp.7-178
$BC_8H_{20}N$ .... $(CH_3)_3N-BC_5H_{11}$ .... B: B Comp.SVol.3/3-189
– .... $(C_2H_5)_2B-N(C_2H_5)_2$ .... B: B Comp.SVol.3/3-148
B: B Comp.SVol.4/3a-226
– .... $(n-C_3H_7)_2B-N(CH_3)_2$ .... B: B Comp.SVol.3/3-149
– .... $(i-C_3H_7)_2B-N(CH_3)_2$ .... B: B Comp.SVol.4/3a-227
B: B Comp.SVol.4/3b-227
– .... $(s-C_4H_9)_2B-NH_2$ .... B: B Comp.SVol.3/3-147
– .... $(t-C_4H_9)_2B-NH_2$ .... B: B Comp.SVol.4/3a-226
$BC_8H_{20}NO_2$ ...... $(CH_3O)_2B-N(C_3H_7-i)_2$ .... B: B Comp.SVol.4/3b-61
$BC_8H_{20}N_3$ .... $[-NH-CH_2CH_2CH_2-N(CH_3)-]B-N(C_2H_5)_2$ ...... B: B Comp.SVol.4/3a-154
$BC_8H_{21}N_3^+$ ...... $[(C_2H_5)_2N-B(-NHCH_3-CH_2-CH_2-NCH_3-)]^+$ ... B: B Comp.SVol.3/3-207
$BC_8H_{21}N_4Si$ ..... $CH_3B[-N(t-C_4H_9)-N=N-NSi(CH_3)_3-]$ .... B: B Comp.SVol.3/3-137/8
– .... $CH_3-B(N_3)-N(C_4H_9-t)-Si(CH_3)_3$ .... B: B Comp.SVol.3/3-110
$BC_8H_{22}N$ .... $H_3B-N(C_3H_7-i)_2-C_2H_5$ .... B: B Comp.SVol.4/3b-10
$BC_8H_{22}NSi$ ...... $1-SiH_3-NC_5H_{10} \cdot B(CH_3)_3$ .... Si: SVol.B4-182
$BC_8H_{22}N_3Si$ ..... $(CH_3)_3SiN(CH_3)B(-NCH_3-CH_2-CH_2-NCH_3-)$ ... B: B Comp.SVol.3/3-96
$BC_8H_{24}NOSi_2$ .... $(CH_3)_2B-N[Si(CH_3)_3]-O-Si(CH_3)_3$ .... B: B Comp.SVol.3/3-153

$BC_8H_{24}NOSi_2$ .... $(CH_3)_2B-N[Si(CH_3)_3]-O-Si(CH_3)_3$ .... B: B Comp.SVol.4/3a-231
$BC_8H_{24}NSi_2$ ..... $(CH_3)_2B-N[Si(CH_3)_3]_2$ .... B: B Comp.SVol.3/3-149
– .... $[(CH_3)_3Si]_2B-N(CH_3)_2$ .... B: B Comp.SVol.3/3-150
$BC_8H_{24}NSn_2$ ..... $(CH_3)_2B-N[Sn(CH_3)_3]_2$ .... B: B Comp.SVol.3/3-150
B: B Comp.SVol.4/3a-227
– .... $[(CH_3)_3Sn]_2B-N(CH_3)_2$ .... B: B Comp.SVol.3/3-150
$BC_8H_{27}N_4Si$ ..... $Si(N(CH_3)_2)_4 \cdot BH_3$ .... Si: SVol.B4-212
$BC_9ClF_2H_{10}O_2$ ... $4-CH_3-C_6H_4-C(O)OCH_3 \cdot BClF_2$ .... B: B Comp.SVol.3/4-64
– .... $4-CH_3-C_6H_4-OC(O)CH_3 \cdot BClF_2$ .... B: B Comp.SVol.3/4-64
– .... $C_6H_5-CH_2-C(O)OCH_3 \cdot BClF_2$ .... B: B Comp.SVol.3/4-64
$BC_9ClF_2H_{10}O_3$ ... $4-CH_3O-C_6H_4-C(O)OCH_3 \cdot BClF_2$ .... B: B Comp.SVol.3/4-64
$BC_9ClF_2H_{13}N$ .... $4-CH_3-C_6H_4-N(CH_3)_2 \cdot BClF_2$ .... B: B Comp.SVol.3/4-65/7
– .... $C_2H_5-NCH_3-C_6H_5 \cdot BClF_2$ .... B: B Comp.SVol.3/4-65/7
$BC_9ClF_2H_{21}N$ .... $N(n-C_3H_7)_3 \cdot BClF_2$ .... B: B Comp.SVol.3/4-65/7
$BC_9ClF_4FeH_6O_3$ .. $[(1-ClC_6H_6)Fe(CO)_3][BF_4]$ .... Fe: Org.Comp.B15-98, 135
– .... $[(2-ClC_6H_6)Fe(CO)_3][BF_4]$ .... Fe: Org.Comp.B15-102, 135
– .... $[(3-ClC_6H_6)Fe(CO)_3][BF_4]$ .... Fe: Org.Comp.B15-105, 135
$BC_9ClF_4H_{12}Se$ ... $[3-Cl-C_6H_4-CH_2Se(CH_3)_2][BF_4]$ .... B: B Comp.SVol.3/4-148
– .... $[4-Cl-C_6H_4-CH_2Se(CH_3)_2][BF_4]$ .... B: B Comp.SVol.3/4-148
BC_9ClH_8 .... $ClB[-CH=CH-(1,2-C_6H_4)-CH_2-]$ .... B: B Comp.SVol.3/4-44
$BC_9ClH_8O_6Re^-$ ... $[fac-(CO)_3Re(CH_2COCOCH_3)(CH_3CO)BCl]^-$ ... Re: Org.Comp.2-418/9
BC_9ClH_9N .... $H_3B \cdot 1-NC_9H_6-4-Cl$ .... B: B Comp.SVol.4/3b-14
$BC_9ClH_9O_6Re$.... $(CO)_3Re[-C(CH_3)O-]_3BCl$ .... Re: Org.Comp.2-379, 380/1
BC_9ClH_{10} .... $1-Cl-1-BC_9H_{10}$ .... B: B Comp.SVol.4/4-48
– .... $1-Cl-2-CH_3-1-BC_8H_7$ .... B: B Comp.SVol.4/4-48
– .... $1-Cl-3-CH_3-1-BC_8H_7$ .... B: B Comp.SVol.4/4-48
– .... $1-Cl-4-CH_3-1-BC_8H_7$ .... B: B Comp.SVol.4/4-48
$BC_9ClH_{12}Si$...... $1,1-(CH_3)_2-3-Cl-1,3-SiBC_7H_6$ .... B: B Comp.SVol.4/4-48
BC_9ClH_{16} .... $(Cl)BC_8H_{13}(CH_3)$ .... B: B Comp.SVol.3/4-44
$BC_9ClH_{19}NSi$ .... $1,2,2-(CH_3)_3-3,4-(C_2H_5)_2-5-Cl-1,2,5-NSiBC_2$ B: B Comp.SVol.4/4-62
$BC_9ClH_{21}NSi$ .... $[2,2,3-(CH_3)_3-4,5-(C_2H_5)_2-1,2,5-NSiBC_2H_2]Cl$ B: B Comp.SVol.4/4-70
$BC_9ClH_{21}N_2PS$ ... $[-N(C_4H_9-t)-BCl-N(C_4H_9-t)-]P(=S)CH_3$....... B: B Comp.SVol.4/4-65
$BC_9ClH_{23}N$ ...... $(CH_3)_3N \cdot HBCl-C(CH_3)_2-C_3H_7-i$ .... B: B Comp.SVol.3/4-38
– .... $(n-C_3H_7)_3N \cdot BH_2Cl$ .... B: B Comp.SVol.4/4-39
$BC_9ClH_{23}NSi$ .... $(C_2H_5)ClBN(t-C_4H_9)Si(CH_3)_3$ .... B: B Comp.SVol.3/4-56
$BC_9ClH_{24}N_2Sn$ ... $N(CH_3)_2-BCl-N(C_4H_9-t)-Sn(CH_3)_3$ .... B: B Comp.SVol.4/4-63
Sn: Org.Comp.18-57/8, 65
$BC_9ClH_{25}NSiSn$ .. $(CH_3)_3SnN(Si(CH_3)_3)BClC_3H_7-i$ .... Sn: Org.Comp.18-77, 81
$BC_9ClH_{25}NSi_2$.... $(i-C_3H_7)ClBN[Si(CH_3)_3]_2$ .... B: B Comp.SVol.3/4-57
$BC_9ClH_{28}N_2Si_3$... $Si(CH_3)_3-NH-BCl-N[Si(CH_3)_3]_2$ .... B: B Comp.SVol.4/4-63
$BC_9Cl_2FH_{10}O_2$ ... $4-CH_3-C_6H_4-C(O)OCH_3 \cdot BCl_2F$ .... B: B Comp.SVol.3/4-64
– .... $4-CH_3-C_6H_4-OC(O)CH_3 \cdot BCl_2F$ .... B: B Comp.SVol.3/4-64
– .... $C_6H_5CH_2-C(O)OCH_3 \cdot BCl_2F$ .... B: B Comp.SVol.3/4-64
$BC_9Cl_2FH_{10}O_3$ ... $4-CH_3O-C_6H_4-C(O)OCH_3 \cdot BCl_2F$ .... B: B Comp.SVol.3/4-64
$BC_9Cl_2FH_{13}N$ .... $4-CH_3-C_6H_4-N(CH_3)_2 \cdot BCl_2F$ .... B: B Comp.SVol.3/4-65/7
– .... $C_2H_5-NCH_3-C_6H_5 \cdot BCl_2F$ .... B: B Comp.SVol.3/4-65/7
$BC_9Cl_2FH_{21}N$ .... $N(n-C_3H_7)_3 \cdot BCl_2F$ .... B: B Comp.SVol.3/4-65/7
$BC_9Cl_2H_{11}$ ...... $2,4,6-(CH_3)_3-C_6H_2-BCl_2$ .... B: B Comp.SVol.4/4-44
– .... $2-CH_3-C_6H_4-CH_2CH_2-BCl_2$.................. B: B Comp.SVol.4/4-44
– .... $n-C_3H_7-2-C_6H_4-BCl_2$ .... B: B Comp.SVol.4/4-43

Formula	Compound	Reference
$BC_9Cl_2H_{11}$	$i-C_3H_7-2-C_6H_4-BCl_2$	B: B Comp.SVol.4/4-43
$BC_9Cl_2H_{13}Si$	$(CH_3)_3Si-2-C_6H_4-BCl_2$	B: B Comp.SVol.4/4-44
–	$(CH_3)_3Si-3-C_6H_4-BCl_2$	B: B Comp.SVol.4/4-44
–	$(CH_3)_3Si-4-C_6H_4-BCl_2$	B: B Comp.SVol.4/4-44
$BC_9Cl_2H_{14}N_3$	$(ClCH_2CH_2)_2NH-BH(CN)(C_4H_4N)$	B: B Comp.SVol.3/3-190
$BC_9Cl_2H_{18}N$	$1-Cl_2B-2,2,6,6-(CH_3)_4-NC_5H_6$	B: B Comp.SVol.3/4-54
		B: B Comp.SVol.4/4-58
$BC_9Cl_2H_{19}Si$	$C_2H_5-BCl-C(C_2H_5)=C(CH_3)-SiCl(CH_3)_2$	B: B Comp.SVol.4/4-49
$BC_9Cl_2H_{21}N_2Si$	$[-N(C_4H_9-t)-BCl-N(C_4H_9-t)-]SiCl-CH_3$	B: B Comp.SVol.4/4-64/5
$BC_9Cl_2H_{22}N$	$(CH_3)_2CHC(CH_3)_2BCl_2 \cdot (CH_3)_3N$	B: B Comp.SVol.3/4-42
$BC_9Cl_3F_4FeH_9O_2PS$	$[C_5H_5Fe(CO)_2CH(SCH_3)PCl_3][BF_4]$	Fe: Org.Comp.B14-133, 140
$BC_9Cl_3H_7InO_3$	$8-In-2-OC_9H_7(=O)_2-1,3 \cdot BCl_3$	In: Org.Comp.1-370
$BC_9Cl_3H_7N$	$C_9H_7N-BCl_3$	B: B Comp.SVol.3/4-32
$BC_9Cl_3H_9N$	$B(CHCClCH_2)_3N$	B: B Comp.SVol.3/3-166
$BC_9Cl_3H_{10}O_2$	$CH_3C(O)O-C_6H_4-4-CH_3 \cdot BCl_3$	B: B Comp.SVol.3/4-52
–	$C_6H_5CH_2-C(O)OCH_3 \cdot BCl_3$	B: B Comp.SVol.3/4-52
$BC_9Cl_3H_{13}N$	$(CH_3)_2N-CH_2C_6H_5 \cdot BCl_3$	B: B Comp.SVol.4/4-33
–	$C_2H_5-NCH_3-C_6H_5 \cdot BCl_3$	B: B Comp.SVol.3/4-66
–	$(Cl)_3NBC_6H_6(C_3H_7-i)$	B: B Comp.SVol.3/4-57
–	$(Cl)_3NBC_6H_6(C_3H_7-n)$	B: B Comp.SVol.3/4-57
$BC_9Cl_4H_{14}N$	$[1-(n-C_4H_9)-NC_5H_5][BCl_4]$	B: B Comp.SVol.4/4-35
BC_9CoH_{10}	$(C_4H_4)Co(C_5BH_6)$	B: B Comp.SVol.3/4-155
$BC_9CrH_8O_3$	$(CO)_3Cr(C_5H_5BCH_3)$	B: B Comp.SVol.3/4-155
$BC_9FH_8O_6Re^-$	$[fac-(CO)_3Re(CH_2COCOCH_3)(CH_3CO)BF]^-$	Re: Org.Comp.2-418/9
$BC_9FH_9O_6Re$	$(CO)_3Re[-C(CH_3)O-]_3BF$	Re: Org.Comp.2-378/9, 380
$BC_9FH_{13}I_2N$	$4-CH_3-C_6H_4-N(CH_3)_2 \cdot BFI_2$	B: B Comp.SVol.3/4-65/7
–	$C_2H_5-NCH_3-C_6H_5 \cdot BFI_2$	B: B Comp.SVol.3/4-65/7
$BC_9FH_{21}I_2N$	$N(n-C_3H_7)_3 \cdot BFI_2$	B: B Comp.SVol.3/4-65/7
$BC_9FH_{21}N_2PS$	$[-BF-N(C_4H_9-t)-P(=S)(CH_3)-N(C_4H_9-t)-]$	B: B Comp.SVol.4/3b-247
$BC_9F_2H_{10}O_6Re$	$fac-(CO)_3Re=C(CH_3)OH[(CH_3CO)_2]BF_2$	Re: Org.Comp.2-315, 317
$BC_9F_2H_{11}$	$2,4,6-(CH_3)_3-C_6H_2-BF_2$	B: B Comp.SVol.4/3b-226
$BC_9F_2H_{13}IN$	$4-CH_3C_6H_4N(CH_3)_2 \cdot BF_2I$	B: B Comp.SVol.3/4-65/7
$BC_9F_2H_{13}Si$	$2-(CH_3)_3Si-C_6H_4-BF_2$	B: B Comp.SVol.4/3b-226
–	$3-(CH_3)_3Si-C_6H_4-BF_2$	B: B Comp.SVol.4/3b-226
–	$4-(CH_3)_3Si-C_6H_4-BF_2$	B: B Comp.SVol.4/3b-226
$BC_9F_2H_{18}N$	$1-BF_2-NC_5H_6-2,2,6,6-(CH_3)_4$	B: B Comp.SVol.4/3b-243
$BC_9F_2H_{21}IN$	$N(n-C_3H_7)_3 \cdot BF_2I$	B: B Comp.SVol.3/4-65/7
$BC_9F_2H_{28}N_3Si_4$	$FB[-NH-Si(CH_3)_2-NSi(CH_3)_3-Si(CH_3)_2-NSi(CH_3)_2F-]$	B: B Comp.SVol.3/3-377
$BC_9F_3FeH_6O_4$	$[2-(F_3B^-O)-C_6H_6{}^+]Fe(CO)_3$	Fe: Org.Comp.B15-103, 144
$BC_9F_3H_{13}N$	$4-CH_3-C_6H_4-N(CH_3)_2 \cdot BF_3$	B: B Comp.SVol.3/4-65/7
–	$C_2H_5-NCH_3-C_6H_5 \cdot BF_3$	B: B Comp.SVol.3/4-65/7
$BC_9F_3H_{18}O_3S$	$(n-C_4H_9)_2B(OSO_2CF_3)$	B: B Comp.SVol.3/4-142
$BC_9F_3H_{21}N$	$N(n-C_3H_7)_3 \cdot BF_3$	B: B Comp.SVol.3/4-65/7
$BC_9F_3H_{21}N_3O_3S$	$[(C_2H_5)_2N-B(-NHCH_3-CH_2CH_2-NCH_3-)]O_3SCF_3$	B: B Comp.SVol.3/3-207
$BC_9F_4FeH_7O_3$	$[(C_6H_7)Fe(CO)_3][BF_4]$	Fe: Org.Comp.B15-54/86
–	$[(C_6H_6D)Fe(CO)_3][BF_4]$	Fe: Org.Comp.B15-56
–	$[(C_6H_5D_2)Fe(CO)_3][BF_4]$	Fe: Org.Comp.B15-56/7

Formula	Compound	Reference
$BC_9F_4FeH_7O_4$	$[(6\text{-}HO\text{-}C_6H_6)Fe(CO)_3][BF_4]$	Fe: Org.Comp.B15-107, 159
$BC_9F_4FeH_8NO_2$	$[(C_5H_5)Fe(CNCH_3)(CO)_2][BF_4]$	Fe: Org.Comp.B15-312, 316/7
$BC_9F_4FeH_9O_2$	$[(C_5H_5)Fe(CO)_2(CH_2{=}CH_2)][BF_4]$	Fe: Org.Comp.B17-4/5, 43/4
–	$[(C_5H_5)(CO)_2Fe{=}CHCH_3][BF_4]$	Fe: Org.Comp.B16a-95
$BC_9F_4FeH_9O_2S$	$[C_5H_5(CO)_2Fe{=}CH(SCH_3)][BF_4]$	Fe: Org.Comp.B16a-113, 122/3
$BC_9F_4FeH_9O_3$	$[(CH_2CHCHCHCH\text{-}CH_3)Fe(CO)_3][BF_4]$	Fe: Org.Comp.B15-13/4, 16, 19, 28/9
–	$[(CH_2CHCHCHCH\text{-}CHD_2)Fe(CO)_3][BF_4]$	Fe: Org.Comp.B15-11
–	$[(C_5H_5)Fe(CO)_2(CH_2{=}CHOH)][BF_4]$	Fe: Org.Comp.B17-20
–	$[(C_5H_5)(CO)_2Fe{=}CCH_3\text{-}OH][BF_4]$	Fe: Org.Comp.B16a-108, 118
$BC_9F_4H_4N_2O_5Re$	$[(CO)_5ReC_4H_4N_2][BF_4]$	Re: Org.Comp.2-152
$BC_9F_4H_6O_5Re$	$[(CO)_5Re(CH_2{=}CHCH{=}CH_2)][BF_4]$	Re: Org.Comp.2-350, 353
$BC_9F_4H_7Te_2$	$[1,2\text{-}Te_2C_3H_2\text{-}3\text{-}C_6H_5][BF_4]$	B: B Comp.SVol.3/4-150/1
$BC_9F_4H_8O_6Re$	$[(CO)_5ReC_4H_8O][BF_4]$	Re: Org.Comp.2-158
$BC_9F_4H_9N_3O_3Re$	$[(CO)_3Re(NCCH_3)_3][BF_4]$	Re: Org.Comp.1-308, 312
$BC_9F_4H_{10}MoNO_4$	$(C_2H_5)_2N\text{-}CMo(CO)_4(FBF_3)$	Mo: Org.Comp.5-97/100
$BC_9F_4H_{11}Te$	$[(\text{-}CH_2\text{-}(1,2\text{-}C_6H_4)\text{-}CH_2\text{-})TeCH_3][BF_4]$	B: B Comp.SVol.3/4-150/1
$BC_9F_4H_{12}NS$	$[NS][BF_4] \cdot 1,3,5\text{-}(CH_3)_3C_6H_3$	B: B Comp.SVol.4/4-167; S: S-N Comp.5-47/9
$BC_9F_4H_{13}N_2O_2S_2$	$[(CH_3)_2NS{=}NS(O)_2C_6H_4\text{-}CH_3\text{-}4][BF_4]$	S: S-N Comp.7-332
$BC_9F_4H_{13}Se$	$[C_6H_5CH_2Se(CH_3)_2][BF_4]$	B: B Comp.SVol.3/4-148
$BC_9F_4H_{13}Sn$	$[(t\text{-}C_4H_9\text{-}C_5H_4)Sn][BF_4]$	B: B Comp.SVol.4/3b-108
$BC_9F_4H_{14}NO_2PRe$	$[(C_5H_5)Re(CO)(NO)(P(CH_3)_3)][BF_4]$	Re: Org.Comp.3-154, 155/6
$BC_9F_4H_{16}N_2O_3Re$	$(CO)_3Re[(CH_3)_2NCH_2CH_2N(CH_3)_2]FBF_3 \cdot 0.5\ H_2O$	Re: Org.Comp.1-157
$BC_9F_4H_{18}N_2O_4Re$	$[(CO)_3Re(H_2O)(N(CH_3)_2C_2H_4\text{-}N(CH_3)_2)][BF_4]$	Re: Org.Comp.1-300, 304, 305
$BC_9F_4H_{21}N_2S$	$[t\text{-}C_4H_9\text{-}N(CH_3)S{=}N\text{-}C_4H_9\text{-}t][BF_4]$	S: S-N Comp.7-333/4
$BC_9F_4H_{21}O_2$	$[((C_2H_5)_2O)_2CH][BF_4]$	B: B Comp.SVol.3/3-267
$BC_9F_4H_{22}N$	$[(n\text{-}C_3H_7)_3NH][BF_4]$	B: B Comp.SVol.4/3b-210
$BC_9F_4H_{27}MnN_2O_{11}P_3$	$[Mn(NO)_2(P(OCH_3)_3)_3][BF_4]$	Mn: MVol.D8-145
$BC_9F_6H_{13}N_2S_2$	$(CH_3)_3N\text{-}BH_2\text{-}N[\text{-}CH{=}C(SCF_3)\text{-}CH{=}C(SCF_3)\text{-}]$	B: B Comp.SVol.4/3b-23/4
$BC_9F_{10}FeH_{13}P_2$	$[(C_5H_5)Fe(CH_2{=}C(CH_3)_2)(PF_3)_2][BF_4]$	Fe: Org.Comp.B16b-7, 10
$BC_9FeH_{10}^-$	$[(C_5H_5)Fe(C_4H_4(BH))]^-$	Fe: Org.Comp.B16b-64
$BC_9Fe_3H_4O_9^-$	$[(CO)_9(H)Fe_3(BH_3)]^-$	Fe: Org.Comp.C6a-298/9
$BC_9Fe_3H_5O_9$	$(CO)_9(H)Fe_3(BH_4)$	Fe: Org.Comp.C6a-295/6
$BC_9GeH_{21}O$	$(C_2H_5)_2Ge(\text{-}CH_2CH_2B(OCH_3)CH_2CH_2\text{-})$	Ge: Org.Comp.3-295
$BC_9H_6NO_8$	$C_5H_5NH[B(\text{-}O\text{-}CO\text{-}CO\text{-}O\text{-})_2]$	B: B Comp.SVol.3/3-227
$BC_9H_8MnO_3$	$(CO)_3Mn(C_5H_5BCH_3)$	B: B Comp.SVol.4/4-180
$BC_9H_9IO_6Re$	$(CO)_3Re[\text{-}C(CH_3)O\text{-}]_3BI$	Re: Org.Comp.2-379
$BC_9H_9N_2O_2$	$H_3B \cdot 1\text{-}NC_9H_6\text{-}6\text{-}NO_2$	B: B Comp.SVol.4/3b-14
$BC_9H_9N_2O_5S$	$HOB[\text{-}C_6H_2(\text{-}O\text{-}CH_2\text{-}O\text{-})\text{-}CH{=}N\text{-}NSO_2CH_3\text{-}]$	B: B Comp.SVol.3/3-164
$BC_9H_9N_3Na$	$Na[HB(NC_4H_4)_2CN] \cdot 3\ C_4H_8O_2$	B: B Comp.SVol.3/3-213
$BC_9H_{10}N$	$H_3B \cdot 1\text{-}NC_9H_7$	B: B Comp.SVol.4/3b-14
–	$H_3B \cdot 2\text{-}NC_9H_7$	B: B Comp.SVol.4/3b-14
$BC_9H_{10}NO$	$2,9,1\text{-}ONBC_9H_{10}$	B: B Comp.SVol.4/3b-66
$BC_9H_{10}N_3O_2$	$C_6H_5B(\text{-}NCH_3\text{-}CO\text{-}NH\text{-}CO\text{-}NH\text{-})$	B: B Comp.SVol.3/3-115
$BC_9H_{11}NS$	$1,3,2\text{-}SNBC_6H_4\text{-}3\text{-}C_3H_7\text{-}i$	B: B Comp.SVol.4/4-146
$BC_9H_{11}N_2O$	$C_6H_5\text{-}B[\text{-}NCH_3\text{-}C(O)\text{-}NCH_3\text{-}]$	B: B Comp.SVol.3/3-112

$BC_9H_{11}N_2O$ ...... C_6H_5-B[-O-N=CH-N(C_2H_5)-] .... B: B Comp.SVol.3/3-163

$BC_9H_{11}N_2OS$ ..... [-B(C_6H_5)-N(CH_3)-N(CH_3)-C(=S)-O-] .... B: B Comp.SVol.4/3b-64
B: B Comp.SVol.4/4-152

$BC_9H_{11}N_2O_2$ ..... [-N(CH_3)-C(=O)-O-B(C_6H_5)-N(CH_3)-] .... B: B Comp.SVol.4/3b-64

$BC_9H_{11}N_2S_2$ ..... S=C[-S-B(C_6H_5)-N(CH_3)-N(CH_3)-] .... B: B Comp.SVol.4/3a-251
B: B Comp.SVol.4/4-145/6

$BC_9H_{11}O_2$ .... C_6H_5-B[-O-$CH_2CH_2CH_2$-O-] .... B: B Comp.SVol.4/3b-69

– .... C_6H_5-B[-O-CH_2-CH(CH_3)-O-] .... B: B Comp.SVol.4/3b-69

$BC_9H_{11}O_2S$ ...... 2-(C_2H_5-O)-3,1,2-$OSBC_7H_6$ .... B: B Comp.SVol.3/4-120

$BC_9H_{11}S_2$ .... 2-C_6H_5-1,3,2-$S_2BC_3H_6$ .... B: B Comp.SVol.4/4-130/1

$BC_9H_{12}NS_2$ ...... 2-$(CH_3)_2$N-5-CH_3-1,3,2-$S_2BC_6H_3$ .... B: B Comp.SVol.3/4-127

$BC_9H_{13}N_2$ .... 1,2,3-$(CH_3)_3$-1,3,2-$N_2BC_6H_4$ .... B: B Comp.SVol.3/3-117

– .... 2-C_6H_5-1,3,2-$N_2BC_3H_8$ .... B: B Comp.SVol.4/3a-175

$BC_9H_{13}N_4$ .... [-B(CH_3)-N(CH_3)-N=N-N(CH_2-C_6H_5)-] .... B: B Comp.SVol.4/3a-199

$BC_9H_{13}O_2S$ ...... $C_6H_5SCH_2B(OCH_3)_2$ .... B: B Comp.SVol.3/4-121

$BC_9H_{13}S_2$ .... 2-C_6H_5-1,3-$S_2C_3H_5$ · BH_3 .... B: B Comp.SVol.4/4-134

$BC_9H_{14}NO$ ...... (C_6H_5)(CH_3O)B-N$(CH_3)_2$ .... B: B Comp.SVol.3/3-154

$BC_9H_{14}NOS$ ..... $(CH_3)_2$B-N(C_6H_5)-S(O)CH_3 .... B: B Comp.SVol.3/3-154
B: B Comp.SVol.3/4-134

$BC_9H_{14}NO_2$ ...... $(CH_3)_3$N-BH[-O-(1,2-C_6H_4)-O-] .... B: B Comp.SVol.3/3-221

$BC_9H_{14}NO_5$ ...... $(CH_3COO)_2$B[-NH=CCH_3-CH=CCH_3-O-] ...... B: B Comp.SVol.3/3-218

$BC_9H_{14}NS$ .... 2,2-$(CH_3)_2$-1,3-SNC_7H_5 · BH_3 .... B: B Comp.SVol.4/3b-7
B: B Comp.SVol.4/4-147/8

– .... 2-C_6H_5-1,3-SNC_3H_6 · BH_3 .... B: B Comp.SVol.4/3b-7
B: B Comp.SVol.4/4-147/8

$BC_9H_{14}N_2PS$ ..... CH_3-B[-N(CH_3)-P(C_6H_5)(=S)-N(CH_3)-] .... B: B Comp.SVol.4/3a-207/8

$BC_9H_{15}N_2$ .... 9-(NC-NH)-[3.3.1]-9-BC_8H_{14} .... B: B Comp.SVol.4/3a-235

$BC_9H_{15}N_3^+$ ...... [(-$(CH)_5$-)N-B(-NCH_3-CH_2-CH_2-NCH_3-)]$^+$ ... B: B Comp.SVol.3/3-207

$BC_9H_{16}N$ .... H_3B-N$(CH_3)_2$-CH_2-C_6H_5 .... B: B Comp.SVol.4/3b-10

$BC_9H_{16}NO$ ...... H_3B-NH_2-CH(CH_3)-CH(OH)-C_6H_5 .... B: B Comp.SVol.3/3-176
B: B Comp.SVol.4/3b-5

$BC_9H_{16}NOSe$ .... 2,3-$(CH_3)_2$-5,6-$(C_2H_5)_2$-1,3,2-$SeNBC_3$(=O)-4. . B: B Comp.SVol.4/4-172

$BC_9H_{16}N_3$ .... (1-H_4C_4N)B(-NCH_3-CH_2-CH_2-CH_2-NCH_3-) ... B: B Comp.SVol.3/3-95

$BC_9H_{17}NP$ .... $(C_2H_5)_2$N-B(-CH=CH-PCH_3-CH=CH-) .... B: B Comp.SVol.3/3-151

$BC_9H_{17}N_2S$ ...... 2-[(n-$C_3H_7)_2$B-NH]-1,3-SNC_3H_2 .... B: B Comp.SVol.4/3a-232

$BC_9H_{17}N_4$ .... 1-[B(-NCH_3-$CH_2CH_2CH_2$-NCH_3-)]-3-CH_3-1,2-$N_2C_3H_2$
B: B Comp.SVol.3/3-96

– .... 1-[B(-NCH_3-CH_2CH_2-NCH_3-)]-3,5-$(CH_3)_2$-1,2-N_2C_3H
B: B Comp.SVol.3/3-96

$BC_9H_{17}OSi$ ...... 1-CH_3O-2-$(CH_3)_3$Si-BC_5H_5 .... B: B Comp.SVol.4/3b-26

$BC_9H_{17}S$ .... [3.3.1]-9-BC_8H_{14}-9-SCH_3 .... B: B Comp.SVol.4/4-132/3

$BC_9H_{18}N$ .... 3-NH_2-7-CH_3-[3.3.1]-3-BC_8H_{13} .... B: B Comp.SVol.3/3-150

– .... 9-(CH_3-NH)-[3.3.1]-9-BC_8H_{14} .... B: B Comp.SVol.4/3a-235

– .... C_5H_{10}B-NC_4H_8 .... B: B Comp.SVol.3/3-156

– .... H_3B-N$(CH_2$-CH=$CH_2)_3$ .... B: B Comp.SVol.4/3b-10

– .... [-B(-$CH_2CH_2CH_2$-$)_3$N-] .... B: B Comp.SVol.4/3a-259

$BC_9H_{18}NO_3$ ...... B[-O-$CHCH_3$-CH_2-$]_3$N .... B: B Comp.SVol.3/3-219

$BC_9H_{18}N_3O_2$ ..... CH_3B[-N(C_2H_5)C(O)-N(C_2H_5)-C(O)N(C_2H_5)-] . . B: B Comp.SVol.3/3-114/5

$BC_9H_{18}N_3S_2$ ..... CH_3B[-N(C_2H_5)C(S)-N(C_2H_5)-C(S)N(C_2H_5)-] . . B: B Comp.SVol.3/3-116
B: B Comp.SVol.3/4-131

Formula	Compound	Reference
$BC_9H_{19}KNSi$	$K[4,5-(C_2H_5)_2-2,2,3-(CH_3)_3-1,2,5-NSiBC_2]$	B: B Comp.SVol.3/3-162
		B: B Comp.SVol.4/3a-250
$BC_9H_{19}NNaSi$	$Na[4,5-(C_2H_5)_2-2,2,3-(CH_3)_3-1,2,5-NSiBC_2]$	B: B Comp.SVol.3/3-162
		B: B Comp.SVol.4/3a-250
$BC_9H_{19}N_2$	$(-CH_2CH_2-)NH-B(C_3H_7-i)_2-CN$	B: B Comp.SVol.4/3b-28
$BC_9H_{19}N_2O_4$	$(CH_3O)_2B[-NH-C(O-C_2H_5)=CH-C(O-C_2H_5)=NH-]$	B: B Comp.SVol.3/3-122/3
$BC_9H_{19}N_2S$	$n-C_4H_9-B[-NH-C(S)-N(C_4H_9-t)-]$	B: B Comp.SVol.3/3-112
		B: B Comp.SVol.3/4-131
$BC_9H_{19}N_2S_2$	$[-B(C_2H_5)_2-NHC(SCH_3)CHC(SCH_3)NH-]$	B: B Comp.SVol.3/3-122/3
		B: B Comp.SVol.4/4-148
$BC_9H_{19}N_4$	$[-N(CH_3)-CH_2CH_2-N(CH_3)-]B-N=C=N-C_4H_9-t$	B: B Comp.SVol.4/3a-154
$BC_9H_{19}N_6$	$BC_2H_2N_6(CH_3)(C_3H_7-n)_2$	B: B Comp.SVol.4/3a-178
$BC_9H_{20}N$	$C_4H_9N-BC_5H_{11}$	B: B Comp.SVol.3/3-189
–	$C_5H_{10}B-N(C_2H_5)_2$	B: B Comp.SVol.3/3-150
–	$C_5H_{10}B-NH-C_4H_9-t$	B: B Comp.SVol.3/3-156
–	$n-C_5H_{11}-B=N-C_4H_9-t$	B: B Comp.SVol.4/3a-214
$BC_9H_{20}NO$	$(C_2H_5)_2NH-B(OCH_3)[-CH_2-CH=CH-CH_2-]$	B: B Comp.SVol.4/3b-70
$BC_9H_{20}NOSi$	$[-NCH_3-B(OC_2H_5)-CCH_3=CCH_3-Si(CH_3)_2-]$	B: B Comp.SVol.4/3b-64
$BC_9H_{20}NO_2Si$	$[-NCH_3-B(O-OC_2H_5)-CCH_3=CCH_3-Si(CH_3)_2-]$	B: B Comp.SVol.4/3b-64
$BC_9H_{20}NSi$	$4,5-(C_2H_5)_2-2,2,3-(CH_3)_3-1,2,5-NSiBC_2H$	B: B Comp.SVol.4/3a-250
$BC_9H_{20}N_3$	$2-CH_3-1,2,4-N_3C_2H_2 \cdot B(C_2H_5)_3$	B: B Comp.SVol.4/3b-28
$BC_9H_{20}N_3O$	$[-N(CH_3)-CH_2CH_2-N(CH_3)-]B-O-1-NC_5H_{10}$	B: B Comp.SVol.4/3b-53
$BC_9H_{21}Li_2N_2$	$t-C_4H_9-NLi-B(CH_3)-NLi-C_4H_9-t$	B: B Comp.SVol.4/3a-169
$BC_9H_{21}NSi^+$	$[2,2,3-(CH_3)_3-4,5-(C_2H_5)_2-1,2,5-NSiBC_2H_2]^+$	B: B Comp.SVol.4/4-70
$BC_9H_{21}N_2$	$(CH_3)_2NH-B(C_3H_7-i)_2-CN$	B: B Comp.SVol.4/3b-27
–	$(n-C_4H_9)_2NH-BH_2-CN$	B: B Comp.SVol.4/3b-22
–	$(i-C_4H_9)_2NH-BH_2-CN$	B: B Comp.SVol.4/3b-22
–	$t-C_4H_9N=C(CH_3)-BCH_3-N(CH_3)_2$	B: B Comp.SVol.3/3-154
$BC_9H_{21}N_2O_2$	$CH_3-O-CH_2-O-CH=CH-CH_2-B[N(CH_3)_2]_2$	B: B Comp.SVol.3/3-108
		B: B Comp.SVol.4/3a-168
–	$t-C_4H_9-OC(O)-B[N(CH_3)_2]_2$	B: B Comp.SVol.3/3-110
$BC_9H_{21}N_2S$	$(n-C_4H_9)_2NH-BH_2-N=C=S$	B: B Comp.SVol.4/3b-23
–	$(i-C_4H_9)_2NH-BH_2-N=C=S$	B: B Comp.SVol.4/3b-23
$BC_9H_{21}N_4O_2$	$(CH_3O)_2B[-NH-CN(CH_3)_2=CH-CN(CH_3)_2=NH-]$	B: B Comp.SVol.3/3-122/3
$BC_9H_{21}S$	$(n-C_3H_7)_2B-S-C_3H_7-n$	B: B Comp.SVol.4/4-129
$BC_9H_{22}N$	$2,2,6,6-(CH_3)_4C_5H_7N \cdot BH_3$	B: B Comp.SVol.4/3b-7
–	$(C_2H_5)_2NH \cdot BC_5H_{11}$	B: B Comp.SVol.3/3-188
–	$(i-C_3H_7)_2B-NH-C_3H_7-i$	B: B Comp.SVol.4/3a-230
–	$(t-C_4H_9)_2B-NH-CH_3$	B: B Comp.SVol.4/3a-230
–	$t-C_4H_9-B(CH_3)-NH-C_4H_9-t$	B: B Comp.SVol.4/3a-234
–	$t-C_4H_9-NH_2 \cdot BC_5H_{11}$	B: B Comp.SVol.3/3-188
–	$(n-C_3H_7)_2B[-CH_2CH_2CH_2-NH_2-]$	B: B Comp.SVol.4/3a-254
$BC_9H_{22}NO_2$	$(n-C_4H_9)_2NH-BH_2-COOH$	B: B Comp.SVol.4/3b-24
–	$(i-C_4H_9)_2NH-BH_2-COOH$	B: B Comp.SVol.4/3b-24
$BC_9H_{22}N_3$	$(i-C_3H_7)_2N-BH-NH-N=C(CH_3)_2$	B: B Comp.SVol.3/3-107
–	$[-NH-CH_2CH_2CH_2-N(C_2H_5)-]B-N(C_2H_5)_2$	B: B Comp.SVol.4/3a-154
$BC_9H_{22}N_3Sn$	$(CH_3)_3SnC_6H_{13}BN_3$	Sn: Org.Comp.18-84, 96
$BC_9H_{23}N$	$(CH_3)_3N-BH-C(CH_3)_2-C_3H_7-i$, radical	B: B Comp.SVol.4/3b-25
$BC_9H_{23}N_2$	$t-C_4H_9-NH-B(CH_3)-NH-C_4H_9-t$	B: B Comp.SVol.4/3a-169

$BC_9H_{23}N_2$ .... $[-N(C_3H_7-i)-CH_2-N(C_3H_7-i)-CH_2CH_2-] \cdot BH_3$ B: B Comp.SVol.4/3b-12
$BC_9H_{23}N_2Si_2$..... $CH_3B[-NSi(CH_3)_3-CH=CH-NSi(CH_3)_3-]$ ...... B: B Comp.SVol.3/3-112/3
$BC_9H_{23}N_4Si$ ..... $C_2H_5-B(N_3)-N(C_4H_9-t)-Si(CH_3)_3$ .... B: B Comp.SVol.3/3-110
– .... $C_2H_5-B[-N(C_4H_9-t)-N=N-N(Si(CH_3)_3)-]$ ...... B: B Comp.SVol.3/3-137/8
– .... $i-C_3H_7-B(N_3)-N(C_3H_7-i)-Si(CH_3)_3$ .... B: B Comp.SVol.3/3-110
– .... $i-C_3H_7-B[-N(C_3H_7-i)-N=N-N(Si(CH_3)_3)-]$..... B: B Comp.SVol.3/3-137/8
$BC_9H_{24}N$........ $(CH_3)_3N-BH_2-C(CH_3)_2-C_3H_7-i$ .... B: B Comp.SVol.4/3b-25
$BC_9H_{24}NOSi$..... $[(CH_3)_3Si][t-C_4H_9O]B-N(CH_3)_2$ .... B: B Comp.SVol.3/3-155
$BC_9H_{24}NSi$ ...... $(i-C_3H_7)_2B-NH-Si(CH_3)_3$ .... B: B Comp.SVol.4/3a-230
$BC_9H_{24}N_3$ .... $B(NH-C_3H_7-i)_3$ .... B: B Comp.SVol.4/3a-153
$BC_9H_{26}N_3Si_2$..... $[-N(C_2H_5)-Si(CH_3)_2-Si(CH_3)_2-N(C_2H_5)-]B-NH-CH_3$
B: B Comp.SVol.4/3a-154
$BC_9H_{27}N_2Si_2$..... $CH_3B[N(CH_3)_2]N[Si(CH_3)_3]_2$................ B: B Comp.SVol.3/3-105
$BC_9H_{27}N_5OP$..... $[(CH_3)_2N]_2BN(CH_3)P(O)[N(CH_3)_2]_2$........... B: B Comp.SVol.3/3-96
$BC_9H_{28}NOSi_2$ .... $(CH_3)_2HN-B[Si(CH_3)_3]_2OCH_3$............... B: B Comp.SVol.3/3-216
$BC_9H_{29}N_2Si_3$..... $[(CH_3)_3Si]_2N-BH-NH[Si(CH_3)_3]$ .... B: B Comp.SVol.3/3-105
$BC_9H_{30}N_3Si_3$..... $(CH_3)_3Si-NH-B(NH_2)-N[Si(CH_3)_3]_2$ .... B: B Comp.SVol.4/3a-154
$BC_{10}ClF_2H_{15}N$ ... $C_6H_5CH_2-NCH_3-C_2H_5 \cdot BClF_2$ .... B: B Comp.SVol.3/4-65/7
– .... $C_6H_5-N(C_2H_5)_2 \cdot BClF_2$.................. B: B Comp.SVol.3/4-65/7
$BC_{10}ClF_4FeH_6O_4$ $[(1-Cl-7-O=C_7H_6)Fe(CO)_3][BF_4]$ .... Fe: Org.Comp.B15-212
$BC_{10}ClF_4FeH_{10}O_2$ $[(C_5H_5)Fe(CO)_2(CH_2=CHCH_2Cl)][BF_4]$........ Fe: Org.Comp.B17-23
$BC_{10}ClFeH_{10}O_2$ .. $(C_5H_5)Fe[C_5H_3(Cl)-B(OH)_2]$ .... Fe: Org.Comp.A9-287/8
– .... $[(HO)_2B-C_5H_4]Fe(C_5H_4-Cl)$ .... Fe: Org.Comp.A9-277, 283
$BC_{10}ClGeH_{24}N_2$ .. $[-N(C_4H_9-t)-BCl-N(C_4H_9-t)-]Ge(CH_3)_2$....... B: B Comp.SVol.4/4-64/5
$BC_{10}ClH_{15}N$ ..... $C_6H_5-BCl-N(C_2H_5)_2$.................... B: B Comp.SVol.3/4-55
– .... $C_6H_5-BCl-NH-C_4H_9-t$ .... B: B Comp.SVol.3/4-56
$BC_{10}ClH_{16}Si$..... $CH_3-BCl-C_6H_4-2-Si(CH_3)_3$ .... B: B Comp.SVol.4/4-49
$BC_{10}ClH_{17}NSi$.... $C_6H_5-BCl-N(CH_3)-Si(CH_3)_3$................ B: B Comp.SVol.4/4-60
$BC_{10}ClH_{18}$ ...... $(c-C_5H_9)_2BCl$ .... B: B Comp.SVol.3/4-43
B: B Comp.SVol.4/4-46
$BC_{10}ClH_{21}N$ ..... $1-Cl-BC_4H_6 \cdot NH(C_3H_7-i)_2$................ B: B Comp.SVol.4/4-69
$BC_{10}ClH_{22}$ ...... $(n-C_5H_{11})_2BCl$ .... B: B Comp.SVol.3/4-43
– .... $[i-C_3H_7-CH(CH_3)]_2BCl$..................... B: B Comp.SVol.4/4-46
$BC_{10}ClH_{23}N_2P$ ... $[-N(C_4H_9-t)-BCl-N(C_4H_9-t)-]P-C_2H_5$ .... B: B Comp.SVol.4/4-64/5
$BC_{10}ClH_{24}N_2Si$... $[-N(C_4H_9-t)-BCl-N(C_4H_9-t)-]Si(CH_3)_2$ .... B: B Comp.SVol.4/4-64/5
$BC_{10}ClH_{25}NSi$.... $(n-C_3H_7)ClBN(t-C_4H_9)Si(CH_3)_3$ .... B: B Comp.SVol.3/4-56
$BC_{10}ClH_{25}NSn$ ... $(CH_3)_3SnN(C_4H_9-t)BClC_3H_7-i$ .... Sn: Org.Comp.18-57, 64/5
$BC_{10}ClH_{27}NSi_2$... $t-C_4H_9-BCl-N[Si(CH_3)_3]_2$.................. B: B Comp.SVol.3/4-57
B: B Comp.SVol.4/4-60
$BC_{10}ClH_{29}NSi_3$... $[(CH_3)_3SiCH_2]ClBN[Si(CH_3)_3]_2$.............. B: B Comp.SVol.3/4-57
$BC_{10}Cl_2FH_{15}N$ ... $C_6H_5CH_2-NCH_3-C_2H_5 \cdot BCl_2F$ .... B: B Comp.SVol.3/4-65/7
– .... $C_6H_5-N(C_2H_5)_2 \cdot BCl_2F$.................. B: B Comp.SVol.3/4-65/7
$BC_{10}Cl_2F_4FeH_8Hg_2$
$[Fe(C_5H_4-HgCl)_2][BF_4]$.................... Fe: Org.Comp.A10-133
$BC_{10}Cl_2H_{10}N$ .... $Cl_2BN(CH_2C_6H_5)(CH_2C\equiv CH)$ .... B: B Comp.SVol.3/4-54
$BC_{10}Cl_2H_{10}O_6Re$ cis-$(CO)_4Re[(CH_3CO)(CH(CH_3)_2CO)]BCl_2$..... Re: Org.Comp.2-346
$BC_{10}Cl_2H_{15}$...... $(CH_3)_5C_5-BCl_2$ .... B: B Comp.SVol.4/4-44
$BC_{10}Cl_2H_{15}O_4$ ... $[B(-O=CCH_3-CH=CCH_3-O-)_2][HCl_2]$ .... B: B Comp.SVol.3/4-52
$BC_{10}Cl_2H_{15}Si$ .... $(CH_3)_3SiCH_2-2-C_6H_4-BCl_2$ .... B: B Comp.SVol.4/4-44
$BC_{10}Cl_2H_{17}$...... $2,7,7-(CH_3)_3-[3.1.1]-C_7H_8-3-BCl_2$........... B: B Comp.SVol.4/4-44

$BC_{10}Cl_2H_{18}N$ .... $(Cl_2BCH_2)C_8H_{13}N(CH_3)$ .... B: B Comp.SVol.3/4-55
$BC_{10}Cl_2H_{23}S$..... $(CH_3)_2S-B(C_8H_{17}-n)Cl_2$ .... B: B Comp.SVol.3/4-137
$BC_{10}Cl_2H_{25}N_2Si$ .. $Cl_2B-N(C_4H_9-t)-Si(CH_3)_2-NH-C_4H_9-t$........ B: B Comp.SVol.4/4-59
$BC_{10}Cl_2H_{28}N_2PSi_2$ $[(CH_3)_3Si]_2N-B(C_4H_9-t)-NH-PCl_2$ .... B: B Comp.SVol.4/3a-170
$BC_{10}Cl_3GaH_{15}N_2$ $C_6H_5B(-NCH_3-CH_2-CH_2-NCH_3-) \cdot GaCl_3$ .... B: B Comp.SVol.3/3-113
$BC_{10}Cl_3H_{11}N$ .... $Cl_2B[-NH(CH_2C_6H_5)-CH_2-CCl=CH-]$......... B: B Comp.SVol.3/4-55
$BC_{10}Cl_3H_{14}O$ .... $C_6H_5-C(CH_3)_2-O-CH_3 \cdot BCl_3$ .... B: B Comp.SVol.4/4-34
$BC_{10}Cl_3H_{14}Si$ .... $CH_3(C_6H_5)SiCl(CH_2)_3BCl_2$ .... B: B Comp.SVol.3/4-42
$BC_{10}Cl_3H_{15}N$ .... $C_6H_5CH_2N(CH_3)C_2H_5 \cdot BCl_3$ .... B: B Comp.SVol.3/4-66
– .... $(Cl)_3NBC_6H_6(n-C_4H_9)$.................... B: B Comp.SVol.3/4-57
$BC_{10}Cl_3H_{23}N_2P$ .. $(i-C_3H_7)_2N-BCl-N(C_4H_9-t)-PCl_2$............ B: B Comp.SVol.4/4-63
$BC_{10}Cl_3H_{24}Si_2$ ... $[(CH_3)_2SiCl(CH_2)_3]_2BCl$ .... B: B Comp.SVol.3/4-44
$BC_{10}CoH_{12}$ ...... $(C_4H_4)Co(C_5H_5BCH_3)$.................... B: B Comp.SVol.4/4-180
– .... $(C_5H_5)Co(C_4H_4BCH_3)$.................... B: B Comp.SVol.4/4-180
$BC_{10}FH_{15}I_2N$..... $(C_6H_5CH_2)(CH_3)(C_2H_5)N-BFI_2$ .... B: B Comp.SVol.3/4-65/7, 101
$BC_{10}FH_{15}N$ ...... $C_6H_5-BF-N(C_2H_5)_2$....................... B: B Comp.SVol.3/3-376
– .... $C_6H_5-BF-NH-C_4H_9-t$....................... B: B Comp.SVol.3/3-376
$BC_{10}FH_{17}NOSi$ ... $C_6H_5-BF-N(OCH_3)-Si(CH_3)_3$ .... B: B Comp.SVol.4/3b-245
$BC_{10}FH_{18}$ .... $(c-C_5H_9)_2BF$ .... B: B Comp.SVol.4/3b-227
$BC_{10}FH_{21}N$ ...... $1-[CH_3-BF]-NC_5H_6-2,2,6,6-(CH_3)_4$ .... B: B Comp.SVol.4/3b-245
$BC_{10}FH_{24}N_2Si$.... $[-BF-N(C_4H_9-t)-Si(CH_3)_2-N(C_4H_9-t)-]$ .... B: B Comp.SVol.4/3b-247
$BC_{10}FH_{27}NSi_2$.... $t-C_4H_9-BF-N[Si(CH_3)_3]_2$ .... B: B Comp.SVol.4/3b-244
$BC_{10}F_2FeH_{10}O_3^-$ $[C_5H_5Fe(CH_2C_2O_2BF_2(CH_3))CO]^-$ .... Fe: Org.Comp.B17-156
$BC_{10}F_2FeH_{13}N_2O$ $(C_5H_5)Fe((CHNCH_3)_2BF_2)CO$ .... Fe: Org.Comp.B16b-127, 138
$BC_{10}F_2H_9O_2$ ..... $[-B(F)_2-OC(CH_3)CHC(C_6H_5)O-]$............. B: B Comp.SVol.3/3-368
B: B Comp.SVol.4/3b-232/4
$BC_{10}F_2H_9O_6Re^-$ .. $[(CO)_4ReC_3(CH_3)_3(-O-BF_2-O-)]^-$ .... Re: Org.Comp.2-389, 393/4
– .... $[(CO)_4ReC_3H_2(C_3H_7-i)(-O-BF_2-O-)]^-$........ Re: Org.Comp.2-389
$BC_{10}F_2H_{10}O_6Re$ .. $(CO)_4Re[(CH_3CO)CH(CH_3)_2CO]BF_2$.......... Re: Org.Comp.2-345/6, 348
$BC_{10}F_2H_{15}$ ...... $C_5(CH_3)_5BF_2$............................ B: B Comp.SVol.3/3-363
$BC_{10}F_2H_{15}IN$..... $(C_6H_5CH_2)(CH_3)(C_2H_5)N-BF_2I$ .... B: B Comp.SVol.3/4-101
$BC_{10}F_2H_{15}N_6NiO_2S$
$Ni[-BF_2-O-N=C(CH_3)-C(CH_3)=N-NC(S-CH_3)N-N=C(CH_3)-C(CH_3)=N-O-]$ .... B: B Comp.SVol.4/3b-124
$BC_{10}F_2H_{17}$ ...... $2,7,7-(CH_3)_3-[3.1.1]-C_7H_8-3-BF_2$ .... B: B Comp.SVol.4/3b-226
$BC_{10}F_3H_{10}O$ ..... $F_3B \cdot C_{10}H_{10}(=O)$...................... B: B Comp.SVol.4/3b-145
$BC_{10}F_3H_{12}O$ ..... $i-C_3H_7-C_7H_5(=O) \cdot BF_3$.................. B: B Comp.SVol.4/3b-145
$BC_{10}F_3H_{15}N$ ..... $C_6H_5CH_2-NCH_3-C_2H_5 \cdot BF_3$.............. B: B Comp.SVol.3/4-65/7
– .... $C_6H_5-N(C_2H_5)_2 \cdot BF_3$ .... B: B Comp.SVol.3/4-65/7
$BC_{10}F_3H_{15}N_3O_3S$ $[(-(CH)_5-)N-B(-NCH_3-CH_2-CH_2-NCH_3-)][O_3SCF_3]$
B: B Comp.SVol.3/3-207
$BC_{10}F_3H_{17}K$ ..... $K[2,7,7-(CH_3)_3-(3.1.1)-C_7H_8-3-BF_3]$......... B: B Comp.SVol.4/3b-228
$BC_{10}F_3H_{17}N_3O_3S$ $[NC_5H_5-1-B(N(CH_3)_2)_2][CF_3-SO_3]$........... B: B Comp.SVol.4/3b-45
$BC_{10}F_3H_{20}N_2O_2$ .. $[(C_2H_5)_2N]_2BOC(O)CF_3$ .... B: B Comp.SVol.3/3-106
$BC_{10}F_4FeH_7O_3$ ... $[(C_7H_7)Fe(CO)_3][BF_4]$..................... Fe: Org.Comp.B15-217, 230
$BC_{10}F_4FeH_7O_4$ ... $[(1-O=CH-C_6H_6)Fe(CO)_3][BF_4]$ .... Fe: Org.Comp.B15-101
– .... $[(6-(O=)C_7H_7)Fe(CO)_3][BF_4]$ .... Fe: Org.Comp.B15-209, 223/5
$BC_{10}F_4FeH_7O_5$ ... $[(1-HOOC-C_6H_6)Fe(CO)_3][BF_4]$............. Fe: Org.Comp.B15-101, 136
– .... $[(2-HOOC-C_6H_6)Fe(CO)_3][BF_4]$............. Fe: Org.Comp.B15-104, 136

Formula	Compound	Reference
$BC_{10}F_4FeH_7O_5$	$[(3\text{-}HOOC\text{-}C_6H_6)Fe(CO)_3][BF_4]$	Fe: Org.Comp.B15-106, 136
–	$[(6\text{-}HOOC\text{-}C_6H_6)Fe(CO)_3][BF_4]$	Fe: Org.Comp.B15-109
–	$[(CH_3COO\text{-}C_5H_4)Fe(CO)_3][BF_4]$	Fe: Org.Comp.B15-51
$BC_{10}F_4FeH_8NO_2$	$[(C_5H_5)Fe(CO)_2(CH_2{=}CHCN)][BF_4]$	Fe: Org.Comp.B17-30
$BC_{10}F_4FeH_9O_2$	$[(C_5H_5)Fe(CO)_2(CH_2{=}C{=}CH_2)][BF_4]$	Fe: Org.Comp.B17-111, 116
$BC_{10}F_4FeH_9O_3$	$[(1\text{-}CH_3\text{-}C_6H_6)Fe(CO)_3][BF_4]$	Fe: Org.Comp.B15-98/9, 135
–	$[(2\text{-}CH_3\text{-}C_6H_6)Fe(CO)_3][BF_4]$	Fe: Org.Comp.B15-103, 144, 155/6
–	$[(3\text{-}CH_3\text{-}C_6H_6)Fe(CO)_3][BF_4]$	Fe: Org.Comp.B15-106, 157/8
–	$[(CH_3\text{-}C_6H_6)Fe(CO)_3][BF_4]$	Fe: Org.Comp.B15-96
–	$[(C_5H_5)Fe(CO)_2(CH_2{=}CHCHO)][BF_4]$	Fe: Org.Comp.B17-29/30
–	$[(C_7H_9)Fe(CO)_3][BF_4]$	Fe: Org.Comp.B15-191/9
–	$[(C_7H_8D)Fe(CO)_3][BF_4]$	Fe: Org.Comp.B15-192/3
$BC_{10}F_4FeH_9O_4$	$[(1\text{-}CH_3O\text{-}C_6H_6)Fe(CO)_3][BF_4]$	Fe: Org.Comp.B15-98, 135
–	$[(2\text{-}CH_3O\text{-}C_6H_6)Fe(CO)_3][BF_4]$	Fe: Org.Comp.B15-103, 139/54
–	$[(3\text{-}CH_3O\text{-}C_6H_6)Fe(CO)_3][BF_4]$	Fe: Org.Comp.B15-105, 157
–	$[(CH_3O\text{-}C_6H_6)Fe(CO)_3][BF_4]$	Fe: Org.Comp.B15-96
$BC_{10}F_4FeH_{10}NO_2$	$[(C_6H_7)Fe(CN\text{-}CH_3)(CO)_2][BF_4]$	Fe: Org.Comp.B15-350
$BC_{10}F_4FeH_{11}N_2O$	$[(C_5H_5)Fe(CN\text{-}CH_3)(CO)(NC\text{-}CH_3)][BF_4]$	Fe: Org.Comp.B15-300
–	$[(C_5H_5)Fe(CN\text{-}CH_3)_2(CO)][BF_4]$	Fe: Org.Comp.B15-333, 340/1
$BC_{10}F_4FeH_{11}O$	$[C_5H_5Fe(CH_2{=}CHCH{=}CH_2)CO][BF_4]$	Fe: Org.Comp.B17-186
$BC_{10}F_4FeH_{11}O_2$	$[(C_5H_5)Fe(CO)_2(CH_2{=}CHCH_3)][BF_4]$	Fe: Org.Comp.B17-6/7, 45/6
–	$[(C_5H_5)Fe(CO)_2(CH_2{=}CDCH_3)][BF_4]$	Fe: Org.Comp.B17-8
–	$[(C_5H_5)Fe(CO)_2(CDH{=}CHCH_3)][BF_4]$	Fe: Org.Comp.B17-8
–	$[(C_5H_5)Fe(CO)_2(CD_2{=}CHCD_3)][BF_4]$	Fe: Org.Comp.B17-8
–	$[(C_5H_5)(CO)_2Fe{=}C(CH_3)_2][BF_4]$	Fe: Org.Comp.B16a-91, 99
–	$[(C_5H_5)(CO)_2Fe{=}CH\text{-}C_2H_5][BF_4]$	Fe: Org.Comp.B16a-87, 95
–	$[(C_5H_5)(CO)_2Fe{=}CD\text{-}C_2H_5][BF_4]$	Fe: Org.Comp.B16a-95
$BC_{10}F_4FeH_{11}O_2S_2$	$[(C_5H_5)(CO)_2Fe{=}C(SCH_3)_2][BF_4]$	Fe: Org.Comp.B16a-152/4
$BC_{10}F_4FeH_{11}O_3$	$[(CH_2C(CH_3)CHC(CH_3)CH_2)Fe(CO)_3][BF_4]$	Fe: Org.Comp.B15-25, 31
–	$[(CH_2(CH)_4\text{-}C_2H_5)Fe(CO)_3][BF_4]$	Fe: Org.Comp.B15-13, 19
–	$[(CH_3\text{-}(CH)_5\text{-}CH_3)Fe(CO)_3][BF_4]$	Fe: Org.Comp.B15-13/6, 21
–	$[(C_5H_5)Fe(CO)_2(CH_2{=}C(CH_3)\text{-}OH)][BF_4]$	Fe: Org.Comp.B17-67
–	$[(C_5H_5)Fe(CO)_2(CH_2{=}CHCH_2\text{-}OH)][BF_4]$	Fe: Org.Comp.B17-24
–	$[(C_5H_5)Fe(CO)_2(CH_2{=}CH\text{-}OCH_3)][BF_4]$	Fe: Org.Comp.B17-21
–	$[(C_5H_5)(CO)_2Fe{=}C(CH_3)\text{-}OCH_3][BF_4]$	Fe: Org.Comp.B16a-108, 118/9
$BC_{10}F_4FeH_{11}O_4$	$[(C_5H_5)Fe(CO)_2(HOCH{=}CHOCH_3)][BF_4]$	Fe: Org.Comp.B17-68/9
$BC_{10}F_4FeH_{13}O_2S$	$[(C_5H_5)Fe(CO)_2CH_2S(CH_3)_2][BF_4]$	Fe: Org.Comp.B14-133/4, 142
$BC_{10}F_4FeH_{14}NO_2$	$[(C_5H_5)Fe(CO)_2CH_2CH_2NH_2CH_3][BF_4]$	Fe: Org.Comp.B14-149, 155
$BC_{10}F_4GeH_{15}$	$[GeC_5(CH_3)_5][BF_4]$	Ge: Org.Comp.3-383/4
$BC_{10}F_4H_5NO_5Re$	$[(CO)_5ReC_5H_5N][BF_4]$	Re: Org.Comp.2-151/2
$BC_{10}F_4H_7NO_7Re$	$[(CO)_5Re(CNCH_2COOC_2H_5)][BF_4]$	Re: Org.Comp.2-252, 254
$BC_{10}F_4H_8NS$	$[NS][BF_4] \cdot C_{10}H_8$	B: B Comp.SVol.4/4-167 S: S-N Comp.5-47/9
$BC_{10}F_4H_9NO_5Re$	$[(CO)_5Re(CNC_4H_9\text{-}t)][BF_4]$	Re: Org.Comp.2-252, 253
$BC_{10}F_4H_{10}IN_2$	$[I(NC_5H_5)_2][BF_4]$	B: B Comp.SVol.4/3b-167/8
$BC_{10}F_4H_{10}MoNO_5$	$[(C_2H_5)_2N\text{-}CMo(CO)_5][BF_4]$	Mo: Org.Comp.5-108, 113
$BC_{10}F_4H_{10}O_5Re$	$[(CO)_5Re(CH_2{=}CHCH_2CH_2CH_3)][BF_4]$	Re: Org.Comp.2-351

$BC_{10}F_4H_{11}MoN_4O_4W$
$[C_5H_5Mo(NO)_2(H)W(NO)_2C_5H_5][BF_4]$ Mo:Org.Comp.6-57/8
$BC_{10}F_4H_{11}MoO_2$ . . $(C_5H_5)Mo(CO)_2(FBF_3)(CH_2{=}CH{-}CH_3)$ Mo:Org.Comp.8-187, 198
$BC_{10}F_4H_{11}S_2$ . . . . $[-CH_2CH_2CH_2{-}SC(C_6H_5)S-][BF_4]$ . . . . B: B Comp.SVol.4/4-163
$BC_{10}F_4H_{12}Se_4$ . . . $[(CH_3)_2Se_2C_3{=}C_3Se_2(CH_3)_2][BF_4]$ . . . . B: B Comp.SVol.4/4-175
$BC_{10}F_4H_{13}NO_3Re$ $[(C_5H_5)Re(CO)(NO)(OC_4H_8)][BF_4]$ . . . . Re: Org.Comp.3-155
$BC_{10}F_4H_{14}MoO_2P$ $[(C_5H_5)Mo(CO)_2(P(CH_3)_3){-}FBF_3]$ Mo:Org.Comp.7-57, 62
$BC_{10}F_4H_{16}MoO_3P$ $[(C_5H_5)Mo(CO)_2(P(CH_3)_3)(H_2O)][BF_4]$ Mo:Org.Comp.7-283, 294
$BC_{10}F_4H_{16}N$ . . . . . $[N(CH_3)_3CH_2C_6H_5][BF_4]$ . . . . B: B Comp.SVol.3/3-343
$BC_{10}F_4H_{19}IMoN_3O$
$[C_5H_5Mo(NO)(N(CH_3)(C_2H_5)N(CH_3)_2)I][BF_4]$ . . . Mo:Org.Comp.6-40
$BC_{10}F_4H_{23}N_2S$ . . . $[t{-}C_4H_9{-}N(C_2H_5)S{=}N{-}C_4H_9{-}t][BF_4]$ . . . . S: S-N Comp.7-334
$BC_{10}F_4H_{24}N$ . . . . . $[(C_2H_5)_3N{-}C_4H_9{-}n][BF_4]$ . . . . B: B Comp.SVol.4/3b-208
$BC_{10}F_4H_{30}N_3Si_5$ . . $[-N(BF_2){-}Si(CH_3)_2{-}N(SiF(CH_3)_2){-}Si(CH_3)_2{-}N$
$(SiF(CH_3)_2){-}Si(CH_3)_2-]$ . . . . B: B Comp.SVol.4/3b-244
$BC_{10}F_5H_9N$ . . . . . . $C_6F_5{-}B{=}N{-}C_4H_9{-}t$ . . . . B: B Comp.SVol.4/3a-214
$BC_{10}F_9H_{12}N_2S_3$ . . $(CH_3)_3N{-}BH_2{-}N[-CH{=}C(SCF_3){-}C(SCF_3){=}C(SCF_3)-]$
B: B Comp.SVol.4/3b-23/4
– . . . . $(CH_3)_3N{-}BH_2{-}N[-C(SCF_3){=}CH{-}C(SCF_3){=}C(SCF_3)-]$
B: B Comp.SVol.4/3b-23/4
$BC_{10}F_{11}H_8Te$ . . . . $[C_6H_5(CH_3)TeC_3F_7][BF_4]$ . . . . B: B Comp.SVol.3/4-150
$BC_{10}FeH_{12}^-$ . . . . . $[C_5H_5FeC_4H_4BCH_3]^-$ . . . . Fe: Org.Comp.B17-222
$BC_{10}FeH_{12}Na$ . . . . $Na[C_5H_5FeC_4H_4BCH_3]$ . . . . Fe: Org.Comp.B17-222
$BC_{10}FeH_{13}$ . . . . . . $C_5H_5Fe(H)C_4H_4BCH_3$ . . . . Fe: Org.Comp.B17-222, 226/7
$BC_{10}FeH_{15}N_2O$ . . . $(C_5H_5)Fe((CHNCH_3)_2BH_2)CO$ . . . . Fe: Org.Comp.B16b-127, 138
$BC_{10}Fe_2H_9O_6S_2$ . . $BFe_2S_2(CO)_6(C_4H_9{-}n)$ . . . . B: B Comp.SVol.4/4-129/30
$BC_{10}Fe_3H_2O_{10}^-$ . . $[(CO)_{10}Fe_3(BH_2)]^-$ . . . . Fe: Org.Comp.C6b-70
$BC_{10}Fe_3H_3O_{10}$ . . . $(CO)_{10}(H)Fe_3(BH_2)$ . . . . Fe: Org.Comp.C6b-70
$BC_{10}Fe_3H_6O_9^-$ . . . $[(CO)_9(H)Fe_3(BH_2{-}CH_3)]^-$ . . . . Fe: Org.Comp.C6a-299
$BC_{10}Fe_3H_7O_9$ . . . . $(CO)_9(H)Fe_3(BH_3{-}CH_3)$ . . . . Fe: Org.Comp.C6a-295/7
$BC_{10}H_8O_4V$ . . . . . . $(CO)_4V(C_5H_5BCH_3)$ . . . . B: B Comp.SVol.3/4-155
B: B Comp.SVol.4/4-180
$BC_{10}H_{10}I_2O_6Re$ . . cis-$(CO)_4Re[(CH_3CO)CH(CH_3)_2CO]BI_2$ . . . . Re: Org.Comp.2-346, 348
$BC_{10}H_{10}N$ . . . . $[=CH{-}CH{=}CH{-}CH{=}CH-]N{-}B[=CH{-}CH{=}CH{-}CH{=}CH-]$
B: B Comp.SVol.4/3b-26
$BC_{10}H_{10}N_3$ . . . . . . $C_5H_5N{-}BH(CN)(C_4H_4N)$ . . . . B: B Comp.SVol.3/3-190
$BC_{10}H_{11}N_3O_4$ . . . . $(CH_3COO)B[-O{-}CH{=}NNH_2-][-O{-}(1,2{-}C_6H_4){-}CH{=}]N$
B: B Comp.SVol.3/3-220
$BC_{10}H_{12}N$ . . . . $(-CH{=}CH{-}CH{=}CH-)B(NH_3){-}C_6H_5$ . . . . B: B Comp.SVol.4/3b-29
– . . . . $H_3B \cdot 1{-}NC_9H_6{-}2{-}CH_3$ . . . . B: B Comp.SVol.4/3b-14
– . . . . $H_3B \cdot 1{-}NC_9H_6{-}3{-}CH_3$ . . . . B: B Comp.SVol.4/3b-14
– . . . . $H_3B \cdot 1{-}NC_9H_6{-}4{-}CH_3$ . . . . B: B Comp.SVol.4/3b-14
– . . . . $H_3B \cdot 1{-}NC_9H_6{-}6{-}CH_3$ . . . . B: B Comp.SVol.4/3b-14
– . . . . $H_3B \cdot 1{-}NC_9H_6{-}8{-}CH_3$ . . . . B: B Comp.SVol.4/3b-14
$BC_{10}H_{12}NO$ . . . . . . $H_3B \cdot 1{-}NC_9H_6{-}6{-}OCH_3$ . . . . B: B Comp.SVol.4/3b-14
$BC_{10}H_{12}NOS_2$ . . . . $(SC_4H_3{-}2)_2B[-NH_2{-}CH_2CH_2{-}O-]$ . . . . B: B Comp.SVol.4/3b-71
B: B Comp.SVol.4/4-152
$BC_{10}H_{12}N_3O_2$ . . . . 1,5-$(CH_3)_2$-2-C_6H_5-1,3,5,2-$N_3BC_2H(=O)_2$-4,6 B: B Comp.SVol.4/3a-176
– . . . . 1,5-$(CH_3)_2$-3-C_6H_5-1,3,5,2-$N_3BC_2H(=O)_2$-4,6 B: B Comp.SVol.4/3a-176
$BC_{10}H_{12}O_7Re$ . . . . fac-$(CO)_3Re[(COCH_3)_3]BOCH_3$ . . . . Re: Org.Comp.2-379

$BC_{10}H_{13}N_2$ ...... $(NC_4H_4-1)_2B(C_2H_5)$ ...... B: B Comp.SVol.4/3a-169
$BC_{10}H_{13}N_2O$ ..... $C_6H_5B[-O-N=CH-N(n-C_3H_7)-]$ ...... B: B Comp.SVol.3/3-163
$BC_{10}H_{13}O_2$ ...... $2-CH_3-C_6H_4-B[-O-CH_2-CH(CH_3)-O-]$ ...... B: B Comp.SVol.4/3b-69
$BC_{10}H_{13}O_2S$ ..... $i-C_3H_7OB[-O-CH_2-(1,2-C_6H_4)-S-]$ ...... B: B Comp.SVol.3/4-120
$BC_{10}H_{14}N$ .... $C_6H_5-B=N-C_4H_9-t$ ...... B: B Comp.SVol.4/3a-214
$BC_{10}H_{14}NO$ ...... $(CH_3)NBOC_9H_{11}$ ...... B: B Comp.SVol.3/3-219
$BC_{10}H_{14}NO_2$ ..... $(i-C_3H_7O)B[-NH-(1,2-C_6H_4)-CH_2-O-]$ ...... B: B Comp.SVol.3/3-164
$BC_{10}H_{15}I_2$ .... $C_5(CH_3)_5-BI_2$ ...... B: B Comp.SVol.4/4-113
$BC_{10}H_{15}N_2$ ...... $C_6H_5B(-NCH_3-CH_2-CH_2-NCH_3-)$ ...... B: B Comp.SVol.3/3-113
$BC_{10}H_{15}S_2$ ...... $2-CH_3-2-C_6H_5-1,3-S_2C_3H_4 \cdot BH_3$ ...... B: B Comp.SVol.4/4-134
– ...... $C_6H_5-B(S-C_2H_5)_2$ ...... B: B Comp.SVol.3/4-110/1
$BC_{10}H_{16}N$ .... $CH_2=CH-CH_2-NCH_3-C_6H_5 \cdot BH_3$ ...... B: B Comp.SVol.3/3-178
– ...... $(CH_3C{\equiv}C)_2B-N(C_2H_5)_2$ ...... B: B Comp.SVol.3/3-149
– ...... $C_5H_5N \cdot BC_5H_{11}$ ...... B: B Comp.SVol.3/3-189
$BC_{10}H_{16}NO$ ...... $O(CH_2-CH_2-)_2NC_6H_5-BH_3$ ...... B: B Comp.SVol.3/3-177
$BC_{10}H_{16}NO_2$ ..... $(CH_3)_3N-BH_2-COO-C_6H_5$ ...... B: B Comp.SVol.4/3b-25
$BC_{10}H_{16}NO_3$ ..... $(HO)_2(C_6H_5)B-HN(-CH_2-CH_2-)_2O$ ...... B: B Comp.SVol.3/3-217
$BC_{10}H_{16}NO_8$ ..... $(C_2H_5)_3NH[B(-O-CO-CO-O-)_2]$ ...... B: B Comp.SVol.3/3-227
$BC_{10}H_{16}NS$ ...... $C_2H_5SB(C_6H_5)N(CH_3)_2$ ...... B: B Comp.SVol.3/4-122
$BC_{10}H_{16}N_3$ ...... $C_5H_{11}N \cdot NC_4H_4-1-BH-CN$ ...... B: B Comp.SVol.3/3-190
– ...... $[-NH-CH_2CH_2CH_2-N(CH_3)-]B-NH-C_6H_5$ ...... B: B Comp.SVol.4/3a-154
$BC_{10}H_{17}N_2$ ...... $(CH_3)_2N-B(C_6H_5)-N(CH_3)_2$ ...... B: B Comp.SVol.4/3a-168
– ...... $(C_2H_5)_2B[-NC_5H_4-CH_2-NH-]$ ...... B: B Comp.SVol.3/3-124
– ...... $C_6H_5-B[N(CH_3)_2]_2$ ...... B: B Comp.SVol.3/3-104
$BC_{10}H_{17}N_2OS$ .... $(CH_3)_2N-B(C_6H_5)-NCH_3-S(O)CH_3$ ...... B: B Comp.SVol.3/3-106
B: B Comp.SVol.3/4-134
$BC_{10}H_{17}N_2Si$ ..... $CH_3B[-NH-(1,2-C_6H_4)-NSi(CH_3)_3-]$ ...... B: B Comp.SVol.3/3-117
$BC_{10}H_{17}N_2Si_2$ .... $C_6H_5B(NH_2)-2,5-(CH_3)_2-1,2,5-NSi_2C_2H_4$ ...... B: B Comp.SVol.3/3-109
$BC_{10}H_{17}N_6$ ...... $[-NCH_3-(CH_2)_2-NCH_3-]B[N_2C_3H_3]_2H$ ...... B: B Comp.SVol.3/3-201
$BC_{10}H_{18}Li_2N$ ..... $Li_2[1-(i-C_3H_7)_2N-BC_4H_4]$ ...... B: B Comp.SVol.3/3-214
B: B Comp.SVol.4/3b-49
$BC_{10}H_{18}N$ .... $1-(i-C_3H_7)_2B-NC_4H_4$ ...... B: B Comp.SVol.4/3a-233
– ...... $2,6-(C_2H_5)_2C_6H_3-NH_2-BH_3$ ...... B: B Comp.SVol.3/3-173
– ...... $H_3B-N(C_2H_5)_2-C_6H_5$ ...... B: B Comp.SVol.4/3b-5, 10
$BC_{10}H_{18}NO$ ...... $2,7-(CH_3)_2-2-NC_8H_9(=O)-6 \cdot BH_3$ ...... B: B Comp.SVol.4/3b-12
– ...... $C_6H_5-CHOH-CHCH_3-NHCH_3 \cdot BH_3$ ...... B: B Comp.SVol.3/3-176
$BC_{10}H_{18}NOS$ ..... $2-CH_3-3,5,6-(C_2H_5)_3-1,3,2-SNBC_3(=O)-4$ ...... B: B Comp.SVol.4/4-151/2
$BC_{10}H_{18}NOSe$ .... $2-CH_3-3,5,6-(C_2H_5)_3-1,3,2-SeNBC_3(=O)-4$ .... B: B Comp.SVol.4/4-172
$BC_{10}H_{18}NO_2$ ..... $2,7-(CH_3)_2-4a-HO-2-NC_8H_8(=O)-6 \cdot BH_3$ .... B: B Comp.SVol.4/3b-12
– ...... $2,7-(CH_3)_2-7a-HO-2-NC_8H_8(=O)-6 \cdot BH_3$ .... B: B Comp.SVol.4/3b-12
$BC_{10}H_{18}N_3$ ...... $2-[2,5-(CH_3)_2-1-NC_4H_2]-1,3,2-N_2BC_2H_4-(CH_3)_2-1,3$
B: B Comp.SVol.3/3-95
– ...... $2-[(i-C_3H_7)_2B-NH]-1,3-N_2C_4H_3$ ...... B: B Comp.SVol.4/3a-232
– ...... $C_{10}H_{15}-N_3-BH_3$, radical ...... B: B Comp.SVol.3/3-172
$BC_{10}H_{19}NS_2$ ..... $n-C_4H_9B[-S-CCH_3-C(C_3H_7-N)-S-]$ ...... B: B Comp.SVol.3/4-114
$BC_{10}H_{19}N_2$ ...... $H_3B-N(CH_3)_2-C_6H_4-4-N(CH_3)_2$ ...... B: B Comp.SVol.4/3b-11
$BC_{10}H_{19}N_4$ ...... $1-[B(-NCH_3-CH_2CH_2CH_2-NCH_3-)]-3,5-$
$(CH_3)_2-1,2-N_2C_3H$ ...... B: B Comp.SVol.3/3-96
$BC_{10}H_{19}N_6O_4$ .... $[-NCH_3-CO-NCH_3-CO-NCH_3-]B[$
$-NHCH_3-CO-NCH_3-CO-NCH_3-]$ ...... B: B Comp.SVol.3/3-124

$BC_{10}H_{19}S$ .... $[3.3.1]-9-BC_8H_{14}-9-SC_2H_5$ .... B: B Comp.SVol.4/4-132/3
$BC_{10}H_{19}Se_2$ ..... $4,5-(C_2H_5)_2-3-(C_4H_9-n)-1,2,3-Se_2BC_2$ .... B: B Comp.SVol.4/4-171/2
$BC_{10}H_{20}N$ .... $(C_2H_5-CH=CH)_2B-N(CH_3)_2$ .... B: B Comp.SVol.3/3-149
– .... $(i-C_3H_7)_2N-B[-CH=CH-CH_2-CH_2-]$ .... B: B Comp.SVol.4/3a-228
– .... $(i-C_3H_7)_2N-B[-CH_2-CH=CH-CH_2-]$ .... B: B Comp.SVol.3/3-150
B: B Comp.SVol.4/3a-228
– .... $C_5H_{10}B-NC_5H_{10}$ .... B: B Comp.SVol.3/3-156
– .... $[3.3.1]-9-BC_8H_{14}-9-N(CH_3)_2$ .... B: B Comp.SVol.3/3-150
– .... $[H_3C-CH=C(CH_3)]_2B-N(CH_3)_2$ .... B: B Comp.SVol.3/3-149
$BC_{10}H_{20}NO$ ...... $5-CH_3-1,2-ONC_3H_2 \cdot B(C_2H_5)_3$ .... B: B Comp.SVol.4/3b-28
$BC_{10}H_{20}NS$ ...... $4-CH_3-1,3-SNC_3H_2 \cdot B(C_2H_5)_3$ .... B: B Comp.SVol.4/3b-28
$BC_{10}H_{20}NS_2$ ..... $(C_2H_5)_2N-B[-S-C(C_2H_5)=C(C_2H_5)-S-]$ .... B: B Comp.SVol.3/4-126
– .... $(C_2H_5)_2N-B[-S-CCH_3=C(C_3H_7-n)-S-]$ .... B: B Comp.SVol.3/4-126
– .... $t-C_4H_9-NH-B[-S-CCH_3=C(C_3H_7-i)-S-]$ .... B: B Comp.SVol.3/4-126
$BC_{10}H_{21}N^+$ ...... $[(-C(CH_3)_2-(CH_2)_3-C(CH_3)_2-)N=BCH_3]^+$ .... B: B Comp.SVol.3/3-205
$BC_{10}H_{21}N_2$ ...... $[3.3.1.1^{3,7}]-1-BC_9H_{15} \cdot CH_3-NH-NH_2$ .... B: B Comp.SVol.4/3b-30
– .... $[(C_2H_5)_2N]_2B-C{\equiv}CH$ .... B: B Comp.SVol.3/3-106
B: B Comp.SVol.4/3a-170
$BC_{10}H_{21}N_2O$ ..... $(CH_3)_2N-B(O-C_6H_9-c)-N(CH_3)_2$ .... B: B Comp.SVol.4/3b-51
$BC_{10}H_{21}S_2$ ...... $n-C_4H_9B[-S-CHCH_3-CH(C_3H_7-n)-S-]$ .... B: B Comp.SVol.3/4-112
$BC_{10}H_{22}N$ .... $(C_2H_5)_2B-N(C_3H_7-i)-C(CH_3)=CH_2$ .... B: B Comp.SVol.3/3-151
– .... $(C_2H_5)_2N-B[-CH(CH_3)-CH_2-CH_2-CH(CH_3)-]$ .. B: B Comp.SVol.4/3a-228
– .... $C_5H_{11}B-NC_5H_{11}$ .... B: B Comp.SVol.3/3-189
– .... $n-C_5H_{11}-B=N-C_5H_{11}-n$ .... B: B Comp.SVol.3/3-141
$BC_{10}H_{22}NO$ ...... $2-(i-C_4H_9)-4,4,6-(CH_3)_3-1,3,2-ONBC_3H_4$ .... B: B Comp.SVol.4/3b-66
$BC_{10}H_{22}NO_2$ ..... $[CH(CH_3)_2C(CH_3)_2]B[-O-CH_2-CH_2-]_2NH$ .... B: B Comp.SVol.3/3-218
$BC_{10}H_{22}NO_6$ ..... $[(CH_3)_4N][HB(OC(O)-CH_3)_3]$ .... B: B Comp.SVol.4/3b-77
$BC_{10}H_{22}NSi$ ..... $4,5-(C_2H_5)_2-1,2,2,3-(CH_3)_4-1,2,5-NSiBC_2$ .... B: B Comp.SVol.3/3-162
B: B Comp.SVol.4/3a-250
$BC_{10}H_{22}N_3$ ...... $(i-C_3H_7)_2B(-NH-CCH_3=N-CCH_3=NH-)$ .... B: B Comp.SVol.3/3-122
– .... $(n-C_5H_{11})_2B-N_3$ .... B: B Comp.SVol.3/3-144
$BC_{10}H_{23}I_2S$ ...... $(CH_3)_2S-BI_2-CH(C_2H_5)-C_5H_{11}-n$ .... B: B Comp.SVol.3/4-137
– .... $(CH_3)_2S-BI_2-C_8H_{17}-n$ .... B: B Comp.SVol.3/4-137
$BC_{10}H_{23}LiNSi_2$ ... $Li[1,3-((CH_3)_3Si)_2-2-CH_3-1,2-NBC_3H_2]$ .... B: B Comp.SVol.4/3b-48
$BC_{10}H_{23}N_2$ ...... $(n-C_3H_7)_3N-BH_2-CN$ .... B: B Comp.SVol.4/3b-23
– .... $t-C_4H_9-N=B-N(C_3H_7-i)_2$ .... B: B Comp.SVol.4/3a-160/5
$BC_{10}H_{23}N_2O$ ..... $(CH_3)_2B-NCH_3-C(O)N(i-C_3H_7)_2$ .... B: B Comp.SVol.3/3-155
– .... $(CH_3)_2B-N(C_3H_7-i)-C(O)N(C_2H_5)_2$ .... B: B Comp.SVol.3/3-156
$BC_{10}H_{23}N_2O_2$ .... $[(C_2H_5)_2N]_2BOC(O)CH_3$ .... B: B Comp.SVol.3/3-106
$BC_{10}H_{23}N_2S$ ..... $(n-C_3H_7)_3N-BH_2-N=C=S$ .... B: B Comp.SVol.4/3b-23
$BC_{10}H_{23}N_2Si$ ..... $1,3,3,4-(CH_3)_4-5,6-(C_2H_5)_2-1,2,3,6-N_2SiBC_2H$ B: B Comp.SVol.4/3a-252
– .... $1-(CH_3-NH)-4,5-(C_2H_5)_2-2,2,3-(CH_3)_3-1,2,5-NSiBC_2$
B: B Comp.SVol.4/3a-250
– .... $2,3,3,4-(CH_3)_4-5,6-(C_2H_5)_2-1,2,3,6-N_2SiBC_2H$ B: B Comp.SVol.4/3a-252
$BC_{10}H_{23}S$ .... $(n-C_3H_7)_2B-S-C_4H_9-n$ .... B: B Comp.SVol.3/4-111
B: B Comp.SVol.4/4-129
$BC_{10}H_{24}N$ .... $(C_2H_5)_2N-CH_3 \cdot BC_5H_{11}$ .... B: B Comp.SVol.3/3-189
– .... $(i-C_3H_7)_2B-N(C_2H_5)_2$ .... B: B Comp.SVol.4/3a-227
– .... $(i-C_3H_7)_2B-NH-C_4H_9-t$ .... B: B Comp.SVol.4/3a-230
– .... $(t-C_4H_9)_2B-N(CH_3)_2$ .... B: B Comp.SVol.3/3-149

$BC_{10}H_{24}N$ .... $(t-C_4H_9)_2B-N(CH_3)_2$ .... B: B Comp.SVol.4/3a-227
$BC_{10}H_{24}NO$ .... $(i-C_3H_7)(t-C_4H_9O)B-NH(i-C_3H_7)$ .... B: B Comp.SVol.3/3-155
$BC_{10}H_{24}NOSi$ .... $CH_3-NH_2-B(C_2H_5)[-O-Si(CH_3)_2-C(CH_3)=C(C_2H_5)-]$
B: B Comp.SVol.4/3b-71
$BC_{10}H_{24}NSi_2$ ..... $2-CH_3-1,3-[(CH_3)_3Si]_2-1,2-NBC_3H_3$ .... B: B Comp.SVol.4/3a-249
$BC_{10}H_{24}N_2OP$ .... $CH_3-B[-N(C_4H_9-t)-P(CH_3)(=O)-N(C_4H_9-t)-]$ ... B: B Comp.SVol.4/3a-207/8
$BC_{10}H_{24}N_2P$ ..... $CH_3-B[-N(C_4H_9-t)-P(CH_3)-N(C_4H_9-t)-]$ ...... B: B Comp.SVol.4/3a-207/8
$BC_{10}H_{24}N_2PS$ .... $CH_3-B[-N(C_4H_9-t)-P(CH_3)(=S)-N(C_4H_9-t)-]$ ... B: B Comp.SVol.4/3a-207/8
$BC_{10}H_{24}N_3$ ...... $(i-C_3H_7)_2N-B(-NCH_3-CH_2-CH_2-NCH_3-)$ ...... B: B Comp.SVol.3/3-95
– .... $t-C_4H_9-N=C(CH_3)-B[N(CH_3)_2]_2$ .... B: B Comp.SVol.3/3-109
$BC_{10}H_{25}N_2$ ...... $i-C_3H_7-NH-B(C_3H_7-i)-N(C_2H_5)_2$ .... B: B Comp.SVol.3/3-107
– .... $i-C_3H_7-NH-B(C_3H_7-i)-NH-C_4H_9-t$ .... B: B Comp.SVol.3/3-107
– .... $[-N(C_3H_7-i)-CH(CH_3)-N(C_3H_7-i)-CH_2CH_2-] \cdot BH_3$
B: B Comp.SVol.4/3b-12
$BC_{10}H_{25}N_2OSi$ ... $(CH_3)_3Si-O-CH=CH-CH_2-B[N(CH_3)_2]_2$ .... B: B Comp.SVol.4/3a-168
$BC_{10}H_{25}N_4Si$ ..... $(CH_3)_3Si-N(C_4H_9-t)-B(N_3)-C_3H_7-i$ .... B: B Comp.SVol.3/3-110
B: B Comp.SVol.4/3a-171
– .... $(CH_3)_3Si-N(C_4H_9-t)-B(N_3)-C_3H_7-n$ .... B: B Comp.SVol.3/3-110
– .... $n-C_3H_7-B[-N(C_4H_9-t)-N=N-N(Si(CH_3)_3)-]$ .... B: B Comp.SVol.3/3-137/8
– .... $i-C_3H_7-B[-N(C_4H_9-t)-N=N-N(Si(CH_3)_3)-]$ .... B: B Comp.SVol.3/3-137/8
$BC_{10}H_{26}N_2O$ ..... $(i-C_3H_7)_2NH-BH_2-N(O)-C_4H_9-t$, radical ...... B: B Comp.SVol.4/3b-26
$BC_{10}H_{26}N_3OS$ .... $[(C_2H_5)_2N]_2B-NCH_3-S(O)CH_3$ .... B: B Comp.SVol.3/3-96
B: B Comp.SVol.3/4-134
$BC_{10}H_{26}N_4^+$ ..... $[(-NCH_3-(CH_2)_2-NCH_3-)B(-N(CH_3)_2-(CH_2)_2-N(CH_3)_2-)]^+$
B: B Comp.SVol.3/3-210/1
$BC_{10}H_{27}N_2Si_2$ .... $(CH_3)_3Si-N=B-N(C_4H_9-t)-Si(CH_3)_3$ .... B: B Comp.SVol.4/3a-160/5
– .... $t-C_4H_9-N=B-N[Si(CH_3)_3]_2$ .... B: B Comp.SVol.4/3a-160/5
$BC_{10}H_{28}NOSi_2$ ... $(C_2H_5)_2B-N[Si(CH_3)_3]-O-Si(CH_3)_3$ .... B: B Comp.SVol.3/3-153
B: B Comp.SVol.4/3a-231
$BC_{10}H_{28}NO_2Si_2$ .. $(C_2H_5-O)_2B-N[Si(CH_3)_3]_2$ .... B: B Comp.SVol.4/3b-61
$BC_{10}H_{28}NSi_2$ ..... $t-C_4H_9-BH-N[Si(CH_3)_3]_2$ .... B: B Comp.SVol.4/3a-224
$BC_{10}H_{28}NSn_2$ .... $(C_2H_5)_2B-N[Sn(CH_3)_3]_2$ .... B: B Comp.SVol.4/3a-227
– .... $[(CH_3)_3Sn]_2B-N(C_2H_5)_2$ .... B: B Comp.SVol.3/3-150
$BC_{10}H_{28}N_2PSn$ ... $(CH_3)_2Sn[-N(C_4H_9-t)-PH(BH_3)-N(C_4H_9-t)-]$ ... Sn: Org.Comp.19-77, 80
$BC_{10}H_{28}N_3Sn_2$ ... $[(CH_3)_3Sn]_2NB(-NCH_3-CH_2-CH_2-NCH_3-)$ ..... B: B Comp.SVol.3/3-96
$BC_{10}H_{29}N_2Si_2$ .... $t-C_4H_9-B(NH_2)-N[Si(CH_3)_3]_2$ .... B: B Comp.SVol.4/3a-169
$BC_{11}ClF_4FeH_{10}$ ... $[(C_6H_6)Fe(C_5H_4-Cl)][BF_4]$ .... Fe: Org.Comp.B18-142/6, 151, 169/70, 188
– .... $[(Cl-C_6H_5)Fe(C_5H_5)][BF_4]$ .... Fe: Org.Comp.B18-142/6, 197, 200, 201, 245/6, 277/80
$BC_{11}ClH_{12}O_6Re^-$ $[(CO)_3Re(CH_2COCO(C_3H_7-i))(CH_3CO)BCl]^-$ ... Re: Org.Comp.2-418/9
$BC_{11}ClH_{13}O_6Re$ .. fac-$(CO)_3Re[(COCH_3)_2COCH(CH_3)_2]BCl$ ...... Re: Org.Comp.2-379, 381/2
$BC_{11}ClH_{15}N$ ..... $1-(C_6H_5-BCl)-NC_5H_{10}$ .... B: B Comp.SVol.3/4-56
– .... $C_2H_5-CH=CH-BCl-NCH_3-C_6H_5$ .... B: B Comp.SVol.3/4-56
$BC_{11}ClH_{18}$ ...... $CH_3-BCl-C_5(CH_3)_5$ .... B: B Comp.SVol.4/4-49
$BC_{11}ClH_{19}N$ ..... $C_5H_5N \cdot BCl(C_3H_7-i)_2$ .... B: B Comp.SVol.4/4-51
$BC_{11}ClH_{21}N_3S$ ... $[(n-C_3H_7)_2B(-NH=CCH_3-NH-(2,3-(1,3-C_3H_2NS))-)]Cl$
B: B Comp.SVol.3/4-62
$BC_{11}ClH_{22}$ ...... $[(CH_3)_2CHC(CH_3)_2](c-C_5H_9)BCl$ .... B: B Comp.SVol.3/4-44

Formula	Compound	Reference
$BC_{11}ClH_{24}$	n-C_5H_{11}-BCl-C$(CH_3)_2$-C_3H_7-i	B: B Comp.SVol.3/4-44
–	n-C_5H_{11}-BCl-C_6H_{13}-n	B: B Comp.SVol.3/4-42
$BC_{11}ClH_{24}N_2O$	1-[$(CH_3)_2$N-BCl-O]-2,2,6,6-$(CH_3)_4$-NC_5H_6	B: B Comp.SVol.4/4-61
$BC_{11}ClH_{25}N_2P$	[-N(C_4H_9-t)-BCl-N(C_4H_9-t)-]P-C_3H_7-i	B: B Comp.SVol.4/4-64/5
$BC_{11}ClH_{27}NSi$	n-C_4H_9-BCl-N(C_4H_9-t)-Si$(CH_3)_3$	B: B Comp.SVol.3/4-57
–	i-C_4H_9-BCl-N(C_4H_9-t)-Si$(CH_3)_3$	B: B Comp.SVol.3/4-57
–	s-C_4H_9-BCl-N(C_4H_9-t)-Si$(CH_3)_3$	B: B Comp.SVol.4/4-60
–	t-C_4H_9-BCl-N(C_4H_9-t)-Si$(CH_3)_3$	B: B Comp.SVol.3/4-57
$BC_{11}ClH_{28}N_2Si$	(s-$C_4H_9)_2$N-BCl-NH-Si$(CH_3)_3$	B: B Comp.SVol.3/4-59
–	t-C_4H_9-NH-BCl-N(C_4H_9-t)-Si$(CH_3)_3$	B: B Comp.SVol.4/4-63
$BC_{11}Cl_2F_4FeH_9$	[(1,2-Cl_2-C_6H_4)Fe(C_5H_5)][BF_4]	Fe: Org.Comp.B19-1, 4/5, 17/8, 87/90
–	[(1,3-Cl_2-C_6H_4)Fe(C_5H_5)][BF_4]	Fe: Org.Comp.B19-1, 4/5, 18, 87/90
–	[(1,4-Cl_2-C_6H_4)Fe(C_5H_5)][BF_4]	Fe: Org.Comp.B19-1, 4/5, 18, 87/90
$BC_{11}Cl_2H_{16}N$	Cl_2B-N(C_4H_9-t)-CH_2-C_6H_5	B: B Comp.SVol.4/4-59
$BC_{11}Cl_2H_{19}O_2$	Cl_2B[-O=C(C_4H_9-t)-CH=C(C_4H_9-t)-O-]	B: B Comp.SVol.3/4-52
$BC_{11}Cl_2H_{21}Si$	C_2H_5-BCl-C(C_2H_5)=C[CCH_3=CH_2]-SiCl$(CH_3)_2$	B: B Comp.SVol.4/4-49
$BC_{11}Cl_2H_{26}N_2P$	(i-$C_3H_7)_2$N-BCl-N(C_4H_9-t)-PCl-CH_3	B: B Comp.SVol.4/4-63
$BC_{11}Cl_3H_{17}N$	C_6H_5-CH_2-NH-C_4H_9-t · BCl_3	B: B Comp.SVol.4/4-33
$BC_{11}Cl_3H_{27}N_2SbSi$	Si$(CH_3)_3$-N(C_4H_9-t)-BCl-N(C_4H_9-t)-$SbCl_2$	B: B Comp.SVol.4/4-63
$BC_{11}Cl_4GeH_{27}N_2Si$	Si$(CH_3)_3$-N(C_4H_9-t)-BCl-N(C_4H_9-t)-$GeCl_3$	B: B Comp.SVol.4/4-63
$BC_{11}Cl_4H_{27}N_2SiSn$	Si$(CH_3)_3$-N(C_4H_9-t)-BCl-N(C_4H_9-t)-$SnCl_3$	B: B Comp.SVol.4/4-63
$BC_{11}Cl_4H_{27}N_2Si_2$	Si$(CH_3)_3$-N(C_4H_9-t)-BCl-N(C_4H_9-t)-$SiCl_3$	B: B Comp.SVol.4/4-63
$BC_{11}CoH_{17}N$	(C_5H_5)Co[$(CH_2$=CH$)_2$B-N$(CH_3)_2$]	B: B Comp.SVol.4/3a-236
$BC_{11}FH_{12}O_6Re^-$	[$(CO)_3$Re($CH_2COCOC_3H_7$-i)(CH_3CO)BF]$^-$	Re: Org.Comp.2-418/9
$BC_{11}FH_{13}O_6Re$	$(CO)_3$Re[$(COCH_3)_2COC_3H_7$-i]BF	Re: Org.Comp.2-379, 381/2
$BC_{11}F_2H_9O_3Te$	[-BF_2-O-$TeC_9H_3(CH_3)(OCH_3)$-O-]	B: B Comp.SVol.4/4-177
$BC_{11}F_2H_{11}NO_2$	F_2B[-O-CC_6H_5=CH-CNC_2H_5=O-]	B: B Comp.SVol.3/3-368
$BC_{11}F_2H_{16}N$	F_2B-N(C_4H_9-t)-CH_2-C_6H_5	B: B Comp.SVol.4/3b-243
$BC_{11}F_2H_{17}N_6NiO_2S$	Ni[-BF_2-O-N=C(CH_3)-C(CH_3)=N-NC(S-C_2H_5)N-N=C(CH_3)-C(CH_3)=N-O-]	B: B Comp.SVol.4/3b-124
$BC_{11}F_3H_{15}IO_3S$	[$(CH_3)_5C_5$BI][CF_3SO_3]	B: B Comp.SVol.3/4-99
$BC_{11}F_3H_{18}O_3S$	(c-$C_5H_9)_2$B(OSO_2CF_3)	B: B Comp.SVol.3/4-142
$BC_{11}F_3H_{26}N_4O_3S$	[(-NCH_3-$(CH_2)_2$-NCH_3-)B(-N$(CH_3)_2$-$(CH_2)_2$-N$(CH_3)_2$-)][O_3SCF_3]	B: B Comp.SVol.3/3-210/1
$BC_{11}F_3H_{33}N_3Si_5$	[-N(BF_2)-Si$(CH_3)_2$-N(Si$(CH_3)_3$)-Si$(CH_3)_2$-N(SiF$(CH_3)_2$)-Si$(CH_3)_2$-]	B: B Comp.SVol.4/3b-244
$BC_{11}F_4FeH_9O_3$	[(6-(CH_2=)C_7H_7)Fe$(CO)_3$][BF_4]	Fe: Org.Comp.B15-209, 225
–	[(CH_3-C_7H_6)Fe$(CO)_3$][BF_4]	Fe: Org.Comp.B15-219, 232
–	[(C_8H_9)Fe$(CO)_3$][BF_4]	Fe: Org.Comp.B15-214, 227/9, 247/8
$BC_{11}F_4FeH_9O_4$	[(1-CH_3-7-O=C_7H_6)Fe$(CO)_3$][BF_4]	Fe: Org.Comp.B15-212
–	[(C_5H_5)Fe$(CO)_2$(H-CC-COO-CH_3)][BF_4]	Fe: Org.Comp.B17-121
$BC_{11}F_4FeH_9O_5$	[(1-$CH_3OC(O)$-C_6H_6)Fe$(CO)_3$][BF_4]	Fe: Org.Comp.B15-101, 136/8

$BC_{11}F_4FeH_9O_5$ . . . $[(2-CH_3OC(O)-C_6H_6)Fe(CO)_3][BF_4]$ . . . . Fe: Org.Comp.B15-105, 136/8
– . . . . $[(3-CH_3OC(O)-C_6H_6)Fe(CO)_3][BF_4]$ . . . . Fe: Org.Comp.B15-107, 136/8
– . . . . $[(6-CH_3C(O)O-C_6H_6)Fe(CO)_3][BF_4]$ . . . . Fe: Org.Comp.B15-107, 159
– . . . . $[(6-CH_3OC(O)-C_6H_6)Fe(CO)_3][BF_4]$ . . . . Fe: Org.Comp.B15-109, 136/8
$BC_{11}F_4FeH_{10}NO_2$ $[(C_5H_5)Fe(CO)_2(CH_2{=}C(CH_3)-CN)][BF_4]$ . . . . . . Fe: Org.Comp.B17-63
– . . . . $[(C_5H_5)Fe(CO)_2(CH_3CH{=}CH-CN)][BF_4]$ . . . . Fe: Org.Comp.B17-59
$BC_{11}F_4FeH_{10}NO_3$ $[(C_5H_5)Fe(CO)_2(NCCH{=}CH-OCH_3)][BF_4]$ . . . . . . Fe: Org.Comp.B17-66, 79
$BC_{11}F_4FeH_{11}$ . . . . $[(C_6H_6)Fe(C_5H_5)][BF_4]$ . . . . Fe: Org.Comp.B18-142/6, 150/4, 157/8, 174/81
$BC_{11}F_4FeH_{11}O_2$ . . $[(C_5H_5)Fe(CO)_2(CH_2{=}C{=}CHCH_3)][BF_4]$ . . . . Fe: Org.Comp.B17-111/2, 117
– . . . . $[(C_5H_5)Fe(CO)_2(CH_2{=}CHCH{=}CH_2)][BF_4]$ . . . . . . Fe: Org.Comp.B17-37, 51/2
– . . . . $[(C_5H_5)Fe(CO)_2(C_4H_6\text{-c})][BF_4]$ . . . . Fe: Org.Comp.B17-86
– . . . . $[(C_5H_5)Fe(CO)_2(H_3C-CC-CH_3)][BF_4]$ . . . . Fe: Org.Comp.B17-122
$BC_{11}F_4FeH_{11}O_2S_2$ $[C_5H_5(CO)_2Fe{=}C(SCH_2)_2CH_2][BF_4]$ . . . . Fe: Org.Comp.B16a-144, 156
$BC_{11}F_4FeH_{11}O_3$ . . $[(1,3-(CH_3)_2C_6H_5)Fe(CO)_3][BF_4]$ . . . . Fe: Org.Comp.B15-112, 161
– . . . . $[(1,4-(CH_3)_2C_6H_5)Fe(CO)_3][BF_4]$ . . . . Fe: Org.Comp.B15-122, 175
– . . . . $[(1-C_2H_5-C_6H_6)Fe(CO)_3][BF_4]$ . . . . Fe: Org.Comp.B15-99, 135/6
– . . . . $[(6,6-(CH_3)_2C_6H_5)Fe(CO)_3][BF_4]$ . . . . Fe: Org.Comp.B15-126
– . . . . $[(6-CH_3-C_7H_8)Fe(CO)_3][BF_4]$ . . . . Fe: Org.Comp.B15-206, 222
– . . . . $[(C_5H_5)Fe(CO)_2(CH_2{=}C{=}CHCH_2OH)][BF_4]$ . . . . . Fe: Org.Comp.B17-112
– . . . . $[(C_5H_5)Fe(CO)_2(CH_2{=}C{=}CDCH_2OD)][BF_4]$ . . . . . Fe: Org.Comp.B17-112
– . . . . $[(C_5H_5)Fe(CO)_2(CH_2{=}CH-C(O)CH_3)][BF_4]$ . . . . . Fe: Org.Comp.B17-30, 50/1
– . . . . $[(C_5H_5)Fe(CO)_2(C_4H_6O)][BF_4]$ . . . . Fe: Org.Comp.B17-93/4
– . . . . $[(C_5H_5)Fe(CO)_2(H_3C-CC-CH_2OH)][BF_4]$ . . . . . . Fe: Org.Comp.B17-122
– . . . . $[(C_5H_5)Fe(CO)_2(H-CC-CH_2CH_2OH)][BF_4]$ . . . . . Fe: Org.Comp.B17-121
– . . . . $[(C_5H_5)(CO)_2Fe{=}C_4H_6O][BF_4]$ . . . . Fe: Org.Comp.B16a-113, 122
– . . . . $[(C_5H_5)(CO)_2Fe{=}C_4H_5DO][BF_4]$ . . . . Fe: Org.Comp.B16a-113, 122
– . . . . $[(C_8H_{11})Fe(CO)_3][BF_4]$ . . . . Fe: Org.Comp.B15-236/7, 255/6
$BC_{11}F_4FeH_{11}O_4$ . . $[(1-CH_3O-3-CH_3-C_6H_5)Fe(CO)_3][BF_4]$ . . . . Fe: Org.Comp.B15-111, 160
– . . . . $[(1-CH_3O-4-CH_3-C_6H_5)Fe(CO)_3][BF_4]$ . . . . Fe: Org.Comp.B15-96/7, 113
– . . . . $[(1-CH_3O-5-CH_3-C_6H_5)Fe(CO)_3][BF_4]$ . . . . Fe: Org.Comp.B15-96/7, 123, 160
– . . . . $[(1-CH_3-3-CH_3O-C_6H_5)Fe(CO)_3][BF_4]$ . . . . Fe: Org.Comp.B15-112, 160/1
– . . . . $[(1-CH_3-4-CH_3O-C_6H_5)Fe(CO)_3][BF_4]$ . . . . Fe: Org.Comp.B15-113, 161/9
– . . . . $[(1-C_2H_5O-C_6H_6)Fe(CO)_3][BF_4]$ . . . . Fe: Org.Comp.B15-98
– . . . . $[(2-CH_3O-3-CH_3-C_6H_5)Fe(CO)_3][BF_4]$ . . . . Fe: Org.Comp.B15-123, 175/6
– . . . . $[(2-CH_3O-4-CH_3-C_6H_5)Fe(CO)_3][BF_4]$ . . . . Fe: Org.Comp.B15-91
– . . . . $[(2-CH_3-3-CH_3O-C_6H_5)Fe(CO)_3][BF_4]$ . . . . Fe: Org.Comp.B15-123, 175/6
– . . . . $[(C_5H_5)Fe(CO)_2(C_4H_6O_2)][BF_4]$ . . . . Fe: Org.Comp.B17-95, 105
$BC_{11}F_4FeH_{11}O_5$ . . $[(1,4-(CH_3O)_2C_6H_5)Fe(CO)_3][BF_4]$ . . . . Fe: Org.Comp.B15-96/7, 113
$BC_{11}F_4FeH_{12}N$ . . . $[(C_6H_6)Fe(C_5H_4-NH_2)][BF_4]$ . . . . Fe: Org.Comp.B18-142/6, 167, 188
– . . . . $[(NH_2-C_6H_5)Fe(C_5H_5)][BF_4]$ . . . . Fe: Org.Comp.B18-142/6, 257, 281
$BC_{11}F_4FeH_{12}NO_2$ $[(C_6H_7)Fe(CN-C_2H_5)(CO)_2][BF_4]$ . . . . Fe: Org.Comp.B15-350, 352
– . . . . $[(C_7H_9)Fe(CN-CH_3)(CO)_2][BF_4]$ . . . . Fe: Org.Comp.B15-351
$BC_{11}F_4FeH_{13}$ . . . . $C_5H_5Fe(C_6H_8)FBF_3$ . . . . Fe: Org.Comp.B17-189/90, 196
$BC_{11}F_4FeH_{13}N_2$ . . $[(NH_2NH-C_6H_5)Fe(C_5H_5)][BF_4]$ . . . . Fe: Org.Comp.B18-142/6, 261

$BC_{11}F_4FeH_{13}O_2$ . . $[(C_5H_5)Fe(CO)_2(CH_2{=}C(CH_3)_2)][BF_4]$ . . . . Fe: Org.Comp.B17-61/2, 73/8
– . . . . $[(C_5H_5)Fe(CO)_2(CD_2{=}C(CH_3)_2)][BF_4]$ . . . . Fe: Org.Comp.B17-62
– . . . . $[(C_5D_5)Fe(CO)_2(CH_2{=}C(CH_3)_2)][BF_4]$ . . . . Fe: Org.Comp.B17-130
– . . . . $[(C_5H_5)Fe(CO)_2(CH_2{=}CH{-}C_2H_5)][BF_4]$ . . . . Fe: Org.Comp.B17-9/10
– . . . . $[(C_5H_5)Fe(CO)_2(CH_3CH{=}CHCH_3)][BF_4]$ . . . . Fe: Org.Comp.B17-56/7, 72
$BC_{11}F_4FeH_{13}O_3$ . . $[(CH_2C(CH_3)C(CH_3)C(CH_3)CH_2)Fe(CO)_3][BF_4]$ Fe: Org.Comp.B15-24
– . . . . $[(CH_2(CH)_4{-}C_3H_7{-}i)Fe(CO)_3][BF_4]$ . . . . Fe: Org.Comp.B15-13, 20
– . . . . $[(CH_3{-}(CH)_4C(CH_3)_2)Fe(CO)_3][BF_4]$ . . . . Fe: Org.Comp.B15-6
– . . . . $[(CH_3{-}(CH)_5{-}C_2H_5)Fe(CO)_3][BF_4]$ . . . . Fe: Org.Comp.B15-13, 22
– . . . . $[(C_5H_5)Fe(CO)_2(CH_2{=}C(CH_3){-}OCH_3)][BF_4]$ . . . . Fe: Org.Comp.B17-67, 79
– . . . . $[(C_5H_5)Fe(CO)_2(CH_2{=}CHCH_2{-}OCH_3)][BF_4]$ . . . . Fe: Org.Comp.B17-25
– . . . . $[(C_5H_5)Fe(CO)_2(CH_2{=}CH{-}O{-}C_2H_5)][BF_4]$ . . . . . . Fe: Org.Comp.B17-21/2, 49
– . . . . $[(C_5H_5)Fe(CO)_2(CH_3CH{=}CH{-}OCH_3)][BF_4]$ . . . . . Fe: Org.Comp.B17-64
$BC_{11}F_4FeH_{13}O_4$ . . $[(C_5H_5)Fe(CO)_2(CH_2{=}C(OCH_3)_2)][BF_4]$ . . . . Fe: Org.Comp.B17-69
– . . . . $[(C_5H_5)Fe(CO)_2(H_3C{-}O{-}CH{=}CH{-}O{-}CH_3)][BF_4]$ Fe: Org.Comp.B17-69, 80/1
$BC_{11}F_4FeH_{13}O_4S$ $[(C_5H_5)Fe(CO)_2(CH_2{=}CHCH_2SO_2CH_3)][BF_4]$ . . . Fe: Org.Comp.B17-26
$BC_{11}F_4FeH_{14}NO_2$ $[(C_5H_5)Fe(CO)_2(CH_2{=}CH{-}N(CH_3)_2)][BF_4]$ . . . . . Fe: Org.Comp.B17-22/3, 49/50
– . . . . $[(C_5H_5)Fe(CO)_2{-}CH_2CH{-}N(CH_3)_2][BF_4]$ . . . . Fe: Org.Comp.B14-127/8
$BC_{11}F_4FeH_{14}NO_5S_2$
$[(C_5H_5)Fe(CO)_2(CH_2{=}CHCH_2S({=}O)NHSO_2CH_3)][BF_4]$
Fe: Org.Comp.B17-25/6
$BC_{11}F_4FeH_{14}N_3$ . . $[(C_5H_5)Fe(CNCH_3)_3][BF_4]$ . . . . Fe: Org.Comp.B15-345
$BC_{11}F_4FeH_{16}NO_2$ $[C_5H_5Fe(CO)_2CH_2CH_2NH(CH_3)_2][BF_4]$ . . . . Fe: Org.Comp.B14-149, 155
$BC_{11}F_4FeH_{16}O_4P$ $[C_5H_5(CO)(P(OCH_3)_3)Fe{=}C{=}CH_2][BF_4]$ . . . . Fe: Org.Comp.B16a-207/8, 210
$BC_{11}F_4FeH_{16}O_5P$ $[C_5H_5Fe(CO)_2CH_2P(OCH_3)_3][BF_4]$ . . . . Fe: Org.Comp.B14-138
$BC_{11}F_4FeH_{18}OP$ . . $[C_5H_5(CO)(P(CH_3)_3)Fe{=}CH(CH_3)][BF_4]$ . . . . Fe: Org.Comp.B16a-18, 24/5
$BC_{11}F_4FeH_{18}O_5P$ $[(C_5H_5)Fe(CH_2{=}CHOH)(CO)P(OCH_3)_3][BF_4]$ . . . Fe: Org.Comp.B16b-86
$BC_{11}F_4H_4O_5ReS_4$ $[(CO)_5ReC_3H_2S_2{=}C_3H_2S_2][BF_4]$ . . . . Re: Org.Comp.2-158
$BC_{11}F_4H_8O_5Re$ . . . $[(CO)_5Re(C_6H_8{-}1,4{-}c)][BF_4]$ . . . . Re: Org.Comp.2-351
$BC_{11}F_4H_{10}O_5Re$ . . $[(CO)_5Re(CH_2{=}CH{-}CH_2CH_2{-}CH{=}CH_2)][BF_4]$ . . . Re: Org.Comp.2-351, 353
– . . . . $[(CO)_5Re(C_6H_{10}{-}c)][BF_4]$ . . . . Re: Org.Comp.2-351
$BC_{11}F_4H_{11}MoN_2O_2$
$[(C_5H_5)Mo(CO)_2(NC{-}CH_3)_2][BF_4]$ . . . . Mo: Org.Comp.7-283, 284, 296/7
$BC_{11}F_4H_{11}MoO_2$ . . $[(C_5H_5)Mo(CO)_2((CH_2)_3C)][BF_4]$ . . . . Mo: Org.Comp.8-305/6, 310, 316
– . . . . $[(C_5H_5)Mo(CO)_2(CH_2{=}CHCH{=}CH_2)][BF_4]$ . . . . . . Mo: Org.Comp.8-305/6, 316
– . . . . $[(C_5H_5)Mo(CO)_2(CH_2{=}C_3H_4{-}c)][BF_4]$ . . . . Mo: Org.Comp.8-186
– . . . . $[(C_5H_5)Mo(CO)_2(CH_3{-}CC{-}CH_3)][BF_4]$ . . . . Mo: Org.Comp.8-186
$BC_{11}F_4H_{13}IMoN_3O$
$[C_5H_5Mo(NO)(NH_2NHC_6H_5)I][BF_4]$ . . . . Mo: Org.Comp.6-39, 43
$BC_{11}F_4H_{13}Ti$ . . . . . $[((C_5H_5)_2Ti(CH_3))_n][BF_4]_n$ . . . . Ti: Org.Comp.5-353/4
$BC_{11}F_4H_{14}MoNO_2S$
$[(C_5H_5)Mo(CO)_2({-}C(N(CH_3)_2){=}S(CH_3){-})][BF_4]$ . . Mo: Org.Comp.8-129
$BC_{11}F_4H_{14}MoN_3O_2$
$[(CH_2CHCH_2)Mo(CO)_2(NC{-}CH_3)_3][BF_4]$ . . . . Mo: Org.Comp.5-200
$BC_{11}F_4H_{14}NTe$ . . . $[3{-}C_2H_5{-}2,5{-}(CH_3)_2{-}1,3{-}TeNC_7H_3][BF_4]$ . . . . . . B: B Comp.SVol.4/4-177

$BC_{11}F_4H_{17}MoO_2S_2$
$[(C_5H_5)Mo(CO)_2(CH_3\text{-}S\text{-}CH_3)_2][BF_4]$ Mo:Org.Comp.7-283, 296
$BC_{11}F_4H_{17}MoO_2Se_2$
$[(C_5H_5)Mo(CO)_2(CH_3\text{-}Se\text{-}CH_3)_2][BF_4]$ Mo:Org.Comp.7-283, 296
$BC_{11}F_4H_{17}MoO_2Te_2$
$[(C_5H_5)Mo(CO)_2(CH_3\text{-}Te\text{-}CH_3)_2][BF_4]$ Mo:Org.Comp.7-283, 296
$BC_{11}F_4H_{23}MoOP_2$ $[C_5H_5Mo(P(CH_3)_3)_2O][BF_4]$ Mo:Org.Comp.6-14
$BC_{11}F_5FeH_{10}$ . . . . $[(F\text{-}C_6H_5)Fe(C_5H_5)][BF_4]$ . . . . Fe: Org.Comp.B18-142/6, 200, 201, 244
$BC_{11}F_6FeH_9$ . . . . . $[(1,2\text{-}F_2\text{-}C_6H_4)Fe(C_5H_5)][BF_4]$ . . . . Fe: Org.Comp.B19-1, 17
– . . . . $[(1,3\text{-}F_2\text{-}C_6H_4)Fe(C_5H_5)][BF_4]$ . . . . Fe: Org.Comp.B19-1, 17
– . . . . $[(1,4\text{-}F_2\text{-}C_6H_4)Fe(C_5H_5)][BF_4]$ . . . . Fe: Org.Comp.B19-1, 17
$BC_{11}F_6FeH_{16}N_2O_2P$
$[C_5H_5Fe(CO)_2C{\equiv}NBH_2N(CH_3)_3][PF_6]$ . . . . Fe: Org.Comp.B14-152, 156
$BC_{11}F_6H_{12}N$ . . . . . $(CH_3)_2N\text{-}BC_7H_6(CF_3)_2$ . . . . B: B Comp.SVol.4/3a-229
$BC_{11}F_{12}H_{11}N_2S_4$. . $(CH_3)_3N\text{-}BH_2\text{-}N[\text{-}C(SCF_3)=C(SCF_3)\text{-}C(SCF_3)=C(SCF_3)\text{-}]$ B: B Comp.SVol.4/3b-23/4
$BC_{11}FeH_{13}$ . . . . . . $C_5H_5FeC_5H_5BCH_3\text{-}1$. . . . . Fe: Org.Comp.B17-323, 325
$BC_{11}FeH_{16}IN_2O_2$. $[C_5H_5Fe(CO)_2C{\equiv}NBH_2N(CH_3)_3]I$. . . . . Fe: Org.Comp.B14-152, 156
$BC_{11}FeH_{16}N_2O_2^+$ $[C_5H_5Fe(CO)_2C{\equiv}NBH_2N(CH_3)_3]^+$ . . . . Fe: Org.Comp.B14-152, 156
$BC_{11}Fe_3H_4O_{10}^-$ . . $[(CO)_{10}Fe_3(BH\text{-}CH_3)]^-$ . . . . Fe: Org.Comp.C6b-70
$BC_{11}Fe_3H_5O_{10}$ . . . $(CO)_{10}(H)Fe_3(BH\text{-}CH_3)$. . . . . Fe: Org.Comp.C6b-70
$BC_{11}Fe_3H_8O_9^-$ . . . $[(CO)_9(H)Fe_3(BH_2\text{-}C_2H_5)]^-$. . . . . Fe: Org.Comp.C6a-299
$BC_{11}Fe_3H_9O_9$ . . . . $(CO)_9(H)Fe_3(BH_3\text{-}C_2H_5)$ . . . . Fe: Org.Comp.C6a-297/8
$BC_{11}GeH_{24}N$. . . . . $1\text{-}(t\text{-}C_4H_9)\text{-}2\text{-}CH_3\text{-}3\text{-}(CH_3)_3Ge\text{-}1,2\text{-}NBC_3H_3$ . . B: B Comp.SVol.4/3a-249
$BC_{11}GeH_{25}O$. . . . . $(C_3H_7)_2Ge(\text{-}CH_2CH_2B(OCH_3)CH_2CH_2\text{-})$ Ge:Org.Comp.3-295/6
$BC_{11}GeH_{27}O_2$. . . . $Ge(C_2H_5)_3CH_2CH_2CH_2B(OCH_3)_2$ Ge:Org.Comp.2-147
$BC_{11}H_8NO_8$. . . . . . $C_7H_7NH[B(\text{-}O\text{-}CO\text{-}CO\text{-}O\text{-})_2]$. . . . . B: B Comp.SVol.3/3-227
$BC_{11}H_9N_4$ . . . . $4\text{-}CN\text{-}C_5H_4N\text{-}BH(CN)(C_4H_4N)$ . . . . B: B Comp.SVol.3/3-190
$BC_{11}H_{11}N_2$ . . . . . . $CH_3B[\text{-}NH\text{-}(1,8\text{-}C_{10}H_6)NH\text{-}]$ . . . . B: B Comp.SVol.3/3-117
$BC_{11}H_{12}N$ . . . . $2\text{-}C_6H_5\text{-}C_5H_4N\text{-}BH_3$. . . . . B: B Comp.SVol.3/3-177
$BC_{11}H_{14}N$ . . . . $(CH_2=CH)_2B\text{-}N(CH_3)C_6H_5$ . . . . B: B Comp.SVol.3/3-151
$BC_{11}H_{14}NO$. . . . . . $1\text{-}C_6H_5\text{-}2,7,1\text{-}ONBC_5H_9$. . . . . B: B Comp.SVol.4/3b-65
– . . . . $6\text{-}HO\text{-}5,6\text{-}NBC_{11}H_{13}$ . . . . B: B Comp.SVol.3/3-163
$BC_{11}H_{14}NO_4$ . . . . . $[3,4\text{-}(OCH_2O)\text{-}C_6H_3]B[\text{-}O\text{-}CH_2\text{-}CH_2\text{-}]_2NH$ . . . . B: B Comp.SVol.3/3-218
$BC_{11}H_{14}N_3O_2$ . . . . $1,2,5\text{-}(CH_3)_3\text{-}3\text{-}C_6H_5\text{-}1,3,5,2\text{-}N_3BC_2(=O)_2\text{-}4,6$ B: B Comp.SVol.4/3a-176
– . . . . $1,3,5\text{-}(CH_3)_3\text{-}2\text{-}C_6H_5\text{-}1,3,5,2\text{-}N_3BC_2(=O)_2\text{-}4,6$ B: B Comp.SVol.3/3-114/5
$BC_{11}H_{14}N_3S_2$ . . . . $1,3,5\text{-}(CH_3)_3\text{-}2\text{-}C_6H_5\text{-}1,3,5,2\text{-}N_3BC_2(=S)_2\text{-}4,6$ B: B Comp.SVol.3/3-116
B: B Comp.SVol.3/4-131
$BC_{11}H_{14}O_7Re$. . . . $(CO)_3Re[(COCH_3)_3]BOC_2H_5$. Re:Org.Comp.2-379
$BC_{11}H_{15}LiN_7$. . . . . $Li[(1,2\text{-}N_2C_3H_3\text{-}1)_3B\text{-}N(CH_3)_2]$ . . . . B: B Comp.SVol.4/3b-40
$BC_{11}H_{15}MoN_2O_2$. $(C_5H_5)Mo(CO)_2[CHN(CH_3)\text{-}BH_2\text{-}N(CH_3)CH]$ . . . Mo:Org.Comp.8-197
$BC_{11}H_{15}N_2O$ . . . . . $C_6H_5\text{-}B[\text{-}O\text{-}N=CH\text{-}N(C_4H_9\text{-}n)\text{-}]$ . . . . B: B Comp.SVol.3/3-163
$BC_{11}H_{15}N_2S$ . . . . . $C_6H_5\text{-}B[\text{-}NH\text{-}C(S)\text{-}N(C_4H_9\text{-}t)\text{-}]$. . . . . B: B Comp.SVol.3/3-112
B: B Comp.SVol.3/4-131
$BC_{11}H_{15}O_2$ . . . . . . $C_6H_5\text{-}B[\text{-}O\text{-}CH_2\text{-}C(CH_3)_2\text{-}CH_2\text{-}O\text{-}]$ . . . . B: B Comp.SVol.4/3b-69
$BC_{11}H_{16}NO_2$. . . . . $(n\text{-}C_4H_9O)B[\text{-}NH\text{-}(1,2\text{-}C_6H_4)\text{-}CH_2\text{-}O\text{-}]$. . . . . . . B: B Comp.SVol.3/3-164
$BC_{11}H_{16}NO_2S$. . . . $(C_2H_5)_2NB[\text{-}O\text{-}(1,2\text{-}C_6H_3\text{-}5\text{-}OCH_3)\text{-}S\text{-}]$ . . . . . . B: B Comp.SVol.3/4-133
$BC_{11}H_{16}NO_4$ . . . . . $(CH_3)_3N\text{-}BH_2\text{-}C(O)\text{-}O\text{-}C(O)\text{-}O\text{-}C_6H_5$ . . . . B: B Comp.SVol.4/3b-25
$BC_{11}H_{16}N_3O$ . . . . . $(C_2H_5)_2N\text{-}B(\text{-}NH\text{-}CC_6H_5=N\text{-}O\text{-})$ . . . . B: B Comp.SVol.3/3-113

Formula	Compound	Reference
$BC_{11}H_{16}N_7$	$(CH_3)_2NH\text{-}B(1,2\text{-}C_3H_3N_2)_3$	B: B Comp.SVol.3/3-94, 190
$BC_{11}H_{17}N_2$	$(CH_3)_3N\text{-}BH_2\text{-}CH(C_6H_5)\text{-}CN$	B: B Comp.SVol.4/3b-25
$BC_{11}H_{17}N_2O$	$2\text{-}(C_2H_5)_2N\text{-}3,1,2\text{-}ONBC_7H_7$	B: B Comp.SVol.4/3b-66
$BC_{11}H_{17}N_4$	$CH_3B[\text{-}N(t\text{-}C_4H_9)\text{-}N{=}N\text{-}NC_6H_5\text{-}]$	B: B Comp.SVol.3/3-137/8
–	$[(CH_3)_2NH_2][NC\text{-}BH(1\text{-}NC_4H_4)_2]$	B: B Comp.SVol.3/3-213
$BC_{11}H_{17}N_4^-$	$[(C_{10}H_{15})NNN\text{-}BH_2CN]^-$, radical anion	B: B Comp.SVol.3/3-184
$BC_{11}H_{17}S_2$	$2,4,6\text{-}(CH_3)_3\text{-}C_6H_2B(SCH_3)_2$	B: B Comp.SVol.3/4-111
$BC_{11}H_{18}N$	$2\text{-}CH_3\text{-}C_5H_4N\text{-}BH[C_5H_{10}]$	B: B Comp.SVol.3/3-189
$BC_{11}H_{18}NO_2$	$(CH_3)_3N\text{-}BH_2\text{-}C(O)O\text{-}CH_2\text{-}C_6H_5$	B: B Comp.SVol.4/3b-25
–	$H_3B\text{-}N(CH_3)_2\text{-}CH_2CH_2\text{-}OC(O)\text{-}C_6H_5$	B: B Comp.SVol.4/3b-15
$BC_{11}H_{19}N$	$2,4,7\text{-}(CH_3)_3\text{-}2\text{-}NC_8H_7 \cdot BH_3$, radical	B: B Comp.SVol.4/3b-12
$BC_{11}H_{19}N_2$	$1,3\text{-}(CH_3)_2\text{-}2\text{-}C_6H_5\text{-}1,3\text{-}N_2C_3H_5 \cdot BH_3$	B: B Comp.SVol.4/3b-12
–	$2\text{-}[(n\text{-}C_3H_7)_2B\text{-}NH]\text{-}NC_5H_4$	B: B Comp.SVol.3/3-152 B: B Comp.SVol.4/3a-231
–	$2\text{-}[(i\text{-}C_3H_7)_2B\text{-}NH]\text{-}NC_5H_4$	B: B Comp.SVol.4/3a-231
–	$(C_2H_5)_2B[\text{-}NC_5H_4\text{-}CH_2CH_2\text{-}NH\text{-}]$	B: B Comp.SVol.3/3-124
$BC_{11}H_{19}N_3^+$	$[(\text{-}CCH_3\text{-}(CH)_3\text{-}CCH_3\text{-})N\text{-}B(\text{-}NCH_3\text{-}CH_2\text{-}CH_2\text{-}NCH_3\text{-})]^+$	B: B Comp.SVol.3/3-207
$BC_{11}H_{19}N_6$	$[\text{-}NCH_3\text{-}(CH_2)_3\text{-}NCH_3\text{-}]B[\text{-}N_2C_3H_3\text{-}H\text{-}N_2C_3H_3\text{-}]$	B: B Comp.SVol.3/3-201
$BC_{11}H_{20}N$	$NC_5H_5 \cdot B(C_2H_5)_3$	B: B Comp.SVol.4/3b-28
$BC_{11}H_{20}NO$	$[HOCH(C_6H_5)\text{-}CH(CH_3)](CH_3)_2N\text{-}BH_3$	B: B Comp.SVol.3/3-176
$BC_{11}H_{20}NOSe$	$2\text{-}CH_3\text{-}3\text{-}(n\text{-}C_3H_7)\text{-}5,6\text{-}(C_2H_5)_2\text{-}1,3,2\text{-}SeNBC_3({=}O)\text{-}4$	B: B Comp.SVol.4/4-172
–	$2\text{-}CH_3\text{-}3\text{-}(i\text{-}C_3H_7)\text{-}5,6\text{-}(C_2H_5)_2\text{-}1,3,2\text{-}SeNBC_3({=}O)\text{-}4$	B: B Comp.SVol.4/4-172/3
$BC_{11}H_{20}N_3OS$	$BC_4H_3N_3S({=}O)(CH_3)(C_3H_7\text{-}n)_2$	B: B Comp.SVol.4/3a-180
$BC_{11}H_{21}$	$7,7\text{-}(CH_3)_2\text{-}2\text{-}C_2H_5\text{-}[3.1.1]\text{-}C_7H_8\text{-}3\text{-}BH_2$	B: B Comp.SVol.4/3b-124
$BC_{11}H_{21}NNaSi$	$Na[4,5\text{-}(C_2H_5)_2\text{-}3\text{-}CH_2{=}CCH_3\text{-}2,2\text{-}(CH_3)_2\text{-}1,2,5\text{-}NSiBC_2]$	B: B Comp.SVol.4/3a-251
$BC_{11}H_{21}NP$	$(C_2H_5)_2N\text{-}B(\text{-}CH{=}CCH_3\text{-}PCH_3\text{-}CCH_3{=}CH\text{-})$	B: B Comp.SVol.3/3-151
$BC_{11}H_{21}N_2S$	$2\text{-}[(n\text{-}C_4H_9)_2B\text{-}NH]\text{-}1,3\text{-}SNC_3H_2$	B: B Comp.SVol.4/3a-232
$BC_{11}H_{21}N_2Si$	$(CH_3)_3Si\text{-}NH\text{-}B(C_6H_5)\text{-}N(CH_3)_2$	B: B Comp.SVol.3/3-108 B: B Comp.SVol.4/3a-168
$BC_{11}H_{21}N_3^+$	$[(\text{-}CCH_3\text{-}(CH)_3\text{-}CCH_3\text{-})N\text{-}B(N(CH_3)_2)_2]^+$	B: B Comp.SVol.3/3-207
$BC_{11}H_{21}N_6$	$[\text{-}NCH_3\text{-}(CH_3)_2\text{-}NCH_3\text{-}]B[\text{-}(CH_3)C_3H_2N_2\text{-}H\text{-}N_2C_3H_3\text{-}]$	B: B Comp.SVol.3/3-201
$BC_{11}H_{21}O_2S$	$2\text{-}[CH_2{=}CHCH(S\text{-}C_2H_5)]\text{-}4,4,5,5\text{-}(CH_3)_4\text{-}1,3,2\text{-}O_2BC_2$	B: B Comp.SVol.4/4-138
$BC_{11}H_{22}N$	$9\text{-}(i\text{-}C_3H_7\text{-}NH)\text{-}[3.3.1]\text{-}9\text{-}BC_8H_{14}$	B: B Comp.SVol.4/3a-235
–	$t\text{-}C_4H_9\text{-}CC\text{-}B(CH_3)\text{-}NH\text{-}C_4H_9\text{-}t$.	B: B Comp.SVol.4/3a-234
–	$[3.3.1.1^{3,7}]\text{-}1\text{-}BC_9H_{15} \cdot C_2H_5\text{-}NH_2$	B: B Comp.SVol.4/3b-30
$BC_{11}H_{22}NO_2$	$8\text{-}[CH_3COO\text{-}CHCH_3]\text{-}[2.2.2]\text{-}1\text{-}NC_7H_{12} \cdot BH_3$	B: B Comp.SVol.4/3b-13
–	$[\text{-}O\text{-}CH_2CH_2\text{-}O\text{-}]B\text{-}1\text{-}NC_5H_6\text{-}2,2,6,6\text{-}(CH_3)_4$	B: B Comp.SVol.4/3b-63
$BC_{11}H_{22}NSe$	$2\text{-}(C_4H_9\text{-}i)\text{-}3\text{-}CH_3\text{-}4,5\text{-}(C_2H_5)_2\text{-}1,2,3\text{-}SeNBC_2$	B: B Comp.SVol.4/4-172/3
–	$2\text{-}(C_4H_9\text{-}t)\text{-}3\text{-}CH_3\text{-}4,5\text{-}(C_2H_5)_2\text{-}1,2,3\text{-}SeNBC_2$	B: B Comp.SVol.4/4-172/3
–	$3\text{-}(C_4H_9\text{-}i)\text{-}2\text{-}CH_3\text{-}4,5\text{-}(C_2H_5)_2\text{-}1,3,2\text{-}SeNBC_2$	B: B Comp.SVol.4/3a-252
–	$3\text{-}(C_4H_9\text{-}t)\text{-}2\text{-}CH_3\text{-}4,5\text{-}(C_2H_5)_2\text{-}1,3,2\text{-}SeNBC_2$	B: B Comp.SVol.4/3a-252
$BC_{11}H_{23}LiN$	$Li[(i\text{-}C_3H_7)_2N\text{-}B(CH_3)(\text{-}CH{=}CH\text{-}CH_2CH_2\text{-})]$	B: B Comp.SVol.4/3b-49
$BC_{11}H_{23}LiNSi$	$Li[1\text{-}(t\text{-}C_4H_9)\text{-}2\text{-}CH_3\text{-}3\text{-}(CH_3)_3Si\text{-}1,2\text{-}NBC_3H_2]$	B: B Comp.SVol.4/3b-48

Formula	Compound	Reference
$BC_{11}H_{23}N_2$	CH_3-B[-N(C_4H_9-t)-CH=CH-N(C_4H_9-t)-]	B: B Comp.SVol.3/3-112/3
–	N(-CH_2-CH_2-)$_3$N-BC_5H_{11}	B: B Comp.SVol.3/3-189
–	N(-CH_2-CH_2-)$_3$N-BH_2-C_5H_9-c	B: B Comp.SVol.3/3-188
–	[(C_2H_5)$_2$N]$_2$B-C≡C-CH_3	B: B Comp.SVol.3/3-106
$BC_{11}H_{23}N_2O$	[-N(BH_3)=C(CH_3)-C(CH_3)$_2$-N(OH)-]C[-(CH_2)$_5$-]	B: B Comp.SVol.4/3b-12
$BC_{11}H_{23}N_2O_2$	(C_2H_5)$_2$B(-NH-COC_2H_5=CH-COC_2H_5=NH-)	B: B Comp.SVol.3/3-122/3
$BC_{11}H_{23}N_6$	$BC_2H_2N_6$(CH_3)(C_4H_9-n)$_2$	B: B Comp.SVol.4/3a-178
–	$BC_2H_2N_6$(C_3H_7-n)$_3$	B: B Comp.SVol.4/3a-178
$BC_{11}H_{24}N$	1-(CH_3)$_2$B-NC_5H_6-2,2,6,6-(CH_3)$_4$	B: B Comp.SVol.4/3b-243
–	(C_2H_5)$_2$B-N(C_3H_7-i)-C(C_2H_5)=CH_2	B: B Comp.SVol.3/3-151
–	(C_2H_5)$_2$B-N(C_3H_7-i)-CCH_3=$CHCH_3$	B: B Comp.SVol.3/3-152
–	[3.3.1]-9-BC_8H_{15} · N(CH_3)$_3$	B: B Comp.SVol.3/3-189
		B: B Comp.SVol.4/3b-29
–	[-(CH_2)$_5$-]N[-BH_2-$CHCH_3$-$CH_2CH_2CH_2CH_2$-]	B: B Comp.SVol.4/3a-255
–	[-(CH_2)$_5$-]N[-BH_2-CH(C_2H_5)-$CH_2CH_2CH_2$-]	B: B Comp.SVol.4/3a-255
–	[-(CH_2)$_5$-]N[-BH_2-CH(C_3H_7-n)-CH_2CH_2-]	B: B Comp.SVol.4/3a-255
$BC_{11}H_{24}NO_2$	[CH(CH_3)$_2$C(CH_3)$_2$]B[-O-CH_2-CH_2-]$_2$$NCH_3$	B: B Comp.SVol.3/3-218
$BC_{11}H_{24}NPb$	1-(t-C_4H_9)-2-CH_3-3-(CH_3)$_3$Pb-1,2-NBC_3H_3	B: B Comp.SVol.4/3a-249
$BC_{11}H_{24}NSi$	1-(t-C_4H_9)-2-CH_3-3-(CH_3)$_3$Si-1,2-NBC_3H_3	B: B Comp.SVol.4/3a-249
$BC_{11}H_{24}NSn$	1-(t-C_4H_9)-2-CH_3-3-(CH_3)$_3$Sn-1,2-NBC_3H_3	B: B Comp.SVol.4/3a-249
$BC_{11}H_{24}N_2$	N(-CH_2-CH_2-)$_3$N-BH[$CHCH_3$-CH(CH_3)$_2$]	B: B Comp.SVol.3/3-189
$BC_{11}H_{24}N_2^+$	[(CH_3)$_2$N=B=N(-C(CH_3)$_2$-(CH_2)$_3$-C(CH_3)$_2$-)]$^+$	B: B Comp.SVol.3/3-204/5
$BC_{11}H_{25}KNSi$	(C_2H_5)$_2$B[-C(C_2H_5)=CCH_3-Si(CH_3)$_2$-NH(K)-]	B: B Comp.SVol.3/3-166
		B: B Comp.SVol.4/3a-254
$BC_{11}H_{25}NNaSi$	(C_2H_5)$_2$B[-C(C_2H_5)=CCH_3-Si(CH_3)$_2$-NH(Na)-]	B: B Comp.SVol.3/3a-166
		B: B Comp.SVol.4/3a-254
$BC_{11}H_{25}NSi_2$	BC_2H_5[-CC_2H_5=CCH_3-Si(CH_3)$_2$-NSi(CH_3)$_2$-]	B: B Comp.SVol.3/3-162
$BC_{11}H_{25}N_2$	t-C_4H_9-N=B-N(C_3H_7-i)-C_4H_9-t	B: B Comp.SVol.4/3a-160/5
–	N(-CH_2-CH_2-)$_3$N-BH_2-CH(CH_3)-C_3H_7-i	B: B Comp.SVol.3/3-188
$BC_{11}H_{25}N_2O_2$	CH_3-O-C(CH_3)$_2$-O-CH=CH-CH_2-B[N(CH_3)$_2$]$_2$	B: B Comp.SVol.3/3-108
		B: B Comp.SVol.4/3a-168
$BC_{11}H_{25}N_4$	(C_2H_5)$_2$B[-NH-CN(CH_3)$_2$=CH-CN(CH_3)$_2$=NH-]	B: B Comp.SVol.3/3-122/3
$BC_{11}H_{25}O_2S$	CH_3-S-CH_2CH_2-B(O-C_4H_9-n)$_2$	B: B Comp.SVol.4/4-138
$BC_{11}H_{26}N$	(C_2H_5)$_3$N-BC_5H_{11}	B: B Comp.SVol.3/3-189
–	(i-C_3H_7)$_2$NH-BC_5H_{11}	B: B Comp.SVol.3/3-188
–	(t-C_4H_9)$_2$B-NH-C_3H_7-i	B: B Comp.SVol.4/3a-230
$BC_{11}H_{26}NSi_2$	[-CH=C(n-C_3H_7)-]B-N[Si(CH_3)$_3$]$_2$	B: B Comp.SVol.3/3-150
$BC_{11}H_{26}NSn$	(CH_3)$_2$Sn[-C(CH_3)=C(C_2H_5)-B(C_2H_5)$_2$-NH_2-]	B: B Comp.SVol.4/3a-255
		Sn: Org.Comp.19-48/9
$BC_{11}H_{26}N_2OP$	C_2H_5-B[-N(C_4H_9-t)-P(CH_3)(=O)-N(C_4H_9-t)-]	B: B Comp.SVol.4/3a-207/8
$BC_{11}H_{26}N_2PS$	C_2H_5-B[-N(C_4H_9-t)-P(CH_3)(=S)-N(C_4H_9-t)-]	B: B Comp.SVol.4/3a-207/8
$BC_{11}H_{27}N_2$	(CH_3)$_2$N-CH_2-CH_2-N(CH_3)$_2$-BH[C_5H_{10}]	B: B Comp.SVol.3/3-189
$BC_{11}H_{27}N_2Si$	1,3-(t-C_4H_9)$_2$-2,2,4-(CH_3)$_3$-1,3,2,4-N_2SiB	B: B Comp.SVol.4/3a-205/6
–	(CH_3)$_3$Si-N(C_4H_9-t)-B=N-C_4H_9-t	B: B Comp.SVol.4/3a-160/5
		B: B Comp.SVol.4/3b-245/6
$BC_{11}H_{27}N_4$	t-C_4H_9N=C[N(CH_3)$_2$]B[N(CH_3)$_2$]$_2$	B: B Comp.SVol.3/3-109
$BC_{11}H_{27}N_4Si$	(CH_3)$_3$Si-N(C_4H_9-t)-B(N_3)-C_4H_9-n	B: B Comp.SVol.3/3-110
		B: B Comp.SVol.4/3a-171
–	(CH_3)$_3$Si-N(C_4H_9-i)-B(N_3)-C_4H_9-i	B: B Comp.SVol.3/3-110

Formula	Compound	Reference
$BC_{11}H_{27}N_4Si$	$(CH_3)_3Si-N(C_4H_9-s)-B(N_3)-C_4H_9-s$	B: B Comp.SVol.3/3-110
–	$(CH_3)_3Si-N(C_4H_9-t)-B(N_3)-C_4H_9-t$	B: B Comp.SVol.3/3-110
–	$(CH_3)_3Si-N[-B(C_4H_9-s)-N(C_4H_9-i)-N=N-]$	B: B Comp.SVol.3/3-137/8
–	$(CH_3)_3Si-N[-B(C_4H_9-t)-N(C_4H_9-i)-N=N-]$	B: B Comp.SVol.4/3a-199
–	$(CH_3)_3Si-N[-B(C_4H_9-n)-N(C_4H_9-n)-N=N-]$	B: B Comp.SVol.3/3-137/8
–	$(CH_3)_3Si-N[-B(C_4H_9-i)-N(C_4H_9-i)-N=N-]$	B: B Comp.SVol.3/3-137/8
$BC_{11}H_{28}NOSi_2$	$3-CH_3-BC_4H_7-1-N[Si(CH_3)_3]-O-Si(CH_3)_3$	B: B Comp.SVol.4/3a-235
$BC_{11}H_{28}NSi$	$(t-C_4H_9)_2B-NH-Si(CH_3)_3$	B: B Comp.SVol.3/3-152
		B: B Comp.SVol.4/3a-230
$BC_{11}H_{28}NSn$	$(t-C_4H_9)_2B-NH-Sn(CH_3)_3$	B: B Comp.SVol.4/3a-231
$BC_{11}H_{28}NSn_2$	$[-(CH_2)_2-CHCH_3-CH_2-]B-N[Sn(CH_3)_3]_2$	B: B Comp.SVol.3/3-150
$BC_{11}H_{29}N_2Si$	$HB[NHSi(CH_3)_3]N(s-C_4H_9)_2$	B: B Comp.SVol.3/3-107
$BC_{11}H_{29}N_2Sn$	$(CH_3)_3SnB[N(C_2H_5)_2]_2$	B: B Comp.SVol.3/3-106
$BC_{11}H_{30}NSi_2$	$t-C_4H_9-B(CH_3)-N[Si(CH_3)_3]_2$	B: B Comp.SVol.4/3a-235
$BC_{11}H_{31}N_2Si_2$	$i-C_3H_7B[N(CH_3)_2]N[Si(CH_3)_3]_2$	B: B Comp.SVol.3/3-108
$BC_{12}ClF_2H_{27}N$	$N(n-C_4H_9)_3 \cdot BClF_2$	B: B Comp.SVol.3/4-65/7
$BC_{12}ClF_4FeH_{12}$	$[(2-Cl-C_6H_4-CH_3)Fe(C_5H_5)][BF_4]$	Fe: Org.Comp.B19-1, 4, 5, 33, 90/2
–	$[(3-Cl-C_6H_4-CH_3)Fe(C_5H_5)][BF_4]$	Fe: Org.Comp.B19-1, 4/6, 33, 90/2
–	$[(4-Cl-C_6H_4-CH_3)Fe(C_5H_5)][BF_4]$	Fe: Org.Comp.B19-1, 4/6, 34, 90/2
$BC_{12}ClF_4FeH_{12}O$	$[(3-Cl-C_6H_4-OCH_3)Fe(C_5H_5)][BF_4]$	Fe: Org.Comp.B19-1, 5, 68
$BC_{12}ClF_4H_{10}$	$[(C_6H_5)_2Cl][BF_4]$	B: B Comp.SVol.3/3-354
$BC_{12}ClH_8$	$5-Cl-5-BC_{12}H_8$	B: B Comp.SVol.4/4-49
$BC_{12}ClH_{10}$	$(C_6H_5)_2BCl$	B: B Comp.SVol.3/4-42/3
		B: B Comp.SVol.4/4-45/6
$BC_{12}ClH_{14}$	$1-Cl-1-BC_{10}H_8-2,3-(CH_3)_2$	B: B Comp.SVol.4/4-48
$BC_{12}ClH_{15}N$	$(C_6H_5)ClBN(CH_2CH=CH_2)_2$	B: B Comp.SVol.3/4-56
$BC_{12}ClH_{16}$	$1-Cl-5,7-(C_2H_5)_2-1-BC_8H_6$	B: B Comp.SVol.4/4-48
$BC_{12}ClH_{16}Si$	$2-Cl-1-Si(CH_3)_3-2-BC_9H_7$	B: B Comp.SVol.4/4-48
$BC_{12}ClH_{17}N$	$C_6H_5(Cl)B-NC_5H_9(CH_3)$	B: B Comp.SVol.3/4-56
$BC_{12}ClH_{19}N$	$C_6H_5-BCl-N(C_3H_7-i)_2$	B: B Comp.SVol.3/4-56
–	$C_6H_5-BCl-N(C_3H_7-n)_2$	B: B Comp.SVol.3/4-56
$BC_{12}ClH_{20}S$	$(CH_3)_5C_5-BCl-S-C_2H_5$	B: B Comp.SVol.4/4-161
$BC_{12}ClH_{21}N$	$(CH_3)_5C_5-BCl-N(CH_3)_2$	B: B Comp.SVol.4/4-60
$BC_{12}ClH_{21}NSi_2$	$(C_6H_5)ClBN[-Si(CH_3)_2-(CH_2)_2-Si(CH_3)_2-]$	B: B Comp.SVol.3/4-57
$BC_{12}ClH_{22}$	$(1-CH_3-c-C_5H_8)_2BCl$	B: B Comp.SVol.4/4-46
–	$(C_2H_5-CH=C-C_2H_5)_2BCl$	B: B Comp.SVol.3/4-43
–	$(t-C_4H_9-CH=CH)_2BCl$	B: B Comp.SVol.3/4-44
–	$(c-C_6H_{11})_2BCl$	B: B Comp.SVol.3/4-43
		B: B Comp.SVol.4/4-46
$BC_{12}ClH_{24}$	$i-C_3H_7-C(CH_3)_2-BCl-C(C_2H_5)=CH-C_2H_5$	B: B Comp.SVol.3/4-44
–	$i-C_3H_7-C(CH_3)_2-BCl-CH=CH-C_4H_9-n$	B: B Comp.SVol.3/4-44
–	$i-C_3H_7-C(CH_3)_2-BCl-CH=CH-C_4H_9-t$	B: B Comp.SVol.3/4-44
$BC_{12}ClH_{26}$	$i-C_3H_7-C(CH_3)_2-BCl-C_6H_{13}-n$	B: B Comp.SVol.3/4-45
–	$(n-C_6H_{13})_2BCl$	B: B Comp.SVol.3/4-43
		B: B Comp.SVol.4/4-47
–	$(s-C_6H_{13})_2BCl$	B: B Comp.SVol.3/4-43
$BC_{12}ClH_{27}N_2P$	$[-N(C_4H_9-t)-BCl-N(C_4H_9-t)-]P-C_4H_9-t$	B: B Comp.SVol.4/4-64/5

Formula	Compound	Reference
$BC_{12}ClH_{28}N_2$	$ClB[N(i\text{-}C_3H_7)_2]_2$	B: B Comp.SVol.3/4-58
$BC_{12}ClH_{29}NSi$	$n\text{-}C_5H_{11}\text{-}BCl\text{-}N(C_4H_9\text{-}t)\text{-}Si(CH_3)_3$	B: B Comp.SVol.4/4-60
$BC_{12}ClH_{32}N_2Si_4$	$ClB[N(\text{-}Si(CH_3)_2\text{-}CH_2\text{-}CH_2\text{-}Si(CH_3)_2)]_2$	B: B Comp.SVol.3/4-59
$BC_{12}Cl_2ErH_{28}O_3$	$ErCl_2[BH_4] \cdot 3\ C_4H_8O$	Sc: MVol.C11b-490
$BC_{12}Cl_2FH_{27}N$	$N(n\text{-}C_4H_9)_3 \cdot BCl_2F$	B: B Comp.SVol.3/4-65/7
$BC_{12}Cl_2H_{10}N$	$Cl_2BN(C_6H_5)_2$	B: B Comp.SVol.3/4-54
$BC_{12}Cl_2H_{11}N_2$	$Cl_2B(NC_{12}H_{11}N)$	B: B Comp.SVol.3/4-55
$BC_{12}Cl_2H_{17}$	$2,4,6\text{-}(C_2H_5)_3\text{-}C_6H_2\text{-}BCl_2$	B: B Comp.SVol.4/4-44
$BC_{12}Cl_2H_{20}NSn$	$(CH_3)_2Sn(Cl)\text{-}N(C_4H_9\text{-}t)\text{-}B(Cl)\text{-}C_6H_5$	Sn: Org.Comp.19-118, 123
$BC_{12}Cl_2H_{25}O$	$ClB[C(Cl)(s\text{-}C_4H_9)(n\text{-}C_5H_{11})]OC_2H_5$	B: B Comp.SVol.3/4-52
$BC_{12}Cl_2H_{29}N_2Si$	$(i\text{-}C_3H_7)_2N\text{-}BCl\text{-}N(C_4H_9\text{-}t)\text{-}SiCl(CH_3)_2$	B: B Comp.SVol.4/4-63
$BC_{12}Cl_3H_{18}NO$	$ON\text{-}BCl_3 \cdot [C_6(CH_3)_6]$	B: B Comp.SVol.4/4-33
$BC_{12}CsH_8O_4$	$Cs[B(\text{-}O\text{-}C_6H_4\text{-}O\text{-})_2]$	B: B Comp.SVol.3/3-227
$BC_{12}FH_8$	$5\text{-}F\text{-}5\text{-}BC_{12}H_8$	B: B Comp.SVol.4/3b-228
$BC_{12}FH_{16}N$	$F(C_6H_5)B\text{-}N[\text{-}CH_2\text{-}CCH_3\text{-}(CH_2)_3\text{-}]$	B: B Comp.SVol.3/3-376
$BC_{12}FH_{17}N$	$F(C_6H_5)B\text{-}N[\text{-}CHCH_3\text{-}(CH_2)_4\text{-}]$	B: B Comp.SVol.3/3-376
$BC_{12}FH_{19}N$	$C_6H_5\text{-}BF\text{-}N(C_3H_7\text{-}i)_2$	B: B Comp.SVol.3/3-376
–	$C_6H_5\text{-}BF\text{-}N(C_3H_7\text{-}n)_2$	B: B Comp.SVol.3/3-376
$BC_{12}FH_{25}N$	$1\text{-}[i\text{-}C_3H_7\text{-}BF]\text{-}NC_5H_6\text{-}2,2,6,6\text{-}(CH_3)_4$	B: B Comp.SVol.4/3b-245
$BC_{12}FH_{27}I_2N$	$N(n\text{-}C_4H_9)_3 \cdot BFI_2$	B: B Comp.SVol.3/4-65/7
$BC_{12}FH_{27}NSi$	$1\text{-}[(CH_3)_3Si\text{-}BF]\text{-}NC_5H_6\text{-}2,2,6,6\text{-}(CH_3)_4$	B: B Comp.SVol.4/3b-243
$BC_{12}FH_{36}N_2Si_4$	$[(CH_3)_3Si]_2N\text{-}BF\text{-}N[Si(CH_3)_3]_2$	B: B Comp.SVol.4/3b-245
$BC_{12}FH_{38}N_4Si_5$	$[(CH_3)_3Si]_2N\text{-}BF\text{-}N[\text{-}Si(CH_3)_2\text{-}NH\text{-}]_2Si(CH_3)_2$	B: B Comp.SVol.3/3-376
$BC_{12}F_2FeH_{13}O_3$	$[\text{-}BF_2\text{-}OC(CH_3)Fe(CO)(C_5H_5)C(CH{=}CHCH_3)O\text{-}]$	Fe: Org.Comp.B16b-127
–	$[\text{-}BF_2\text{-}OC(CH_3)Fe(CO)(C_5H_5)C(CCH_3{=}CH_2)O\text{-}]$	Fe: Org.Comp.B16b-127, 139/40
$BC_{12}F_2FeH_{14}O_3^-$	$[C_5H_5Fe(CH_2C_2O_2BF_2(C_3H_7\text{-}i))CO]^-$	Fe: Org.Comp.B17-156, 158
$BC_{12}F_2H_{11}N_2$	$1\text{-}F_2B\text{-}1,10\text{-}N_2C_{12}H_{11}$	B: B Comp.SVol.3/3-375, 377/8
$BC_{12}F_2H_{17}NO_6Re$	$[N(CH_3)_4][cis\text{-}(CO)_4ReCH_2COCO(CH_3)BF_2]$	Re: Org.Comp.2-388
$BC_{12}F_2H_{27}IN$	$N(n\text{-}C_4H_9)_3 \cdot BF_2I$	B: B Comp.SVol.3/4-65/7
$BC_{12}F_2H_{36}N_3Si_5$	$F_2B\text{-}N[\text{-}Si(CH_3)_2\text{-}N(Si(CH_3)_3)\text{-}]_2Si(CH_3)_2$	B: B Comp.SVol.3/3-375
$BC_{12}F_3H_{11}N$	$(C_6H_5)_2HN\text{-}BF_3$	B: B Comp.SVol.3/3-294
$BC_{12}F_3H_{19}N_3O_3S$	$[(\text{-}CCH_3\text{-}(CH)_3\text{-}CCH_3\text{-})N\text{-}B(\text{-}NCH_3\text{-}CH_2\text{-}CH_2\text{-}NCH_3\text{-})][O_3SCF_3]$	B: B Comp.SVol.3/3-207, 210
$BC_{12}F_3H_{21}N_3O_3S$	$[2,6\text{-}(CH_3)_2\text{-}NC_5H_3\text{-}1\text{-}B(N(CH_3)_2)_2][CF_3\text{-}SO_3]$	B: B Comp.SVol.3/3-207
		B: B Comp.SVol.4/3b-45
$BC_{12}F_3H_{22}N_2$	$H_{10}C_5N\text{-}CH{=}CH\text{-}NC_5H_{10} \cdot BF_3$	B: B Comp.SVol.3/3-379
$BC_{12}F_3H_{27}N$	$N(n\text{-}C_4H_9)_3 \cdot BF_3$	B: B Comp.SVol.3/4-65/7
$BC_{12}F_3H_{27}OP$	$(C_4H_9)_3PO \cdot BF_3$	B: B Comp.SVol.3/3-252
$BC_{12}F_3H_{29}NOSi_2$	$CF_3\text{-}CH_2\text{-}O\text{-}B(C_4H_9\text{-}t)\text{-}N[Si(CH_3)_3]_2$	B: B Comp.SVol.4/3b-61
$BC_{12}F_3H_{30}Si_3$	$[(CH_3)_3Si][(CH_3)_2(t\text{-}C_4H_9)Si][(CH_3)_2SiF]CBF_2$	B: B Comp.SVol.3/3-363
$BC_{12}F_4$	$C_{12}[BF_4]$	B: B Comp.SVol.4/3b-107
$BC_{12}F_4FeH_7O_3$	$[(C_9H_7)Fe(CO)_3][BF_4]$	Fe: Org.Comp.B15-53
$BC_{12}F_4FeH_9O_3$	$[(C_9H_9)Fe(CO)_3][BF_4]$	Fe: Org.Comp.B15-253/4
$BC_{12}F_4FeH_{11}O_2$	$[(C_5H_5)Fe(CO)_2(C_5H_6\text{-}c)][BF_4]$	Fe: Org.Comp.B17-87
$BC_{12}F_4FeH_{11}O_4$	$[(6\text{-}CH_3C(O)\text{-}C_7H_8)Fe(CO)_3][BF_4]$	Fe: Org.Comp.B15-208, 223
$BC_{12}F_4FeH_{11}O_5$	$[(1\text{-}CH_3OC(O)CH_2\text{-}C_6H_6)Fe(CO)_3][BF_4]$	Fe: Org.Comp.B15-100, 136

Formula	Compound	Reference
$BC_{12}F_4FeH_{11}O_5$	$[(1-CH_3OC(O)-C_7H_8)Fe(CO)_3][BF_4]$	Fe: Org.Comp.B15-201
–	$[(6-CH_3OC(O)CH_2-C_6H_6)Fe(CO)_3][BF_4]$	Fe: Org.Comp.B15-108, 159
$BC_{12}F_4FeH_{11}O_6$	$[(1-CH_3OC(O)-2-CH_3O-C_6H_5)Fe(CO)_3][BF_4]$	Fe: Org.Comp.B15-110, 159
–	$[(1-CH_3OC(O)-3-CH_3O-C_6H_5)Fe(CO)_3][BF_4]$	Fe: Org.Comp.B15-112
–	$[(1-CH_3O-3-CH_3OC(O)-C_6H_5)Fe(CO)_3][BF_4]$	Fe: Org.Comp.B15-111
–	$[(2-CH_3O-3-CH_3OC(O)-C_6H_5)Fe(CO)_3][BF_4]$	Fe: Org.Comp.B15-123
$BC_{12}F_4FeH_{12}$	$[(CH_2C_6H_5)Fe(C_5H_5)][BF_4]$	Fe: Org.Comp.B18-270
$BC_{12}F_4FeH_{12}NO_2$	$[(C_8H_9)Fe(CNCH_3)(CO)_2][BF_4]$	Fe: Org.Comp.B15-351
$BC_{12}F_4FeH_{13}$	$[(CH_3-C_6H_5)Fe(C_5H_5)][BF_4]$	Fe: Org.Comp.B18-142/6, 197, 200, 201, 205/6, 269/72
–	$[(C_6H_6)Fe(C_6H_7)][BF_4]$	Fe: Org.Comp.B18-142/6, 153, 172, 189/90
–	$[(c-C_7H_8)Fe(C_5H_5)][BF_4]$	Fe: Org.Comp.B19-340/1, 343
$BC_{12}F_4FeH_{13}O$	$[(CH_3O-C_6H_5)Fe(C_5H_5)][BF_4]$	Fe: Org.Comp.B18-142/6, 197, 198, 201, 249/50
$BC_{12}F_4FeH_{13}O_2$	$[(C_5H_5)Fe(CO)_2(CH_2=C=C(CH_3)_2)][BF_4]$	Fe: Org.Comp.B17-113
–	$[(C_5H_5)Fe(CO)_2(CH_2=CHC(CH_3)=CH_2)][BF_4]$	Fe: Org.Comp.B17-37
–	$[(C_5H_5)Fe(CO)_2(CH_2=CHCH=CHCH_3)][BF_4]$	Fe: Org.Comp.B17-38
–	$[(C_5H_5)Fe(CO)_2(CH_2=CH-C_3H_5-c)][BF_4]$	Fe: Org.Comp.B17-15
–	$[(C_5H_5)Fe(CO)_2(CH_2=CD-C_3H_5-c)][BF_4]$	Fe: Org.Comp.B17-15
–	$[(C_5H_5)Fe(CO)_2(CD_2=CH-C_3H_5-c)][BF_4]$	Fe: Org.Comp.B17-15
–	$[(C_5H_5)Fe(CO)_2(C_5H_8-c)][BF_4]$	Fe: Org.Comp.B17-87/8, 103
–	$[(C_5H_5)Fe(CO)_2(H_3CCH=C=CHCH_3)][BF_4]$	Fe: Org.Comp.B17-114
–	$[(C_5H_5)Fe(CO)_2(H-CC-C_3H_7-n)][BF_4]$	Fe: Org.Comp.B17-121
–	$[(C_5H_5)(CO)_2Fe=CH(CH=C(CH_3)_2)][BF_4]$	Fe: Org.Comp.B16a-88, 95
–	$[(C_5H_5)(CO)_2Fe=CH(CH=CH-C_2H_5)][BF_4]$	Fe: Org.Comp.B16a-95
$BC_{12}F_4FeH_{13}O_3$	$[(1,3,5-(CH_3)_3C_6H_4)Fe(CO)_3][BF_4]$	Fe: Org.Comp.B15-129, 181/2
–	$[(C_5H_5)Fe(CO)_2(2-CH_3-OC_4H_5)][BF_4]$	Fe: Org.Comp.B17-94
–	$[(C_5H_5)Fe(CO)_2(CH_2=C=CH-CH_2CH_2OH)][BF_4]$	Fe: Org.Comp.B17-112
–	$[(C_5H_5)Fe(CO)_2(CH_2=C_4H_6O)][BF_4]$	Fe: Org.Comp.B17-70
–	$[(C_5H_5)(CO)_2Fe=C(C_3H_5-c)-OCH_3][BF_4]$	Fe: Org.Comp.B16a-111
$BC_{12}F_4FeH_{13}O_4$	$[(1,5-(CH_3)_2-3-CH_3O-C_6H_4)Fe(CO)_3][BF_4]$	Fe: Org.Comp.B15-129, 181
–	$[(1-CH_3O-3,5-(CH_3)_2C_6H_4)Fe(CO)_3][BF_4]$	Fe: Org.Comp.B15-96/7, 129
–	$[(C_5H_5)Fe(CO)_2(5-CH_3-1,4-O_2C_4H_5)][BF_4]$	Fe: Org.Comp.B17-96
–	$[(C_5H_5)Fe(CO)_2(6-CH_3-1,4-O_2C_4H_5)][BF_4]$	Fe: Org.Comp.B17-96
–	$[(C_5H_5)Fe(CO)_2(CH_2=CHCH_2-COO-CH_3)][BF_4]$	Fe: Org.Comp.B17-33
–	$[(C_5H_5)Fe(CO)_2(CH_2=CHCH_2-OOCCH_3)][BF_4]$	Fe: Org.Comp.B17-25
$BC_{12}F_4FeH_{14}NO_2$	$[(C_5H_5)Fe(CN-C_4H_9-t)(CO)_2][BF_4]$	Fe: Org.Comp.B15-313, 318
$BC_{12}F_4FeH_{15}N_2O$	$[(C_5H_5)Fe(CN-C_2H_5)_2(CO)][BF_4]$	Fe: Org.Comp.B15-335
$BC_{12}F_4FeH_{15}O_2$	$[(C_5H_5)Fe(CO)_2(CH_2=CH-C_3H_7-i)][BF_4]$	Fe: Org.Comp.B17-12
–	$[(C_5H_5)Fe(CO)_2(CH_2=CH-C_3H_7-n)][BF_4]$	Fe: Org.Comp.B17-12, 47
–	$[(C_5H_5)Fe(CO)_2(CH_3CH=C(CH_3)_2)][BF_4]$	Fe: Org.Comp.B17-85
–	$[(C_5H_5)Fe(CO)_2(CH_3CH=CH-C_2H_5)][BF_4]$	Fe: Org.Comp.B17-57/8
–	$[(C_5H_5)(CO)_2Fe=CH-C_4H_9-t][BF_4]$	Fe: Org.Comp.B16a-88, 96
$BC_{12}F_4FeH_{15}O_3$	$[(CH_2(CH)_4-C_4H_9-t)Fe(CO)_3][BF_4]$	Fe: Org.Comp.B15-13, 20
–	$[(CH_3(CH)_5-C_3H_7-i)Fe(CO)_3][BF_4]$	Fe: Org.Comp.B15-13, 22
–	$[(C_5H_5)Fe(CO)_2(CH_2=C(CH_3)-O-C_2H_5)][BF_4]$	Fe: Org.Comp.B17-67, 79/80
–	$[(C_5H_5)Fe(CO)_2(CH_3CH=CH-O-C_2H_5)][BF_4]$	Fe: Org.Comp.B17-65, 78
$BC_{12}F_4FeH_{15}O_3Si$	$[(1-(CH_3)_3Si-C_6H_6)Fe(CO)_3][BF_4]$	Fe: Org.Comp.B15-102, 138

Formula	Compound	Reference
$BC_{12}F_4FeH_{15}O_3Si$	$[(2\text{-}(CH_3)_3Si\text{-}C_6H_6)Fe(CO)_3][BF_4]$	Fe: Org.Comp.B15-105, 138
–	$[(3\text{-}(CH_3)_3Si\text{-}C_6H_6)Fe(CO)_3][BF_4]$	Fe: Org.Comp.B15-107, 158
$BC_{12}F_4FeH_{15}O_4Si$	$[(1\text{-}(CH_3)_3SiO\text{-}C_6H_6)Fe(CO)_3][BF_4]$	Fe: Org.Comp.B15-91/2
–	$[(2\text{-}(CH_3)_3SiO\text{-}C_6H_6)Fe(CO)_3][BF_4]$	Fe: Org.Comp.B15-91/2
$BC_{12}F_4FeH_{16}NO_2$	$[(C_5H_5)Fe(CO)_2CH(\text{-}CH_2CH_2CH_2\text{-}NH_2\text{-}CH_2\text{-})][BF_4]$	Fe: Org.Comp.B14-151, 155
–	$[(C_5H_5)Fe(CO)_2(CH_2{=}C(CH_3)\text{-}N(CH_3)_2)][BF_4]$	Fe: Org.Comp.B17-68
–	$[(C_5H_5)Fe(CO)_2CH_2CH(\text{-}CH_2\text{-}CH_2\text{-}CH_2\text{-}NH_2\text{-})][BF_4]$	Fe: Org.Comp.B14-151, 155/6
$BC_{12}F_4FeH_{17}O_2Si$	$[((CH_3)_3Si\text{-}C_5H_4)Fe(CO)_2(CH_2{=}CH_2)][BF_4]$	Fe: Org.Comp.B17-132
–	$[(C_5H_5)Fe(CO)_2(CH_2{=}CH\text{-}Si(CH_3)_3)][BF_4]$	Fe: Org.Comp.B17-23
$BC_{12}F_4FeH_{18}NO_2$	$[(C_5H_5)Fe(CO)_2\text{-}CH_2CH(CH_3)\text{-}NH(CH_3)_2][BF_4]$	Fe: Org.Comp.B14-150
–	$[(C_5H_5)Fe(CO)_2\text{-}CH_2CH_2\text{-}N(CH_3)_3][BF_4]$	Fe: Org.Comp.B14-149, 155
$BC_{12}F_4FeH_{18}O_4P$	$[(C_5H_5)(CO)(P(OCH_3)_3)Fe{=}C{=}CHCH_3][BF_4]$	Fe: Org.Comp.B16a-208, 210
$BC_{12}F_4FeH_{20}O_6P$	$[(C_5H_5)(CO)(P(OCH_3)_3)Fe{=}COH\text{-}CH_2OCH_3][BF_4]$	Fe: Org.Comp.B16a-42, 56
$BC_{12}F_4H_5NO_5Re$	$[(CO)_5Re(CNC_6H_5)][BF_4]$	Re: Org.Comp.2-253, 254
$BC_{12}F_4H_6O_6Re$	$[(CO)_5ReOC(C_6H_5)H][BF_4]$	Re: Org.Comp.2-157
$BC_{12}F_4H_{10}I$	$[(C_6H_5)_2I][BF_4]$	B: B Comp.SVol.3/3-354/5 B: B Comp.SVol.4/3b-219 B: B Comp.SVol.4/4-115
$BC_{12}F_4H_{10}IO$	$[(C_6H_5)_2IO][BF_4]$	B: B Comp.SVol.3/3-355
$BC_{12}F_4H_{11}NO_5Re$	$[(CO)_5Re(CNC_6H_{11})][BF_4]$	Re: Org.Comp.2-253/4
$BC_{12}F_4H_{11}N_3PS_2$	$[5,5\text{-}(C_6H_5)_2\text{-}1,3,2,4,6,5\text{-}S_2N_3PH][BF_4]$	B: B Comp.SVol.4/3b-166
$BC_{12}F_4H_{12}O_4Re$	$[C_8H_{12}Re(CO)_4][BF_4]$	Re: Org.Comp.2-420
$BC_{12}F_4H_{13}MoN_2O_3$	$[C_5H_5Mo(NO)(NCCD_3)CH{=}C(\text{-}C_2H_4OC({=}O)\text{-})][BF_4]$	Mo: Org.Comp.6-178
$BC_{12}F_4H_{13}MoN_2O_5$	$[(C_5H_5)Mo(CO)_2(\text{-}C(OH){=}C(COO\text{-}C_2H_5)\text{-}NH\text{-}NH\text{-})][BF_4]$	Mo: Org.Comp.8-139
$BC_{12}F_4H_{13}MoO$	$[C_5H_5Mo(CO)(H\text{-}CC\text{-}CH_3)_2][BF_4]$	Mo: Org.Comp.6-316, 322
$BC_{12}F_4H_{13}MoO_2$	$[(C_5H_5)Mo(CO)_2(CH_2{=}C(CH_3)\text{-}CH{=}CH_2)][BF_4]$	Mo: Org.Comp.8-305/6, 307, 308, 316
$BC_{12}F_4H_{15}MoO$	$[C_5H_5Mo(CO)(H_3C\text{-}CC\text{-}CH_3)H_2C{=}CH_2][BF_4]$	Mo: Org.Comp.6-315, 322
$BC_{12}F_4H_{18}IMoN_6$	$[(I)Mo(CN\text{-}CH_3)_6][BF_4]$	Mo: Org.Comp.5-71, 73
$BC_{12}F_4H_{18}NS$	$[NS][BF_4] \cdot (CH_3)_6C_6$	B: B Comp.SVol.4/4-167 S: S-N Comp.5-47/9
$BC_{12}F_4H_{24}N_3O_3S$	$[(1,4\text{-}ONC_4H_8\text{-}4)_3S][BF_4]$	S: S-N Comp.8-222/4, 228
$BC_{12}F_4H_{28}N$	$[(n\text{-}C_3H_7)_4N][BF_4]$	B: B Comp.SVol.4/3b-210
–	$[(n\text{-}C_4H_9)_3NH][BF_4]$	B: B Comp.SVol.4/3b-210
$BC_{12}F_{10}N_3$	$(C_6F_5)_2BN_3$	B: B Comp.SVol.3/3-145
$BC_{12}FeH_{13}O_4$	$(HO)_2B\text{-}C_5H_4FeC_5H_4OC(O)CH_3$	Fe: Org.Comp.A9-279, 283/4
$BC_{12}FeH_{19}O_3SeSi$	$BC_2FeSeSi(CO)_3(C_2H_5)_2(CH_3)_3$	B: B Comp.SVol.4/4-180
$BC_{12}Fe_2H_5O_6S_2$	$BFe_2S_2(CO)_6(C_6H_5)$	B: B Comp.SVol.4/4-129/30
$BC_{12}GeH_{28}NSi$	$1\text{-}(CH_3)_3Ge\text{-}4,5\text{-}(C_2H_5)_2\text{-}2,2,3\text{-}(CH_3)_3\text{-}1,2,5\text{-}NSiBC_2$	B: B Comp.SVol.4/3a-250/1
$BC_{12}H_8I$	$5\text{-}I\text{-}5\text{-}BC_{12}H_8$	B: B Comp.SVol.4/4-113/4
$BC_{12}H_8O_4^-$	$[B(\text{-}O\text{-}1\text{-}C_6H_4\text{-}2\text{-}O\text{-})_2]^-$	B: B Comp.SVol.4/3b-77
$BC_{12}H_8O_4Rb$	$Rb[B(\text{-}O\text{-}C_6H_4\text{-}O\text{-})_2]$	B: B Comp.SVol.3/3-227

Formula	Compound	Reference
$BC_{12}H_{10}NO_2$	$C_6H_5OB[-NH-(1,2-C_6H_4)-O-]$	B: B Comp.SVol.3/3-164
$BC_{12}H_{10}NO_3$	$[C_6H_4-1,2-(O)_2]B[-NH_2-(1,2-C_6H_4)-O-]$	B: B Comp.SVol.3/3-219
$BC_{12}H_{10}N_3$	$(C_6H_5)_2B-N_3$	B: B Comp.SVol.3/3-144/5
		B: B Comp.SVol.4/3a-218
$BC_{12}H_{10}N_6^-$	$[(1,2,3-N_3C_6H_4-1)_2BH_2]^-$	B: B Comp.SVol.4/3b-50
$BC_{12}H_{10}N_6O_3Re$	$(CO)_3Re(1,2-N_2C_3H_3)_3BH$	Re: Org.Comp.1-115, 117
$BC_{12}H_{11}I_2N_2$	$I_2BN(CH_2)_3C_9H_5N$	B: B Comp.SVol.3/4-100/1
$BC_{12}H_{12}NO$	$H_3B \cdot NC_5H_4-2-C(O)C_6H_5$	B: B Comp.SVol.4/3b-13
$BC_{12}H_{12}NO_4$	$NH_4[B(-O-C_6H_4-O-)_2]$	B: B Comp.SVol.3/3-226/7
$BC_{12}H_{12}N_3$	$(NC_4H_4-1)_3B$	B: B Comp.SVol.4/3a-153
$BC_{12}H_{13}N_3^-$	$[HB(NC_4H_4)_3]^-$	B: B Comp.SVol.3/3-213
$BC_{12}H_{14}N$	$(C_6H_5)_2NH-BH_3$	B: B Comp.SVol.3/3-174
		B: B Comp.SVol.4/3b-5, 10
$BC_{12}H_{14}NO$	$H_3B \cdot NC_5H_4-2-[CH(OH)-C_6H_5]$	B: B Comp.SVol.4/3b-13
$BC_{12}H_{15}N_2$	$C_6H_5C\equiv C-B(-NCH_3-CH_2-CH_2-NCH_3-)$	B: B Comp.SVol.3/3-113
$BC_{12}H_{15}N_2O_5S$	$HOB[-C_6H_2(-OCH_2O-)-CH=N-NSO_2(C_4H_9-n)-]$	B: B Comp.SVol.3/3-164
$BC_{12}H_{15}S_2$	$n-C_4H_9B(-S-CH=CC_6H_5-S-)$	B: B Comp.SVol.3/4-114
$BC_{12}H_{16}LiS_2$	$Li[B(C_2H_5)_2(SC_4H_3)_2]$	B: B Comp.SVol.3/4-117
$BC_{12}H_{16}N$	$(CH_3)_2B-N(CH_2C_6H_5)-CH_2-C\equiv CH$	B: B Comp.SVol.3/3-151
–	$NC_5H_5 \cdot C_5H_5-B(CH_3)_2$	B: B Comp.SVol.4/3b-28
$BC_{12}H_{16}NO$	$1-C_6H_5-2,8,1-ONBC_6H_{11}$	B: B Comp.SVol.4/3b-65
$BC_{12}H_{16}NS$	$CH_3SBCH_3N(CH_2C_6H_5)(CH_2C\equiv CH)$	B: B Comp.SVol.3/4-122
$BC_{12}H_{16}NS_2$	$(C_2H_5)_2N-B[-S-CH=C(C_6H_5)-S-]$	B: B Comp.SVol.3/4-126
–	$t-C_4H_9-NH-B[-S-CH=C(C_6H_5)-S-]$	B: B Comp.SVol.3/4-126
$BC_{12}H_{16}N_3O_2$	$1-C_2H_5-3,5-(CH_3)_2-2-C_6H_5-1,3,5,2-N_3BC_2(=O)_2-4,6$	B: B Comp.SVol.4/3a-176
$BC_{12}H_{16}O_7Re$	$fac-(CO)_3Re[(COCH_3)_3]BOCH_2CH_2CH_3$	Re: Org.Comp.2-379
$BC_{12}H_{17}LiN$	$Li[(CH_3)_2N-B(C_6H_5)(-CH_2-CH=CH-CH_2-)]$	B: B Comp.SVol.4/3b-49
$BC_{12}H_{17}N_4$	$[H_2C=CH-CH_2-NH_3][BH(NC_4H_4)_2CN]$	B: B Comp.SVol.3/3-213
$BC_{12}H_{17}S_2$	$n-C_4H_9B(-S-CH_2-CHC_6H_5-S-)$	B: B Comp.SVol.3/4-112
$BC_{12}H_{18}N$	$1-NC_9H_{10}[-1-BH_2-CH_2CH_2CH_2-1-]$	B: B Comp.SVol.4/3a-255
–	$(-CH_2-CH=CH-CH_2-)B(C_6H_5)-NH-(CH_3)_2$	B: B Comp.SVol.4/3b-29
$BC_{12}H_{18}NO_4$	$(CH_3)_3N-BH_2-C(O)-O-C(O)-O-CH_2-C_6H_5$	B: B Comp.SVol.4/3b-25
$BC_{12}H_{19}N_4$	$[-B(C_2H_5)-N(C_4H_9-t)-N=N-N(C_6H_5)-]$	B: B Comp.SVol.3/3-137/8
–	$[-B(C_3H_7-i)-N(C_3H_7-i)-N=N-N(C_6H_5)-]$	B: B Comp.SVol.3/3-137/8
–	$[-B(C_4H_9-t)-N(CH_3)-N=N-N(CH_2-C_6H_5)-]$	B: B Comp.SVol.4/3a-199
–	$[(CH_3)_3NH][(NC_4H_4-1)_2BH-CN]$	B: B Comp.SVol.3/3-213
$BC_{12}H_{19}O_3RuSeSi$	$BC_2RuSeSi(CO)_3(C_2H_5)_2(CH_3)_3$	B: B Comp.SVol.4/4-180
$BC_{12}H_{19}S_2$	$2-(CH_3)_5C_5-1,3,2-S_2BC_2H_4$	B: B Comp.SVol.4/4-130
$BC_{12}H_{20}N$	$2-C_2H_5-C_5H_4N-BC_5H_{11}$	B: B Comp.SVol.3/3-189
–	$(i-C_3H_7)_2B-NH-C_6H_5$	B: B Comp.SVol.4/3a-230
$BC_{12}H_{20}NO$	$(C_6H_5)(C_2H_5O)B-N(C_2H_5)_2$	B: B Comp.SVol.3/3-154
$BC_{12}H_{20}NO_2$	$H_3B-N(CH_3)_2-CH_2CH_2-OC(O)-CH_2-C_6H_5$	B: B Comp.SVol.4/3b-15
$BC_{12}H_{20}NO_3$	$H_3B-N(CH_3)_2-CH_2CH_2-OC(O)-CH_2-O-C_6H_5$	B: B Comp.SVol.4/3b-15
$BC_{12}H_{20}NO_8$	$(C_2H_5)_3NH[B(-O-CO-CH_2-CO-O-)_2]$	B: B Comp.SVol.3/3-227
$BC_{12}H_{20}N_3$	$(n-C_3H_7)_2B[-(1,2-NC_5H_4)-N=CH-NH-]$	B: B Comp.SVol.3/3-123
$BC_{12}H_{20}N_3O$	$(n-C_3H_7)_2B[-(1,2-NC_5H_4)-NH-CO-NH-]$	B: B Comp.SVol.3/3-123
$BC_{12}H_{20}NaO_3Se$	$Na[C_6H_5-Se-B(O-C_2H_5)_3]$	B: B Comp.SVol.4/4-174
$BC_{12}H_{21}N_2$	$(C_2H_5)_2N-B(C_6H_5)-N(CH_3)_2$	B: B Comp.SVol.3/3-105

$BC_{12}H_{21}N_2$ ...... $(C_3H_7)_2B[-NC_5H_4-CH_2-NH-]$ .... B: B Comp.SVol.3/3-124
– .... $n-C_4H_9-NH-B(C_6H_5)-N(CH_3)_2$ .... B: B Comp.SVol.3/3-108
– .... $i-C_4H_9-NH-B(C_6H_5)-N(CH_3)_2$ .... B: B Comp.SVol.3/3-108
– .... $s-C_4H_9-NH-B(C_6H_5)-N(CH_3)_2$ .... B: B Comp.SVol.3/3-108
– .... $t-C_4H_9-NH-B(C_6H_5)-N(CH_3)_2$ .... B: B Comp.SVol.3/3-108
B: B Comp.SVol.4/3a-169
– .... $[-NCH_3-C(CH_3)(C_6H_5)-NCH_3-CH_2CH_2-] \cdot BH_3$
B: B Comp.SVol.4/3b-12
$BC_{12}H_{21}N_2OS$.... $(C_2H_5)_2N-B(C_6H_5)-NCH_3-S(O)CH_3$ .... B: B Comp.SVol.3/3-106
B: B Comp.SVol.3/4-134
$BC_{12}H_{21}N_2Si_2$.... $C_6H_5B[N(CH_3)_2]-2,5-(CH_3)_2-1,2,5-NSi_2C_2H_4$ .. B: B Comp.SVol.3/3-109
$BC_{12}H_{21}N_6$ ...... $[-NCH_3-(CH_2)_2-NCH_3-]B[-(CH_3)C_3H_2N_2-H$
$-(CH_3)C_3H_2N_2-]$ .... B: B Comp.SVol.3/3-201
– .... $[-NCH_3-(CH_2)_2-NCH_3-]B[-N_2C_3H_3-H-C_3HN_2(CH_3)_2-]$
B: B Comp.SVol.3/3-201
$BC_{12}H_{22}I$......... $n-C_4H_9-CC-(CH_2)_5-BI-CH_3$ .... B: B Comp.SVol.4/4-113
$BC_{12}H_{22}N$ .... $1-(t-C_4H_9)_2B-NC_4H_4$ .... B: B Comp.SVol.4/3a-233
$BC_{12}H_{22}NOSe$.... $2-CH_3-3-(n-C_4H_9)-5,6-(C_2H_5)_2-1,3,2-SeNBC_3(=O)-4$
B: B Comp.SVol.4/4-172/3
– .... $3-CH_3-2-(n-C_4H_9)-5,6-(C_2H_5)_2-1,3,2-SeNBC_3(=O)-4$
B: B Comp.SVol.4/4-172/3
$BC_{12}H_{22}NO_2Si_2$ .. $1,3,2-O_2BC_6H_4-2-N[Si(CH_3)_3]_2$ .... B: B Comp.SVol.4/3b-63
$BC_{12}H_{22}N_3S$ ..... $(n-C_4H_9)_2B[-NH-CH=N-(2,3-(1,3-C_3H_2NS))-]$ B: B Comp.SVol.3/4-132
$BC_{12}H_{23}$ .... $1-(C_4H_9)_2B-2-(CH_2=)-C_3H_3$ .... B: B Comp.SVol.4/3b-124
$BC_{12}H_{23}NP$ ...... $(C_2H_5)_2N-B(-CH=CCH_3-PC_2H_5-CCH_3=CH-)$... B: B Comp.SVol.3/3-151
$BC_{12}H_{23}N_2O_2$.... $2,2,6,6-(CH_3)_4-NC_5H_6-1-B[-NH-C(=O)-CH(CH_3)-O-]$
B: B Comp.SVol.4/3b-56
$BC_{12}H_{23}N_2Sn$.... $(CH_3)_3SnN(CH_3)B(C_6H_5)N(CH_3)_2$ .... Sn: Org.Comp.18-57, 62
$BC_{12}H_{23}N_4O_3$.... $CH_3B[-NC_2H_5-(CO-NC_2H_5-)_3]$ .... B: B Comp.SVol.3/3-116
$BC_{12}H_{24}LiN_2$..... $[(C_6H_5BH_3)(CH_3)_2N-CH_2-CH_2-N(CH_3)_2]Li$ .... B: B Comp.SVol.3/3-213
$BC_{12}H_{24}N$ .... $1-(BC_5H_{10}-1)-NC_5H_8-(CH_3)_2-2,6$ .... B: B Comp.SVol.3/3-156
– .... $9-(t-C_4H_9-NH)-[3.3.1]-9-BC_8H_{14}$ .... B: B Comp.SVol.4/3a-235
– .... $[3.3.1]-9-BC_8H_{15} \cdot NC_4H_9$ .... B: B Comp.SVol.3/3-189
– .... $[3.3.1.1^{3,7}]-1-BC_9H_{15} \cdot n-C_3H_7-NH_2$ .... B: B Comp.SVol.3/3-191
B: B Comp.SVol.4/3b-30
– .... $[3.3.1.1^{3,7}]-1-BC_9H_{15} \cdot N(CH_3)_3$ .... B: B Comp.SVol.4/3b-30
– .... $[-B(-CH_2-CH(CH_3)-CH_2-)_3N-]$ .... B: B Comp.SVol.4/3a-259
$BC_{12}H_{24}NO_2$..... $(CH_3)_2B-N(C_4H_9-i)-C(O)C(O)-C_4H_9-i$......... B: B Comp.SVol.3/3-155
– .... $(CH_3)_2B-N(C_4H_9-t)-C(O)C(O)-C_4H_9-t$......... B: B Comp.SVol.3/3-155
$BC_{12}H_{24}NO_8$..... $[t-C_4H_9-NH_3][B(OC(O)-CH_3)_4]$ .... B: B Comp.SVol.4/3b-77
$BC_{12}H_{24}NSe$..... $2-(C_4H_9-t)-3,4,5-(C_2H_5)_3-1,2,3-SeNBC_2$ ..... B: B Comp.SVol.4/4-172/3
– .... $3-(C_4H_9-t)-2,4,5-(C_2H_5)_3-1,3,2-SeNBC_2$ ..... B: B Comp.SVol.4/3a-252
$BC_{12}H_{24}NSi$ ..... $4,5-(C_2H_5)_2-3-[CH_2=C(CH_3)]-1,2,2-(CH_3)_3-1,2,5-NSiBC_2$
B: B Comp.SVol.4/3a-251
$BC_{12}H_{24}NSn$..... $(C_2H_5)_2N-B[-CH=CCH_3-Sn(CH_3)_2-CCH_3=CH-]$ B: B Comp.SVol.3/3-151
$BC_{12}H_{24}N_3$ ...... $B[N(CH_2)_4]_3$ .... B: B Comp.SVol.3/3-94
$BC_{12}H_{24}N_3O_2$.... $CH_3B[-N(C_3H_7-i)-C(O)-N(C_3H_7-i)-C(O)-N(C_3H_7-i)-]$
B: B Comp.SVol.3/3-114/5
– .... $CH_3B[-N(C_4H_9-n)-C(O)-NCH_3-C(O)-N(C_4H_9-n)-]$
B: B Comp.SVol.3/3-115

$BC_{12}H_{25}N_2$ ...... 1-(i-C_3H_7-N=B)-2,2,6,6-$(CH_3)_4$-NC_5H_6 ...... B: B Comp.SVol.4/3a-160/5
– .... N(-CH_2-CH_2-$)_3$N-BH_2-C_6H_{11}-c .... B: B Comp.SVol.3/3-188
$BC_{12}H_{25}N_2O_2$ .... OC_5H_9-2-O-CH=CH-CH_2-B[N$(CH_3)_2]_2$ .... B: B Comp.SVol.4/3a-168
$BC_{12}H_{26}N$ .... BC_5H_{11} · NC_5H_9-2,6-$(CH_3)_2$ .... B: B Comp.SVol.3/3-189
– .... [3.3.1]-9-BC_8H_{15} · HN$(C_2H_5)_2$ .... B: B Comp.SVol.3/3-188
– .... [3.3.1]-9-BC_8H_{15} · NH_2-C_4H_9-t .... B: B Comp.SVol.3/3-188
$BC_{12}H_{26}NOSi$ .... CH_3NH_2-B(C_2H_5)[-O-Si$(CH_3)_2$-C(CCH_3=CH_2)=C(C_2H_5)-]
B: B Comp.SVol.4/3b-71
$BC_{12}H_{26}N_3$ ...... (i-$C_3H_7)_2$B[-NH-C(C_2H_5)=N-C(C_2H_5)=NH-] ... B: B Comp.SVol.3/3-122
– .... (n-$C_4H_9)_2$B(-NH-CCH_3=N-CCH_3=NH-) .... B: B Comp.SVol.3/3-122
$BC_{12}H_{26}N_3O_2$ .... [NC_5H_{10}-1-O-$]_2$B-N$(CH_3)_2$ .... B: B Comp.SVol.4/3b-61
$BC_{12}H_{26}PS$ ...... [3.3.1]-9-BC_8H_{14}-9-SCH_3 · P$(CH_3)_3$ .... B: B Comp.SVol.4/4-132
$BC_{12}H_{27}N_2$ ...... t-C_4H_9-N=B-N$(C_4H_9$-t$)_2$ .... B: B Comp.SVol.4/3a-160/5
– .... N(-CH_2-CH_2-$)_3$N-BH_2-C$(CH_3)_2$-C_3H_7-i ...... B: B Comp.SVol.3/3-188
$BC_{12}H_{27}N_2O$ ..... $(CH_3)_2$B-N(i-C_3H_7)-C(O)N(i-$C_3H_7)_2$ .... B: B Comp.SVol.3/3-156
$BC_{12}H_{27}N_2Si$ ..... 1-[$(CH_3)_3$Si-N=B]-2,2,6,6-$(CH_3)_4$-NC_5H_6 ..... B: B Comp.SVol.4/3a-160/5
– .... 3-$(CH_3)_3$Si-1,3-$N_2C_3H_3$ · B$(C_2H_5)_3$ .... B: B Comp.SVol.4/3b-28
$BC_{12}H_{27}S$ .... (n-$C_4H_9)_2$BS(n-C_4H_9) .... B: B Comp.SVol.3/4-111
$BC_{12}H_{28}N$ .... BC_5H_{11} · $(C_2H_5)_2$N-C_3H_7-i .... B: B Comp.SVol.3/3-189
– .... (t-$C_4H_9)_2$B-N$(C_2H_5)_2$ .... B: B Comp.SVol.4/3a-227
– .... (n-$C_4H_9)_2$B-NH-C_4H_9-t .... B: B Comp.SVol.4/3a-230
– .... (t-$C_4H_9)_2$B-NH-C_4H_9-t .... B: B Comp.SVol.4/3a-230
– .... n-C_4H_9-B(C_4H_9-t)-NH-C_4H_9-t .... B: B Comp.SVol.4/3a-234
$BC_{12}H_{28}NO$ ...... (n-C_4H_9)(t-C_4H_9O)B-NH(t-C_4H_9) .... B: B Comp.SVol.3/3-155
$BC_{12}H_{28}NSi$ ..... $(CH_3)_3$Si-N(C_4H_9-t)-B=CH-C_4H_9-t .... B: B Comp.SVol.4/3a-216
– .... $(C_2H_5)_2$B[-C(C_2H_5)=CCH_3-Si$(CH_3)_2$-$NHCH_3$-] B: B Comp.SVol.4/3a-254
$BC_{12}H_{28}NSiSn$ ... 1-$(CH_3)_3$Sn-4,5-$(C_2H_5)_2$-2,2,3-$(CH_3)_3$-1,2,5-$NSiBC_2$
B: B Comp.SVol.4/3a-250/1
Sn: Org.Comp.18-83/4, 95
$BC_{12}H_{28}NSi_2$ ..... 1-$(CH_3)_3$Si-4,5-$(C_2H_5)_2$-2,2,3-$(CH_3)_3$-1,2,5-$NSiBC_2$
B: B Comp.SVol.4/3a-250/1
– .... [-C(C_2H_5)=C(C_2H_5)-]B-N[Si$(CH_3)_3]_2$ .... B: B Comp.SVol.3/3-150
– .... [-CH=C(n-C_4H_9)-]B-N[Si$(CH_3)_3]_2$ .... B: B Comp.SVol.3/3-150
– .... [-CCH_3=C(n-C_3H_7)-]B-N[Si$(CH_3)_3]_2$ .... B: B Comp.SVol.3/3-150
$BC_{12}H_{28}NSn$ ..... $(CH_3)_2$B[-CCH_3=CCH_3-Sn$(CH_3)_2$-N$(C_2H_5)_2$-] .. B: B Comp.SVol.4/3a-255
Sn: Org.Comp.19-48/9
– .... $(C_2H_5)_2$B[-C(C_2H_5)=CCH_3-Sn$(CH_3)_2$-$NHCH_3$-] B: B Comp.SVol.4/3a-255
Sn: Org.Comp.19-48/9
$BC_{12}H_{28}N_2^+$ ..... [(i-$C_3H_7)_2$N=B=N(i-$C_3H_7)_2]^+$ .... B: B Comp.SVol.3/3-205
$BC_{12}H_{28}N_3$ ...... (i-$C_3H_7)_2$N-B=N-N(i-$C_3H_7)_2$ .... B: B Comp.SVol.3/3-100
– .... i-C_3H_7-N=C(C_4H_9-n)-B[N$(CH_3)_2]_2$ .... B: B Comp.SVol.3/3-109
$BC_{12}H_{28}N_5$ ...... B[N(i-$C_3H_7)_2]_2N_3$ .... B: B Comp.SVol.3/3-97
$BC_{12}H_{29}N_2$ ...... $(C_2H_5)_2$N-B(C_4H_9-t)-N$(C_2H_5)_2$ .... B: B Comp.SVol.4/3a-168
– .... i-C_3H_7-NH-B(C_3H_7-i)-N$(C_3H_7$-i$)_2$ .... B: B Comp.SVol.3/3-107
– .... n-C_4H_9-B[NH-C_4H_9-t$]_2$ .... B: B Comp.SVol.3/3-107
– .... t-C_4H_9-NH-B(C_4H_9-n)-N$(C_2H_5)_2$ .... B: B Comp.SVol.3/3-107
$BC_{12}H_{29}N_2OSi$ ... $(CH_3)_3SiCH_2CH_2$-O-CH=$CHCH_2$-B[N$(CH_3)_2]_2$ .. B: B Comp.SVol.4/3a-168
$BC_{12}H_{29}N_2Si$ ..... 1-(t-C_4H_9)-2,2-$(CH_3)_2$-3-(i-$C_3H_7)_2$N-1,2,3-NSiB
B: B Comp.SVol.4/3a-205
$BC_{12}H_{29}N_2Sn$ .... 1,3-(t-$C_4H_9)_2$-2,4,4-$(CH_3)_3$-1,3,4,2-N_2SnBCH_2 B: B Comp.SVol.4/3a-175

Formula	Compound	Reference
$BC_{12}H_{29}N_2Sn$	1,3-(t-C_4H_9)$_2$-2,4,4-(CH_3)$_3$-1,3,4,2-N_2SnBCH_2	Sn: Org.Comp.19-56, 58/9
$BC_{12}H_{30}LiN_2Si$	$LiN(C_4H_9$-t)-B(CH_3)-N(C_4H_9-t)-Si(CH_3)$_3$ · (CH_3)$_2$N-CH_2CH_2-N(CH_3)$_2$	B: B Comp.SVol.4/3a-169
$BC_{12}H_{30}N$	H_3N-B(C_4H_9-n)$_3$	B: B Comp.SVol.4/3b-27
$BC_{12}H_{30}NO_6$	H_3N-BH_3 · [-O(CH_2CH_2O)$_5CH_2CH_2$-]	B: B Comp.SVol.3/3-171
$BC_{12}H_{30}N_3$	B[N(C_2H_5)$_2$]$_3$	B: B Comp.SVol.3/3-94
$BC_{12}H_{30}O_9P_3$	[(C_2H_5O)$_2$PO]$_3$B	B: B Comp.SVol.3/3-251
$BC_{12}H_{31}N_2Si$	(CH_3)$_2$B-N(C_4H_9-t)-Si(CH_3)$_2$-NH-C_4H_9-t	B: B Comp.SVol.4/3a-231
$BC_{12}H_{32}NOSi_2$	(n-C_3H_7)$_2$B-N[Si(CH_3)$_3$]-O-Si(CH_3)$_3$	B: B Comp.SVol.3/3-153
–	(i-C_3H_7)$_2$B-N[Si(CH_3)$_3$]-O-Si(CH_3)$_3$	B: B Comp.SVol.3/3-154
$BC_{12}H_{32}NSn_2$	(i-C_3H_7)$_2$B-N[Sn(CH_3)$_3$]$_2$	B: B Comp.SVol.4/3a-227
$BC_{12}H_{33}N_2Si_2$	(CH_3)$_2$N-B(C_4H_9-t)-N[Si(CH_3)$_3$]$_2$	B: B Comp.SVol.3/3-108
–	i-C_3H_7-NH-B(C_3H_7-i)-N[Si(CH_3)$_3$]$_2$	B: B Comp.SVol.3/3-107
$BC_{12}H_{35}N_2Si_3$	(CH_3)$_3$SiCH_2B[N(CH_3)$_2$]N[Si(CH_3)$_3$]$_2$	B: B Comp.SVol.3/3-108
$BC_{12}H_{36}NSi_5$	[(CH_3)$_3$Si]$_3$Si-B=N-Si(CH_3)$_3$	B: B Comp.SVol.3/3-142/3 B: B Comp.SVol.4/3a-215
$BC_{13}ClF_4FeH_{14}$	[(Cl-C_6H_5)Fe(C_5H_4-C_2H_5)][BF_4]	Fe: Org.Comp.B18-142/6, 197/8, 246
$BC_{13}ClF_5H_8NO$	(C_6F_5)ClBNH(4-$CH_3OC_6H_4$)	B: B Comp.SVol.3/4-56
$BC_{13}ClH_{16}$	2-Cl-4-(n-C_4H_9)-2-BC_9H_7	B: B Comp.SVol.3/4-44
–	2-Cl-4-(t-C_4H_9)-2-BC_9H_7	B: B Comp.SVol.3/4-44
$BC_{13}ClH_{19}N$	9-Cl-[3.3.1]-9-BC_8H_{14} · NC_5H_5	B: B Comp.SVol.4/4-69
$BC_{13}ClH_{20}NO_6Re$	[N(CH_3)$_4$][(CO)$_3$Re($CH_2COCOCH_3$)(CH_3CO)BCl]	Re: Org.Comp.2-418/9
$BC_{13}ClH_{23}N$	(n-C_4H_9)$_2$BCl · NC_5H_5	B: B Comp.SVol.4/4-69
–	(t-C_4H_9)$_2$BCl · NC_5H_5	B: B Comp.SVol.4/4-51
$BC_{13}ClH_{23}NSi$	C_6H_5-BCl-N(C_4H_9-t)-Si(CH_3)$_3$	B: B Comp.SVol.4/4-60
$BC_{13}ClH_{25}N$	(C_2H_5)$_2NC_9H_{15}$B(Cl)	B: B Comp.SVol.3/4-57
$BC_{13}ClH_{27}N$	1-(t-C_4H_9-BCl)-2,2,6,6-(CH_3)$_4$-NC_5H_6	B: B Comp.SVol.4/4-60
$BC_{13}ClH_{28}N_2$	1-(t-C_4H_9-NH-BCl)-2,2,6,6-(CH_3)$_4$-NC_5H_6	B: B Comp.SVol.4/4-63
–	1-[(C_2H_5)$_2$N-BCl]-2,2,6,6-(CH_3)$_4$-NC_5H_6	B: B Comp.SVol.3/4-59
$BC_{13}ClH_{31}NSi$	n-C_6H_{13}-BCl-N(C_4H_9-t)-Si(CH_3)$_3$	B: B Comp.SVol.4/4-60
$BC_{13}Cl_2GeH_{33}N_2Si$	Si(CH_3)$_3$-N(C_4H_9-t)-BCl-N(C_4H_9-t)-GeCl(CH_3)$_2$	B: B Comp.SVol.4/4-63
$BC_{13}Cl_2H_{17}$	t-C_4H_9-CH=C($CH_2C_6H_5$)-BCl_2	B: B Comp.SVol.3/4-42
–	$C_6H_5CH_2$-BCl-CH=CCl-C_4H_9-t	B: B Comp.SVol.3/4-44
$BC_{13}Cl_2H_{22}N$	[H_2N(i-C_3H_7)$_2$][1,1-Cl_2-1-BC_7H_6]	B: B Comp.SVol.4/4-45
$BC_{13}Cl_2H_{24}N$	(CH_3)$_3$N · BCl_2-C_5(CH_3)$_5$	B: B Comp.SVol.4/4-70
$BC_{13}Cl_2H_{27}HgN_2$	2,2,6,6-(CH_3)$_4$-NC_5H_6-1-BCl-N(C_4H_9-t)-HgCl	B: B Comp.SVol.4/4-63
$BC_{13}Cl_2H_{27}N_2Pd$	[-C(CH_3)$_2$-$CH_2CH_2CH_2$-C(CH_3)$_2$-]N[-BCl -N(C_4H_9-t)-PdCl-]	B: B Comp.SVol.4/4-65
$BC_{13}Cl_2H_{27}N_2S$	2,2,6,6-(CH_3)$_4$-NC_5H_6-1-BCl-N(C_4H_9-t)-SCl	B: B Comp.SVol.4/4-63
$BC_{13}Cl_2H_{27}N_2S_2$	2,2,6,6-(CH_3)$_4$-NC_5H_6-1-BCl-N(C_4H_9-t)-SS-Cl	B: B Comp.SVol.4/4-63
$BC_{13}Cl_2H_{31}N_2Si$	(i-C_3H_7)$_2$N-BCl-N(C_4H_9-t)-SiCl(CH_3)-C_2H_5	B: B Comp.SVol.4/4-63
$BC_{13}Cl_2H_{33}N_2SiSn$	Si(CH_3)$_3$-N(C_4H_9-t)-BCl-N(C_4H_9-t)-SnCl(CH_3)$_2$	B: B Comp.SVol.4/4-63 Sn: Org.Comp.19-117/8

$BC_{13}Cl_2H_{33}N_2Si_2$ $Si(CH_3)_3$-$N(C_4H_9$-t)-BCl-$N(C_4H_9$-t)-$SiCl(CH_3)_2$ B: B Comp.SVol.4/4-63

$BC_{13}Cl_3F_3H_{27}O_3SSi$ $(n$-$C_4H_9)_3SiO(BCl_3)SO_2CF_3$ . . . . B: B Comp.SVol.3/4-32

$BC_{13}Cl_3H_{11}N$ . . . . C_6H_5-CH=N-C_6H_5 · BCl_3 . . . . B: B Comp.SVol.3/4-32

$BC_{13}Cl_3H_{11}NO$ . . . 4-(C_6H_5-CO)-$C_6H_4NH_2$-BCl_3 . . . . B: B Comp.SVol.3/4-32

$BC_{13}Cl_3H_{27}N_2P$ . . 2,2,6,6-$(CH_3)_4$-NC_5H_6-1-BCl-$N(C_4H_9$-t)-PCl_2 B: B Comp.SVol.4/4-63

$BC_{13}Cl_3H_{27}N_2Sb$. [-$C(CH_3)_2$-$CH_2CH_2CH_2$-$C(CH_3)_2$-]N[-BCl -$N(C_4H_9$-t)-$SbCl_2$-] . . . . B: B Comp.SVol.4/4-65

$BC_{13}Cl_3H_{28}N_2Si$ . . 1-[Cl_3Si-$N(C_4H_9$-t)-BH]-2,2,6,6-$(CH_3)_4$-NC_5H_6 B: B Comp.SVol.4/3a-225

$BC_{13}Cl_4H_{27}N_2Si$ . . 2,2,6,6-$(CH_3)_4$-NC_5H_6-1-BCl-$N(C_4H_9$-t)-$SiCl_3$ B: B Comp.SVol.4/4-63

$BC_{13}Cl_4H_{27}N_2Sn$. [-$C(CH_3)_2$-$CH_2CH_2CH_2$-$C(CH_3)_2$-]N[-BCl -$N(C_4H_9$-t)-$SnCl_3$-] . . . . B: B Comp.SVol.4/4-65

$BC_{13}Cl_4H_{27}N_2Ti$ . . [-$C(CH_3)_2$-$CH_2CH_2CH_2$-$C(CH_3)_2$-]N[-BCl -$N(C_4H_9$-t)-$TiCl_3$-]. . . . . B: B Comp.SVol.4/4-65

$BC_{13}Cl_5H_{27}N_2Nb$ [-$C(CH_3)_2$-$CH_2CH_2CH_2$-$C(CH_3)_2$-]N[-BCl -$N(C_4H_9$-t)-$NbCl_4$-] . . . . B: B Comp.SVol.4/4-65

$BC_{13}Cl_5H_{27}N_2Ta$. . [-$C(CH_3)_2$-$CH_2CH_2CH_2$-$C(CH_3)_2$-]N[-BCl -$N(C_4H_9$-t)-$TaCl_4$-] . . . . B: B Comp.SVol.4/4-65

$BC_{13}Cl_6H_{27}N_2Si_2$ 2,2,6,6-$(CH_3)_4$-NC_5H_6-1-BCl-$N(C_4H_9$-t)-Si_2Cl_5 B: B Comp.SVol.4/4-63

$BC_{13}CoH_{10}O_2$ . . . . $(CO)_2Co(C_5H_5BC_6H_5)$ . . . . B: B Comp.SVol.4/4-180

$BC_{13}FH_{20}NO_6Re$. . [$N(CH_3)_4$][fac-$(CO)_3Re(CH_2COCOCH_3)(CH_3CO)BF$] Re: Org.Comp.2-418/9

$BC_{13}FH_{28}N_2$ . . . . . 2,2,6,6-$(CH_3)_4$-NC_5H_6-1-BF-$N(C_2H_5)_2$ . . . . B: B Comp.SVol.3/3-376

– . . . . 2,2,6,6-$(CH_3)_4$-NC_5H_6-1-BF-NH-C_4H_9-t . . . . . B: B Comp.SVol.4/3b-246

$BC_{13}FH_{29}NSi$ . . . . 1-[$(CH_3)_3Si$-CH_2-BF]-NC_5H_6-2,2,6,6-$(CH_3)_4$ . . B: B Comp.SVol.4/3b-245

$BC_{13}FH_{32}N_2Si$. . . . $(CH_3)_3Si$-$N(C_4H_9$-t)-BF-$N(C_3H_7$-i$)_2$ . . . . B: B Comp.SVol.4/3b-246

$BC_{13}FH_{36}N_2Si_3$. . . $(CH_3)_3Si$-$N(C_4H_9$-t)-BF-N[$Si(CH_3)_3]_2$ . . . . B: B Comp.SVol.4/3b-246

$BC_{13}FH_{39}NSi_5$. . . . [$(CH_3)_3Si]_3Si$-BF-$N(CH_3)$-$Si(CH_3)_3$ . . . . B: B Comp.SVol.4/3b-245

$BC_{13}F_2H_7O_2$ . . . . . $C_{13}H_7$[-O-BF_2-O-]. . . . . B: B Comp.SVol.4/3b-234

$BC_{13}F_2H_{13}N_2Na_2O_6S_2$ $Na_2[(CH_3)_4BC_9F_2HN_2(SO_3)_2]$ . . . . B: B Comp.SVol.4/3b-250

$BC_{13}F_2H_{14}N_2NaO_3S$ $Na[(CH_3)_4BC_9F_2H_2N_2(SO_3)]$. . . . . B: B Comp.SVol.4/3b-250

$BC_{13}F_2H_{15}N_2$ . . . . $N_2BC_9H_3(F)_2(CH_3)_4$ . . . . B: B Comp.SVol.3/3-378; B: B Comp.SVol.4/3b-250

$BC_{13}F_3H_{27}N_2Sb$ . . 2,2,6,6-$(CH_3)_4$-NC_5H_6-1-BF-$N(C_4H_9$-t)-SbF_2 . B: B Comp.SVol.4/3b-246

$BC_{13}F_3H_{28}N_2O_3S$ [(i-$C_3H_7)_2N]_2B$-O-S(=O$)_2$-CF_3 . . . . B: B Comp.SVol.4/3b-51; B: B Comp.SVol.4/4-167

$BC_{13}F_4FeH_{11}O_2$ . . [$(C_5H_5)Fe(CO)_2(CH_2$=C=CH-CC-$CH_3)$][BF_4] . . . Fe: Org.Comp.B17-112, 117

– . . . . [$(C_9H_7)Fe(CO)_2(CH_2$=$CH_2)$][BF_4] . . . . Fe: Org.Comp.B17-133, 136

$BC_{13}F_4FeH_{11}O_3$ . . [(5-CH_3-$C_9H_8)Fe(CO)_3$][BF_4]. . . . . Fe: Org.Comp.B15-254

– . . . . [(9-CH_3-$C_9H_8)Fe(CO)_3$][BF_4]. . . . . Fe: Org.Comp.B15-254

– . . . . [$(C_{10}H_{11})Fe(CO)_3$][BF_4] . . . . Fe: Org.Comp.B15-111

$BC_{13}F_4FeH_{11}O_4$ . . [(1,4-$(HOOC)_2$-$C_6H_4)Fe(C_5H_5)$][BF_4]. . . . . Fe: Org.Comp.B19-16

$BC_{13}F_4FeH_{12}NO_2$ [$(C_5H_5)Fe(CO)_2CH_2$-1-NC_5H_5][BF_4] . . . . Fe: Org.Comp.B14-136

– . . . . [$(C_5H_5)Fe(CO)_2CH_2$-2-NC_5H_5][BF_4] . . . . Fe: Org.Comp.B14-151

– . . . . [$(C_5H_5)Fe(CO)_2CH_2$-3-NC_5H_5][BF_4] . . . . Fe: Org.Comp.B14-151, 156

$BC_{13}F_4FeH_{13}$ .... $[(c-C_8H_8)Fe(C_5H_5)][BF_4]$ .... Fe: Org.Comp.B19-340, 342, 345

$BC_{13}F_4FeH_{13}O$ ... $[C_5H_5Fe(C_7H_8)CO][BF_4]$ .... Fe: Org.Comp.B17-188, 192

$BC_{13}F_4FeH_{13}O_2$ .. $[(4-CH_3-C_6H_4-COOH)Fe(C_5H_5)][BF_4]$ .... Fe: Org.Comp.B19-23

– .... $[(CH_3-OOC-C_6H_5)Fe(C_5H_5)][BF_4]$ .... Fe: Org.Comp.B18-142/6, 200, 201, 211

– .... $[(C_5H_5)Fe(CO)_2(C_6H_8-c)][BF_4]$ .... Fe: Org.Comp.B17-90

$BC_{13}F_4FeH_{13}O_3$ .. $[(C_{10}H_{13})Fe(CO)_3][BF_4]$ .... Fe: Org.Comp.B15-124, 176/7

$BC_{13}F_4FeH_{13}O_4$ .. $[(CH_3O-C_9H_{10})Fe(CO)_3][BF_4]$ .... Fe: Org.Comp.B15-127, 179

$BC_{13}F_4FeH_{13}O_5$ .. $[(1-CH_3OC(O)-CH(CH_3)-C_6H_6)Fe(CO)_3][BF_4]$ .. Fe: Org.Comp.B15-100, 136

– .... $[(1-CH_3OC(O)-CH_2-C_7H_8)Fe(CO)_3][BF_4]$ ..... Fe: Org.Comp.B15-205, 222

$BC_{13}F_4FeH_{13}O_6$ .. $[(1-CH_3OC(O)-CH_2-4-CH_3O-C_6H_5)Fe(CO)_3][BF_4]$ Fe: Org.Comp.B15-115, 171

$BC_{13}F_4FeH_{14}NO$ .. $[(CH_3-C(=O)NH-C_6H_5)Fe(C_5H_5)][BF_4]$ .... Fe: Org.Comp.B18-142/6, 261

– .... $[(C_6H_6)Fe(C_5H_4-NH-C(=O)CH_3)][BF_4]$ .... Fe: Org.Comp.B18-142/6, 168

$BC_{13}F_4FeH_{15}$ .... $[(1,2-(CH_3)_2-C_6H_4)Fe(C_5H_5)][BF_4]$ .... Fe: Org.Comp.B19-1, 4, 5, 7/8

– .... $[(1,3-(CH_3)_2-C_6H_4)Fe(C_5H_5)][BF_4]$ .... Fe: Org.Comp.B19-1, 4/5, 9/10

– .... $[(1,4-(CH_3)_2-C_6H_4)Fe(C_5H_5)][BF_4]$ .... Fe: Org.Comp.B19-1, 4/5, 11/2, 81/7

– .... $[(C_2H_5-C_6H_5)Fe(C_5H_5)][BF_4]$ .... Fe: Org.Comp.B18-142/6, 200, 201, 213

– .... $[(C_6H_6)Fe(C_5H_4-C_2H_5)][BF_4]$ .... Fe: Org.Comp.B18-142/6, 151, 163, 187

$BC_{13}F_4FeH_{15}O$ ... $[(2-CH_3-C_6H_4-OCH_3)Fe(C_5H_5)][BF_4]$ .... Fe: Org.Comp.B19-1, 5, 46

– .... $[(3-CH_3-C_6H_4-OCH_3)Fe(C_5H_5)][BF_4]$ .... Fe: Org.Comp.B19-1, 5, 47

– .... $[(4-CH_3-C_6H_4-OCH_3)Fe(C_5H_5)][BF_4]$ .... Fe: Org.Comp.B19-1/2, 5, 47

– .... $[(C_2H_5-O-C_6H_5)Fe(C_5H_5)][BF_4]$ .... Fe: Org.Comp.B18-142/6, 197, 198, 250

– .... $[(C_6H_6)Fe(C_5H_4-O-C_2H_5)][BF_4]$ .... Fe: Org.Comp.B18-142/6, 153, 169

– .... $[(C_6H_7)Fe(C_6H_8)(CO)][BF_4]$ .... Fe: Org.Comp.B17-244

– .... $[(C_7H_9)Fe(C_4H_3-CH_3)(CO)][BF_4]$ .... Fe: Org.Comp.B17-244

$BC_{13}F_4FeH_{15}O_2$ .. $[(1,4-(CH_3O)_2-C_6H_4)Fe(C_5H_5)][BF_4]$ .... Fe: Org.Comp.B19-1, 5, 19, 87

– .... $[(C_5H_5)Fe(CO)_2(CH_2=CHCH_2CH_2CH=CH_2)][BF_4]$ Fe: Org.Comp.B17-40

– .... $[(C_5H_5)Fe(CO)_2(CH_2=C_5H_8-c)][BF_4]$ .... Fe: Org.Comp.B17-70

– .... $[(C_5H_5)Fe(CO)_2(C_2H_5-CC-C_2H_5)][BF_4]$ .... Fe: Org.Comp.B17-122/3

– .... $[(C_5H_5)Fe(CO)_2(C_6H_{10}-c)][BF_4]$ .... Fe: Org.Comp.B17-90, 104

– .... $[(C_5H_5)Fe(CO)_2(H-CC-C_4H_9-n)][BF_4]$ .... Fe: Org.Comp.B17-121

$BC_{13}F_4FeH_{15}O_3$ .. $[(1-CH_3-4-CH_2=C(CH_3)-C_6H_7)Fe(CO)_3][BF_4]$ .. Fe: Org.Comp.B15-258

– .... $[(1-(i-C_3H_7)-4-CH_3-C_6H_5)Fe(CO)_3][BF_4]$ ..... Fe: Org.Comp.B15-122, 175

– .... $[(2,4,6,6-(CH_3)_4C_6H_3)Fe(CO)_3][BF_4]$ .... Fe: Org.Comp.B15-91

– .... $[(2-(n-C_4H_9)-C_6H_6)Fe(CO)_3][BF_4]$ .... Fe: Org.Comp.B15-104

– .... $[(3-(n-C_4H_9)-C_6H_6)Fe(CO)_3][BF_4]$ .... Fe: Org.Comp.B15-106

– .... $[(CH_2(CH)_4CH_2CH_2CH_2CH=CH_2)Fe(CO)_3][BF_4]$ Fe: Org.Comp.B15-25

– .... $[((CH_3)_5C_5)Fe(CO)_3][BF_4]$ .... Fe: Org.Comp.B15-52/3

– .... $[(C_5H_5)Fe(CO)_2(2,5-(CH_3)_2-OC_4H_4)][BF_4]$ ..... Fe: Org.Comp.B17-94/5

– .... $[(C_5H_5)Fe(CO)_2(2-C_2H_5-OC_4H_5)][BF_4]$ .... Fe: Org.Comp.B17-94

$BC_{13}F_4FeH_{15}O_3$ .. $[(C_5H_5)Fe(CO)_2(5,5\text{-}(CH_3)_2\text{-}OC_4H_4)][BF_4]$ ..... Fe: Org.Comp.B17-95

– .... $[(C_5H_5)Fe(CO)_2(CH_2{=}C{=}CHC(CH_3)_2OH)][BF_4]$ .. Fe: Org.Comp.B17-112

– .... $[(C_5H_5)Fe(CO)_2(CH_2{=}CHCH_2CH_2\text{-}C({=}O)CH_3)][BF_4]$ Fe: Org.Comp.B17-35

$BC_{13}F_4FeH_{15}O_4$ .. $[(C_5H_5)Fe(CO)_2(2\text{-}CH_2{=}CHCH_2\text{-}1,3\text{-}O_2C_3H_5)][BF_4]$ Fe: Org.Comp.B17-29

– .... $[(C_5H_5)Fe(CO)_2(5,6\text{-}(CH_3)_2\text{-}1,4\text{-}O_2C_4H_4)][BF_4]$ Fe: Org.Comp.B17-96, 105/6

– .... $[(C_5H_5)Fe(CO)_2(CH_3CH{=}CH\text{-}COO\text{-}C_2H_5)][BF_4]$ Fe: Org.Comp.B17-59

$BC_{13}F_4FeH_{17}O_2$ .. $[(C_5(CH_3)_5)(CO)_2Fe{=}CH_2][BF_4]$ .... Fe: Org.Comp.B16a-103/4

– .... $[(C_5H_5)Fe(CO){=}C(O\text{-}C_2H_5)CH_2CH_2CH{=}CH_2][BF_4]$ Fe: Org.Comp.B17-167

– .... $[(C_5H_5)Fe(CO)_2(CH_2{=}CH\text{-}C_4H_9\text{-}n)][BF_4]$ ...... Fe: Org.Comp.B17-13, 47

– .... $[(C_5H_5)Fe(CO)_2(CH_2{=}CH\text{-}CH(CH_3)C_2H_5)][BF_4]$ Fe: Org.Comp.B17-13

– .... $[(C_5H_5)Fe(CO)_2(CH_2{=}CH\text{-}C_4H_9\text{-}t)][BF_4]$ ...... Fe: Org.Comp.B17-12

$BC_{13}F_4FeH_{17}O_4$ .. $[(C_5H_5)Fe(CO)_2(C_2H_5OCH{=}CHOC_2H_5)][BF_4]$ ... Fe: Org.Comp.B17-69/70, 81/2

$BC_{13}F_4FeH_{19}O_2Si$ $[((CH_3)_3SiC_5H_3CH_3)Fe(CO)_2(CH_2{=}CH_2)][BF_4]$ .. Fe: Org.Comp.B17-132/3

$BC_{13}F_4FeH_{19}S$ ... $[C_5H_5Fe(C_6H_8)S(CH_3)_2][BF_4]$ .... Fe: Org.Comp.B17-190, 196

$BC_{13}F_4FeH_{19}Se$ .. $[C_5H_5Fe(C_6H_8)Se(CH_3)_2][BF_4]$ .... Fe: Org.Comp.B17-190, 196

$BC_{13}F_4FeH_{19}Te$ .. $[C_5H_5Fe(C_6H_8)Te(CH_3)_2][BF_4]$ .... Fe: Org.Comp.B17-190, 196

$BC_{13}F_4FeH_{20}NO_2$ $[C_5H_5Fe(CO)_2CH_2CH(CH_3)N(CH_3)_3][BF_4]$ ..... Fe: Org.Comp.B14-150

$BC_{13}F_4FeH_{20}O_4P$ $[C_5H_5(CO)(P(OCH_3)_3)Fe{=}C{=}C(CH_3)_2][BF_4]$ .... Fe: Org.Comp.B16a-208, 211

$BC_{13}F_4FeH_{20}O_5P$ $[C_5H_5Fe(CO)_2C(CH_3)_2P(OCH_3)_3][BF_4]$ .... Fe: Org.Comp.B14-138, 143/4

$BC_{13}F_4FeH_{21}O_2$ .. $[(C_5H_5)Fe(CH_2{=}CHCH_3)(CO)(C_2H_5OC_2H_5)][BF_4]$ Fe: Org.Comp.B16b-82, 89/90

$BC_{13}F_4FeH_{27}O_6P_2$ $[(C_5H_5)Fe(CH_2{=}CH_2)(P(OCH_3)_3)_2][BF_4]$ .... Fe: Org.Comp.B16b-4/5, 8/9

$BC_{13}F_4H_8O_6Re$ ... $[(CO)_5ReOC(CH_3)C_6H_5][BF_4]$ .... Re: Org.Comp.2-157

$BC_{13}F_4H_{10}N_2O_2Re$ $[(C_5H_5)Re(CO)_2\text{-}N{=}N\text{-}C_6H_5][BF_4]$ .... Re: Org.Comp.3-193/7, 203

$BC_{13}F_4H_{10}N_2O_4Re$ $[(CO)_3Re(NC_5H_4\text{-}C_5H_4N)(H_2O)][BF_4]$ .... Re: Org.Comp.1-301

$BC_{13}F_4H_{12}O_5Re$ .. $[(CO)_5Re(CH_2{=}C_4H(CH_3)_3)][BF_4]$ .... Re: Org.Comp.2-351, 353/4

– .... $[(CO)_5Re(C_8H_{12}\text{-}c)][BF_4]$ .... Re: Org.Comp.2-351, 353

$BC_{13}F_4H_{13}MoO_2$ .. $[(C_5H_5)Mo(CO)_2(C_6H_8)][BF_4]$ .... Mo: Org.Comp.8-305/6, 313, 317/8

$BC_{13}F_4H_{15}MoO_2$ .. $[(C_5H_5)Mo(CO)_2((CH_2)_2CC(CH_3)_2)][BF_4]$ ...... Mo: Org.Comp.8-305/6, 311

– .... $[(C_5H_5)Mo(CO)_2(CH_2{=}CH\text{-}C(CH_3){=}CH\text{-}CH_3)][BF_4]$ Mo: Org.Comp.8-305/6, 309

– .... $[(C_5H_5)Mo(CO)_2(CH_3\text{-}CHC(CH_2)CH\text{-}CH_3)][BF_4]$ Mo: Org.Comp.8-305/6, 311, 312

$BC_{13}F_4H_{19}MoN_2O_2$ $[C_5H_5Mo(NO)_2(C_8H_{14}\text{-}c)][BF_4]$ .... Mo: Org.Comp.6-156

$BC_{13}F_4H_{20}MoO_2P$ $[(C_5H_5)Mo(CO)_2(P(C_2H_5)_3)\text{-}FBF_3]$ .... Mo: Org.Comp.7-57, 63, 103

$BC_{13}F_4H_{20}MoO_3P$ $[(C_5H_5)Mo(CO)_2(P(CH_3)_3)(O{=}C(CH_3)_2)][BF_4]$ ... Mo: Org.Comp.7-283, 294

$BC_{13}F_4H_{22}MoO_3P$ $[(C_5H_5)Mo(CO)_2(P(C_2H_5)_3)(H_2O)][BF_4]$ .... Mo: Org.Comp.7-283, 294

$BC_{13}F_4H_{23}MoO_2P_2$ $[(C_5H_5)Mo(CO)_2(P(CH_3)_3)_2][BF_4]$ .... Mo: Org.Comp.7-283, 286

$BC_{13}F_4H_{27}N_2Si$ ... 2,2,6,6-$(CH_3)_4\text{-}NC_5H_6\text{-}1\text{-}BF\text{-}N(C_4H_9\text{-}t)\text{-}SiF_3$ .. B: B Comp.SVol.4/3b-246

$BC_{13}F_6FeH_{16}P$ ... $[(CH_3C_6H_5)Fe(BC_5H_5\text{-}1\text{-}CH_3)][PF_6]$ .... Fe: Org.Comp.B18-142/6, 209

$BC_{13}F_{10}FeH_{13}P_2$ . . $[(C_5H_5)Fe(CH_2{=}CHC_6H_5)(PF_3)_2][BF_4]$ . . . . Fe: Org.Comp.B16b-7
$BC_{13}F_{10}FeH_{21}P_2$ . . $[(C_5H_5)Fe(CH_2{=}CHC_6H_{13}\text{-}n)(PF_3)_2][BF_4]$ . . . . . . Fe: Org.Comp.B16b-6
$BC_{13}FeH_9O_3$ . . . . . $(CO)_3Fe(C_4H_4BC_6H_5)$ . . . . B: B Comp.SVol.4/4-180
$BC_{13}FeH_{16}^+$ . . . . . $[(CH_3C_6H_5)Fe(BC_5H_5\text{-}1\text{-}CH_3)]^+$ . . . . Fe: Org.Comp.B18-142/6, 209
$BC_{13}FeH_{18}NO_2$ . . . $(C_5H_5)Fe[C_5H_3(B(OH)_2)CH_2N(CH_3)_2]$ . . . . Fe: Org.Comp.A9-287, 288
$BC_{13}FeH_{20}N$ . . . . . $[H_3B\text{-}N(CH_3)_2\text{-}CH_2\text{-}C_5H_4]Fe(C_5H_5)$ . . . . B: B Comp.SVol.4/3b-15
$BC_{13}GeH_{27}O$ . . . . . $Ge(CH_3)_3C(B(C_4H_9)_2){=}C{=}O$ . . . . Ge: Org.Comp.2-7
$BC_{13}GeH_{31}S_2$ . . . . $Ge(C_2H_5)_3CH_2CH_2CH_2B(SC_2H_5)_2$ . . . . Ge: Org.Comp.2-147, 153
$BC_{13}H_9O_3Ru$ . . . . . $(CO)_3Ru(C_4H_4BC_6H_5)$ . . . . B: B Comp.SVol.3/4-155
B: B Comp.SVol.4/4-180
$BC_{13}H_{11}NS$ . . . . . . 1,3,2-$SNBC_6H_4\text{-}3\text{-}CH_2\text{-}C_6H_5$ . . . . B: B Comp.SVol.4/4-146
$BC_{13}H_{11}N_2O_2$ . . . . $O_2C_7H_5\text{-}N_2BC_6H_6$ . . . . B: B Comp.SVol.3/3-117
– . . . . $O_2N_2BC_7H_6(C_6H_5)$ . . . . B: B Comp.SVol.3/3-117
$BC_{13}H_{11}O_2S$ . . . . . $C_6H_5\text{-}O\text{-}B[\text{-}O\text{-}CH_2\text{-}(1,2\text{-}C_6H_4)\text{-}S\text{-}]$ . . . . B: B Comp.SVol.3/4-120
$BC_{13}H_{11}O_2S_2$ . . . . $C_6H_5\text{-}S\text{-}B[\text{-}O\text{-}(1,2\text{-}C_6H_3\text{-}4\text{-}OCH_3)\text{-}S\text{-}]$ . . . . . . B: B Comp.SVol.3/4-120/1
$BC_{13}H_{11}O_3S$ . . . . . $C_6H_5\text{-}O\text{-}B[\text{-}O\text{-}(1,2\text{-}C_6H_3\text{-}4\text{-}OCH_3)\text{-}S\text{-}]$ . . . . . . B: B Comp.SVol.3/4-120
$BC_{13}H_{11}S_2$ . . . . . . $CH_3B[\text{-}S\text{-}(1,2\text{-}(C_6H_3\text{-}5\text{-}C_6H_5))\text{-}S\text{-}]$ . . . . B: B Comp.SVol.3/4-114
– . . . . $C_6H_5\text{-}B[\text{-}S\text{-}(1,2\text{-}(C_6H_3\text{-}5\text{-}CH_3))\text{-}S\text{-}]$ . . . . B: B Comp.SVol.3/4-114
$BC_{13}H_{12}N$ . . . . $CH_3B[\text{-}(1,2\text{-}C_6H_4)\text{-}(1,2\text{-}C_6H_4)\text{-}NH\text{-}]$ . . . . B: B Comp.SVol.3/3-163
$BC_{13}H_{12}NO$ . . . . . . $C_6H_5\text{-}B[\text{-}NH\text{-}(1,2\text{-}C_6H_4)\text{-}CH_2\text{-}O\text{-}]$ . . . . B: B Comp.SVol.3/3-164
– . . . . $HB[\text{-}N(CH_2C_6H_5)\text{-}(1,2\text{-}C_6H_4)\text{-}O\text{-}]$ . . . . B: B Comp.SVol.3/3-164
$BC_{13}H_{12}NO_8$ . . . . . $C_7H_7NH[B(\text{-}O\text{-}CO\text{-}CH_2\text{-}CO\text{-}O\text{-})_2]$ . . . . B: B Comp.SVol.3/3-227
$BC_{13}H_{12}NS$ . . . . . . 1-C_6H_5-4-CH_3-5,2,1-$SNBC_6H_4$ . . . . B: B Comp.SVol.3/4-130
– . . . . 2-C_6H_5-1,3-SNC_7H_4 · BH_3 . . . . B: B Comp.SVol.4/4-147/8
$BC_{13}H_{13}N_4$ . . . . . . $[\text{-}B(C_6H_5)\text{-}N(CH_3)\text{-}N{=}N\text{-}N(C_6H_5)\text{-}]$ . . . . B: B Comp.SVol.4/3a-199
$BC_{13}H_{14}N$ . . . . $(C_6H_5)_2B\text{-}NH\text{-}CH_3$ . . . . B: B Comp.SVol.4/3a-231
$BC_{13}H_{14}NS$ . . . . . . 2-C_6H_5-1,3-SNC_7H_6 · BH_3 . . . . B: B Comp.SVol.4/3b-7
B: B Comp.SVol.4/4-147/8
$BC_{13}H_{16}N$ . . . . $H_3B\text{-}N(C_6H_5)_2\text{-}CH_3$ . . . . B: B Comp.SVol.4/3b-10
– . . . . $H_3B\text{-}NH(C_6H_5)\text{-}CH_2\text{-}C_6H_5$ . . . . B: B Comp.SVol.4/3b-5, 10
– . . . . $H_3B\text{-}NH_2\text{-}CH(C_6H_5)_2$ . . . . B: B Comp.SVol.4/3b-10
$BC_{13}H_{16}N_3$ . . . . . . $CH_3(C_6H_5)CH\text{-}NH_2\text{-}BH(CN)(C_4H_4N)$ . . . . B: B Comp.SVol.3/3-190
$BC_{13}H_{18}N$ . . . . $C_6H_5\text{-}CC\text{-}B(CH_3)\text{-}NH\text{-}C_4H_9\text{-}t$ . . . . B: B Comp.SVol.4/3a-234
$BC_{13}H_{18}NO$ . . . . . . 1-C_6H_5-2,9,1-$ONBC_7H_{13}$ . . . . B: B Comp.SVol.4/3b-66
$BC_{13}H_{18}NO_2$ . . . . . $[\text{-}B(C_2H_5)_2\text{-}O\text{-}C({=}O)\text{-}CH_2\text{-}N({=}CH\text{-}C_6H_5)\text{-}]$ . . . . B: B Comp.SVol.4/3b-70
$BC_{13}H_{19}NO_2$ . . . . . $C_6H_5\text{-}B[\text{-}O\text{-}CH_2CH_2\text{-}O\text{-}]$ · NC_5H_{10} . . . . B: B Comp.SVol.4/3b-69
$BC_{13}H_{19}N_4O$ . . . . . $[O(CH_2CH_2)_2NH_2][BH(NC_4H_4)_2CN]$ . . . . B: B Comp.SVol.3/3-213
$BC_{13}H_{20}N$ . . . . 7-$(i\text{-}C_3H_7)_2N$-7-BC_7H_6 . . . . B: B Comp.SVol.4/3a-229
– . . . . $(\text{-}CH{=}CH\text{-}CH_2\text{-}CH_2\text{-})B(C_6H_5)\text{-}N(CH_3)_3$ . . . . B: B Comp.SVol.4/3b-29
– . . . . $(\text{-}CH_2\text{-}CH{=}CH\text{-}CH_2\text{-})B(C_6H_5)\text{-}N(CH_3)_3$ . . . . B: B Comp.SVol.3/3-192
B: B Comp.SVol.4/3b-29
– . . . . $t\text{-}C_4H_9\text{-}B{=}N\text{-}C_6H_2\text{-}2,4,6\text{-}(CH_3)_3$ . . . . B: B Comp.SVol.4/3a-214
– . . . . [3.3.1]-9-BC_8H_{15} · NC_5H_5 . . . . B: B Comp.SVol.3/3-189
B: B Comp.SVol.4/3b-29
$BC_{13}H_{20}NO$ . . . . . . $C_6H_5\text{-}B(OCH_3)\text{-}N[\text{-}CHCH_3\text{-}(CH_2)_4\text{-}]$ . . . . B: B Comp.SVol.3/3-156
– . . . . $C_6H_5\text{-}B(OCH_3)\text{-}N[\text{-}CH_2\text{-}CHCH_3\text{-}(CH_2)_3\text{-}]$ . . . . . B: B Comp.SVol.3/3-156
$BC_{13}H_{20}NO_2$ . . . . . $[\text{-}B(C_2H_5)_2\text{-}O\text{-}C({=}O)\text{-}CH(CH_2\text{-}C_6H_5)\text{-}NH_2\text{-}]$ . . . B: B Comp.SVol.4/3b-70
$BC_{13}H_{20}N_5$ . . . . . . $BC_6H_6N_3({=}N\text{-}CN)(C_3H_7\text{-}i)_2$ . . . . B: B Comp.SVol.4/3a-257
$BC_{13}H_{21}N_2$ . . . . . . $C_6H_5B[N(CH_3)_2]\text{-}NC_5H_{10}$ . . . . B: B Comp.SVol.3/3-109
$BC_{13}H_{21}N_4$ . . . . . . $n\text{-}C_3H_7\text{-}B[\text{-}N(C_4H_9\text{-}t)\text{-}N{=}N\text{-}N(C_6H_5)\text{-}]$ . . . . B: B Comp.SVol.3/3-137/8

Formula	Compound	Reference
$BC_{13}H_{21}N_4$	$i-C_3H_7-B[-N(C_4H_9-t)-N=N-N(C_6H_5)-]$	B: B Comp.SVol.3/3-137/8
$BC_{13}H_{22}N$	$2-i-C_3H_7-C_5H_4N-BH[C_5H_{10}]$	B: B Comp.SVol.3/3-189
$BC_{13}H_{22}NO$	$CH_3O-B(C_6H_5)-N(C_3H_7-i)_2$	B: B Comp.SVol.3/3-154
–	$i-C_3H_7-O-B(C_6H_5)-N(C_2H_5)_2$	B: B Comp.SVol.3/3-154
$BC_{13}H_{22}N_3S_2$	$[-B(C_3H_7-n)_2-N(CH_2-CH=CH_2)-C(=S)-NH-C_3H_2NS-]$	B: B Comp.SVol.4/3a-179 B: B Comp.SVol.4/4-149/50
$BC_{13}H_{23}NO_5Re$	$[(C_2H_5)_4N][(CO)_5ReBH_3]$	Re: Org.Comp.2-168
$BC_{13}H_{23}N_2$	$2-[(n-C_4H_9)_2B-NH]-NC_5H_4$	B: B Comp.SVol.4/3a-232
$BC_{13}H_{23}N_6$	$[-NCH_3-(CH_2)_2-NCH_3-]B[-(CH_3)C_3H_2N_2-H$ $-C_3HN_2(CH_3)_2-]$	B: B Comp.SVol.3/3-201
$BC_{13}H_{24}I$	$n-C_4H_9-CC-(CH_2)_6-BI-CH_3$	B: B Comp.SVol.4/4-113
$BC_{13}H_{24}N$	$1-(t-C_4H_9)-2-CH_3-3-(CH_2=CHCH_2CH_2CH_2)-1,2-NBC_3H_3$	B: B Comp.SVol.4/3a-249
–	$(CH_3)_2N[-CH_2CH_2-BC_9H_{14}-]$	B: B Comp.SVol.4/3a-259
$BC_{13}H_{24}NO$	$(C_{10}H_{15})(CH_3)B(-NH_2-CH_2-CH_2-O-)$	B: B Comp.SVol.3/3-217
$BC_{13}H_{24}NSi_2$	$C_6H_5(CH_3)B-N[-Si(CH_3)_2-CH_2-CH_2-Si(CH_3)_2-]$	B: B Comp.SVol.3/3-156
$BC_{13}H_{24}N_3OS$	$BC_4H_3N_3S(=O)(CH_3)(C_4H_9-n)_2$	B: B Comp.SVol.4/3a-180
$BC_{13}H_{25}N_2$	$N(-CH_2-CH_2-)_3N-BH_2(C_7H_{11})$	B: B Comp.SVol.3/3-188
$BC_{13}H_{25}N_2Si$	$C_6H_5B[NHSi(CH_3)_3]N(C_2H_5)_2$	B: B Comp.SVol.3/3-108
$BC_{13}H_{25}N_2Si_2$	$1,3-[(CH_3)_2SiH]_2-2-C_6H_5-1,3,2-N_2BC_3H_6$	B: B Comp.SVol.4/3a-175
$BC_{13}H_{25}O_2S$	$2-[CH_2=CHCH(S-C_4H_9-t)]-4,4,5,5-(CH_3)_4-1,3,2-O_2BC_2$	B: B Comp.SVol.4/4-138
$BC_{13}H_{26}LiN_2$	$[(C_6H_5CH_2BH_3)(CH_3)_2N-CH_2-CH_2-N(CH_3)_2]Li$	B: B Comp.SVol.3/3-213
$BC_{13}H_{26}N$	$9-(t-C_4H_9-CH_2-NH)-[3.3.1]-9-BC_8H_{14}$	B: B Comp.SVol.4/3a-235
$BC_{13}H_{27}NNaSi$	$Na[-B(C_2H_5)_2-NH-Si(CH_3)_2-C(C(CH_3)=CH_2)=C(C_2H_5)-]$	B: B Comp.SVol.4/3b-48
$BC_{13}H_{27}N_2$	$1-(t-C_4H_9-N=B)-2,2,6,6-(CH_3)_4-NC_5H_6$	B: B Comp.SVol.3/3-100 B: B Comp.SVol.4/3a-160/5
$BC_{13}H_{27}N_2O_2S$	$2,2,6,6-(CH_3)_4-NC_5H_6-1-B[-N(C_4H_9-t)-S(=O)-O-]$	B: B Comp.SVol.4/3b-55
$BC_{13}H_{28}IN_2$	$(C_2H_5)_2N-BI-NC_5H_6(CH_3)_4$	B: B Comp.SVol.3/4-101
$BC_{13}H_{28}N$	$1-(t-C_4H_9-BH)-2,2,6,6-(CH_3)_4-NC_5H_6$	B: B Comp.SVol.4/3a-225
$BC_{13}H_{28}N_2^+$	$[(C_2H_5)_2N=B=N(-C(CH_3)_2(CH_2)_3-C(CH_3)_2-)]^+$	B: B Comp.SVol.3/3-205
$BC_{13}H_{28}N_3O$	$[-N(CH_3)-CH_2CH_2-N(CH_3)-]B-O$ $-1-NC_5H_6-2,2,6,6-(CH_3)_4$	B: B Comp.SVol.4/3b-53
$BC_{13}H_{29}N_2OSi$	$[(C_2H_5)_2N]_2BC[Si(CH_3)_3]=C=O$	B: B Comp.SVol.3/3-105
$BC_{13}H_{29}N_2O_2$	$t-C_4H_9OC(O)B[N(C_2H_5)_2]_2$	B: B Comp.SVol.3/3-110
$BC_{13}H_{29}N_2Sn$	$(CH_3)_3Sn-CC-B[N(C_2H_5)_2]_2$	B: B Comp.SVol.4/3a-170
$BC_{13}H_{30}N$	$(t-C_4H_9)_2B-N(CH_3)-C_4H_9-t$	B: B Comp.SVol.4/3a-231
$BC_{13}H_{30}N_3$	$1-[t-C_4H_9-NH-B(NH_2)]-2,2,6,6-(CH_3)_4-NC_5H_6$	B: B Comp.SVol.4/3a-155
$BC_{13}H_{31}N_2O_3$	$(i-C_3H_7O)_3B(-NCH_3-CH_2-CH_2-NCH_3-)$	B: B Comp.SVol.3/3-113
$BC_{13}H_{31}N_2SSeSi_2$	$[(CH_3)_3Si]_2N-S-N[-C(C_2H_5)=C(C_2H_5)-Se-B(CH_3)-]$	B: B Comp.SVol.4/3a-252
$BC_{13}H_{31}N_2Si$	$1-(t-C_4H_9)-2-CH_3-2-C_2H_5-3-(i-C_3H_7)_2N-1,2,3-NSiB$	B: B Comp.SVol.4/3a-205
$BC_{13}H_{32}NSi_2$	$(i-C_3H_7)_2N-B=C[Si(CH_3)_3]_2$	B: B Comp.SVol.4/3a-215/6
$BC_{13}H_{32}NSi_3$	$1,3,3-[(CH_3)_3Si]_3-2-CH_3-1,2-NBC_3H_2$	B: B Comp.SVol.4/3a-250
$BC_{13}H_{33}N_2O_3Si$	$(i-C_3H_7O)_3SiB[N(CH_3)_2]_2$	B: B Comp.SVol.3/3-110

$BC_{13}H_{33}N_2Si$ ..... $i-C_3H_7B[NH(i-C_3H_7)]N(t-C_4H_9)Si(CH_3)_3$ ...... B: B Comp.SVol.3/3-107
$BC_{13}H_{34}NO_7$ ..... $H_3N-BH_3 \cdot [-O(CH_2CH_2O)_5CH_2CH_2-] \cdot CH_3OH$
B: B Comp.SVol.3/3-171
$BC_{13}H_{36}NSi_4$ ..... $[(CH_3)_3Si]_3C-B=N-Si(CH_3)_3$ ..... B: B Comp.SVol.3/3-142/4
B: B Comp.SVol.4/3a-215
– ..... $[(CH_3)_3Si]_3Si-B=N-C_4H_9-t$ ..... B: B Comp.SVol.4/3a-215
$BC_{13}H_{38}NOSi_4$ ... $[(CH_3)_3Si]_3C-B(OH)-NH-Si(CH_3)_3$ ..... B: B Comp.SVol.4/3b-61
$BC_{13}H_{39}N_2Si_4$ .... $[(CH_3)_3Si]_3SiB[N(CH_3)_2]_2$ ..... B: B Comp.SVol.3/3-105
$BC_{14}ClF_4H_{14}Se$ .. $[2-Cl-C_6H_4CH_2-SeCH_3-C_6H_5][BF_4]$ ..... B: B Comp.SVol.3/4-148
– ..... $[4-Cl-C_6H_4CH_2-SeCH_3-C_6H_5][BF_4]$ ..... B: B Comp.SVol.3/4-148
$BC_{14}ClH_9NS_2$ .... $4-C_6H_5-5-Cl-2,6,4,5-S_2NBC_8H_4$ ..... B: B Comp.SVol.3/4-57
B: B Comp.SVol.4/4-62
$BC_{14}ClH_{10}MnN_6O_4$
$Mn[(6-NO_2-1,2-N_2C_7H_4-1)_2BH_2]Cl$ ..... Mn: MVol.D8-16/8
$BC_{14}ClH_{11}O_6Re$ .. $cis-(CO)_4Re[(CH_3CO)_2]B(C_6H_5)Cl$ ..... Re: Org.Comp.2-344
$BC_{14}ClH_{12}MnN_4$ .. $Mn[(1,2-N_2C_7H_5-1)_2BH_2]Cl$ ..... Mn: MVol.D8-16/8
$BC_{14}ClH_{12}NO_5Re$ $cis-(CO)_4Re[(CH_3CO)(CH_3CNH)]B(C_6H_5)Cl$ ... Re: Org.Comp.2-343
$BC_{14}ClH_{12}N_2$ .... $1-Cl-2-C_6H_5-3-CH_3-2,4,1-N_2BC_7H_4$ ..... B: B Comp.SVol.4/4-62
$BC_{14}ClH_{14}$ ...... $(2-CH_3-C_6H_4)_2BCl$ ..... B: B Comp.SVol.3/4-44
$BC_{14}ClH_{15}N$ ..... $(C_6H_5)ClBN(C_2H_5)C_6H_5$ ..... B: B Comp.SVol.3/4-56
$BC_{14}ClH_{23}N$ ..... $C_6H_5-BCl-N(C_4H_9-n)_2$ ..... B: B Comp.SVol.3/4-56
– ..... $C_6H_5-BCl-N(C_4H_9-s)_2$ ..... B: B Comp.SVol.3/4-56
$BC_{14}ClH_{23}N_2P$ ... $[-N(C_4H_9-t)-BCl-N(C_4H_9-t)-]P-C_6H_5$ ..... B: B Comp.SVol.4/4-64/5
$BC_{14}ClH_{24}$ ...... $t-C_4H_9-BCl-C_5(CH_3)_5$ ..... B: B Comp.SVol.4/4-49
$BC_{14}ClH_{26}$ ...... $(2-CH_3-c-C_6H_{10})_2BCl$ ..... B: B Comp.SVol.4/4-47
$BC_{14}ClH_{26}O_2$ .... $(n-C_3H_7)ClB[-O=C(C_4H_9-t)-CH=C(C_4H_9-t)-O-]$
B: B Comp.SVol.3/4-52
$BC_{14}ClH_{28}N_6$ .... $[((C_2H_5)_2NH)_2B(1,2-N_2C_3H_3)_2]Cl$ ..... B: B Comp.SVol.4/4-70
$BC_{14}ClH_{30}$ ...... $[(CH_3)_2CHC(CH_3)_2](n-C_8H_{17})BCl$ ..... B: B Comp.SVol.3/4-45
$BC_{14}ClH_{36}N_2Si_2$ .. $ClB[N(C_4H_9-s)_2][N(Si(CH_3)_3)_2]$ ..... B: B Comp.SVol.3/4-59
$BC_{14}ClH_{38}Si_4$ .... $[((CH_3)_3Si)_2CH]_2BCl$ ..... B: B Comp.SVol.3/4-44
$BC_{14}Cl_2H_{10}NS_2$ .. $3-(3-SC_4H_3)-4-[BCl_2-N(C_6H_5)]-SC_4H_2$ ..... B: B Comp.SVol.3/4-54
B: B Comp.SVol.4/4-59
$BC_{14}Cl_2H_{10}O_6Re$ $cis-(CO)_4Re[(CH_3CO)(CH_2(C_6H_5)CO)]BCl_2$ .... Re: Org.Comp.2-345
$BC_{14}Cl_3F_3H_{13}O_3SSi$
$(CH_3)(C_6H_5)_2SiO(BCl_3)SO_2CF_3$ ..... B: B Comp.SVol.3/4-32
$BC_{14}Cl_3H_{12}MoN_6O_2$
$(CH_2CHCH_2)Mo(CO)_2[-N_2C_3H_2(Cl)-]_3BH$ ..... Mo: Org.Comp.5-226, 227
$BC_{14}Cl_3H_{14}N_2$ .... $CH_3C(=N-C_6H_5)-NH-C_6H_5 \cdot BCl_3$ ..... B: B Comp.SVol.4/4-33
$BC_{14}Cl_3H_{30}N_2Si$ .. $2,2,6,6-(CH_3)_4-NC_5H_6-1-BCl-N(C_4H_9-t)-SiCl_2-CH_3$
B: B Comp.SVol.4/4-63
$BC_{14}CoH_{24}SeSi$ .. $BC_2CoSeSi(C_5H_5)(C_2H_5)_2(CH_3)_3$ ..... B: B Comp.SVol.4/4-180
$BC_{14}Co_2H_{18}NO_6$ .. $t-C_4H_9-B[-N(C_4H_9-t)-Co(CO)_3-Co(CO)_3-]$ .... B: B Comp.SVol.4/3a-249
$BC_{14}CrH_9O_4$ ..... $(CO)_4Cr(C_4H_4BC_6H_5)$ ..... B: B Comp.SVol.4/4-180
$BC_{14}FH_{23}N$ ...... $C_6H_5-BF-N(C_4H_9-n)_2$ ..... B: B Comp.SVol.3/3-376
– ..... $C_6H_5-BF-N(C_4H_9-s)_2$ ..... B: B Comp.SVol.3/3-376
$BC_{14}FH_{36}N_2Si_2$ ... $(CH_3)_3Si-N(C_4H_9-t)-BF-N(C_4H_9-t)-Si(CH_3)_3$ .. B: B Comp.SVol.4/3b-245
– ..... $[(CH_3)_3Si]_2N-BF-N(C_4H_9-t)_2$ ..... B: B Comp.SVol.4/3b-246
$BC_{14}FH_{39}NSi_4$ .... $[(CH_3)_3Si]_3C-BF-N(CH_3)-Si(CH_3)_3$ ..... B: B Comp.SVol.4/3b-245
$BC_{14}FH_{41}NSi_5$ .... $[(CH_3)_3Si]_3Si-BF-N(C_2H_5)-Si(CH_3)_3$ ..... B: B Comp.SVol.4/3b-245

$BC_{14}F_2FeH_{22}NO_3$ $[N(CH_3)_4][C_5H_5Fe(CH_2C_2O_2BF_2(CH_3))CO]$ .... Fe: Org.Comp.B17-156
$BC_{14}F_2H_{10}O_6Re$ .. cis-$(CO)_4Re[(CH_3CO)(CH_2(C_6H_5)CO)]BF_2$ .... Re: Org.Comp.2-345
$BC_{14}F_2H_{16}NO_3$ ... $BC_{14}F_2H_{16}NO_3$ .... B: B Comp.SVol.4/3b-234
$BC_{14}F_2H_{20}O_2$ .... $F_2B[$-O-(1,2-C_6H_2-3,6-(t-$C_4H_9)_2$)-O-], radical B: B Comp.SVol.3/3-369/70
$BC_{14}F_2H_{21}NO_6Re$ $[N(CH_3)_4]$[cis-$(CO)_4ReCH_2COCO(CH(CH_3)_2)BF_2]$
Re: Org.Comp.2-389
$BC_{14}F_3H_{19}NO_3S$ .. NC_5H_5 · [3.3.1]-9-BC_8H_{14}-9-[O-S(O)$_2$-CF_3] .. B: B Comp.SVol.4/3b-69
$BC_{14}F_3H_{23}NO_3S$ .. (n-$C_4H_9)_2$B-O-SO_2-CF_3 · NC_5H_5 .... B: B Comp.SVol.4/4-165
$BC_{14}F_3H_{28}N_2O_2$ .. $CF_3OC(O)B[N(i\text{-}C_3H_7)_2]_2$ .... B: B Comp.SVol.3/3-110
$BC_{14}F_3H_{28}N_2O_3S$ $(C_2H_5)_2$N-B[O-S(O)$_2$-CF_3]-1-NC_5H_6-2,2,6,6-$(CH_3)_4$
B: B Comp.SVol.4/3b-52
– .... t-C_4H_9-NH-B[O-S(O)$_2$-CF_3]-1-NC_5H_6-2,2,6,6-$(CH_3)_4$
B: B Comp.SVol.4/3b-52
$BC_{14}F_3H_{36}N_4P_2$ .. [-P(N$(CH_3)_2)_2$C(CH_3)P(N$(CH_3)_3)_2$C$(CH_3)(BF_3)$-] B: B Comp.SVol.4/3b-124
$BC_{14}F_4FeH_8O_7ReS_2$
$[C_5H_5(CO)_2Fe{=}C(SCH_3)SRe(CO)_5][BF_4]$ ...... Fe: Org.Comp.B16a-142, 154
$BC_{14}F_4FeH_9O_3$ ... $[(C_{11}H_9)Fe(CO)_3][BF_4]$ .... Fe: Org.Comp.B15-220
$BC_{14}F_4FeH_{11}O_2$ .. $[C_5H_5(CO)_2Fe{=}CH(C_6H_5)][BF_4]$ .... Fe: Org.Comp.B16a-97/9
$BC_{14}F_4FeH_{11}O_3$ .. $[(CH_2(CH)_4C_6H_5)Fe(CO)_3][BF_4]$ .... Fe: Org.Comp.B15-13, 21
$BC_{14}F_4FeH_{13}O_2$ .. $[(C_5H_5)Fe(CO)_2((2.2.1)\text{-}C_7H_8)][BF_4]$ .... Fe: Org.Comp.B17-98
– .... $[(C_5H_5)Fe(CO)_2((3.2.0)\text{-}C_7H_8)][BF_4]$ .... Fe: Org.Comp.B17-96
– .... $[(C_5H_5)Fe(CO)_2(C_7H_8\text{-}c)][BF_4]$ .... Fe: Org.Comp.B17-91
– .... $[(C_9H_7)Fe(CO)_2(CH_2{=}CHCH_3)][BF_4]$ .... Fe: Org.Comp.B17-133/4, 137
$BC_{14}F_4FeH_{13}O_3$ .. $[(8\text{-}C_2H_5CH{=}C_8H_7)Fe(CO)_3][BF_4]$ .... Fe: Org.Comp.B15-241/2
$BC_{14}F_4FeH_{13}O_4$ .. $[(CH_3O\text{-}C_{10}H_{10})Fe(CO)_3][BF_4]$ .... Fe: Org.Comp.B15-128
$BC_{14}F_4FeH_{15}O_2$ .. $[(C_5H_5)Fe(CO)_2(CH_2{=}CHCH{=}CHCH_2CH{=}CH_2)][BF_4]$
Fe: Org.Comp.B17-41
– .... $[(C_5H_5)Fe(CO)_2(C_6H_7\text{-}CH_3\text{-}5)][BF_4]$ .... Fe: Org.Comp.B17-91
– .... $[(C_5H_5)Fe(CO)_2(C_7H_{10})][BF_4]$ .... Fe: Org.Comp.B17-91, 99, 106/7, 115, 118
$BC_{14}F_4FeH_{15}O_3$ .. $[C_5H_5Fe(CH_2{=}C(CO_2CH_3)HC{=}C{=}CHCH_3)CO][BF_4]$
Fe: Org.Comp.B17-188, 191/2
$BC_{14}F_4FeH_{15}O_4$ .. $[(CH_3O\text{-}C_{10}H_{12})Fe(CO)_3][BF_4]$ .... Fe: Org.Comp.B15-127, 179
$BC_{14}F_4FeH_{15}O_5$ .. $[(3\text{-}CH_3\text{-}6\text{-}CH_3C(O)OCH_2CH_2\text{-}C_6H_5)Fe(CO)_3][BF_4]$
Fe: Org.Comp.B15-125, 178
– .... $[(6\text{-}CH_3C(O)OCH(CH_3)CH_2\text{-}C_6H_6)Fe(CO)_3][BF_4]$
Fe: Org.Comp.B15-108, 159
$BC_{14}F_4FeH_{15}O_6$ .. $[(1\text{-}CH_3C(O)OCH_2CH_2\text{-}4\text{-}CH_3O\text{-}C_6H_5)Fe(CO)_3][BF_4]$
Fe: Org.Comp.B15-118
– .... $[(3\text{-}CH_3O\text{-}6\text{-}CH_3C(O)OCH_2CH_2\text{-}C_6H_5)Fe(CO)_3][BF_4]$
Fe: Org.Comp.B15-125, 178
$BC_{14}F_4FeH_{17}$ .... $[(1,2,3\text{-}(CH_3)_3\text{-}C_6H_3)Fe(C_5H_5)][BF_4]$ .... Fe: Org.Comp.B19-99, 121
– .... $[(1,2,4\text{-}(CH_3)_3\text{-}C_6H_3)Fe(C_5H_5)][BF_4]$ .... Fe: Org.Comp.B19-99, 121
– .... $[(1,3,5\text{-}(CH_3)_3\text{-}C_6H_3)Fe(C_5H_5)][BF_4]$ .... Fe: Org.Comp.B19-99, 103/4, 133/6
– .... $[(1,4\text{-}(CH_3)_2\text{-}C_6H_4)Fe(C_5H_4\text{-}CH_3)][BF_4]$ ...... Fe: Org.Comp.B19-2, 13
– .... $[(CH_3\text{-}C_6H_5)Fe(C_7H_9)][BF_4]$ .... Fe: Org.Comp.B18-142/6, 209
– .... $[(C_6H_6)Fe(C_5H_4\text{-}C_3H_7\text{-}i)][BF_4]$ .... Fe: Org.Comp.B18-142/6, 151, 164, 187

$BC_{14}F_4FeH_{17}$ .... $[(C_6H_6)Fe(C_5H_4-C_3H_7-n)][BF_4]$ .... Fe: Org.Comp.B18-142/6, 151, 164, 187
$BC_{14}F_4FeH_{17}O$ ... $[(2,6-(CH_3)_2-C_6H_3-OCH_3)Fe(C_5H_5)][BF_4]$ .... Fe: Org.Comp.B19-99, 126
– .... $[(C_6H_7)Fe(CO)(C_7H_{10})][BF_4]$ .... Fe: Org.Comp.B17-244
– .... $[(C_7H_9)Fe(CO)(C_6H_8)][BF_4]$ .... Fe: Org.Comp.B17-244
$BC_{14}F_4FeH_{17}O_2$ .. $[(CH_3O-C_6H_6)Fe(CO)(C_6H_8)][BF_4]$ .... Fe: Org.Comp.B17-244
– .... $[(C_5H_5)Fe(CO)_2(CH_2=CHCH_2CH_2C(CH_3)=CH_2)][BF_4]$ Fe: Org.Comp.B17-40
– .... $[(C_5H_5)Fe(CO)_2(CH_2=C_6H_{10}-c)][BF_4]$ .... Fe: Org.Comp.B17-72, 82
– .... $[(C_5H_5)Fe(CO)_2((CH_3)_2C=C=C(CH_3)_2)][BF_4]$ ... Fe: Org.Comp.B17-114, 118
– .... $[(C_5H_5)Fe(CO)_2(C_6H_9-CH_3-1)][BF_4]$ .... Fe: Org.Comp.B17-91, 104
– .... $[(C_5H_5)Fe(CO)_2(C_7H_{12}-c)][BF_4]$ .... Fe: Org.Comp.B17-91/2, 104
– .... $[(C_5H_5)(CO)_2Fe=CH-(CH_2)_4-CH=CH_2][BF_4]$ ... Fe: Org.Comp.B16a-88, 95
$BC_{14}F_4FeH_{17}O_5$ .. $[(C_5H_5)Fe(CO)_2(CH_2=C(OC_2H_5)CO_2C_2H_5)][BF_4]$ Fe: Org.Comp.B17-68, 80
$BC_{14}F_4FeH_{19}O_2$ .. $[(C_5H_5)Fe(CO)_2(CH_2=CH-C_5H_{11}-n)][BF_4]$ .... Fe: Org.Comp.B17-13/4
– .... $[(C_5H_5)Fe(CO)_2(CH_3CH=CH-C_4H_9-n)][BF_4]$ ... Fe: Org.Comp.B17-58
$BC_{14}F_4FeH_{19}O_3$ .. $[(C_5(CH_3)_5)(CO)_2Fe=CH-OCH_3][BF_4]$ .... Fe: Org.Comp.B16a-129
– .... $[(C_5H_5)(CO)_2Fe=C(C_4H_9-t)-O-C_2H_5][BF_4]$ .... Fe: Org.Comp.B16a-111
– .... $[(C_5H_5)Fe(CO)_2(C_2H_5-CH=C(C_2H_5)-OCH_3)][BF_4]$ Fe: Org.Comp.B17-85
$BC_{14}F_4FeH_{19}O_4$ .. $[(C_5H_5)Fe(CO)_2(CH_2=CHCH(O-C_2H_5)_2)][BF_4]$ .. Fe: Org.Comp.B17-29
– .... $[(C_5H_5)Fe(CO)_2(CH_3CH=CH-O-CH(CH_3)CH(CH_3)OH)][BF_4]$ .... Fe: Org.Comp.B17-65, 78
$BC_{14}F_4FeH_{20}N$ ... $[C_5H_5FeC_4((CH_3)_4-2,3,4,5)NCH_3][BF_4]$ .... Fe: Org.Comp.B17-225, 235/6
$BC_{14}F_4FeH_{20}O_5P$ $[C_5H_5(CO)(P(OCH_3)_3)Fe=C(CH_3)OCH_2-CCH][BF_4]$ Fe: Org.Comp.B16a-38, 53
$BC_{14}F_4FeH_{21}O_2Si$ $[(n-C_3H_7Si(CH_3)_2C_5H_4)Fe(CO)_2(CH_2=CH_2)][BF_4]$ Fe: Org.Comp.B17-132
$BC_{14}F_4FeH_{22}NO_2$ $[C_5H_5Fe(CO)_2CH_2N(C_2H_5)_3][BF_4]$ .... Fe: Org.Comp.B14-136
$BC_{14}F_4FeH_{22}O_4P$ $[(C_5H_5)(CO)(P(OCH_3)_3)Fe=CHCH=C(CH_3)_2][BF_4]$ Fe: Org.Comp.B16a-20, 26
– .... $[(C_5H_5)Fe(c-C_4H_5-CH_3)(CO)(P(OCH_3)_3)][BF_4]$ Fe: Org.Comp.B16b-91/2
$BC_{14}F_4FeH_{22}O_5P$ $[C_5H_5(CO)(P(OCH_3)_3)Fe=C(CH_3)OCH_2CH=CH_2][BF_4]$ Fe: Org.Comp.B16a-38, 53
$BC_{14}F_4FeH_{24}OP$ .. $[C_5H_5(CO)(P(C_2H_5)_3)Fe=CH(CH_3)][BF_4]$ .... Fe: Org.Comp.B16a-18, 25
$BC_{14}F_4FeH_{29}O_6P_2$ $[(CH_3-C_6H_5)Fe(P(OCH_3)_3)_2-CH_3][BF_4]$ .... Fe: Org.Comp.B18-21
– .... $[(C_5H_5)Fe(CH_2=CHCH_3)(P(OCH_3)_3)_2][BF_4]$ .... Fe: Org.Comp.B16b-5/6, 8/9
$BC_{14}F_4H_8N_2O_2$ ... $BC_{14}F_4H_8N_2O_2$ .... B: B Comp.SVol.4/3b-251
$BC_{14}F_4H_9NO_7ReS$ $[(CO)_5Re(CNCH_2SO_2C_6H_4CH_3)][BF_4]$ .... Re: Org.Comp.2-252, 254
$BC_{14}F_4H_{11}MoO$ .. $[C_9H_7Mo(CO)(H-CC-H)_2][BF_4]$ .... Mo: Org.Comp.6-315
$BC_{14}F_4H_{12}I$ ...... $[C_6H_5-CH=CH-I-C_6H_5][BF_4]$ .... B: B Comp.SVol.4/3b-219
$BC_{14}F_4H_{12}N_2O_2Re$ $[(C_5H_5)Re(CO)_2-N=N-C_6H_4-4-CH_3][BF_4]$ .... Re: Org.Comp.3-193/7, 203
$BC_{14}F_4H_{12}N_2O_3Re$ $[(C_5H_5)Re(CO)_2-N=N-C_6H_4-2-OCH_3][BF_4]$ .... Re: Org.Comp.3-193/7, 203
– .... $[(C_5H_5)Re(CO)_2-N=N-C_6H_4-4-OCH_3][BF_4]$ .... Re: Org.Comp.3-193/7, 203, 209
$BC_{14}F_4H_{13}Te$ .... $[(-CH_2-(1,2-C_6H_4)-CH_2-)TeC_6H_5][BF_4]$ .... B: B Comp.SVol.3/4-150/1

$BC_{14}F_4H_{14}N_2O_3Re$
$[(C_5H_5)Re(CO)_2\text{-}NHNH\text{-}C_6H_4\text{-}4\text{-}OCH_3][BF_4]$ .. Re: Org.Comp.3-193/7, 204
$BC_{14}F_4H_{15}Se$ .... $[(C_6H_5CH_2)(CH_3)(C_6H_5)Se][BF_4]$ .... B: B Comp.SVol.3/4-147/8
$BC_{14}F_4H_{17}MoO$ .. $[C_5H_5Mo(CO)(H_3C\text{-}CC\text{-}CH_3)_2][BF_4]$ .... Mo: Org.Comp.6-318, 324/5
$BC_{14}F_4H_{18}MoNO_2$ $[C_5H_5Mo(CO)(NO)C_8H_{13}\text{-}c][BF_4]$............ Mo: Org.Comp.6-350, 354
$BC_{14}F_4H_{29}NP_2Re$ $[(CH_3)_2Re(P(CH_3)_3)_2{=}N\text{-}C_6H_5][BF_4]$.......... Re: Org.Comp.1-18/9
$BC_{14}F_4H_{30}InO_2$... $[(i\text{-}C_3H_7)_2In(OC_4H_8)_2][BF_4]$ .... In: Org.Comp.1-336/7
$BC_{14}F_4H_{30}NP_2Re$ $[(CH_3)_2Re(P(CH_3)_3)_2(NH\text{-}C_6H_5)][BF_4]$ .... Re: Org.Comp.1-19
$BC_{14}F_4H_{32}MoO_{11}P_3$
$[(CH_2CHCH_2)Mo(CO)_2(P(OCH_3)_3)_3][BF_4]$...... Mo: Org.Comp.5-200, 201
$BC_{14}F_4H_{34}MoP_3$ .. $[C_5H_5Mo(P(CH_3)_3)_3H_2][BF_4]$............... Mo: Org.Comp.6-13
$BC_{14}F_6FeH_{21}NP$ .. $[(C_6H_6)Fe(2\text{-}CH_3\text{-}1\text{-}(t\text{-}C_4H_9)\text{-}1,2\text{-}NBC_3H_3)][PF_6]$
Fe: Org.Comp.B18-99
$BC_{14}F_7H_9N_2O_2Re$ $[(C_5H_5)Re(CO)_2\text{-}N{=}N\text{-}C_6H_4\text{-}2\text{-}CF_3][BF_4]$ ..... Re: Org.Comp.3-193/7, 203, 208
$BC_{14}FeH_{20}NO_2$... $(C_5H_5)Fe[C_5H_3(B(OH)_2)CH_2CH_2N(CH_3)_2]$ ..... Fe: Org.Comp.A9-287, 288
$BC_{14}FeH_{21}INO_2$ .. $[C_5H_5FeC_5H_3(B(OH)_2)CH_2N(CH_3)_3]I$ .... Fe: Org.Comp.A9-288
$BC_{14}FeH_{21}N^+$.... $[(C_6H_6)Fe(2\text{-}CH_3\text{-}1\text{-}(t\text{-}C_4H_9)\text{-}1,2\text{-}NBC_3H_3)]^+$.. Fe: Org.Comp.B18-99/101
$BC_{14}FeH_{21}NO_2{}^+$.. $[C_5H_5FeC_5H_3(B(OH)_2)CH_2N(CH_3)_3]^+$......... Fe: Org.Comp.A9-288
$BC_{14}FeH_{21}O_3SeSi$ $BC_2FeSeSi(CO)_3(C_2H_5)_2(CH_3)_2[C(CH_3){=}CH_2]$ .. B: B Comp.SVol.4/4-180
$BC_{14}GeH_{31}$ ...... $(CH_3)_2Ge[\text{-}CH(CH_3)CH_2\text{-}B(C(CH_3)_2C_3H_7\text{-}i)$
$\text{-}CH_2CH(CH_3)\text{-}]$ .... Ge: Org.Comp.3-296
$BC_{14}H_{10}NOS_2$.... $C_6H_5B(\text{-}NOH\text{-}C_4H_2S\text{-}C_4H_2S\text{-})$.............. B: B Comp.SVol.3/3-164
$BC_{14}H_{10}NS_2$ ..... $(C_4H_2S)_2(\text{-}BC_6H_5\text{-}NH\text{-})$ .... B: B Comp.SVol.3/3-163
$BC_{14}H_{10}N_2NaS_2$ .. $Na[B(C_6H_5)_2(SCN)_2]$.................... B: B Comp.SVol.3/4-132
$BC_{14}H_{11}N_2O_4$.... $(O_2C_7H_5)B[\text{-}NH\text{-}(O_2C_7H_4)\text{-}NH\text{-}]$ .... B: B Comp.SVol.3/3-117
$BC_{14}H_{12}MoN_9O_8$. $(CH_2CHCH_2)Mo(CO)_2[\text{-}N_2C_3H_2(NO_2)\text{-}]_3BH$.... Mo: Org.Comp.5-226, 227
$BC_{14}H_{12}N$....... $2\text{-}BC_9H_7 \cdot NC_5H_5$ .... B: B Comp.SVol.4/3b-26
$BC_{14}H_{12}NO_2$..... $1\text{-}CH_3\text{-}2\text{-}C_6H_5\text{-}3,1,2\text{-}ONBC_7H_4({=}O)\text{-}4$ .... B: B Comp.SVol.4/3b-67
$BC_{14}H_{12}NO_6$..... $NH_4[B(O\text{-}C_6H_4\text{-}2\text{-}COO)_2]$ .... B: B Comp.SVol.3/3-227
$BC_{14}H_{12}NO_8$..... $NH_4[B(O\text{-}C_6H_3\text{-}2\text{-}COO\text{-}4\text{-}OH)_2]$ .... B: B Comp.SVol.3/3-227
$BC_{14}H_{12}NS_2$ ..... $(C_6H_5)_2N\text{-}B(\text{-}S\text{-}CH{=}CH\text{-}S\text{-})$................ B: B Comp.SVol.3/4-126
– .... $C_6H_5\text{-}B(\text{-}C_4H_4S\text{-}C_4H_2S\text{-}NH\text{-})$ .... B: B Comp.SVol.3/3-163
$BC_{14}H_{14}LiS_3$..... $Li[BC_2H_5(SC_4H_3)_3]$...................... B: B Comp.SVol.3/4-117
$BC_{14}H_{14}N$ .... $9\text{-}(CH_3)_2B\text{-}9\text{-}NC_{12}H_8$...................... B: B Comp.SVol.4/3a-233
$BC_{14}H_{14}NS_2$ ..... $(C_4H_4S)_2(\text{-}BC_6H_5\text{-}NH\text{-})$ .... B: B Comp.SVol.3/3-163
$BC_{14}H_{14}N_3$ ...... $(2\text{-}CH_3C_6H_4)_2BN_3$..................... B: B Comp.SVol.3/3-145
$BC_{14}H_{15}MoN_6O_2$. $(CH_2CHCH_2)Mo(CO)_2[\text{-}N_2C_3H_3\text{-}]_3BH$ .... Mo: Org.Comp.5-225/7
$BC_{14}H_{15}N_2$ ...... $C_6H_5B[\text{-}NCH_3\text{-}(1,2\text{-}C_6H_4)\text{-}NCH_3\text{-}]$ .... B: B Comp.SVol.3/3-117
$BC_{14}H_{15}N_4$ ...... $[\text{-}B(C_6H_5)\text{-}N(CH_3)\text{-}N{=}N\text{-}N(CH_2\text{-}C_6H_5)\text{-}]$...... B: B Comp.SVol.4/3a-199
$BC_{14}H_{16}N$ .... $(C_6H_5)_2B\text{-}N(CH_3)_2$ .... B: B Comp.SVol.3/3-148/9
– .... $(C_6H_5)_2B\text{-}NH\text{-}C_2H_5$ .... B: B Comp.SVol.4/3a-231
$BC_{14}H_{16}NO_4$..... $NH_4[B(O\text{-}C_6H_4\text{-}2\text{-}CH_2O)_2]$................ B: B Comp.SVol.3/3-227
$BC_{14}H_{16}N_5$ ...... $[4\text{-}H_2N\text{-}C_5H_4NH][BH(NC_4H_4)_2CN]$........... B: B Comp.SVol.3/3-213
$BC_{14}H_{17}N_2$ ...... $(C_6H_5)_2B\text{-}NHN(CH_3)_2$..................... B: B Comp.SVol.3/3-153
$BC_{14}H_{18}N$ .... $H_3B\text{-}NH(CH_2\text{-}C_6H_5)_2$ .... B: B Comp.SVol.4/3b-10
$BC_{14}H_{18}NO_4$..... $[\text{-}B(C_2H_5)_2\text{-}O\text{-}C({=}O)\text{-}CH_2\text{-}N({=}CH\text{-}C_6H_4\text{-}2\text{-}COOH)\text{-}]$
B: B Comp.SVol.4/3b-70
$BC_{14}H_{18}N_3$ ...... $(C_2H_5)_2B[\text{-}(1,2\text{-}NC_5H_4)\text{-}N{=}(2,1\text{-}NC_5H_4)\text{-}]$..... B: B Comp.SVol.3/3-123

$BC_{14}H_{18}N_4^+$ ..... $[(NCH_3-CH_2-CH_2-NCH_3-)B(-(1,2-NC_5H_4)$
$-(1,2-NC_5H_4)-)]^+$ ..... B: B Comp.SVol.3/3-210/1
$BC_{14}H_{18}N_5$ ..... $(CH_3)_2NH-B(C_6H_5)(1,2-C_3H_3N_2)_2$ ..... B: B Comp.SVol.3/3-190
$BC_{14}H_{19}NP$ ..... $(C_2H_5)_2N-B(-CH=CH-PC_6H_5-CH=CH-)$ ..... B: B Comp.SVol.3/3-151
$BC_{14}H_{19}N_2O_2$ ..... $(C_2H_5CH=CH)_2B-NH(C_6H_4-4-NO_2)$ ..... B: B Comp.SVol.3/3-153
$BC_{14}H_{19}S$ ..... $[3.3.1]-9-BC_8H_{14}-9-S-C_6H_5$ ..... B: B Comp.SVol.4/4-132
$BC_{14}H_{19}Se$ ..... $[3.3.1]-9-BC_8H_{14}-9-(Se-C_6H_5)$ ..... B: B Comp.SVol.4/4-170
$BC_{14}H_{20}N$ ..... $(CH_3CH=CCH_3)_2B-NH-C_6H_5$ ..... B: B Comp.SVol.3/3-152
– ..... $(C_2H_5-CH=CH)_2B-NH-C_6H_5$ ..... B: B Comp.SVol.3/3-152
– ..... $[3.3.1.1^{3,7}]-1-BC_9H_{15} \cdot NC_5H_5$ ..... B: B Comp.SVol.3/3-191
B: B Comp.SVol.4/3b-30
$BC_{14}H_{20}NS_2$ ..... $(n-C_3H_7)_2NB(-S-CH=CC_6H_5-S-)$ ..... B: B Comp.SVol.3/4-126
$BC_{14}H_{20}NSe_2$ ..... $3-[2,6-(CH_3)_2-C_6H_3-NH]-4,5-(C_2H_5)_2-1,2,3-Se_2BC_2$
B: B Comp.SVol.4/4-174
$BC_{14}H_{21}LiN$ ..... $Li[(C_2H_5)_2N-B(C_6H_5)(-CH_2-CH=CH-CH_2-)]$ ..... B: B Comp.SVol.4/3b-49
$BC_{14}H_{21}NO_2$ ..... $C_6H_5-B[-O-CH_2CH_2CH_2-O-] \cdot NC_5H_{10}$ ..... B: B Comp.SVol.4/3b-69
– ..... $C_6H_5-B[-O-CH_2-CH(CH_3)-O-] \cdot NC_5H_{10}$ ..... B: B Comp.SVol.4/3b-69
$BC_{14}H_{21}N_3NaO_4$ ..... $Na[HB(NC_4H_4)(CN)_2] \cdot 2\ OC_4H_8O$ ..... B: B Comp.SVol.3/3-213
$BC_{14}H_{21}N_4$ ..... $BC_6H_6N_3(=CH-CN)(C_3H_7-i)_2$ ..... B: B Comp.SVol.4/3a-181
– ..... $[C_5H_{12}N][(NC_4H_4-1)_2BH-CN]$ ..... B: B Comp.SVol.3/3-213
$BC_{14}H_{21}N_6$ ..... $BC_2H_2N_6(C_6H_5)(C_3H_7-n)_2$ ..... B: B Comp.SVol.4/3a-178
$BC_{14}H_{21}O_3RuSeSi$ $BC_2RuSeSi(CO)_3(C_2H_5)_2(CH_3)_2[C(CH_3)=CH_2]$ ..... B: B Comp.SVol.4/4-180
$BC_{14}H_{22}N$ ..... $9-[1-(C_5H_4N(CH_3)-2)]-[3.3.1]-9-BC_8H_{15}$ ..... B: B Comp.SVol.3/3-189
– ..... $(-CH_2-CH=CH-CH_2-)B(C_6H_4-2-CH_3)-N(CH_3)_3$ B: B Comp.SVol.4/3b-29
– ..... $C_2H_5-CH=C(C_2H_5)-BCH_3-NCH_3-C_6H_5$ ..... B: B Comp.SVol.3/3-155
$BC_{14}H_{22}NS$ ..... $2-CH_3-NC_5H_9-1-B(C_6H_5)-S-C_2H_5$ ..... B: B Comp.SVol.4/4-139
– ..... $3-CH_3-NC_5H_9-1-B(C_6H_5)-S-C_2H_5$ ..... B: B Comp.SVol.4/4-139
– ..... $[3.3.1]-9-BC_8H_{14}-9-SCH_3 \cdot NC_5H_5$ ..... B: B Comp.SVol.4/4-132
$BC_{14}H_{23}N_2$ ..... $C_6H_5B[N(CH_3)_2]-2-CH_3-NC_5H_9$ ..... B: B Comp.SVol.3/3-109
$BC_{14}H_{23}N_4$ ..... $1-CH_3-2-[(CH_3)_2N]_2B-5-C_6H_5-1,3-N_2C_3H_3$ ..... B: B Comp.SVol.3/3-109
– ..... $[-B(C_4H_9-n)-N(C_4H_9-n)-N=N-N(C_6H_5)-]$ ..... B: B Comp.SVol.3/3-137/8
– ..... $[-B(C_4H_9-n)-N(C_4H_9-t)-N=N-N(C_6H_5)-]$ ..... B: B Comp.SVol.3/3-137/8
– ..... $[-B(C_4H_9-i)-N(C_4H_9-i)-N=N-N(C_6H_5)-]$ ..... B: B Comp.SVol.3/3-137/8
– ..... $[-B(C_4H_9-s)-N(C_4H_9-s)-N=N-N(C_6H_5)-]$ ..... B: B Comp.SVol.3/3-137/8
– ..... $[-B(C_4H_9-s)-N(C_4H_9-t)-N=N-N(C_6H_5)-]$ ..... B: B Comp.SVol.4/3a-199
– ..... $[-B(C_4H_9-t)-N(C_4H_9-t)-N=N-N(C_6H_5)-]$ ..... B: B Comp.SVol.3/3-137/8
– ..... $[-B(C_4H_9-t)-N(C_6H_5)-N=N-N(C_4H_9-i)-]$ ..... B: B Comp.SVol.4/3a-199
$BC_{14}H_{24}N$ ..... $(t-C_4H_9)_2B-NH-C_6H_5$ ..... B: B Comp.SVol.4/3a-230
$BC_{14}H_{24}NO_2$ ..... $(C_{10}H_{15})B[-O-CH_2-CH_2-]_2NH$ ..... B: B Comp.SVol.3/3-218
$BC_{14}H_{24}NO_3S$ ..... $2-(CH_2=CH-CH_2)-3-[CH_3-S(O)_2]-7,8,8-(CH_3)_3$
$-1,3,2-ONBC_7H_7$ ..... B: B Comp.SVol.4/3b-65
$BC_{14}H_{24}NS$ ..... $C_2H_5SB(C_6H_5)N(C_3H_7-i)_2$ ..... B: B Comp.SVol.3/4-122
$BC_{14}H_{24}NSi_2$ ..... $[-CH=CC_6H_5-]B-N[Si(CH_3)_3]_2$ ..... B: B Comp.SVol.3/3-150
$BC_{14}H_{24}N_3$ ..... $(n-C_4H_9)_2B[-(1,2-NC_5H_4)-N=CH-NH-]$ ..... B: B Comp.SVol.3/3-123
$BC_{14}H_{24}N_3O$ ..... $BC_6H_5N_3(O-C_2H_5)(C_3H_7-i)_2$ ..... B: B Comp.SVol.4/3a-257
$BC_{14}H_{24}N_3S$ ..... $[-B(C_3H_7-n)_2-N(CH_2-CH=CH_2)-C(=S)-NH-C_4H_4N-]$
B: B Comp.SVol.4/4-149/50
$BC_{14}H_{25}N_2$ ..... $2-[(n-C_4H_9)_2B-N(CH_3)]-NC_5H_4$ ..... B: B Comp.SVol.4/3a-232
– ..... $(CH_3)_5C_5-B[-N(CH_3)-CH_2CH_2-N(CH_3)-]$ ..... B: B Comp.SVol.4/3a-174
– ..... $n-C_4H_9-NH-B(C_6H_5)-N(C_2H_5)_2$ ..... B: B Comp.SVol.3/3-108

$BC_{14}H_{25}N_2$ — $i\text{-}C_4H_9\text{-}NH\text{-}B(C_6H_5)\text{-}N(C_2H_5)_2$ — B: B Comp.SVol.3/3-108
– $s\text{-}C_4H_9\text{-}NH\text{-}B(C_6H_5)\text{-}N(C_2H_5)_2$ — B: B Comp.SVol.3/3-108
– $t\text{-}C_4H_9\text{-}NH\text{-}B(C_6H_5)\text{-}N(C_2H_5)_2$ — B: B Comp.SVol.3/3-108
– $C_6H_5\text{-}B[NH\text{-}C_4H_9\text{-}t]_2$ — B: B Comp.SVol.3/3-108
– $C_6H_5\text{-}B[N(C_2H_5)_2]_2$ — B: B Comp.SVol.3/3-105
$BC_{14}H_{25}N_6$ — $[\text{-}NCH_3\text{-}(CH_2)_2\text{-}NCH_3\text{-}]B[\text{-}(CH_3)_2C_3HN_2\text{-}H\text{-}(CH_3)_2C_3HN_2\text{-}]$ — B: B Comp.SVol.3/3-201
$BC_{14}H_{25}S_2$ — $(CH_3)_5C_5\text{-}B(SC_2H_5)_2$ — B: B Comp.SVol.4/4-129
$BC_{14}H_{26}N$ — $(CH_3)_2N[\text{-}CH_2CH_2CH_2\text{-}BC_9H_{14}\text{-}]$ — B: B Comp.SVol.4/3a-259; B: B Comp.SVol.4/3b-30
– $[3.3.1.1^{3,7}]\text{-}1\text{-}BC_9H_{15} \cdot NC_5H_{11}$ — B: B Comp.SVol.3/3-191; B: B Comp.SVol.4/3b-30
– $[3.3.1.1^{3,7}]\text{-}C_{10}H_{15}\text{-}BCH_3\text{-}NH\text{-}C_3H_7\text{-}i$ — B: B Comp.SVol.3/3-155
$BC_{14}H_{26}NRh_2$ — $(C_2H_4)Rh[C_4H_4BN(i\text{-}C_3H_7)_2]Rh(C_2H_4)$ — B: B Comp.SVol.4/4-180
$BC_{14}H_{27}N_2$ — $(CH_3)_5C_5\text{-}B[N(CH_3)_2]_2$ — B: B Comp.SVol.4/3a-170
– $N(\text{-}CH_2\text{-}CH_2\text{-})_3N\text{-}9\text{-}[3.3.1]\text{-}9\text{-}BC_8H_{15}$ — B: B Comp.SVol.3/3-189
$BC_{14}H_{27}N_2OS$ — $2,2,6,6\text{-}(CH_3)_4NC_5H_6\text{-}1\text{-}B[\text{-}N(C_4H_9\text{-}t)\text{-}C(O)S\text{-}]$ — B: B Comp.SVol.4/4-152
$BC_{14}H_{27}N_2O_2$ — $2,2,6,6\text{-}(CH_3)_4NC_5H_6\text{-}1\text{-}B[\text{-}N(C_4H_9\text{-}t)\text{-}C(O)O\text{-}]$ — B: B Comp.SVol.4/3b-55
– $2,2,6,6\text{-}(CH_3)_4NC_5H_6\text{-}1\text{-}B[\text{-}NH\text{-}C(O)\text{-}CH(C_3H_7\text{-}i)\text{-}O\text{-}]$ — B: B Comp.SVol.4/3b-56
$BC_{14}H_{27}N_2Se_2$ — $2\text{-}[2,2,6,6\text{-}(CH_3)_4NC_5H_6\text{-}1]\text{-}3\text{-}(C_4H_9\text{-}t)\text{-}1,3,2\text{-}SeNBC(=Se)\text{-}4$ — B: B Comp.SVol.4/4-171
$BC_{14}H_{27}N_2Sn$ — $BC_5H_3N_2Sn\text{-}(CH_3)_3\text{-}(C_2H_5)_3$ — B: B Comp.SVol.4/3a-258; Sn: Org.Comp.19-47/8, 52
$BC_{14}H_{28}N$ — $9\text{-}[1\text{-}(1\text{-}(CH_3)NC_5H_{10})]\text{-}BH[C_8H_{14}]$ — B: B Comp.SVol.3/3-189
$BC_{14}H_{28}NO_2Si$ — $2,4,7\text{-}(CH_3)_3\text{-}4a\text{-}[(CH_3)_3Si\text{-}O]\text{-}2\text{-}NC_8H_7(=O)\text{-}6 \cdot BH_3$ — B: B Comp.SVol.4/3b-13
$BC_{14}H_{28}N_3$ — $2,6\text{-}(CH_3)_2C_5H_8N\text{-}B=N\text{-}NC_5H_8(CH_3)_2\text{-}2,6$ — B: B Comp.SVol.3/3-100
– $C_7H_{14}N\text{-}B[\text{-}NH\text{-}NC_5H_8(CH_3)\text{-}CH_2\text{-}]$ — B: B Comp.SVol.3/3-117
$BC_{14}H_{28}N_3Sn$ — $BC_4HN_3Sn\text{-}(CH_3)_4\text{-}(C_2H_5)_3$ — B: B Comp.SVol.4/3a-258; Sn: Org.Comp.19-56, 59
$BC_{14}H_{28}N_5$ — $N_3B[NC_5H_8(CH_3)_2\text{-}2,6]_2$ — B: B Comp.SVol.3/3-97
$BC_{14}H_{28}N_6^+$ — $[((C_2H_5)_2NH)_2B(1,2\text{-}N_2C_3H_3)_2]^+$ — B: B Comp.SVol.4/4-70
$BC_{14}H_{29}N_2O$ — $2,2,6,6\text{-}(CH_3)_4\text{-}NC_5H_6\text{-}1\text{-}B[\text{-}N(C_4H_9\text{-}t)\text{-}CH_2\text{-}O\text{-}]$ — B: B Comp.SVol.4/3b-55
$BC_{14}H_{30}N$ — $(i\text{-}C_3H_7)_2NH\text{-}[BHC_8H_{14}]$ — B: B Comp.SVol.3/3-189
$BC_{14}H_{31}N_2$ — $1\text{-}[(CH_3)_2N\text{-}B(C_3H_7\text{-}i)]\text{-}2,2,6,6\text{-}(CH_3)_4\text{-}NC_5H_6$ — B: B Comp.SVol.4/3a-169
– $(CH_3)_2NCH_2CH_2N(CH_3)_2\text{-}9\text{-}[3.3.1]\text{-}9\text{-}BC_8H_{15}$ — B: B Comp.SVol.3/3-189
$BC_{14}H_{31}N_2O$ — $t\text{-}C_4H_9\text{-}NH\text{-}B(OCH_3)\text{-}1\text{-}NC_5H_6\text{-}2,2,6,6\text{-}(CH_3)_4$ — B: B Comp.SVol.4/3b-52
$BC_{14}H_{31}N_2O_2$ — $CH_3OC(O)B[N(i\text{-}C_3H_7)_2]_2$ — B: B Comp.SVol.3/3-110
$BC_{14}H_{31}N_2SSeSi$ — $(CH_3)_3Si\text{-}N(C_4H_9\text{-}t)\text{-}S\text{-}N[\text{-}C(C_2H_5)=C(C_2H_5)\text{-}Se\text{-}B(CH_3)\text{-}]$ — B: B Comp.SVol.4/3a-252
$BC_{14}H_{32}NSi_2$ — $1\text{-}(t\text{-}C_4H_9)\text{-}3,3\text{-}[(CH_3)_3Si]_2\text{-}2\text{-}CH_3\text{-}1,2\text{-}NBC_3H_2$ — B: B Comp.SVol.4/3a-250
– $[\text{-}C(n\text{-}C_3H_7)=C(n\text{-}C_3H_7)\text{-}]B\text{-}N[Si(CH_3)_3]_2$ — B: B Comp.SVol.3/3-150
$BC_{14}H_{32}N_5$ — $(i\text{-}C_3H_7)_2N\text{-}B[\text{-}N(C_4H_9\text{-}t)\text{-}N=N\text{-}N(C_4H_9\text{-}i)\text{-}]$ — B: B Comp.SVol.4/3a-199
$BC_{14}H_{33}N_2$ — $t\text{-}C_4H_9\text{-}NH\text{-}B(C_4H_9\text{-}t)\text{-}N(C_3H_7\text{-}i)_2$ — B: B Comp.SVol.3/3-107
– $i\text{-}C_4H_9\text{-}NH\text{-}B(C_4H_9\text{-}i)\text{-}N(C_3H_7\text{-}i)_2$ — B: B Comp.SVol.3/3-107
$BC_{14}H_{33}N_2OSi$ — $2,2,6,6\text{-}(CH_3)_4\text{-}NC_5H_6\text{-}1\text{-}O\text{-}B(CH_3)\text{-}N(CH_3)\text{-}Si(CH_3)_3$ — B: B Comp.SVol.4/3b-61

$BC_{14}H_{33}N_2SSeSi_2$ $[(CH_3)_3Si]_2N\text{-}S\text{-}N[\text{-}C(C_2H_5)=C(C_2H_5)\text{-}Se\text{-}B(C_2H_5)\text{-}]$ B: B Comp.SVol.4/3a-252

$BC_{14}H_{34}NO_4Si$ . . . $[(i\text{-}C_3H_7O)_3Si][i\text{-}C_3H_7O]B\text{-}N(CH_3)_2$ . . . . B: B Comp.SVol.3/3-155

$BC_{14}H_{35}N_2OSi$ . . . $t\text{-}C_4H_9\text{-}NH\text{-}B(O\text{-}C_3H_7\text{-}i)\text{-}N(C_4H_9\text{-}t)\text{-}Si(CH_3)_3$ B: B Comp.SVol.4/3b-51

$BC_{14}H_{35}N_2Si_2$ . . . . $t\text{-}C_4H_9\text{-}CH_2\text{-}C(CH_3)_2\text{-}N=B\text{-}N[Si(CH_3)_3]_2$ . . . . . B: B Comp.SVol.4/3a-160/5

$BC_{14}H_{36}IN_2Si_2$ . . . $[(CH_3)_3Si\text{-}N(C_4H_9\text{-}t)=B=N(C_4H_9\text{-}t)\text{-}Si(CH_3)_3]I$ . . B: B Comp.SVol.4/4-114/5

$BC_{14}H_{36}NOSi_2$ . . . $(n\text{-}C_4H_9)_2B\text{-}N[Si(CH_3)_3]\text{-}O\text{-}Si(CH_3)_3$ . . . . B: B Comp.SVol.4/3a-231

– . . . . $(i\text{-}C_4H_9)_2B\text{-}N[Si(CH_3)_3]\text{-}O\text{-}Si(CH_3)_3$ . . . . B: B Comp.SVol.3/3-154

$BC_{14}H_{36}NSi_3$ . . . . . $[(CH_3)_3Si]_3C\text{-}B=N\text{-}C_4H_9\text{-}t$ . . . . B: B Comp.SVol.4/3a-215

$BC_{14}H_{36}NSn_2$ . . . . $(CH_3)_2B\text{-}N[Sn(C_2H_5)_3]_2$ . . . . B: B Comp.SVol.4/3a-227

– . . . . $(n\text{-}C_4H_9)_2B\text{-}N[Sn(CH_3)_3]_2$ . . . . B: B Comp.SVol.4/3a-227

– . . . . $(t\text{-}C_4H_9)_2B\text{-}N[Sn(CH_3)_3]_2$ . . . . B: B Comp.SVol.4/3a-227

$BC_{14}H_{36}N_5Si_2$ . . . . $(CH_3)_3Si\text{-}N(C_4H_9\text{-}t)\text{-}B(N_3)\text{-}N(C_4H_9\text{-}t)\text{-}Si(CH_3)_3$ B: B Comp.SVol.4/3a-156

$BC_{14}H_{37}N_2Si_2$ . . . . $(s\text{-}C_4H_9)_2N\text{-}BH\text{-}N[Si(CH_3)_3]_2$ . . . . B: B Comp.SVol.3/3-107

– . . . . $t\text{-}C_4H_9\text{-}NH\text{-}B(C_4H_9\text{-}n)\text{-}N[Si(CH_3)_3]_2$ . . . . B: B Comp.SVol.3/3-107

$BC_{14}H_{38}NSi_3$ . . . . . $[t\text{-}C_4H_9][(CH_3)_3SiCH_2]B\text{-}N[Si(CH_3)_3]_2$ . . . . B: B Comp.SVol.3/3-154

$BC_{14}H_{40}NOSi_4$ . . . $[(CH_3)_3Si]_3C\text{-}B(OCH_3)\text{-}NH\text{-}Si(CH_3)_3$ . . . . B: B Comp.SVol.4/3b-61

$BC_{14}H_{40}N_4PSi_2$ . . . $[(CH_3)_3Si]_2N\text{-}B(C_4H_9\text{-}t)\text{-}NH\text{-}P[N(CH_3)_2]_2$ . . . . . B: B Comp.SVol.4/3a-170

$BC_{15}ClF_{12}H_{11}NO_2$ $ClB[OC(CF_3)_2]_2NC_5H_2(C_4H_9\text{-}t)\text{-}5$ . . . . B: B Comp.SVol.3/4-61

$BC_{15}ClH_{12}MoN_8O_2$ $ClCMo(CO)_2[\text{-}N_2C_3H_3\text{-}]_3B\text{-}N_2C_3H_3$ . . . . Mo: Org.Comp.5-94/5

$BC_{15}ClH_{16}N_2O$ . . . $(C_6H_5)ClBN(C_6H_5)C(O)N(CH_3)_2$ . . . . B: B Comp.SVol.3/4-55/6

$BC_{15}ClH_{17}N$ . . . . . $(C_6H_5)ClBNH(2,4,6\text{-}(CH_3)_3C_6H_2)$ . . . . B: B Comp.SVol.3/4-56

$BC_{15}ClH_{19}NSn$ . . . $(CH_3)_2Sn(Cl)\text{-}N(CH_3)\text{-}B(C_6H_5)_2$ . . . . Sn: Org.Comp.19-117

$BC_{15}ClH_{23}N$ . . . . . 9-Cl-[3.3.1]-9-BC_8H_{14} · 2,4-$(CH_3)_2NC_5H_3$ . . . B: B Comp.SVol.4/4-69

– . . . . 9-Cl-[3.3.1]-9-BC_8H_{14} · 2,6-$(CH_3)_2NC_5H_3$ . . . B: B Comp.SVol.4/4-69

$BC_{15}ClH_{23}NO$ . . . . 1-$(C_6H_5\text{-}BCl\text{-}O)$-2,2,6,6-$(CH_3)_4NC_5H_6$ . . . . B: B Comp.SVol.4/4-55

$BC_{15}ClH_{24}NO_6Re$ $[N(CH_3)_4][fac\text{-}(CO)_3Re(CH_2COCO(C_3H_7\text{-}i))(CH_3CO)BCl$ Re: Org.Comp.2-418/9

$BC_{15}ClH_{27}N$ . . . . . $(n\text{-}C_4H_9)_2BCl$ · NC_5H_3-2,4$(CH_3)_2$ . . . . B: B Comp.SVol.4/4-69

– . . . . $(n\text{-}C_4H_9)_2BCl$ · NC_5H_3-2,6$(CH_3)_2$ . . . . B: B Comp.SVol.4/4-69

$BC_{15}ClH_{27}NSi$ . . . . 2-$(CH_3)_3Si\text{-}C_6H_4\text{-}BCl\text{-}N(C_3H_7\text{-}i)_2$ . . . . B: B Comp.SVol.4/4-60

$BC_{15}ClH_{31}N_2O_2P$ 2,2,6,6-$(CH_3)_4\text{-}NC_5H_6$-1-$B[\text{-}N(C_4H_9\text{-}t)\text{-}P(Cl)$ $\text{-}O\text{-}CH_2CH_2\text{-}O\text{-}]$ . . . . B: B Comp.SVol.4/3b-54

$BC_{15}ClH_{33}N_3$ . . . . $ClB[N(i\text{-}C_3H_7)NH(i\text{-}C_3H_7)]\text{-}NC_5H_6(CH_3)_4$ . . . . . B: B Comp.SVol.3/4-59

$BC_{15}ClH_{34}N_2Si$ . . . 1-$[(CH_3)_2SiCl\text{-}N(C_4H_9\text{-}t)\text{-}BH]$-2,2,6,6-$(CH_3)_4\text{-}NC_5H_6$ B: B Comp.SVol.4/3a-225

$BC_{15}Cl_2H_{20}N$ . . . . NC_5H_5 · $BCl_2\text{-}C_5(CH_3)_5$ . . . . B: B Comp.SVol.4/4-70

$BC_{15}Cl_2H_{33}N_2Si$ . . 2,2,6,6-$(CH_3)_4\text{-}NC_5H_6$-1-$BCl\text{-}N(C_4H_9\text{-}t)\text{-}SiCl(CH_3)_2$ B: B Comp.SVol.4/4-63

$BC_{15}Cl_3H_{16}Si$ . . . . $(C_6H_5)_2SiCl(CH_2)_3BCl_2$ . . . . B: B Comp.SVol.3/4-42

$BC_{15}CoH_{14}$ . . . . . . $(C_5H_5)Co(C_4H_4BC_6H_5)$ . . . . B: B Comp.SVol.4/4-180

$BC_{15}CoH_{17}N$ . . . . . $(C_5H_5)Co[(CH_2=CH)_2B\text{-}NH\text{-}C_6H_5]$ . . . . B: B Comp.SVol.4/3a-236

$BC_{15}FH_{19}NSi$ . . . . $C_6H_5\text{-}BF\text{-}N(C_6H_5)\text{-}Si(CH_3)_3$ . . . . B: B Comp.SVol.4/3b-245

$BC_{15}FH_{24}NO_6Re$ . . $[N(CH_3)_4][fac\text{-}(CO)_3Re(CH_2COCO(C_3H_7\text{-}i))(CH_3CO)BF$ Re: Org.Comp.2-418/9

$BC_{15}FH_{31}N_2O_2P$ . . 2,2,6,6-$(CH_3)_4\text{-}NC_5H_6$-1-$B[\text{-}N(C_4H_9\text{-}t)\text{-}P(F)$ $\text{-}O\text{-}CH_2CH_2\text{-}O\text{-}]$ . . . . B: B Comp.SVol.4/3b-54

$BC_{15}FH_{43}NSi_5$ . . . . $[(CH_3)_3Si]_3Si\text{-}BF\text{-}N(C_3H_7\text{-}i)\text{-}Si(CH_3)_3$ . . . . B: B Comp.SVol.4/3b-245

$BC_{15}FH_{45}NSi_6$.... $[(CH_3)_3Si]_3Si-BF-N[Si(CH_3)_3]_2$ .... B: B Comp.SVol.4/3b-245

$BC_{15}F_2H_9O_2Te$ ... $[-BF_2-O-TeC_9H_4(C_6H_5)-O-]$ .... B: B Comp.SVol.4/4-177

$BC_{15}F_2H_{10}NO_4$ ... $F_2B[-O-CC_6H_5=CH-C(4-C_6H_4NO_2)=O-]$ ...... B: B Comp.SVol.3/3-368

$BC_{15}F_2H_{11}O_2$ .... $F_2B[-O-C(C_6H_5)=CH-C(C_6H_5)=O-]$ .... B: B Comp.SVol.3/3-368
B: B Comp.SVol.4/3b-233/4

$BC_{15}F_2H_{19}N_2$ .... $BC_9F_2HN_2(CH_3)_6$ .... B: B Comp.SVol.3/3-378
B: B Comp.SVol.4/3b-250

$BC_{15}F_2H_{26}NSi$.... $(CH_3)_3Si-N(BF_2)-C_6H_3-2,6-(C_3H_7-i)_2$ .... B: B Comp.SVol.4/3b-244

$BC_{15}F_3Fe_2H_{10}O_3S_3$
$C_5H_5(CO)_2Fe=C(-SC(=SBF_3)Fe(CO)(C_5H_5)S-)$ Fe: Org.Comp.B16a-185

$BC_{15}F_3H_{18}N_4O_3S$ $[(-NCH_3-CH_2-CH_2-NCH_3-)B(-(1,2-NC_5H_4)$
$-(1,2-NC_5H_4)-)][O_3SCF_3]$................ B: B Comp.SVol.3/3-210/1

$BC_{15}F_3H_{28}N_2O_2$ .. $t-C_4H_9-NH-B[O-C(O)-CF_3]-1-NC_5H_6-2,2,6,6-(CH_3)_4$
B: B Comp.SVol.4/3b-52

$BC_{15}F_3H_{30}N_2O_3S$ $t-C_4H_9-N(CH_3)-B[O-S(O)_2-CF_3]-1-NC_5H_6-2,2,6,6-(CH_3)_4$
B: B Comp.SVol.4/3b-52

$BC_{15}F_3H_{36}Si_3$ .... $(CH_3)_2SiF-C[Si(CH_3)_3][BF_2]-Si(C_4H_9-t)_2-CH_3$ B: B Comp.SVol.4/3b-226

$BC_{15}F_4FeH_8O_7Re$ $[(C_5H_5)Fe(CO)_2(H_3C-CC-Re(CO)_5)][BF_4]$ ..... Fe: Org.Comp.B17-123

$BC_{15}F_4FeH_{11}O_2$ .. $[(C_5H_5)(CO)_2Fe=C=CH-C_6H_5][BF_4]$ .... Fe: Org.Comp.B16a-212/3

– .... $[(C_5H_5)Fe(CO)_2(H-CC-C_6H_5)][BF_4]$ .... Fe: Org.Comp.B17-122

$BC_{15}F_4FeH_{11}O_3$ .. $[(1-C_6H_5-C_6H_6)Fe(CO)_3][BF_4]$ .... Fe: Org.Comp.B15-101, 136

– .... $[(1-C_6H_5-C_6H_4D_2)Fe(CO)_3][BF_4]$ .... Fe: Org.Comp.B15-136

– .... $[(2-C_6H_5-C_6H_6)Fe(CO)_3][BF_4]$ .... Fe: Org.Comp.B15-104, 156

– .... $[(3-C_6H_5-C_6H_6)Fe(CO)_3][BF_4]$ .... Fe: Org.Comp.B15-106, 156

$BC_{15}F_4FeH_{13}$ .... $[(C_{10}H_8)Fe(C_5H_5)][BF_4]$ .... Fe: Org.Comp.B19-216,
219/20, 233, 302/5

$BC_{15}F_4FeH_{13}O_2$ .. $[(C_5H_5)(CO)_2Fe=C(CH_3)-C_6H_5][BF_4]$ .... Fe: Org.Comp.B16a-91, 99

– .... $[(C_5H_5)Fe(CO)_2(CH_2=CH-C_6H_5)][BF_4]$ .... Fe: Org.Comp.B17-18, 47/8

– .... $[(C_5H_5)Fe(CO)_2(CHD=CH-C_6H_5)][BF_4]$ .... Fe: Org.Comp.B17-18/9

– .... $[(C_5H_5)Fe(CO)_2(CH_2=C_7H_6-c)][BF_4]$ .... Fe: Org.Comp.B17-72

– .... $[(C_5H_5)Fe(CO)_2(C_8H_8-c)][BF_4]$............... Fe: Org.Comp.B17-92

– .... $[(C_5H_5)Fe(CO)_2-CH_2C_7H_6][BF_4]$ .... Fe: Org.Comp.B14-129/31

$BC_{15}F_4FeH_{13}O_3$ .. $[(CH_3-(CH)_5-C_6H_5)Fe(CO)_3][BF_4]$ .... Fe: Org.Comp.B15-13/4, 22

– .... $[(C_5H_5)(CO)_2Fe=CH-C_6H_4-OCH_3-4][BF_4]$ ..... Fe: Org.Comp.B16a-91, 97/9

$BC_{15}F_4FeH_{13}O_3Ru$
$[(C_5H_5)(CO)Fe(CO)(CH=CH_2)Ru(CO)(C_5H_5)][BF_4]$
Fe: Org.Comp.B16b-148/9

– .... $[(C_5H_5)(CO)Fe(CO)(CCH_3)Ru(CO)(C_5H_5)][BF_4]$ Fe: Org.Comp.B16b-147

$BC_{15}F_4FeH_{15}O_2$ .. $[(C_9H_7)Fe(CO)_2(CH_2=C(CH_3)_2)][BF_4]$......... Fe: Org.Comp.B17-134

$BC_{15}F_4FeH_{15}O_2S$ $[C_5H_5Fe(CO)_2CH_2S(CH_3)C_6H_5][BF_4]$ .... Fe: Org.Comp.B14-133, 135,
142

$BC_{15}F_4FeH_{15}O_6$ .. $[(C_5H_5)Fe(CO)_2(CH_2=(3-C_4H_2O(=O-2)$
$(CO_2CH_3-4)CH_3-5))][BF_4]$ .... Fe: Org.Comp.B17-70

$BC_{15}F_4FeH_{17}$ .... $[(C_{10}H_{12})Fe(C_5H_5)][BF_4]$ .... Fe: Org.Comp.B19-219/20,
225

$BC_{15}F_4FeH_{17}N_2O_4$ $[(C_5H_5)(CO)_2Fe=C(-N(CH_3)-C(=O)-N(CH_3)$
$-C(O-C_2H_5)=CH-)][BF_4]$ .... Fe: Org.Comp.B16a-117, 127

$BC_{15}F_4FeH_{17}O$ ... $[(C_6H_6)Fe(C_9H_{10}-4-OH)][BF_4]$.............. Fe: Org.Comp.B18-142/6,
153, 171, 189

$BC_{15}F_4FeH_{17}O_2$ .. $[(C_5H_5)(CO)_2Fe=CH-C_7H_{11}][BF_4]$............ Fe: Org.Comp.B16a-89, 96

$BC_{15}F_4FeH_{17}O_2$ .. $[(C_5H_5)Fe(CO)_2(CH_2{=}CHCH_2{-}CC{-}C_3H_7{-}n)][BF_4]$ Fe: Org.Comp.B17-42
– .... $[(C_5H_5)Fe(CO)_2(CH_2{=}CH{-}C_6H_9{-}c)][BF_4]$ ...... Fe: Org.Comp.B17-17
– .... $[(C_5H_5)Fe(CO)_2(C_8H_{12})][BF_4]$ .... Fe: Org.Comp.B17-92/3, 101, 107
– .... $[(C_5H_5)(CO)_2Fe{=}C_8H_{12}][BF_4]$ .... Fe: Org.Comp.B16a-92, 100/1
$BC_{15}F_4FeH_{17}O_3$ .. $[(6{-}(n{-}C_4H_9{-}CH{=})C_7H_7)Fe(CO)_3][BF_4]$ .... Fe: Org.Comp.B15-221, 233
– .... $[(C_5H_5)Fe(CO)_2(CH_2{=}CH{-}C_6H_9({=}O){-}2)][BF_4]$... Fe: Org.Comp.B17-16
– .... $[(C_5H_5)Fe(CO)_2(CH_2{=}CH{-}C_6H_9O)][BF_4]$....... Fe: Org.Comp.B17-16
– .... $[(C_5H_5)Fe(CO)_2(C_8H_{12}O)][BF_4]$ .... Fe: Org.Comp.B17-97
– .... $[(C_{12}H_{17})Fe(CO)_3][BF_4]$ .... Fe: Org.Comp.B15-258/9
$BC_{15}F_4FeH_{17}O_4$ .. $[(CH_3O{-}C_{10}H_{11}{-}CH_3)Fe(CO)_3][BF_4]$ .... Fe: Org.Comp.B15-127, 180
$BC_{15}F_4FeH_{17}O_6$ .. $[(3{-}CH_3O{-}6{-}CH_3C(O)OCH_2CH_2{-}6{-}CH_3{-}C_6H_4)Fe(CO)_3][BF_4]$......................... Fe: Org.Comp.B15-130, 182
$BC_{15}F_4FeH_{18}N_3O_3$ $[C_5H_5(CO)_2Fe{=}C({-}NCH_3C({=}O)NCH_3C(N(CH_3)_2)CH{-})][BF_4]$ .... Fe: Org.Comp.B16a-118
$BC_{15}F_4FeH_{19}$ .... $[(1,2,3,5{-}(CH_3)_4{-}C_6H_2)Fe(C_5H_5)][BF_4]$........ Fe: Org.Comp.B19-142, 148
– .... $[(1,2,4,5{-}(CH_3)_4{-}C_6H_2)Fe(C_5H_5)][BF_4]$........ Fe: Org.Comp.B19-142/3, 149, 190/2
$BC_{15}F_4FeH_{19}O$ ... $[(C_7H_9)Fe(CO)(C_7H_{10})][BF_4]$................ Fe: Org.Comp.B17-244/5
– .... $[(C_8H_{11})Fe(CO)(C_6H_8)][BF_4]$................ Fe: Org.Comp.B17-244/5
$BC_{15}F_4FeH_{19}O_2$ .. $[(C_5H_5)Fe(CO)_2(CH_2{=}CH{-}(CH_2)_4{-}CH{=}CH_2)][BF_4]$ Fe: Org.Comp.B17-41
– .... $[(C_5H_5)Fe(CO)_2(CH_2{=}CH{-}C_6H_{11}{-}c)][BF_4]$ ..... Fe: Org.Comp.B17-15
– .... $[(C_5H_5)Fe(CO)_2(C_8H_{14})][BF_4]$............... Fe: Org.Comp.B17-93
– .... $[(C_9H_{11})Fe(CO)_2(CH_2{=}C(CH_3)_2)][BF_4]$ .... Fe: Org.Comp.B17-134
$BC_{15}F_4FeH_{19}O_3$ .. $[(C_5H_5)Fe(CO)_2(C_7H_{11}(OCH_3{-}1))][BF_4]$ .... Fe: Org.Comp.B17-92
$BC_{15}F_4FeH_{19}O_4$ .. $[(C_5H_5)(CO)_2Fe{=}C(OCH_3){-}O{-}C_6H_{11}{-}c][BF_4]$... Fe: Org.Comp.B16a-136, 149
– .... $[(C_5H_5)Fe(CO)_2(CH_2{=}CH{-}C_6H_9(OH)_2{-}3,4)][BF_4]$ Fe: Org.Comp.B17-16
$BC_{15}F_4FeH_{21}O_2$ .. $[((CH_3)_5C_5)Fe(CO)_2(CH_2{=}CHCH_3)][BF_4]$ ...... Fe: Org.Comp.B17-131
– .... $[(C_5H_5)Fe(CO)_2(CH_2{=}CH{-}C_6H_{13}{-}n)][BF_4]$ ..... Fe: Org.Comp.B17-14
– .... $[(C_5H_5)Fe(CO)_2(CH_3CH{=}CH{-}C_5H_{11}{-}n)][BF_4]$... Fe: Org.Comp.B17-58
– .... $[(C_5H_5)Fe(CO)_2(n{-}C_3H_7{-}CH{=}CH{-}C_3H_7{-}n)][BF_4]$ Fe: Org.Comp.B17-58
$BC_{15}F_4FeH_{21}O_3$ .. $[(C_5H_5)Fe(CO)_2(CH_2{=}C(CH_3){-}O{-}CH_2C_4H_9{-}t)][BF_4]$ Fe: Org.Comp.B17-68
– .... $[(C_5H_5)Fe(CO)_2(n{-}C_4H_9{-}CH{=}CH{-}O{-}C_2H_5)][BF_4]$ Fe: Org.Comp.B17-66
$BC_{15}F_4FeH_{21}Si$... $[((CH_3)_3Si{-}CH_2{-}C_6H_5)Fe(C_5H_5)][BF_4]$ .... Fe: Org.Comp.B18-142/6, 199, 219
$BC_{15}F_4FeH_{22}O_4P$ $[(C_5H_5)Fe((CH_3)_2C{=}CH_2)(CO)(P({-}OCH_2{-})_3C{-}CH_3)][BF_4]$ Fe: Org.Comp.B16b-86
– .... $[(C_5H_5)Fe((CH_3)_2C{=}CD_2)(CO)(P({-}OCH_2{-})_3C{-}CH_3)][BF_4]$ Fe: Org.Comp.B16b-86
– .... $[(C_5H_5)Fe(CH_3CH{=}CHCH_3)(CO)(P({-}OCH_2{-})_3C{-}CH_3)][BF_4]$ Fe: Org.Comp.B16b-85
$BC_{15}F_4FeH_{23}O_3Si_2$ $[(3,6{-}((CH_3)_3Si)_2C_6H_5)Fe(CO)_3][BF_4]$......... Fe: Org.Comp.B15-91

$BC_{15}F_4FeH_{24}N_3Pd$ $[(C_5H_5)Fe(C_5H_3(Pd(NH_2\text{-}CH_2CH_2\text{-}NH_2))\text{-}CH_2$
$\text{-}N(CH_3)_2)][BF_4]$ . . . . Fe: Org.Comp.A10-187/8
$BC_{15}F_4FeH_{24}OP$ . . $[C_5H_5(CO)(P(CH_3)_3)Fe{=}C{=}CHC_4H_9\text{-}t][BF_4]$ . . . . Fe: Org.Comp.B16a-208, 210
$BC_{15}F_4FeH_{24}O_4P$ $[(C_5H_5)Fe(c\text{-}C_5H_7CH_3)(CO)P(OCH_3)_3][BF_4]$ . . . Fe: Org.Comp.B16b-91/2
$BC_{15}F_4FeH_{24}O_5P$ $[(C_5H_5)(CO)(P(OCH_3)_3)Fe{=}CCH_3\text{-}O$
$\text{-}CHCH_3\text{-}CH{=}CH_2][BF_4]$ . . . . Fe: Org.Comp.B16a-38, 53
– . . . . $[(C_5H_5)(CO)(P(OCH_3)_3)Fe{=}CCH_3\text{-}O$
$\text{-}CH_2CH{=}CHCH_3][BF_4]$ . . . . Fe: Org.Comp.B16a-38, 53
– . . . . $[(C_5H_5)(CO)(P(OCH_3)_3)Fe{=}C(C_2H_5)\text{-}O\text{-}CH_2CH{=}CH_2][BF_4]$
Fe: Org.Comp.B16a-40, 55
– . . . . $[(C_5H_5)Fe(CO)_2CH_2CH_2P(O\text{-}C_2H_5)_3][BF_4]$ . . . . . Fe: Org.Comp.B14-153, 157
$BC_{15}F_4FeH_{25}O_2Si_2$
$[(1,3\text{-}((CH_3)_3Si)_2C_5H_3)Fe(CO)_2(CH_2{=}CH_2)][BF_4]$
Fe: Org.Comp.B17-133
$BC_{15}F_4FeH_{25}S_2$ . . $[(C_5H_5)Fe(c\text{-}C_6H_8(S(CH_3)_2)_2)][BF_4]$ . . . . Fe: Org.Comp.B16b-8, 10
$BC_{15}F_4FeH_{25}Se_2$ $[(C_5H_5)Fe(c\text{-}C_6H_8(Se(CH_3)_2)_2)][BF_4]$ . . . . Fe: Org.Comp.B16b-8, 10
$BC_{15}F_4FeH_{25}Te_2$. $[(C_5H_5)Fe(c\text{-}C_6H_8(Te(CH_3)_2)_2)][BF_4]$ . . . . Fe: Org.Comp.B16b-8, 10
$BC_{15}F_4FeH_{31}O_6P_2$ $[(C_5H_5)Fe(CH_2{=}CHC_2H_5)(P(OCH_3)_3)_2][BF_4]$ . . . . Fe: Org.Comp.B16b-6
$BC_{15}F_4Fe_2H_{11}O_4S_2$
$[C_5H_5(CO)_2Fe{=}C(SH)SFe(CO)_2C_5H_5][BF_4]$ . . . . Fe: Org.Comp.B16a-140, 152
$BC_{15}F_4H_{11}N_3O_3Re$
$[(CO)_3Re(NC_5H_4\text{-}C_5H_4N)NCCH_3][BF_4]$ . . . . Re: Org.Comp.1-294
$BC_{15}F_4H_{13}MoN_2O_2$
$[(C_9H_7)Mo(CO)_2(NC\text{-}CH_3)_2][BF_4]$. Mo:Org.Comp.7-283, 284, 297
$BC_{15}F_4H_{13}MoO_2$. . $[(C_9H_7)Mo(CO)_2(CH_2{=}CHCH{=}CH_2)][BF_4]$. Mo:Org.Comp.8-305/7
$BC_{15}F_4H_{14}I$ . . . . . . $[C_6H_5\text{-}CH_2\text{-}CH{=}CH\text{-}I\text{-}C_6H_5][BF_4]$ . . . . B: B Comp.SVol.4/3b-219
$BC_{15}F_4H_{14}N_2O_2Re$
$[(C_5H_5)Re(CO)_2\text{-}N{=}N\text{-}C_6H_3(CH_3)_2\text{-}2,4][BF_4]$. . . Re: Org.Comp.3-193/7, 204
– . . . . $[(C_5H_5)Re(CO)_2\text{-}N{=}N\text{-}C_6H_3(CH_3)_2\text{-}2,6][BF_4]$. . . Re: Org.Comp.3-193/7, 204
– . . . . $[(C_5H_5)Re(CO)_2\text{-}N{=}N\text{-}C_6H_3(CH_3)_2\text{-}3,5][BF_4]$. . . Re: Org.Comp.3-193/7, 204
$BC_{15}F_4H_{15}MoO_4$. . $[(C_5H_5)Mo(CO)_2(CH_2{=}CH\text{-}C({=}C{=}CHCH_3)COO\text{-}CH_3)][BF_4]$
Mo:Org.Comp.8-305/6, 310
$BC_{15}F_4H_{15}N_3O_2Re$
$[(C_5H_5)Re(CO)(NC\text{-}CH_3)\text{-}N{=}N\text{-}C_6H_4\text{-}OCH_3\text{-}4][BF_4]$
Re: Org.Comp.3-141
$BC_{15}F_4H_{16}MoN_3O_2$
$[(CH_2CHCH_2)Mo(CO)_2(NH_3)(NC_5H_4\text{-}2\text{-}(2\text{-}C_5H_4N))][BF_4]$
Mo:Org.Comp.5-200, 201
$BC_{15}F_4H_{16}N_2O_2Re$
$[(C_5H_5)Re(CO)_2\text{-}NHN(CH_3)\text{-}C_6H_4\text{-}4\text{-}CH_3][BF_4]$ Re: Org.Comp.3-193/7, 204, 209
$BC_{15}F_4H_{16}N_2O_3Re$
$[(C_5H_5)Re(CO)_2\text{-}NHN(CH_3)\text{-}C_6H_4\text{-}4\text{-}OCH_3][BF_4]$
Re: Org.Comp.3-193/7, 204/5
$BC_{15}F_4H_{16}O_5Re$ . . $[(CO)_5Re(C_{10}H_{16})][BF_4]$ . . . . Re: Org.Comp.2-351, 354
$BC_{15}F_4H_{17}MoO_2$. . $[(C_5H_5)Mo(CO)_2(C_8H_{12})][BF_4]$ Mo:Org.Comp.8-305, 315
$BC_{15}F_4H_{17}Se$ . . . . $[4\text{-}CH_3\text{-}C_6H_4CH_2Se(CH_3)C_6H_5][BF_4]$. . . . . B: B Comp.SVol.3/4-148
$BC_{15}F_4H_{20}MoN$. . . $[C_5H_5Mo(NCCH_3)(H_3C\text{-}CC\text{-}CH_3)_2][BF_4]$ Mo:Org.Comp.6-136/7, 145/6
$BC_{15}F_4H_{24}MnN_3S_6$
$[Mn((SC({=}S))NC_4H_8)_3][BF_4]$ Mn:MVol.D7-178/9

$BC_{15}F_4H_{27}MoP_2$ .. $[C_5H_5Mo(P(CH_3)_2CH_2CH_2P(CH_3)_2)H_3C\text{-}CC\text{-}CH_3][BF_4]$ Mo:Org.Comp.6-118

$BC_{15}F_4H_{29}MoO_6P_2$
$[C_5H_5Mo(P(OCH_3)_3)_2H_3C\text{-}CC\text{-}CH_3][BF_4]$ Mo:Org.Comp.6-118, 128

$BC_{15}F_4H_{29}MoP_2$ .. $[C_5H_5Mo(P(CH_3)_3)_2H_3C\text{-}CC\text{-}CH_3][BF_4]$ Mo:Org.Comp.6-118

$BC_{15}F_4H_{30}MnN_3S_6$
$[Mn(SC(=S)N(C_2H_5)_2)_3][BF_4]$ Mn:MVol.D7-162/3

$BC_{15}F_4H_{34}MoO_{11}P_3$
$[(CH_2C(CH_3)CH_2)Mo(CO)_2(P(OCH_3)_3)_3][BF_4]$ .. Mo:Org.Comp.5-200, 203

$BC_{15}F_5FeH_{12}$ .... $[(2\text{-}F\text{-}C_{10}H_7)Fe(C_5H_5)][BF_4]$ .... Fe: Org.Comp.B19-216, 241

– .... $[(6\text{-}F\text{-}C_{10}H_7)Fe(C_5H_5)][BF_4]$ .... Fe: Org.Comp.B19-216, 241

$BC_{15}F_6FeH_{23}NP$ .. $[(CH_3\text{-}C_6H_5)Fe(2\text{-}CH_3\text{-}1\text{-}(t\text{-}C_4H_9)\text{-}1,2\text{-}NBC_3H_3)][PF_6]$
Fe: Org.Comp.B18-99/100

$BC_{15}F_{10}H_9N_4Si$... $C_6F_5\text{-}B(N_3)\text{-}N(C_6F_5)\text{-}Si(CH_3)_3$.............. B: B Comp.SVol.3/3-110

– .... $C_6F_5\text{-}B[\text{-}N(C_6F_5)\text{-}N{=}N\text{-}N(Si(CH_3)_3)\text{-}]$ .... B: B Comp.SVol.3/3-137/8

$BC_{15}FeH_{14}^-$ ..... $[C_5H_5FeC_4H_4BC_6H_5]^-$.................... Fe: Org.Comp.B17-223

$BC_{15}FeH_{14}NO_2$... $(C_5H_5)Fe[C_5H_3(B(OH)_2)(2\text{-}C_5H_4N)]$ .... Fe: Org.Comp.A9-287, 288

$BC_{15}FeH_{14}Na$.... $Na[C_5H_5FeC_4H_4BC_6H_5]$ .... Fe: Org.Comp.B17-223

$BC_{15}FeH_{15}$ ...... $C_5H_5Fe(H)C_4H_4BC_6H_5$ .... Fe: Org.Comp.B17-200, 226/7

$BC_{15}FeH_{18}N$ ..... $[1\text{-}BH_3\text{-}2,3\text{-}(CH_3)_2\text{-}1\text{-}NC_8H_4]Fe(C_5H_5)$....... Fe: Org.Comp.B19-279

$BC_{15}FeH_{23}INO_2$ .. $[C_5H_5FeC_5H_3(B(OH)_2)CH_2CH_2N(CH_3)_3]I$ ...... Fe: Org.Comp.A9-288

$BC_{15}FeH_{23}N^+$.... $[(CH_3\text{-}C_6H_5)Fe(2\text{-}CH_3\text{-}1\text{-}(t\text{-}C_4H_9)\text{-}1,2\text{-}NBC_3H_3)]^+$
Fe: Org.Comp.B18-99/102

$BC_{15}FeH_{23}NO_2^+$.. $[C_5H_5FeC_5H_3(B(OH)_2)CH_2CH_2N(CH_3)_3]^+$ ..... Fe: Org.Comp.A9-288

$BC_{15}FeH_{28}O_2P$ ... $C_5(CH_3)_5(CO)(P(CH_3)_3)Fe{=}C(H)OBH_3$ .... Fe: Org.Comp.B16a-176/7

$BC_{15}GeH_{33}Sn_2$ ... $(CH_3)_2Ge(\text{-}C(Sn(CH_3)_3){=}C(B(CH_3)_2)$
$\text{-}C(CH_3){=}C(Sn(CH_3)_3)\text{-})$ Ge:Org.Comp.3-278/9

$BC_{15}H_{12}NOS_2$.... $C_6H_5B(\text{-}NOCH_3\text{-}C_4H_2S\text{-}C_4H_2S\text{-})$............ B: B Comp.SVol.3/3-164

$BC_{15}H_{17}MoN_6O_2$. $[CH_2C(CH_3)CH_2]Mo(CO)_2[\text{-}N_2C_3H_3\text{-}]_3BH$ Mo:Org.Comp.5-225/6, 228, 231, 232

$BC_{15}H_{17}N_2O_7$ .... $(CH_3COO)_2B[\text{-}N(C_6H_4\text{-}4\text{-}NO_2){=}CCH_3\text{-}CH{=}CCH_3\text{-}O\text{-}]$
B: B Comp.SVol.3/3-218

$BC_{15}H_{17}S_2$ ...... $2,2\text{-}(C_6H_5)_2\text{-}1,3\text{-}S_2C_3H_4 \cdot BH_3$............. B: B Comp.SVol.4/4-134

$BC_{15}H_{18}N$....... $(C_6H_5)_2B\text{-}NCH_3\text{-}C_2H_5$ .... B: B Comp.SVol.3/3-153

– .... $(C_6H_5)_2B\text{-}NH\text{-}C_3H_7\text{-}i$.................... B: B Comp.SVol.3/3-153

$BC_{15}H_{18}NO_5$..... $(CH_3COO)_2B[\text{-}NC_6H_5{=}CCH_3\text{-}CH{=}CCH_3\text{-}O\text{-}]$... B: B Comp.SVol.3/3-218

$BC_{15}H_{19}N_4Si$..... $(C_6H_5)B[\text{-}NC_6H_5\text{-}N{=}N\text{-}NSi(CH_3)_3\text{-}]$.......... B: B Comp.SVol.3/3-137/8

$BC_{15}H_{20}NO_4$..... $[\text{-}B(C_2H_5)_2\text{-}O\text{-}C({=}O)\text{-}CH_2\text{-}N({=}CH\text{-}C_6H_4\text{-}2\text{-}COOCH_3)\text{-}]$
B: B Comp.SVol.4/3b-70

$BC_{15}H_{20}NSi$ ..... $(C_6H_5)_2B\text{-}NHSi(CH_3)_3$.................... B: B Comp.SVol.3/3-153

$BC_{15}H_{21}MoN_4O_2$. $(CH_2CHCH_2)Mo(CO)_2[\text{-}N_2C_3H(CH_3)_2\text{-}]_2BH_2$ ... Mo:Org.Comp.5-213, 214, 218

– .... $(CH_2CHCH_2)Mo(CO)_2[\text{-}N_2C_3H_3\text{-}]_2B(C_2H_5)_2$ ... Mo:Org.Comp.5-213, 214/5, 218/9

$BC_{15}H_{21}N_2$ ...... $NC_5H_4\text{-}2\text{-}[CH(CH_3)\text{-}NH(BH_3)\text{-}CH(CH_3)\text{-}C_6H_5]$ B: B Comp.SVol.4/3b-7

$BC_{15}H_{21}O_2S$..... $2\text{-}(C_6H_5\text{-}S\text{-}CH_2\text{-}CH{=}CH)\text{-}4,4,5,5\text{-}(CH_3)_4\text{-}1,3,2\text{-}O_2BC_2$
B: B Comp.SVol.4/4-138

$BC_{15}H_{22}N$....... $2\text{-}BC_9H_7 \cdot N(C_2H_5)_3$ .... B: B Comp.SVol.4/3b-26

– .... $2\text{-}CH_3\text{-}[3.3.1.1^{3,7}]\text{-}1\text{-}BC_9H_{14} \cdot NC_5H_5$........ B: B Comp.SVol.4/3b-30

– .... $3\text{-}BC_{10}H_{17} \cdot NC_5H_5$..................... B: B Comp.SVol.3/3-191

– .... $(C_2H_5\text{-}CH{=}CH)_2B\text{-}NCH_3\text{-}C_6H_5$ .... B: B Comp.SVol.3/3-152

$BC_{15}H_{22}N$ .... $[3.3.1.1^{3,7}]$-1-BC_9H_{15} · NC_5H_4-2-CH_3 .... B: B Comp.SVol.4/3b-30

– .... $[3.3.1.1^{3,7}]$-1-BC_9H_{15} · NC_5H_4-3-CH_3 .... B: B Comp.SVol.4/3b-30

– .... $[3.3.1.1^{3,7}]$-1-BC_9H_{15} · NH_2-C_6H_5 .... B: B Comp.SVol.3/3-191

– .... $[CH_3CH{=}C(CH_3)]_2B$-NCH_3-C_6H_5 .... B: B Comp.SVol.3/3-152

$BC_{15}H_{22}N_5$ .... $BC_3H_3N_5(C_6H_5)(C_3H_7$-n$)_2$ .... B: B Comp.SVol.4/3a-179

– .... [-$B(C_6H_5)$-$N(C_4H_9$-t)-N=N-$N(CH_2CH_2$-CH_2CH_2-CN)-] B: B Comp.SVol.4/3a-199

$BC_{15}H_{23}N^+$ .... $[(-C(CH_3)_2-(CH_2)_3-C(CH_3)_2-)N{=}BC_6H_5]^+$ .... B: B Comp.SVol.3/3-205

$BC_{15}H_{23}NO_2$ .... 2-CH_3-C_6H_4-B[-O-CH_2-$CH(CH_3)$-O-] · NC_5H_{10} B: B Comp.SVol.4/3b-69

$BC_{15}H_{23}N_2$ .... $C_6H_5B(1$-$NC_4H_8)$-NC_5H_{10} .... B: B Comp.SVol.3/3-109

$BC_{15}H_{23}N_6$ .... $BC_2H_2N_6(C_6H_4$-2-$CH_3)(C_3H_7$-n$)_2$ .... B: B Comp.SVol.4/3a-178

– .... $BC_2H_2N_6(C_6H_4$-4-$CH_3)(C_3H_7$-n$)_2$ .... B: B Comp.SVol.4/3a-178

$BC_{15}H_{24}N$ .... 2-NC_9H_9 · $B(C_2H_5)_3$ .... B: B Comp.SVol.4/3b-29

– .... 9-[1-$(C_5H_4N(C_2H_5)$-2)]-[3.3.1]-9-BC_8H_{15} .... B: B Comp.SVol.3/3-189

$BC_{15}H_{24}NO$ .... 1-[C_6H_5-O-BH]-NC_5H_6-2,2,6,6-$(CH_3)_4$ .... B: B Comp.SVol.4/3b-61

$BC_{15}H_{24}NSi$ .... 1-C_6H_5-4,5-$(C_2H_5)_2$-2,2,3-$(CH_3)_3$-1,2,5-$NSiBC_2$ B: B Comp.SVol.4/3a-250

– .... NC_5H_5 · $(CH_3)_2B$-C_5H_4-$Si(CH_3)_3$ .... B: B Comp.SVol.4/3b-28

$BC_{15}H_{25}N_2$ .... $(C_2H_5)_2N$-$B(C_6H_5)$-1-NC_5H_{10} .... B: B Comp.SVol.3/3-109

– .... s-C_4H_9-NH-$B(C_6H_5)$-1-NC_5H_{10} .... B: B Comp.SVol.3/3-108

– .... t-C_4H_9-NH-$B(C_6H_5)$-1-NC_5H_{10} .... B: B Comp.SVol.3/3-108

$BC_{15}H_{25}N_2O$ .... $BC_7H_6N_2(O$-$C_2H_5)(C_3H_7$-i$)_2$ .... B: B Comp.SVol.4/3a-257

– .... $BC_7H_6N_2(O$-$C_2H_5)(C_3H_7$-n$)_2$ .... B: B Comp.SVol.4/3a-257

$BC_{15}H_{25}N_2Si$ .... 1-(2-NH_2-C_6H_4)-4,5-$(C_2H_5)_2$-2,2,3-$(CH_3)_3$-1,2,5-$NSiBC_2$ B: B Comp.SVol.4/3a-250

– .... 1-C_6H_5-5,6-$(C_2H_5)_2$-3,3,4-$(CH_3)_3$-1,2,3,6-N_2SiBC_2H B: B Comp.SVol.4/3a-252

– .... 1-(C_6H_5-NH)-4,5-$(C_2H_5)_2$-2,2,3-$(CH_3)_3$-1,2,5-$NSiBC_2$ B: B Comp.SVol.4/3a-250/1

– .... 2-C_6H_5-5,6-$(C_2H_5)_2$-3,3,4-$(CH_3)_3$-1,2,3,6-N_2SiBC_2H B: B Comp.SVol.4/3a-252

$BC_{15}H_{26}NO$ .... C_6H_5-$B(OCH_3)$-$N(C_4H_9$-n$)_2$ .... B: B Comp.SVol.3/3-154

– .... C_6H_5-$B(OCH_3)$-$N(C_4H_9$-s$)_2$ .... B: B Comp.SVol.3/3-154

$BC_{15}H_{26}N_2PS$ .... C_6H_5-B[-$N(C_4H_9$-t)-$P(CH_3)(=S)$-$N(C_4H_9$-t)-] .... B: B Comp.SVol.4/3a-207/8

$BC_{15}H_{26}N_3$ .... $C_6H_5N{=}C(C_4H_9$-t$)B[N(CH_3)_2]_2$ .... B: B Comp.SVol.3/3-109

$BC_{15}H_{26}N_3O$ .... $BC_6H_5N_3(=O)(CH_3)(C_4H_9$-n$)_2$ .... B: B Comp.SVol.4/3a-180/1

$BC_{15}H_{26}N_3S_2$ .... [-$B(C_4H_9$-n$)_2$-$N(CH_2$-$CH{=}CH_2)$-C(=S)-NH-C_3H_2NS-] B: B Comp.SVol.4/3a-179; B: B Comp.SVol.4/4-149/50

$BC_{15}H_{27}N_2$ .... [-$N(C_3H_7$-i)-$CH(C_6H_5)$-$N(C_3H_7$-i)-CH_2CH_2-] · BH_3 B: B Comp.SVol.4/3b-12

$BC_{15}H_{28}N$ .... $(C_2H_5)_2B$-$N(C_5H_7$-c)-C_6H_{11}-c .... B: B Comp.SVol.3/3-152

– .... [3.3.1]-9-BC_8H_{15} · [2.2.2]-1-NC_7H_{13} .... B: B Comp.SVol.4/3b-29

– .... [4.2.1]-9-BC_8H_{15} · [2.2.2]-1-NC_7H_{13} .... B: B Comp.SVol.4/3b-30

$BC_{15}H_{28}NSn$ .... NC_4H_4-1-$B(C_2H_5)$-$C(C_2H_5){=}C(CH_3)$-$Sn(CH_3)_2$-C_2H_5 B: B Comp.SVol.4/3a-235

$BC_{15}H_{29}N_2$ .... C_5H_5-$B[N(C_3H_7$-i$)_2]$-NH-C_4H_9-t .... B: B Comp.SVol.4/3a-170

$BC_{15}H_{29}N_2O_2$ .... 2,2,6,6-$(CH_3)_4NC_5H_6$-1-B[-NHC(O)-$CH(C_4H_9$-i)-O-] B: B Comp.SVol.4/3b-56

$BC_{15}H_{29}N_2O_2$ .... $2,2,6,6-(CH_3)_4NC_5H_6-1-B[-NHC(O)-CH(C_4H_9-s)-O-]$
B: B Comp.SVol.4/3b-56
$BC_{15}H_{29}N_2Si_2$.... $1,3-[(CH_3)_3Si]_2-2-C_6H_5-1,3,2-N_2BC_3H_6$ ...... B: B Comp.SVol.4/3a-175
$BC_{15}H_{29}N_2Sn$ .... $BC_5H_2N_2Sn-(CH_3)_4-(C_2H_5)_3$ .... B: B Comp.SVol.4/3a-258
Sn: Org.Comp.19-56, 59
$BC_{15}H_{30}N$ .... $[3.3.1]-9-BC_8H_{15} \cdot NC_5H_9(CH_3)_2-2,6$ .... B: B Comp.SVol.3/3-189
– .... $[3.3.1.1^{3,7}]-1-BC_9H_{15} \cdot (n-C_3H_7)_2NH$ .... B: B Comp.SVol.4/3b-30
– .... $[3.3.1.1^{3,7}]-1-BC_9H_{15} \cdot N(C_2H_5)_3$ .... B: B Comp.SVol.4/3b-30
$BC_{15}H_{30}N_3S_2$ .... $CH_3B[-N(C_4H_9-n)-C(S)-N(C_4H_9-n)-C(S)-N(C_4H_9-n)-]$
B: B Comp.SVol.3/3-116
B: B Comp.SVol.3/4-131
$BC_{15}H_{31}NO_2P$.... $9-[(C_2H_5-O)_2P-N(C_3H_7-i)]-[3.3.1]-9-BC_8H_{14}$.. B: B Comp.SVol.4/3a-236
$BC_{15}H_{31}N_2O_2$ .... $t-C_4H_9-NH-B[O-C(O)-CH_3]-1-NC_5H_6-2,2,6,6-(CH_3)_4$
B: B Comp.SVol.4/3b-52
$BC_{15}H_{32}N_3$ ...... $1-[(i-C_3H_7)_2N-N=B]-2,2,6,6-(CH_3)_4-NC_5H_6$ ... B: B Comp.SVol.4/3a-160/5
– .... $[-B(NC_5H_6(CH_3)_4)-N(i-C_3H_7)-N(i-C_3H_7)-]$ .... B: B Comp.SVol.3/3-135
$BC_{15}H_{32}N_3O_2$ .... $t-C_4H_9-NH-B[O-C(O)-CH_2-NH_2]$
$-1-NC_5H_6-2,2,6,6-(CH_3)_4$ .... B: B Comp.SVol.4/3b-52
$BC_{15}H_{33}N_2OSi$ ... $2-[(CH_3)_3Si-N(C_4H_9-t)]-3-(t-C_4H_9)$
$-4-CH_2=CCH_3-1,3,2-ONBCH$. .... B: B Comp.SVol.4/3b-55
– .... $2-[(CH_3)_3Si-N(C_4H_9-t)]-3-(t-C_4H_9)$
$-4-CH_3CH=CH-1,3,2-ONBCH$ .... B: B Comp.SVol.4/3b-55
$BC_{15}H_{34}NSn$..... $(C_2H_5)_2B[-C(C_2H_5)=CCH_3-Sn(CH_3)_2-N(C_2H_5)_2-]$
B: B Comp.SVol.4/3a-255
Sn: Org.Comp.19-47/8, 50
– .... $(C_2H_5)_2B[-C(C_2H_5)=CCH_3-Sn(CH_3)_2-NH(C_4H_9-n)-]$
B: B Comp.SVol.4/3a-255
Sn: Org.Comp.19-48/50
$BC_{15}H_{34}N_3$ ...... $1-[(CH_3)_2N-B(NH-C_4H_9-t)]-2,2,6,6-(CH_3)_4-NC_5H_6$
B: B Comp.SVol.4/3a-155
$BC_{15}H_{35}N_2$ ...... $(C_2H_5)_2N-CH_2-CH_2-N(C_2H_5)_2-BH[C_5H_{10}]$ .... B: B Comp.SVol.3/3-189
$BC_{15}H_{35}N_2Si$..... $t-C_4H_9-CH_2-C(CH_3)_2-N=B-N(C_4H_9-t)-Si(CH_3)_3$
B: B Comp.SVol.4/3a-160/5
$BC_{15}H_{35}N_4$ ...... $1-[(CH_3)_2N-NH-B(NH-C_4H_9-t)]-2,2,6,6-(CH_3)_4-NC_5H_6$
B: B Comp.SVol.4/3a-155
– .... $1-[CH_3NH-N(CH_3)-B(NH-C_4H_9-t)]-2,2,6,6-(CH_3)_4-NC_5H_6$
B: B Comp.SVol.4/3a-155
$BC_{15}H_{36}IN_2Si$ .... $[(t-C_4H_9)_2N=B=N(C_4H_9-t)-Si(CH_3)_3]I$. .... B: B Comp.SVol.4/4-114/5
$BC_{15}H_{36}LiN_2Si$ ... $LiN(C_4H_9-t)-B(C_4H_9-n)-N(C_4H_9-t)-Si(CH_3)_3$
$\cdot (CH_3)_2N-CH_2CH_2-N(CH_3)_2$ .... B: B Comp.SVol.4/3a-169
– .... $LiN(C_4H_9-t)-B(C_4H_9-t)-N(C_4H_9-t)-Si(CH_3)_3$
$\cdot (CH_3)_2N-CH_2CH_2-N(CH_3)_2$ .... B: B Comp.SVol.4/3a-169
$BC_{15}H_{36}N_3Si$..... $[(CH_3)_2N]_2BN(i-C_3H_7)C[Si(CH_3)_3]CHC_3H_7$..... B: B Comp.SVol.3/3-96
$BC_{15}H_{37}N_2Si$..... $i-C_4H_9B[NH(i-C_4H_9)]N(t-C_4H_9)Si(CH_3)_3$ ...... B: B Comp.SVol.3/3-107
$BC_{15}H_{38}N_3Si$..... $(CH_3)_3Si-N(C_4H_9-t)-B(NH-C_4H_9-t)_2$ .... B: B Comp.SVol.4/3a-154
$BC_{16}ClFeH_{14}N_2$ .. $HNC_6H_4HNB-C_5H_4FeC_5H_4-Cl$ .... Fe: Org.Comp.A9-280, 284
$BC_{16}ClH_{10}MnN_2O_4$
$Mn[(1,3-(O=)_2-2-NC_8H_4-2)_2BH_2]Cl$. .... Mn: MVol.D8-7/8
$BC_{16}ClH_{14}O_6$ .... $C_6H_5(ClO_4)B[-O=CC_6H_5-CH=CCH_3-O-]$ ...... B: B Comp.SVol.3/4-52
$BC_{16}ClH_{15}NS$ .... $ClB[-NC_6H_5-(3,4-C_4H_2S)-(1,2-C_6H_8)-]$. ...... B: B Comp.SVol.3/4-57

$BC_{16}ClH_{15}O_6Re$ .. cis-$(CO)_4Re[(CH_3CO)(CH(CH_3)_2CO)]B(C_6H_5)Cl$ Re: Org.Comp.2-346
$BC_{16}ClH_{16}$ ...... 1-Cl-2,5-$(C_6H_5)_2$-BC_4H_6 .... B: B Comp.SVol.4/4-47
$BC_{16}ClH_{18}$ ...... $(C_6H_5$-$CH_2CH_2)_2BCl$ .... B: B Comp.SVol.4/4-47
$BC_{16}ClH_{18}O$ ..... t-C_4H_9-BCl-O-C_6H_4-2-C_6H_5 .... B: B Comp.SVol.4/4-55
$BC_{16}ClH_{19}N$ ..... $(C_6H_5)ClBN(s$-$C_4H_9)C_6H_5$ .... B: B Comp.SVol.3/4-56
$BC_{16}ClH_{20}N_2$ .... [B(-$NHCH_3$-CH_2-(1,2-C_6H_4)-$)_2$]Cl .... B: B Comp.SVol.3/4-62
$BC_{16}ClH_{23}N$ ..... 2-($ClCH_2$-CH_2)-[3.3.1.$1^{3,7}$]-1-BC_9H_{14} · NC_5H_5
B: B Comp.SVol.4/3b-30
$BC_{16}ClH_{23}N_3S$ ... [(n-$C_3H_7)_2$B(-NH=CC_6H_5-NH-(2,3-(1,3-C_3H_2NS))))]Cl
B: B Comp.SVol.3/4-62
$BC_{16}ClH_{25}N$ ..... 1-(C_6H_5-CH_2-BCl)-2,2,6,6-$(CH_3)_4$-NC_5H_6 .... B: B Comp.SVol.4/4-60
$BC_{16}ClH_{27}N$ ..... $(C_6H_5)ClBN(i$-$C_5H_{11})_2$ .... B: B Comp.SVol.3/4-56
$BC_{16}ClH_{29}NSi$ .... t-C_4H_9-BCl-N[$Si(CH_3)_3$]-C_6H_2-2,4,6-$(CH_3)_3$ .. B: B Comp.SVol.4/4-60
$BC_{16}ClH_{33}N_2PSi_2$ [$(CH_3)_3Si]_2$N-B(C_4H_9-t)-NH-PCl-C_6H_5 .... B: B Comp.SVol.4/3a-170
$BC_{16}ClH_{34}$ ...... (n-$C_8H_{17})_2BCl$ .... B: B Comp.SVol.3/4-44
$BC_{16}ClH_{39}N$ ..... [$(C_4H_9)_4$N][H_3BCl] .... B: B Comp.SVol.4/4-39
$BC_{16}ClH_{46}N_3PSi_4$ [$(CH_3)_3Si]_2$N-B(C_4H_9-t)-NH-PCl-N[$Si(CH_3)_3]_2$ B: B Comp.SVol.4/3a-170
$BC_{16}Cl_2H_{30}N$ .... $(C_2H_5)_3N$ · BCl_2-$C_5(CH_3)_5$ .... B: B Comp.SVol.4/4-70
$BC_{16}Cl_2H_{34}N_2P$ .. 2,2,6,6-$(CH_3)_4$-NC_5H_6-1-BCl-N(C_4H_9-t)-PCl-C_3H_7-i
B: B Comp.SVol.4/4-63
$BC_{16}Cl_3F_4H_{31}MnN_3S_6$
[Mn(SC(=S)N$(C_2H_5)_2)_3$]BF_4 · $CHCl_3$ .... Mn: MVol.D7-162/3
$BC_{16}Cl_3H_{37}N$ .... [(n-$C_4H_9)_4$N][$HBCl_3$] .... B: B Comp.SVol.3/4-38
$BC_{16}CoH_{26}SeSi$ .. $BC_2CoSeSi(C_5H_5)(C_2H_5)_2(CH_3)_2[C(CH_3)=CH_2]$ B: B Comp.SVol.4/4-180
$BC_{16}Co_2H_{16}NO_6$ .. [CH_3C≡C][$CH_3C_2Co_2(CO)_6$]B-N$(C_2H_5)_2$ .... B: B Comp.SVol.3/3-156
$BC_{16}FH_{27}N$ ...... F(C_6H_5)B-N(i-$C_5H_{11})_2$ .... B: B Comp.SVol.3/3-376
$BC_{16}FH_{29}NSi$ .... CH_3-BF-N[$Si(CH_3)_3$]-C_6H_3-2,6-(C_3H_7-i$)_2$ .... B: B Comp.SVol.4/3b-245
$BC_{16}FH_{36}N_2Si$ .... [$(CH_3)_4C_5H_6$N]BF-N[$Si(CH_3)_3$][t-C_4H_9] .... B: B Comp.SVol.3/3-376
$BC_{16}FH_{38}NPSi_2$ .. 2,2,6,6-$(CH_3)_4$-NC_5H_6-1-BF-PH-CH[$Si(CH_3)_3]_2$
B: B Comp.SVol.4/3b-243
$BC_{16}FH_{41}NSi_3$ .... [$(CH_3)_3Si]_3$C-BF-N(C_3H_7-i$)_2$ .... B: B Comp.SVol.4/3b-244
$BC_{16}FH_{43}NSi_4$ .... [$(CH_3)_3Si]_3$C-BF-N(C_3H_7-i)-$Si(CH_3)_3$ .... B: B Comp.SVol.4/3b-245
$BC_{16}FH_{45}NSi_5$ .... [$(CH_3)_3Si]_3$C-BF-N[$Si(CH_3)_3]_2$ .... B: B Comp.SVol.4/3b-245
– .... [$(CH_3)_3Si]_3$Si-BF-N(C_4H_9-t)-$Si(CH_3)_3$ .... B: B Comp.SVol.4/3b-245
$BC_{16}F_2FeH_{15}O_3$ .. $F_2BO_2C_4H_4$-$C_5H_4FeC_5H_4$-$COCH_3$ .... Fe: Org.Comp.A9-280
$BC_{16}F_2FeH_{15}O_4$ .. [F_2BOC=CHC(CH_3)=O]$C_5H_4FeC_5H_4$-C(O)OCH_3 Fe: Org.Comp.A9-280
$BC_{16}F_2FeH_{23}O_3$ .. (C_5H_5)Fe(C(C_4H_9-t)O$)_2BF_2$(CO) .... Fe: Org.Comp.B16b-127, 139
$BC_{16}F_2FeH_{26}NO_3$ [N$(CH_3)_4$][C_5H_5Fe($CH_2C_2O_2BF_2$(C_3H_7-i))CO] .. Fe: Org.Comp.B17-156, 158
$BC_{16}F_2H_{11}O_3Te$ .. [-BF_2-O-$TeC_9H_3(C_6H_5)(OCH_3)$-O-] .... B: B Comp.SVol.4/4-177
$BC_{16}F_2H_{12}N_3O$ ... F_2B[-O-C$(C_5H_4N)_2$-C_5H_4N-] .... B: B Comp.SVol.3/3-378
$BC_{16}F_2H_{13}O_2$ .... [-B(F$)_2$OC(C_6H_5)C(C_6H_5)C(CH_3)O-] .... B: B Comp.SVol.4/3b-232/3
$BC_{16}F_2H_{14}NO_2$ ... $(CH_3)_2$B-N(C_6H_4-4-F)-C(O)C(O)(C_6H_4-4-F) ... B: B Comp.SVol.3/3-155
$BC_{16}F_2H_{19}N_6NiO_2S$
Ni[-BF_2-O-N=C(CH_3)-C(CH_3)=N-NC(S-CH_2-
C_6H_5)N-N=C(CH_3)-C(CH_3)=N-O-] .... B: B Comp.SVol.4/3b-124
$BC_{16}F_3H_{23}NO_3S$ .. 2,4-$(CH_3)_2$-NC_5H_3 · [3.3.1]-9-BC_8H_{14}-9-[O-S(O$)_2$-CF_3]
B: B Comp.SVol.4/3b-69
– .... [9-(2,6-$(CH_3)_2$-NC_5H_3-1)-(3.3.1)-9-BC_8H_{14}][CF_3-SO_3]
B: B Comp.SVol.4/3b-45

$BC_{16}F_3H_{23}NSi$.... 1-(2-CF_3-C_6H_4)-4,5-$(C_2H_5)_2$-2,2,3-$(CH_3)_3$-1,2,5-$NSiBC_2$
B: B Comp.SVol.4/3a-250

$BC_{16}F_3H_{23}NSn$ ... 1-(2-CF_3-C_6H_4)-4,5-$(C_2H_5)_2$-2,2,3-$(CH_3)_3$-1,2,5-$NSnBC_2$ .... B: B Comp.SVol.4/3a-251
Sn: Org.Comp.19-55, 58

$BC_{16}F_3H_{27}NO_3S$ .. 2,4-$(CH_3)_2$-NC_5H_3 · $B(C_4H_9\text{-}n)_2$-O-$S(O)_2$-CF_3
B: B Comp.SVol.4/4-165

– .... [2,6-$(CH_3)_2$-NC_5H_3-1-$B(C_4H_9\text{-}n)_2$][CF_3-SO_3] B: B Comp.SVol.4/3b-45
B: B Comp.SVol.4/4-165

$BC_{16}F_3H_{44}NOsSi_4^-$
[$N(C_4H_9\text{-}n)_4$][$((CH_3)_3Si\text{-}CH_2)_4OsN\text{-}BF_3$] ...... Os: Org.Comp.A1-32

$BC_{16}F_3H_{50}N_6Si_8$ .. $FB[N(\text{-}Si(CH_3)_2\text{-}N(SiF(CH_3)_2)\text{-}Si(CH_3)_2\text{-}NH\text{-}Si(CH_3)_2\text{-})]_2$
B: B Comp.SVol.3/3-376

$BC_{16}F_4$ .... $C_{16}BF_4$ .... B: B Comp.SVol.3/3-251

$BC_{16}F_4FeH_{11}O_3$ .. [$(C_{13}H_{11})Fe(CO)_3$][BF_4] .... Fe: Org.Comp.B15-254/5

$BC_{16}F_4FeH_{11}O_4$ .. [$(1\text{-}C_6H_5\text{-}7\text{-}O{=}C_7H_6)Fe(CO)_3$][$BF_4$] .... Fe: Org.Comp.B15-212

$BC_{16}F_4FeH_{13}O_2$ .. [$(C_5H_5)Fe(CO)_2(CH_2{=}C{=}CH\text{-}C_6H_5)$][$BF_4$] ...... Fe: Org.Comp.B17-113, 117/8

– .... [$(C_5H_5)Fe(CO)_2(C_9H_8)$][BF_4] .... Fe: Org.Comp.B17-97/8

$BC_{16}F_4FeH_{14}NO_2$ [$(C_5H_5)Fe(CN\text{-}C_6H_3(CH_3)_2\text{-}2{,}6)(CO)_2$][$BF_4$] .... Fe: Org.Comp.B15-315, 318

$BC_{16}F_4FeH_{15}$ .... [$(1\text{-}CH_3\text{-}C_{10}H_7)Fe(C_5H_5)$][$BF_4$] .... Fe: Org.Comp.B19-216, 220, 237, 305/6

– .... [$(5\text{-}CH_3\text{-}C_{10}H_7)Fe(C_5H_5)$][$BF_4$] .... Fe: Org.Comp.B19-216, 220, 237, 305/6

– .... [$(C_5H_5\text{-}C_6H_5)Fe(C_5H_5)$][$BF_4$] .... Fe: Org.Comp.B18-142/6, 200, 219

$BC_{16}F_4FeH_{15}O_2$ .. [$(C_5H_5)Fe(CO)_2(CH_2{=}CHCH_2\text{-}C_6H_5)$][$BF_4$] ..... Fe: Org.Comp.B17-31

– .... [$(C_5H_5)Fe(CO)_2(CH_3CH{=}CH\text{-}C_6H_5)$][$BF_4$] ..... Fe: Org.Comp.B17-60

$BC_{16}F_4FeH_{15}O_3$ .. [$(CH_3\text{-}CHC(CH_3)CHCHCH\text{-}C_6H_5)Fe(CO)_3$][$BF_4$] Fe: Org.Comp.B15-25

– .... [$(C_5H_5)Fe(CO)_2(CH_2{=}CHCH_2\text{-}O\text{-}C_6H_5)$][$BF_4$] .. Fe: Org.Comp.B17-25

$BC_{16}F_4FeH_{15}O_3Ru$
[$(C_5H_5)(CO)Fe(CO)(CH{=}CHCH_3)Ru(CO)(C_5H_5)$][$BF_4$]
Fe: Org.Comp.B16b-149/50

$BC_{16}F_4FeH_{15}O_4$ .. [$(CH_3(CH)_5\text{-}C_6H_4\text{-}4\text{-}OCH_3)Fe(CO)_3$][$BF_4$] ..... Fe: Org.Comp.B15-22

$BC_{16}F_4FeH_{16}NO_3$ [$(C_5H_5)Fe(CO)_2CH(C(O)CH_3)CH_2\text{-}NC_5H_5$][$BF_4$] Fe: Org.Comp.B14-151

$BC_{16}F_4FeH_{17}OS$ .. [$C_5H_5Fe(CH_2S(C_6H_5)CH_2CH{=}CH_2)CO$][$BF_4$] ... Fe: Org.Comp.B17-166/7

$BC_{16}F_4FeH_{17}O_2$ .. [$(C_5H_5)Fe(CO)_2(CH_2{=}CH\text{-}C_7H_9)$][$BF_4$] .... Fe: Org.Comp.B17-19, 99

$BC_{16}F_4FeH_{17}O_2S$ [$C_5H_5Fe(CO)_2CH(CH_3)S(CH_3)C_6H_5$][$BF_4$] ..... Fe: Org.Comp.B14-133, 135, 143

$BC_{16}F_4FeH_{17}O_3$ .. [$(8\text{-}(n\text{-}C_4H_9\text{-}CH{=})C_8H_7)Fe(CO)_3$][$BF_4$] .... Fe: Org.Comp.B15-242/3

– .... [$(8\text{-}(n\text{-}C_4H_9\text{-}CD{=})C_8H_7)Fe(CO)_3$][$BF_4$] .... Fe: Org.Comp.B15-242/3

$BC_{16}F_4FeH_{17}O_4$ .. [$(C_5H_5)Fe(CO)_2(C_9H_{12}O_2)$][$BF_4$] .... Fe: Org.Comp.B17-71, 82

$BC_{16}F_4FeH_{18}NO_2$ [$C_5H_5Fe(CO)_2CH_2CH_2NH_2CH_2C_6H_5$][$BF_4$] ..... Fe: Org.Comp.B14-150, 155

$BC_{16}F_4FeH_{18}OP$ .. [$C_5H_5(CO)(P(CH_3)_2C_6H_5)Fe{=}C{=}CH_2$][$BF_4$] ..... Fe: Org.Comp.B16a-207, 209

$BC_{16}F_4FeH_{18}O_2P$ [$C_5H_5Fe(CO)_2CH_2P(CH_3)_2C_6H_5$][$BF_4$] .... Fe: Org.Comp.B14-137

$BC_{16}F_4FeH_{19}O$ ... [$(1{,}3{,}5\text{-}(CH_3)_3\text{-}C_6H_3)Fe(C_5H_4\text{-}C({=}O)\text{-}CH_3)$][$BF_4$]
Fe: Org.Comp.B19-99, 107

$BC_{16}F_4FeH_{19}O_2$ .. [$(C_5H_5)(CO)_2Fe{=}CH\text{-}C_8H_{13}$][$BF_4$] .... Fe: Org.Comp.B16a-89, 96/7

– .... [$(C_5H_5)Fe(CO)_2(C_9H_{14})$][$BF_4$] .... Fe: Org.Comp.B17-101, 107, 115

– .... [$(C_9H_{11})Fe(CO)_2(CH_2{=}C(CH_3)CH{=}CH_2)$][$BF_4$] .. Fe: Org.Comp.B17-134/5

$BC_{16}F_4FeH_{19}O_3$ . . $[(C_5H_5)Fe(CO)_2(CH_2{=}CH{-}C_6H_8{-}CH_3{-}1{-}({=}O){-}2)][BF_4]$
Fe: Org.Comp.B17-16, 64
– . . . . $[(C_5H_5)Fe(CO)_2(CH_2{=}CH{-}C_6H_8{-}CH_3{-}3{-}({=}O){-}2)][BF_4]$
Fe: Org.Comp.B17-17, 47
– . . . . $[(C_5H_5)Fe(CO)_2(CH_2{=}CH{-}C_6H_8{-}CH_3{-}6{-}({=}O){-}2)][BF_4]$
Fe: Org.Comp.B17-17
– . . . . $[(C_5H_5)Fe(CO)_2(CH_2{=}CH{-}C_6H_8{-}OCH_3{-}2)][BF_4]$ Fe: Org.Comp.B17-17, 47
– . . . . $[(C_5H_5)Fe(CO)_2(CH_3CH{=}CH{-}C_6H_9({=}O))][BF_4]$ . . Fe: Org.Comp.B17-60
– . . . . $[(C_5H_5)Fe(CO)_2(CH_3{-}C_8H_{11}O)][BF_4]$ . . . . Fe: Org.Comp.B17-97
$BC_{16}F_4FeH_{19}O_6$ . . $[(1{-}CH_3OC(O){-}(CH_2)_4{-}4{-}CH_3O{-}C_6H_5)Fe(CO)_3][BF_4]$
Fe: Org.Comp.B15-116
$BC_{16}F_4FeH_{20}N$ . . . $[(C_6H_6)Fe(1{-}C_5H_4{-}NC_5H_{10})][BF_4]$ . . . . Fe: Org.Comp.B18-142/6, 153, 168
– . . . . $[(NC_5H_{10}{-}1{-}C_6H_5)Fe(C_5H_5)][BF_4]$ . . . . Fe: Org.Comp.B18-142/6, 198, 263
$BC_{16}F_4FeH_{20}O_2P$ $[C_5H_5(CO)(P(CH_3)_2C_6H_5)Fe{=}C(CH_3)OH][BF_4]$ . . Fe: Org.Comp.B16a-35, 51
$BC_{16}F_4FeH_{21}$ . . . . $[(1,3,5{-}(CH_3)_3{-}C_6H_3)Fe(C_5H_4{-}C_2H_5)][BF_4]$ . . . . Fe: Org.Comp.B19-99, 106
– . . . . $[((CH_3)_5C_6H)Fe(C_5H_5)][BF_4]$. . . . . Fe: Org.Comp.B19-142/3, 151, 192
– . . . . $[(C_6H_6)Fe(C_5(CH_3)_5)][BF_4]$ . . . . Fe: Org.Comp.B18-142/6, 170, 188/9
$BC_{16}F_4FeH_{23}N_2O$ $[(C_5H_5)Fe(CN{-}C_4H_9{-}t)_2CO][BF_4]$. . . . . Fe: Org.Comp.B15-337, 341
$BC_{16}F_4FeH_{23}O$ . . . $[(CH_3)_5C_5Fe(CH_2{=}CHC(CH_3){=}CH_2)CO][BF_4]$ . . . Fe: Org.Comp.B17-189
$BC_{16}F_4FeH_{23}O_3$ . . $[(C_5H_5)Fe(CH_2{=}CH_2)({=}C(OCH_3)OC_6H_{11}{-}c)(CO)][BF_4]$
Fe: Org.Comp.B17-138
$BC_{16}F_4FeH_{24}O_4P$ $[(CH_3{-}C_5H_4)Fe((CH_3)_2C{=}CH_2)(CO)$ $(P({-}OCH_2{-})_3C{-}CH_3)][BF_4]$ . . . . Fe: Org.Comp.B16b-87
– . . . . $[(CH_3{-}C_5H_4)Fe(CH_3CH{=}CHCH_3)(CO)$ $(P({-}OCH_2{-})_3C{-}CH_3)][BF_4]$ . . . . Fe: Org.Comp.B16b-87
– . . . . $[(C_5H_5)Fe(C_2H_5{-}CH{=}CHCH_3)(CO)$ $(P({-}OCH_2{-})_3C{-}CH_3)][BF_4]$ . . . . Fe: Org.Comp.B16b-86
$BC_{16}F_4FeH_{25}NO_4PS$
$[C_5H_5(CO)(P(OCH_3)_3)Fe{=}C_5H_5NS(CH_3)_2][BF_4]$ Fe: Org.Comp.B16a-49, 60
$BC_{16}F_4FeH_{26}N_3Pd$ $[(C_5H_5)Fe(C_5H_3(Pd(NH_2{-}CH_2CH_2{-}NH_2))$ ${-}CH(CH_3){-}N(CH_3)_2)][BF_4]$ . . . . Fe: Org.Comp.A10-188, 189
$BC_{16}F_4FeH_{26}O_4P$ $[(C_5H_5)Fe(c{-}C_6H_9CH_3)(CO)P(OCH_3)_3][BF_4]$ . . . Fe: Org.Comp.B16b-91/2
$BC_{16}F_4FeH_{26}O_5P$ $[C_5H_5(CO)(P(OCH_3)_3)Fe{=}C(C_2H_5)OCH_2CH{=}CHCH_3][BF_4]$
Fe: Org.Comp.B16a-40, 55
$BC_{16}F_4FeH_{26}Si_2$ . . $[Fe(C_5H_4Si(CH_3)_3)_2][BF_4]$. . . . . Fe: Org.Comp.A9-303/4
$BC_{16}F_4FeH_{28}O_2P$ $[C_5(CH_3)_5(CO)(P(CH_3)_3)Fe{=}CH(OCH_3)][BF_4]$. . . Fe: Org.Comp.B16a-61
$BC_{16}F_4FeH_{29}P$ . . . $[C_5H_7Fe(C_5H_7)P(C_2H_5)_3][BF_4]$ . . . . Fe: Org.Comp.B17-179
$BC_{16}F_4FeH_{31}O_6P_2$ $[(C_5H_5)Fe(CH_2{=}CHC(CH_3){=}CH_2)(P(OCH_3)_3)_2][BF_4]$
Fe: Org.Comp.B16b-7, 9
$BC_{16}F_4FeH_{33}P_3Si$ $[(C_6H_6)Fe(((CH_3)_2P{-}CH_2)_3Si{-}CH_3)][BF_4] \cdot 0.25\ CH_3CN$
Fe: Org.Comp.B18-5/6
$BC_{16}F_4FeH_{35}O_6P_2Si$
$[(C_5H_5)Fe(CH_2{=}CHSi(CH_3)_3)(P(OCH_3)_3)_2][BF_4]$ Fe: Org.Comp.B16b-7
$BC_{16}F_4Fe_2H_{13}O_5$ $[C_5H_5(CO)_2Fe{=}C(CH_3)OFe(CO)_2C_5H_5][BF_4]$ . . . Fe: Org.Comp.B16a-109, 119
$BC_{16}F_4H_8N_2O_4Re$ $[(CO)_4Re(1,10{-}N_2C_{12}H_8)][BF_4]$. . . . . Re: Org.Comp.1-475
$BC_{16}F_4H_{14}NS$ . . . . $[SN][BF_4] \cdot 9,10{-}(CH_3)_2C_{14}H_8$. . . . . B: B Comp.SVol.4/4-167

$BC_{16}F_4H_{14}NS$ .... $[SN][BF_4] \cdot 9,10-(CH_3)_2C_{14}H_8$ .... S: S-N Comp.5-47/9
$BC_{16}F_4H_{15}MoO$ .. $[C_9H_7Mo(CO)(H-CC-CH_3)_2][BF_4]$ Mo:Org.Comp.6-316
$BC_{16}F_4H_{15}MoO_2$ .. $[(C_9H_7)Mo(CO)_2(CH_2=C(CH_3)-CH=CH_2)][BF_4]$ .. Mo:Org.Comp.8-305/6, 307, 308
$BC_{16}F_4H_{17}MoO$ .. $[C_9H_7Mo(CO)(H_3C-CC-CH_3)H_2C=CH_2][BF_4]$ ... Mo:Org.Comp.6-315
$BC_{16}F_4H_{18}MoN_3O_2$
$[(CH_2C(CH_3)CH_2)Mo(CO)_2(NC_5H_4-2-(2-C_5H_4N))(NH_3)][BF_4]$ Mo:Org.Comp.5-200, 203
$BC_{16}F_4H_{21}MoO$ .. $[(C_5H_5)Mo(CO)(H_3C-CC-C_2H_5)_2][BF_4]$ Mo:Org.Comp.6-319, 325
– .... $[(C_5H_5)Mo(CO)(H-CC-C_3H_7-i)_2][BF_4]$ Mo:Org.Comp.6-316, 323
$BC_{16}F_4H_{21}MoO_2$ .. $[((CH_3)_5C_5)Mo(CO)_2((CH_2)_3C)][BF_4]$ Mo:Org.Comp.8-305/6, 310/1, 316/7
– .... $[((CH_3)_5C_5)Mo(CO)_2(CH_2=CHCH=CH_2)][BF_4]$ .. Mo:Org.Comp.8-305/6, 307
$BC_{16}F_4H_{22}I$ ...... $[4-(t-C_4H_9)-c-C_6H_8-I-C_6H_5][BF_4]$ .... B: B Comp.SVol.4/3b-219
$BC_{16}F_4H_{24}I$ ...... $[n-C_8H_{17}-CH=CH-I-C_6H_5][BF_4]$ .... B: B Comp.SVol.4/3b-219
$BC_{16}F_4H_{25}MoNO_6P$
$[(C_5H_5)Mo(CO)_2(P(OCH_3)_3)=C(CH_3)-4-(1,4-ONC_4H_8)][BF_4]$ Mo:Org.Comp.8-110
$BC_{16}F_4H_{26}MoOP$ $[C_5H_5Mo(CO)(P(C_2H_5)_3)H_3C-CC-CH_3][BF_4]$ ... Mo:Org.Comp.6-286
$BC_{16}F_4H_{30}O_{10}P_2Re$
trans-$[(CO)_4Re(P(OC_2H_5)_3)_2][BF_4]$ .... Re: Org.Comp.1-481/2
$BC_{16}F_4H_{31}MoO_6P_2$
$[(C_5H_5)Mo(P(OCH_3)_3)_2(H_3C-CC-C_2H_5)][BF_4]$ .. Mo:Org.Comp.6-119, 129
– .... $[(C_5H_5)Mo(P(OCH_3)_3)_2(H-CC-C_3H_7-i)][BF_4]$ ... Mo:Org.Comp.6-116, 128
– .... $[(C_5H_5)Mo(P(OCH_3)_3)_2(D-CC-C_3H_7-i)][BF_4]$ ... Mo:Org.Comp.6-116, 128
$BC_{16}F_4H_{33}MoO_6P_2Si$
$[(C_5H_5)Mo(P(OCH_3)_3)_2(H-CC-Si(CH_3)_3)][BF_4]$ Mo:Org.Comp.6-116
$BC_{16}F_4H_{36}N$ ..... $[(n-C_4H_9)_4N][BF_4]$ .... B: B Comp.SVol.3/3-344/5
B: B Comp.SVol.4/3b-209, 210
$BC_{16}F_6FeH_{25}NP$ .. $[(1,2-(CH_3)_2C_6H_4)Fe(2-CH_3-1-(t-C_4H_9)-1,2-NBC_3H_3)][PF_6]$ .... Fe: Org.Comp.B18-99, 100
– .... $[(1,4-(CH_3)_2C_6H_4)Fe(2-CH_3-1-(t-C_4H_9)-1,2-NBC_3H_3)][PF_6]$ .... Fe: Org.Comp.B18-99/101
$BC_{16}F_{20}H_{36}NO_4Te_4$
$[(n-C_4H_9)_4N][B(OTeF_5)_4]$ .... B: B Comp.SVol.4/4-176
$BC_{16}FeH_{15}$ ...... $(C_5H_5)Fe[BC_5H_5(C_6H_5)]$ .... Fe: Org.Comp.B17-323, 325
– .... $(C_6H_6)Fe[BC_4H_4-1-C_6H_5]$ .... Fe: Org.Comp.B18-106
$BC_{16}FeH_{25}N^+$ .... $[(1,2-(CH_3)_2C_6H_4)Fe(2-CH_3-1-(t-C_4H_9)-1,2-NBC_3H_3)]^+$ Fe: Org.Comp.B18-99, 100
– .... $[(1,4-(CH_3)_2C_6H_4)Fe(2-CH_3-1-(t-C_4H_9)-1,2-NBC_3H_3)]^+$ Fe: Org.Comp.B18-99/101
$BC_{16}FeH_{26}N$ ..... $1,4-(CH_3)_2C_6H_5FeC_3H_3B(CH_3)NC(CH_3)_3$ ...... Fe: Org.Comp.B17-183/4
$BC_{16}GaH_{36}N_2$ .... $[-C(CH_3)_2-CH_2CH_2CH_2-C(CH_3)_2-]N[-Ga(CH_3)_2-N(C_4H_9-t)-B(CH_3)-]$ .... B: B Comp.SVol.4/3a-206/7
$BC_{16}GeH_{35}O_2$ .... $Ge(C_4H_9)_3CH_2CH_2B(-OCH_2CH_2O-)$ Ge:Org.Comp.3-19
$BC_{16}H_9Mn_2O_6$ ... $(CO)_3Mn(C_4H_4BC_6H_5)Mn(CO)_3$ .... B: B Comp.SVol.3/4-155
B: B Comp.SVol.4/4-180
$BC_{16}H_{12}LiS_4$ ..... $Li[B(SC_4H_3)_4]$ .... B: B Comp.SVol.3/4-117
$BC_{16}H_{12}NS_2$ ..... $C_6H_5B[-C(2-C_4H_3S)=CH-(2,3-C_4H_2S)-NH-]$ ... B: B Comp.SVol.3/4-130

$BC_{16}H_{13}MoO_4$ . . . $(1\text{-}C_6H_5\text{-}BC_6H_8)Mo(CO)_4$ Mo:Org.Comp.5-312, 314, 316
$BC_{16}H_{13}S$ . . . . $HB(\text{-}CH{=}CC_6H_5\text{-}S\text{-}CC_6H_5{=}CH\text{-})$ . . . . . . B: B Comp.SVol.3/4-115
$BC_{16}H_{14}NS_2$ . . . . . $C_6H_5B[\text{-}C(C_4H_3S){=}CH\text{-}(C_4H_4S)\text{-}NH\text{-}]$ . . . . . B: B Comp.SVol.3/3-163
$BC_{16}H_{15}MoN_6O_2$. $(C_5H_5)Mo(CO)_2[\text{-}N_2C_3H_3\text{-}BH(N_2C_3H_3)\text{-}N_2C_3H_3\text{-}]$
Mo:Org.Comp.7-156, 158, 178
$BC_{16}H_{15}Ru$ . . . . . . $(C_6H_6)Ru(C_4H_4BC_6H_5)$ . . . . B: B Comp.SVol.3/4-155
$BC_{16}H_{16}NO_2$ . . . . . $(CH_3)_2B\text{-}N(C_6H_5)\text{-}C(O)C(O)C_6H_5$ . . . . . . B: B Comp.SVol.3/3-155
$BC_{16}H_{16}N_3O_2$ . . . . $1,5\text{-}(CH_3)_2\text{-}2,3\text{-}(C_6H_5)_2\text{-}1,3,5,2\text{-}N_3BC_2({=}O)_2\text{-}4,6$
B: B Comp.SVol.4/3a-176
$BC_{16}H_{17}MoN_6O_2$. $(c\text{-}C_5H_7)Mo(CO)_2[\text{-}N_2C_3H_3\text{-}]_3BH$ Mo:Org.Comp.5-225/6, 229
$BC_{16}H_{18}N$ . . . . $5\text{-}(C_2H_5)_2N\text{-}5\text{-}BC_{12}H_8$ . . . . B: B Comp.SVol.4/3a-229
$BC_{16}H_{18}NO$ . . . . . . $[\text{-}CH(C_6H_5)\text{-}CH(CH_3)\text{-}N(CH_3)\text{-}B(C_6H_5)\text{-}O\text{-}]$ . . . B: B Comp.SVol.4/3b-64
$BC_{16}H_{19}N_2O$ . . . . . $(n\text{-}C_4H_9)B[\text{-}(1,2\text{-}NC_5H_5)\text{-}N{=}CC_6H_5\text{-}O{=}]$ . . . . . . B: B Comp.SVol.3/3-219
$BC_{16}H_{19}N_2O_6$ . . . . $(CH_3COO)_2B[\text{-}NH{=}CCH_3\text{-}C(CONHC_6H_5){=}CCH_3\text{-}O\text{-}]$
B: B Comp.SVol.3/3-218
$BC_{16}H_{19}N_4$ . . . . . . $[\text{-}B(C_6H_5)\text{-}N(C_4H_9\text{-}t)\text{-}N{=}N\text{-}N(C_6H_5)\text{-}]$ . . . . . B: B Comp.SVol.4/3a-199
$BC_{16}H_{19}S$ . . . . $(C_6H_5)_2B\text{-}S\text{-}C_4H_9\text{-}n$ . . . . B: B Comp.SVol.4/4-129
$BC_{16}H_{20}N$ . . . . $(C_6H_5)_2B\text{-}N(C_2H_5)_2$ . . . . B: B Comp.SVol.3/3-149
– . . . . $(C_6H_5)_2B\text{-}NH\text{-}C_4H_9\text{-}n$ . . . . B: B Comp.SVol.3/3-153
– . . . . $(C_6H_5)_2B\text{-}NH\text{-}C_4H_9\text{-}s$ . . . . B: B Comp.SVol.3/3-153
– . . . . $(C_6H_5)_2B\text{-}NH\text{-}C_4H_9\text{-}t$ . . . . B: B Comp.SVol.3/3-153
– . . . . $C_6H_5\text{-}C{\equiv}N\text{-}BC_9H_{15}$ . . . . B: B Comp.SVol.3/3-191
$BC_{16}H_{20}NO_2$ . . . . . $C_6H_5\text{-}O\text{-}B(C_6H_5)[\text{-}O\text{-}CH_2CH_2\text{-}N(CH_3)_2\text{-}]$ . . . . . B: B Comp.SVol.4/3b-70
$BC_{16}H_{20}N_5$ . . . . . . $[4\text{-}(CH_3)_2N\text{-}C_5H_4NH][BH(NC_4H_4)_2CN]$ . . . . B: B Comp.SVol.3/3-213
$BC_{16}H_{21}LiNO$ . . . . $C_6H_5\text{-}CH(CH_3)\text{-}NH(BH_3)\text{-}CH(C_6H_5)\text{-}CH_2\text{-}OLi$ . . B: B Comp.SVol.4/3b-7
$BC_{16}H_{21}N_2O$ . . . . . $C_6H_5\text{-}NH\text{-}B(C_6H_5)[\text{-}O\text{-}CH_2CH_2\text{-}N(CH_3)_2\text{-}]$ . . . . B: B Comp.SVol.4/3b-56
$BC_{16}H_{21}N_3^+$ . . . . . $[(C_6H_5)_2N\text{-}B(\text{-}NHCH_3\text{-}CH_2\text{-}CH_2\text{-}NCH_3\text{-})]^+$ . . . B: B Comp.SVol.3/3-207/9
$BC_{16}H_{22}N$ . . . . $BC_{16}H_{22}N$ . . . . B: B Comp.SVol.4/3a-259
B: B Comp.SVol.4/3b-30
– . . . . $H_3B\text{-}NH[\text{-}CH(CH_3)\text{-}C_6H_5]_2$ . . . . B: B Comp.SVol.4/3b-7
$BC_{16}H_{22}NO$ . . . . . . $C_6H_5\text{-}CH(CH_3)\text{-}NH(BH_3)\text{-}CH(C_6H_5)\text{-}CH_2\text{-}OH$ . . B: B Comp.SVol.4/3b-7
$BC_{16}H_{22}NSn$ . . . . . $(CH_3)_3SnN(CH_3)B(C_6H_5)_2$ . . . . Sn: Org.Comp.18-57, 61
$BC_{16}H_{22}N_3$ . . . . . . $6\text{-}N(C_3H_7\text{-}i)_2\text{-}5,7,6\text{-}N_2BC_{10}H_8$ . . . . B: B Comp.SVol.4/3a-156
– . . . . $(n\text{-}C_3H_7)_2B[\text{-}NC_5H_4\text{-}N{=}NC_5H_4\text{-}]$ . . . . . . B: B Comp.SVol.3/3-123
$BC_{16}H_{22}N_3OS$ . . . . $[\text{-}B(C_3H_7\text{-}n)_2\text{-}N(C_6H_5)\text{-}C({=}O)\text{-}NH\text{-}SNC_3H_2\text{-}]$ . . B: B Comp.SVol.4/3a-179/80
B: B Comp.SVol.4/4-152/3
– . . . . $[\text{-}B(C_3H_7\text{-}n)_2\text{-}N(C_6H_5)\text{-}C({=}O)\text{-}SNC_3H_2{=}NH\text{-}]$ . . B: B Comp.SVol.4/3a-180
– . . . . $[\text{-}B(C_3H_7\text{-}n)_2\text{-}O\text{-}C(NH\text{-}C_6H_5){=}N\text{-}SNC_3H_2\text{-}]$ . . . B: B Comp.SVol.4/4-152/3
$BC_{16}H_{22}N_3S_2$ . . . . $BN_2S_2C_4H_2\text{-}(C_3H_7\text{-}n)_2\text{-}(NH\text{-}C_6H_5)$ . . . . . . B: B Comp.SVol.4/4-149/50
– . . . . $[\text{-}B(C_3H_7\text{-}n)_2\text{-}N(C_6H_5)\text{-}C(S)NH\text{-}C_3H_2NS\text{-}]$ . . . . B: B Comp.SVol.4/3a-179
B: B Comp.SVol.4/4-149/50
$BC_{16}H_{23}MoN_4O_2$. $[CH_2C(CH_3)CH_2]Mo(CO)_2[\text{-}N_2C_3H(CH_3)_2\text{-}]_2BH_2$
Mo:Org.Comp.5-213, 215
$BC_{16}H_{23}NOP$ . . . . . $(C_2H_5)_2N\text{-}B[\text{-}CH{=}CCH_3\text{-}P(O)(C_6H_5)\text{-}CCH_3{=}CH\text{-}]$
B: B Comp.SVol.3/3-151
$BC_{16}H_{23}NP$ . . . . . . $(C_2H_5)_2N\text{-}B(\text{-}CH{=}CCH_3\text{-}PC_6H_5\text{-}CCH_3{=}CH\text{-})$ . . . B: B Comp.SVol.3/3-151
$BC_{16}H_{23}NPS$ . . . . . $(C_2H_5)_2N\text{-}B[\text{-}CH{=}CCH_3\text{-}P(S)(C_6H_5)\text{-}CCH_3{=}CH\text{-}]$
B: B Comp.SVol.3/3-151
$BC_{16}H_{23}NPSe$ . . . . $(C_2H_5)_2N\text{-}B[\text{-}CH{=}CCH_3\text{-}P(Se)(C_6H_5)\text{-}CCH_3{=}CH\text{-}]$
B: B Comp.SVol.3/3-151

$BC_{16}H_{23}N_4$ ...... $[HC(CH_2CH_2)_3NH][BH(NC_4H_4)_2CN]$ .... B: B Comp.SVol.3/3-213

$BC_{16}H_{23}O_2S$ ..... 2-(C_6H_5-CH_2SCH_2CH=CH)-4,4,5,5-$(CH_3)_4$-1,3,2-O_2BC_2 B: B Comp.SVol.4/4-138

– .... 2-[C_6H_5-S-$CHCH_3$-CH=CH]-4,4,5,5-$(CH_3)_4$-1,3,2-O_2BC_2 B: B Comp.SVol.4/4-138

$BC_{16}H_{23}O_3S$ ..... 2-[C_6H_5-S(O)-$CHCH_3$-CH=CH]-4,4,5,5-$(CH_3)_4$-1,3,2-O_2BC_2. .... B: B Comp.SVol.4/4-138

$BC_{16}H_{24}N$ .... 2,2-$(CH_3)_2$-[3.3.1.1^{3,7}]-1-BC_9H_{13} · NC_5H_5 ... B: B Comp.SVol.4/3b-30

– .... 3-$BC_{10}H_{16}$-4-CH_3 · NC_5H_5 .... B: B Comp.SVol.3/3-191

– .... $[CH_3CH=C(CH_3)]_2B$-NH-C_6H_3-2,6-$(CH_3)_2$ .... B: B Comp.SVol.3/3-153

$BC_{16}H_{24}N_5$ ...... $BC_3H_3N_5(C_6H_4$-2-$CH_3)(C_3H_7$-n$)_2$ .... B: B Comp.SVol.4/3a-179

– .... $BC_3H_3N_5(C_6H_4$-4-$CH_3)(C_3H_7$-n$)_2$ .... B: B Comp.SVol.4/3a-179

$BC_{16}H_{25}LiN$ ..... Li[(i-$C_3H_7)_2$N-B(C_6H_5)(-CH=CH-CH_2CH_2-)]... B: B Comp.SVol.4/3b-49

– .... Li[(i-$C_3H_7)_2$N-B(C_6H_5)(-CH_2-CH=CH-CH_2-)].. B: B Comp.SVol.4/3b-49

$BC_{16}H_{25}NO_2$ .... C_6H_5-B[-O-CH_2-C$(CH_3)_2$-CH_2-O-] · NC_5H_{10} B: B Comp.SVol.4/3b-69

$BC_{16}H_{25}N_6$ ...... $BC_2H_2N_6(C_6H_5)(C_4H_9$-n$)_2$ .... B: B Comp.SVol.4/3a-178

$BC_{16}H_{26}N$ .... 9-[1-($C_5H_4N(C_3H_7$-i)-2)]-BHC_8H_{14}. .... B: B Comp.SVol.3/3-189

$BC_{16}H_{26}NSi$ ..... 1-(C_6H_5-CH_2)-4,5-$(C_2H_5)_2$-2,2,3-$(CH_3)_3$-1,2,5-$NSiBC_2$ B: B Comp.SVol.4/3a-250

$BC_{16}H_{27}N_2$ ...... 1-[$(C_2H_5)_2$N-B(C_6H_5)]-2-CH_3-NC_5H_9 .... B: B Comp.SVol.3/3-109

– .... 1-[t-C_4H_9-NH-B(C_6H_5)]-2-CH_3-NC_5H_9 ...... B: B Comp.SVol.3/3-108

– .... (i-$C_3H_7)_2$B-N=C(C_6H_5)NH-C_3H_7-i. .... B: B Comp.SVol.3/3-156

$BC_{16}H_{27}N_4$ ...... (n-C_5H_{11})B[-N(n-C_5H_{11})-N=N-NC_6H_5-]...... B: B Comp.SVol.3/3-137/8

$BC_{16}H_{28}N$ .... (t-$C_4H_9C{\equiv}C)_2$B-N$(C_2H_5)_2$. .... B: B Comp.SVol.3/3-149

$BC_{16}H_{28}NS$ ...... $C_2H_5SB(C_6H_5)N(C_4H_9$-s$)_2$ .... B: B Comp.SVol.3/4-122

$BC_{16}H_{28}N_3S$ ..... [-B(C_4H_9-n$)_2$-N(CH_2-CH=CH_2)-C(=S)-NH-C_4H_4N-] B: B Comp.SVol.4/4-149/50

$BC_{16}H_{29}N_2$ ...... $(CH_3)_2$N-B(C_6H_5)-N(C_4H_9-s$)_2$ .... B: B Comp.SVol.3/3-108

– .... $(C_2H_5)_2$N-B(C_6H_5)-N(C_3H_7-n$)_2$ .... B: B Comp.SVol.3/3-109

– .... t-C_4H_9-NH-B(C_6H_5)-N(C_3H_7-i$)_2$ .... B: B Comp.SVol.3/3-108

$BC_{16}H_{29}N_3^+$ ..... [(-C$(CH_3)_2$-$(CH_2)_3$-C$(CH_3)_2$-)N-BN$(CH_3)_2$(C_5H_5N)]$^+$ B: B Comp.SVol.3/3-207

$BC_{16}H_{30}IO_4$. ..... CH_3-COO-$(CH_2)_6$-BI-$(CH_2)_6$-OC(=O)-CH_3. ... B: B Comp.SVol.4/4-113

$BC_{16}H_{30}N$ .... $(C_2H_5)_2$B-N(C_6H_9-c)-C_6H_{11}-c. .... B: B Comp.SVol.3/3-152

– .... $(C_2H_5)_2$N-C_6H_5 · BH_2-C$(CH_3)_2$-C_3H_7-i ...... B: B Comp.SVol.3/3-188

$BC_{16}H_{30}N_2$ ...... N(-CH_2-CH_2-$)_3$N-BH_2($C_{10}H_{16}$) .... B: B Comp.SVol.3/3-188

$BC_{16}H_{30}S^-$ ...... $[(SC_4H_3)B(C_4H_9$-n$)_3]^-$. .... B: B Comp.SVol.3/4-117

– .... $[(SC_4H_3)B(C_4H_9$-i$)_3]^-$ .... B: B Comp.SVol.3/4-117

– .... $[(SC_4H_3)B(C_4H_9$-s$)_3]^-$. .... B: B Comp.SVol.3/4-117

$BC_{16}H_{31}N_2Si$. .... $(CH_3)_2$N-B(CH_2-C_6H_5)-N(C_4H_9-t)-Si$(CH_3)_3$ ... B: B Comp.SVol.4/3a-169

$BC_{16}H_{31}N_4$ ...... 1-[1,2-$N_2C_3H_3$-1-B(NH-C_4H_9-t)]-2,2,6,6-$(CH_3)_4$-NC_5H_6 B: B Comp.SVol.4/3a-155

– .... 1-[1,3-$N_2C_3H_3$-1-B(NH-C_4H_9-t)]-2,2,6,6-$(CH_3)_4$-NC_5H_6 B: B Comp.SVol.4/3a-155

$BC_{16}H_{32}N$ .... 9,10-(C_4H_9-n$)_2$-[3.3.2]-9,10-NBC_8H_{14} .... B: B Comp.SVol.4/3a-252

$BC_{16}H_{33}NO_2P$.... 9-[$(C_2H_5$-O$)_2$P-N(C_4H_9-t)]-[3.3.1]-9-BC_8H_{14}.. B: B Comp.SVol.4/3a-236

$BC_{16}H_{33}N_2Si$. .... 2-(C_4H_9-t)-3-[$(CH_3)_3$Si-N(C_4H_9-t)]-[2.2.1]-2,3-NBC_5H_6 B: B Comp.SVol.4/3a-174

– .... $(CH_3)_3$Si-N(C_4H_9-i)-B(C_5H_5)-NH-C_4H_9-t ..... B: B Comp.SVol.4/3a-170

$BC_{16}H_{33}N_4Si_3$ .... $[-N(Si(CH_3)_3)-CH_2CH_2-N(Si(CH_3)_3)-]$
$Si[-N(CH_3)-B(C_6H_5)-N(CH_3)-]$ .... B: B Comp.SVol.4/3a-206

$BC_{16}H_{34}N_2O_2P$ ... 2,2,6,6-$(CH_3)_4$-NC_5H_6-1-B[-$N(C_4H_9$-t)
-$P(CH_3)-O-CH_2CH_2-O-$] .... B: B Comp.SVol.4/3b-54

$BC_{16}H_{34}N_2O_3P$ ... 1,3,2-$O_2PC_2H_4$-2-[$N(C_4H_9$-t)-$B(OCH_3)$-
1-NC_5H_6-2,2,6,6-$(CH_3)_4$] .... B: B Comp.SVol.4/3b-52

$BC_{16}H_{34}N_2O_3Sb$ .. 1,3,2-$O_2SbC_2H_4$-2-[$N(C_4H_9$-t)-$B(OCH_3)$-
1-NC_5H_6-2,2,6,6-$(CH_3)_4$] .... B: B Comp.SVol.4/3b-52

$BC_{16}H_{34}N_3O$ ..... 1-[CH_3-C(=O)-$N(CH_3)$-B(NH-C_4H_9-t)]-
2,2,6,6-$(CH_3)_4$-NC_5H_6 .... B: B Comp.SVol.4/3a-155

$BC_{16}H_{34}N_3O_2$ .... t-C_4H_9-NH-B[O-C(O)-$CH(NH_2)-CH_3$]-
1-NC_5H_6-2,2,6,6-$(CH_3)_4$ .... B: B Comp.SVol.4/3b-52

$BC_{16}H_{35}N_2O$ ..... t-C_4H_9-NH-B(O-C_3H_7-i)-1-NC_5H_6-2,2,6,6-$(CH_3)_4$
B: B Comp.SVol.4/3b-52

$BC_{16}H_{36}IN_2Si$ .... [2,2,6,6-$(CH_3)_4$-NC_5H_6-1-(=B=$N(C_4H_9$-t)-$Si(CH_3)_3$)]I
B: B Comp.SVol.4/4-114/5

$BC_{16}H_{36}NSi_2$..... [-CH=CH-$CHSi(CH_3)_3$-$CHSi(CH_3)_3$-]B-N(i-$C_3H_7)_2$
B: B Comp.SVol.3/3-150

$BC_{16}H_{36}N_2O_3Sb$ .. $(CH_3O)_2Sb$-$N(C_4H_9$-t)-$B(OCH_3)$-1-NC_5H_6-2,2,6,6-$(CH_3)_4$
B: B Comp.SVol.4/3b-52

$BC_{16}H_{36}N_2P$ ..... n-C_4H_9-B[-$N(C_4H_9$-t)-$P(C_4H_9$-n)-$N(C_4H_9$-t)-] B: B Comp.SVol.4/3a-207/8

$BC_{16}H_{36}N_2Si^+$ ... [(t-C_4H_9)($(CH_3)_3Si$)N=B=N(-$C(CH_3)_2$-$(CH_2)_3$-$C(CH_3)_2$-)]$^+$
B: B Comp.SVol.3/3-205

$BC_{16}H_{36}N_3$ ...... 1-[t-C_4H_9-NH-B(NH-C_3H_7-i)]-2,2,6,6-$(CH_3)_4$-NC_5H_6
B: B Comp.SVol.4/3a-155

$BC_{16}H_{37}N_2O_3$ .... (t-$C_4H_9O)_3$B(-NCH_3-CH_2-CH_2-NCH_3-) ...... B: B Comp.SVol.3/3-113

$BC_{16}H_{38}N_3Si$..... $[(CH_3)_2N]_2BN(n$-$C_4H_9)C[Si(CH_3)_3]CHC_3H_7$ .... B: B Comp.SVol.3/3-96

$BC_{16}H_{39}LiN_3Si_2$ .. [Li($(CH_3)_2N$-CH_2CH_2-$N(CH_3)_2$)]
[1,3-$((CH_3)_3Si)_2$-2-CH_3-1,2-NBC_3H_2]....... B: B Comp.SVol.4/3b-48

$BC_{16}H_{39}N_2O_3Si$ .. (t-$C_4H_9O)_3SiB[N(CH_3)_2]_2$ .... B: B Comp.SVol.3/3-110

$BC_{16}H_{40}NSn_2$ .... $(C_2H_5)_2B$-$N[Sn(C_2H_5)_3]_2$ .... B: B Comp.SVol.4/3a-227

$BC_{16}H_{41}N_4Si_3$.... 3,5,5-$[(CH_3)_3Si]_3$-4-(i-$C_3H_7)_2$N-1,2,3,4-N_3BC B: B Comp.SVol.4/3a-250

$BC_{16}H_{45}N_4Si_4$.... [-$N(C_2H_5)$-$Si(CH_3)_2$-$Si(CH_3)_2$-$N(C_2H_5)$-]B
-$N(C_2H_5)$-$Si(CH_3)_2$-$Si(CH_3)_2$-NH-C_2H_5 ..... B: B Comp.SVol.4/3a-154/5

$BC_{16.5}F_4FeH_{33.75}N_{0.25}P_3Si$
$[(C_6H_6)Fe(((CH_3)_2P$-$CH_2)_3Si$-$CH_3)][BF_4]$ · 0.25 CH_3CN
Fe: Org.Comp.B18-5/6

$BC_{17}ClF_4H_{20}N_3O_2Re$
$[(C_5H_5)Re(CO)_2$-N=N-C_6H_4-4-$N(C_2H_5)_2][BF_4]$ · HCl
Re: Org.Comp.3-193/7, 204

$BC_{17}ClH_{13}N$ ..... 5-Cl-5-$BC_{12}H_8$ · NC_5H_5................. B: B Comp.SVol.4/4-69

$BC_{17}Cl_2H_{33}O_2S_2$.. $C_6H_{13}C(S_2C_2H_4)B[O(CH_2)_4Cl]_2$ .... B: B Comp.SVol.3/4-112

$BC_{17}Cl_4GaH_{13}N$ .. [5-(NC_5H_5-1)-5-$BC_{12}H_8$][$GaCl_4$]............ B: B Comp.SVol.4/3b-45

$BC_{17}Cl_9H_{14}NO_3P$ CCl_3-CH(OH)-P[-$CH(CCl_3)$-O-$B(C_6H_5)$
(1-NC_5H_5)-O-$CH(CCl_3)$-] .... B: B Comp.SVol.4/3b-73

– .... $[NC_5H_6][P(-CH(CCl_3)-O-)_3B-C_6H_5]$ .... B: B Comp.SVol.4/3b-73

$BC_{17}CoH_{28}N_2O_4$.. [(2,2,6,6-$(CH_3)_4$-NC_5H_6-1)=B=NH-C_4H_9-t][$Co(CO)_4$]
B: B Comp.SVol.4/3b-45

$BC_{17}FH_{45}NSi_4$.... $[(CH_3)_3Si]_3C$-BF-$N(C_4H_9$-t)-$Si(CH_3)_3$ .... B: B Comp.SVol.4/3b-245

$BC_{17}F_2FeH_{15}O_3$ .. $(C_5H_5)Fe(C(CCH_3=CH_2)OBF_2OCC_6H_5)CO$ ..... Fe: Org.Comp.B16b-130, 141

$BC_{17}F_2FeH_{19}O_3$ .. $(C_5H_5)Fe(C(C_7H_8CH_3)OBF_2OC(CH_3))CO$ ...... Fe: Org.Comp.B16b-129/30
$BC_{17}F_2FeH_{21}O_3$ .. $(C_5H_5)Fe(C(c-C_6H_7(CH_3)_2)OBF_2OCCH_3)CO$ ... Fe: Org.Comp.B16b-128, 139/40
$BC_{17}F_2H_{15}O_4$ .... $[-B(F)_2OC(C_6H_4-4-OCH_3)CHC(C_6H_4-4-OCH_3)O-]$ B: B Comp.SVol.4/3b-233
$BC_{17}F_2H_{35}N_2Si_3$ .. $[(CH_3)_3Si]_2N-BF-N[C_6H_2-2,4,6-(CH_3)_3]-SiF(CH_3)_2$ B: B Comp.SVol.3/3-376
$BC_{17}F_3Fe_3H_{17}NO_{11}$
$[NH(C_2H_5)_3][(H)Fe_3(CO)_{11}-BF_3]$ .... Fe: Org.Comp.C6b-91
$BC_{17}F_3H_{14}O_5S$ ... $C_6H_5(CF_3SO_3)B[-O=CC_6H_5-CH=CCH_3-O-]$ ... B: B Comp.SVol.3/4-52
$BC_{17}F_3H_{21}N_3O_3S$ $[(C_6H_5)_2N-B(-NHCH_3-CH_2-CH_2-NCH_3-)]O_3SCF_3$ B: B Comp.SVol.3/3-207/9
$BC_{17}F_3H_{29}N_3O_3S$ $[(-C(CH_3)_2-(CH_2)_3-C(CH_3)_2-)N-BN(CH_3)_2-1-NC_5H_5][O_3SCF_3]$ .... B: B Comp.SVol.3/3-207
$BC_{17}F_3H_{36}N_2O_3SSi$
$[(2,2,6,6-(CH_3)_4-NC_5H_6-1)=B=N(C_4H_9-t)-Si(CH_3)_3][CF_3-SO_3]$ .... B: B Comp.SVol.3/3-205
B: B Comp.SVol.4/3b-45
$BC_{17}F_4FeH_{12}NO_5$ $[(O_2N-C_6H_4-C_8H_8)Fe(CO)_3][BF_4]$ .... Fe: Org.Comp.B15-214, 229
$BC_{17}F_4FeH_{13}O_3$ .. $[(CH_3-C_{13}H_{10})Fe(CO)_3][BF_4]$ .... Fe: Org.Comp.B15-255
– .... $[(C_6H_5-CH=C_7H_7)Fe(CO)_3][BF_4]$ .... Fe: Org.Comp.B15-210, 226
$BC_{17}F_4FeH_{13}O_4$ .. $[(6-C_6H_5C(O)-C_7H_8)Fe(CO)_3][BF_4]$ .... Fe: Org.Comp.B15-208, 223
$BC_{17}F_4FeH_{15}$ .... $[(C_6H_6)Fe(C_5H_4-C_6H_5)][BF_4]$ .... Fe: Org.Comp.B18-142/6, 151, 165
– .... $[(C_6H_5-C_6H_5)Fe(C_5H_5)][BF_4]$ .... Fe: Org.Comp.B18-142/6, 197, 200, 201, 223/4, 274/6
$BC_{17}F_4FeH_{15}O$ ... $[(C_6H_5-O-C_6H_5)Fe(C_5H_5)][BF_4]$ .... Fe: Org.Comp.B18-142/6, 201, 251/2
$BC_{17}F_4FeH_{15}O_3$ .. $[(6-C_7H_7-C_7H_8)Fe(CO)_3][BF_4]$ .... Fe: Org.Comp.B15-207, 223
– .... $[(C_5H_5)Fe(CO)_2(OC_4H_5-C_6H_5)][BF_4]$ .... Fe: Org.Comp.B17-95
– .... $[(C_{14}H_{15})Fe(CO)_3][BF_4]$ .... Fe: Org.Comp.B15-111
$BC_{17}F_4FeH_{17}$ .... $[(C_{10}H_8)Fe(C_5H_4-C_2H_5)][BF_4]$ .... Fe: Org.Comp.B19-216, 235
$BC_{17}F_4FeH_{17}O_2$ .. $[(C_5H_5)Fe(CO)_2(CH_2=CHCH_2CH_2-C_6H_5)][BF_4]$ Fe: Org.Comp.B17-34
– .... $[(C_5H_5)Fe(CO)_2(CH_2=CHCH_2-C_7H_7-c)][BF_4]$ ... Fe: Org.Comp.B17-32
– .... $[(C_5H_5)Fe(CO)_2(C_6H_5-CH=C(CH_3)_2)][BF_4]$ ..... Fe: Org.Comp.B17-85
– .... $[(C_5H_5)Fe(CO)_2(C_{10}H_{12})][BF_4]$ .... Fe: Org.Comp.B17-98, 100
– .... $[(C_5H_5)(CO)_2Fe=CH-C(CH_3)_2-C_6H_5][BF_4]$ ..... Fe: Org.Comp.B16a-88
$BC_{17}F_4FeH_{17}O_2S$ $[(C_5H_5)Fe(CO)_2CH_2S(C_6H_5)CH_2CH=CH_2][BF_4]$ Fe: Org.Comp.B14-133, 135, 142/3
$BC_{17}F_4FeH_{17}O_3Ru$
$[(C_5H_5)(CO)Fe(CO)(C(CH_3)=CHCH_3)Ru(CO)(C_5H_5)][BF_4]$ Fe: Org.Comp.B16b-149/50
$BC_{17}F_4FeH_{17}O_4$ .. $[(C_5H_5)Fe(CO)_2(CH_2=CHCH_2C_6H_3(OH-4)OCH_3-3)][BF_4]$ Fe: Org.Comp.B17-32
$BC_{17}F_4FeH_{19}OS$ .. $[C_5H_5Fe(CH_2S(C_6H_5)CH_2CH_2CH=CH_2)CO][BF_4]$ Fe: Org.Comp.B17-167
$BC_{17}F_4FeH_{19}O_2$ .. $[(C_5H_5)Fe(CO)_2(C_{10}H_{14})][BF_4]$ .... Fe: Org.Comp.B17-100

$BC_{17}F_4FeH_{19}O_2S$ $[C_5H_5Fe(CO)_2CH(CH_3)S(C_2H_5)C_6H_5][BF_4]$ ..... Fe: Org.Comp.B14-133, 135, 143

$BC_{17}F_4FeH_{19}O_2Si$ $[(C_6H_5Si(CH_3)_2C_5H_4)Fe(CO)_2(CH_2{=}CH_2)][BF_4]$ Fe: Org.Comp.B17-132

$BC_{17}F_4FeH_{19}O_3$ .. $[(3-CH_3-8-(n-C_4H_9-CH{=})C_8H_6)Fe(CO)_3][BF_4]$ Fe: Org.Comp.B15-243

– .... $[(4-CH_3-8-(n-C_4H_9-CH{=})C_8H_6)Fe(CO)_3][BF_4]$ Fe: Org.Comp.B15-243/4

– .... $[(C_5H_5)Fe(CO)_2(CH_2{=}CCH_3-C_6H_6(CH_3)({=}O))][BF_4]$ Fe: Org.Comp.B17-64

$BC_{17}F_4FeH_{19}O_6$ .. $[(C_5H_5)Fe(CO)_2(C_5H_7(CH(CO_2CH_3)_2-3))][BF_4]$ Fe: Org.Comp.B17-89, 103/4

$BC_{17}F_4FeH_{20}NO_2$ $[(C_5H_5)Fe(CO)_2-CH(C_6H_5)CH_2NH(CH_3)_2][BF_4]$ Fe: Org.Comp.B14-150

– .... $[(C_5H_5)Fe(CO)_2-CH_2-CHCH_3-NH_2CH_2C_6H_5][BF_4]$ Fe: Org.Comp.B14-150, 155

$BC_{17}F_4FeH_{21}O$ ... $[(1,2,4,5-(CH_3)_4-C_6H_2)Fe(C_5H_4-CO-CH_3)][BF_4]$ Fe: Org.Comp.B19-150

$BC_{17}F_4FeH_{21}O_2$ .. $[((CH_3)_5C_5)Fe(CO)_2(C_5H_6-c)][BF_4]$ .... Fe: Org.Comp.B17-131

– .... $[(C_5H_5)Fe(CO)_2(CH_2{=}C_9H_{14})][BF_4]$ .... Fe: Org.Comp.B17-72

$BC_{17}F_4FeH_{21}O_3$ .. $[(C_5H_5)(CO)_2Fe{=}C(C_7H_{11})-O-C_2H_5][BF_4]$ ..... Fe: Org.Comp.B16a-112, 122

– .... $[(C_5H_5)Fe(CO)_2(CH_2{=}CCH_3-C_6H_8(CH_3)({=}O))][BF_4]$ Fe: Org.Comp.B17-64

– .... $[(C_5H_5)Fe(CO)_2(CH_3CH{=}CH-C_6H_8(CH_3)({=}O))][BF_4]$ Fe: Org.Comp.B17-60

$BC_{17}F_4FeH_{22}$ .... $[((CH_2)C_6(CH_3)_5)Fe(C_5H_5)][BF_4]$ .... Fe: Org.Comp.B19-178

$BC_{17}F_4FeH_{23}$ .... $[((CH_3)_6C_6)Fe(C_5H_5)][BF_4]$ .... Fe: Org.Comp.B19-142/3, 156, 193/7

$BC_{17}F_4FeH_{23}O$ ... $[(CH_3)_5C_5Fe(C_6H_8)CO][BF_4]$ .... Fe: Org.Comp.B17-189, 195/6

$BC_{17}F_4FeH_{23}O_2$ .. $[(C_5H_5)Fe(CO)_2(CH_2{=}CHCH(CH_3)CH_2CH_2CH{=}C(CH_3)_2)][BF_4]$ .... Fe: Org.Comp.B17-40/1

$BC_{17}F_4FeH_{25}O_2$ .. $[(C_5H_5)Fe(CO)_2(n-C_4H_9CH{=}CHC_4H_9-n)][BF_4]$ .. Fe: Org.Comp.B17-58

$BC_{17}F_4FeH_{25}O_4P$ $[(CH_3C_5H_4)Fe(CH_3CH{=}CC_2H_5)(CO)P(OCH_2)_3CCH_3][BF_4]$ Fe: Org.Comp.B16b-88

$BC_{17}F_4FeH_{27}O_4Si$ $[(C_5H_5)Fe(CO)_2(CH_3CH{=}CHOCH(CH_3)CH(CH_3)OSi(CH_3)_3)][BF_4]$ .... Fe: Org.Comp.B17-65, 78

$BC_{17}F_4FeH_{28}O_4P$ $[(C_5H_5)Fe(CH_3CH{=}C(CH_2)_5)(CO)P(OCH_3)_3][BF_4]$ Fe: Org.Comp.B16b-93

$BC_{17}F_4FeH_{30}OP$ .. $[C_5(CH_3)_5(CO)(P(CH_3)_3)Fe{=}C(CH_3)_2][BF_4]$ ..... Fe: Org.Comp.B16a-29/30

$BC_{17}F_4Fe_2H_{15}O_5$ $[C_5H_4CH_3(CO)_2Fe{=}C(CH_3)OFe(CO)_2C_5H_5][BF_4]$ Fe: Org.Comp.B16a-127

$BC_{17}F_4H_{15}MoO_2$.. $[(C_9H_7)Mo(CO)_2(C_6H_8)][BF_4]$ Mo:Org.Comp.8-305/6, 313

$BC_{17}F_4H_{15}MoO_3$.. $[(C_9H_7)Mo(CO)_2(C_6H_7-1-OH)][BF_4]$.......... Mo:Org.Comp.8-305/6, 314

$BC_{17}F_4H_{17}MoO_2$.. $[(C_9H_7)Mo(CO)_2(CH_2{=}CH-C(CH_3){=}CH-CH_3)][BF_4]$ Mo:Org.Comp.8-305/6, 310

$BC_{17}F_4H_{18}I$ ...... $[C_6H_5-CH_2CH_2-C(CH_3){=}CH-I-C_6H_5][BF_4]$ .... B: B Comp.SVol.4/3b-219

$BC_{17}F_4H_{19}N_3O_2Re$ $[(C_5H_5)Re(CO)_2-N{=}N-C_6H_4-4-N(C_2H_5)_2][BF_4]$ Re: Org.Comp.3-193/7, 203/4

$BC_{17}F_4H_{24}MoN$... $[C_5H_5Mo(NCCH_3)(H-CC-C_3H_7-i)_2][BF_4]$ Mo:Org.Comp.6-136

$BC_{17}F_4H_{25}MoO_8P_2$ $[C_9H_7Mo(CO)_2(P(OCH_3)_3)_2][BF_4]$............ Mo:Org.Comp.7-289

$BC_{17}F_4H_{28}MoOP$ $[C_5H_5Mo(CO)(P(C_2H_5)_3)H_3C-CC-C_2H_5][BF_4]$ .. Mo:Org.Comp.6-287

$BC_{17}F_4H_{33}MoO_6P_2$ $[(C_5H_5)Mo(P(OCH_3)_3)_2(H_3C-CC-C_3H_7-i)][BF_4]$ Mo:Org.Comp.6-119, 129

– .... $[(C_5H_5)Mo(P(OCH_3)_3)_2(H-CC-C_4H_9-t)][BF_4]$... Mo:Org.Comp.6-116, 128

Formula	Compound	Reference
$BC_{17}F_4H_{33}MoP_2$	$[(C_5H_5)Mo(P(CH_3)_3)_2(H-CC-C_4H_9-t)][BF_4]$	Mo:Org.Comp.6-116
$BC_{17}F_4H_{35}MoO_6P_2$	$[(P(OCH_3)_3)_2(H)(C_5H_5)MoC-CH_2C_4H_9-t][BF_4]$	Mo:Org.Comp.6-74/5, 80
$BC_{17}F_5FeH_{14}$	$[(F-4-C_6H_4-C_6H_5)Fe(C_5H_5)][BF_4]$	Fe: Org.Comp.B18-142/6, 222
$BC_{17}F_6FeH_{27}NP$	$[(1,3,5-(CH_3)_3C_6H_3)Fe(2-CH_3-1-(t-C_4H_9)-1,2-NBC_3H_3)][PF_6]$	Fe: Org.Comp.B18-99, 100
$BC_{17}FeH_{16}O^+$	$[C_6H_7FeC_4H_4BC_6H_5CO]^+$	Fe: Org.Comp.B17-245
$BC_{17}FeH_{17}$	$(C_5H_5)Fe[BC_5H_4(CH_3)(C_6H_5)]$	B: B Comp.SVol.3/4-155 Fe: Org.Comp.B17-323
$BC_{17}FeH_{24}NO_2$	$(C_5H_5)Fe[C_5H_3(B(OH)_2)-CH_2-1-NC_5H_9-2-CH_3]$	Fe: Org.Comp.A9-287, 288
$BC_{17}FeH_{27}N^+$	$[(1,3,5-(CH_3)_3C_6H_3)Fe(2-CH_3-1-(t-C_4H_9)-1,2-NBC_3H_3)]^+$	Fe: Org.Comp.B18-99, 100
$BC_{17}FeH_{28}N$	$1,3,5-(CH_3)_3C_6H_4FeC_3H_3B(CH_3)NC(CH_3)_3$	Fe: Org.Comp.B17-184
$BC_{17}Fe_3H_{13}O_9P^-$	$[(CO)_9Fe_3(BH_2)(P(CH_3)_2-C_6H_5)]^-$	Fe: Org.Comp.C6a-295, 299
$BC_{17}H_{12}MoN_9O_2$	$(CH_2CHCH_2)Mo(CO)_2[-N_2C_3H_2(CN)-]_3BH$	Mo:Org.Comp.5-226, 227
$BC_{17}H_{15}NO$	$C_6H_5B[-O-(1,2-C_{10}H_6)-CH_2-NH_2-]$	B: B Comp.SVol.3/3-220
$BC_{17}H_{16}N_5O_6$	$[CH_3C(O)O]_2B(-NC_6H_5-N{=}CNO_2-N{=}NC_6H_5-)$	B: B Comp.SVol.3/3-121/2
$BC_{17}H_{17}MoN_8O_2$	$(CH_2CHCH_2)Mo(CO)_2(N_2C_3H_3)_4B$	Mo:Org.Comp.5-225/6, 228
$BC_{17}H_{17}N_4O_4$	$[CH_3C(O)O]_2B(-NC_6H_5-N{=}CH-N{=}NC_6H_5-)$	B: B Comp.SVol.3/3-121
$BC_{17}H_{19}N_2$	$CH_3B[-NC_6H_5-CCH_3{=}CCH_3-NC_6H_5-]$	B: B Comp.SVol.3/3-112/3
$BC_{17}H_{19}N_2S_2$	$[-B(C_6H_5)_2-NHC(SCH_3)CHC(SCH_3)NH-]$	B: B Comp.SVol.3/3-122/3 B: B Comp.SVol.4/4-148/9
$BC_{17}H_{19}N_3^+$	$[C_{13}H_9N-B(-NCH_3-CH_2-CH_2-NCH_3-)]^+$	B: B Comp.SVol.3/3-207/9
$BC_{17}H_{20}NO_2$	$(CH_3O)(C_6H_5)B[-O-(1,2-C_6H_4)-CH{=}NC_3H_7-]$	B: B Comp.SVol.3/3-219/20
$BC_{17}H_{21}MoN_4O_2$	$(C_5H_5)Mo(CO)_2[-N_2C_3H_3-B(C_2H_5)_2-N_2C_3H_3-]$	Mo:Org.Comp.7-156, 159, 178
$BC_{17}H_{21}N_4$	$[-B(C_6H_5)-N(C_4H_9-t)-N{=}N-N(CH_2-C_6H_5)-]$	B: B Comp.SVol.4/3a-199
–	$[C_6H_5CH(CH_3)NH_3][(NC_4H_4-1)_2BH-CN]$	B: B Comp.SVol.3/3-213
$BC_{17}H_{22}MoN_7O_2S$	$HB(3,5-(CH_3)_2C_3HN_2)_3Mo(NS)(CO)_2$	S: S-N Comp.5-51, 58
$BC_{17}H_{22}N$	$2-CH_3-C_6H_4-C{\equiv}N-BC_9H_{15}$	B: B Comp.SVol.3/3-191
–	$4-CH_3-C_6H_4-C{\equiv}N-BC_9H_{15}$	B: B Comp.SVol.3/3-191
–	$(C_6H_5)_2B-NHCH_2-C_4H_9-t$	B: B Comp.SVol.3/3-153
$BC_{17}H_{22}N_7O_2SW$	$HB(3,5-(CH_3)_2C_3HN_2)_3W(NS)(CO)_2$	S: S-N Comp.5-51, 59
$BC_{17}H_{23}N_2$	$1,3-(C_6H_5CH_2)_2-1,3-N_2C_3H_6 \cdot BH_3$	B: B Comp.SVol.4/3b-12
–	$2-[(n-C_3H_7)_2B-N(C_6H_5)]-NC_5H_4$	B: B Comp.SVol.4/3a-232
$BC_{17}H_{23}N_4Si$	$2-CH_3-C_6H_4B[-N(C_6H_4-2-CH_3)-N{=}N-NSi(CH_3)_3-]$	B: B Comp.SVol.3/3-137/8
–	$2-CH_3-C_6H_4-B(N_3)-N(C_6H_4-CH_3-2)-Si(CH_3)_3$	B: B Comp.SVol.3/3-110
$BC_{17}H_{24}NO$	$C_6H_5-CHCH_3-NH(BH_3)-CHCH_3-C_6H_4-2-OCH_3$	B: B Comp.SVol.4/3b-7
–	$C_6H_5-CHCH_3-NH(BH_3)-CH(C_6H_5)-CH_2OCH_3$	B: B Comp.SVol.4/3b-7
$BC_{17}H_{25}MoN_4O_2$	$(CH_2CHCH_2)Mo(CO)_2[-N_2C_3(CH_3)_3-]_2BH_2$	Mo:Org.Comp.5-213, 214
$BC_{17}H_{26}N$	$3-BC_{10}H_{16}-4-C_2H_5 \cdot NC_5H_5$	B: B Comp.SVol.3/3-191
–	$C_8H_{14}B[-N(CH_3)_2-CH_2-(1,2-C_6H_4)-]$	B: B Comp.SVol.3/3-166
$BC_{17}H_{26}N_5$	$BC_3H_3N_5(C_6H_5)(C_4H_9-n)_2$	B: B Comp.SVol.4/3a-179
$BC_{17}H_{27}N_6$	$BC_2H_2N_6(C_6H_4-2-CH_3)(C_4H_9-n)_2$	B: B Comp.SVol.4/3a-178
–	$BC_2H_2N_6(C_6H_4-4-CH_3)(C_4H_9-n)_2$	B: B Comp.SVol.4/3a-178
$BC_{17}H_{28}N$	$2-[H_3B-N(CH_3)(C_6H_5)]-1,7,7-(CH_3)_3-[2.2.1]-C_7H_8$	B: B Comp.SVol.4/3b-10
$BC_{17}H_{29}N_2$	$1-[(i-C_3H_7)_2N-B(C_6H_5)]-NC_5H_{10}$	B: B Comp.SVol.3/3-109
–	$(i-C_3H_7)_2B-N{=}C(C_6H_5)NH-C_4H_9-t$	B: B Comp.SVol.3/3-156

$BC_{17}H_{29}N_2O$ ..... $BC_7H_6N_2(O-C_2H_5)(C_4H_9-i)_2$ .... B: B Comp.SVol.4/3a-257

$BC_{17}H_{29}N_4$ ...... 2,4,6-$(CH_3)_3-C_6H_2-N[-B(C_4H_9-t)-N(C_4H_9-i)-N=N-]$
B: B Comp.SVol.4/3a-199

– .... $[-B(C_6H_{13}-n)-N(C_4H_9-t)-N=N-N(CH_2C_6H_5)-]$.. B: B Comp.SVol.4/3a-199

$BC_{17}H_{31}N_2O_2$ .... 1-$[CH_3-O-C(=O)-CC-B(NH-C_4H_9-t)]$-
2,2,6,6-$(CH_3)_4-NC_5H_6$ .... B: B Comp.SVol.4/3a-170

$BC_{17}H_{32}N$ .... $(C_2H_5)_2B-N(C_6H_{11}-c)-C_7H_{11}-c$ .... B: B Comp.SVol.3/3-152

– .... $NC_5H_5 \cdot B(C_4H_9-n)_3$ .... B: B Comp.SVol.4/3b-28

$BC_{17}H_{32}N_3$ ...... 1-$[t-C_4H_9-NH-B(1-NC_4H_4)]$-2,2,6,6-$(CH_3)_4-NC_5H_6$
B: B Comp.SVol.4/3a-155

$BC_{17}H_{32}N_5Si$..... $(CH_3)_3Si-N(C_4H_9-t)-B[-N(C_4H_9-t)-N=N-N(C_6H_5)-]$
B: B Comp.SVol.4/3a-199

$BC_{17}H_{33}N_3^+$ ..... $[(i-C_3H_7)_2N-BN(i-C_3H_7)_2C_5H_5N]^+$........... B: B Comp.SVol.3/3-207

$BC_{17}H_{34}N_3O$ ..... 1-$[2-(O=)-NC_4H_6-1-B(NH-C_4H_9-t)]$-
2,2,6,6-$(CH_3)_4-NC_5H_6$ .... B: B Comp.SVol.4/3a-155

$BC_{17}H_{35}NO_2P$.... 9-$[(C_2H_5-O)_2P-N(CH_2-C_4H_9-t)]$-[3.3.1]-9-$BC_8H_{14}$
B: B Comp.SVol.4/3a-236

$BC_{17}H_{35}N_2$ ...... 1-$[t-C_4H_9-CH_2-C(CH_3)_2-N=B]$-2,2,6,6-$(CH_3)_4-NC_5H_6$
B: B Comp.SVol.4/3a-160/5

$BC_{17}H_{36}N$ .... 1-$[t-C_4H_9-B(C_4H_9-i)]$-2,2,6,6-$(CH_3)_4-NC_5H_6$.. B: B Comp.SVol.4/3a-233

$BC_{17}H_{36}NSi$ ..... 1-$(n-C_8H_{17})$-4,5-$(C_2H_5)_2$-2,2,3-$(CH_3)_3$-1,2,5-$NSiBC_2$
B: B Comp.SVol.4/3a-250

$BC_{17}H_{36}N_2P$ ..... 1-$[(t-C_4H_9)_2P-N=B]$-2,2,6,6-$(CH_3)_4-NC_5H_6$ ... B: B Comp.SVol.4/3a-160/5

$BC_{17}H_{36}N_3O_2$ .... 1-$[C_2H_5-OC(O)-CH_2-NH-B(NH-C_4H_9-t)]$-
2,2,6,6-$(CH_3)_4-NC_5H_6$ .... B: B Comp.SVol.4/3a-155

$BC_{17}H_{37}N_2O$ ..... $t-C_4H_9-NH-B(O-C_4H_9-t)-1-NC_5H_6$-2,2,6,6-$(CH_3)_4$
B: B Comp.SVol.4/3b-52

$BC_{17}H_{37}N_2O_2$ .... $t-C_4H_9OC(O)B[N(i-C_3H_7)_2]_2$ .... B: B Comp.SVol.3/3-110

$BC_{17}H_{37}N_3O_2P$ ... 1-[1,3,2-$O_2PC_2H_4$-2-$N(C_4H_9-t)-B(N(CH_3)_2)]$-
2,2,6,6-$(CH_3)_4-NC_5H_6$ .... B: B Comp.SVol.4/3a-155

$BC_{17}H_{37}N_3O_2Sb$.. 1-[1,3,2-$O_2SbC_2H_4$-2-$N(C_4H_9-t)-B(N(CH_3)_2)]$-
2,2,6,6-$(CH_3)_4-NC_5H_6$ .... B: B Comp.SVol.4/3a-155

$BC_{17}H_{38}NSn$ ..... $(C_2H_5)_2B[-C(C_2H_5)=CCH_3-Sn(CH_3)_2-N(C_3H_7-i)_2-]$
B: B Comp.SVol.4/3a-255
Sn: Org.Comp.19-47/8, 50

$BC_{17}H_{38}N_3$ ...... 1-$[t-C_4H_9-NH-B(NH-C_4H_9-t)]$-2,2,6,6-$(CH_3)_4-NC_5H_6$
B: B Comp.SVol.4/3a-155

$BC_{17}H_{39}N_2OSi$ ... $(CH_3)_3Si-N(C_4H_9-t)-B(OCH_3)-1-NC_5H_6$-2,2,6,6-$(CH_3)_4$
B: B Comp.SVol.4/3b-52

$BC_{17}H_{39}O_4Sn$.... $(C_4H_9)_2Sn(OC_3H_7-i)OB(OC_3H_7-i)_2$ .... Sn: Org.Comp.16-192

$BC_{17}H_{42}NSi_2Sn$ .. $[(CH_3)_3Si]_2N-Sn(CH_3)_2-C(CH_3)=C(C_2H_5)-B(C_2H_5)_2$
B: B Comp.SVol.4/3a-255
Sn: Org.Comp.19-47/8, 51

$BC_{17}H_{45}N_3PSi_3$... $[(CH_3)_3Si]_2N-P[-N(C_4H_9-t)-B(C_4H_9-t)-N(Si(CH_3)_3)-]$
B: B Comp.SVol.4/3a-207/8

$BC_{18}ClH_{12}O$ ..... 6-Cl-5,6-$OBC_{12}H_7$-4-C_6H_5 .... B: B Comp.SVol.4/4-55

$BC_{18}ClH_{20}$ ...... $C_6H_5-BCl-C(C_6H_5)=CH-C_4H_9-n$ .... B: B Comp.SVol.3/4-43, 45

– .... $C_6H_5-BCl-C(C_6H_5)=CH-C_4H_9-t$............ B: B Comp.SVol.3/4-43, 45

– .... $C_6H_5-BCl-CH=C(C_6H_5)-C_4H_9-n$ .... B: B Comp.SVol.3/4-45

– .... $C_6H_5-BCl-CH=C(C_6H_5)-C_4H_9-t$............ B: B Comp.SVol.3/4-45

$BC_{18}ClH_{22}$ ...... $(2\text{-}CH_3\text{-}C_6H_4\text{-}CH_2CH_2)_2BCl$ ...... B: B Comp.SVol.4/4-47

$BC_{18}ClH_{22}MoN_6O_2$
$ClCMo(CO)_2[\text{-}N_2C_3H(CH_3)_2\text{-}]_3BH$ Mo:Org.Comp.5-94

$BC_{18}ClH_{26}O$ ..... $2,7,7\text{-}(CH_3)_3\text{-}[3.1.1]\text{-}C_7H_8\text{-}3\text{-}BCl\text{-}OCH(CH_3)\text{-}C_6H_5$
B: B Comp.SVol.4/4-55

$BC_{18}ClH_{36}N_2$ .... $ClB[NC_5H_6(CH_3)_4]_2$ ...... B: B Comp.SVol.3/4-59

$BC_{18}ClH_{45}NSi_4$ ... $2,2,6,6\text{-}(CH_3)_4\text{-}NC_5H_6\text{-}1\text{-}BCl\text{-}Si[Si(CH_3)_3]_3$ ... B: B Comp.SVol.4/4-61

$BC_{18}Cl_2H_{16}P$ ..... $(C_6H_5)_3P \cdot BHCl_2$ ...... B: B Comp.SVol.4/4-39

$BC_{18}Cl_2H_{30}N$ .... $Cl_2BNH[C_6H_2\text{-}2,4,6\text{-}(t\text{-}C_4H_9)_3]$ ...... B: B Comp.SVol.3/4-54

$BC_{18}Cl_3H_{15}N$ .... $H_3N\text{-}B(C_6H_4\text{-}4\text{-}Cl)_3$ ...... B: B Comp.SVol.3/3-190

$BC_{18}Cl_3H_{15}P$ ..... $(C_6H_5)_3P \cdot BCl_3$ ...... B: B Comp.SVol.4/4-34

$BC_{18}Cl_4GaH_{36}N_2$ $[(\text{-}C(CH_3)_2\text{-}(CH_2)_3\text{-}C(CH_3)_2\text{-})N{=}B{=}$
$N(\text{-}C(CH_3)_2\text{-}(CH_2)_3\text{-}C(CH_3)_2\text{-})][GaCl_4]$ ...... B: B Comp.SVol.3/3-205

$BC_{18}CrFeH_{14}O_3$ .. $(C_5H_5)Fe(C_4H_4BC_6H_5)Cr(CO)_3$ ...... B: B Comp.SVol.4/4-180

$BC_{18}CrFeH_{14}O_3^-$ $[C_5H_5FeC_4H_4BC_6H_5Cr(CO)_3]^-$ ...... Fe: Org.Comp.B17-200, 227

$BC_{18}FH_{22}$ .... $[2,4,6\text{-}(CH_3)_3\text{-}C_6H_2]_2BF$ ...... B: B Comp.SVol.3/3-364
B: B Comp.SVol.4/3b-228

$BC_{18}FH_{34}N_2Si$ .... $C_6H_5\text{-}CH_2\text{-}N(C_4H_9\text{-}t)\text{-}BF\text{-}N(C_4H_9\text{-}t)\text{-}Si(CH_3)_3$ B: B Comp.SVol.4/3b-246

$BC_{18}FH_{36}N_2$ ..... $FB[1\text{-}NC_5H_6\text{-}2,2,6,6\text{-}(CH_3)_4]_2$ ...... B: B Comp.SVol.3/3-376

– ...... $FB[N{=}C(C_4H_9\text{-}t)_2]_2$ ...... B: B Comp.SVol.3/3-376

$BC_{18}FH_{49}N_3Si_4$ ... $[(CH_3)_3Si]_2N\text{-}BF\text{-}N[Si(CH_3)_2\text{-}C_4H_9\text{-}t]\text{-}NH$
$\text{-}Si(CH_3)_2\text{-}C_4H_9\text{-}t$ ...... B: B Comp.SVol.4/3b-246

$BC_{18}FH_{54}N_4Si_7$ ... $[(CH_3)_3Si]_2N\text{-}BF\text{-}N[Si(CH_3)_3]\text{-}Si(CH_3)_2\text{-}$
$N[\text{-}Si(CH_3)_2\text{-}N(Si(CH_3)_3)\text{-}Si(CH_3)_2\text{-}]$ ...... B: B Comp.SVol.4/3b-247

$BC_{18}F_2FeH_{23}O_3$ .. $(C_5H_5)Fe(C(c\text{-}C_6H_6(CH_3)_3)OBF_2OCCH_3)CO$ ... Fe: Org.Comp.B16b-128/9, 141

$BC_{18}F_2H_{30}N$ ..... $F_2B\text{-}NH[C_6H_2\text{-}2,4,6\text{-}(t\text{-}C_4H_9)_3]$ ...... B: B Comp.SVol.3/3-375

$BC_{18}F_3H_{19}N_3O_3S$ $[C_{13}H_9N\text{-}B(\text{-}NCH_3\text{-}CH_2CH_2\text{-}NCH_3\text{-})]O_3SCF_3$ .. B: B Comp.SVol.3/3-207/9

$BC_{18}F_3H_{29}NSn$ ... $(C_2H_5)_2B[\text{-}C(C_2H_5){=}CCH_3\text{-}Sn(CH_3)_2\text{-}NH(C_6H_4\text{-}2\text{-}CF_3)\text{-}]$
B: B Comp.SVol.4/3a-255
Sn: Org.Comp.19-48, 50

$BC_{18}F_3H_{33}N_3O_3S$ $[NC_5H_5\text{-}1\text{-}B(N(C_3H_7\text{-}i)_2)_2][CF_3\text{-}SO_3]$ ...... B: B Comp.SVol.4/3b-45

$BC_{18}F_4FeH_{13}O_3$ .. $[(9\text{-}C_6H_5\text{-}C_9H_8)Fe(CO)_3][BF_4]$ ...... Fe: Org.Comp.B15-254

$BC_{18}F_4FeH_{14}$ .... $[(C_{13}H_9)Fe(C_5H_5)][BF_4]$ ...... Fe: Org.Comp.B19-251

$BC_{18}F_4FeH_{15}$ .... $[(C_{13}H_{10})Fe(C_5H_5)][BF_4]$ ...... Fe: Org.Comp.B19-220, 249/50, 306/7

$BC_{18}F_4FeH_{15}O_2$ .. $[(C_5H_5)Fe(CO)_2(CH_2{=}C{=}C{=}C(CH_3)C_6H_5)][BF_4]$ Fe: Org.Comp.B17-116, 119/20

$BC_{18}F_4FeH_{15}O_3$ .. $[(C_6H_5\text{-}CH{=}CH\text{-}C_7H_8)Fe(CO)_3][BF_4]$ ...... Fe: Org.Comp.B15-207

– ...... $[(C_7H_7\text{-}C_8H_8)Fe(CO)_3][BF_4]$ ...... Fe: Org.Comp.B15-215

$BC_{18}F_4FeH_{15}O_4W$ $[C_5H_5(CO)_2Fe{=}CO_2W(C_5H_5)_2][BF_4]$ ...... Fe: Org.Comp.B16a-136, 150

$BC_{18}F_4FeH_{16}$ .... $[(C_6H_5CHC_6H_5)Fe(C_5H_5)][BF_4]$ ...... Fe: Org.Comp.B18-142/6, 231

$BC_{18}F_4FeH_{17}$ .... $[(C_6H_6)Fe(C_5H_4\text{-}CH_2\text{-}C_6H_5)][BF_4]$ ...... Fe: Org.Comp.B18-142/6, 151, 166, 188

– ...... $[(C_6H_5\text{-}CH_2\text{-}C_6H_5)Fe(C_5H_5)][BF_4]$ ...... Fe: Org.Comp.B18-142/6, 200, 201, 230

$BC_{18}F_4FeH_{19}HgO_6$
$[(C_5H_5)Fe(CO)_2(C_7H_8(HgO_2CCH_3)(O_2CCH_3))][BF_4]$
Fe: Org.Comp.B17-99

$BC_{18}F_4FeH_{19}O_2$ .. $[(C_9H_{11})Fe(CO)_2(C_7H_8)][BF_4]$ ...... Fe: Org.Comp.B17-135/6

$BC_{18}F_4FeH_{21}O$ ... $[(OC_{13}H_{16})Fe(C_5H_5)][BF_4]$ ... Fe: Org.Comp.B19-216, 222/3, 300

$BC_{18}F_4FeH_{21}O_2$ .. $[C_5H_5(CO)_2Fe{=}CHC_{10}H_{15}][BF_4]$ ... Fe: Org.Comp.B16a-89, 96/7

$BC_{18}F_4FeH_{21}O_3$ .. $[C_5H_5(CO)_2Fe{=}CHC_6H_8({=}O)(C_2H_4CH{=}CH_2)\text{-}c][BF_4]$ Fe: Org.Comp.B16a-89, 97

$BC_{18}F_4FeH_{22}NO_2$ $[(C_5H_5)Fe(CO)_2CH(CH_3)CH(CH_3)NH_2CH_2C_6H_5][BF_4]$ Fe: Org.Comp.B14-150/1, 155

$BC_{18}F_4FeH_{23}O_2$ .. $[((CH_3)_6C_6)Fe(C_5H_4\text{-}COOH)][BF_4]$ ... Fe: Org.Comp.B19-142, 160/1, 202

– ... $[(C_5H_5)(CO)_2Fe{=}CH\text{-}CH_2CH_2CH_2\text{-}C_6H_9{=}CH_2][BF_4]$ Fe: Org.Comp.B16a-88, 95/6

– ... $[(C_5H_5)(CO)_2Fe{=}CH\text{-}CH_2CH_2CH_2\text{-}C_6H_8\text{-}CH_3][BF_4]$ Fe: Org.Comp.B16a-89, 97

– ... $[(C_5H_5)(CO)_2Fe{=}C_3(C_4H_9\text{-}t)_2][BF_4]$ ... Fe: Org.Comp.B16a-91

$BC_{18}F_4FeH_{23}O_3$ .. $[(C_5H_5)(CO)_2Fe{=}C(C_8H_{13})\text{-}O\text{-}C_2H_5][BF_4]$ ... Fe: Org.Comp.B16a-112

– ... $[(C_5H_5)(CO)_2Fe{=}CH\text{-}C_6H_8({=}O)(C_4H_9)][BF_4]$ ... Fe: Org.Comp.B16a-89, 97

– ... $[(C_5H_5)Fe(CO)_2{=}C(O\text{-}C_2H_5)\text{-}C_8H_{13}][BF_4]$ ... Fe: Org.Comp.B13-47

$BC_{18}F_4FeH_{24}$ .... $[((CH_3)_6C_6)Fe(C_6H_6)][BF_4]$ ... Fe: Org.Comp.B19-405

$BC_{18}F_4FeH_{25}$ .... $[((CH_3)_6C_6)Fe(C_5H_4\text{-}CH_3)][BF_4]$ ... Fe: Org.Comp.B19-159

$BC_{18}F_4FeH_{25}O$ ... $[(CH_3)_5C_5Fe(C_6H_7CH_3)CO][BF_4]$ ... Fe: Org.Comp.B17-189

$BC_{18}F_4FeH_{25}O_2$ .. $[(CH_3)_5C_5Fe(C_6H_7OCH_3)CO][BF_4]$ ... Fe: Org.Comp.B17-189

$BC_{18}F_4FeH_{25}O_4$ .. $[(C_5H_5)Fe(CO)_2(CH_2{=}CHCH_2C_3H_4O_2(C_5H_{11}\text{-}n))][BF_4]$ Fe: Org.Comp.B17-34

$BC_{18}F_4FeH_{33}O_6P_2$ $[(C_5H_5)Fe(C_7H_{10}\text{-}c)(P(OCH_3)_3)_2][BF_4]$ ... Fe: Org.Comp.B16b-8

$BC_{18}F_4H_{15}S$ ..... $[(C_6H_5)_3S][BF_4]$ ... B: B Comp.SVol.3/4-140; B: B Comp.SVol.4/4-163

$BC_{18}F_4H_{15}Se$ .... $[(C_6H_5)_3Se][BF_4]$ ... B: B Comp.SVol.3/4-147

$BC_{18}F_4H_{15}Te$ .... $[(C_6H_5)_3Te][BF_4]$ ... B: B Comp.SVol.3/4-150; B: B Comp.SVol.4/4-176

$BC_{18}F_4H_{16}MoNO_5$ $[(C_5H_5)Mo(CO)_2O_2CC({=}CHC_6H_5)NH_2C(O)CH_3][BF_4]$ Mo:Org.Comp.7-190

$BC_{18}F_4H_{17}MoO_2$.. $[(C_9H_7)Mo(CO)_2(C_6H_7\text{-}1\text{-}CH_3)][BF_4]$ ... Mo:Org.Comp.8-305/6, 313, 317/8

– ... $[(C_9H_7)Mo(CO)_2(C_6H_7\text{-}2\text{-}CH_3)][BF_4]$ ... Mo:Org.Comp.8-305/6, 313, 317/8

$BC_{18}F_4H_{17}MoO_3$.. $[(C_9H_7)Mo(CO)_2(C_6H_7\text{-}1\text{-}OCH_3)][BF_4]$ ... Mo:Org.Comp.8-305/6, 314, 317/8

$BC_{18}F_4H_{19}MoO$ .. $[C_9H_7Mo(CO)(H_3C\text{-}CC\text{-}CH_3)_2][BF_4]$ ... Mo:Org.Comp.6-318, 325

$BC_{18}F_4H_{22}MoN_2O_5P$ $[(CH_2CHCH_2)Mo(CO)_2(NC_5H_4\text{-}2\text{-}(2\text{-}C_5H_4N))(P(OCH_3)_3)][BF_4]$ ... Mo:Org.Comp.5-200, 202

$BC_{18}F_4H_{22}N_2O_3Re$ $[(C_5H_5)Re(CO)_2\text{-}NHN(C_4H_9\text{-}n)\text{-}C_6H_4\text{-}4\text{-}OCH_3][BF_4]$ Re: Org.Comp.3-193/7, 205

$BC_{18}F_4H_{25}MoO$ .. $[C_5H_5Mo(CO)(H\text{-}CC\text{-}C_4H_9\text{-}t)_2][BF_4]$ ... Mo:Org.Comp.6-316/7, 323

$BC_{18}F_4H_{25}MoO_2$.. $[((CH_3)_5C_5)Mo(CO)_2(CH_3\text{-}CHC(CH_2)CH\text{-}CH_3)][BF_4]$ Mo:Org.Comp.8-305/6, 311, 312

$BC_{18}F_4H_{26}MoN$... $[C_5H_5Mo(NCC_4H_9\text{-}t)(H_3C\text{-}CC\text{-}CH_3)_2][BF_4]$ ... Mo:Org.Comp.6-137

$BC_{18}F_4H_{29}MoO_6P_2$
$[C_9H_7Mo(P(OCH_3)_3)_2H{-}CC{-}CH_3][BF_4]$ Mo:Org.Comp.6-121

$BC_{18}F_4H_{30}MnN_3S_6$
$[Mn((SC(=S))NC_5H_{10})_3][BF_4]$ Mn:MVol.D7-178/9

$BC_{18}F_4H_{35}MoO_6P_2$
$[C_5H_5Mo(P(OCH_3)_3)_2H_3C{-}CC{-}C_4H_9{-}t][BF_4]$ ... Mo:Org.Comp.6-119, 129

$BC_{18}F_4H_{35}MoO_6P_2S$
$[C_5H_5Mo(P(OCH_3)_3)_2H_3CS{-}CC{-}C_4H_9{-}t][BF_4]$ .. Mo:Org.Comp.6-117

$BC_{18}F_5H_{34}N_4O_7Re_2$
$[(CO)_3ReF(N(CH_3)_2CH_2CH_2N(CH_3)_2)]_2H \cdot HOBF_3$
Re: Org.Comp.1-158, 177, 178

$BC_{18}F_6H_{14}NO_2$ ... $(CH_3)_2B{-}N(C_6H_4{-}3{-}CF_3){-}C(O)C(O)(C_6H_4{-}3{-}CF_3)$
B: B Comp.SVol.3/3-155

$BC_{18}F_9H_{10}Se$ .... $[(C_6H_5)_2Se(C_6F_5)][BF_4]$ .... B: B Comp.SVol.3/4-147

$BC_{18}F_9H_{10}Te$ .... $[(C_6H_5)_2TeC_6F_5][BF_4]$ .... B: B Comp.SVol.3/4-150

$BC_{18}F_{10}H_5N_4$ .... $C_6F_5B({-}NC_6F_5{-}N{=}N{-}NC_6H_5{-})$ .... B: B Comp.SVol.3/3-137/8

$BC_{18}F_{20}H_{24}O_4Te_4Tl$
$[(1,3,5{-}(CH_3)_3{-}C_6H_3)_2Tl][B(OTeF_5)_4]$ .... B: B Comp.SVol.4/4-176

$BC_{18}FeH_{24}O_3PSi$ $BC_2FePSi(C_2H_5)_2(CH_3)_3(C_6H_5)(CO)_3$ .... B: B Comp.SVol.4/4-179

$BC_{18}FeH_{24}O_4{}^+$ ... $[6{-}(n{-}C_4H_9)_2BO{-}C_7H_6Fe(CO)_3]^+$ .... Fe: Org.Comp.B15-218, 230/1

$BC_{18}Fe_3H_{15}O_9P^-$ $[(CO)_9Fe_3(H{-}B{-}CH_3)(P(CH_3)_2{-}C_6H_5)]^-$ .... Fe: Org.Comp.C6a-299

$BC_{18}GeH_{39}Sn_2$ ... $(CH_3)_2Ge[{-}C(Sn(CH_3)_3){=}C(B(C_2H_5)_2)C(C_2H_5){=}$
$C(Sn(CH_3)_3){-}]$ Ge:Org.Comp.3-279

$BC_{18}H_{13}Mn_2O_6$ ... $(CO)_3Mn[(C_2H_5)C_4H_3BC_6H_5]Mn(CO)_3$ .... B: B Comp.SVol.3/4-155

$BC_{18}H_{13}N_9{}^-$ ..... $[(1,2,3{-}N_3C_6H_4{-}1)_3BH]^-$ .... B: B Comp.SVol.4/3b-50

$BC_{18}H_{14}N$ .... $C_6H_5B[{-}(1,2{-}C_6H_4){-}(1,2{-}C_6H_4){-}NH{-}]$ .... B: B Comp.SVol.3/3-163

$BC_{18}H_{15}N_4$ ...... $C_6H_5B({-}NC_6H_5{-}N{=}N{-}NC_6H_5{-})$ .... B: B Comp.SVol.3/3-137/8

$BC_{18}H_{15}Se_3$ ..... $B(SeC_6H_5)_3$ .... B: B Comp.SVol.3/4-147

$BC_{18}H_{16}NO_3$ ..... $(C_6H_5O)_2B[{-}NH_2{-}(1,2{-}C_6H_4){-}O{-}]$ .... B: B Comp.SVol.3/3-219

$BC_{18}H_{17}MoN_6O_2$. $(c{-}C_7H_7)Mo(CO)_2[{-}N_2C_3H_3{-}BH(N_2C_3H_3){-}N_2C_3H_3{-}]$
Mo:Org.Comp.7-156, 159

– .... $(c{-}C_7H_7)Mo(CO)_2[{-}N_2C_3H_3{-}]_3BH$ Mo:Org.Comp.5-226, 229

$BC_{18}H_{18}N$ .... $(C_6H_5)_3N{-}BH_3$ .... B: B Comp.SVol.3/3-178

– .... $H_3N{-}B(C_6H_5)_3$ .... B: B Comp.SVol.3/3-190
B: B Comp.SVol.4/3b-27

$BC_{18}H_{19}MoN_8O_2$. $[CH_2C(CH_3)CH_2]Mo(CO)_2(N_2C_3H_3)_4B$ Mo:Org.Comp.5-225/6, 228

$BC_{18}H_{19}N_2O_2$ .... $3,4{-}(C_6H_5)_2{-}2,4,6,3{-}ON_2BC_6H_9({=}O){-}5$ .... B: B Comp.SVol.4/3b-67/8

$BC_{18}H_{19}N_4O_4$ .... $[CH_3C(O)O]_2B[{-}N(C_6H_5)NC(CH_3)NN(C_6H_5){-}]$ .. B: B Comp.SVol.3/3-121/2

– .... $[CH_3OC(O)]_2B[{-}N(C_6H_5)NC(C_6H_5)NN(CH_3){-}]$ .. B: B Comp.SVol.4/3b-57

$BC_{18}H_{19}N_4O_5$ .... $[CH_3OC(O)]_2B[{-}N(C_6H_5)NC(C_6H_5)NN(OCH_3){-}]$ B: B Comp.SVol.4/3b-57

$BC_{18}H_{19}O_4W_2$.... $W_2[(CH_3)CB(H)C_2H_5](CO)_4(C_5H_5)_2$ .... B: B Comp.SVol.3/4-155

$BC_{18}H_{19}S_2$ ...... $n{-}C_4H_9B({-}S{-}CC_6H_5{=}CC_6H_5{-}S{-})$ .... B: B Comp.SVol.3/4-114

$BC_{18}H_{20}NO_2$ ..... $(CH_3)_2B{-}N(CH_2C_6H_5){-}C(O){-}C(O)CH_2C_6H_5$ .... B: B Comp.SVol.3/3-155

$BC_{18}H_{20}NS_2$ ..... $(C_2H_5)_2NB[{-}C(C_6H_5){=}C(C_6H_5){-}S{-}S{-}]$ .... B: B Comp.SVol.3/4-126

– .... $(C_2H_5)_2NB[{-}S{-}C(C_6H_5){=}C(C_6H_5){-}S{-}]$ .... B: B Comp.SVol.3/4-126

$BC_{18}H_{21}S_2$ ...... $n{-}C_4H_9B({-}S{-}CHC_6H_5{-}CHC_6H_5{-}S{-})$.......... B: B Comp.SVol.3/4-112

$BC_{18}H_{22}IMoN_6O_2$ $ICMo(CO)_2[{-}N_2C_3H(CH_3)_2{-}]_3BH$ Mo:Org.Comp.5-94

$BC_{18}H_{22}IS_2$...... $IB[SC_6H_2{-}2,4,6{-}(CH_3)_3]_2$ .... B: B Comp.SVol.3/4-99

$BC_{18}H_{22}N$ .... $1{-}[(C_6H_5)_2B]{-}2{-}CH_3{-}NC_5H_9$ .... B: B Comp.SVol.3/3-156

– .... $9{-}(i{-}C_3H_7)_2B{-}9{-}NC_{12}H_8$ .... B: B Comp.SVol.4/3a-233

$BC_{18}H_{22}N$ .... $C_9H_7N\text{-}BC_9H_{15}$ .... B: B Comp.SVol.3/3-191
$BC_{18}H_{22}NS$ ...... $1\text{-}(1\text{-}C_5H_5N)\text{-}2\text{-}(2\text{-}C_4H_3S)\text{-}1\text{-}BC_9H_{14}$ .... B: B Comp.SVol.3/3-191
$BC_{18}H_{22}N_3$ ...... $[2,4,6\text{-}(CH_3)_3C_6H_2]_2BN_3$ .... B: B Comp.SVol.3/3-145
$BC_{18}H_{22}N_6O_3Re$ .. $(CO)_3Re[1,2\text{-}N_2C_3H(CH_3)_2\text{-}3,5]_3BH$ .... Re: Org.Comp.1-115, 117
$BC_{18}H_{23}MoN_6O_2$. $(CH_2CHCH_2)Mo(CO)_2[\text{-}N_2C_3H_3\text{-}]_3B\text{-}C_4H_9\text{-}n$ .. Mo: Org.Comp.5-225/7
$BC_{18}H_{24}N$ .... $2\text{-}(i\text{-}C_3H_7)_2N\text{-}2\text{-}BC_{12}H_{10}$ .... B: B Comp.SVol.4/3a-229
– .... $(i\text{-}C_3H_7)_2B\text{-}N(C_6H_5)_2$ .... B: B Comp.SVol.4/3a-227
– .... $(C_6H_5)_2B\text{-}N(C_3H_7\text{-}i)_2$ .... B: B Comp.SVol.3/3-149
– .... $(C_6H_5)_2B\text{-}N(C_3H_7\text{-}n)_2$ .... B: B Comp.SVol.3/3-149
– .... $[2,4,6\text{-}(CH_3)_3\text{-}C_6H_2]_2B\text{-}NH_2$ .... B: B Comp.SVol.3/3-149
B: B Comp.SVol.4/3a-226
B: B Comp.SVol.4/3b-228
$BC_{18}H_{24}NO_2$ ..... $C_6H_5\text{-}O\text{-}B(C_6H_5)[\text{-}O\text{-}CH_2CH_2\text{-}N(C_2H_5)_2\text{-}]$ .... B: B Comp.SVol.4/3b-70
$BC_{18}H_{24}N_3O$ ..... $BC_6H_5N_3(=O)(C_6H_5)(C_3H_7\text{-}n)_2$ .... B: B Comp.SVol.4/3a-180/1
$BC_{18}H_{24}O_3PRuSi$ $BC_2PRuSi(C_2H_5)_2(CH_3)_3(C_6H_5)(CO)_3$ .... B: B Comp.SVol.4/4-179
$BC_{18}H_{25}MoN_4O_2$. $(c\text{-}C_6H_9)Mo(CO)_2[\text{-}N_2C_3H(CH_3)_2\text{-}]_2BH_2$ ...... Mo: Org.Comp.5-213, 216
– .... $(c\text{-}C_6H_9)Mo(CO)_2[\text{-}N_2C_3H_3\text{-}]_2B(C_2H_5)_2$ ...... Mo: Org.Comp.5-213, 216/7
$BC_{18}H_{25}MoN_6O_2$. $(CH_2CHCH_2)(C_3H_4N_2)Mo(CO)_2[\text{-}N_2C_3H_3\text{-}]_2B(C_2H_5)_2$
Mo: Org.Comp.5-224, 225
– .... $(CH_2CHCH_2)Mo(CO)_2[\text{-}N_2C_3H_3\text{-}]_2B(C_2H_5)_2 \cdot N_2C_3H_4$
Mo: Org.Comp.5-218
$BC_{18}H_{25}N_2$ ...... $2\text{-}[(n\text{-}C_3H_7)_2B\text{-}N(CH_2\text{-}C_6H_5)]\text{-}NC_5H_4$ .... B: B Comp.SVol.4/3a-232
– .... $[\text{-}N(CH_2C_6H_5)\text{-}CHCH_3\text{-}N(CH_2C_6H_5)\text{-}CH_2CH_2\text{-}] \cdot BH_3$
B: B Comp.SVol.4/3b-12
$BC_{18}H_{25}N_2O_2$ .... $NC_5H_5 \cdot (i\text{-}C_3H_7)_2B\text{-}N(C_6H_5)\text{-}COOH$ .... B: B Comp.SVol.4/3b-28
$BC_{18}H_{26}MoN_3O_3$. $(C_5H_5)Mo(CO)_2\text{-}C(=O)\text{-}C[N(C_2H_5)_2]=CH\text{-}2\text{-}$
$[1,3,2\text{-}N_2BC_2H_4\text{-}1,3\text{-}(CH_3)_2]$ .... Mo: Org.Comp.8-202, 204
$BC_{18}H_{26}N$ .... $C_6H_5\text{-}CH(CH_3)\text{-}N(BH_3)(C_3H_7\text{-}i)\text{-}CH_2\text{-}C_6H_5$ ... B: B Comp.SVol.4/3b-10
$BC_{18}H_{26}NO$ ...... $(C_8H_{14}BNOC_9H_9)CH_3$ .... B: B Comp.SVol.3/3-219
$BC_{18}H_{26}N_3$ ...... $(n\text{-}C_4H_9)_2B[\text{-}(1,2\text{-}NC_5H_4)\text{-}N=(2,1\text{-}NC_5H_4)\text{-}]$ ... B: B Comp.SVol.3/3-123
$BC_{18}H_{26}N_3OS$ .... $BC_4H_3N_3S(=O)(C_6H_5)(C_4H_9\text{-}n)_2$ .... B: B Comp.SVol.4/3a-180
– .... $[\text{-}B(C_4H_9\text{-}n)_2\text{-}O\text{-}C(NH\text{-}C_6H_5)=N\text{-}SNC_3H_2\text{-}]$ ... B: B Comp.SVol.4/4-152/3
$BC_{18}H_{26}N_4O_8P_2Re$
$[CH_3C(\text{-}O\text{-}CH_2\text{-})_3P]_2Re(CO)_2[\text{-}(1,2\text{-}N_2C_3H_3)$
$\text{-}BH_2\text{-}(1,2\text{-}N_2C_3H_3)\text{-}]$ .... Re: Org.Comp.1-69, 80
$BC_{18}H_{27}N_2O_2$ .... $2,2,6,6\text{-}(CH_3)_4\text{-}NC_5H_6\text{-}1\text{-}B[\text{-}NH\text{-}C(=O)$
$\text{-}CH(CH_2\text{-}C_6H_5)\text{-}O\text{-}]$ .... B: B Comp.SVol.4/3b-56
$BC_{18}H_{28}N$ .... $C_5H_5N\text{-}3\text{-}BC_{10}H_{16}\text{-}4\text{-}(n\text{-}C_3H_7)$ .... B: B Comp.SVol.3/3-191
$BC_{18}H_{28}NSn_2$ .... $(C_6H_5)_2B\text{-}N[Sn(CH_3)_3]_2$ .... B: B Comp.SVol.4/3a-227
$BC_{18}H_{28}N_5$ ...... $BC_3H_3N_5(C_6H_4\text{-}2\text{-}CH_3)(C_4H_9\text{-}n)_2$ .... B: B Comp.SVol.4/3a-179
– .... $BC_3H_3N_5(C_6H_4\text{-}4\text{-}CH_3)(C_4H_9\text{-}n)_2$ .... B: B Comp.SVol.4/3a-179
$BC_{18}H_{29}N_2$ ...... $1\text{-}[2,4,6\text{-}(CH_3)_3C_6H_2\text{-}N=B]\text{-}2,2,6,6\text{-}(CH_3)_4NC_5H_6$
B: B Comp.SVol.3/3-100
B: B Comp.SVol.4/3a-160/5
$BC_{18}H_{30}N$ .... $1\text{-}[i\text{-}C_3H_7\text{-}B(C_6H_5)]\text{-}2,2,6,6\text{-}(CH_3)_4\text{-}NC_5H_6$ ... B: B Comp.SVol.4/3a-233
– .... $i\text{-}C_3H_7\text{-}C(CH_3)_2\text{-}B=N\text{-}C_6H_3\text{-}2,6\text{-}(C_3H_7\text{-}i)_2$ .... B: B Comp.SVol.4/3a-214
$BC_{18}H_{30}NO$ ...... $(C_2H_5)_2B[\text{-}N(C_3H_7\text{-}i)=C(C_2H_5)CH_2CH(C_6H_5)\text{-}O\text{-}]$
B: B Comp.SVol.3/3-218
– .... $(C_2H_5)_2B[\text{-}N(C_3H_7\text{-}i)=CCH_3\text{-}CHCH_3\text{-}CH(C_6H_5)\text{-}O\text{-}]$
B: B Comp.SVol.3/3-218

$BC_{18}H_{30}NSi$ ..... $C_5H_5N-BC_9H_{14}-2-CH_2Si(CH_3)_3$ .... B: B Comp.SVol.3/3-191
$BC_{18}H_{31}N_2$ ...... $C_6H_5B[N(C_3H_7-i)_2]-2-CH_3-NC_5H_9$ .... B: B Comp.SVol.3/3-109
$BC_{18}H_{32}NO$...... $i-C_3H_7-C(CH_3)_2-B(OH)-NH-C_6H_3-2,6-(C_3H_7-i)_2$
B: B Comp.SVol.4/3b-61
$BC_{18}H_{33}N_2$ ...... $1-[t-C_4H_9-NH-B(C_5H_5)]-2,2,6,6-(CH_3)_4-NC_5H_6$
B: B Comp.SVol.4/3a-169
$BC_{18}H_{33}N_2OSi$ ... $2-[(CH_3)_3Si-N(C_4H_9-t)]-3-(t-C_4H_9)-4-C_6H_5-$
1,3,2-ONBCH .... B: B Comp.SVol.4/3b-55
$BC_{18}H_{33}N_2O_2Si$ .. $(t-C_4H_9O)_2(C_6H_5)SiB(-NCH_3-CH_2-CH_2-NCH_3-)$
B: B Comp.SVol.3/3-113
$BC_{18}H_{34}N$ .... $9-(C_4H_9-n)-10-(C_6H_{11}-c)-[3.3.2]-9,10-NBC_8H_{14}$
B: B Comp.SVol.4/3a-252
– .... $10-(C_4H_9-n)-9-(C_6H_{11}-c)-[3.3.2]-9,10-NBC_8H_{14}$
B: B Comp.SVol.4/3a-252
– .... $(C_2H_5)_2B-N(C_6H_{11}-c)-C_8H_{13}-c$ .... B: B Comp.SVol.3/3-152
$BC_{18}H_{34}N_3$ ...... $B(NH_2)_2[HNC_6H_2-2,4,6-(t-C_4H_9)_3]$.......... B: B Comp.SVol.3/3-95
$BC_{18}H_{35}N_2$ ...... $N(-CH_2-CH_2-)_3N-BH(c-C_6H_{11})_2$ .... B: B Comp.SVol.3/3-189
$BC_{18}H_{35}N_2Si_2$.... $2,6-(i-C_3H_7)_2-C_6H_3-N=B-N[Si(CH_3)_3]_2$....... B: B Comp.SVol.4/3a-160/5
$BC_{18}H_{36}N$ .... $(n-C_4H_9)_2B-N(i-C_4H_9)(c-C_6H_9)$............. B: B Comp.SVol.3/3-152
$BC_{18}H_{36}NSn$..... $10-[(C_2H_5)_2N-Sn(CH_3)_2-CCH_3=]-9-CH_3-$
$[3.3.2]-9-BC_9H_{14}$ .... B: B Comp.SVol.4/3a-257
Sn: Org.Comp.19-48, 51
$BC_{18}H_{36}N_2^+$ ..... $[(-C(CH_3)_2-(CH_2)_3-C(CH_3)_2-)N=B=$
$N(-C(CH_3)_2-(CH_2)_3-C(CH_3)_2-)]^+$.......... B: B Comp.SVol.3/3-205
$BC_{18}H_{36}N_3$ ...... $1-[2,2,6,6-(CH_3)_4-NC_5H_6-1-N=B]-2,2,6,6-(CH_3)_4-NC_5H_6$
B: B Comp.SVol.4/3a-160/5
$BC_{18}H_{38}N_3$ ...... $[N(n-C_4H_9)_4][H_2B(CN)_2]$.................. B: B Comp.SVol.3/3-213
$BC_{18}H_{38}N_3O_2$ .... $t-C_4H_9-NH-B[OOC-CH(NH_2)-C_3H_7-i]-$
$1-NC_5H_6-2,2,6,6-(CH_3)_4$ .... B: B Comp.SVol.4/3b-52
– .... $[C_2H_5-OOC-CHCH_3-NH-B(NH-C_4H_9-t)]-$
$1-NC_5H_6-2,2,6,6-(CH_3)_4$ .... B: B Comp.SVol.4/3a-155
$BC_{18}H_{39}IN_2P$..... $[2,2,6,6-(CH_3)_4-NC_5H_6-1-B=N-P(C_4H_9-t)_2CH_3]I$
B: B Comp.SVol.4/4-115
$BC_{18}H_{39}N_2Si$..... $1-[(i-C_3H_7)_3Si-N=B]-2,2,6,6-(CH_3)_4-NC_5H_6$ ... B: B Comp.SVol.4/3a-160/5
$BC_{18}H_{40}NSn$ ..... $(i-C_3H_7)_2B[-C(C_3H_7-i)=C(CH_3)-Sn(CH_3)_2-N(C_2H_5)_2-]$
B: B Comp.SVol.4/3a-255
$BC_{18}H_{44}NSn_2$ .... $(i-C_3H_7)_2B-N[Sn(C_2H_5)_3]_2$ .... B: B Comp.SVol.4/3a-227
$BC_{18}H_{46}LiN_4Si$ ... $LiN(C_4H_9-t)-B(CH_3)-N(C_4H_9-t)-Si(CH_3)_3$
$\cdot (CH_3)_2N-CH_2CH_2-N(CH_3)_2$ .... B: B Comp.SVol.4/3a-169
$BC_{18}H_{48}N_3Si_4$.... $1,2-[t-C_4H_9-Si(CH_3)_2]_2-3-[(CH_3)_3Si]_2N-N_2B$ .. B: B Comp.SVol.3/3-135
B: B Comp.SVol.4/3a-195
B: B Comp.SVol.4/3b-246
$BC_{18}H_{58}N_3Si_6Th$.. $[(CH_3)_3Si-N-Si(CH_3)_3]_3Th[BH_4]$............. Th: SVol.D4-210
$BC_{18.5}Cl_{1.5}F_4H_{30.5}MnN_3S_6$
$[Mn((SC(=S))NC_5H_{10})_3]BF_4 \cdot 0.5\ Cl_3CH$ ...... Mn: MVol.D7-178/9
$BC_{19}ClF_4H_{46}N_2P_3Re$
$[Re(CNC_4H_9-t)(P(CH_3)_3)_3(CNHC_4H_9-t)Cl][BF_4]$ Re: Org.Comp.2-231/2, 240
$BC_{19}ClH_{17}N$ ..... $5-Cl-5-BC_{12}H_8 \cdot 2,4-(CH_3)_2-NC_5H_3$ .... B: B Comp.SVol.4/4-69
– .... $5-Cl-5-BC_{12}H_8 \cdot 2,6-(CH_3)_2-NC_5H_3$ .... B: B Comp.SVol.4/4-69
$BC_{19}ClH_{26}$ ...... $2,4,6-(CH_3)_3-C_6H_2-BCl-C_5(CH_3)_5$........... B: B Comp.SVol.4/4-50

$BC_{19}ClH_{27}NSi_2$ . . . 9-$[(CH_3)_3Si]_2N$-BCl-$C_{13}H_9$ . . . . B: B Comp.SVol.4/4-61

$BC_{19}ClH_{31}N_2O_2P$ 2,2,6,6-$(CH_3)_4$-NC_5H_6-1-B[-N(C_4H_9-t)-P(Cl) -O-(1,2-C_6H_4)-O-] . . . . B: B Comp.SVol.4/3b-54

$BC_{19}ClH_{45}NSi_3$ . . . 2,2,6,6-$(CH_3)_4$-NC_5H_6-1-BCl-C$[Si(CH_3)_3]_3$. . . . B: B Comp.SVol.4/4-61

$BC_{19}Cl_2H_{22}N$ . . . . 1-NC_9H_7 · BCl_2-$C_5(CH_3)_5$. . . . . B: B Comp.SVol.4/4-70

– . . . . 2-NC_9H_7 · BCl_2-$C_5(CH_3)_5$. . . . . B: B Comp.SVol.4/4-70

$BC_{19}Cl_3H_{19}P$. . . . . $[CH_3P(C_6H_5)_3][HBCl_3]$ . . . . B: B Comp.SVol.3/4-38

$BC_{19}CoFeH_{14}O_4$ . . $(C_5H_5)(CO)Fe(CO)_2Co(C_4H_4BC_6H_5)CO$ . . . . Fe: Org.Comp.B16b-145/6

$BC_{19}CrH_{24}O_4PSi$ $BC_2CrPSi(CO)_4(C_2H_5)_2(CH_3)_3(C_6H_5)$ . . . . B: B Comp.SVol.4/4-179

$BC_{19}FH_{27}NSi_2$. . . . $C_{13}H_9$-9-BF-N$[Si(CH_3)_3]_2$ . . . . B: B Comp.SVol.4/3b-244

$BC_{19}FH_{31}N_2O_2P$ . . 2,2,6,6-$(CH_3)_4$-NC_5H_6-1-B[-N(C_4H_9-t)-P(F) -O-(1,2-C_6H_4)-O-] . . . . B: B Comp.SVol.4/3b-54

$BC_{19}FH_{31}N_2O_2Sb$ 1-[1,3,2-$O_2SbC_6H_4$-2-N(C_4H_9-t)-BF] -NC_5H_6-2,2,6,6-$(CH_3)_4$ . . . . B: B Comp.SVol.4/3b-246

$BC_{19}FH_{35}NSi$ . . . . t-C_4H_9-BF-N$[Si(CH_3)_3]$-C_6H_3-2,6-$(C_3H_7\text{-}i)_2$ . . B: B Comp.SVol.4/3b-245

$BC_{19}FH_{45}NSi_2$. . . . $[(CH_3)_3Si]_2N$-BF-$C(C_4H_9\text{-}t)_3$ . . . . B: B Comp.SVol.4/3b-243

$BC_{19}FH_{49}N_3Si_3$ . . . $(CH_3)_3Si$-N(C_4H_9-t)-BF-N$[Si(CH_3)_2$-C_4H_9-t] -NH-$Si(CH_3)_2$-C_4H_9-t . . . . B: B Comp.SVol.4/3b-246

$BC_{19}F_2H_{29}NO_6Re$ $[NC_5H_8(CH_3)_4][(CO)_4ReC_3(CH_3)_3(\text{-O-}BF_2\text{-O-})]$ Re: Org.Comp.2-389, 393/4

– . . . . $[NC_5H_8(CH_3)_4][(CO)_4ReC_3H_2(C_3H_7\text{-}i)(\text{-O-}BF_2\text{-O-})]$ Re: Org.Comp.2-389

$BC_{19}F_3H_{26}N_2O_3S$ $[BC_{10}H_8N_2(C_4H_9\text{-}n)_2][CF_3\text{-}SO_3]$ . . . . B: B Comp.SVol.4/3b-46

$BC_{19}F_3H_{36}N_2O_3S$ [(-$C(CH_3)_2$-$(CH_2)_3$-$C(CH_3)_2$-)N=B= N(-$C(CH_3)_2$-$(CH_2)_3$-$C(CH_3)_2$-)]$[O_3SCF_3]$ . . . . B: B Comp.SVol.3/3-205

$BC_{19}F_4FeH_{13}O_2$ . . $[(C_5H_5)Fe(CO)_2(C_{12}H_8)][BF_4]$. . . . . Fe: Org.Comp.B17-98, 106

– . . . . $[(C_5D_5)Fe(CO)_2(C_{12}H_8)][BF_4]$. . . . . Fe: Org.Comp.B17-130

$BC_{19}F_4FeH_{14}NO_2$ $[(2\text{-}NC_8H_4(C_6H_5\text{-}2)(=O)_2\text{-}1,3)Fe(C_5H_5)][BF_4]$ . . Fe: Org.Comp.B18-142/6, 198, 263

$BC_{19}F_4FeH_{15}$ . . . . $[(C_{14}H_{10})Fe(C_5H_5)][BF_4]$ . . . . Fe: Org.Comp.B19-219/20, 270, 311/2

$BC_{19}F_4FeH_{17}$ . . . . $[(C_6H_5CH{=}CHC_6H_5)Fe(C_5H_5)][BF_4]$ . . . . Fe: Org.Comp.B18-142/6, 233

$BC_{19}F_4FeH_{19}O_4$ . . $[(C_5H_5)Fe(CO)_2(CH_2{=}CHCH_2C_3H_4O_2(C_6H_5))][BF_4]$ Fe: Org.Comp.B17-34, 51

$BC_{19}F_4FeH_{20}NO_6$ $[(C_5H_5)Fe(CO)_2(C_5H_5(CO_2C_2H_5)_2\text{-}3,4(CN\text{-}4))][BF_4]$ Fe: Org.Comp.B17-89

$BC_{19}F_4FeH_{21}O_2$ . . $[(C_5H_5)Fe(CO)_2(n\text{-}C_4H_9CH{=}CHC_6H_5)][BF_4]$. . . . Fe: Org.Comp.B17-61

$BC_{19}F_4FeH_{21}O_3$ . . $[C_5H_5(CO)_2Fe{=}C(C(CH_3)_2C_6H_5)OC_2H_5][BF_4]$. . . Fe: Org.Comp.B16a-111

$BC_{19}F_4FeH_{21}O_9$ . . $[(6\text{-}(CH_3OC(O))_2CHCH_2CH(CH_2OC(O)CH_3)\text{-}C_6H_6)Fe(CO)_3][BF_4]$ . . . . Fe: Org.Comp.B15-109, 159

$BC_{19}F_4FeH_{22}O_5P$ $[C_5H_5Fe(CO)_2C(=CHC_6H_5)CH_2P(OCH_3)_3][BF_4]$ Fe: Org.Comp.B14-153

$BC_{19}F_4FeH_{23}O_2$ . . $[(C_9H_{11})Fe(CO)_2(CH_2{=}CHC_6H_9)][BF_4]$ . . . . Fe: Org.Comp.B17-135

$BC_{19}F_4FeH_{23}O_6$ . . $[(C_5H_5)Fe(CO)_2(C_5H_7(CH(CO_2C_2H_5)_2\text{-}3))][BF_4]$ Fe: Org.Comp.B17-89, 104

$BC_{19}F_4FeH_{26}NO_2$ $[(C_5H_5)Fe(CO)_2CH_2CH_2\text{-}C_6H_9(=NC_4H_8)\text{-}2][BF_4]$ Fe: Org.Comp.B14-152, 156

$BC_{19}F_4FeH_{27}$ . . . . $[((CH_3)_6C_6)Fe(C_7H_9)][BF_4]$. . . . . Fe: Org.Comp.B19-172

$BC_{19}F_4FeH_{27}O$ . . . $[(CH_3)_5C_5Fe(C_8H_{12})CO][BF_4]$. . . . . Fe: Org.Comp.B17-189

$BC_{19}F_4FeH_{29}O_4$ . . $[(C_5H_5)Fe(CO)_2(i\text{-}C_5H_{11}\text{-}OCH{=}CHO\text{-}C_5H_{11}\text{-}i)][BF_4]$ Fe: Org.Comp.B17-70

$BC_{19}F_4FeH_{31}O_6P_2$ $[(C_5H_5)Fe(CH_2{=}CHC_6H_5)(P(OCH_3)_3)_2][BF_4]$. . . . Fe: Org.Comp.B16b-7, 9/10

$BC_{19}F_4FeH_{39}O_6P_2$ $[(C_5H_5)Fe(CH_2{=}CH_2)(P(O\text{-}C_2H_5)_3)_2][BF_4]$ . . . . . Fe: Org.Comp.B16b-5

$BC_{19}F_4FeH_{39}O_6P_2$ $[(C_5H_5)Fe(CH_2=CH-C_6H_{13}-n)(P(OCH_3)_3)_2][BF_4]$
Fe: Org.Comp.B16b-6
$BC_{19}F_4Fe_2H_{17}O_2$ $[C_5H_5Fe(CO)_2CH_2CHC_5H_4FeC_5H_5][BF_4]$ ...... Fe: Org.Comp.B14-128
$BC_{19}F_4Fe_2H_{17}O_3$ $[C_5H_5Fe(CO)_2CH_2C(OH)C_5H_4FeC_5H_5][BF_4]$.... Fe: Org.Comp.B14-128
$BC_{19}F_4H_{15}MoO_2$.. $[(C_9H_7)Mo(CO)_2(C_8H_8)][BF_4]$ Mo:Org.Comp.8-305, 315
$BC_{19}F_4H_{17}MoO_4$.. $[(C_9H_7)Mo(CO)_2(C_6H_7-1-COO-CH_3)][BF_4]$ Mo:Org.Comp.8-305/6, 314, 317/8
$BC_{19}F_4H_{19}MoO_2$.. $[(C_9H_7)Mo(CO)_2(C_6H_6-1,4-(CH_3)_2)][BF_4]$ Mo:Org.Comp.8-305/6, 314, 317/8
– .... $[(C_9H_7)Mo(CO)_2(C_6H_6-2,5-(CH_3)_2)][BF_4]$ Mo:Org.Comp.8-305/6, 314
$BC_{19}F_4H_{21}MoN_2O_3$
$[(CH_2C(CH_3)CH_2)Mo(CO)_2(NC_5H_4-2-(2-C_5H_4N))(O=C(CH_3)_2)][BF_4]$ Mo:Org.Comp.5-200, 203
$BC_{19}F_4H_{22}MoN$... $[C_9H_7Mo(NCCH_3)(H_3C-CC-CH_3)_2][BF_4]$ Mo:Org.Comp.6-137, 146
$BC_{19}F_4H_{23}P_2$ .... $[2,4,6-(CH_3)_3-C_6H_2-P=CH=P-C_6H_2-2,4,6-(CH_3)_3][BF_4]$
B: B Comp.SVol.4/3b-166
$BC_{19}F_4H_{28}MoN$... $[C_5H_5Mo(NCCH_3)(H-CC-C_4H_9-t)_2][BF_4]$ Mo:Org.Comp.6-136, 145
$BC_{19}F_4H_{29}MoO_6P_2$
$[C_5H_5Mo(P(OCH_3)_3)_2H-CC-C_6H_5][BF_4]$....... Mo:Org.Comp.6-116
$BC_{19}F_4H_{31}MoO_6P_2$
$[C_9H_7Mo(P(OCH_3)_3)_2H_3C-CC-CH_3][BF_4]$ Mo:Org.Comp.6-122, 130
$BC_{19}F_4H_{31}MoP_2$.. $[C_9H_7Mo(P(CH_3)_3)_2H_3C-CC-CH_3][BF_4]$....... Mo:Org.Comp.6-121, 129/30
$BC_{19}F_4H_{35}MoO_2P_2$
$[(C_5H_5)Mo(CO)_2(P(C_2H_5)_3)_2][BF_4]$ Mo:Org.Comp.7-283, 287
$BC_{19}FeH_{14}NbO_4$.. $C_5H_5FeC_4H_4BC_6H_5Nb(CO)_4$................ Fe: Org.Comp.B17-224
$BC_{19}FeH_{14}O_4Ta$.. $C_5H_5FeC_4H_4BC_6H_5Ta(CO)_4$................ Fe: Org.Comp.B17-224
$BC_{19}FeH_{14}O_4V$... $C_5H_5FeC_4H_4BC_6H_5V(CO)_4$................. Fe: Org.Comp.B17-223
$BC_{19}FeH_{22}NO_2$... $(C_5H_5)Fe[C_5H_3(B(OH)_2)-2-C_5H_3N-(C_4H_9-n)-6]$ Fe: Org.Comp.A9-287
$BC_{19}GaH_{42}N_2$.... $[-C(CH_3)_2-CH_2CH_2CH_2-C(CH_3)_2-]N[-Ga(C_2H_5)_2-N(C_4H_9-t)-B(C_2H_5)-]$........ B: B Comp.SVol.4/3a-206/7
$BC_{19}H_{15}KNS$..... $K[B(C_6H_5)_3SCN]$........................ B: B Comp.SVol.3/4-132
$BC_{19}H_{16}NO_2$..... $C_6H_5OB[-NCH_2C_6H_5-(1,2-C_6H_4)-O-]$........ B: B Comp.SVol.3/3-164
$BC_{19}H_{16}NO_2S$.... $(C_6H_5)_2NB[-O-(1,2-C_6H_3-5-OCH_3)-S-]$ ...... B: B Comp.SVol.3/4-133
$BC_{19}H_{16}N_3$ ...... $2,3,5-(C_6H_5)_3-1,2,4,3-N_3BCH$............... B: B Comp.SVol.4/3a-174/5
$BC_{19}H_{17}MoN_8O_2$. $(C_5H_5)Mo(CO)_2[-N_2C_3H_3-B(N_2C_3H_3)_2-N_2C_3H_3-]$
Mo:Org.Comp.7-156, 159/60, 178/80
$BC_{19}H_{19}MoN_8O_2$. $(c-C_5H_7)Mo(CO)_2(N_2C_3H_3)_4B$............... Mo:Org.Comp.5-225/6, 229
$BC_{19}H_{19}N_4O_5$.... $[CH_3C(O)O]_2B[-N(C_6H_5)NC(C(O)CH_3)NN(C_6H_5)-]$
B: B Comp.SVol.3/3-121/2
$BC_{19}H_{20}N$....... $CH_3-NH_2-B(C_6H_5)_3$ .... B: B Comp.SVol.4/3b-27
$BC_{19}H_{20}N_3$...... $(H_2N)_2C=NH-B(C_6H_5)_3$.................... B: B Comp.SVol.4/3b-27
$BC_{19}H_{23}MoN_4O_2$. $(c-C_7H_7)Mo(CO)_2[-N_2C_3H(CH_3)_2-]_2BH_2$ Mo:Org.Comp.5-213, 217, 221/2
– .... $(c-C_7H_7)Mo(CO)_2[-N_2C_3H_3-]_2B(C_2H_5)_2$ Mo:Org.Comp.5-213, 217, 222
$BC_{19}H_{23}N_2O_2$.... $7-[(C_2H_5-O)_2B-N=C(C_6H_5)]-1-NC_8H_8$........ B: B Comp.SVol.4/3b-124
– .... $(C_6H_5)_2B[-NH-C(O-C_2H_5)=CH-C(O-C_2H_5)=NH-]$
B: B Comp.SVol.3/3-122/3
$BC_{19}H_{24}MoO_4PSi$ $BC_2MoPSi(CO)_4(C_2H_5)_2(CH_3)_3(C_6H_5)$ .... B: B Comp.SVol.4/4-179
$BC_{19}H_{24}N$....... $C_5H_5N-B[-C(n-C_3H_7)=C(n-C_3H_7)-][C_6H_5]$ .... B: B Comp.SVol.3/3-192

$BC_{19}H_{24}O_4PSiW$ . . $BC_2PSiW(CO)_4(C_2H_5)_2(CH_3)_3(C_6H_5)$ . . . . B: B Comp.SVol.4/4-179
$BC_{19}H_{25}N_4$ . . . . . . $(C_6H_5)_2B[-NH-CN(CH_3)_2=CH-CN(CH_3)_2=NH-]$ B: B Comp.SVol.3/3-122/3
$BC_{19}H_{25}S$ . . . . $[2,4,6-(CH_3)_3-C_6H_2]_2BSCH_3$ . . . . B: B Comp.SVol.3/4-111
$BC_{19}H_{26}N$ . . . . $(1-c-C_6H_9)_2B-NCH_3-C_6H_5$. . . . . B: B Comp.SVol.3/3-153
– . . . . $[2,4,6-(CH_3)_3C_6H_2]_2B-NHCH_3$ . . . . B: B Comp.SVol.3/3-153
$BC_{19}H_{27}N_6$ . . . . . . $2,2,6,6-(CH_3)_4-NC_5H_6-1-B[-C(CN)_2-C(CN)_2-N(C_4H_9-t)-]$
B: B Comp.SVol.4/3a-174
$BC_{19}H_{28}N$ . . . . $BC_{14}H_{23} \cdot NC_5H_5$. . . . . B: B Comp.SVol.4/3b-30
$BC_{19}H_{28}NO_2$ . . . . . $[(-O-(CH_2)_3-O-)B(C_6H_5)_2][(C_2H_5)_2NH_2]$ . . . . . . B: B Comp.SVol.3/3-215
$BC_{19}H_{28}NSi$ . . . . . $(C_6H_5)_2B-N(C_4H_9-t)-Si(CH_3)_3$ . . . . B: B Comp.SVol.4/3a-231
$BC_{19}H_{29}MoN_4O_2$. $(CH_2CHCH_2)Mo(CO)_2[-N_2C_3H(C_2H_5)_2-]_2BH_2$ . . Mo:Org.Comp.5-213, 214
$BC_{19}H_{30}N$ . . . . $[C_2H_5CH=C(C_2H_5)]_2B-N(CH_3)C_6H_5$ . . . . B: B Comp.SVol.3/3-153
$BC_{19}H_{30}NSn$ . . . . . $1-NC_8H_6-1-B(C_2H_5)-C(C_2H_5)=C(CH_3)-Sn(CH_3)_2-C_2H_5$
B: B Comp.SVol.4/3a-235
$BC_{19}H_{30}NTi$ . . . . . $t-C_4H_9-B[-N(C_4H_9-t)-Ti(C_5H_5)_2-CH_2-]$ . . . . . . B: B Comp.SVol.4/3a-249
$BC_{19}H_{31}N_2Sn$ . . . . $BC_9H_4N_2Sn-(CH_3)_4-(C_2H_5)_3$ . . . . B: B Comp.SVol.4/3a-258
Sn: Org.Comp.19-56, 59
$BC_{19}H_{32}N$ . . . . $1-[CH_3C_6H_4-B(C_3H_7-i)]-2,2,6,6-(CH_3)_4NC_5H_6$ B: B Comp.SVol.4/3a-233
– . . . . $1-[i-C_3H_7-B(CH_2-C_6H_5)]-2,2,6,6-(CH_3)_4NC_5H_6$
B: B Comp.SVol.4/3a-233
$BC_{19}H_{33}N_2O$ . . . . . $t-C_4H_9-NH-B(O-C_6H_5)-1-NC_5H_6-2,2,6,6-(CH_3)_4$
B: B Comp.SVol.4/3b-52
$BC_{19}H_{34}NOSi$ . . . . $2-[(CH_3)_3Si-N(C_4H_9-t)]-3-(C_4H_9-t)-4-C_6H_5-1,2-OBC_2H_2$
B: B Comp.SVol.4/3b-62
$BC_{19}H_{34}N_3$ . . . . . . $1-[t-C_4H_9-NH-B(NH-C_6H_5)]-2,2,6,6-(CH_3)_4-NC_5H_6$
B: B Comp.SVol.4/3a-155
$BC_{19}H_{35}N_2Si$. . . . . $2,6-(i-C_3H_7)_2-C_6H_3-N=B-N(C_4H_9-t)-Si(CH_3)_3$ B: B Comp.SVol.4/3a-160/5
$BC_{19}H_{36}N_3OSi$ . . . $(CH_3)_3Si-N(C_4H_9-t)-B[-O-N(CH_3)-CH(C_6H_5)$
$-N(C_4H_9-t)-]$. . . . . B: B Comp.SVol.4/3b-53
$BC_{19}H_{36}S^-$ . . . . . . $[(H_3C_4S)B(C_5H_{11}-c)_3]^-$. . . . . B: B Comp.SVol.3/4-117
$BC_{19}H_{38}N$ . . . . $(n-C_4H_9)_2B-N(i-C_5H_{11})(c-C_6H_9)$ . . . . B: B Comp.SVol.3/3-152
$BC_{19}H_{38}NSn$ . . . . . $10-[(C_2H_5)_2N-Sn(CH_3)_2-CCH_3=]-9-C_2H_5-$
$[3.3.2]-9-BC_9H_{14}$ . . . . B: B Comp.SVol.4/3a-257
Sn: Org.Comp.19-48, 52
$BC_{19}H_{40}N_3O_2$ . . . . $t-C_4H_9-NH-B[OOC-CH(NH_2)-C_4H_9-i]-$
$1-NC_5H_6-2,2,6,6-(CH_3)_4$ . . . . B: B Comp.SVol.4/3b-52
– . . . . $t-C_4H_9-NH-B[OOC-CH(NH_2)-C_4H_9-s]-$
$1-NC_5H_6-2,2,6,6-(CH_3)_4$ . . . . B: B Comp.SVol.4/3b-52
$BC_{19}H_{41}N_2$ . . . . . . $n-C_{16}H_{33}-N(CH_3)_2-BH_2-CN$ . . . . B: B Comp.SVol.4/3b-23
$BC_{19}H_{42}NO_2$ . . . . . $n-C_{16}H_{33}-N(CH_3)_2-BH_2-COOH$. . . . . B: B Comp.SVol.4/3b-25
$BC_{19}H_{43}NSi_4$. . . . . $[((CH_3)_3Si)_2CH]_2BNC_5H_5$ . . . . B: B Comp.SVol.3/3-211
$BC_{19}H_{44}N_3Si$. . . . . $1-[(CH_3)_3Si-N(C_4H_9-t)-B(NH-C_3H_7-i)]-$
$2,2,6,6-(CH_3)_4-NC_5H_6$ . . . . B: B Comp.SVol.4/3a-155
$BC_{19}H_{45}N_4SSi_2$. . . $2,2,6,6-(CH_3)_4-NC_5H_6-1-B[-N(C_4H_9-t)$
$-S(=NSi(CH_3)_3)-N(Si(CH_3)_3)-]$. . . . . B: B Comp.SVol.4/3a-155/6
$BC_{19}H_{45}N_5Sb$ . . . . $1-[((CH_3)_2N)_2Sb-N(C_4H_9-t)-B(N(CH_3)_2)]-$
$2,2,6,6-(CH_3)_4-NC_5H_6$ . . . . B: B Comp.SVol.4/3a-155
$BC_{19}H_{47}OOsP_2$. . . $(H)(CO)(BH_4)Os[P(C_3H_7-i)_3]_2$. . . . . Os: Org.Comp.A1-125/6, 131
– . . . . $(H)(CO)(BH_4)Os[P(C_4H_9-t)_2-CH_3]_2$ . . . . Os: Org.Comp.A1-126, 131

$BC_{19}H_{48}N_3Si_3$ .... 1,2-[t-C_4H_9-Si$(CH_3)_2$]$_2$-3-[$(CH_3)_3$Si-N(C_4H_9-t)]-N_2B
B: B Comp.SVol.4/3a-195
B: B Comp.SVol.4/3b-246

$BC_{20}ClH_{15}O_6Re$ .. cis-$(CO)_4$Re[(CH_3CO)(CH_2(C_6H_5)CO)]B(C_6H_5)Cl
Re: Org.Comp.2-345

$BC_{20}ClH_{23}N_2^+$ ... [(2-(NC_5H_4-2)-NC_5H_4)BCl-$C_5(CH_3)_5$]$^+$ .... B: B Comp.SVol.4/4-70

$BC_{20}ClH_{27}NSi$.... 9-[$(CH_3)_3$Si-N(C_4H_9-t)-BCl]-$C_{13}H_9$ .... B: B Comp.SVol.4/4-61

$BC_{20}ClH_{30}$ ...... [$(CH_3)_5C_5$]$_2$BCl .... B: B Comp.SVol.4/4-47

$BC_{20}Cl_2F_4GaH_{16}N_4$
[$GaCl_2$(2-(NC_5H_4-2)-NC_5H_4)$_2$][BF_4] .... Ga: SVol.D1-265/6

$BC_{20}Cl_2H_{23}N_2$.... [(-NC_5H_4-C_5H_4N-)BCl-$C_5(CH_3)_5$]Cl .... B: B Comp.SVol.4/4-70

$BC_{20}Cl_3H_{16}N_4$.... (NH_2)$N_3C_2H_2$-B(C_6H_4-4-Cl)$_3$.............. B: B Comp.SVol.3/3-191

$BC_{20}Cl_3H_{19}N$ .... $(CH_3)_2$NH-B(C_6H_4-4-Cl)$_3$................ B: B Comp.SVol.4/3b-27

$BC_{20}CrH_{24}O_5PSi$ BC_2CrPSi$(CO)_5$(C_2H_5)$_2$(CH_3)$_3$(C_6H_5) .... B: B Comp.SVol.4/4-180

$BC_{20}FH_{30}$ .... [$(CH_3)_5C_5$]$_2$BF......................... B: B Comp.SVol.4/3b-227

$BC_{20}F_2H_{30}I$ ...... [(($C_6H_5CH_2$)(CH_3)(C_2H_5))$_2BF_2$]I .... B: B Comp.SVol.3/4-101

$BC_{20}F_2H_{30}IN_2$.... [(($C_6H_5CH_2$)(C_2H_5)(CH_3)N)$_2BF_2$]I .... B: B Comp.SVol.3/3-379

$BC_{20}F_2H_{30}N_2^+$ ... [(($C_6H_5CH_2$)(C_2H_5)(CH_3)N)$_2BF_2$]$^+$ .... B: B Comp.SVol.3/3-379

$BC_{20}F_2H_{36}N_3SSi$.. [((CH_3)$_2$N)$_3$S][1-(CH_3)$_2$B(F)-8-(CH_3)$_2$Si(F)-$C_{10}H_6$]
B: B Comp.SVol.4/3b-228/9

$BC_{20}F_2H_{41}N_2Si_3$.. [$(CH_3)_3$Si]$_2$N-BF-N(C_4H_9-t)-SiF(C_6H_5)C_4H_9-t B: B Comp.SVol.3/3-376

$BC_{20}F_3H_{37}N_3O_3S$ [2,6-$(CH_3)_2$-NC_5H_3-1-B(N(C_3H_7-i)$_2$)$_2$][CF_3-SO_3]
B: B Comp.SVol.4/3b-45

$BC_{20}F_4FeH_{10}O_7Re$
[(C_5H_5)Fe$(CO)_2$(C_6H_5-CC-Re$(CO)_5$)][BF_4] .... Fe: Org.Comp.B17-123/4

$BC_{20}F_4FeH_{15}O_2$ .. [(C_5H_5)Fe$(CO)_2$($C_{13}H_{10}$)][BF_4].............. Fe: Org.Comp.B17-100

$BC_{20}F_4FeH_{15}O_3$ .. [(C_6H_5-$(CH)_5$-C_6H_5)Fe$(CO)_3$][BF_4] .... Fe: Org.Comp.B15-23, 29/30

– .... [(C_6H_5-CHCDCHCHCH-C_6H_5)Fe$(CO)_3$][BF_4]... Fe: Org.Comp.B15-29/30

$BC_{20}F_4FeH_{17}O_2S$ [C_5H_5Fe$(CO)_2$$CH_2$S($C_6H_5$)$_2$][$BF_4$]............ Fe: Org.Comp.B14-133, 135, 143

$BC_{20}F_4FeH_{19}O$ ... [(CH_3C(=O)-4-C_6H_4-CH_2-C_6H_5)Fe(C_5H_5)][BF_4]
Fe: Org.Comp.B18-142/6, 234/5

$BC_{20}F_4FeH_{23}O_2$ .. [(C_9H_{11})Fe$(CO)_2$(C_9H_{12})][BF_4].............. Fe: Org.Comp.B17-135, 136

$BC_{20}F_4FeH_{23}O_2S$ [(C_5H_5)Fe$(CO)_2$$CH_2$S($C_6H_5$)-$(CH_2)_4$-CH=$CH_2$][$BF_4$]
Fe: Org.Comp.B14-133, 135

$BC_{20}F_4FeH_{23}O_8$ .. [((CH_3O)(CH_3)C_6H_4-C_5H_7(OC(O)CH_3)$_2$)Fe$(CO)_3$][BF_4]
Fe: Org.Comp.B15-131

$BC_{20}F_4FeH_{25}O_3$ .. [$C_5H_5$$(CO)_2$Fe=CH$C_6H_8$(=O)($C_2H_4$CH=C$(CH_3)_2$)-c][$BF_4$]
Fe: Org.Comp.B16a-90, 97

$BC_{20}F_4FeH_{25}O_7$ .. [((CH_3O)(CH_3)C_6H_4-C_5H_7(CH_2OCH_3)OC(O)CH_3)Fe$(CO)_3$][BF_4] .... Fe: Org.Comp.B15-131, 182

$BC_{20}F_4FeH_{27}N_2O$ [(C_5H_5)Fe(CN-C_6H_{11}-c)$_2$CO][BF_4]........... Fe: Org.Comp.B15-338

$BC_{20}F_4FeH_{27}O_2Si_2$
[(1-(C_6H_5Si$(CH_3)_2$)(3-$(CH_3)_3$Si)C_5H_3)Fe$(CO)_2$(CH_2=CH_2)][BF_4]....................... Fe: Org.Comp.B17-133

$BC_{20}F_4FeH_{29}$ .... [(1,2,4,5-$(CH_3)_4$-C_6H_2)Fe(C_6H_3-$(CH_3)_4$-1,2,4,5)][BF_4]
Fe: Org.Comp.B19-144, 150

$BC_{20}F_4FeH_{31}O_2$ .. [((CH_3)$_5C_5$)Fe$(CO)_2$(CH_2=CHC_6H_{13}-n)][BF_4]... Fe: Org.Comp.B17-131

$BC_{20}F_4FeH_{32}N_3$ .. [(C_5H_5)Fe(CN-C_4H_9-t)$_3$][BF_4] .... Fe: Org.Comp.B15-345

$BC_{20}F_4Fe_2H_{10}O_9ReS_2$
$[C_5H_5(CO)_2Fe{=}C(SRe(CO)_5)SFe(CO)_2C_5H_5][BF_4]$ Fe: Org.Comp.B16a-143, 154/5

$BC_{20}F_4Fe_2H_{19}O_3$ $[C_5H_5Fe(CO)_2CH_2C(OCH_3)C_5H_4FeC_5H_5][BF_4]$. Fe: Org.Comp.B14-129

$BC_{20}F_4GeH_{31}Si$ .. $[GeC_5(CH_3)_4Si(CH_3)_2C_5H(CH_3)_4][BF_4]$........ Ge: Org.Comp.3-385

$BC_{20}F_4H_{16}N_2O_7Re$
$[(CO)_5Re(CNC(COOCH_3){=}C(-N(CH_2C_6H_5)CH_2CH_2CH_2-))][BF_4]$........................ Re: Org.Comp.2-253

$BC_{20}F_4H_{16}O_3PW$.. $[(C_5H_5)(CO)_3WPH(C_6H_5)_2][BF_4]$............. B: B Comp.SVol.4/3b-124

$BC_{20}F_4H_{17}MoN_4O_3$
$[C_5H_5Mo(CO)(2\text{-}NNC_6H_4OCH_2CH_2OC_6H_4NN\text{-}2)][BF_4]$ Mo: Org.Comp.6-235

$BC_{20}F_4H_{18}MoN_3O_2$
$[(CH_2CHCH_2)Mo(CO)_2(NC_5H_4\text{-}2\text{-}(2\text{-}C_5H_4N))(NC_5H_5)][BF_4]$.......................... Mo: Org.Comp.5-200, 201/2, 204/5

$BC_{20}F_4H_{18}N_2O_3Re$
$[(C_5H_5)Re(CO)_2\text{-}NHN(C_6H_5)\text{-}C_6H_4\text{-}4\text{-}OCH_3][BF_4]$ Re: Org.Comp.3-193/7, 205

$BC_{20}F_4H_{21}MoO_2$.. $[(C_9H_7)Mo(CO)_2(C_6H_6\text{-}5\text{-}C_2H_5\text{-}2\text{-}CH_3)][BF_4]$.. Mo: Org.Comp.8-305/6, 314

$BC_{20}F_4H_{24}Se_8$ ... $[((CH_3)_2Se_2C_3{=}C_3Se_2(CH_3)_2)_2][BF_4]$......... B: B Comp.SVol.4/4-175

$BC_{20}F_4H_{27}MoOSi_2$ $[C_9H_7Mo(CO)(H\text{-}CC\text{-}Si(CH_3)_3)_2][BF_4]$........ Mo: Org.Comp.6-317/8

$BC_{20}F_4H_{27}MoO_2$.. $[((CH_3)_5C_5)Mo(CO)_2(C_4(CH_3)_4)][BF_4]$......... Mo: Org.Comp.8-305, 312

$BC_{20}F_4H_{28}MoOP$ $[C_9H_7Mo(CO)(P(C_2H_5)_3)H_3C\text{-}CC\text{-}CH_3][BF_4]$... Mo: Org.Comp.6-286, 289

$BC_{20}F_4H_{29}MoO$ .. $[C_5H_5Mo(CO)(H_3C\text{-}CC\text{-}C_4H_9\text{-}t)_2][BF_4]$....... Mo: Org.Comp.6-319

$BC_{20}F_4H_{29}MoOS_2$ $[C_5H_5Mo(CO)(t\text{-}C_4H_9\text{-}CC\text{-}SCH_3)_2][BF_4]$...... Mo: Org.Comp.6-320/1

$BC_{20}F_4H_{30}Sb$ .... $[((CH_3)_5C_5)_2Sb][BF_4]$..................... B: B Comp.SVol.3/3-252

$BC_{20}F_4H_{31}MoO_6P_2$
$[C_5H_5Mo(P(OCH_3)_3)_2H_3C\text{-}CC\text{-}C_6H_5][BF_4]$..... Mo: Org.Comp.6-120, 129

$BC_{20}F_4H_{31}MoO_6P_2S$
$[C_5H_5Mo(P(OCH_3)_3)_2H_3CS\text{-}CC\text{-}C_6H_5][BF_4]$ ... Mo: Org.Comp.6-117

$BC_{20}F_4H_{35}MoO_6P_2Si$
$[C_9H_7Mo(P(OCH_3)_3)_2H\text{-}CC\text{-}Si(CH_3)_3][BF_4]$.... Mo: Org.Comp.6-121, 129

$BC_{20}FeH_{26}O_3PSi$ $BC_2FePSi(CO)_3(C_2H_5)_2(CH_3)_2(C_6H_5)[C(CH_3){=}CH_2]$ B: B Comp.SVol.4/4-179

$BC_{20}H_{16}N_3O_2$.... $3\text{-}(4\text{-}HOOC\text{-}C_6H_4)\text{-}2,5\text{-}(C_6H_5)_2\text{-}1,2,4,3\text{-}N_3BCH$ B: B Comp.SVol.4/3a-174/5

$BC_{20}H_{18}N$....... $CH_3\text{-}CN\text{-}B(C_6H_5)_3$...................... B: B Comp.SVol.4/3b-28

$BC_{20}H_{18}NO_2$..... $(C_6H_5)_2B[-O\text{-}(1,2\text{-}C_6H_4)\text{-}CCH_3{=}NOH-]$....... B: B Comp.SVol.3/3-219/20

$BC_{20}H_{18}N_3$...... $3\text{-}(4\text{-}CH_3\text{-}C_6H_4)\text{-}2,5\text{-}(C_6H_5)_2\text{-}1,2,4,3\text{-}N_3BCH$.. B: B Comp.SVol.4/3a-174/5

$BC_{20}H_{18}N_3O$..... $3\text{-}(4\text{-}CH_3O\text{-}C_6H_4)\text{-}2,5\text{-}(C_6H_5)_2\text{-}1,2,4,3\text{-}N_3BCH$ B: B Comp.SVol.4/3a-174/5

$BC_{20}H_{19}MoN_6O_2$. $(CH_2CHCH_2)Mo(CO)_2[-N_2C_3H_3-]_3B\text{-}C_6H_5$.... Mo: Org.Comp.5-227, 230

– .... $[CH_2C(C_6H_5)CH_2]Mo(CO)_2[-N_2C_3H_3-]_3BH$.... Mo: Org.Comp.5-225/6, 228

$BC_{20}H_{19}N_2$...... $(C_6H_5)_2B[-NH\text{-}(1,2\text{-}C_6H_4)\text{-}CH{=}NCH_3-]$....... B: B Comp.SVol.3/3-123

$BC_{20}H_{19}N_4$...... $2\text{-}CH_3C_6H_4\text{-}B[-N(C_6H_4CH_3\text{-}2)\text{-}N{=}N\text{-}N(C_6H_5)-]$ B: B Comp.SVol.3/3-137/8

– .... $(C_6H_5)_3B \cdot 4\text{-}H_2N\text{-}1,2,4\text{-}N_3C_2H_2$........... B: B Comp.SVol.3/3-191

$BC_{20}H_{20}N$....... $(C_6H_5C{\equiv}C)_2B\text{-}N(C_2H_5)_2$.................. B: B Comp.SVol.3/3-149

$BC_{20}H_{20}NO_2$..... $(C_6H_5)(C_6H_5O)B[-O\text{-}CH(C_6H_5)\text{-}CH_2\text{-}NH_2-]$... B: B Comp.SVol.3/3-217

$BC_{20}H_{21}MoN_8O_2$. $(c-C_6H_9)Mo(CO)_2(N_2C_3H_3)_4B$ ... Mo: Org.Comp.5-225/6, 229
$BC_{20}H_{21}N_2O_6$ ... $[-B(C_2H_5)_2-O-C(=O)-CH_2-N(=CH$
$-C_6H_4-2-(COO-C_6H_4-4-NO_2))-]$ ... B: B Comp.SVol.4/3b-70
$BC_{20}H_{21}N_4O_6$ ... $[CH_3C(O)O]_2B(-NC_6H_5-N=CCOOC_2H_5-N=NC_6H_5-)$
B: B Comp.SVol.3/3-121/2
$BC_{20}H_{21}S$ ... $(CH_3)_2S-B(C_6H_5)_3$ ... B: B Comp.SVol.3/4-116
$BC_{20}H_{22}N$ ... $(CH_3)_2NH-B(C_6H_5)_3$ ... B: B Comp.SVol.4/3b-27
$BC_{20}H_{22}NO_2$ ... $[(-O-(CH_2)_3-O-)B(C_6H_5)_2][C_5H_5NH]$ ... B: B Comp.SVol.3/3-215
$BC_{20}H_{24}N_3OS$ ... $BC_4H_3N_3S(=O)(2-C_{10}H_7)(C_3H_7-n)_2$ ... B: B Comp.SVol.4/3a-180
$BC_{20}H_{24}O_5PSiW$ .. $BC_2PSiW(CO)_5(C_2H_5)_2(CH_3)_3(C_6H_5)$ ... B: B Comp.SVol.4/4-180
$BC_{20}H_{26}N$ ... $1-(C_2H_5)_2N-2,5-(C_6H_5)_2-BC_4H_6$ ... B: B Comp.SVol.4/3a-228
– ... $9-(t-C_4H_9)_2B-9-NC_{12}H_8$ ... B: B Comp.SVol.4/3a-233
– ... $(n-C_3H_7)_2B-N(C_6H_5)-C(C_6H_5)=CH_2$ ... B: B Comp.SVol.3/3-152
$BC_{20}H_{26}NSi$ ... $(CH_3)_3Si-N(C_4H_9-t)-B[9-(=C_{13}H_8)]$ ... B: B Comp.SVol.4/3a-216
$BC_{20}H_{26}N_3$ ... $(i-C_3H_7)_2B(-NH-CC_6H_5=N-CC_6H_5=NH-)$ ... B: B Comp.SVol.3/3-122
$BC_{20}H_{26}O_3PRuSi$ $BC_2PRuSi(CO)_3(C_2H_5)_2(CH_3)_2(C_6H_5)[C(CH_3)=CH_2]$
B: B Comp.SVol.4/4-179
$BC_{20}H_{27}MoN_6O_2$. $(CH_2CHCH_2)Mo(CO)_2[-N_2C_3H(CH_3)_2-]_3BH$ ... Mo: Org.Comp.5-226, 227
$BC_{20}H_{27}S$ ... $[2,4,6-(CH_3)_3-C_6H_2]_2BSC_2H_5$ ... B: B Comp.SVol.3/4-111
$BC_{20}H_{28}N$ ... $(t-C_4H_9)_2B-N(C_6H_5)_2$ ... B: B Comp.SVol.4/3a-227
– ... $[2,4,6-(CH_3)_3C_6H_2]_2B-N(CH_3)_2$ ... B: B Comp.SVol.3/3-149
$BC_{20}H_{28}NNi_2$ ... $(C_5H_5)Ni[C_4H_4BN(i-C_3H_7)_2]Ni(C_5H_5)$ ... B: B Comp.SVol.4/4-180
$BC_{20}H_{28}N_3$ ... $BC_6H_5N_2(=N-C_6H_5)(C_4H_9-n)_2$ ... B: B Comp.SVol.4/3a-180
$BC_{20}H_{29}N_6$ ... $BC_2H_2N_6(C_6H_5)(C_6H_{11}-c)_2$ ... B: B Comp.SVol.4/3a-178
$BC_{20}H_{29}N_8$ ... $[(CH_3)_2C_3HN_2]_3B-C_3H_2N_2(CH_3)_2$ ... B: B Comp.SVol.3/3-201
$BC_{20}H_{31}MoN_4O_2$. $[CH_2C(CH_3)CH_2]Mo(CO)_2[-N_2C_3H(C_2H_5)_2-]_2BH_2$
Mo: Org.Comp.5-213, 215
$BC_{20}H_{32}N$ ... $(n-C_4H_9)_2B-N(C_6H_5)(c-C_6H_9)$ ... B: B Comp.SVol.3/3-152
$BC_{20}H_{32}N_3S$ ... $2,2,6,6-(CH_3)_4-NC_5H_6-1-B[-N(C_4H_9-t)-C(=S)-N(C_6H_5)-]$
B: B Comp.SVol.4/3a-155/6
$BC_{20}H_{33}N_2S_2Sn$ .. $(C_2H_5)_2B-C(C_2H_5)=C(CH_3)-Sn(CH_3)[-S$
$-CH=C(CH_3)-N=C(CH_3)-CH=C_3HNS(CH_3)-]$ .. Sn: Org.Comp.19-148/9
$BC_{20}H_{34}I$ ... $2,7,7-(CH_3)_3-[3.1.1]-C_7H_8-3-BI-$
$3-[3.1.1]-C_7H_8-2,7,7-(CH_3)_3$ ... B: B Comp.SVol.4/4-113
$BC_{20}H_{34}N_2O_2P$ ... $2,2,6,6-(CH_3)_4-NC_5H_6-1-B[-N(C_4H_9-t)$
$-P(CH_3)-O-(1,2-C_6H_4)-O-]$ ... B: B Comp.SVol.4/3b-54
$BC_{20}H_{38}N$ ... $(C_2H_5)_2B-N(c-C_6H_{11})(c-C_{10}H_{17})$ ... B: B Comp.SVol.3/3-152
$BC_{20}H_{38}NOSi_2$ ... $2-(i-C_3H_7)_2N-3,3-[(CH_3)_3Si]_2-4-C_6H_5-1,2-OBC_2H$
B: B Comp.SVol.4/3b-63
$BC_{20}H_{39}S$ ... $(CH_3)_2S-B(C_6H_{11})_3$ ... B: B Comp.SVol.3/4-116
$BC_{20}H_{42}N_3O_2$ ... $1-[C_2H_5-OC(O)-CH(C_3H_7-i)-NH-B(NH$
$-C_4H_9-t)]-2,2,6,6-(CH_3)_4-NC_5H_6$ ... B: B Comp.SVol.4/3a-155
$BC_{20}H_{45}N_2P_2$ ... $t-C_4H_9-B[-N(C_4H_9-t)-P(C_4H_9-t)-P(C_4H_9-t)-N(C_4H_9-t)-]$
B: B Comp.SVol.4/3a-208
$BC_{20}H_{45}N_3P$ ... $t-C_4H_9-N=P(C_4H_9-t)[-N(C_4H_9-t)-B(C_4H_9-t)-N(C_4H_9-t)-]$
B: B Comp.SVol.4/3a-207/8
$BC_{20}H_{45}O_4Sn$ ... $(C_4H_9)_2Sn(OC_4H_9)OB(OC_4H_9)_2$ ... Sn: Org.Comp.16-192
$BC_{20}H_{46}N$ ... $H_3B-N(CH_3)_2-C_{18}H_{37}-n$ ... B: B Comp.SVol.4/3b-10
$BC_{20}H_{46}N_3Sn$ ... $1-[(CH_3)_3Sn-N(C_4H_9-t)-B(N(C_2H_5)_2)]-$
$2,2,6,6-(CH_3)_4-NC_5H_6$ ... B: B Comp.SVol.4/3a-155

$BC_{20}H_{57}N_3PSi_5$ . . . $[(CH_3)_3Si]_2N-B(C_4H_9-t)-NH-P[CH_2-Si(CH_3)_3]$
$-N[Si(CH_3)_3]_2$ . . . . B: B Comp.SVol.4/3a-170

$BC_{20}H_{60}NSi_8$. . . . . $[((CH_3)_3Si)_3Si]_2B-N(CH_3)_2$ . . . . B: B Comp.SVol.3/3-150

$BC_{21}ClF_4FeH_{19}O_2PS$
$[C_5H_5Fe(CO)_2CH(SCH_3)PCl(C_6H_5)_2][BF_4]$ . . . . . Fe: Org.Comp.B14-133, 140

$BC_{21}ClF_4H_{46}N_3P_2Re$
$[Re(CNC_4H_9-t)_2(P(CH_3)_3)_2(\equiv CN(H)C_4H_9-t)Cl][BF_4]$
Re: Org.Comp.2-258/9, 264/5

$BC_{21}ClH_{13}MnN_9O_6$
$Mn[(6-NO_2-1,2-N_2C_7H_4-1)_3BH]Cl$. . . . . Mn: MVol.D8-16/8

$BC_{21}ClH_{16}MnN_6$ . . $Mn[(1,2-N_2C_7H_5-1)_3BH]Cl$ . . . . Mn: MVol.D8-16/8

$BC_{21}ClH_{16}O_6$ . . . . $C_6H_5(ClO_4)B[-O=CC_6H_5-CH=CC_6H_5-O-]$ . . . . . B: B Comp.SVol.3/4-52

$BC_{21}Cl_2H_{35}N_2S$ . . $1-[2,6-(i-C_3H_7)_2C_6H_3-N(SCl)-BCl]-$
$2,2,6,6-(CH_3)_4-NC_5H_6$ . . . . B: B Comp.SVol.4/4-64

$BC_{21}Cl_2H_{35}N_2S_2$. . $1-[2,6-(i-C_3H_7)_2C_6H_3-N(SS-Cl)-BCl]-$
$2,2,6,6-(CH_3)_4-NC_5H_6$ . . . . B: B Comp.SVol.4/4-64

$BC_{21}Cl_3H_{35}N_2Sb$. $1-[2,6-(i-C_3H_7)_2C_6H_3-N(SbCl_2)-BCl]-$
$2,2,6,6-(CH_3)_4-NC_5H_6$ . . . . B: B Comp.SVol.4/4-64

$BC_{21}Cl_4GeH_{35}N_2$ $1-[2,6-(i-C_3H_7)_2C_6H_3-N(GeCl_3)-BCl]-$
$2,2,6,6-(CH_3)_4-NC_5H_6$ . . . . B: B Comp.SVol.4/4-64

$BC_{21}Cl_4H_{35}N_2Si$ . . $1-[2,6-(i-C_3H_7)_2C_6H_3-N(SiCl_3)-BCl]-$
$2,2,6,6-(CH_3)_4-NC_5H_6$ . . . . B: B Comp.SVol.4/4-64

$BC_{21}Cl_4H_{35}N_2Ti$ . . $1-[2,6-(i-C_3H_7)_2C_6H_3-N(TiCl_3)-BCl]-$
$2,2,6,6-(CH_3)_4-NC_5H_6$ . . . . B: B Comp.SVol.4/4-64

$BC_{21}Cl_6H_{35}N_2Si_2$ $1-[2,6-(i-C_3H_7)_2C_6H_3-N(SiCl_2SiCl_3)-BCl]-$
$2,2,6,6-(CH_3)_4-NC_5H_6$ . . . . B: B Comp.SVol.4/4-64

$BC_{21}FH_{31}NSi$ . . . . $C_6H_5-BF-N[Si(CH_3)_3]-C_6H_3-2,6-(C_3H_7-i)_2$. . . . B: B Comp.SVol.4/3b-245

$BC_{21}FH_{39}N_3SSi$ . . $[((CH_3)_2N)_3S][1-(CH_3)_2B(F)-8-(CH_3)_3Si-C_{10}H_6]$
B: B Comp.SVol.4/3b-228/9

$BC_{21}FH_{44}N_2Si_3$. . . $[(CH_3)_3Si]_2N-BF-N[Si(CH_3)_3]-C_6H_3-2,6-(C_3H_7-i)_2$
B: B Comp.SVol.4/3b-246

$BC_{21}F_2FeH_{34}NO_3$ $[C_9H_{20}N][C_5H_5Fe(CH_2C_2O_2BF_2(C_3H_7-i))CO]$ . . Fe: Org.Comp.B17-156

$BC_{21}F_3H_{28}N_2O_3S$ $[2,2,6,6-(CH_3)_4-NC_5H_6-1-B(C_6H_5)-1-NC_5H_5][CF_3-SO_3]$
B: B Comp.SVol.4/3b-45

$BC_{21}F_3H_{35}N_2Sb$ . . $2,2,6,6-(CH_3)_4-NC_5H_6-1-BF-N(SbF_2)-C_6H_3-$
$2,6-(C_3H_7-i)_2$ . . . . B: B Comp.SVol.4/3b-246

$BC_{21}F_4FeH_{15}O_2$ . . $[(C_5H_5)Fe(CO)_2(C_6H_5-CC-C_6H_5)][BF_4]$ . . . . Fe: Org.Comp.B17-123

$BC_{21}F_4FeH_{17}O_2$ . . $[(C_5H_5)Fe(CO)_2(C_6H_5CH=CHC_6H_5)][BF_4]$. . . . . . Fe: Org.Comp.B17-60/1

$BC_{21}F_4FeH_{21}$ . . . . $[(-CH_2CH_2-(1,4-C_6H_4)-CH_2CH_2-(1,4-C_6H_4)-)$
$Fe(C_5H_5)][BF_4]$ . . . . Fe: Org.Comp.B19-329/30, 336/7

$BC_{21}F_4FeH_{23}O_2$ . . $[(C_9H_{11})Fe(CO)_2(C_{10}H_{12})][BF_4]$ . . . . Fe: Org.Comp.B17-136

$BC_{21}F_4FeH_{25}O_2S$ $[(C_5H_5)Fe(CO)_2CH(-(CH_2)_4-CH=CH_2)S(CH_3)C_6H_5][BF_4]$
Fe: Org.Comp.B14-133, 136

$BC_{21}F_4FeH_{25}O_8$ . . $[((CH_3O)(CH_3)C_6H_4-C_6H_9(OC(O)CH_3)_2)Fe(CO)_3][BF_4]$
Fe: Org.Comp.B15-131, 183

$BC_{21}F_4Fe_2H_{21}O_3$ $[C_5H_5Fe(CO)_2CH_2C(OC_2H_5)C_5H_4FeC_5H_5][BF_4]$ Fe: Org.Comp.B14-129

$BC_{21}F_4H_{15}MoN_2O_2$
$[(C_9H_7)Mo(CO)_2(NC_5H_4-2-)_2][BF_4]$ . . . . Mo: Org.Comp.7-269

$BC_{21}F_4H_{17}MoN_4O_4$
$[(C_5H_5)Mo(CO)_2(N_2C_6H_4OCH_2CH_2OC_6H_4N_2)][BF_4]$ Mo:Org.Comp.7-269

$BC_{21}F_4H_{20}MoN_3O_2$
$[(CH_2C(CH_3)CH_2)Mo(CO)_2(NC_5H_4$-2-$(2$-$C_5H_4N))(NC_5H_5)][BF_4]$ Mo:Org.Comp.5-200, 203/4

$BC_{21}F_4H_{21}O_3Se$ . . $[(4$-OCH_3-$C_6H_4)_3Se][BF_4]$ . . . . B: B Comp.SVol.3/4-147

$BC_{21}F_4H_{28}MoN_2O_5P$
$[(CH_2CHCH_2)Mo(CO)_2(NC_5H_4$-2-$(2$-$C_5H_4N))(P(O$-$C_2H_5)_3)][BF_4]$. Mo:Org.Comp.5-200, 202

$BC_{21}F_4H_{31}MnN_3O_6P_2$
$[Mn(NO)_2(C_6H_5$-$P(OCH_3)_2)_2(t$-C_4H_9-$NC)][BF_4]$ Mn:MVol.D8-119

$BC_{21}F_4H_{35}MoO_6P_2$
$[C_9H_7Mo(P(OCH_3)_3)_2H$-CC-C_4H_9-$t][BF_4]$ Mo:Org.Comp.6-121, 129

$BC_{21}F_4H_{35}N_2Ti$. . . 2,2,6,6-$(CH_3)_4$-NC_5H_6-1-BF-$N(TiF_3)$-C_6H_3-2,6-$(C_3H_7$-$i)_2$ B: B Comp.SVol.4/3b-246

$BC_{21}F_4H_{41}MoP_2$ . . $[C_5H_5Mo(P(C_2H_5)_3)_2H_3C$-$CC$-$CH_3][BF_4]$ Mo:Org.Comp.6-118

$BC_{21}F_4H_{42}MnN_3S_6$
$[Mn(SC(=S)N(C_3H_7$-$i)_2)_3][BF_4]$. Mn:MVol.D7-162/3

$BC_{21}F_4H_{45}O_{12}P_3Re$
mer-$[(CO)_3Re(P(OC_2H_5)_3)_3][BF_4]$ . . . . Re: Org.Comp.1-310

$BC_{21}FeH_{30}LiN_2$. . . $[Li(CH_3)_2NCH_2CH_2N(CH_3)_2][C_5H_5FeC_4H_4BC_6H_5]$ Fe: Org.Comp.B17-223

$BC_{21}H_{17}N_6O_9$ . . . . $[CH_3$-$OC(O)]_2B[$-$N(C_6H_5)NC(2$-OC_4H_2-5-$NO_2)NN(C_6H_4$-4-$NO_2)$-$]$ . . . . B: B Comp.SVol.4/3b-57

$BC_{21}H_{18}NO_3$ . . . . . $(C_6H_5)_2B[$-O-$(1,2$-$C_6H_4)$-$CH=N(CH_2COOH)$-$]$. . B: B Comp.SVol.3/3-219/20

$BC_{21}H_{18}N_5O_7$ . . . . $[CH_3$-$OOC]_2B[$-$N(C_6H_5)NC(2$-$OC_4H_3)NN(C_6H_4$-4-$NO_2)$-$]$ B: B Comp.SVol.4/3b-57

– . . . . $[CH_3$-$OOC]_2B[$-$N(C_6H_5)NC(2$-OC_4H_2-5-$NO_2)NN(C_6H_5)$-$]$ B: B Comp.SVol.4/3b-57

$BC_{21}H_{19}MoN_8O_2$. $(c$-$C_7H_7)Mo(CO)_2(N_2C_3H_3)_4B$. Mo:Org.Comp.5-225/6, 230, 231, 233

$BC_{21}H_{19}N_4O_5$ . . . . $[CH_3$-$OC(O)]_2B[$-$N(C_6H_5)NC(2$-$OC_4H_3)NN(C_6H_5)$-$]$ B: B Comp.SVol.4/3b-57

$BC_{21}H_{20}NO_2$ . . . . . $(C_6H_5)_2B[$-O-$(1,2$-$C_6H_4)$-$CH=N(CH_2CH_2OH)$-$]$ B: B Comp.SVol.3/3-219/20

$BC_{21}H_{21}NOP$. . . . . $(C_6H_5)_2B[$-$NH=CCH_3$-PC_6H_5-CH_2-O-$]$ . . . . B: B Comp.SVol.3/3-218

$BC_{21}H_{24}N$ . . . . H_3B-$N(CH_2$-$C_6H_5)_3$ . . . . B: B Comp.SVol.4/3b-10

– . . . . H_3N-$B(C_6H_4$-2-$CH_3)_3$ . . . . B: B Comp.SVol.3/3-191

– . . . . H_3N-$B(C_6H_4$-3-$CH_3)_3$ . . . . B: B Comp.SVol.3/3-191

– . . . . H_3N-$B(C_6H_4$-4-$CH_3)_3$ . . . . B: B Comp.SVol.3/3-191

$BC_{21}H_{24}NO_3$ . . . . . H_3N-$B(C_6H_4$-4-$OCH_3)_3$ . . . . B: B Comp.SVol.3/3-191

$BC_{21}H_{24}N_3$ . . . . . . $B(NHC_6H_4$-2-$CH_3)_3$ . . . . B: B Comp.SVol.3/3-94

$BC_{21}H_{25}MoN_4O_2$. $[CH_2C(C_6H_5)CH_2]Mo(CO)_2[$-$N_2C_3H(CH_3)_2$-$]_2BH_2$ Mo:Org.Comp.5-213, 216

– . . . . $[CH_2C(C_6H_5)CH_2]Mo(CO)_2[$-$N_2C_3H_3$-$]_2B(C_2H_5)_2$ Mo:Org.Comp.5-213, 216, 220/1

$BC_{21}H_{26}N$ . . . . $(t$-$C_4H_9)_2N$-$B=C_{13}H_8$ . . . . B: B Comp.SVol.4/3a-216

$BC_{21}H_{28}N$ . . . . $(C_6H_5)_2B$-$N=C(t$-$C_4H_9)_2$ . . . . B: B Comp.SVol.3/3-156

$BC_{21}H_{29}MoN_6O_2$. $[CH_2C(CH_3)CH_2]Mo(CO)_2[$-$N_2C_3H(CH_3)_2$-$]_3BH$ Mo:Org.Comp.5-226, 228, 231

$BC_{21}H_{29}NPSi$ .... 1-$(C_6H_5)_2$P-4,5-$(C_2H_5)_2$-2,2,3-$(CH_3)_3$-1,2,5-$NSiBC_2$ B: B Comp.SVol.4/3a-250/1
$BC_{21}H_{30}N$ .... [2,4,6-$(CH_3)_3C_6H_2]_2$B-NH-C_3H_7-i .... B: B Comp.SVol.3/3-153
– .... (c-$C_7H_{11})_2$B-NCH_3-C_6H_5 .... B: B Comp.SVol.3/3-153
$BC_{21}H_{30}N_3O$ ..... $BC_6H_5N_3$(=O)$(C_6H_4$-2-$CH_3)(C_4H_9$-n$)_2$ .... B: B Comp.SVol.4/3a-180/1
– .... $(C_2H_5)_2$N-B(C_6H_5)-N(C_6H_5)-C(O)NH-C_4H_9-i .. B: B Comp.SVol.3/3-109
– .... t-C_4H_9-NH-B(C_6H_5)-N(C_6H_5)-C(O)N$(C_2H_5)_2$ .. B: B Comp.SVol.3/3-109
$BC_{21}H_{31}N_4Si$..... 2,4,6-$(CH_3)_3C_6H_2$B[-N$(C_6H_2$-2,4,6-$(CH_3)_3)$ -N=N-NSi$(CH_3)_3$-] .... B: B Comp.SVol.3/3-137/8
$BC_{21}H_{32}NO$...... (c-$C_6H_{11})_2$B[-N$(CH_3)(C_6H_5)$-CH_2-CO-] ...... B: B Comp.SVol.3/3-165
$BC_{21}H_{32}NO_2$..... [(-O-$(CH_2)_3$-O-)B$(C_6H_5)_2$][$(C_2H_5)_3$NH] .... B: B Comp.SVol.3/3-215
$BC_{21}H_{33}N_3NaO_6$.. [HB$(NC_4H_4)_2$CN]Na · 3 $C_4H_8O_2$ .... B: B Comp.SVol.3/3-213
$BC_{21}H_{34}KN_6$..... K[(3-(t-C_4H_9)-1,2-$N_2C_3H_2$-1$)_3$BH] .... B: B Comp.SVol.4/3b-40
$BC_{21}H_{34}N_6Tl$..... Tl[(3-(t-C_4H_9)-1,2-$N_2C_3H_2$-1$)_3$BH] .... B: B Comp.SVol.4/3b-40
$BC_{21}H_{35}N_2$ ...... 1-[2,6-(i-$C_3H_7)_2$-C_6H_3-N=B]-2,2,6,6-$(CH_3)_4$-NC_5H_6 B: B Comp.SVol.4/3a-160/5
$BC_{21}H_{39}N_4Si$..... 2,6-(i-$C_3H_7)_2$-C_6H_3-N[Si$(CH_3)_3$]-B(N_3)-C$(CH_3)_2$-C_3H_7-i B: B Comp.SVol.4/3a-171
$BC_{21}H_{39}O_5PRe$... [(n-$C_4H_9)_4$P][$(CO)_5ReBH_3$] .... Re: Org.Comp.2-168/9
$BC_{21}H_{40}NO$...... 1-(i-C_4H_9)-2-(n-C_4H_9-O)-3,3-(n-$C_3H_7)_2$-1,2-NBC_7H_8 B: B Comp.SVol.4/3b-65
$BC_{21}H_{40}N_2O$..... $(CH_3)_3$N-BH_2-N(O)-C_6H_2-2,4,6-$(C_4H_9$-t$)_3$, radical B: B Comp.SVol.4/3b-26
$BC_{21}H_{44}N_3O_2$.... 1-[C_2H_5-OC(O)-CH$(C_4H_9$-i)-NH-B(NH -C_4H_9-t)]-2,2,6,6-$(CH_3)_4$-NC_5H_6 .... B: B Comp.SVol.4/3a-155
$BC_{21}H_{45}N_2$ ...... n-$C_{18}H_{37}$-N$(CH_3)_2$-BH_2-CN .... B: B Comp.SVol.4/3b-23
$BC_{21}H_{46}NO_2$..... n-$C_{18}H_{37}$-N$(CH_3)_2$-BH_2-COOH .... B: B Comp.SVol.4/3b-25
$BC_{21}H_{46}NSn$..... (i-$C_4H_9)_2$B[-C$(C_4H_9$-i)=CCH_3-Sn$(CH_3)_2$-N$(C_2H_5)_2$-] B: B Comp.SVol.4/3a-255 Sn: Org.Comp.19-48, 51
$BC_{21}H_{49}N_2Si_5$.... [$(CH_3)_3$Si]$_3$Si-B[NH-Si$(CH_3)_3$]-NH-C_6H_2-2,4,6-$(CH_3)_3$ B: B Comp.SVol.4/3a-170
$BC_{21}H_{52}LiN_4Si$... LiN$(C_4H_9$-t)-B$(C_4H_9$-n)-N$(C_4H_9$-t)-Si$(CH_3)_3$ · $(CH_3)_2$N-CH_2CH_2-N$(CH_3)_2$ .... B: B Comp.SVol.4/3a-169
– .... LiN$(C_4H_9$-t)-B$(C_4H_9$-t)-N$(C_4H_9$-t)-Si$(CH_3)_3$ · $(CH_3)_2$N-CH_2CH_2-N$(CH_3)_2$ .... B: B Comp.SVol.4/3a-169
$BC_{22}ClH_{16}$ ...... 4-Cl-4-$BC_{22}H_{16}$ .... B: B Comp.SVol.4/4-49
$BC_{22}ClH_{20}$ ...... 2,4,6-$(CH_3)_3$-C_6H_2-BCl-9-$C_{13}H_9$ .... B: B Comp.SVol.4/4-50
$BC_{22}ClH_{23}N_2^+$ ... [(1,10-$N_2C_{12}H_8$)BCl-$C_5(CH_3)_5]^+$ .... B: B Comp.SVol.4/4-70
$BC_{22}ClH_{27}N$ ..... 1-$(C_{13}H_9$-9-BCl)-2,2,6,6-$(CH_3)_4$-NC_5H_6 ...... B: B Comp.SVol.4/4-61
$BC_{22}ClH_{28}Si$..... (C_6H_5)[n-C_4H_9($(CH_3)_3$Si)CC=CC_6H_5]BCl ...... B: B Comp.SVol.3/4-45
$BC_{22}ClH_{29}N$ ..... 1-[$(C_6H_5)_2$CH-BCl]-2,2,6,6-$(CH_3)_4$-NC_5H_6 .... B: B Comp.SVol.4/4-60
$BC_{22}ClH_{32}N_2$ .... Cl-B[N$(C_4H_9$-t)-CH_2-$C_6H_5]_2$ .... B: B Comp.SVol.4/4-63
$BC_{22}Cl_2H_{23}N_2$.... [(1,10-$N_2C_{12}H_8$)BCl-$C_5(CH_3)_5$]Cl .... B: B Comp.SVol.4/4-70
$BC_{22}Cl_3H_{19}N_3$.... (-N=CCH_3-CH=CNH_2-)NH-B$(C_6H_4$-4-Cl$)_3$ .... B: B Comp.SVol.3/3-192
$BC_{22}Cl_3H_{38}N_2Sn$. 1-[2,6-(i-$C_3H_7)_2C_6H_3$-N$(SnCl_2$-$CH_3)$-BCl]- 2,2,6,6-$(CH_3)_4$-NC_5H_6 .... B: B Comp.SVol.4/4-64
$BC_{22}CoH_{31}PSi$ ... $BC_2CoPSi(C_6H_5)(C_5H_5)(C_2H_5)_2(CH_3)_2$[C$(CH_3)$=$CH_2$] B: B Comp.SVol.4/4-180
$BC_{22}CoH_{34}N_7S$... HB[-$N_2C_3H_2(C_4H_9$-t)-]$_3$Co-NCS .... B: B Comp.SVol.4/3b-40

$BC_{22}Co_2H_{28}NO_6$ . . $[(t-C_4H_9)C\equiv C][(t-C_4H_9)C_2Co_2(CO)_6]B-N(C_2H_5)_2$
B: B Comp.SVol.3/3-156

$BC_{22}FH_{27}N$ . . . . . . $1-(C_{13}H_9-9-BF)-NC_5H_6-2,2,6,6-(CH_3)_4$ . . . . . . B: B Comp.SVol.4/3b-245

$BC_{22}FH_{29}N$ . . . . . . $1-[(C_6H_5)_2CH-BF]-NC_5H_6-2,2,6,6-(CH_3)_4$ . . . . . B: B Comp.SVol.4/3b-245

$BC_{22}FH_{32}N_2$ . . . . . $C_6H_5-CH_2-N(C_4H_9-t)-BF-N(C_4H_9-t)-CH_2-C_6H_5$
B: B Comp.SVol.4/3b-246

$BC_{22}FH_{41}N_3OSSi$ $[((CH_3)_2N)_3S][1-(CH_3)_2B(F)-8-(CH_3)_2Si(O-C_2H_5)-C_{10}H_6]$
B: B Comp.SVol.4/3b-228/9

$BC_{22}F_2FeH_{21}O_3$ . . $(C_5H_5)Fe(C(C_7H_8CH_3)OBF_2OCC_6H_5)CO$ . . . . . . Fe: Org.Comp.B16b-131

$BC_{22}F_2FeH_{23}O_3$ . . $(C_5H_5)Fe(C(c-C_6H_7(CH_3)_2)OBF_2OCC_6H_5)CO$ . . Fe: Org.Comp.B16b-130

$BC_{22}F_3H_{16}O_5S$ . . . $C_6H_5(CF_3SO_3)B[-O=CC_6H_5-CH=CC_6H_5-O-]$ . . . B: B Comp.SVol.3/4-52

$BC_{22}F_3H_{34}N_2Si_2$ . . $FB[N(C_6H_2-2,4,6-(CH_3)_3)(Si(CH_3)_2F)]_2$ . . . . B: B Comp.SVol.3/3-375

$BC_{22}F_4FeH_{15}O_2$ . . $[C_5H_5(CO)_2Fe=C_3(C_6H_5)_2-c][BF_4]$ . . . . Fe: Org.Comp.B16a-92, 99/100

$BC_{22}F_4FeH_{19}N_2O_6$ $[((HO)(CH_3O)_2C_{16}H_{12}N-CN)Fe(CO)_3][BF_4]$ . . . . Fe: Org.Comp.B15-132

$BC_{22}F_4FeH_{23}O_3$ . . $[C_5H_5(CO)_2Fe=CHC_6H_8(=O)(C_2H_4C_6H_5)-c][BF_4]$
Fe: Org.Comp.B16a-90, 97

$BC_{22}F_4FeH_{30}NO_2$ $[(C_5H_5)Fe(CO)_2-C_5H_8-C_6H_9(=NC_4H_8)][BF_4]$ . . . Fe: Org.Comp.B14-152, 156

$BC_{22}F_4FeH_{31}O_2$ . . $[(C_5H_5)Fe(CO)_2(CH_2=CHCH(CH_3)CH_2CH_2$
$CH=C(CH_3)CH_2CH_2CH=C(CH_3)_2)][BF_4]$ . . . . . . Fe: Org.Comp.B17-41/2

$BC_{22}F_4FeH_{33}N_2O_2$ $[C_5H_5(CO)_2Fe=C_3(N(C_3H_7-i)_2)_2-c][BF_4]$ . . . . . . Fe: Org.Comp.B16a-99/100

$BC_{22}F_4FeH_{44}P_2$ . . $[(CH_2CHCHCH=CH_2)_2Fe(P(C_2H_5)_3)_2][BF_4]$ . . . . . Fe: Org.Comp.B17-179

$BC_{22}F_4Fe_3H_{15}O_6S_2$
$[C_5H_5(CO)_2Fe=C(SFe(CO)_2C_5H_5)_2][BF_4]$ . . . . . . Fe: Org.Comp.B16a-143, 154

$BC_{22}F_4H_{17}MoO$ . . $[C_5H_5Mo(CO)(H-CC-C_6H_5)_2][BF_4]$ . . . . Mo: Org.Comp.6-317, 322

$BC_{22}F_4H_{18}MoN_3O_2$
$[(CH_2CHCH_2)Mo(CO)_2(1,10-N_2C_{12}H_8)(NC_5H_5)][BF_4]$
Mo: Org.Comp.5-200, 202/3

$BC_{22}F_4H_{21}OSSe$ . . $[((C_6H_5)_2Se)((CH_3)_2S)C(COC_6H_5)][BF_4]$ . . . . B: B Comp.SVol.3/4-148

$BC_{22}F_4H_{25}MoO_2$ . . $[(C_9H_7)Mo(CO)_2(C_5H_{11}-CH=CH-C(=CH_2)-CH=CH_2)][BF_4]$
Mo: Org.Comp.8-305/6, 308

$BC_{22}F_4H_{25}O_2P_2W$ $[(C_5H_5)(CO)_2(P(CH_3)_3)WPH(C_6H_5)_2][BF_4]$ . . . . . B: B Comp.SVol.4/3b-124

$BC_{22}F_4H_{27}MoO$ . . $[C_9H_7Mo(CO)(H-CC-C_4H_9-t)_2][BF_4]$ . . . . Mo: Org.Comp.6-317

$BC_{22}F_4H_{37}MoO_6P_2Si$
$[C_5H_5Mo(P(OCH_3)_3)_2C_6H_5-CC-Si(CH_3)_3][BF_4]$ Mo: Org.Comp.6-121

$BC_{22}H_{18}NO_2$ . . . . . $HB[-O-(1,2-C_{10}H_6)-CH_2-]_2NH$ . . . . B: B Comp.SVol.3/3-220/1

$BC_{22}H_{18}N_3$ . . . . . . $(C_6H_5)_2B[-NC_5H_4-N=NC_5H_4-]$ . . . . B: B Comp.SVol.3/3-123
B: B Comp.SVol.4/3a-181

$BC_{22}H_{18}N_3S_2$ . . . . $BN_2S_2C_4H_2-(C_6H_5)_2-(NH-C_6H_5)$ . . . . B: B Comp.SVol.4/4-149/50

– . . . . $[-B(C_6H_5)_2-N(C_6H_5)-C(=S)-NH-C_3H_2NS-]$ . . . . B: B Comp.SVol.4/3a-179
B: B Comp.SVol.4/4-149/50

$BC_{22}H_{18}N_5O_9$ . . . . $[CH_3-OC(O)]_2B[-N(C_6H_5)NC$
$(2-OC_4H_2-5-NO_2)NN(C_6H_4-4-COOH)-]$ . . . . . B: B Comp.SVol.4/3b-57

$BC_{22}H_{19}N_4O_7$ . . . . $[CH_3-OC(O)]_2B[-N(C_6H_5)NC(2-OC_4H_3)NN$
$(C_6H_4-4-COOH)-]$ . . . . B: B Comp.SVol.4/3b-57

$BC_{22}H_{20}N_3O_2$ . . . . $3-[4-(C_2H_5-O-C(=O))-C_6H_4]-2,5-(C_6H_5)_2-1,2,4,3-N_3BCH$
B: B Comp.SVol.4/3a-174/5

$BC_{22}H_{22}N_3$ . . . . . . $(-N=CCH_3-CH=CNH_2-)NH-B(C_6H_5)_3$ . . . . B: B Comp.SVol.3/3-192

$BC_{22}H_{22}N_3O_8$ . . . . $(CH_3COO)_2B[-N(C_6H_4-4-NO_2)=CCH_3-$
$C(CONHC_6H_5)=CCH_3-O-]$ . . . . B: B Comp.SVol.3/3-218

$BC_{22}H_{23}N_2O_6$ . . . . $(CH_3COO)_2B[-NC_6H_5=CCH_3-C(CONHC_6H_5)=CCH_3-O-]$ B: B Comp.SVol.3/3-218

$BC_{22}H_{24}N$ . . . . $(C_6H_5)_2B[-N(CH_3)_2-CH(CH_3)-(1,2-C_6H_4)-]$ . . . . B: B Comp.SVol.4/3a-256

$BC_{22}H_{26}N$ . . . . $1-(C_{13}H_8=B)-2,2,6,6-(CH_3)_4-NC_5H_6$ . . . . B: B Comp.SVol.4/3a-216/7

– . . . . $CH_3-NH_2-B(C_6H_4-3-CH_3)_3$ . . . . B: B Comp.SVol.4/3b-27

– . . . . $CH_3-NH_2-B(C_6H_4-4-CH_3)_3$ . . . . B: B Comp.SVol.4/3b-27

$BC_{22}H_{26}NO_3$ . . . . . $CH_3-NH_2-B(C_6H_4-2-OCH_3)_3$ . . . . B: B Comp.SVol.4/3b-27

$BC_{22}H_{26}N_3$ . . . . . . $(1-NC_8H_6-1)_2B-N(C_3H_7-i)_2$ . . . . B: B Comp.SVol.4/3a-156

– . . . . $(H_2N)_2C=NH-B(C_6H_4-2-CH_3)_3$ . . . . B: B Comp.SVol.4/3b-27

$BC_{22}H_{27}MoO_2P_2$ . . $(C_5H_5)Mo(CO)_2[P(CH_3)_3]-P(C_6H_5)_2 \cdot BH_3$ Mo:Org.Comp.7-130/1

$BC_{22}H_{28}N$ . . . . $1-(C_{13}H_9-BH)-2,2,6,6-(CH_3)_4-NC_5H_6$ . . . . B: B Comp.SVol.4/3a-225

– . . . . $1-[(C_6H_5)_2C=B]-2,2,6,6-(CH_3)_4-NC_5H_6$ . . . . B: B Comp.SVol.4/3a-217

$BC_{22}H_{28}NO_2$ . . . . . $(CH_3)_2B-N[C_6H_2-2,4,6-(CH_3)_3]-C(O)C(O)-$
$C_6H_2-2,4,6-(CH_3)_3$ . . . . B: B Comp.SVol.3/3-155

$BC_{22}H_{30}N$ . . . . $1-[(C_6H_5)_2CH-BH]-2,2,6,6-(CH_3)_4-NC_5H_6$ . . . . B: B Comp.SVol.4/3a-225

– . . . . $(n-C_4H_9)_2B-N(C_6H_5)-C(C_6H_5)=CH_2$ . . . . B: B Comp.SVol.3/3-152

$BC_{22}H_{32}N$ . . . . $[2,4,6-(CH_3)_3C_6H_2]_2B-NH(t-C_4H_9)$ . . . . B: B Comp.SVol.3/3-153

$BC_{22}H_{32}N_2^+$ . . . . . $[(t-C_4H_9)(C_6H_5CH_2)N=B=N(CH_2C_6H_5)(t-C_4H_9)]^+$
B: B Comp.SVol.3/3-205

$BC_{22}H_{33}LiNO$ . . . . $[(2,4,6-(CH_3)_3-C_6H_2)_2B-NH-Li(O(C_2H_5)_2)]_2$ . . . B: B Comp.SVol.4/3a-231

$BC_{22}H_{33}N_2$ . . . . . . $C_6H_5CH_2-N(C_4H_9-t)-BH-N(C_4H_9-t)-CH_2C_6H_5$ B: B Comp.SVol.3/3-107
B: B Comp.SVol.4/3a-169

$BC_{22}H_{33}N_2Si$. . . . . $1,3-[2,6-(CH_3)_2-C_6H_3]_2-4-(t-C_4H_9)-$
$2,2-(CH_3)_2-1,3,2,4-N_2SiB$ . . . . B: B Comp.SVol.4/3a-205/6

$BC_{22}H_{34}N$ . . . . $(c-C_6H_{11})_2B[-N(CH_3)(C_6H_5)-CH_2-CH=CH-]$ . . . B: B Comp.SVol.3/3-165

$BC_{22}H_{38}N_2PSi_2$. . . $[(CH_3)_3Si]_2N-B(C_4H_9-t)-NH-P(C_6H_5)_2$ . . . . B: B Comp.SVol.4/3a-170

$BC_{22}H_{38}N_3O_2$ . . . . $t-C_4H_9-NH-B[O-C(O)-CH(NH_2)-CH_2-C_6H_5]-$
$1-NC_5H_6-2,2,6,6-(CH_3)_4$ . . . . B: B Comp.SVol.4/3b-52

$BC_{22}H_{40}NO$. . . . . . $i-C_3H_7-C(CH_3)_2-B(O-C_4H_9-t)-NH-C_6H_3-2,6-(C_3H_7-i)_2$
B: B Comp.SVol.4/3b-61

– . . . . $[7,7,2-(CH_3)_3-(3.1.1)-C_7H_8-3]_2B[-O-CH_2CH_2-NH_2-]$
B: B Comp.SVol.4/3b-71

$BC_{22}H_{41}N_2$ . . . . . . $i-C_3H_7-C(CH_3)_2-B(NH-C_4H_9-t)-NH-C_6H_3-2,6-(C_3H_7-i)_2$
B: B Comp.SVol.4/3a-170

$BC_{22}H_{42}N$ . . . . $(C_2H_5)_2B-N(C_6H_{11}-c)-C_{12}H_{21}-c$ . . . . B: B Comp.SVol.3/3-152

– . . . . $(C_2H_5)_2B-N(C_6H_9-c)-C_{12}H_{23}-c$ . . . . B: B Comp.SVol.3/3-152

$BC_{22}H_{42}S^-$ . . . . . . $[(H_3C_4S)B(C_6H_{13})_3]^-$ . . . . B: B Comp.SVol.3/4-117

$BC_{22}H_{49}N_2Si_4$. . . . $[(CH_3)_3Si]_3Si-B(NH-C_4H_9-t)-NH-C_6H_2-2,4,6-(CH_3)_3$
B: B Comp.SVol.4/3a-170

$BC_{23}ClH_{24}N^+$ . . . . $[(5-NC_{13}H_9)BCl-C_5(CH_3)_5]^+$ . . . . B: B Comp.SVol.4/4-70

– . . . . $[(10-NC_{13}H_9)BCl-C_5(CH_3)_5]^+$ . . . . B: B Comp.SVol.4/4-70

$BC_{23}Cl_2H_{24}N$ . . . . $[(5-NC_{13}H_9)BCl-C_5(CH_3)_5]Cl$. . . . . B: B Comp.SVol.4/4-70

– . . . . $[(10-NC_{13}H_9)BCl-C_5(CH_3)_5]Cl$ . . . . B: B Comp.SVol.4/4-70

$BC_{23}Cl_3H_{20}N_2$. . . . $(-N=CCH_3-CH=CCH_3-)NH-B(C_6H_4-4-Cl)_3$ . . . . B: B Comp.SVol.3/3-191

$BC_{23}FH_{42}NP$ . . . . . $(t-C_4H_9)_2N-BF-PH-C_6H_2-2,4,6-(C_3H_7-i)_3$ . . . . B: B Comp.SVol.4/3b-243

$BC_{23}F_2FeH_{25}O_3$ . . $(C_5H_5)Fe(C(c-C_6H_6(CH_3)_3)OBF_2OCC_6H_5)CO$ . . Fe: Org.Comp.B16b-130

$BC_{23}F_2H_{13}N_2$ . . . . $F_2B(NC_{11}H_6)_2CH$. . . . . B: B Comp.SVol.3/3-378

$BC_{23}F_2H_{15}O_2$ . . . . $[-B(F)_2OC(2-C_{10}H_7)CHC(2-C_{10}H_7)O-]$ . . . . B: B Comp.SVol.4/3b-233

$BC_{23}F_2H_{19}N_2O_2$ . . $1-NC_8H_6-3-CH_2-CH[-NH_2-B(C_6H_4-3-F)_2-O-C(=O)-]$
B: B Comp.SVol.4/3b-71

$BC_{23}F_3H_{17}N$ ..... $NC_5H_5 \cdot B(C_6H_4\text{-}4\text{-}F)_3$ ..... B: B Comp.SVol.4/3b-28
$BC_{23}F_3H_{32}N_2O_3S$ $C_6H_5\text{-}CH_2\text{-}N(C_4H_9\text{-}t)\text{-}B[O\text{-}S(O)_2\text{-}CF_3]$
$\text{-}N(C_4H_9\text{-}t)\text{-}CH_2\text{-}C_6H_5$ ..... B: B Comp.SVol.4/3b-52
$BC_{23}F_4FeH_{15}O_2$ .. $[(C_5H_5)Fe(CO)_2(CH_2{=}C{=}C{=}C_{13}H_8)][BF_4]$ ..... Fe: Org.Comp.B17-116, 119/20
$BC_{23}F_4FeH_{17}O_2$ .. $[(C_5H_5)Fe(CO)_2(CH_2{=}C{=}C{=}C(C_6H_5)_2)][BF_4]$ ..... Fe: Org.Comp.B17-116, 119/20
$BC_{23}F_4FeH_{20}MoNO_4$
$[(C_5H_5)Mo(CO)_2(C_6H_7\text{-}4\text{-}CN\text{-}4\text{-}CH{=}CH_2)Fe$
$(CO)_2(C_5H_5)][BF_4]$ ..... Mo: Org.Comp.8-272
$BC_{23}F_4FeH_{35}$ .... $[(1,3,5\text{-}(C_4H_9\text{-}t)_3\text{-}C_6H_3)Fe(C_5H_5)][BF_4]$ ..... Fe: Org.Comp.B19-99, 122
– ..... $[((C_2H_5)_6C_6)Fe(C_5H_5)][BF_4]$ ..... Fe: Org.Comp.B19-173, 205/9
$BC_{23}F_4FeH_{36}O_6P$ $[(C_5H_5)Fe(CO)_2(CH_2{=}CHCH(CH_3)CH_2CH_2$
$CH{=}C(CH_3)CH_2OP(O)(O\text{-}C_3H_7\text{-}i)_2)][BF_4]$ .... Fe: Org.Comp.B17-41
$BC_{23}F_4Fe_3H_{17}O_7$ $[C_5H_5(CO)_2Fe{=}C(CH_2Fe(CO)_2C_5H_5)OFe(CO)_2C_5H_5][BF_4]$
Fe: Org.Comp.B16a-110, 121
$BC_{23}F_4H_{15}O_5PRe$ $[(CO)_5ReP(C_6H_5)_3][BF_4]$ ..... Re: Org.Comp.2-155
$BC_{23}F_4H_{20}NOPRe$ $[(C_5H_5)Re(NO)(P(C_6H_5)_3)][BF_4] \cdot CH_2Cl_2$ ..... Re: Org.Comp.3-29, 31, 36
$BC_{23}F_4H_{26}O_9P_2PtReW$
$[(CO)_5ReW(CO)_4Pt(P(CH_3)_3)_2(C(H)C_6H_4CH_3\text{-}4)][BF_4]$
Re: Org.Comp.2-195/6, 205
$BC_{23}F_4H_{33}MoO_2$ .. $[((CH_3)_5C_5)Mo(CO)_2(C_5H_{11}\text{-}CH{=}CH\text{-}C({=}CH_2)$
$\text{-}CH{=}CH_2)][BF_4]$ ..... Mo: Org.Comp.8-305/6, 309
$BC_{23}FeH_{26}Rh$ .... $C_5H_5FeC_4H_4BC_6H_5RhC_8H_{12}$ ..... Fe: Org.Comp.B17-225
$BC_{23}GeH_{27}O_2$ .... $Ge(C_6H_5)_3CH_2CH_2CH_2B(OCH_3)_2$ ..... Ge: Org.Comp.3-94
$BC_{23}H_{20}N$ ..... $NC_5H_5 \cdot B(C_6H_5)_3$ ..... B: B Comp.SVol.4/3b-28
$BC_{23}H_{20}NO_3$ ..... $(C_6H_5\text{-}O)_3B \cdot NC_5H_5$ ..... B: B Comp.SVol.4/3b-69
$BC_{23}H_{20}N_5O_6$ .... $[CH_3\text{-}OC(O)]_2B[\text{-}N(C_6H_5)NC(C_6H_4\text{-}4\text{-}NO_2)NN(C_6H_5)\text{-}]$
B: B Comp.SVol.4/3b-57
$BC_{23}H_{21}MoN_4O_2$. $(CH_2CHCH_2)Mo(CO)_2[\text{-}N_2C_3H_3\text{-}]_2B(C_6H_5)_2$ ... Mo: Org.Comp.5-213, 215
$BC_{23}H_{21}MoN_8O_2$. $[CH_2C(C_6H_5)CH_2]Mo(CO)_2(N_2C_3H_3)_4B$ ..... Mo: Org.Comp.5-225/6, 229
$BC_{23}H_{23}N_2$ ..... $(\text{-}N{=}CCH_3\text{-}CH{=}CCH_3\text{-})NH\text{-}B(C_6H_5)_3$ ..... B: B Comp.SVol.3/3-191
$BC_{23}H_{24}N_3$ ..... $(\text{-}N{=}CCH_3\text{-}CH{=}CNH_2\text{-})NCH_3\text{-}B(C_6H_5)_3$ ..... B: B Comp.SVol.3/3-192
$BC_{23}H_{25}N_4$ ..... $(NH_2)N_3C_2H_2\text{-}B(C_6H_4\text{-}4\text{-}CH_3)_3$ ..... B: B Comp.SVol.3/3-191
$BC_{23}H_{25}N_4O_3$ .... $(NH_2)N_3C_2H_2\text{-}B(C_6H_4\text{-}2\text{-}OCH_3)_3$ ..... B: B Comp.SVol.3/3-191
– ..... $(NH_2)N_3C_2H_2\text{-}B(C_6H_4\text{-}4\text{-}OCH_3)_3$ ..... B: B Comp.SVol.3/3-191
$BC_{23}H_{27}N_2$ ..... $[\text{-}N(CH_2\text{-}C_6H_5)\text{-}CH(C_6H_5)\text{-}N(CH_2\text{-}C_6H_5)$
$\text{-}CH_2CH_2\text{-}] \cdot BH_3$ ..... B: B Comp.SVol.4/3b-12
$BC_{23}H_{28}N$ ..... $(CH_3)_2NH\text{-}B(C_6H_4\text{-}2\text{-}CH_3)_3$ ..... B: B Comp.SVol.4/3b-27
– ..... $(CH_3)_2NH\text{-}B(C_6H_4\text{-}4\text{-}CH_3)_3$ ..... B: B Comp.SVol.4/3b-27
$BC_{23}H_{28}NO_3$ ..... $(CH_3)_2NH\text{-}B(C_6H_4\text{-}2\text{-}OCH_3)_3$ ..... B: B Comp.SVol.4/3b-27
$BC_{23}H_{28}N_3O$ ..... $(C_{10}H_{15})BCH_3[\text{-}NC_6H_5\text{-}CO\text{-}(1,2\text{-}NC_5H_4)\text{-}NH\text{-}]$ B: B Comp.SVol.3/3-123
$BC_{23}H_{30}NO$ ..... $CH_3O\text{-}B(C_{13}H_9)\text{-}1\text{-}NC_5H_6\text{-}2,2,6,6\text{-}(CH_3)_4$ ..... B: B Comp.SVol.4/3b-61
$BC_{23}H_{32}N$ ..... $[2,4,6\text{-}(CH_3)_3C_6H_2]_2B\text{-}N(\text{-}CH_2\text{-})_5$ ..... B: B Comp.SVol.3/3-156
$BC_{23}H_{32}N_3$ ..... $(i\text{-}C_3H_7)_2B[\text{-}NH{=}CC_6H_5\text{-}N(i\text{-}C_3H_7)\text{-}C({=}NC_6H_5)\text{-}]$
B: B Comp.SVol.3/3-165
$BC_{23}H_{33}N_2$ ..... $2,6\text{-}(i\text{-}C_3H_7)_2\text{-}C_6H_3\text{-}N{=}B\text{-}N(C_4H_9\text{-}t)\text{-}CH_2\text{-}C_6H_5$
B: B Comp.SVol.4/3a-160/5
$BC_{23}H_{35}N_2Si$ ..... $8\text{-}(C_3H_7\text{-}i)\text{-}4\text{-}CH_3\text{-}1,2\text{-}NBC_8H_7\text{-}$
$2\text{-}N[Si(CH_3)_3]\text{-}C_6H_3\text{-}2,6\text{-}(CH_3)_2$ ..... B: B Comp.SVol.4/3a-176

$BC_{23}H_{36}NO_2$ ..... $HO-CH_2CH_2CH_2-O-B(C_6H_5)_2-NH_2-C(CH_3)_2-CH_2-C_4H_9-t$
B: B Comp.SVol.4/3b-69

$BC_{23}H_{37}N_2$ ...... $1-(i-C_4H_9)-2-(C_6H_5-NH)-3,3-(n-C_3H_7)_2-1,3-NBC_7H_8$
B: B Comp.SVol.3/3-117

– .... $1-C_6H_5-2-(n-C_4H_9-NH)-3,3-(n-C_3H_7)_2-1,3-NBC_7H_8$
B: B Comp.SVol.3/3-117

$BC_{23}H_{45}N_2$ ...... $1-(i-C_4H_9)-2-(t-C_4H_9-NH)-3,3-(n-C_4H_9)_2-1,3-NBC_7H_8$
B: B Comp.SVol.3/3-117

$BC_{24}Cl_2F_4H_{22}NOPRe$
$[(C_5H_5)Re(NO)(P(C_6H_5)_3)(CH_2Cl_2)][BF_4]$ ...... Re: Org.Comp.3-29, 31, 36

$BC_{24}Cl_4H_{20}Sb$.... $[Sb(C_6H_5)_4][BCl_4]$ .... B: B Comp.SVol.3/4-33/4

$BC_{24}FH_{36}N_2$ ..... $2,6-(i-C_3H_7)_2-C_6H_3-NH-BF-NH-C_6H_3-2,6-(C_3H_7-i)_2$
B: B Comp.SVol.4/3b-246

$BC_{24}FH_{42}NP$ ..... $2,2,6,6-(CH_3)_4-NC_5H_6-1-[BF-PH-C_6H_2-2,4,6-(C_3H_7-i)_3]$
B: B Comp.SVol.4/3b-243

$BC_{24}FH_{62}OSi_6$ ... $FB[O(CH_2)_4C(Si(CH_3)_3)_3][C(Si(CH_3)_3)_3]$ .... B: B Comp.SVol.3/3-367

$BC_{24}F_4$ .... $C_{24}[BF_4]$ .... B: B Comp.SVol.4/3b-107

$BC_{24}F_4FeH_{20}$ .... $[(C_5H_5)Fe(C(C_6H_5)_3)][BF_4]$ .... Fe: Org.Comp.B18-142/6, 238

$BC_{24}F_4FeH_{22}MoNO_4$
$[(C_5H_5)Mo(CO)_2(C_7H_9-4-CN-4-CH=CH_2)Fe$
$(CO)_2(C_5H_5)][BF_4]$ .... Mo: Org.Comp.8-287

$BC_{24}F_4FeH_{24}NO_7$ $[((CH_3COO)(CH_3O)_2C_{16}H_{12}N-CH_3)Fe(CO)_3][BF_4]$
Fe: Org.Comp.B15-132

$BC_{24}F_4FeH_{29}O_2S$ $[(C_5H_5)Fe(CO)_2CH(S(CH_3)C_6H_5)CH_2$
$CH_2-c-C_6H_9(=CH_2)-2][BF_4]$............... Fe: Org.Comp.B14-133, 136

$BC_{24}F_4FeH_{37}$ .... $[(1,2,4,5-(CH_3)_4-C_6H_2)Fe(1,2,4,5-$
$(CH_3)_4-C_6H_2-6-C_4H_9-t)][BF_4]$............. Fe: Org.Comp.B19-144, 150/1

$BC_{24}F_4Fe_3H_{19}O_7$ $[(C_5H_5)(CO)_2Fe=C(CH_2Fe(CO)_2(C_5H_4CH_3))-O$
$-Fe(CO)_2(C_5H_5)][BF_4]$ .... Fe: Org.Comp.B16a-111, 121

– .... $[(C_5H_5)(CO)_2Fe=C(CH_2Fe(CO)_2(C_5H_5))-O$
$-Fe(CO)_2(C_5H_4CH_3)][BF_4]$ .... Fe: Org.Comp.B16a-110, 121

$BC_{24}F_4H_{17}O_4PRe$ $[(2-(CH_2=CH)C_6H_4P(C_6H_5)_2)Re(CO)_4][BF_4]$.... Re: Org.Comp.2-327, 328/9

$BC_{24}F_4H_{20}NO_2PRe$
$[(C_5H_5)Re(CO)(NO)(P(C_6H_5)_3)][BF_4]$ .... Re: Org.Comp.3-153/4, 156, 157

$BC_{24}F_4H_{21}MoO$ .. $[C_5H_5Mo(CO)(H_3C-CC-C_6H_5)_2][BF_4]$......... Mo: Org.Comp.6-319, 325

$BC_{24}F_4H_{21}MoOS_2$ $[C_5H_5Mo(CO)(C_6H_5-CC-SCH_3)_2][BF_4]$ .... Mo: Org.Comp.6-321

$BC_{24}F_4H_{22}NOPRe$ $[(C_5H_5)Re(NO)(P(C_6H_5)_3)=CH_2][BF_4]$......... Re: Org.Comp.3-109/11, 112

$BC_{24}F_4H_{23}INOPRe$
$[(C_5H_5)Re(NO)(P(C_6H_5)_3)-I-CH_3][BF_4]$ .... Re: Org.Comp.3-29, 31/2, 37

$BC_{24}F_4H_{33}MnN_2O_2P_3$
$[Mn(NO)_2(P(C_6H_5)(CH_3)_2)_3][BF_4]$............ Mn: MVol.D8-64

$BC_{24}F_4H_{33}MnN_2O_8P_3$
$[Mn(NO)_2(C_6H_5-P(OCH_3)_2)_3][BF_4]$........... Mn: MVol.D8-118/9

$BC_{24}F_4H_{34}MoN_3O_2$
$[(CH_2CHCH_2)Mo(CO)_2(NC_5H_5)$
$(c-C_6H_{11}-N=CHCH=N-C_6H_{11}-c)][BF_4]$ ...... Mo: Org.Comp.5-263, 266

$BC_{24}F_4H_{36}MoNO_2$ $[2-i-C_3H_7-5-CH_3-c-C_6H_9C_5H_4Mo(CO)$
$(NO)C_8H_{13}-c][BF_4]$..................... Mo: Org.Comp.6-351

Formula	Compound	Reference
$BC_{24}F_4H_{72}LiN_{12}O_4P_4$	$Li[BF_4] \cdot 4\ O{=}P[N(CH_3)_2]_3$	B: B Comp.SVol.4/3b-124
$BC_{24}F_5H_{24}N_2$	$[2,4,6\text{-}(CH_3)_3C_6H_2NH]_2BC_6F_5$	B: B Comp.SVol.3/4-41
$BC_{24}FeH_{46}O_2P$	$C_5(CH_3)_5(CO)(P(C_4H_9\text{-}n)_3)Fe{=}C(H)OBH_3$	Fe: Org.Comp.B16a-176/7
$BC_{24}H_{15}N_4$	$(NC)_2C{=}C(CN)_2\text{-}B(C_6H_5)_3$	B: B Comp.SVol.3/3-192
$BC_{24}H_{16}N_{12}^-$	$[(1,2,3\text{-}N_3C_6H_4\text{-}1)_4B]^-$	B: B Comp.SVol.4/3b-50
$BC_{24}H_{21}MoN_6O_2$	$(c\text{-}C_7H_7)Mo(CO)_2[\text{-}N_2C_3H_3\text{-}]_3B\text{-}C_6H_5$	Mo:Org.Comp.5-230, 231, 232
$BC_{24}H_{23}MoN_4O_2$	$[CH_2C(CH_3)CH_2]Mo(CO)_2[\text{-}N_2C_3H_3\text{-}]_2B(C_6H_5)_2$	Mo:Org.Comp.5-213, 215/6, 219/20
$BC_{24}H_{26}N_3$	$8\text{-}N(C_3H_7\text{-}i)_2\text{-}7,9,8\text{-}N_2BC_{18}H_{12}$	B: B Comp.SVol.4/3a-156
$BC_{24}H_{27}N_4$	$2,4,6\text{-}(CH_3)_3C_6H_2B[\text{-}N(C_6H_2\text{-}2,4,6\text{-}(CH_3)_3)\text{-}N{=}N\text{-}NC_6H_5\text{-}]$	B: B Comp.SVol.3/3-137/8
$BC_{24}H_{27}S$	$[2,4,6\text{-}(CH_3)_3\text{-}C_6H_2]_2BSC_6H_5$	B: B Comp.SVol.3/4-111
$BC_{24}H_{28}N$	$[2,4,6\text{-}(CH_3)_3\text{-}C_6H_2]_2B\text{-}NH\text{-}C_6H_5$	B: B Comp.SVol.3/3-153
		B: B Comp.SVol.4/3a-232
		B: B Comp.SVol.4/3b-228
$BC_{24}H_{28}N_3O$	$BC_6H_4N_3({=}O)(C_6H_5)_2(C_3H_7\text{-}i)_2$	B: B Comp.SVol.4/3a-180/1
$BC_{24}H_{29}MoN_6O_2$	$(C_7H_7)Mo(CO)_2[\text{-}N_2C_3H(CH_3)_2\text{-}BH(N_2C_3H(CH_3)_2)\text{-}N_2C_3H(CH_3)_2\text{-}]$	Mo:Org.Comp.7-156, 159
$BC_{24}H_{29}N_2$	$[2,4,6\text{-}(CH_3)_3C_6H_2NH]_2BC_6H_5$	B: B Comp.SVol.3/3-109
$BC_{24}H_{29}N_4$	$BC_6H_4N_3(C_6H_5)(NH\text{-}C_6H_5)(C_3H_7\text{-}i)_2$	B: B Comp.SVol.4/3a-180
–	$BC_6H_5N_3({=}N\text{-}C_6H_5)(C_6H_5)(C_3H_7\text{-}i)_2$	B: B Comp.SVol.4/3a-181, 257
$BC_{24}H_{30}N$	$(C_2H_5)_3N\text{-}B(C_6H_5)_3$	B: B Comp.SVol.3/3-191
$BC_{24}H_{30}NOSi_2$	$6\text{-}[(CH_3)_3Si]_2N\text{-}4\text{-}C_6H_5\text{-}5,6\text{-}OBC_{12}H_7$	B: B Comp.SVol.4/3b-67
$BC_{24}H_{30}N_3$	$(1\text{-}NC_9H_8\text{-}1)_2B\text{-}N(C_3H_7\text{-}i)_2$	B: B Comp.SVol.4/3a-156
$BC_{24}H_{33}N_2$	$1\text{-}[(CH_3)_2N\text{-}B(9\text{-}C_{13}H_9)]\text{-}2,2,6,6\text{-}(CH_3)_4\text{-}NC_5H_6$	B: B Comp.SVol.4/3a-170
$BC_{24}H_{34}N_3$	$(i\text{-}C_3H_7)_2B[\text{-}NH{=}CC_6H_5\text{-}N(t\text{-}C_4H_9)\text{-}C({=}NC_6H_5)\text{-}]$	B: B Comp.SVol.3/3-165
$BC_{24}H_{39}N_2Ti$	$2,2,6,6\text{-}(CH_3)_4\text{-}NC_5H_6\text{-}1\text{-}B[\text{-}N(C_4H_9\text{-}t)\text{-}Ti(C_5H_5)_2\text{-}CH_2\text{-}]$	B: B Comp.SVol.4/3a-249
$BC_{24}H_{43}N_4S$	$(n\text{-}C_4H_9)_2B[\text{-}NC_3H_2S\text{-}N{=}CNH(c\text{-}C_6H_{11})\text{-}N(c\text{-}C_6H_{11})\text{-}]$	B: B Comp.SVol.3/3-123
$BC_{24}H_{44}N_2O_3Sb$	$2,6\text{-}(i\text{-}C_3H_7)_2\text{-}C_6H_3\text{-}N[Sb(OCH_3)_2]\text{-}B(OCH_3)\text{-}1\text{-}NC_5H_6\text{-}2,2,6,6\text{-}(CH_3)_4$	B: B Comp.SVol.4/3b-52
$BC_{25}ClF_4H_{25}NOPRe$	$[(C_5H_5)Re(NO)(P(C_6H_5)_3)(Cl\text{-}C_2H_5)][BF_4]$	Re: Org.Comp.3-29, 31
$BC_{25}ClF_6FeH_{27}IrO_3P_2$	$[(C_5H_5)(CO)Fe(CO)(CF_2)IrCl(CO)(P(CH_3)_2C_6H_5)_2][BF_4]$	Fe: Org.Comp.B16b-148
$BC_{25}ClH_{37}NSi$	$1\text{-}[Si(CH_3)_3\text{-}CH(C_6H_5)\text{-}4\text{-}C_6H_4\text{-}BCl]\text{-}2,2,6,6\text{-}(CH_3)_4\text{-}NC_5H_6$	B: B Comp.SVol.4/4-61
$BC_{25}Cl_2H_{37}N_2Si$	$1\text{-}[(C_6H_5)_2SiCl\text{-}N(C_4H_9\text{-}t)\text{-}BCl]\text{-}2,2,6,6\text{-}(CH_3)_4\text{-}NC_5H_6$	B: B Comp.SVol.4/4-63
$BC_{25}Cl_3H_{18}N_2$	$C_7H_6N_2\text{-}B(C_6H_4\text{-}4\text{-}Cl)_3$	B: B Comp.SVol.3/3-192
$BC_{25}FH_{34}OSi_3$	$HO\text{-}BF\text{-}C[Si(CH_3)_2\text{-}C_6H_5]_3$	B: B Comp.SVol.4/3b-232
$BC_{25}FH_{35}NSi$	$1\text{-}[9\text{-}(CH_3)_3Si\text{-}C_{13}H_8\text{-}9\text{-}BF]\text{-}NC_5H_6\text{-}2,2,6,6\text{-}(CH_3)_4$	B: B Comp.SVol.4/3b-245

$BC_{25}FH_{37}NSi$ .... 1-[$(CH_3)_3Si$-$C(C_6H_5)_2$-BF]-NC_5H_6-2,2,6,6-$(CH_3)_4$
B: B Comp.SVol.4/3b-245
$BC_{25}F_2H_{33}Si_3$ .... [$(CH_3)_2Si(C_6H_5)$]$_3$C-BF_2 .... B: B Comp.SVol.4/3b-226
$BC_{25}F_4FeH_{21}O$ ... [(C_6H_5-C(=O)-4-C_6H_4-CH_2-C_6H_5)Fe(C_5H_5)][BF_4]
Fe: Org.Comp.B18-142/6, 197, 240
$BC_{25}F_4FeH_{22}OP$ .. [C_5H_5(CO)(P$(C_6H_5)_3$)Fe=CH_2][BF_4] .... Fe: Org.Comp.B16a-18, 24
$BC_{25}F_4FeH_{51}O_6P_2$ [(C_5H_5)Fe(CH_2=CH_2)(P$(OC_3H_7$-i$)_3)_2$][BF_4]..... Fe: Org.Comp.B16b-5
$BC_{25}F_4H_{20}MoO_2P$ [(C_5H_5)Mo$(CO)_2$(P$(C_6H_5)_3$)-FBF_3] Mo:Org.Comp.7-57, 68/9, 105
$BC_{25}F_4H_{20}MoO_5P$ [(C_5H_5)Mo$(CO)_2$(P(O-$C_6H_5)_3$)(FBF_3)]......... Mo:Org.Comp.7-57, 88, 114
$BC_{25}F_4H_{22}MoO_3P$ [(C_5H_5)Mo$(CO)_2$(P$(C_6H_5)_3$)(H_2O)][BF_4] Mo:Org.Comp.7-283, 294
$BC_{25}F_4H_{22}MoO_6P$ [(C_5H_5)Mo$(CO)_2$(P(O-$C_6H_5)_3$)(H_2O)][BF_4] Mo:Org.Comp.7-295
$BC_{25}F_4H_{23}N_2OPRe$
[(C_5H_5)Re(NO)(P$(C_6H_5)_3$)(NC-CH_3)][BF_4] ..... Re: Org.Comp.3-29/30, 33
$BC_{25}F_4H_{24}MoN$... [C_5H_5Mo($NCCH_3$)(C_6H_5-CC-$CH_3)_2$][BF_4] Mo:Org.Comp.6-137, 146
$BC_{25}F_4H_{25}INOPRe$
[(C_5H_5)Re(NO)(P$(C_6H_5)_3$)-I-C_2H_5][BF_4] ...... Re: Org.Comp.3-29, 32, 37
$BC_{25}F_4H_{27}NO_2P$ .. [$(C_6H_5)_3$PC($COOCH_3$)=C(CH_3)-NH-C_2H_5][BF_4] B: B Comp.SVol.4/3b-210
$BC_{25}F_4H_{33}MoO_6P_2$
[C_5H_5Mo(P$(OCH_3)_3)_2C_6H_5$-CC-C_6H_5][BF_4].... Mo:Org.Comp.6-120, 129
$BC_{25}F_4H_{45}MoN_6O$ [Mo(CN-C_4H_9-t$)_5$(NO)][BF_4] Mo:Org.Comp.5-56
$BC_{25}F_6FeH_{20}OP$ .. [C_5H_5(CO)(P$(C_6H_5)_3$)Fe=CF_2][BF_4] .... Fe: Org.Comp.B16a-65, 69/70
$BC_{25}GeH_{33}N_2O_2$ .. Ge$(C_6H_5)_3CH_2CH_2CH_2$B$(OCH_2CH_2NH_2)_2$ Ge:Org.Comp.3-94
$BC_{25}H_{17}N^+$ ...... [$C_{13}H_9$N-$BC_{12}H_8$]$^+$ .... B: B Comp.SVol.3/3-207
$BC_{25}H_{20}N$ .... C_5H_5N-B$(C_6H_5)_2$C≡CC_6H_5 .... B: B Comp.SVol.3/3-192
$BC_{25}H_{20}NO_2$ ..... $(C_6H_5)_2$B[-O-(1,2-C_6H_4)-CH=N(C_6H_4-2-OH)-] B: B Comp.SVol.3/3-219/20
$BC_{25}H_{21}N_2$ ...... $C_7H_6N_2$-B$(C_6H_5)_3$ .... B: B Comp.SVol.3/3-192
$BC_{25}H_{23}N_4O_6$ .... [CH_3-OOC]$_2$B[-N(C_6H_5)NC(C_6H_4-2-OOC-CH_3)NN(C_6H_5)-] .... B: B Comp.SVol.4/3b-57
– .... [CH_3-OOC]$_2$B[-N(C_6H_5)NC(C_6H_4-4-OOC-CH_3)NN(C_6H_5)-] .... B: B Comp.SVol.4/3b-57
– .... [CH_3-OOC]$_2$B[-N(C_6H_4-4-OOC-CH_3)NC(C_6H_5)NN(C_6H_5)-] .... B: B Comp.SVol.4/3b-57
$BC_{25}H_{24}N$ .... 2-C_2H_5-NC_5H_4 · B$(C_6H_5)_3$ .... B: B Comp.SVol.4/3b-28
$BC_{25}H_{24}NO$...... $(C_6H_5)_2$B[-NH($CH_3C_{10}H_6$)-CH_2-CH_2-O-] ..... B: B Comp.SVol.3/3-217
$BC_{25}H_{24}NO_3$ ..... (C_6H_5-O$)_3$B · NC_5H_4-2-C_2H_5 .... B: B Comp.SVol.4/3b-69
$BC_{25}H_{25}NO_2PS$... $HOCH_2$-P(S)(C_6H_5)-CH_2O-B$(C_6H_5)_2$-NC_5H_5
= [HNC_5H_5][OCH_2-P(S)(C_6H_5)-CH_2O-B$(C_6H_5)_2$]
B: B Comp.SVol.3/3-217
B: B Comp.SVol.4/3b-78
B: B Comp.SVol.4/4-153
$BC_{25}H_{25}NO_2PSe$.. $HOCH_2$-P(Se)(C_6H_5)-CH_2O-B$(C_6H_5)_2$-NC_5H_5
= [HNC_5H_5][OCH_2-P(Se)(C_6H_5)-CH_2O-B$(C_6H_5)_2$]
B: B Comp.SVol.4/3b-78
B: B Comp.SVol.4/4-174
$BC_{25}H_{28}N_3$ ...... [-N=CCH_3-CH=C(NH_2)-]NH-B(C_6H_4-2-$CH_3)_3$ B: B Comp.SVol.3/3-192
– .... [-N=CCH_3-CH=C(NH_2)-]NH-B(C_6H_4-4-$CH_3)_3$ B: B Comp.SVol.3/3-192
$BC_{25}H_{28}N_3O_3$ .... [-N=CCH_3-CH=C(NH_2)-]NH-B(C_6H_4-2-$OCH_3)_3$
B: B Comp.SVol.3/3-192

$BC_{25}H_{28}N_3O_3$ .... $[-N{=}CCH_3-CH{=}C(NH_2)-]NH-B(C_6H_4-4-OCH_3)_3$
B: B Comp.SVol.3/3-192

$BC_{25}H_{29}S$ .... $[2,4,6-(CH_3)_3-C_6H_2]_2BSC_6H_4-4-CH_3$ .... B: B Comp.SVol.3/4-111

$BC_{25}H_{30}N$ .... $[2,4,6-(CH_3)_3C_6H_2]_2B-NHCH_2C_6H_5$ .... B: B Comp.SVol.3/3-153

$BC_{25}H_{30}N_3O$ ..... $BC_6H_4N_3(=O)(C_6H_5)(C_6H_4-2-CH_3)(C_3H_7-i)_2$ ... B: B Comp.SVol.4/3a-180/1

$BC_{25}H_{32}N$ .... $CH_3-NH_2-B[C_6H_3-2,5-(CH_3)_2]_3$ .... B: B Comp.SVol.4/3b-27

$BC_{25}H_{32}NO_3$ ..... $CH_3-NH_2-B(C_6H_4-2-O-C_2H_5)_3$ .... B: B Comp.SVol.4/3b-27

$BC_{25}H_{32}N_3O_3$ .... $(H_2N)_2C{=}NH-B(C_6H_4-2-OC_2H_5)_3$ .... B: B Comp.SVol.4/3b-27

$BC_{25}H_{33}MoN_4O_2$. $[CH_2C(C_6H_5)CH_2]Mo(CO)_2[-N_2C_3H(C_2H_5)_2-]_2BH_2$
Mo:Org.Comp.5-213, 216

$BC_{25}H_{33}N_2$ ...... $C_6H_5NH-B[-NC_6H_5-C_2(-(CH_2)_4-)-C(n-C_3H_7)_2-]$
B: B Comp.SVol.3/3-117

$BC_{25}H_{33}N_2O_2S$ ... $CH_3-O-C(O)-B(C_3H_7-i)[-NHC(CH_2-C_6H_5)$
$C(C_6H_5)C(S-C_4H_9-n)NH-]$ .... B: B Comp.SVol.4/3b-56

$BC_{25}H_{35}N_2S$ ..... $[-B(C_3H_7-n)_2-NHC(CH_2C_6H_5)C(C_6H_5)C(S-C_3H_7-n)NH-]$
B: B Comp.SVol.4/3a-178
B: B Comp.SVol.4/4-149

$BC_{25}H_{35}N_2Si$..... $1-[1-(CH_3)_3Si-1,2-NBC_{13}H_8-2]-2,2,6,6-(CH_3)_4-NC_5H_6$
B: B Comp.SVol.4/3a-174

$BC_{25}H_{38}NOSi$ .... $2-[(CH_3)_3Si-N(C_4H_9-t)]-3-(t-C_4H_9)-$
$4,4-(C_6H_5)_2-1,2-OBC_2H$. .... B: B Comp.SVol.4/3b-62

$BC_{25}H_{39}N_2O_8$ .... $2-[2,2,6,6-(CH_3)_4-NC_5H_6-1]-3-(t-C_4H_9)-$
$4,5,6-[CH_3-O-C(O)]_3-6a-CH_3O-1,3,2-ONBC_5$
B: B Comp.SVol.4/3b-65

$BC_{25}H_{40}NO$...... $1-C_6H_5-2-(C_6H_{13}-O)-3,3-(n-C_3H_7)_2-1,2-NBC_7H_8$
B: B Comp.SVol.4/3b-65

$BC_{25}H_{40}NO_2$ ..... $HO-CH_2CH_2CH_2-O-B(C_6H_5)_2-NH_2-C_{10}H_{21}-n$ B: B Comp.SVol.4/3b-69

$BC_{25}H_{41}N_2$ ...... $1-(i-C_4H_9)-2-(C_6H_5-NH)-3,3-(n-C_4H_9)_2-1,2-NBC_7H_8$
B: B Comp.SVol.3/3-117

– .... $1-C_6H_5-2-(t-C_4H_9-NH)-3,3-(n-C_4H_9)_2-1,2-NBC_7H_8$
B: B Comp.SVol.3/3-117

$BC_{25}H_{44}N$ .... $[2,7,7-(CH_3)_3-(3.1.1)-C_7H_8-3]_2BH \cdot NC_5H_9$ ... B: B Comp.SVol.4/3b-26

$BC_{25}H_{45}N_2O$ ..... $2,2,6,6-(CH_3)_4-NC_5H_6-1-O-B(C_4H_9-t)-NH$
$-C_6H_3-2,6-(C_3H_7-i)_2$ .... B: B Comp.SVol.4/3b-61

$BC_{25}H_{45}N_2O_2$ .... $2,2,6,6-(CH_3)_4-NC_5H_6-1-O-B(O-C_4H_9-t)-NH$
$-C_6H_3-2,6-(C_3H_7-i)_2$ .... B: B Comp.SVol.4/3b-61

$BC_{25}H_{55}S$ .... $(CH_3)_2S-B(C_5H_{11}-n)(C_6H_{13}-n)(C_{12}H_{25}-n)$ .... B: B Comp.SVol.3/4-137

$BC_{26}ClF_4FeH_{23}OP$ $[C_5H_5(CO)(P(C_6H_5)_3)Fe{=}C(CH_3)Cl][BF_4]$ ...... Fe: Org.Comp.B16a-34

$BC_{26}ClH_{34}N_4$ .... $[(C_6H_5-NH)(C_6H_5)BC_6H_5N_3(C_4H_9-n)_2]Cl$...... B: B Comp.SVol.4/3b-46

$BC_{26}Cl_3F_6FeH_{29}IrO_3P_2$
$[(C_5H_5)(CO)Fe(CO)(CF_2)IrCl(CO)$
$(P(CH_3)_2C_6H_5)_2][BF_4] \cdot CH_2Cl_2$ .... Fe: Org.Comp.B16b-148/9

$BC_{26}Cl_3H_{24}Si_2$ ... $(C_6H_5)_2SiClCH_2Si(C_6H_5)_2CH_2BCl_2$........... B: B Comp.SVol.3/4-42

$BC_{26}F_4FeH_{19}O_3$ .. $[(C_6H_5)_2C_{11}H_9Fe(CO)_3][BF_4]$ .... Fe: Org.Comp.B15-256/7

$BC_{26}F_4FeH_{21}O_2$ .. $[(C_6H_5-C(=O)O-C(C_6H_5){=}CHC_6H_5)Fe(C_5H_5)][BF_4]$
Fe: Org.Comp.B18-142/6, 199, 241

$BC_{26}F_4FeH_{22}OP$ .. $[C_5H_5(CO)(P(C_6H_5)_3)Fe{=}C{=}CH_2][BF_4]$ .... Fe: Org.Comp.B16a-207, 209/10

$BC_{26}F_4FeH_{22}O_2P$ $[C_5H_5Fe(CO)_2CH_2P(C_6H_5)_3][BF_4]$............ Fe: Org.Comp.B14-137, 143

$BC_{26}F_4FeH_{24}OP$ .. $[(C_5H_5)Fe(CH_2{=}CH_2)(CO)P(C_6H_5)_3][BF_4]$...... Fe: Org.Comp.B16b-80, 88/9
$BC_{26}F_4FeH_{24}OPS$ $[C_5H_5(CO)(P(C_6H_5)_3)Fe{=}C(CH_3)SH][BF_4]$ ..... Fe: Org.Comp.B16a-46, 58
$BC_{26}F_4FeH_{24}O_2P$ $[(C_5H_5)(CO)(P(C_6H_5)_3)Fe{=}CCH_3\text{-}OH][BF_4]$ .... Fe: Org.Comp.B16a-35, 51
– .... $[(C_5H_5)(CO)(P(C_6H_5)_3)Fe{=}CH\text{-}OCH_3][BF_4]$ .... Fe: Org.Comp.B16a-34, 51
$BC_{26}F_4FeH_{24}O_4P$ $[(C_5H_5)Fe(CH_2{=}CH_2)(CO)P(OC_6H_5)_3][BF_4]$ .... Fe: Org.Comp.B16b-81
$BC_{26}F_4FeH_{29}NO_4P$
$[C_5H_5(CO)(P(OCH_3)_3)Fe{=}C(\text{-}NCH_3CHC_6H_5$
$C({=}CHC_6H_5)\text{-})][BF_4]$..................... Fe: Org.Comp.B16a-49, 60
$BC_{26}F_4FeH_{40}OP$ .. $[C_5H_5(CO)(P(C_6H_{11}\text{-}c)_3)Fe{=}C{=}CH_2][BF_4]$ ..... Fe: Org.Comp.B16a-207, 209
$BC_{26}F_4FeH_{42}O_2P$ $[C_5H_5(CO)(P(C_6H_{11}\text{-}c)_3)Fe{=}C(CH_3)OH][BF_4]$... Fe: Org.Comp.B16a-35, 51
$BC_{26}F_4H_{22}MoOP$ $[C_5H_5Mo(CO)(P(C_6H_5)_3)H\text{-}CC\text{-}H][BF_4]\cdot 0.33\ CH_2Cl_2$
Mo:Org.Comp.6-285
$BC_{26}F_4H_{27}INOPRe$
$[(C_5H_5)Re(NO)(P(C_6H_5)_3)\text{-}I\text{-}C_3H_7\text{-}n][BF_4]$..... Re: Org.Comp.3-29, 32, 37
$BC_{26}F_4H_{30}MoOP$ $[C_5H_5Mo(CO)(P(C_2H_5)_3)C_6H_5\text{-}CC\text{-}C_6H_5][BF_4]$ Mo:Org.Comp.6-288, 289
$BC_{26}F_4H_{35}MoO_6P_2$
$[C_5H_5Mo(P(OCH_3)_3)_2C_6H_5\text{-}CC\text{-}CH_2C_6H_5][BF_4]$ Mo:Org.Comp.6-120, 129
$BC_{26}F_4H_{39}MoO_6P_2Si$
$[C_9H_7Mo(P(OCH_3)_3)_2C_6H_5\text{-}CC\text{-}Si(CH_3)_3][BF_4]$ Mo:Org.Comp.6-122, 130
$BC_{26}F_5H_{18}N_4O_7Re_2$
$[(CO)_3Re(NC_5H_4\text{-}2\text{-}(2\text{-}C_5H_4N))F]_2H\cdot HOBF_3$ Re: Org.Comp.1-157/8
$BC_{26}FeH_{26}NO_4$... $2,2,6,6\text{-}(CH_3)_4\text{-}NC_5H_6\text{-}1\text{-}B[\text{-}Fe(CO)_4\text{-}9\text{-}C_{13}H_8\text{-}9\text{-}]$
B: B Comp.SVol.4/3a-237
$BC_{26}H_{22}NO$...... $(C_6H_5)_2B[\text{-}O\text{-}(1,2\text{-}C_6H_4)\text{-}CCH_3{=}NC_6H_5\text{-}]$ ..... B: B Comp.SVol.3/3-219/20
$BC_{26}H_{23}NOP$..... $(C_6H_5)_2B[\text{-}NH{=}CC_6H_5\text{-}PC_6H_5\text{-}CH_2\text{-}O\text{-}]$ ...... B: B Comp.SVol.3/3-218
$BC_{26}H_{23}N_2O$..... $(C_6H_5)_2B[\text{-}O\text{-}(1,2\text{-}C_6H_4)\text{-}CCH_3{=}N(NHC_6H_5)\text{-}]$ B: B Comp.SVol.3/3-219/20
$BC_{26}H_{25}MoN_6O_2$. $(CH_2CHCH_2)(C_3H_4N_2)Mo(CO)_2[\text{-}N_2C_3H_3\text{-}]_2B(C_6H_5)_2$
Mo:Org.Comp.5-224/5
$BC_{26}H_{26}N$....... $C_{10}H_6[\text{-}1\text{-}B(C_6H_5)_2\text{-}N(CH_3)_2\text{-}CHCH_3\text{-}2\text{-}]$..... B: B Comp.SVol.4/3a-256
– .... $C_{10}H_6[\text{-}1\text{-}CHCH_3\text{-}N(CH_3)_2\text{-}B(C_6H_5)_2\text{-}2\text{-}]$..... B: B Comp.SVol.4/3a-256
$BC_{26}H_{29}N_2$...... $(\text{-}N{=}CCH_3\text{-}CH{=}CCH_3\text{-})NH\text{-}B(C_6H_4\text{-}2\text{-}CH_3)_3$... B: B Comp.SVol.3/3-191
– .... $(\text{-}N{=}CCH_3\text{-}CH{=}CCH_3\text{-})NH\text{-}B(C_6H_4\text{-}3\text{-}CH_3)_3$... B: B Comp.SVol.3/3-191
– .... $(\text{-}N{=}CCH_3\text{-}CH{=}CCH_3\text{-})NH\text{-}B(C_6H_4\text{-}4\text{-}CH_3)_3$... B: B Comp.SVol.3/3-191
$BC_{26}H_{29}N_2O_3$.... $(\text{-}N{=}CCH_3\text{-}CH{=}CCH_3\text{-})NH\text{-}B(C_6H_4\text{-}2\text{-}OCH_3)_3$ B: B Comp.SVol.3/3-191
– .... $(\text{-}N{=}CCH_3\text{-}CH{=}CCH_3\text{-})NH\text{-}B(C_6H_4\text{-}4\text{-}OCH_3)_3$ B: B Comp.SVol.3/3-191
$BC_{26}H_{30}NO_2$..... $2\text{-}[2,2,6,6\text{-}(CH_3)_4\text{-}NC_5H_6\text{-}1]\text{-}4\text{-}CH_3O\text{-}$
$4\text{-}(CH{\equiv}C)\text{-}1,2\text{-}OBC_{14}H_8$.................. B: B Comp.SVol.4/3b-63
$BC_{26}H_{30}N_3O_3$.... $[\text{-}N{=}CCH_3\text{-}CH{=}C(NH_2)\text{-}]NCH_3\text{-}B(C_6H_4\text{-}2\text{-}OCH_3)_3$
B: B Comp.SVol.3/3-192
– .... $[\text{-}N{=}CCH_3\text{-}CH{=}C(NH_2)\text{-}]NCH_3\text{-}B(C_6H_4\text{-}4\text{-}OCH_3)_3$
B: B Comp.SVol.3/3-192
$BC_{26}H_{31}MoN_6O_2$. $[CH_2C(C_6H_5)CH_2]Mo(CO)_2[\text{-}N_2C_3H(CH_3)_2\text{-}]_3BH$
Mo:Org.Comp.5-226, 229
$BC_{26}H_{31}N_4$...... $(NH_2)N_3C_2H_2\text{-}B[C_6H_3\text{-}2,5\text{-}(CH_3)_2]_3$.......... B: B Comp.SVol.3/3-191
$BC_{26}H_{31}N_4O_3$.... $(NH_2)N_3C_2H_2\text{-}B(C_6H_4\text{-}4\text{-}OC_2H_5)_3$........... B: B Comp.SVol.3/3-191
$BC_{26}H_{33}N_2O_2Sn$.. $(C_2H_5)_2B\text{-}C(C_2H_5){=}C(CH_3)\text{-}Sn(CH_3)[\text{-}O$
$\text{-}C_6H_4\text{-}N{=}C(CH_3)\text{-}CH{=}C_7H_4NO\text{-}]$........... Sn: Org.Comp.19-147/9
$BC_{26}H_{33}N_4$...... $BC_6H_4N_3(C_6H_5)(NH\text{-}C_6H_5)(C_4H_9\text{-}n)_2$......... B: B Comp.SVol.4/3a-180
$BC_{26}H_{35}LiN$..... $Li[2,2,6,6\text{-}(CH_3)_4\text{-}NC_5H_6\text{-}1\text{-}B(C_4H_9\text{-}t){=}C_{13}H_8]$ B: B Comp.SVol.4/3b-47

$BC_{26}H_{35}NO_2PS$... $HOCH_2P(S)(C_6H_5)-CH_2O-B(C_6H_5)_2-N(C_2H_5)_3$ = $[HN(C_2H_5)_3][OCH_2-P(S)(C_6H_5)-CH_2O-B(C_6H_5)_2]$
B: B Comp.SVol.4/3b-78
B: B Comp.SVol.4/4-153

$BC_{26}H_{35}NO_2PSe$.. $HOCH_2P(Se)(C_6H_5)-CH_2O$ $-B(C_6H_5)_2-N(C_2H_5)_3$ = $[HN(C_2H_5)_3]$ $[OCH_2-P(Se)(C_6H_5)-CH_2O-B(C_6H_5)_2]$ .... B: B Comp.SVol.4/3b-78
B: B Comp.SVol.4/4-174

$BC_{26}H_{35}NO_3P$.... $(C_2H_5)_3N-B(C_6H_5)_2-OCH_2-P(=O)(C_6H_5)$ $-CH_2OH$ = $[(C_2H_5)_3NH][B(C_6H_5)_2-OCH_2-$ $P(O)(C_6H_5)-CH_2O]$ .... B: B Comp.SVol.4/3b-78

$BC_{26}H_{36}N$ .... $1-[C_{13}H_9-9-B(C_4H_9-t)]-2,2,6,6-(CH_3)_4-NC_5H_6$ B: B Comp.SVol.4/3a-234

$BC_{26}H_{37}N_2S$ ..... $[-B(C_3H_7-n)_2-NHC(CH_2-C_6H_5)C(C_6H_5)C(S-C_4H_9-n)NH-]$
B: B Comp.SVol.4/4-149
B: B Comp.SVol.4/3a-178

$BC_{26}H_{38}NO_4Sn$... $(C_4H_9)_3SnN(C_6H_5)COOB(OCH_2C_6H_4O)$ .... Sn: Org.Comp.18-181, 188

$BC_{26}H_{39}Se$ ...... $[2,7,7-(CH_3)_3-[3.3.1]-C_7H_8-3-]_2B-Se-C_6H_5$ ... B: B Comp.SVol.4/4-170

$BC_{26}H_{45}N_2$ ...... $N(-CH_2-CH_2-)_3N-BH[C_{10}H_{16}]_2$ .... B: B Comp.SVol.3/3-190

$BC_{26.33}Cl_{0.67}F_4H_{22.67}MoOP$
$[C_5H_5Mo(CO)(P(C_6H_5)_3)H-CC-H][BF_4]$ · 0.33 CH_2Cl_2
Mo:Org.Comp.6-285

$BC_{27}Cl_3H_{40}N_2Si$ .. $1-[2,6-(i-C_3H_7)_2C_6H_3-N(SiCl_2-C_6H_5)-BCl]-$ $2,2,6,6-(CH_3)_4-NC_5H_6$ .... B: B Comp.SVol.4/4-64

$BC_{27}Cl_6H_{23}NO_3P$ NC_5H_5 · $B(C_6H_5)_2-O-CH(CCl_3)-P(O)(C_6H_5)-CHOH-CCl_3$
B: B Comp.SVol.4/3b-78

– .... $[NC_5H_6][-B(C_6H_5)_2-O-CH(CCl_3)-P(O)(C_6H_5)$ $-CH(CCl_3)-O-]$ .... B: B Comp.SVol.4/3b-78

$BC_{27}CrFeH_{28}NO_3$ $[(CH_3)_3NC_6H_5][C_5H_5FeC_4H_4BC_6H_5Cr(CO)_3]$ ... Fe: Org.Comp.B17-200, 227

$BC_{27}FH_{39}N_2O_2Sb$ $1-[1,3,2-O_2SbC_6H_4-2-N(C_6H_3-2,6-(C_3H_7-i)_2)$ $-BF]-NC_5H_6-2,2,6,6-(CH_3)_4$. .... B: B Comp.SVol.4/3b-246

$BC_{27}FH_{44}N_2Si$.... $2,6-(i-C_3H_7)_2-C_6H_3-NH-BF-N[Si(CH_3)_3]$ $-C_6H_3-2,6-(C_3H_7-i)_2$ .... B: B Comp.SVol.4/3b-246

$BC_{27}FH_{48}NP$ ..... $2,2,6,6-(CH_3)_4-NC_5H_6-1-[BF-PH-C_6H_2-2,4,6-(C_4H_9-t)_3]$
B: B Comp.SVol.4/3b-243

$BC_{27}F_4FeH_{24}N_2O_2S$
$[(C_5H_5)Fe(CN-C_6H_4-OCH_3-3)_2S-C_6H_5][BF_4]$ .. Fe: Org.Comp.B15-328, 329

$BC_{27}F_4FeH_{24}OP$ .. $[(C_5H_5)Fe(CH_2=C=CH_2)(CO)P(C_6H_5)_3][BF_4]$.... Fe: Org.Comp.B16b-96

$BC_{27}F_4FeH_{24}O_2P$ $[(C_5H_5)Fe(CO)_2-CH(CH_3)P(C_6H_5)_3][BF_4]$...... Fe: Org.Comp.B14-133, 138

– .... $[(C_5H_5)Fe(CO)_2-CH_2CH_2P(C_6H_5)_3][BF_4]$ ...... Fe: Org.Comp.B14-152, 157

$BC_{27}F_4FeH_{24}O_2PS$
$[(C_5H_5)Fe(CO)_2-CH(SCH_3)P(C_6H_5)_3][BF_4]$..... Fe: Org.Comp.B14-133, 140, 144

$BC_{27}F_4FeH_{24}O_4P$ $[(C_5H_5)Fe(CH_2=C=CH_2)(CO)P(OC_6H_5)_3][BF_4]$ .. Fe: Org.Comp.B16b-96/7

$BC_{27}F_4FeH_{26}OP$ .. $[(C_5H_5)(CO)(P(C_6H_5)_3)Fe=C(CH_3)_2][BF_4]$...... Fe: Org.Comp.B16a-22, 28

– .... $[(C_5H_5)Fe(CH_2=CHCH_3)(CO)P(C_6H_5)_3][BF_4]$ ... Fe: Org.Comp.B16b-82

$BC_{27}F_4FeH_{26}OPS$ $[(C_5H_5)(CO)(P(C_6H_5)_3)Fe=C(CH_3)-SCH_3][BF_4]$ Fe: Org.Comp.B16a-46

$BC_{27}F_4FeH_{26}O_2P$ $[(C_5H_5)(CO)(P(C_6H_5)_3)Fe=C(CH_3)-OCH_3][BF_4]$ Fe: Org.Comp.B16a-37, 52

– .... $[(C_5H_5)(CO)(P(C_6H_5)_3)Fe=C(CH_2D)-OCH_3][BF_4]$
Fe: Org.Comp.B16a-40

$BC_{27}F_4FeH_{26}O_4P$ $[(C_5H_5)Fe(CH_2=CHCH_3)(CO)P(OC_6H_5)_3][BF_4]$ .. Fe: Org.Comp.B16b-83

$BC_{27}F_4FeH_{27}NOP$ $[C_5H_5(CO)(P(C_6H_5)_3)Fe{=}C(CH_3)NHCH_3][BF_4]$ .. Fe: Org.Comp.B16a-46, 58

$BC_{27}F_4FeH_{39}O_6$ .. $[(C_5H_5)Fe(CO)_2(CH_2{=}(3\text{-}C_4H_2O({=}O\text{-}2)(CO_2CH_3\text{-}4)n\text{-}C_{13}H_{27}\text{-}5))][BF_4]$ .... Fe: Org.Comp.B17-71

$BC_{27}F_4FeH_{44}O_2P$ $[C_5H_5(CO)(P(C_6H_{11}\text{-}c)_3)Fe{=}C(CH_3)OCH_3][BF_4]$ Fe: Org.Comp.B16a-36, 52

$BC_{27}F_4H_{28}NOPRe$ $[(C_5H_5)Re(NO)(P(C_6H_5)_3){=}CH\text{-}C_3H_7\text{-}i][BF_4]$ ... Re: Org.Comp.3-109/11, 115

$BC_{27}F_4H_{31}INOPReSi$
$[(C_5H_5)Re(NO)(P(C_6H_5)_3)\text{-}I\text{-}CH_2\text{-}Si(CH_3)_3][BF_4] \cdot 0.5\ CH_2Cl_2$ .... Re: Org.Comp.3-29, 32, 37

$BC_{27}F_4H_{37}MoO_6P_2$
$[C_5H_5Mo(P(OCH_3)_3)_2(4\text{-}CH_3C_6H_4\text{-}CC\text{-}C_6H_4CH_3\text{-}4)][BF_4]$
Mo: Org.Comp.6-120

$BC_{27}F_{10}FeH_{18}N_2S$ $[(C_5H_5)Fe(CN\text{-}C_6H_4CF_3\text{-}3)_2S\text{-}C_6H_5][BF_4]$ ..... Fe: Org.Comp.B15-328, 329

$BC_{27}GeH_{34}NSi$ ... $1\text{-}(C_6H_5)_3Ge\text{-}4,5\text{-}(C_2H_5)_2\text{-}2,2,3\text{-}(CH_3)_3\text{-}1,2,5\text{-}NSiBC_2$
B: B Comp.SVol.4/3a-250/1

$BC_{27}H_{21}N_4$ ...... $(NC)_2C{=}C(CN)_2\text{-}B(C_6H_4\text{-}2\text{-}CH_3)_3$ .... B: B Comp.SVol.3/3-192

$BC_{27}H_{22}KN_6$ ..... $K[(3\text{-}C_6H_5\text{-}1,2\text{-}N_2C_3H_2\text{-}1)_3BH]$ .... B: B Comp.SVol.4/3b-40

$BC_{27}H_{22}N_6Tl$ ..... $Tl[(3\text{-}C_6H_5\text{-}1,2\text{-}N_2C_3H_2\text{-}1)_3BH]$ .... B: B Comp.SVol.4/3b-40

$BC_{27}H_{23}MoN_4O_2$. $(c\text{-}C_7H_7)Mo(CO)_2[\text{-}N_2C_3H_3\text{-}]_2B(C_6H_5)_2$ ...... Mo: Org.Comp.5-213, 217

$BC_{27}H_{25}Th$ ...... $(C_9H_7)_3Th[BH_4]$ .... Th: SVol.D4-177

$BC_{27}H_{26}N_3$ ...... $n\text{-}C_3H_7\text{-}B[\text{-}NH\text{-}C({=}C(C_6H_5)\text{-}CN)\text{-}C(C_6H_5){=}C(CH_2\text{-}C_6H_5)\text{-}NH\text{-}]$ .... B: B Comp.SVol.4/3a-175/6

$BC_{27}H_{27}N_2$ ...... $(CH_3)_2C{=}N\text{-}NH(C_6H_5)\text{-}B(C_6H_5)_3$ .... B: B Comp.SVol.3/3-191

$BC_{27}H_{29}N_3O_3Re$ .. $[(CO)_3Re(NH_3)_3][B(C_6H_5)_4]$ .... Re: Org.Comp.1-307

$BC_{27}H_{33}S_3$ ...... $B(SC_6H_2\text{-}2,4,6\text{-}(CH_3)_3)_3$ .... B: B Comp.SVol.3/4-99

$BC_{27}H_{34}N$ .... $[2,4,6\text{-}(CH_3)_3\text{-}C_6H_2]_2B\text{-}NH\text{-}C_6H_2\text{-}2,4,6\text{-}(CH_3)_3$ B: B Comp.SVol.4/3a-232

$BC_{27}H_{34}NO_3$ ..... $(C_6H_5\text{-}O)_3B \cdot NC_5H_7\text{-}2,2,6,6\text{-}(CH_3)_4$ .... B: B Comp.SVol.4/3b-69

$BC_{27}H_{36}N$ .... $H_3N\text{-}B[C_6H_2\text{-}2,4,6\text{-}(CH_3)_3]_3$ .... B: B Comp.SVol.4/3b-27

$BC_{27}H_{36}NO_3$ ..... $H_3N\text{-}B[C_6H_4(OC_3H_7\text{-}i)\text{-}2]_3$ .... B: B Comp.SVol.3/3-191

$BC_{27}H_{37}N_2$ ...... $C_6H_5NH\text{-}B[\text{-}NC_6H_5\text{-}C_2(\text{-}(CH_2)_4\text{-})\text{-}C(n\text{-}C_4H_9)_2\text{-}]$
B: B Comp.SVol.3/3-117

$BC_{27}H_{38}N$ .... $1\text{-}[9\text{-}CH_3\text{-}C_{13}H_8\text{-}9\text{-}B(C_4H_9\text{-}t)]\text{-}2,2,6,6\text{-}(CH_3)_4\text{-}NC_5H_6$
B: B Comp.SVol.4/3a-234

$BC_{27}H_{41}N_2SiTi$ ... $[\text{-}N(C_4H_9\text{-}t)\text{-}Ti(C_5H_5)_2\text{-}(1,2\text{-}C_6H_4)\text{-}]B\text{-}N(C_4H_9\text{-}t)\text{-}Si(CH_3)_3$ .... B: B Comp.SVol.4/3a-175

$BC_{27}H_{41}N_2SiZr$ ... $[\text{-}N(C_4H_9\text{-}t)\text{-}Zr(C_5H_5)_2\text{-}(1,2\text{-}C_6H_4)\text{-}]B\text{-}N(C_4H_9\text{-}t)\text{-}Si(CH_3)_3$ .... B: B Comp.SVol.4/3a-175

$BC_{27}H_{44}NO$ ...... $1\text{-}C_6H_5\text{-}2\text{-}(C_6H_{13}\text{-}O)\text{-}3,3\text{-}(n\text{-}C_4H_9)_2\text{-}1,2\text{-}NBC_7H_8$
B: B Comp.SVol.4/3b-65

$BC_{27}H_{47}NO$ ...... $(C_{10}H_{17})_2BN(\text{-}C(CH_3)_2\text{-}CH_2\text{-}O\text{-}CC_2H_5\text{-})$ ..... B: B Comp.SVol.3/3-156

$BC_{27}H_{47}NO^+$ .... $[(\text{-}C(CH_3)_2\text{-}(CH_2)_3\text{-}C(CH_3)_2\text{-})N{=}B\text{-}O\text{-}C_6H_2\text{-}2,4,6\text{-}(t\text{-}C_4H_9)_3]^+$ .... B: B Comp.SVol.3/3-203/5

$BC_{27}H_{47}N_2$ ...... $1\text{-}[2,4,6\text{-}(t\text{-}C_4H_9)_3\text{-}C_6H_2\text{-}N{=}B]\text{-}2,2,6,6\text{-}(CH_3)_4\text{-}NC_5H_6$
B: B Comp.SVol.3/3-100
B: B Comp.SVol.4/3a-160/5

$BC_{27}H_{53}N_5Sb$ .... $1\text{-}[((CH_3)_2N)_2Sb\text{-}N(C_6H_3\text{-}2,6\text{-}(C_3H_7\text{-}i)_2)\text{-}B(N(CH_3)_2)]\text{-}2,2,6,6\text{-}(CH_3)_4\text{-}NC_5H_6$ .... B: B Comp.SVol.4/3a-155

$BC_{27}H_{54}N_3$ ...... $B[N{=}C(t\text{-}C_4H_9)_2]_3$ .... B: B Comp.SVol.3/3-94

$BC_{27.5}ClF_4H_{32}INOPReSi$
$[(C_5H_5)Re(NO)(P(C_6H_5)_3)\text{-}I\text{-}CH_2\text{-}Si(CH_3)_3][BF_4] \cdot 0.5\ CH_2Cl_2$ .... Re: Org.Comp.3-29, 32, 37

$BC_{28}ClFeH_{23}$ .... $(Cl\text{-}C_5H_4)Fe[C_5H_4\text{-}B(C_6H_5)_3]$ .... Fe: Org.Comp.A9-282
$BC_{28}ClFeH_{23}^-$ .... $[ClC_5H_4FeC_5H_4B(C_6H_5)_3]^-$ .... Fe: Org.Comp.A9-277, 282
$BC_{28}ClH_{16}MnN_{12}O_8$
$Mn[(6\text{-}NO_2\text{-}1,2\text{-}N_2C_7H_4\text{-}1)_4B]Cl$ .... Mn: MVol.D8-16/8
$BC_{28}ClH_{20}MnN_8$ .. $Mn[(1,2\text{-}N_2C_7H_5\text{-}1)_4B]Cl$ .... Mn: MVol.D8-16/8
$BC_{28}F_4FeH_{24}O_2P$ $[C_5H_5Fe(CO)_2C(=CH_2)CH_2P(C_6H_5)_3][BF_4]$ ..... Fe: Org.Comp.B14-153
$BC_{28}F_4FeH_{25}O_2P$ $[(C_5H_5)Fe(CO)_2(CH_2=CHCH_2P(C_6H_5)_3)][BF_4]$ .. Fe: Org.Comp.B17-27
$BC_{28}F_4FeH_{26}OP$ .. $[(C_5H_5)(CO)(P(C_6H_5)_3)Fe=C=C(CH_3)_2][BF_4]$.... Fe: Org.Comp.B16a-208, 210/1
– .... $[(C_5H_5)Fe(CH_3\text{-}CC\text{-}CH_3)(CO)P(C_6H_5)_3][BF_4]$ .. Fe: Org.Comp.B16b-98
$BC_{28}F_4FeH_{26}O_2P$ $[C_5H_5Fe(CO)_2CH_2CH(CH_3)P(C_6H_5)_3][BF_4]$..... Fe: Org.Comp.B14-153
$BC_{28}F_4FeH_{26}O_4P$ $[(C_5H_5)Fe(CH_3\text{-}CC\text{-}CH_3)(CO)P(OC_6H_5)_3][BF_4]$ Fe: Org.Comp.B16b-99
$BC_{28}F_4FeH_{28}OP$ .. $[(C_5H_5)Fe(CH_2=CHC_2H_5)(CO)P(C_6H_5)_3][BF_4] \cdot CH_2Cl_2$
Fe: Org.Comp.B16b-84
$BC_{28}F_4FeH_{28}O_2P$ $[(C_5H_5)(CO)(P(C_6H_5)_3)Fe=CCH_3\text{-}O\text{-}C_2H_5][BF_4]$ Fe: Org.Comp.B16a-37, 53
– .... $[(C_5H_5)(CO)(P(C_6H_5)_3)Fe=C(C_2H_5)\text{-}OCH_3][BF_4]$ Fe: Org.Comp.B16a-40, 54/5
– .... $[(C_5H_5)(CO)(P(C_6H_5)_3)Fe=C(C_3H_7\text{-}i)OH][BF_4]$.. Fe: Org.Comp.B16a-40
$BC_{28}F_4FeH_{28}O_4P$ $[(C_5H_5)Fe(CH_2=CHC_2H_5)(CO)P(O\text{-}C_6H_5)_3][BF_4]$ Fe: Org.Comp.B16b-84
– .... $[(C_5H_5)Fe(CH_3CH=CHCH_3)(CO)P(O\text{-}C_6H_5)_3][BF_4]$
Fe: Org.Comp.B16b-84/5
$BC_{28}F_4FeH_{29}NOP$ $[C_5H_5(CO)(P(C_6H_5)_3)Fe=C(CH_3)N(CH_3)_2][BF_4]$ Fe: Org.Comp.B16a-47, 59
$BC_{28}F_4FeH_{43}S_2$ .. $[(6\text{-}(1,3\text{-}S_2C_4H_7\text{-}2)\text{-}C_6(CH_3)_6)Fe(C_6(CH_3)_6)][BF_4]$
Fe: Org.Comp.B19-143, 168
$BC_{28}F_4FeH_{46}O_2P$ $[C_5H_5(CO)(P(C_6H_{11}\text{-}c)_3)Fe=C(CH_3)OC_2H_5][BF_4]$
Fe: Org.Comp.B16a-37
$BC_{28}F_4H_{26}MoOP$ $[(C_5H_5)Mo(CO)(P(CH_3)_2C_6H_5)(C_6H_5\text{-}CC\text{-}C_6H_5)][BF_4]$
Mo: Org.Comp.6-288, 290
– .... $[(C_5H_5)Mo(CO)(P(C_6H_5)_3)(H_3C\text{-}CC\text{-}CH_3)][BF_4]$ Mo: Org.Comp.6-287, 289/90
$BC_{28}F_4H_{26}MoO_3P$ $[(C_5H_5)Mo(CO)_2(P(C_6H_5)_3)(O=C(CH_3)_2)][BF_4]$.. Mo: Org.Comp.7-283, 295
$BC_{28}F_4H_{28}Os$ .... $[(2\text{-}CH_3\text{-}C_6H_4)_4Os][BF_4]$ .... Os: Org.Comp.A1-36
$BC_{28}F_4H_{33}MoN_4O_3$
$[C_5H_5Mo(CO)(2\text{-}NNC_6H_3(t\text{-}C_4H_9\text{-}4)OCH_2$
$CH_2OC_6H_3(t\text{-}C_4H_9\text{-}4)NN\text{-}2)][BF_4]$.......... Mo: Org.Comp.6-235
$BC_{28}F_4H_{34}MoOP$ $[C_5H_5Mo(CO)(P(C_2H_5)_3)(4\text{-}CH_3C_6H_4\text{-}CC$
$\text{-}C_6H_4CH_3\text{-}4)][BF_4]$ .... Mo: Org.Comp.6-288
$BC_{28}F_4H_{44}MoOP$ $[C_5H_5Mo(CO)(P(C_6H_{11}\text{-}c)_3)H_3C\text{-}CC\text{-}CH_3][BF_4]$ Mo: Org.Comp.6-286
$BC_{28}FeH_{23}$ ...... $[(C_6H_5)_2B\text{-}C_5H_4]Fe(C_5H_4\text{-}C_6H_5)$ .... Fe: Org.Comp.A9-277, 281/2
$BC_{28}H_{22}NO_2$ ..... $C_6H_5B[\text{-}O\text{-}(1,2\text{-}C_{10}H_6)\text{-}CH_2\text{-}]_2NH$........... B: B Comp.SVol.3/3-220/1
$BC_{28}H_{26}N_2O_4Re$ .. $[(CO)_4Re(NH_3)_2][B(C_6H_5)_4]$ .... Re: Org.Comp.1-479
$BC_{28}H_{26}N_3$ ...... $(\text{-}N=CCH_3\text{-}CH=CNH_2\text{-})NC_6H_5\text{-}B(C_6H_5)_3$ ...... B: B Comp.SVol.3/3-192
$BC_{28}H_{27}N_2$ ...... $C_7H_6N_2\text{-}B(C_6H_4\text{-}2\text{-}CH_3)_3$.................. B: B Comp.SVol.3/3-192
– .... $C_7H_6N_2\text{-}B(C_6H_4\text{-}4\text{-}CH_3)_3$.................. B: B Comp.SVol.3/3-192
$BC_{28}H_{27}N_2O_3$ .... $C_7H_6N_2\text{-}B(C_6H_4\text{-}2\text{-}OCH_3)_3$ .... B: B Comp.SVol.3/3-192
– .... $C_7H_6N_2\text{-}B(C_6H_4\text{-}4\text{-}OCH_3)_3$ .... B: B Comp.SVol.3/3-192
$BC_{28}H_{31}N_2$ ...... $1\text{-}(1\text{-}C_6H_5\text{-}1,2\text{-}NBC_{13}H_8\text{-}2)\text{-}2,2,6,6\text{-}(CH_3)_4\text{-}NC_5H_6$
B: B Comp.SVol.4/3a-174
$BC_{28}H_{37}N_2$ ...... $1\text{-}[3\text{-}(C_2H_5)_2N\text{-}1\text{-}BC_{15}H_9\text{-}1]\text{-}2,2,6,6\text{-}(CH_3)_4\text{-}NC_5H_6$
B: B Comp.SVol.4/3a-236

$BC_{28.5}ClF_4FeH_{29}O_2P$
$[C_5H_5(CO)(P(C_6H_5)_3)Fe{=}C(CH_3)OC_2H_5][BF_4]$ · 0.5 CH_2Cl_2 .... Fe: Org.Comp.B16a-38

$BC_{29}Cl_2F_4FeH_{30}OP$
$[(C_5H_5)Fe(CH_2{=}CHC_2H_5)(CO)P(C_6H_5)_3][BF_4]$ · CH_2Cl_2 Fe: Org.Comp.B16b-84

$BC_{29}Cl_3H_{25}N_3O$ .. $(NH_2)C_{11}H_{11}N_2O\text{-}B(C_6H_4\text{-}4\text{-}Cl)_3$ .... B: B Comp.SVol.3/3-192

$BC_{29}F_4FeH_{23}O_3$ .. $[(6\text{-}(C_6H_5)_3C\text{-}C_7H_8)Fe(CO)_3][BF_4]$ .... Fe: Org.Comp.B15-207, 223

$BC_{29}F_4FeH_{25}O_2$ .. $[(C_5H_5)Fe(CO)_2(CH_2{=}CHCH_2C(C_6H_5)_3)][BF_4]$ .. Fe: Org.Comp.B17-32

$BC_{29}F_4FeH_{26}NO_2$ $[(C_5H_5)Fe(C(OC_2H_5)N(C_6H_5)C(CH(C_6H_5)_2))CO][BF_4]$ Fe: Org.Comp.B16b-120

$BC_{29}F_4FeH_{26}N_3O_3$ $[(C_5H_5)Fe(CN\text{-}C_6H_4\text{-}4\text{-}OCH_3)_3][BF_4]$ .... Fe: Org.Comp.B15-346

$BC_{29}F_4FeH_{26}O_2P$ $[C_5H_5Fe(CO)_2C({=}CHCH_3)CH_2P(C_6H_5)_3][BF_4]$ .. Fe: Org.Comp.B14-153

$BC_{29}F_4FeH_{26}O_3P$ $[(C_5H_5)Fe(CH_3\text{-}CC\text{-}CO_2CH_3)(CO)P(C_6H_5)_3][BF_4]$ Fe: Org.Comp.B16b-99

$BC_{29}F_4FeH_{26}O_6P$ $[(C_5H_5)Fe(CH_3\text{-}CC\text{-}CO_2CH_3)(CO)P(OC_6H_5)_3][BF_4]$ Fe: Org.Comp.B16b-100

$BC_{29}F_4FeH_{28}O_2P$ $[C_5H_5(CO)(P(C_6H_5)_3)Fe{=}C(CH_3)OCH_2CH{=}CH_2][BF_4]$ Fe: Org.Comp.B16a-38, 53

$BC_{29}F_4FeH_{28}O_5P$ $[(C_5H_5)Fe(CH_3\text{-}CC\text{-}CH_2OCH_3)(CO)P(OC_6H_5)_3][BF_4]$ Fe: Org.Comp.B16b-99

$BC_{29}F_4FeH_{30}O_2P$ $[C_5H_5(CO)(P(C_6H_5)_3)Fe{=}C(CH_3)OC_3H_7\text{-}i][BF_4]$ Fe: Org.Comp.B16a-37

$BC_{29}F_4H_{25}MoNO_4P$
$[C_5H_5Mo(CO)(NO)(P(C_6H_5)_3CH_2C_4H_3O({=}O))][BF_4]$ Mo:Org.Comp.6-306

$BC_{29}F_4H_{25}MoNO_7P$
$[C_5H_5Mo(CO)(NO)(P(C_6H_5O)_3CH_2C_4H_3O({=}O))][BF_4]$ Mo:Org.Comp.6-306

$BC_{29}F_4H_{25}MoN_3OP$
$[C_5H_5Mo(NO)(N{=}NC_6H_5)P(C_6H_5)_3][BF_4]$ Mo:Org.Comp.6-61

$BC_{29}F_4H_{26}O_4P_2Re$ $[(CO)_3Re(P(C_6H_5)_2CH_2CH_2P(C_6H_5)_2)(H_2O)][BF_4]$ Re: Org.Comp.1-302/3

$BC_{29}F_4H_{28}MoOP$ $[(C_5H_5)Mo(CO)(P(C_6H_5)_3)(H_3C\text{-}CC\text{-}C_2H_5)][BF_4]$ Mo:Org.Comp.6-287

– .... $[(C_5H_5)Mo(CO)(P(C_6H_5)_3)(H\text{-}CC\text{-}C_3H_7\text{-}i)][BF_4]$ Mo:Org.Comp.6-285

$BC_{29}F_4H_{28}MoP$ ... $[(C_5H_5)Mo(H_3C\text{-}CC\text{-}CH_3)P(C_6H_5)_2C_6H_4CH{=}CH_2\text{-}2][BF_4]$ Mo:Org.Comp.6-138, 146

– .... $[(C_5H_5)Mo(H_3C\text{-}CC\text{-}CH_3)P(C_6H_5)_2C_6H_4CD{=}CH_2\text{-}2][BF_4]$ Mo:Org.Comp.6-138, 146

$BC_{29}F_4H_{33}MoN_4O_4$
$[(C_5H_5)Mo(CO)_2(N_2C_6H_3(C_4H_9\text{-}t\text{-}4)OCH_2CH_2OC_6H_3(C_4H_9\text{-}t\text{-}4)N_2)][BF_4]$ Mo:Org.Comp.7-269

$BC_{29}F_4H_{35}MoO_6P_2$
$[C_9H_7Mo(P(OCH_3)_3)_2C_6H_5\text{-}CC\text{-}C_6H_5][BF_4]$ Mo:Org.Comp.6-122

$BC_{29}F_4H_{46}MoO_3P$ $[(C_5H_5)Mo(CO)_2(P(C_6H_{11}\text{-}c)_3){=}C(CH_3)O\text{-}C_2H_5]$ $[BF_4]$ · 0.5 CH_2Cl_2 .... Mo:Org.Comp.8-107

$BC_{29}F_5H_{24}MoN_3OP$
$[(C_5H_5)Mo(NO)(N{=}N\text{-}C_6H_4\text{-}F\text{-}3)P(C_6H_5)_3][BF_4]$ Mo:Org.Comp.6-61

– .... $[(C_5H_5)Mo(NO)(N{=}N\text{-}C_6H_4\text{-}F\text{-}4)P(C_6H_5)_3][BF_4]$ Mo:Org.Comp.6-62

$BC_{29}FeH_{25}$ ...... $[(C_5H_5)Fe(B(C_6H_5)_4)]$ ...... Fe: Org.Comp.B18-146/8, 266, 282
$BC_{29}FeH_{25}^-$ ...... $[(C_5H_5)Fe(B(C_6H_5)_4)]^-$ ...... Fe: Org.Comp.B18-148, 266
$BC_{29}FeH_{25}K$ ...... $K[(C_5H_5)Fe(B(C_6H_5)_4)]$ ...... Fe: Org.Comp.B18-148, 266
$BC_{29}H_{23}NO_5Re$ ... $[(CO)_5ReNH_3][B(C_6H_5)_4]$ ...... Re: Org.Comp.2-150/1
$BC_{29}H_{24}NO_3Si$ ... $[(C_6H_5)_3SiO]B(NC_5H_5)[-O-(1,2-C_6H_4)-O-]$ .... B: B Comp.SVol.3/3-221
$BC_{29}H_{28}N_3O$ ...... $(NH_2)C_{11}H_{11}N_2O-B(C_6H_5)_3$ ...... B: B Comp.SVol.3/3-192
$BC_{29}H_{32}NO$ ...... 2'-[2,2,6,6-$(CH_3)_4$-NC_5H_6-1]-4'-C_6H_5-1',2'-$OBC_{14}H_9$ B: B Comp.SVol.4/3b-63
$BC_{29}H_{35}N_2O_3$ .... (-N=CCH_3-CH=CCH_3-)NH-B(C_6H_4-4-OC_2H_5)$_3$ B: B Comp.SVol.3/3-191
$BC_{29}H_{36}N$ ...... 1-[$(C_6H_5)_2$CH-B($CH_2C_6H_5$)]-2,2,6,6-$(CH_3)_4$-NC_5H_6 B: B Comp.SVol.4/3a-233
– ...... CH_3-CN-B[C_6H_2-2,4,6-$(CH_3)_3$]$_3$ ...... B: B Comp.SVol.4/3b-28
$BC_{29}H_{38}NSi$ ...... 2,4,6-$(CH_3)_3$-C_6H_2-B(9-$C_{13}H_9$)-N(C_4H_9-t)-Si$(CH_3)_3$ B: B Comp.SVol.4/3a-235
$BC_{29}H_{38}N_5$ ...... (-N=CCH_3-CH=CCH_3-)NH-B(C_6H_4-4-N$(CH_3)_2$)$_3$ B: B Comp.SVol.3/3-191
$BC_{29}H_{39}N_2$ ...... 1-[3-$(C_2H_5)_2$N-4-CH_3-1-$BC_{15}H_8$-1]-2,2,6,6-$(CH_3)_4$-NC_5H_6 ...... B: B Comp.SVol.4/3a-236
$BC_{29}H_{40}NO_3$ ...... $(CH_3)_2$NH-B[C_6H_4-2-(O-C_3H_7-n)]$_3$ ...... B: B Comp.SVol.4/3b-27
$BC_{29.5}ClF_4H_{47}MoO_3P$ $[(C_5H_5)Mo(CO)_2(P(C_6H_{11}$-c$)_3)$=C$(CH_3)$O-$C_2H_5$] [$BF_4$] · 0.5 CH_2Cl_2 ...... Mo: Org.Comp.8-107
$BC_{30}FH_{52}N_2Si_2$ ... 2,6-(i-C_3H_7)$_2$-C_6H_3-N[Si$(CH_3)_3$]-BF -N[Si$(CH_3)_3$]-C_6H_3-2,6-(C_3H_7-i)$_2$ ...... B: B Comp.SVol.4/3b-246
$BC_{30}F_4FeH_{23}O_3$ .. [(8-$(C_6H_5)_3$C-C_8H_8)Fe$(CO)_3$][BF_4] ...... Fe: Org.Comp.B15-240
$BC_{30}F_4FeH_{27}O_2$ .. [(C_5H_5)Fe$(CO)_2$(CH_2=$CHCH_2CH_2$C$(C_6H_5)_3$)][BF_4] Fe: Org.Comp.B17-35
$BC_{30}F_4FeH_{28}O_3P$ [(C_5H_5)Fe(CH_3-CC-$CO_2C_2H_5$)(CO)P$(C_6H_5)_3$][BF_4] Fe: Org.Comp.B16b-100
$BC_{30}F_4FeH_{30}OP$ .. [(C_5H_5)Fe(C_2H_5-CC-C_2H_5)(CO)P$(C_6H_5)_3$][BF_4] Fe: Org.Comp.B16b-100
$BC_{30}F_4FeH_{30}O_2P$ [C_5H_5(CO)(P$(C_6H_5)_3$)Fe=CH(CH=C$(CH_3)OC_2H_5$)][BF_4] Fe: Org.Comp.B16a-23, 29
$BC_{30}F_4FeH_{30}O_4P$ [(C_5H_5)Fe(CH_3-CC-C_3H_7-i)(CO)P(O-C_6H_5)$_3$][BF_4] Fe: Org.Comp.B16b-99
– ...... [(C_5H_5)Fe(C_2H_5-CC-C_2H_5)(CO)P(O-C_6H_5)$_3$][BF_4] Fe: Org.Comp.B16b-100
$BC_{30}F_4FeH_{31}N_4O_3$ [C_5H_5(4-$CH_3OC_6H_4$NC)$_2$Fe=C($NHCH_3$)NH $C_6H_4OCH_3$-4][BF_4] ...... Fe: Org.Comp.B16a-169/70
$BC_{30}F_4FeH_{32}O_2P$ [C_5H_5(CO)(P$(C_6H_5)_3$)Fe=C(C_4H_9-n)OCH_3][BF_4] Fe: Org.Comp.B16a-41
$BC_{30}F_4FeH_{43}O_3$ .. [(i-C_6H_{13}CH(CH_3)-$C_{17}H_{20}(CH_3)_2$)Fe$(CO)_3$][BF_4] Fe: Org.Comp.B15-130, 182
$BC_{30}F_4H_{23}MoO_2$ .. [(C_5H_5)Mo$(CO)_2$(CH_3-$C_4(C_6H_5)_3$)][BF_4] ...... Mo: Org.Comp.8-305, 312/3
$BC_{30}F_4H_{28}MoOP$ [C_5H_5Mo(CO)CH(CH_3)=C(CH_3)CH=CHC_6H_4 P$(C_6H_5)_2$-2][BF_4] ...... Mo: Org.Comp.6-361/3
$BC_{30}F_4H_{28}MoO_4P$ [(C_5H_5)Mo$(CO)_2$(P$(C_6H_5)_3$)=C(-CH_2CH_2 CH(OCH_3)-O-)][BF_4] ...... Mo: Org.Comp.8-109/10
$BC_{30}F_4H_{30}MoOP$ [(C_5H_5)Mo(CO)(P$(C_6H_5)_3$)(H_3C-CC-C_3H_7-i)][BF_4] Mo: Org.Comp.6-287, 289
– ...... [(C_5H_5)Mo(CO)(P$(C_6H_5)_3$)(H-CC-C_4H_9-t)][BF_4] Mo: Org.Comp.6-286

$BC_{30}F_4H_{56}IMoN_6$ $[(t\text{-}C_4H_9\text{-}NH\text{-}CC\text{-}NH\text{-}C_4H_9\text{-}t)Mo(I)(CN\text{-}C_4H_9\text{-}t)_4][BF_4]$
Mo:Org.Comp.5-172, 173

$BC_{30}FeH_{27}$ ...... $[(CH_3\text{-}C_5H_4)Fe(B(C_6H_5)_4)]$ .... Fe: Org.Comp.B18-146/8, 266

$BC_{30}FeH_{34}O_2P$ ... $C_5(CH_3)_5(CO)(P(C_6H_5)_3)Fe{=}C(H)OBH_3$ .... Fe: Org.Comp.B16a-176/7

$BC_{30}H_{20}O_6Re$ .... $[Re(CO)_6][B(C_6H_5)_4]$ .... Re: Org.Comp.2-217, 222

$BC_{30}H_{24}N$ .... $H_3N\text{-}B(1\text{-}C_{10}H_7)_3$ .... B: B Comp.SVol.4/3b-27

$BC_{30}H_{24}N_3O$ ..... $BC_6H_4N_3({=}O)(C_6H_5)_4$ .... B: B Comp.SVol.4/3a-180/1

$BC_{30}H_{26}NO_3Si$ ... $[(C_6H_5)_3SiO]B(NH_2C_6H_5)[\text{-}O\text{-}(1,2\text{-}C_6H_4)\text{-}O\text{-}]$ .. B: B Comp.SVol.3/3-221

$BC_{30}H_{27}MoN_6O_2$. $(c\text{-}C_7H_7)(C_3H_4N_2)Mo(CO)_2[\text{-}N_2C_3H_3\text{-}]_2B(C_6H_5)_2$
Mo:Org.Comp.5-225

$BC_{30}H_{32}N$ .... $1\text{-}(3\text{-}C_6H_5\text{-}1\text{-}BC_{15}H_9\text{-}1)\text{-}2,2,6,6\text{-}(CH_3)_4\text{-}NC_5H_6$
B: B Comp.SVol.4/3a-236

– .... $1\text{-}(4\text{-}C_6H_5\text{-}1\text{-}BC_{15}H_9\text{-}1)\text{-}2,2,6,6\text{-}(CH_3)_4\text{-}NC_5H_6$
B: B Comp.SVol.4/3a-236

– .... $[2,4,6\text{-}(CH_3)_3C_6H_2]_2B\text{-}N(C_6H_5)_2$ .... B: B Comp.SVol.3/3-149

$BC_{30}H_{33}N_2$ ...... $(CH_3)_2C{=}N\text{-}NH(C_6H_5)\text{-}B(C_6H_4\text{-}2\text{-}CH_3)_3$ ...... B: B Comp.SVol.3/3-191

– .... $(CH_3)_2C{=}N\text{-}NH(C_6H_5)\text{-}B(C_6H_4\text{-}3\text{-}CH_3)_3$ ...... B: B Comp.SVol.3/3-191

– .... $(CH_3)_2C{=}N\text{-}NH(C_6H_5)\text{-}B(C_6H_4\text{-}4\text{-}CH_3)_3$ ...... B: B Comp.SVol.3/3-191

$BC_{30}H_{33}N_2O_3$ .... $(CH_3)_2C{=}N\text{-}NH(C_6H_5)\text{-}B(C_6H_4\text{-}2\text{-}OCH_3)_3$ ..... B: B Comp.SVol.3/3-191

– .... $(CH_3)_2C{=}N\text{-}NH(C_6H_5)\text{-}B(C_6H_4\text{-}4\text{-}OCH_3)_3$ ..... B: B Comp.SVol.3/3-191

$BC_{30}H_{34}NO_3Si$ ... $[(C_6H_5)_3SiO]B[N(C_2H_5)_3][\text{-}O\text{-}(1,2\text{-}C_6H_4)\text{-}O\text{-}]$ .. B: B Comp.SVol.3/3-221

$BC_{30}H_{38}N$ .... $1\text{-}(9\text{-}BC_{21}H_{20}\text{-}9)\text{-}2,2,6,6\text{-}(CH_3)_4\text{-}NC_5H_6$ ...... B: B Comp.SVol.4/3a-236

$BC_{30}H_{43}NO_2PS$... $C_2H_5\text{-}CH(OH)\text{-}P({=}S)(C_6H_5)\text{-}CH(C_2H_5)\text{-}O$
$\text{-}B(C_6H_5)_2\text{-}N(C_2H_5)_3$ = $[HN(C_2H_5)_3][\text{-}O$
$\text{-}CH(C_2H_5)\text{-}P({=}S)(C_6H_5)\text{-}CH(C_2H_5)\text{-}O\text{-}B(C_6H_5)_2\text{-}]$
B: B Comp.SVol.4/4-153

$BC_{31}Cl_2F_4H_{29}MoP_2$
$[C_5H_5Mo(P(C_6H_5)_2CH_2CH_2P(C_6H_5)_2)Cl_2][BF_4]$ Mo:Org.Comp.6-23

$BC_{31}Cl_3H_{25}N_3$.... $(C_6H_5\text{-}NH)_2C{=}NH\text{-}B(C_6H_4\text{-}4\text{-}Cl)_3$ .... B: B Comp.SVol.3/3-191
B: B Comp.SVol.4/3b-27

$BC_{31}F_4FeH_{23}O_3$ .. $[(c\text{-}C_3(C_6H_5)_3\text{-}6\text{-}C_7H_8)Fe(CO)_3][BF_4]$ .... Fe: Org.Comp.B15-207

$BC_{31}F_4FeH_{27}NOP$ $[C_5H_5(CO)(P(C_6H_5)_3)FeC(NC_5H_5){=}CH_2][BF_4]$ .. Fe: Org.Comp.B16a-215

$BC_{31}F_4FeH_{30}O_2P$ $[C_5H_5(CO)(P(C_6H_5)_3)Fe{=}C(CH_2CH_2\text{-}CCH)OC_2H_5][BF_4]$
Fe: Org.Comp.B16a-42, 55

$BC_{31}F_4FeH_{30}O_4P$ $[C_5H_5Fe(CO)_2CH_2CH_2CH(COOC_2H_5)P(C_6H_5)_3][BF_4]$
Fe: Org.Comp.B14-154, 157/8

$BC_{31}F_4FeH_{32}O_2P$ $[(C_5(CH_3)_5)Fe(CO)_2\text{-}CH_2P(C_6H_5)_3][BF_4]$ ...... Fe: Org.Comp.B14-146

– .... $[(C_5H_5)(CO)(P(C_6H_5)_3)Fe{=}C(O\text{-}C_2H_5)$
$\text{-}CH_2CH_2CH{=}CH_2][BF_4]$ .... Fe: Org.Comp.B16a-42, 55

$BC_{31}F_4FeH_{32}O_4P$ $[(C_5H_5)Fe(c\text{-}C_7H_{12})(CO)P(OC_6H_5)_3][BF_4]$ ..... Fe: Org.Comp.B16b-92/3

$BC_{31}F_4FeH_{34}O_2PSi$
$[C_5H_5(CO)(P(C_6H_5)_3)Fe{=}C(CH{=}CHSi(CH_3)_3)OCH_3][BF_4]$
Fe: Org.Comp.B16a-42, 55/6

$BC_{31}F_4FeH_{43}$ .... $[((CH_3)_6C_6)Fe(C_6(CH_3)_6\text{-}6\text{-}CH_2\text{-}C_6H_5)][BF_4]$ .. Fe: Org.Comp.B19-143, 166

$BC_{31}F_4H_{29}IMoNOP_2$
$[C_5H_5Mo(NO)(P(C_6H_5)_2CH_2CH_2P(C_6H_5)_2)I][BF_4]$
Mo:Org.Comp.6-40

$BC_{31}F_4H_{30}MoO_3PS$
$[(C_5H_5)Mo(CO)_2(P(C_6H_5)_3){=}C(\text{-}CH_2CH_2CH(S$
$\text{-}C_2H_5)\text{-}O\text{-})][BF_4]$ Mo:Org.Comp.8-109/10

$BC_{31}F_4H_{30}MoO_4P$ $[(C_5H_5)Mo(CO)_2(P(C_6H_5)_3)=C(-CH_2CH_2CH(O-C_2H_5)-O-)][BF_4]$ . . . Mo:Org.Comp.8-109/10

$BC_{31}F_4H_{31}MoNP$. $[C_5H_5Mo(NCCH_3)P(C_6H_5)_2C_6H_4CH=CH C(CH_3)=CHCH_3-2][BF_4]$ . . . Mo:Org.Comp.6-195, 201/2

$BC_{31}F_4H_{32}MoP$ . . . $[C_5H_5Mo(H-CC-C_4H_9-t)P(C_6H_5)_2C_6H_4CH=CH_2-2][BF_4]$ Mo:Org.Comp.6-138

$BC_{31}F_4H_{45}P_3Re$ . . $[(CH_2C(CH_3)CHC(CH_3)CH_2)ReH(P(CH_3)_2-C_6H_5)_3][BF_4]$ Re: Org.Comp.3-3/4

$BC_{31}GaH_{42}N_2$ . . . . $[-C(CH_3)_2-CH_2CH_2CH_2-C(CH_3)_2-]N[-Ga(C_6H_5)_2-N(C_4H_9-t)-B(C_6H_5)-]$ . . . B: B Comp.SVol.4/3a-206/7

$BC_{31}H_{26}N$ . . . . $CH_3-NH_2-B(1-C_{10}H_7)_3$ . . . B: B Comp.SVol.4/3b-27

$BC_{31}H_{26}N_3$ . . . . . . $(H_2N)_2C=NH-B(1-C_{10}H_7)_3$ . . . B: B Comp.SVol.4/3b-27

$BC_{31}H_{28}N_3$ . . . . . . $(C_6H_5-NH)_2C=NH-B(C_6H_5)_3$ . . . B: B Comp.SVol.3/3-191; B: B Comp.SVol.4/3b-27

$BC_{31}H_{32}N$ . . . . $9-[2,4,6-(CH_3)_3-C_6H_2-B=]-C_{13}H_8 \cdot NC_5H_4-4-C_4H_9-t$ B: B Comp.SVol.4/3a-217; B: B Comp.SVol.4/3b-26

$BC_{31}H_{32}N_3$ . . . . . . $(-N=CCH_3-CH=CNH_2-)NC_6H_5-B(C_6H_4-4-CH_3)_3$ B: B Comp.SVol.3/3-192

$BC_{31}H_{33}MoP_2$ . . . . $C_5H_5Mo(P(C_6H_5)_2CH_2CH_2P(C_6H_5)_2)H_2BH_2$ Mo:Org.Comp.6-17

$BC_{31}H_{33}N_2O_3$ . . . . $C_7H_6N_2-B(C_6H_4-2-OC_2H_5)_3$ . . . B: B Comp.SVol.3/3-192

$BC_{31}H_{42}InN_2$ . . . . . $[-C(CH_3)_2-CH_2CH_2CH_2-C(CH_3)_2-]N[-In(C_6H_5)_2-N(C_4H_9-t)-B(C_6H_5)-]$ . . . B: B Comp.SVol.4/3a-206/7

$BC_{31}H_{56}NO_2$ . . . . . $(CH_3)_3N-BH_2-C(O)[-O-C_{17}H_{22}(CH_3)_2-CH(CH_3)-CH_2CH_2CH_2-C_3H_7-i]$ . . . B: B Comp.SVol.4/3b-25

$BC_{31}H_{57}N_2O$ . . . . . $t-C_4H_9-NH-B[O-C_6H_2-2,4,6-(C_4H_9-t)_3]-1-NC_5H_6-2,2,6,6-(CH_3)_4$ . . . B: B Comp.SVol.4/3b-52

$BC_{32}F_2FeH_{25}O_2$ . . $[C_5H_5(CO)_2Fe=CF_2][B(C_6H_5)_4]$ . . . Fe: Org.Comp.B16a-135, 148

$BC_{32}F_3H_{80}N_2OsSi_4$ $[N(C_4H_9-n)_4][((CH_3)_3Si-CH_2)_4OsN-BF_3]$ . . . Os: Org.Comp.A1-32

$BC_{32}F_4FeH_{23}O_3$ . . $[(C_6H_5)_3C_{11}H_8Fe(CO)_3][BF_4]$ . . . Fe: Org.Comp.B15-257

$BC_{32}F_4FeH_{26}OP$ . . $[C_5H_5(CO)(P(C_6H_5)_3)Fe=C=CHC_6H_5][BF_4]$ . . . Fe: Org.Comp.B16a-208, 210

$BC_{32}F_4FeH_{29}NOP$ $[C_5H_5(CO)(P(C_6H_5)_3)FeC(NC_5H_4CH_3-4)=CH_2][BF_4]$ Fe: Org.Comp.B16a-215

$BC_{32}F_4FeH_{30}NP_2$ $[(C_5H_5)Fe(CNH)P(C_6H_5)_2CH_2CH_2P(C_6H_5)_2][BF_4]$ Fe: Org.Comp.B15-280

$BC_{32}F_4FeH_{30}OP$ . . $[(C_5H_5)Fe(CH_2=CH_2)(CO)P(C_6H_5)_3][BF_4] \cdot C_6H_6$ Fe: Org.Comp.B16b-81, 89

$BC_{32}F_4FeH_{34}OP$ . . $[(C_5H_5)Fe(c-C_8H_{14})(CO)P(C_6H_5)_3][BF_4] \cdot CH_2Cl_2$ Fe: Org.Comp.B16b-92/3

$BC_{32}F_4FeH_{34}O_4P$ $[(C_5H_5)Fe(c-C_8H_{14})(CO)P(OC_6H_5)_3][BF_4] \cdot 0.5\ CH_2Cl_2$ Fe: Org.Comp.B16b-92/3

$BC_{32}F_4FeH_{35}NOP$ $[C_5H_5(CO)(P(C_6H_5)_3)Fe=CH(CH=C(CH_3)N(C_2H_5)_2)][BF_4]$ Fe: Org.Comp.B16a-24, 29

$BC_{32}F_4H_{26}MoOP$ $[(C_5H_5)Mo(CO)(P(C_6H_5)_3)=C=CH-C_6H_5][BF_4]$ . . Mo:Org.Comp.6-259/60

– . . . . $[(C_5H_5)Mo(CO)(P(C_6H_5)_3)(H-CC-C_6H_5)][BF_4]$ . . Mo:Org.Comp.6-286

$BC_{32}F_4H_{28}MoOP$ $[C_9H_7Mo(CO)(P(C_6H_5)_3)H_3C-CC-CH_3][BF_4]$ . . . Mo:Org.Comp.6-287

$BC_{32}F_4H_{30}NO_2PRe$ $[(C_5H_5)Re(NO)(P(C_6H_5)_3)=C(OCH_3)-CH_2C_6H_5][BF_4]$ Re: Org.Comp.3-109/11, 124

$BC_{32}F_4H_{32}MoO_4P$ $[(C_5H_5)Mo(CO)_2(P(C_6H_5)_3)=C(-CH_2CH_2CH(O$
$-C_3H_7-i)-O-)][BF_4]$ Mo:Org.Comp.8-109/10
– .... $[(C_5H_5)Mo(CO)_2(P(C_6H_5)_3)=C(-CH_2CH_2CH(O$
$-C_3H_7-n)-O-)][BF_4]$.................... Mo:Org.Comp.8-109/10
$BC_{32}F_4H_{35}MoOSi_2$ $[C_9H_7Mo(CO)(C_6H_5-CC-Si(CH_3)_3)_2][BF_4]$ Mo:Org.Comp.6-321/2
$BC_{32}F_4H_{37}MoO_3P_2$
$[C_5H_5Mo(P(OCH_3)_3)P(C_6H_5)_2C_6H_4CH=CH$
$C(CH_3)=CHCH_3-2][BF_4]$ Mo:Org.Comp.6-196
$BC_{32}FeH_{25}O_2S$ ... $[(C_5H_5)Fe(CS)(CO)_2][B(C_6H_5)_4]$ .... Fe: Org.Comp.B15-273
$BC_{32}FeH_{25}O_3$ .... $[(C_5H_5)Fe(CO)_3][B(C_6H_5)_4]$................. Fe: Org.Comp.B15-36/40, 44/7
$BC_{32}H_{24}N$ .... $CH_3-CN-B(1-C_{10}H_7)_3$ .... B: B Comp.SVol.4/3b-28
$BC_{32}H_{28}N$ .... $(CH_3)_2NH-B(1-C_{10}H_7)_3$ .... B: B Comp.SVol.4/3b-27
$BC_{32}H_{32}N_3$ ...... $10-N(C_3H_7-i)_2-9,11,10-N_2BC_{26}H_{18}$ .... B: B Comp.SVol.4/3a-156
$BC_{32}H_{32}N_8Ti_2^+$... $[((C_5H_5)_2Ti)_2((C_3H_3N_2)_4B)]^+$ .... Ti: Org.Comp.5-176
$BC_{32}H_{33}N_2S$ ..... $[-B(C_6H_5)_2-NHC(CH_2C_6H_5)C(C_6H_5)C(S-C_4H_9-n)NH-]$
B: B Comp.SVol.4/3a-178
B: B Comp.SVol.4/4-149
$BC_{32}H_{34}N$ .... $9-[2,3,5,6-(CH_3)_4-C_6H-B=]-C_{13}H_8 \cdot NC_5H_4-4-C_4H_9-t$
B: B Comp.SVol.4/3a-217
B: B Comp.SVol.4/3b-26
$BC_{32}H_{34}N_3O$ ..... $(NH_2)C_{11}H_{11}N_2O-B(C_6H_4-2-CH_3)_3$ .... B: B Comp.SVol.3/3-192
– .... $(NH_2)C_{11}H_{11}N_2O-B(C_6H_4-4-CH_3)_3$ .... B: B Comp.SVol.3/3-192
$BC_{32}H_{34}N_3O_4$ .... $(NH_2)C_{11}H_{11}N_2O-B(C_6H_4-4-OCH_3)_3$......... B: B Comp.SVol.3/3-192
$BC_{32}H_{38}N$ .... $NC_5H_5 \cdot B[C_6H_2-2,4,6-(CH_3)_3]_3$ .... B: B Comp.SVol.4/3b-28
$BC_{32}H_{47}LiNO_2$ ... $[2,4,6-(CH_3)_3-C_6H_2]_2B-N(C_6H_5)-Li[O(C_2H_5)_2]_2$ B: B Comp.SVol.4/3a-232
$BC_{32}H_{47}NO_2PS$... $i-C_3H_7-CHOH-P(S)(C_6H_5)-CH(C_3H_7-i)-O$
$-B(C_6H_5)_2-N(C_2H_5)_3 = [HN(C_2H_5)_3][O$
$-CH(C_3H_7-i)-P(S)(C_6H_5)-CH(C_3H_7-i)-O-B(C_6H_5)_2]$
B: B Comp.SVol.4/3b-78
B: B Comp.SVol.4/4-153
$BC_{32}H_{47}NO_3P$.... $(C_2H_5)_3N-B(C_6H_5)_2-O-CH(C_3H_7-i)$
$-P(=O)(C_6H_5)-CHOH-C_3H_7-i$
$= [(C_2H_5)_3NH][B(C_6H_5)_2-O-CH(C_3H_7-i)$
$-P(O)(C_6H_5)-CH(C_3H_7-i)-O]$ .... B: B Comp.SVol.4/3b-78
$BC_{32}H_{56}NO_4$..... $(CH_3)_3N-BH_2-C(O)-O-C(O)[-O-C_{17}H_{22}(CH_3)_2$
$-CH(CH_3)-CH_2CH_2CH_2-C_3H_7-i]$ .... B: B Comp.SVol.4/3b-25
$BC_{32.5}ClF_4FeH_{35}O_4P$
$[(C_5H_5)Fe(c-C_8H_{14})(CO)P(OC_6H_5)_3][BF_4] \cdot 0.5\ CH_2Cl_2$
Fe: Org.Comp.B16b-92/3
$BC_{33}Cl_2F_4FeH_{36}OP$
$[(C_5H_5)Fe(c-C_8H_{14})(CO)P(C_6H_5)_3][BF_4] \cdot CH_2Cl_2$
Fe: Org.Comp.B16b-92/3
$BC_{33}Cl_2H_{45}N_2Si$.. $1-[2,6-(i-C_3H_7)_2C_6H_3-N(SiCl(C_6H_5)_2)-BCl]-$
$2,2,6,6-(CH_3)_4-NC_5H_6$ .... B: B Comp.SVol.4/4-64
$BC_{33}CoH_{26}$ ...... $3,4,5,6-(C_6H_5)_4-1,3,4,5,6-(C_5H_5)CoC_4BH$ ..... B: B Comp.SVol.3/4-155
$BC_{33}F_4FeH_{26}O_2P$ $[(C_5H_5)Fe(CO)_2-C(P(C_6H_5)_3)=CH-C_6H_5][BF_4]$.. Fe: Org.Comp.B14-141, 144/6
Fe: Org.Comp.B16a-216, 218/9
$BC_{33}F_4FeH_{28}OP$.. $[(C_5H_5)Fe(CH_3-CC-C_6H_5)(CO)P(C_6H_5)_3][BF_4]$. Fe: Org.Comp.B16b-99

$BC_{33}F_4FeH_{28}O_2P$ $[C_5H_5Fe(CO)_2CH_2CH(C_6H_5)P(C_6H_5)_3][BF_4]$ . . . . Fe: Org.Comp.B14-153
$BC_{33}F_4FeH_{28}O_4P$ $[(C_5H_5)Fe(CH_3-CC-C_6H_5)(CO)P(OC_6H_5)_3][BF_4]$ Fe: Org.Comp.B16b-99
$BC_{33}F_4FeH_{29}OP_2$ $[C_5H_5(CO)Fe((C_6H_5)_2PCH_2P(C_6H_5)_2)C=CH_2][BF_4]$ Fe: Org.Comp.B16a-216, 217/8
$BC_{33}F_4FeH_{30}O_2P$ $[C_5H_5(CO)(P(C_6H_5)_3)Fe=C(CH_2C_6H_5)OCH_3][BF_4]$ Fe: Org.Comp.B16a-42, 56
$BC_{33}F_4FeH_{31}NOP$ $[C_5H_5(CO)(P(C_6H_5)_3)Fe=C(CH_3)NHCH_2C_6H_5][BF_4]$ Fe: Org.Comp.B16a-46/7
$BC_{33}F_4Fe_2H_{28}O_4P$ $[C_5H_5(CO)(P(C_6H_5)_3)Fe=C(CH_3)OFe(CO)_2C_5H_5][BF_4]$ Fe: Org.Comp.B16a-39, 54
$BC_{33}F_4H_{28}MoN_2O_2P$ $[(CH_2CHCH_2)Mo(CO)_2(NC_5H_4-2-(2-C_5H_4N))(P(C_6H_5)_3)][BF_4]$ Mo:Org.Comp.5-200, 202
$BC_{33}F_4H_{28}MoOP$ $[C_5H_5Mo(CO)(P(C_6H_5)_3)H_3C-CC-C_6H_5][BF_4]$ . . Mo:Org.Comp.6-287/8, 289
$BC_{33}F_4H_{28}MoO_4P$ $[C_5H_5Mo(CO)(P(OC_6H_5)_3)H_3C-CC-C_6H_5][BF_4]$ Mo:Org.Comp.6-288
$BC_{33}F_4H_{29}MoO_2P_2$ $[(C_5H_5)Mo(CO)_2((C_6H_5)_2P-CH_2CH_2-P(C_6H_5)_2)][BF_4]$ Mo:Org.Comp.7-274
$BC_{33}F_4H_{31}N_2OPRe$ $[(C_5H_5)Re(NO)(P(C_6H_5)_3)(NC-CH(C_6H_5)-C_2H_5)][BF_4]$ Re: Org.Comp.3-29/30, 33
$BC_{33}F_4H_{34}MoO_4P$ $[(C_5H_5)Mo(CO)_2(P(C_6H_5)_3)=C(-CH_2CH_2CH(O-C_4H_9-t)-O-)][BF_4]$ Mo:Org.Comp.8-109/10
$BC_{33}FeH_{27}O_3$ . . . . $[(CH_3-C_5H_4)Fe(CO)_3][B(C_6H_5)_4]$ . . . . Fe: Org.Comp.B15-51
– . . . . $[(c-C_6H_7)Fe(CO)_3][B(C_6H_5)_4]$ . . . . Fe: Org.Comp.B15-54/70
$BC_{33}FeH_{28}N$ . . . . . $(C_6H_5)_2B-C_5H_4FeC_5H_4-C_6H_5 \cdot NC_5H_5$ . . . . Fe: Org.Comp.A9-281/2
$BC_{33}FeH_{29}O_2$ . . . . $[(C_5H_5)Fe(CO)_2(CH_2=CH_2)][B(C_6H_5)_4]$ . . . . Fe: Org.Comp.B17-5
$BC_{33}FeH_{33}$ . . . . . . $[(B(C_6H_4-3-CH_3)_4)Fe(C_5H_5)]$ . . . . Fe: Org.Comp.B19-67
– . . . . $[(B(C_6H_4-4-CH_3)_4)Fe(C_5H_5)]$ . . . . Fe: Org.Comp.B19-67
– . . . . $[(CH_3-C_6H_4)_3B-C_6H_4-CH_3]Fe(C_5H_5)$ . . . . Fe: Org.Comp.B19-6
$BC_{33}FeH_{33}O_4$ . . . . $[(B(C_6H_4-4-O-CH_3)_4)Fe(C_5H_5)]$ . . . . Fe: Org.Comp.B19-77
– . . . . $[(CH_3-O-C_6H_4)_3B-C_6H_4-O-CH_3]Fe(C_5H_5)$ . . . . Fe: Org.Comp.B19-6
$BC_{33}H_{33}N_4$ . . . . . . $(NC)_2C=C(CN)_2-B[C_6H_2-2,4,6-(CH_3)_3]_3$ . . . . B: B Comp.SVol.3/3-192
$BC_{33}H_{33}S$ . . . . $[(2-CH_3C_6H_4)(CH_3)_2S][B(C_6H_5)_4]$ . . . . B: B Comp.SVol.3/4-117
$BC_{33}H_{39}N_2$ . . . . . . $(CH_3)_2C=N-NH(C_6H_5)-B[C_6H_3-2,5-(CH_3)_2]_3$ . . . B: B Comp.SVol.3/3-191
– . . . . $(CH_3)_2C=N-NH(C_6H_5)-B[C_6H_3-3,4-(CH_3)_2]_3$ . . . B: B Comp.SVol.3/3-191
$BC_{33}H_{39}N_2O_3$ . . . . $(CH_3)_2C=N-NH(C_6H_5)-B(C_6H_4-4-OC_2H_5)_3$ . . . . B: B Comp.SVol.3/3-191
$BC_{33}H_{48}N_3$ . . . . . . $B[N(t-C_4H_9)(CH_2C_6H_5)]_3$ . . . . B: B Comp.SVol.3/3-94
$BC_{34}Cl_2F_4FeH_{28}P_2Pt$ $[Fe(C_5H_4-P(C_6H_5)_2)_2PtCl_2][BF_4]$ . . . . Fe: Org.Comp.A10-40
$BC_{34}F_4FeH_{28}O_2P$ $[C_5H_5Fe(CO)_2C(=CHC_6H_5)CH_2P(C_6H_5)_3][BF_4]$. . Fe: Org.Comp.B14-153
$BC_{34}F_4FeH_{31}NOP$ $[C_5H_5(CO)(P(C_6H_5)_3)Fe=CH(CH=C(CH_3)NHC_6H_5)][BF_4]$ Fe: Org.Comp.B16a-23, 29
$BC_{34}F_4FeH_{33}NOP$ $[C_5H_5(CO)(P(C_6H_5)_3)Fe=C(CH_3)NHCH(CH_3)C_6H_5][BF_4]$ Fe: Org.Comp.B16a-47, 58/9
$BC_{34}F_4FeH_{33}OP_2$ $[C_5H_5(CO)(P(C_6H_5)_3)FeC(P(CH_3)_2C_6H_5)=CH_2][BF_4]$ Fe: Org.Comp.B16a-215
$BC_{34}F_4FeH_{33}P_2$ . . $[C_5H_5((C_6H_5)_2PCH_2CH_2P(C_6H_5)_2)Fe=C=CHCH_3][BF_4]$ Fe: Org.Comp.B16a-198, 202

$BC_{34}F_4FeH_{37}NOP$ $[C_5H_5(CO)(P(C_6H_5)_3)Fe{=}CH(CH{=}C(CH_3)NH$ C_6H_{11}-c)][BF_4] .... Fe: Org.Comp.B16a-23/4, 29

$BC_{34}F_4FeH_{40}NPRh$ $[(C_5H_5)Fe(C_5H_3(P(C_6H_5)_2)$-$CH(CH_3)$-$N(CH_3)_2)$ $Rh(C_8H_{12})][BF_4]$ .... Fe: Org.Comp.A10-67, 70, 73/5

$BC_{34}F_4FeH_{49}O_2$ .. $[(C_5H_5)Fe(CO)_2((CH_3)_2C_{17}H_{21}CH(CH_3)$-$C_6H_{13}$-i)][$BF_4$] Fe: Org.Comp.B17-101

$BC_{34}F_4H_{25}MoO$ .. $[C_5H_5Mo(CO)(C_6H_5$-CC-$C_6H_5)_2][BF_4]$ .... Mo: Org.Comp.6-320, 325/6

$BC_{34}F_4H_{30}MoN_2O_2P$ $[(CH_2C(CH_3)CH_2)Mo(CO)_2(NC_5H_4$-2-(2-$C_5H_4N))(P(C_6H_5)_3)][BF_4]$ .... Mo: Org.Comp.5-200, 204

$BC_{34}F_4H_{30}MoN_2O_5P$ $[(CH_2C(CH_3)CH_2)Mo(CO)_2(NC_5H_4$-2-(2-$C_5H_4N))(P(O$-$C_6H_5)_3)][BF_4]$ .... Mo: Org.Comp.5-200, 204

$BC_{34}F_4H_{30}O_2PSe$ $[((C_6H_5)_2Se)((C_6H_5)_3P)C(COOC_2H_5)][BF_4]$ .... B: B Comp.SVol.3/4-148

$BC_{34}F_4H_{37}MoNP$. $[C_5H_5Mo(NCC_4H_9$-t)$P(C_6H_5)_2C_6H_4CH{=}CH$ $C(CH_3){=}CHCH_3$-2][BF_4] .... Mo: Org.Comp.6-196, 201/2

$BC_{34}F_5H_{29}P_2Re$ .. [trans-C_6H_5-C≡C-Re$(P(C_6H_5)_2CH_2CH_2$ $P(C_6H_5)_2)(F)][BF_4]$ .... Re: Org.Comp.1-11

$BC_{34}F_9H_{25}N_3$ .... $(C_6H_5NH)_2C{=}NH$-$B(C_6H_4$-3-$CF_3)_3$ .... B: B Comp.SVol.3/3-191

$BC_{34}FeH_{28}$ ...... C_6H_5-$C_5H_4FeC_5H_4B(C_6H_5)_3$ .... Fe: Org.Comp.A9-282

$BC_{34}FeH_{29}O_3$ .... [(c-C_7H_9)$Fe(CO)_3][B(C_6H_5)_4]$ .... Fe: Org.Comp.B15-191

$BC_{34}FeH_{29}O_4$ .... $[(CH_3O$-2-$C_6H_6)Fe(CO)_3][B(C_6H_5)_4]$ .... Fe: Org.Comp.B15-103, 139/54

$BC_{34}FeH_{30}Mn$ .... $[(C_5H_5)Mn(C_6H_5$-$B(C_6H_5)_2$-$C_6H_5)Fe(C_5H_5)]$ ... Fe: Org.Comp.B18-146/8, 267

$BC_{34}FeH_{31}N_2O$ ... $[(C_5H_5)Fe(CNCH_3)_2CO][B(C_6H_5)_4]$ .... Fe: Org.Comp.B15-333

$BC_{34}FeH_{33}O_2S$ ... $[C_5H_5Fe(CO)_2CH_2S(CH_3)_2][B(C_6H_5)_4]$ .... Fe: Org.Comp.B14-133/5, 142

$BC_{34}FeH_{35}$ ...... $[(CH_3$-$C_6H_4)_3B$-C_6H_4-$CH_3]Fe(C_5H_4$-$CH_3)$ ..... Fe: Org.Comp.B19-6

$BC_{34}FeH_{35}O_4$ .... $[(CH_3$-O-$C_6H_4)_3B$-C_6H_4-O-$CH_3]Fe(C_5H_4$-$CH_3)$ Fe: Org.Comp.B19-6

$BC_{34}H_{34}N_3$ ...... $(C_6H_5$-$NH)_2C{=}NH$-$B(C_6H_4$-2-$CH_3)_3$ .... B: B Comp.SVol.3/3-191; B: B Comp.SVol.4/3b-27

– .... $(C_6H_5$-$NH)_2C{=}NH$-$B(C_6H_4$-3-$CH_3)_3$ .... B: B Comp.SVol.3/3-191; B: B Comp.SVol.4/3b-27

– .... $(C_6H_5$-$NH)_2C{=}NH$-$B(C_6H_4$-4-$CH_3)_3$ .... B: B Comp.SVol.3/3-191; B: B Comp.SVol.4/3b-27

$BC_{34}H_{34}N_3O_3$ .... $(C_6H_5$-$NH)_2C{=}NH$-$B(C_6H_4$-2-$OCH_3)_3$ .... B: B Comp.SVol.3/3-191; B: B Comp.SVol.4/3b-27

– .... $(C_6H_5$-$NH)_2C{=}NH$-$B(C_6H_4$-4-$OCH_3)_3$ .... B: B Comp.SVol.3/3-191; B: B Comp.SVol.4/3b-27

$BC_{35}ClFeH_{30}$ .... $[(Cl$-$C_6H_5)Fe(C_5H_5)][B(C_6H_5)_4]$ .... Fe: Org.Comp.B18-142/6, 197, 201, 246

$BC_{35}ClFeH_{33}N$ ... $[C_5H_5NC_2H_5][(C_6H_5)_3B$-$C_5H_4FeC_5H_4$-Cl] ..... Fe: Org.Comp.A9-282

$BC_{35}F_2H_{45}N_4O_2$ .. $BF_2C_{35}H_{45}N_4O_2$ .... B: B Comp.SVol.3/3-378

$BC_{35}F_4FeH_{34}O_2P$ $[C_5H_5(CO)(P(C_6H_5)_3)Fe{=}C(CH_2CH_2C_6H_5)OC_2H_5][BF_4]$ Fe: Org.Comp.B16a-41

$BC_{35}F_4FeH_{47}OS_2$ $[((CH_3)_6C_6)Fe(6$-(1,3-$S_2C_4H_7$-2)-5-$(C_6H_5$-CO-$CH_2)$-C_6-1,2,3,4,6-$(CH_3)_5)][BF_4]$ .... Fe: Org.Comp.B19-144/5, 172

$BC_{35}F_4H_{28}MoN$ ... $[C_5H_5Mo(NCCH_3)(C_6H_5$-CC-$C_6H_5)_2][BF_4]$ ..... Mo: Org.Comp.6-137, 146

$BC_{35}F_4H_{30}MoN_4P$ $[C_5H_5Mo(N=NC_6H_5)_2P(C_6H_5)_3][BF_4]$ Mo:Org.Comp.6-62
$BC_{35}F_4H_{33}MoP_2$ . . $[C_5H_5Mo(P(C_6H_5)_2CH=CHP(C_6H_5)_2)H_3C-CC-CH_3][BF_4]$ Mo:Org.Comp.6-119, 129
$BC_{35}F_4H_{35}MoP_2$ . . $[(C_5H_5)Mo(P(C_6H_5)_2CH_2CH_2P(C_6H_5)_2)(H_2C=CHCH=CH_2)][BF_4]$. Mo:Org.Comp.6-193, 200
– . . . . $[(C_5H_5)Mo(P(C_6H_5)_2CH_2CH_2P(C_6H_5)_2)(H_3C-CC-CH_3)][BF_4]$ Mo:Org.Comp.6-119, 128
$BC_{35}F_4H_{37}MoP_2$ . . $[C_5H_5Mo(P(C_6H_5)_2CH_3)_2H_3C-CC-CH_3][BF_4]$. . . Mo:Org.Comp.6-118
$BC_{35}F_4H_{53}OP_3Re$ $[(CH_2C(CH_3)CHC(CH_3)CH_2)ReH(P(CH_3)_2-C_6H_5)_3][BF_4] \cdot OC_4H_8$ . . . . Re: Org.Comp.3-3/4
$BC_{35}F_5H_{29}MoN_4P$ $[(C_5H_5)Mo(N=N-C_6H_5)(N=N-C_6H_4F-3)P(C_6H_5)_3][BF_4]$ Mo:Org.Comp.6-62
– . . . . $[(C_5H_5)Mo(N=N-C_6H_5)(N=N-C_6H_4F-4)P(C_6H_5)_3][BF_4]$ Mo:Org.Comp.6-62
$BC_{35}FeH_{29}O_3$ . . . . $[(C_8H_9)Fe(CO)_3][B(C_6H_5)_4]$. . . . . Fe: Org.Comp.B15-214, 227/9
$BC_{35}FeH_{31}$ . . . . . . $[(C_6H_6)Fe(C_5H_5)][B(C_6H_5)_4]$ . . . . Fe: Org.Comp.B18-142/6, 150/2, 154, 158/9
$BC_{35}FeH_{33}O_2$ . . . . $[(C_5H_5)Fe(CO)_2(CH_2=CHC_2H_5)][B(C_6H_5)_4]$. . . . . Fe: Org.Comp.B17-11
$BC_{35}FeH_{36}N_3O$ . . . $[C_5H_5(CH_3NC)(CO)Fe=C(NHCH_3)_2][B(C_6H_5)_4]$. . Fe: Org.Comp.B16a-163, 168
$BC_{35}H_{26}N$ . . . . $NC_5H_5 \cdot B(1-C_{10}H_7)_3$ . . . . B: B Comp.SVol.4/3b-28
$BC_{35}H_{29}MoN_4O_2$. $(CH_2CHCH_2)Mo(CO)_2[-N_2C_3H(C_6H_5)_2-]_2BH_2$ . . Mo:Org.Comp.5-213, 214
$BC_{35}H_{29}N_2$ . . . . . . $(-N=CCH_3-CH=CCH_3-)NH-B[1-C_{10}H_7]_3$ . . . . . . B: B Comp.SVol.3/3-191
$BC_{35}H_{31}MoN_2O_2$. $[(C_5H_5)Mo(CO)_2(NC-CH_3)_2][B(C_6H_5)_4]$ Mo:Org.Comp.7-284
$BC_{35}H_{36}NO$. . . . . . 2'-$[2,2,6,6-(CH_3)_4-NC_5H_6-1]$-4',4'-$(C_6H_5)_2$-1',2'-$OBC_{14}H_8$ B: B Comp.SVol.4/3b-63
$BC_{35}H_{40}N_3O_4$ . . . . $(NH_2)C_{11}H_{11}N_2O-B(C_6H_4-4-OC_2H_5)_3$ . . . . B: B Comp.SVol.3/3-192
$BC_{35}H_{43}MnN_4O_2S$ $[Mn(4-CH_3C_6H_4SO_2)(NH_2CH_2CH_2NH_2)_2](B(C_6H_5)_4)$ Mn:MVol.D7-109/11
$BC_{35}H_{53}LiNO_2$ . . . $[2,4,6-(CH_3)_3-C_6H_2]_2B-N[C_6H_2-2,4,6-(CH_3)_3]-Li[O(C_2H_5)_2]_2$ . . . . B: B Comp.SVol.4/3a-232
$BC_{36}ClH_{26}O_2$ . . . . $ClB[O-C_6H_3-2,6-(C_6H_5)_2]_2$ . . . . B: B Comp.SVol.4/4-55
$BC_{36}ClH_{33}NP_2$ . . . $[(C_6H_5)_3P=N=P(C_6H_5)_3][H_3BCl]$ . . . . B: B Comp.SVol.3/4-38
B: B Comp.SVol.4/4-39/40
$BC_{36}Cl_3F_2H_{46}N_4O_2$
$BF_2C_{35}H_{45}N_4O_2 \cdot CHCl_3$. . . . . B: B Comp.SVol.3/3-378
$BC_{36}Cl_3H_{66}SSn_2$. . $[(c-C_6H_{11})_3Sn]_2S \cdot BCl_3$ . . . . B: B Comp.SVol.4/4-34
$BC_{36}FH_{26}O_2$ . . . . . $2,6-(C_6H_5)_2-C_6H_3-O-FB-O-C_6H_3-2,6-(C_6H_5)_2$ B: B Comp.SVol.4/3b-232
$BC_{36}F_4FeH_{34}O_2P$ $[C_5H_5(CO)(P(C_6H_5)_3)Fe=CH(CH=C(C_6H_4CH_3-4)OC_2H_5)][BF_4]$ . . . . Fe: Org.Comp.B16a-23, 29
$BC_{36}F_4FeH_{35}NOP$ $[C_5H_5(CO)(P(C_6H_5)_3)Fe=C(-NCH_3CHC_6H_5C(CH_3)_2-)][BF_4]$ . . . . Fe: Org.Comp.B16a-47, 59
$BC_{36}F_4FeH_{35}P_2S_2$ $[C_5H_5((C_6H_5)_2PCH_2CH_2P(C_6H_5)_2)Fe=C=C(CH_3)CS_2CH_3][BF_4]$ . . . . Fe: Org.Comp.B16a-201, 204
$BC_{36}F_4FeH_{39}P_2$ . . $[C_5H_5((C_6H_5)_2PCH_2CH_2P(C_6H_5)_2)Fe=CH(C_4H_9-t)][BF_4]$ Fe: Org.Comp.B16a-7
$BC_{36}F_4H_{29}MoO$ . . $[C_5H_5Mo(CO)(C_6H_5CH_2-CC-C_6H_5)_2][BF_4]$ Mo:Org.Comp.6-319
$BC_{36}F_4H_{32}MoN_4P$ $[C_5H_5Mo(N=NC_6H_5)(N=NC_6H_4CH_3-4)P(C_6H_5)_3][BF_4]$ Mo:Org.Comp.6-62
$BC_{36}F_5H_{31}MoN_4P$ $[C_5H_5Mo(N=NC_6H_4CH_3-4)(N=NC_6H_4F-4)P(C_6H_5)_3][BF_4]$ Mo:Org.Comp.6-62

$BC_{36}FeH_{30}N$ ..... $[(C_6H_6)Fe(C_5H_4\text{-}CN)][B(C_6H_5)_4]$ .... Fe: Org.Comp.B18-142/6, 151, 154, 161
$BC_{36}FeH_{33}$ ...... $[(CH_3C_6H_5)Fe(C_5H_5)][B(C_6H_5)_4]$............ Fe: Org.Comp.B18-142/6, 197, 201, 206
$BC_{36}FeH_{33}O$ ..... $[(CH_3O\text{-}C_6H_5)Fe(C_5H_5)][B(C_6H_5)_4]$ .... Fe: Org.Comp.B18-142/6, 198, 201, 250
– .... $[(C_6H_6)Fe(C_5H_4\text{-}OCH_3)][B(C_6H_5)_4]$ .... Fe: Org.Comp.B18-142/6, 169
$BC_{36}H_{39}S$ .... $[(CH_3)(C_6H_5)(2\text{-}C_5H_{11})S][B(C_6H_5)_4]$.......... B: B Comp.SVol.3/4-117
$BC_{36}H_{44}N_3O_3S$ ... $[(1,4\text{-}ONC_4H_8\text{-}4)_3S][B(C_6H_5)_4]$ .... S: S-N Comp.8-222/4, 228/9
$BC_{36}H_{46}N_3O_2S$ ... $[(1,4\text{-}ONC_4H_8\text{-}4\text{-})_2S\text{-}N(C_2H_5)_2][B(C_6H_5)_4]$ .... S: S-N Comp.8-230, 243
$BC_{36}H_{48}N_3OS$.... $[((C_2H_5)_2N)_2S\text{-}4\text{-}1,4\text{-}ONC_4H_8][B(C_6H_5)_4]$ ..... S: S-N Comp.8-230, 241
$BC_{36}H_{50}N_3S$ ..... $[((C_2H_5)_2N)_3S][B(C_6H_5)_4]$ .... S: S-N Comp.8-230, 240
$BC_{37}Cl_2F_4H_{30}NO_2OsP_2$
$[(ON)(CO)(Cl)_2Os(P(C_6H_5)_3)_2][BF_4]$ .... Os: Org.Comp.A1-164, 190/1
$BC_{37}Cl_3H_{35}NP_2$ .. $[(C_6H_5)_3P{=}N{=}P(C_6H_5)_3][H_3BCl] \cdot CH_2Cl_2$ ..... B: B Comp.SVol.4/4-37/8
$BC_{37}F_4FeH_{37}NOP$ $[(C_5H_5)(CO)(P(C_6H_5)_3)Fe{=}C(\text{-}NCH_3 CH(C_6H_4\text{-}CH_3\text{-}3)C(CH_3)_2\text{-})][BF_4]$ .... Fe: Org.Comp.B16a-48, 59
– .... $[(C_5H_5)(CO)(P(C_6H_5)_3)Fe{=}C(\text{-}NCH_3 CH(C_6H_4\text{-}CH_3\text{-}4)C(CH_3)_2\text{-})][BF_4]$ .... Fe: Org.Comp.B16a-48, 59
$BC_{37}F_4FeH_{41}P_2$ .. $[C_5(CH_3)_5((C_6H_5)_2PCH_2CH_2P(C_6H_5)_2)Fe{=}CH_2][BF_4]$
Fe: Org.Comp.B16a-7/8
$BC_{37}F_4H_{31}MoO_2P_2$
$[(C_9H_7)Mo(CO)_2((C_6H_5)_2P\text{-}CH_2CH_2\text{-}P(C_6H_5)_2)][BF_4]$
Mo:Org.Comp.7-274/5
$BC_{37}F_4H_{34}MoN_4P$ $[C_5H_5Mo(N{=}NC_6H_4CH_3\text{-}4)_2P(C_6H_5)_3][BF_4]$ Mo:Org.Comp.6-62
$BC_{37}F_4H_{37}MoO_2P_2S$
$[C_5H_5Mo(P(C_6H_5)_2CH_2CH_2P(C_6H_5)_2)C_6H_8\text{-}c][BF_4] \cdot SO_2$
Mo:Org.Comp.6-194
$BC_{37}F_4H_{37}MoP_2$.. $[C_5H_5Mo(P(C_6H_5)_2CH_2CH_2P(C_6H_5)_2)C_6H_8\text{-}c][BF_4]$
Mo:Org.Comp.6-194
$BC_{37}FeH_{35}$ ...... $[(1,4\text{-}(CH_3)_2\text{-}C_6H_4)Fe(C_5H_5)][B(C_6H_5)_4]$ ...... Fe: Org.Comp.B19-1, 5, 12
$BC_{37}FeH_{35}O$ ..... $[(C_2H_5\text{-}O\text{-}C_6H_5)Fe(C_5H_5)][B(C_6H_5)_4]$ .... Fe: Org.Comp.B18-142/6, 197, 198, 250
– .... $[(C_6H_6)Fe(C_5H_4\text{-}O\text{-}C_2H_5)][B(C_6H_5)_4]$ .... Fe: Org.Comp.B18-142/6, 153, 169
$BC_{37}FeH_{47}O_6P_2$ .. $[(CH_3C_6H_5)Fe(H)(P(OCH_3)_3)_2][B(C_6H_5)_4]$...... Fe: Org.Comp.B18-6
$BC_{37}H_{27}N_2$ ...... $C_7H_6N_2\text{-}B(1\text{-}C_{10}H_7)_3$ .... B: B Comp.SVol.3/3-192
$BC_{37}H_{37}MoNO_4P$ $[(C_5H_5)Mo(CO)_2(1,3,6,2\text{-}O_2NPC_4H_9\text{-}2\text{-}CH{=}CH_2)][B(C_6H_5)_4]$.................. Mo:Org.Comp.7-249, 271, 281
$BC_{37}H_{37}N_4S$ ..... $[\text{-}S\text{-}C(N(CH_3)_2)NC(N(CH_3)_2)\text{-}N{=}C(C_6H_5)\text{-}][B(C_6H_5)_4]$
B: B Comp.SVol.4/4-150
$BC_{37}H_{39}S$ .... $[(i\text{-}C_3H_7)(C_6H_5)(CH_2{=}CHCH_2CH_2)S][B(C_6H_5)_4]$ B: B Comp.SVol.3/4-117
$BC_{37}H_{40}N_3$ ...... $(C_6H_5\text{-}NH)_2C{=}NH\text{-}B[C_6H_3\text{-}2,5\text{-}(CH_3)_2]_3$ ...... B: B Comp.SVol.4/3b-27
$BC_{37}H_{40}N_3O_3$ .... $(C_6H_5\text{-}NH)_2C{=}NH\text{-}B(C_6H_4\text{-}2\text{-}O\text{-}C_2H_5)_3$ ...... B: B Comp.SVol.4/3b-27
– .... $(C_6H_5\text{-}NH)_2C{=}NH\text{-}B(C_6H_4\text{-}4\text{-}O\text{-}C_2H_5)_3$ ...... B: B Comp.SVol.3/3-191
B: B Comp.SVol.4/3b-27
$BC_{37}H_{43}MoO_8P_2$.. $[(C_5H_5)Mo(CO)_2(P(OCH_3)_3)_2][B(C_6H_5)_4]$ Mo:Org.Comp.7-283, 290
$BC_{37}H_{44}N$ .... $9\text{-}[2,4,6\text{-}(C_3H_7\text{-}i)_3\text{-}C_6H_2\text{-}B{=}]\text{-}C_{13}H_8 \cdot NC_5H_4\text{-}4\text{-}C_4H_9\text{-}t$
B: B Comp.SVol.4/3b-26

$BC_{37}H_{45}N$ .... 4-(t-C_4H_9)-NC_5H_4-1-B[9-(=$C_{13}H_8$)]-C_6H_3-2,4,6-(C_3H_7-i)$_3$ .... B: B Comp.SVol.4/3a-217

$BC_{37}H_{46}N_3O_2S$ ... [(1,4-ONC_4H_8-4-)$_2$S-NC_5H_{10}-1][B(C_6H_5)$_4$]. ... S: S-N Comp.8-230, 244

$BC_{37}H_{49}N_4PdS$ ... [Pd(NCS)((C_2H_5)$_2NCH_2CH_2NHCH_2CH_2$N(C_2H_5)$_2$)][B(C_6H_5)$_4$] .... Pd: SVol.B2-307

– .... [Pd(SCN)((C_2H_5)$_2NCH_2CH_2NHCH_2CH_2$N(C_2H_5)$_2$)][B(C_6H_5)$_4$] .... Pd: SVol.B2-307

$BC_{38}ClF_4H_{29}MoN_3O_3Pd_3$
[C_5H_5Mo(CO)$_3$(PdC_9H_5(CH_3)N)$_3$Cl][BF_4] Mo:Org.Comp.6-211/2

$BC_{38}Cl_2H_{41}NO_3P$ (C_2H_5)$_3$N-B(C_6H_5)$_2$-O-CH(C_6H_4-4-Cl)-P(O)(C_6H_5)-CHOH-C_6H_4-4-Cl. .... B: B Comp.SVol.4/3b-78

– .... [(C_2H_5)$_3$NH][(C_6H_5)$_2$B(-O-CH(C_6H_4-4-Cl)-P(O)(C_6H_5)-CH(C_6H_4-4-Cl)-O-)] .... B: B Comp.SVol.4/3b-78

$BC_{38}F_3H_{50}N_2Si_2$ .. FB[N(C_6H_2-2,4,6-(CH_3)$_3$)SiF(C_6H_5)C_4H_9-t]$_2$ ... B: B Comp.SVol.3/3-376

$BC_{38}F_4FeH_{30}O_4P$ [(C_5H_5)Fe(C_6H_5-CC-C_6H_5)(CO)P(OC_6H_5)$_3$][BF_4]
Fe: Org.Comp.B16b-100/1

$BC_{38}F_4FeH_{35}OP_2$ [C_5H_5((C_6H_5)$_2PCH_2CH_2P$(C_6H_5)$_2$)Fe=C(C_6H_5)OH][BF_4]
Fe: Org.Comp.B16a-9

$BC_{38}F_4FeH_{37}NOP$ [C_5H_5(CO)(P(C_6H_5)$_3$)Fe=C(NCH_3CH(CH=CHC_6H_5)C(CH_3)$_2$-)][BF_4]. .... Fe: Org.Comp.B16a-48, 59

$BC_{38}F_4FeH_{40}NOP_2$
[(C_5H_5)Fe(CNC(CH_3)$_2CH_2$C(O)CH_3)P(C_6H_5)$_2$$CH_2CH_2P$($C_6H_5$)$_2$][$BF_4$] .... Fe: Org.Comp.B15-282

$BC_{38}F_4FeH_{41}$ .... [((CH_3)$_6C_6$)Fe(C_7H_8-C(C_6H_5)$_3$)][BF_4]. .... Fe: Org.Comp.B19-172

$BC_{38}F_4H_{27}MoO$ .. [C_9H_7Mo(CO)(C_6H_5-CC-C_6H_5)$_2$][BF_4]. Mo:Org.Comp.6-320

$BC_{38}F_4H_{30}MoOP$ [C_5H_5Mo(CO)(P(C_6H_5)$_3$)C_6H_5-CC-C_6H_5][BF_4] Mo:Org.Comp.6-288, 289

$BC_{38}F_4H_{30}OPSe$ .. [((C_6H_5)$_2$Se)((C_6H_5)$_3$P)C(COC_6H_5)][BF_4] ...... B: B Comp.SVol.3/4-148

$BC_{38}F_4H_{33}MoO$ .. [C_5H_5Mo(CO)(4-$CH_3C_6H_4$-CC-$C_6H_4CH_3$-4)$_2$][BF_4]
Mo:Org.Comp.6-320, 326

$BC_{38}F_4H_{36}MoN_4P$ [$CH_3C_5H_4$Mo(N=$NC_6H_4CH_3$-4)$_2$P(C_6H_5)$_3$][BF_4] Mo:Org.Comp.6-62

$BC_{38}FeH_{35}O_2$ .... [(C_2H_5-OOC-C_6H_5)Fe(C_5H_5)][B(C_6H_5)$_4$] ...... Fe: Org.Comp.B18-142/6, 215

– .... [(C_6H_6)Fe(C_5H_4-COO-C_2H_5)][B(C_6H_5)$_4$] ...... Fe: Org.Comp.B18-142/6, 151, 164

$BC_{38}FeH_{37}$ ...... [(1,3,5-(CH_3)$_3$-C_6H_3)Fe(C_5H_5)][B(C_6H_5)$_4$] ..... Fe: Org.Comp.B19-99, 104, 133/6

– .... [(CH_3-C_6H_5)Fe(C_7H_9)][B(C_6H_5)$_4$]. .... Fe: Org.Comp.B18-142/6, 209

$BC_{38}FeH_{43}$ ...... [(CH_3-C_6H_4)$_3$B-C_6H_4-CH_3]Fe[C_5(CH_3)$_5$]. ..... Fe: Org.Comp.B19-6

$BC_{38}FeH_{43}O_4$ .... [(CH_3-O-C_6H_4)$_3$B-C_6H_4-O-CH_3]Fe[C_5(CH_3)$_5$] Fe: Org.Comp.B19-6

$BC_{38}H_{39}MoNO_4P$ [(C_5H_5)Mo(CO)$_2$(1,3,6,2-$O_2NPC_4H_9$-2-CH_2CH=CH_2)][B(C_6H_5)$_4$]. Mo:Org.Comp.7-249, 271, 281

$BC_{38}H_{41}N_3O_7P$ ... (C_2H_5)$_3$N-B(C_6H_5)$_2$-O-CH(C_6H_4-4-NO_2)-P(O)(C_6H_5)-CHOH-C_6H_4-4-NO_2 .... B: B Comp.SVol.4/3b-78

– .... [(C_2H_5)$_3$NH][(C_6H_5)$_2$B(-O-CH(C_6H_4-4-NO_2)-P(O)(C_6H_5)-CH(C_6H_4-4-NO_2)-O-)] .... B: B Comp.SVol.4/3b-78

$BC_{38}H_{43}NO_3P$.... (C_2H_5)$_3$N-B(C_6H_5)$_2$-O-CH(C_6H_5)-P(O)(C_6H_5)-CHOH-C_6H_5 .... B: B Comp.SVol.4/3b-78

– .... [(C_2H_5)$_3$NH][(C_6H_5)$_2$B(-O-CH(C_6H_5)-P(O)(C_6H_5)-CH(C_6H_5)-O-)] .... B: B Comp.SVol.4/3b-78

$BC_{39}ClF_4H_{30}O_3P_2Re$
[(CO)$_3$ReCl(P(C_6H_5)$_3$)$_2$][BF_4] .... Re: Org.Comp.1-249, 306

$BC_{39}CoH_{30}$ ...... $(C_5H_5)Co[(C_6H_5)_4C_4BC_6H_5]$ .... B: B Comp.SVol.3/4-155
$BC_{39}F_4FeH_{37}OP_2$ $[C_5H_5((C_6H_5)_2PCH_2CH_2P(C_6H_5)_2)Fe{=}C(C_6H_5)OCH_3][BF_4]$ Fe: Org.Comp.B16a-9
$BC_{39}F_4FeH_{37}P_2$ .. $[(C_5H_5)Fe(CH_2{=}CHCH_2CH_2P(C_6H_5)_2)(P(C_6H_5)_3)][BF_4]$ Fe: Org.Comp.B16b-3
$BC_{39}F_4H_{30}MoN$... $[C_9H_7Mo(NCCH_3)(C_6H_5-CC-C_6H_5)_2][BF_4]$..... Mo:Org.Comp.6-137, 146
$BC_{39}F_4H_{66}MnN_3S_6$
$[Mn(SC({=}S)N(C_6H_{11}-c)_2)_3][BF_4]$............. Mn:MVol.D7-162/3
$BC_{39}FeH_{33}$ ...... $[(C_{10}H_8)Fe(C_5H_5)][B(C_6H_5)_4]$ .... Fe: Org.Comp.B19-216, 220, 233/4
$BC_{39}FeH_{37}$ ...... $[(C_{10}H_{12})Fe(C_5H_5)][B(C_6H_5)_4]$ .... Fe: Org.Comp.B19-216, 219/20, 225
$BC_{39}FeH_{39}S$ ..... $[(n-C_4H_9-S-C_6H_5)Fe(C_5H_5)][B(C_6H_5)_4]$....... Fe: Org.Comp.B18-142/6, 198, 253
$BC_{39}H_{44}N_3O_2S$... $[(1,4-ONC_4H_8-4-)_2S-NCH_3-C_6H_5][B(C_6H_5)_4]$.. S: S-N Comp.8-230, 244
$BC_{39}H_{47}N_3O_3Re$.. $[cis-(CO)_3Re(C_4H_9N)_3][B(C_6H_5)_4]$ .... Re: Org.Comp.1-308
$BC_{39}H_{50}N_3S$ ..... $[(1-C_5H_{10}N)_3S][B(C_6H_5)_4]$ .... S: S-N Comp.8-222/4, 228
$BC_{40}F_4Fe_2H_{38}O_2P$ $[C_5H_5(CO)(P(C_6H_5)_3)Fe{=}C(CH_2CH_2C_6H_4FeC_5H_5)OC_2H_5][BF_4]$ .... Fe: Org.Comp.B16a-42
$BC_{40}F_4H_{30}O_{10}P_2Re$
trans-$[(CO)_4Re(P(OC_6H_5)_3)_2][BF_4]$ .... Re: Org.Comp.1-482
$BC_{40}F_4H_{34}MoOP$ $[C_5H_5Mo(CO)(P(C_6H_5)_3)(4-CH_3C_6H_4-CC-C_6H_4CH_3-4)][BF_4]$ Mo:Org.Comp.6-288
$BC_{40}FeH_{35}$ ...... $[(C_5H_5-C_6H_5)Fe(C_5H_5)][B(C_6H_5)_4]$........... Fe: Org.Comp.B18-142/6, 200, 219
$BC_{40}FeH_{38}O_2P$... $[(C_5H_5)Fe(CO)_2CH_2P(CH_3)_2C_6H_5][B(C_6H_5)_4]$... Fe: Org.Comp.B14-136
$BC_{40}FeH_{39}O$ ..... $[(1,3,5-(CH_3)_3-C_6H_3)Fe(C_5H_4-C({=}O)-CH_3)][B(C_6H_5)_4]$ Fe: Org.Comp.B19-99, 107
$BC_{40}FeH_{40}N$ ..... $[(C_6H_6)Fe(1-C_5H_4-NC_5H_{10})][B(C_6H_5)_4]$....... Fe: Org.Comp.B18-142/6, 168
$BC_{40}FeH_{41}$ ...... $[(1,3,5-(CH_3)_3-C_6H_3)Fe(C_5H_4-C_2H_5)][B(C_6H_5)_4]$ Fe: Org.Comp.B19-99, 106
$BC_{40}FeH_{41}O_3$ .... $[((CH_3)_7C_6)Fe(CO)_3][B(C_6H_5)_4]$ .... Fe: Org.Comp.B15-134, 185/6
$BC_{40}Fe_2H_{36}O_4Sb$ $[(CH_3)_2Sb(Fe(CO)_2C_5H_5)_2][B(C_6H_5)_4]$ .... Sb: Org.Comp.5-219
$BC_{40}H_{46}N_3O_3$ .... $(C_6H_5-NH)_2C{=}NH-B[C_6H_4-2-(O-C_3H_7-n)]_3$ ... B: B Comp.SVol.4/3b-27
$BC_{40}H_{46}N_3O_4S_2$ .. $[(1,4-ONC_4H_8-4-)_2S-NCH_3-S(O)_2-C_6H_4-4-CH_3][B(C_6H_5)_4]$...................... S: S-N Comp.8-230, 244
$BC_{40}H_{54}NO_{10}$ .... $H_3B-NH_3 \cdot 2,3,11,12-(4-CH_3O-C_6H_4)_4-1,4,7,10,13,16-O_6C_{12}H_{20}$ .... B: B Comp.SVol.4/3b-2
$BC_{41}ClFeH_{34}O$ ... $[(3-Cl-C_6H_4-O-C_6H_5)Fe(C_5H_5)][B(C_6H_5)_4]$.... Fe: Org.Comp.B18-142/6, 198, 252
$BC_{41}CoGeH_{61}P_3$.. $(CH_3)_2Ge(-CH{=}C(CH_3)C(CH_3){=}CH-)Co(P(CH_3)_3)_3[B(C_6H_5)_4]$................. Ge: Org.Comp.3-282, 286
$BC_{41}F_4FeH_{35}NOP$ $[C_5H_5(CO)(P(C_6H_5)_3)Fe{=}C(-NCH_3CHC_6H_5C({=}CHC_6H_5)-)][BF_4]$.................... Fe: Org.Comp.B16a-49, 60
$BC_{41}F_4H_{35}IMoNOP_2$
$[C_5H_5Mo(NO)(P(C_6H_5)_3)_2I][BF_4]$............. Mo:Org.Comp.6-39
$BC_{41}F_4H_{35}NOP_2Re$
$[(C_5H_5)Re(NO)(P(C_6H_5)_3)_2][BF_4]$ .... Re: Org.Comp.3-29/30, 36
$BC_{41}F_4H_{38}P_2Re$ .. $[(C_5H_5)Re(P(C_6H_5)_3)_2(H)_3][BF_4] \cdot H_2O$ .... Re: Org.Comp.3-12, 13

$BC_{41}F_4H_{43}N_4OOsP_2S_2$
$[(H)(CO)Os(SNN(CH_3)_2)_2(P(C_6H_5)_3)_2][BF_4]$ .... Os: Org.Comp.A1-147

$BC_{41}FeH_{34}N$ ..... $[(NC-C_6H_5)Fe(C_5H_5)][C_{10}H_7-2-CH_2-B(C_6H_5)_3]$ Fe: Org.Comp.B18-142/6, 201, 202/3, 268

$BC_{41}FeH_{35}$ ...... $[(C_6H_6)Fe(C_5H_4-C_6H_5)][B(C_6H_5)_4]$ .... Fe: Org.Comp.B18-142/6, 151, 165

– .... $[(C_6H_5-C_6H_5)Fe(C_5H_5)][B(C_6H_5)_4]$ .... Fe: Org.Comp.B18-142/6, 197, 201, 224/5

$BC_{41}FeH_{35}O$ ..... $[(C_6H_5-O-C_6H_5)Fe(C_5H_5)][B(C_6H_5)_4]$ .... Fe: Org.Comp.B18-142/6, 197, 198, 201, 252

$BC_{41}FeH_{35}S$ ..... $[(C_6H_5-S-C_6H_5)Fe(C_5H_5)][B(C_6H_5)_4]$ .... Fe: Org.Comp.B18-142/6, 198, 253

$BC_{41}FeH_{37}$ ...... $[(CH_3C_6H_5)Fe(C_5H_5)][C_{10}H_7-2-CH_2B(C_6H_5)_3]$. Fe: Org.Comp.B18-142/6, 201, 206/7

$BC_{41}FeH_{43}$ ...... $[((CH_3)_6C_6)Fe(C_5H_5)][B(C_6H_5)_4]$ .... Fe: Org.Comp.B19-142, 156, 193/7

$BC_{41}H_{35}MoN_2O_2$. $[(C_5H_5)Mo(CO)_2(NC_5H_5)_2][B(C_6H_5)_4]$ .... Mo: Org.Comp.7-284

$BC_{41}H_{43}MoO_8P_2$ .. $[(C_5H_5)Mo(CO)_2(P(-OCH_2-)_3CCH_3)_2][B(C_6H_5)_4]$ Mo: Org.Comp.7-283, 291

$BC_{41}H_{46}MoN_4O_2P$ $[(C_5H_5)Mo(CO)_2(C_{10}H_{21}N_4P)][B(C_6H_5)_4]$ ...... Mo: Org.Comp.7-250, 270

$BC_{42}ClFeH_{36}O$ ... $[(3-CH_3-C_6H_4-O-C_6H_4-3-Cl)Fe(C_5H_5)][B(C_6H_5)_4]$ Fe: Org.Comp.B19-2, 49

$BC_{42}F_4H_{35}MoOP_2$ $C_5H_5Mo(CO)(P(C_6H_5)_3)_2FBF_3$ .... Mo: Org.Comp.6-224

$BC_{42}FeH_{35}$ ...... $[(C_{13}H_{10})Fe(C_5H_5)][B(C_6H_5)_4]$ .... Fe: Org.Comp.B19-216, 220, 250

$BC_{42}FeH_{37}O$ ..... $[(2-CH_3-C_6H_4-O-C_6H_5)Fe(C_5H_5)][B(C_6H_5)_4]$ .. Fe: Org.Comp.B19-2, 48

– .... $[(3-CH_3-C_6H_4-O-C_6H_5)Fe(C_5H_5)][B(C_6H_5)_4]$ .. Fe: Org.Comp.B18-142/6, 198, 252
Fe: Org.Comp.B19-2, 48

– .... $[(4-CH_3-C_6H_4-O-C_6H_5)Fe(C_5H_5)][B(C_6H_5)_4]$ .. Fe: Org.Comp.B18-142/6, 198, 252

$BC_{42}FeH_{42}O_2P$ ... $[(C_5H_5)Fe(CO)_2CH_2P(C_2H_5)_2C_6H_5][B(C_6H_5)_4]$ .. Fe: Org.Comp.B14-136

$BC_{42}Fe_2H_{38}O_4Sb$ $[(CH_3)(CH_2=CHCH_2)Sb(Fe(CO)_2C_5H_5)_2][B(C_6H_5)_4]$ Sb: Org.Comp.5-230

$BC_{42}H_{37}Pt$ .... $(C_8H_{12})Pt[(C_6H_5)_4C_4BC_6H_5]$ .... B: B Comp.SVol.3/4-155

$BC_{42}H_{41}MoNO_4P$ $[(C_5H_5)Mo(CO)_2(1,3,6,2-O_2NPC_4H_9-2-CH_2-C_6H_5)][B(C_6H_5)_4]$ .... Mo: Org.Comp.7-249, 271, 281

$BC_{42}H_{41}O_2SSe$ ... $[((C_6H_5)_2Se)((CH_3)_2S)C(COOC_2H_5)][B(C_6H_5)_4]$ B: B Comp.SVol.3/4-148

$BC_{43}Cl_2F_4H_{35}N_2OOsP_2$
$4-Cl-C_6H_4-N=NH-OsCl(CO)[P(C_6H_5)_3]_2-F-BF_3$ Os: Org.Comp.A1-161, 166

– .... $[4-Cl-C_6H_4-N=NH-OsCl(CO)(P(C_6H_5)_3)_2][BF_4]$ Os: Org.Comp.A1-166

$BC_{43}F_4FeH_{39}O_6P_2$ $[(C_5H_5)Fe(CH_2=CH_2)(P(OC_6H_5)_3)_2][BF_4]$ ...... Fe: Org.Comp.B16b-5

$BC_{43}F_4H_{35}MoO_2P_2$
$[(C_5H_5)Mo(CO)_2(P(C_6H_5)_3)_2][BF_4]$ .... Mo: Org.Comp.7-283, 288/9, 298

$BC_{43}F_4H_{35}MoO_2Sb_2$
$[(C_5H_5)Mo(CO)_2(Sb(C_6H_5)_3)_2][BF_4]$ .... Mo: Org.Comp.7-283, 292

$BC_{43}F_4H_{35}MoO_5P_2$
$[(C_5H_5)Mo(CO)_2(P(C_6H_5)_3)(P(O-C_6H_5)_3)][BF_4]$ Mo: Org.Comp.7-283, 293

$BC_{43}F_4H_{35}MoO_8P_2$
$[(C_5H_5)Mo(CO)_2(P(O\text{-}C_6H_5)_3)_2][BF_4]$ Mo:Org.Comp.7-283, 291

$BC_{43}FeH_{37}$ ...... $[(C_{14}H_{12})Fe(C_5H_5)][B(C_6H_5)_4]$ .... Fe: Org.Comp.B19-216, 220, 263/4

$BC_{43}FeH_{37}O_3$ .... $[(CH_3OC(=O)\text{-}3\text{-}C_6H_4\text{-}O\text{-}C_6H_5)Fe(C_5H_5)][B(C_6H_5)_4]$
Fe: Org.Comp.B18-142/6, 198, 252

$BC_{43}FeH_{39}O$ ..... $[(2\text{-}CH_3\text{-}C_6H_4\text{-}O\text{-}C_6H_4\text{-}3\text{-}CH_3)Fe(C_5H_5)][B(C_6H_5)_4]$
Fe: Org.Comp.B19-2, 49

– .... $[(2\text{-}CH_3\text{-}C_6H_4\text{-}O\text{-}C_6H_4\text{-}4\text{-}CH_3)Fe(C_5H_5)][B(C_6H_5)_4]$
Fe: Org.Comp.B19-2, 49

– .... $[(3\text{-}CH_3\text{-}C_6H_4\text{-}O\text{-}C_6H_4\text{-}3\text{-}CH_3)Fe(C_5H_5)][B(C_6H_5)_4]$
Fe: Org.Comp.B19-2, 49

– .... $[(3\text{-}CH_3\text{-}C_6H_4\text{-}O\text{-}C_6H_4\text{-}4\text{-}CH_3)Fe(C_5H_5)][B(C_6H_5)_4]$
Fe: Org.Comp.B19-2, 49

$BC_{43}FeH_{47}$ ...... $[(1,3,5\text{-}(CH_3)_3\text{-}C_6H_3)Fe(C_6H_3\text{-}1,2,3,5\text{-}(CH_3)_4)][B(C_6H_5)_4]$
Fe: Org.Comp.B19-101, 112

$BC_{43}H_{34}N_3$ ...... $(C_6H_5\text{-}NH)_2C{=}NH\text{-}B(1\text{-}C_{10}H_7)_3$ .... B: B Comp.SVol.4/3b-27

$BC_{43}H_{55}MoO_8P_2$ .. $[(C_5H_5)Mo(CO)_2(P(O\text{-}C_2H_5)_3)_2][B(C_6H_5)_4]$ Mo:Org.Comp.7-283, 290

$BC_{44}ClF_4H_{38}N_2OOsP_2$
$4\text{-}CH_3\text{-}C_6H_4\text{-}N{=}NH\text{-}OsCl(CO)[P(C_6H_5)_3]_2\text{-}F\text{-}BF_3$
Os: Org.Comp.A1-161, 167

– .... $[4\text{-}CH_3\text{-}C_6H_4\text{-}N{=}NH\text{-}OsCl(CO)(P(C_6H_5)_3)_2][BF_4]$
Os: Org.Comp.A1-167

$BC_{44}ClF_4H_{38}N_2O_2OsP_2$
$4\text{-}CH_3O\text{-}C_6H_4\text{-}N{=}NH\text{-}OsCl(CO)[P(C_6H_5)_3]_2\text{-}F\text{-}BF_3$
Os: Org.Comp.A1-161, 168

– .... $[4\text{-}CH_3O\text{-}C_6H_4\text{-}N{=}NH\text{-}OsCl(CO)(P(C_6H_5)_3)_2][BF_4]$
Os: Org.Comp.A1-168

$BC_{44}F_4FeH_{35}O_2P_2PtS_2$
$[C_5H_5(CO)_2Fe{=}CS_2Pt(P(C_6H_5)_3)_2][BF_4]$ .... Fe: Org.Comp.B16a-144, 156

$BC_{44}F_4FeH_{37}$ .... $[(1,3,5\text{-}(CH_3)_3\text{-}C_6H_3)Fe(C_5(C_6H_5)_5)][BF_4]$ ..... Fe: Org.Comp.B19-109

$BC_{44}F_4FeH_{37}OP_2$ $[C_5H_5(CO)(P(C_6H_5)_3)FeC(P(C_6H_5)_3){=}CH_2][BF_4]$ Fe: Org.Comp.B16a-215/7

$BC_{44}FeH_{39}O_3$ .... $[(CH_3\text{-}COO\text{-}C_6H_4\text{-}3\text{-}O\text{-}C_6H_4\text{-}2\text{-}CH_3)Fe(C_5H_5)][B(C_6H_5)_4]$ .... Fe: Org.Comp.B19-2, 49

– .... $[(CH_3\text{-}COO\text{-}C_6H_4\text{-}3\text{-}O\text{-}C_6H_4\text{-}3\text{-}CH_3)Fe(C_5H_5)][B(C_6H_5)_4]$ .... Fe: Org.Comp.B19-2, 49

$BC_{44}Fe_2H_{40}O_4Sb$ $[(CH_2{=}CHCH_2)_2Sb(Fe(CO)_2C_5H_5)_2][B(C_6H_5)_4]$ .. Sb: Org.Comp.5-219

$BC_{44}H_{38}MoN_3O_2$. $[(CH_2CHCH_2)Mo(CO)_2(NC_5H_4\text{-}2\text{-}(2\text{-}C_5H_4N))(NC_5H_5)][B(C_6H_5)_4]$ Mo:Org.Comp.5-200, 202

$BC_{45}F_4H_{36}N_2O_4P_2Re$
$[(\text{-}O\text{-}C(C_6H_5){=}NH\text{-}C(O)\text{-})Re(P(C_6H_5)_3)_2(NO)(CO)][BF_4]$
Re: Org.Comp.1-53

$BC_{45}F_4H_{39}MoO_2P_2$
$[(C_5H_5)Mo(CO)_2(P(C_6H_5)_3)\text{-}CH(CH_3)\text{-}P(C_6H_5)_3][BF_4]$
Mo:Org.Comp.8-99

$BC_{45}F_4H_{39}MoP_2$ .. $[C_5H_5Mo(P(C_6H_5)_2CH_2CH_2P(C_6H_5)_2)C_6H_5\text{-}CC\text{-}C_6H_5][BF_4]$ Mo:Org.Comp.6-120

$BC_{45}FeH_{35}O_3$ .... $[((C_6H_5)_2CHC_5H_4)Fe(CO)_3][B(C_6H_5)_4]$ .... Fe: Org.Comp.B15-51/2

$BC_{45}FeH_{40}O_2P$ ... $[(C_5H_5)Fe(CO)_2CH_2P(C_6H_5)_2CH_3][B(C_6H_5)_4]$ ... Fe: Org.Comp.B14-136

$BC_{45}Fe_3H_{34}NO_9P_2$
$[(C_6H_5)_3P{=}N{=}P(C_6H_5)_3][(CO)_9(H)Fe_3(BH_3)]$ .... Fe: Org.Comp.C6a-298/9

$BC_{45}H_{52}MoNO_2$ .. $[2\text{-}i\text{-}C_3H_7\text{-}5\text{-}CH_3\text{-}c\text{-}C_6H_9C_5H_4Mo(CO)(NO)C_3H_3(CH_3)_2\text{-}1,2][B(C_6H_5)_4]$ .... Mo: Org.Comp.6-349

$BC_{45}H_{53}O_7Ti_3$.... $[(C_5H_5)_3Ti_3(OCH_3)_6O][B(C_6H_5)_4]$ .... Ti: Org.Comp.5-280

$BC_{45}H_{69}OOsP_4$... $[(H)(CO)Os((C_2H_5)_2P\text{-}CH_2CH_2\text{-}P(C_2H_5)_2)_2][B(C_6H_5)_4]$ Os: Org.Comp.A1-120

$BC_{46}F_4FeH_{38}Sn_2$ $[Fe(C_5H_4\text{-}Sn(C_6H_5)_3)_2][BF_4]$ .... Fe: Org.Comp.A10-159/60

$BC_{46}F_4H_{39}NO_3P_2Re$
$[(CO)_2Re(P(C_6H_5)_3)_2NH_2C(O)C_6H_4\text{-}4\text{-}CH_3][BF_4]$ Re: Org.Comp.1-75

$BC_{46}FeH_{42}O_2P$... $[(C_5H_5)Fe(CO)_2CH_2P(C_6H_5)_2C_2H_5][B(C_6H_5)_4]$.. Fe: Org.Comp.B14-137

$BC_{46}Fe_3H_{32}NO_{10}P_2$
$[(C_6H_5)_3P{=}N{=}P(C_6H_5)_3][(CO)_{10}Fe_3(BH_2)]$ ..... Fe: Org.Comp.C6b-70

$BC_{46}Fe_3H_{36}NO_9P_2$
$[(C_6H_5)_3P{=}N{=}P(C_6H_5)_3][(CO)_9(H)Fe_3(BH_2\text{-}CH_3)]$ Fe: Org.Comp.C6a-299

$BC_{47}ClF_4H_{42}O_2P_4Re$
$[(CO)_2Re(Cl)(P(C_6H_5)_2CH_2P(C_6H_5)_2)P(C_6H_5)CH_2P(CH_3)(C_6H_5)_2][BF_4]$ .... Re: Org.Comp.1-91

$BC_{47}F_4H_{39}NO_4P_2Re$
$[(CO)_3Re(P(C_6H_5)_3)_2NH_2C(O)C_6H_4\text{-}4\text{-}CH_3][BF_4]$ Re: Org.Comp.1-311

$BC_{47}FeH_{55}$...... $[((C_2H_5)_6C_6)Fe(C_5H_5)][B(C_6H_5)_4]$............ Fe: Org.Comp.B19-142, 173/4, 205/9

$BC_{47}Fe_2H_{40}O_4Sb$ $[(C_6H_5)(CH_2{=}CHCH_2)Sb(Fe(CO)_2C_5H_5)_2][B(C_6H_5)_4]$ Sb: Org.Comp.5-230

$BC_{47}Fe_3H_{34}NO_{10}P_2$
$[(C_6H_5)_3P{=}N{=}P(C_6H_5)_3][(CO)_{10}Fe_3(BH\text{-}CH_3)]$ .. Fe: Org.Comp.C6b-70

$BC_{48}F_4FeH_{43}OP_2$ $[C_5H_5((C_6H_5)_2PCH_2CH_2P(C_6H_5)_2)Fe{=}C{=}C(CH_3)C(O)CH(C_6H_5)_2][BF_4]$ .... Fe: Org.Comp.B16a-201, 204

$BC_{48}F_4H_{50}N_4P_2Re$ $[Re(CNCH(CH_3)_2)_2(P(C_6H_5)_3)_2(NCCH_3)_2][BF_4]$ Re: Org.Comp.2-256

$BC_{48}FeH_{41}O$..... $[(2\text{-}CH_3\text{-}C_6H_4\text{-}O\text{-}C_6H_4\text{-}4\text{-}C_6H_5)Fe(C_5H_5)][B(C_6H_5)_4]$ Fe: Org.Comp.B19-2, 49

$BC_{48}H_{61}MoN_4O$ .. $[(CH_2CHCH_2)Mo(CO)(CN\text{-}C_4H_9\text{-}t)_4][B(C_6H_5)_4]$ Mo: Org.Comp.5-296/9

$BC_{48}H_{65}MoN_2OP_2$ $[(CH_2CHCH_2)Mo(CO)(CN\text{-}C_6H_{11}\text{-}c)_2(P(CH_3)_3)_2][B(C_6H_5)_4]$.................... Mo: Org.Comp.5-296/9

$BC_{49}ClF_2FeH_{47}IrO_3P_2$
$[(C_5H_5)(CO)Fe(CO)(CF_2)IrCl(CO)(P(CH_3)_2C_6H_5)_2][B(C_6H_5)_4]$ .... Fe: Org.Comp.B16b-148

$BC_{49}F_4FeH_{45}O_6P_2$ $[(C_5H_5)Fe(CH_2{=}CH_2)(P(OC_6H_5)_3)_2][BF_4] \cdot C_6H_6$ Fe: Org.Comp.B16b-5

$BC_{49}H_{39}O_4PRe$... $[(2\text{-}(CH_2{=}CHCH_2)C_6H_4P(C_6H_5)_2)Re(CO)_4][B(C_6H_5)_4]$ Re: Org.Comp.2-327, 329

$BC_{49}H_{67}MoO_8P_2$.. $[(C_5H_5)Mo(CO)_2(P(O\text{-}C_3H_7\text{-}i)_3)_2][B(C_6H_5)_4]$ ... Mo: Org.Comp.7-283, 291

– .... $[(C_5H_5)Mo(CO)_2(P(O\text{-}C_3H_7\text{-}n)_3)_2][B(C_6H_5)_4]$... Mo: Org.Comp.7-283, 290

$BC_{49}H_{69}MoN_7^+$ .. $[Mo(CN\text{-}C_4H_9\text{-}t)_6(NC\text{-}B(C_6H_5)_3)][(C_6H_5)_3B\text{-}CN\text{-}B(C_6H_5)_3]$........................ Mo: Org.Comp.5-76, 81/2

$BC_{50}F_4H_{54}N_4P_2Re$ $[Re(CNC_4H_9\text{-}t)_2(P(C_6H_5)_3)_2(NCCH_3)_2][BF_4]$ ... Re: Org.Comp.2-258

$BC_{50}FeH_{42}O_2P$... $[(C_5H_5)Fe(CO)_2CH_2P(C_6H_5)_3][B(C_6H_5)_4]$ ...... Fe: Org.Comp.B14-137, 143

$BC_{50}FeH_{43}NOP$ .. $[(C_5H_5)Fe(CNCH_3)(CO)P(C_6H_5)_3][B(C_6H_5)_4]$ ... Fe: Org.Comp.B15-301

$BC_{50}Fe_2H_{40}O_4Sb$ $[(C_6H_5)_2Sb(Fe(CO)_2C_5H_5)_2][B(C_6H_5)_4]$ .... Sb: Org.Comp.5-219

$BC_{51}F_4H_{41}MoO_2P_2$
$[(C_5H_5)Mo(CO)_2(P(C_6H_5)_3)-C(=CH-C_6H_5)-P(C_6H_5)_3][BF_4]$ Mo:Org.Comp.8-99

$BC_{51}F_4H_{52}N_4P_2Re$ $[Re(CNCH(CH_3)_2)_2(P(C_6H_5)_3)_2(NCCH_3)C_5H_5N][BF_4]$ Re: Org.Comp.2-257

$BC_{51}FeH_{48}N_2OP$ .. $[C_5H_5(CO)(P(C_6H_5)_3)Fe=C(NHCH_3)_2][B(C_6H_5)_4]$ Fe: Org.Comp.B16a-68, 72/3

$BC_{52}FeH_{46}O_2P$ ... $[C_5H_5(CO)(P(C_6H_5)_3)Fe=C_4H_6O-c][B(C_6H_5)_4]$ .. Fe: Org.Comp.B16a-44/5, 57

$BC_{53}F_4H_{56}N_4P_2Re$ $[Re(CNC_4H_9-t)_2(P(C_6H_5)_3)_2(NCCH_3)C_5H_5N][BF_4]$ Re: Org.Comp.2-258

$BC_{53}Fe_3H_{43}NO_9P_3$
$[(C_6H_5)_3P=N=P(C_6H_5)_3][(CO)_9Fe_3(BH_2)(P(CH_3)_2-C_6H_5)]$ Fe: Org.Comp.C6a-299

$BC_{53}H_{46}MoO_3P$ .. $[(C_5H_5)Mo(CO)_2(P(C_6H_5)_3)=C(-CH_2CH_2CH_2$ $-O-)][B(C_6H_5)_4]$ Mo:Org.Comp.8-108

$BC_{54}ClF_4H_{52}NP_4Re$
$[CH_3NH-C{\equiv}ReCl(P(C_6H_5)_2CH_2CH_2P(C_6H_5)_2)_2][BF_4]$ Re: Org.Comp.1-12/3

$BC_{54}F_4H_{48}O_2P_4Re$ trans-$[(CO)_2Re(P(C_6H_5)_2CH_2CH_2P(C_6H_5)_2)_2][BF_4]$ Re: Org.Comp.1-93

$BC_{54}Fe_3H_{45}NO_9P_3$
$[(C_6H_5)_3P=N=P(C_6H_5)_3][(CO)_9Fe_3(H-B(CH_3))$ $(P(CH_3)_2-C_6H_5)]$ .... Fe: Org.Comp.C6a-299

$BC_{54}H_{48}MoO_3P$ .. $[(C_5H_5)Mo(CO)_2(P(C_6H_5)_3)=C(-CH_2CH_2CH_2$ $CH_2-O-)][B(C_6H_5)_4]$ Mo:Org.Comp.8-109

$BC_{54}H_{74}N_6Re$ .... $[Re(CNC_4H_9-t)_6][B(C_6H_5)_4]$ .... Re: Org.Comp.2-292

$BC_{55}F_5H_{45}NO_2P_3Re$
$[(CO)ReF(P(C_6H_5)_3)_3(NO)][BF_4]$ .... Re: Org.Comp.1-47/8

$BC_{55}H_{50}MoO_4P$ .. $[(C_5H_5)Mo(CO)_2(P(C_6H_5)_3)=C(-CH_2CH_2CH(O$ $-C_2H_5)-O-)][B(C_6H_5)_4]$ Mo:Org.Comp.8-109/10

$BC_{55}H_{55}MoO_6P_2$ .. $[(C_5H_5)Mo(CO)_2(C_6H_5-P(O-CH_2CH=CH_2)_2)_2][B(C_6H_5)_4]$ Mo:Org.Comp.7-283, 292

$BC_{55}H_{76}MoN_7$ .... $[(t-C_4H_9-NH-CC-NH-C_4H_9-t)Mo(CN)(CN$ $-C_4H_9-t)_4][B(C_6H_5)_4]$ Mo:Org.Comp.5-172, 174, 177/8

$BC_{55}H_{79}MoO_8P_2$ .. $[(C_5H_5)Mo(CO)_2(P(O-C_4H_9-n)_3)_2][B(C_6H_5)_4]$ ... Mo:Org.Comp.7-283, 291

$BC_{56}F_4H_{54}N_2P_4Re$ $[Re(CNCH_3)_2(P(C_6H_5)_2CH_2CH_2P(C_6H_5)_2)_2][BF_4]$ Re: Org.Comp.2-256, 262

$BC_{56}F_4H_{55}MoN_2P_4$
$[CH_3-NH-CMo(CN-CH_3)(P(C_6H_5)_2CH_2CH_2$ $P(C_6H_5)_2)_2][BF_4]$ Mo:Org.Comp.5-115/8

– .... $[HMo(CN-CH_3)_2(P(C_6H_5)_2CH_2CH_2P(C_6H_5)_2)_2][BF_4]$ Mo:Org.Comp.5-12, 13/4, 19

$BC_{56}F_4H_{57}MoN_2P_4$
$[CH_3NH-CMo(=CH-NHCH_3)(P(C_6H_5)_2CH_2CH_2$ $P(C_6H_5)_2)_2][BF_4] \cdot 0.5\ O(C_2H_5)_2$ Mo:Org.Comp.5-119

$BC_{56}F_4H_{66}N_4P_2Re$ $[Re(CNC_4H_9-t)_4(P(C_6H_5)_3)_2][BF_4]$ .... Re: Org.Comp.2-278, 280

$BC_{56}H_{45}OSe_2$ .... $[((C_6H_5)_2Se)_2C(COC_6H_5)][B(C_6H_5)_4]$ .... B: B Comp.SVol.3/4-148

$BC_{56}H_{69}MoN_4O$ .. $[(CH_2CHCH_2)Mo(CO)(CN-C_6H_{11}-c)_4][B(C_6H_5)_4]$ Mo:Org.Comp.5-296/9

$BC_{57}ClF_4H_{58}NP_4Re$
$[t-C_4H_9-NH-C{\equiv}ReCl(P(C_6H_5)_2CH_2CH_2P(C_6H_5)_2)_2][BF_4]$ Re: Org.Comp.1-13

$BC_{57}F_4H_{45}O_{12}P_3Re$
mer-$[(CO)_3Re(P(OC_6H_5)_3)_3][BF_4]$ . . . . Re: Org.Comp.1-310
$BC_{57}F_4H_{53}IMoNOP_4$
$[C_5H_5Mo(NO)(P(C_6H_5)_2CH_2CH_2P(C_6H_5)_2)_2I][BF_4]$
Mo: Org.Comp.6-39
$BC_{57}F_4H_{53}MoP_4$ . . $[C_5H_5Mo(P(C_6H_5)_2CH_2CH_2P(C_6H_5)_2)_2][BF_4]$ . . . Mo: Org.Comp.6-24
$BC_{57}H_{49}MoO_2P_2$ . . $[(C_5H_5)Mo(CO)_2((C_6H_5)_2P\text{-}CH_2CH_2\text{-}P(C_6H_5)_2)][B(C_6H_5)_4]$
Mo: Org.Comp.7-250, 274
$BC_{58}F_4H_{62}MoN_2O_{0.5}P_4$
$[CH_3NH\text{-}CMo(=CH\text{-}NHCH_3)(P(C_6H_5)_2\text{-}CH_2$
$CH_2\text{-}P(C_6H_5)_2)_2][BF_4] \cdot 0.5\ O(C_2H_5)_2$ . . . . Mo: Org.Comp.5-119
$BC_{58}FeH_{60}NPRh$ . . $[(C_5H_5)Fe(C_5H_3(P(C_6H_5)_2)\text{-}CH(CH_3)$
$\text{-}N(CH_3)_2)Rh(C_8H_{12})][B(C_6H_5)_4]$ . . . . Fe: Org.Comp.A10-67, 70, 73/5
$BC_{60}H_{65}O_2OsP_4$ . . $[O=CH\text{-}Os(CO)(P(C_6H_5)_2\text{-}CH_2\text{-}P(C_6H_5)_2)_2][B(C_2H_5)_4]$
Os: Org.Comp.A1-220/1, 232/3
$BC_{61}ClF_4H_{57}P_4Re$ trans-$[(CH_2C_6H_5)C_2H_2ReCl[P(C_6H_5)_2CH_2CH_2$
$P(C_6H_5)_2]_2][BF_4]$. . . . . Re: Org.Comp.2-304, 306/7
$BC_{61}GeH_{52}P$ . . . . . $Ge(C_6H_5)_3CH_2P(C_6H_5)_3[B(C_6H_5)_4]$. . . . . Ge: Org.Comp.3-67
$BC_{61}H_{55}MoO_4P_2$ . . $[(C_5H_5)Mo(CO)_2(CH_2=CHCH_2\text{-}O\text{-}P(C_6H_5)_2)_2][B(C_6H_5)_4]$
Mo: Org.Comp.7-283, 292
$BC_{62}F_4H_{54}N_2OOsP_3$
$[(H)(CO)Os(NH=N\text{-}C_6H_4\text{-}4\text{-}CH_3)(P(C_6H_5)_3)_3][BF_4]$
Os: Org.Comp.A1-168
$BC_{62}F_4H_{66}N_2P_4Re$ trans-$[Re(CNC_4H_9\text{-}t)_2(P(C_6H_5)_2CH_2CH_2P(C_6H_5)_2)_2][BF_4]$
Re: Org.Comp.2-257
$BC_{63}Cl_2H_{71}O_2OsP_4$
$[O=CH\text{-}Os(CO)(P(C_6H_5)_2\text{-}CH_2CH_2\text{-}P(C_6H_5)_2)_2]$
$[B(C_2H_5)_4] \cdot CH_2Cl_2$ . . . . Os: Org.Comp.A1-220, 233
$BC_{64}H_{59}NO_2OsP_3$ $[(ON)(CO)Os(P(C_6H_5)_2\text{-}CH_3)_3][B(C_6H_5)_4]$ . . . . . Os: Org.Comp.A1-162
$BC_{64}H_{59}NO_5OsP_3$ $[(ON)(CO)Os(P(C_6H_5)_3)_2(P(O\text{-}CH_3)_3)][B(C_6H_5)_4]$
Os: Org.Comp.A1-162
$BC_{66}Cl_2F_4H_{56}N_2P_4Re$
trans-$[Re(CNC_6H_4Cl\text{-}4)_2(P(C_6H_5)_2CH_2CH_2$
$P(C_6H_5)_2)_2][BF_4]$. . . . . Re: Org.Comp.2-261
$BC_{66}Cl_4F_4H_{54}N_2P_4Re$
trans-$[Re(CNC_6H_3Cl_2\text{-}2,6)_2(P(C_6H_5)_2CH_2CH_2$
$P(C_6H_5)_2)_2][BF_4]$. . . . . Re: Org.Comp.2-261
$BC_{66}H_{50}IMoN_6$ . . . $[(I)Mo(CN\text{-}C_6H_5)_6][B(C_6H_5)_4]$. . . . . Mo: Org.Comp.5-75
$BC_{67}FeH_{55}$ . . . . . . $[(1,2\text{-}(CH_3)_2\text{-}C_6H_4)Fe(C_5(C_6H_5)_5)][B(C_6H_5)_4]$ . . Fe: Org.Comp.B19-8
– . . . . $[(1,3\text{-}(CH_3)_2\text{-}C_6H_4)Fe(C_5(C_6H_5)_5)][B(C_6H_5)_4]$ . . Fe: Org.Comp.B19-10
– . . . . $[(1,4\text{-}(CH_3)_2\text{-}C_6H_4)Fe(C_5(C_6H_5)_5)][B(C_6H_5)_4]$ . . Fe: Org.Comp.B19-13/4
$BC_{68}F_4H_{62}MoN_2P_4$
$[Mo(CN\text{-}C_6H_4\text{-}CH_3\text{-}4)_2(P(C_6H_5)_2\text{-}CH_2CH_2\text{-}$
$P(C_6H_5)_2)_2][BF_4]$. . . . . Mo: Org.Comp.5-16, 20
$BC_{68}F_4H_{62}N_2O_2P_4Re$
trans-$[Re(CNC_6H_4OCH_3\text{-}4)_2(P(C_6H_5)_2CH_2$
$CH_2P(C_6H_5)_2)_2][BF_4]$ . . . . Re: Org.Comp.2-261/2

$BC_{68}F_4H_{62}N_2P_4Re$ trans-$[Re(CNC_6H_4CH_3\text{-}4)_2(P(C_6H_5)_2CH_2CH_2P(C_6H_5)_2)_2][BF_4]$ . . . . Re: Org.Comp.2-260

$BC_{68}F_4H_{63}MoN_2P_4$ $[MoH(CN\text{-}C_6H_4\text{-}CH_3\text{-}4)_2(P(C_6H_5)_2\text{-}CH_2CH_2\text{-}P(C_6H_5)_2)_2][BF_4]$ Mo:Org.Comp.5-12, 17

$BC_{68}FeH_{57}$ . . . . . . $[(1,3,5\text{-}(CH_3)_3\text{-}C_6H_3)Fe(C_5(C_6H_5)_5)][B(C_6H_5)_4]$ Fe: Org.Comp.B19-109/10

$BC_{68}H_{57}INP_2Re$ . . $[Re(CNC_6H_4CH_3\text{-}4)(P(C_6H_5)_3)_2I][B(C_6H_5)_4]$ . . . . Re: Org.Comp.2-233, 241

$BC_{69}FeH_{59}$ . . . . . . $[(1,2,4,5\text{-}(CH_3)_4\text{-}C_6H_2)Fe(C_5(C_6H_5)_5)][B(C_6H_5)_4]$ Fe: Org.Comp.B19-150

$BC_{69}H_{57}N_3O_4P_2Re$ $[(CO)Re(P(C_6H_5)_3)_2(NO)(NCO)NH_2C(O)C_6H_5][B(C_6H_5)_4]$ Re: Org.Comp.1-50

$BC_{69}H_{59}NO_2OsP_3$ $[(ON)(CO)Os((C_6H_5)_2P\text{-}CH_2CH_2\text{-}P(C_6H_5)_2)(P(C_6H_5)_3)][B(C_6H_5)_4]$ . . . . Os: Org.Comp.A1-163

$BC_{69}H_{62}O_4P_2ReSi$ $[(CO)_3Re((C_6H_5)_2PCH_2CH_2P(C_6H_5)_2)=C(OC_2H_5)Si(C_6H_5)_2CH_3][B(C_6H_5)_4]$ . . . . Re: Org.Comp.1-193

$BC_{70}F_4FeH_{51}$ . . . . $[(C_5H(C_6H_5)_5)Fe(C_5(C_6H_5)_5)][BF_4]$ . . . . Fe: Org.Comp.B18-142/6, 243

$BC_{70}H_{59}N_3O_4P_2Re$ $[(CO)Re(P(C_6H_5)_3)_2(NO)(NCO)NH_2C(O)C_6H_4CH_3\text{-}4][B(C_6H_5)_4]$ . . . . Re: Org.Comp.1-50

$BC_{71}FeH_{63}$ . . . . . . $[((CH_3)_6C_6)Fe(C_5(C_6H_5)_5)][B(C_6H_5)_4]$ . . . . Fe: Org.Comp.B19-163

$BC_{72}H_{62}N_6Re$ . . . . $[Re(CNC_6H_4CH_3\text{-}4)_6][B(C_6H_5)_4]$ . . . . Re: Org.Comp.2-294, 301

$BC_{77}ClH_{116}OOsP_4$ $[(CO)(Cl)Os((c\text{-}C_6H_{11})_2P\text{-}CH_2CH_2\text{-}P(C_6H_{11}\text{-}c)_2)_2][B(C_6H_5)_4]$ . . . . Os: Org.Comp.A1-123

$BC_{78}ClH_{64}O_4P_4ReRh$ $[(OC)_3Re(Cl)((C_6H_5)_2PCH_2P(C_6H_5)_2)_2Rh(CO)][B(C_6H_5)_4]$ Re: Org.Comp.1-321

$BC_{78}H_{68}O_2P_4Re$ . . cis-$[(CO)_2Re(P(C_6H_5)_2CH_2CH_2P(C_6H_5)_2)_2][B(C_6H_5)_4]$ Re: Org.Comp.1-92

$BC_{79}H_{65}NO_2OsP_3$ $[(ON)(CO)Os(P(C_6H_5)_3)_3][B(C_6H_5)_4]$ . . . . Os: Org.Comp.A1-162

$BC_{83}F_4FeH_{131}O_{60}$ $[(C_6H_6)Fe(C_5H_5)][BF_4] \cdot 2\ C_{36}H_{60}O_{30} \cdot 8\ H_2O$ Fe: Org.Comp.B18-180

$BC_{92}ClH_{82}MoN_2P_4$ $[MoCl(CN\text{-}C_6H_4\text{-}CH_3\text{-}4)_2(P(C_6H_5)_2\text{-}CH_2CH_2\text{-}P(C_6H_5)_2)_2][B(C_6H_5)_4]$ Mo:Org.Comp.5-17, 20/1

$BC_{92}H_{82}MoN_2P_4$ . . $[Mo(CN\text{-}C_6H_4\text{-}CH_3\text{-}4)_2(P(C_6H_5)_2\text{-}CH_2CH_2\text{-}P(C_6H_5)_2)_2][B(C_6H_5)_4] \cdot 0.5\ CH_2Cl_2$ Mo:Org.Comp.5-16, 20

$BC_{92.5}ClH_{83}MoN_2P_4$ $[Mo(CN\text{-}C_6H_4\text{-}CH_3\text{-}4)_2(P(C_6H_5)_2\text{-}CH_2CH_2\text{-}P(C_6H_5)_2)_2][B(C_6H_5)_4] \cdot 0.5\ CH_2Cl_2$ Mo:Org.Comp.5-16, 20

$BCaS_2$ . . . . $Ca[BS_2]$ . . . . B: B Comp.SVol.3/4-108

$BCeO_3$ . . . . $CeBO_3$ . . . . Sc: MVol.C11b-391/416

doped with rare earth ions . . . . Sc: MVol.C11b-410/1

– . . . . $CeBO_3$ solid solutions

$CeBO_3\text{-}TbBO_3$ . . . . Sc: MVol.C11b-417

BCl . . . . BCl . . . . B: B Comp.SVol.3/4-1/2, 6

B: B Comp.SVol.4/4-1/4

– . . . . $^{10}B^{35}Cl$ . . . . B: B Comp.SVol.3/4-2

– . . . . $^{11}B^{35}Cl$ . . . . B: B Comp.SVol.3/4-1/2

– . . . . $^{11}B^{37}Cl$ . . . . B: B Comp.SVol.3/4-1

BCl^+	BCl^+	B: B Comp.SVol.3/4-6
BCl^{2+}	BCl^{2+}	B: B Comp.SVol.3/4-6/7
$BClEu_2O_3$	Eu_2BO_3Cl	Sc: MVol.C11b-450
$BClF_2$	$BClF_2$	B: B Comp.SVol.3/4-64/7
		B: B Comp.SVol.4/4-70/1
$BClF_6$	$[ClF_2][BF_4]$	B: B Comp.SVol.4/3b-218/9
$BClF_{10}$	$[ClF_6][BF_4]$	B: B Comp.SVol.3/3-354
$BClH^+$	$BClH^+$	B: B Comp.SVol.4/4-39
–	$HBCl^+$	B: B Comp.SVol.4/4-39
$BClH_2$	BH_2Cl	B: B Comp.SVol.3/4-37
		B: B Comp.SVol.4/4-39
$BClH_3^-$	$[H_3BCl]^-$	B: B Comp.SVol.3/4-38
		B: B Comp.SVol.4/4-39/40
$BClH_3Li$	$Li[H_3BCl]$	B: B Comp.SVol.4/4-40
$BClI_2$	$BClI_2$	B: B Comp.SVol.3/4-101
		B: B Comp.SVol.4/4-115
BClO	BOCl	B: B Comp.SVol.4/4-54
–	ClBO	B: B Comp.SVol.3/4-48/9
		B: B Comp.SVol.4/4-54
–	$^{35}Cl^{10}B^{16}O$	B: B Comp.SVol.3/4-48/9
–	$^{35}Cl^{10}B^{18}O$	B: B Comp.SVol.3/4-48
–	$^{35}Cl^{11}B^{16}O$	B: B Comp.SVol.3/4-48/9
–	$^{35}Cl^{11}B^{18}O$	B: B Comp.SVol.3/4-48
–	$^{37}Cl^{10}B^{16}O$	B: B Comp.SVol.3/4-48
–	$^{37}Cl^{10}B^{18}O$	B: B Comp.SVol.3/4-48
–	$^{37}Cl^{11}B^{16}O$	B: B Comp.SVol.3/4-48/9
–	$^{37}Cl^{11}B^{18}O$	B: B Comp.SVol.3/4-48
BClS	BSCl	B: B Comp.SVol.3/4-136
		B: B Comp.SVol.4/4-154/6
–	ClBS	B: B Comp.SVol.3/4-135/6
		B: B Comp.SVol.4/4-154/7, 159/60
$BClS^+$	$[ClB=S]^+$	B: B Comp.SVol.3/4-136
		B: B Comp.SVol.4/4-158/60
BCl_2	BCl_2, radical	B: B Comp.SVol.3/4-2/3, 6
		B: B Comp.SVol.4/4-4
BCl_2^+	BCl_2^+	B: B Comp.SVol.3/4-3, 6/7
BCl_2^{2+}	BCl_2^{2+}	B: B Comp.SVol.3/4-6/7
BCl_2ErH_4	$ErCl_2[BH_4] \cdot n\ C_4H_8O$	Sc: MVol.C11b-490
BCl_2F	BCl_2F	B: B Comp.SVol.3/4-64/7
		B: B Comp.SVol.4/4-70/1
BCl_2F_4O	$[Cl_2O][BF_4]$	B: B Comp.SVol.3/3-354
BCl_2H	$HBCl_2$	B: B Comp.SVol.3/4-35/7
		B: B Comp.SVol.4/4-37/8
–	$H^{10}B^{35}Cl_2$	B: B Comp.SVol.4/4-37/8
–	$H^{11}B^{35}Cl_2$	B: B Comp.SVol.4/4-37/8
–	$DBCl_2$	B: B Comp.SVol.4/4-37/8
–	$D^{10}B^{35}Cl_2$	B: B Comp.SVol.4/4-37/8
–	$D^{11}B^{35}Cl_2$	B: B Comp.SVol.4/4-37/8
BCl_2HO	Cl_2BOH	B: B Comp.SVol.3/4-50

$BCl_2H_2^-$ ... $[H_2BCl_2]^-$ ... B: B Comp.SVol.3/4-38
BCl_2H_2N ... $Cl_2B\text{-}NH_2$ ... B: B Comp.SVol.4/4-56/7
BCl_2H_4Yb ... $YbCl_2[BH_4] \cdot n\ C_4H_8O$ ... Sc: MVol.C11b-490
$BCl_2H_6NSi_2$ ... $(SiH_3)_2NBCl_2$ ... Si: SVol.B4-149
BCl_2I ... BCl_2I ... B: B Comp.SVol.3/4-101
B: B Comp.SVol.4/4-115
BCl_3 ... BCl_3
Chemical reactions ... B: B Comp.SVol.3/4-13/31
B: B Comp.SVol.4/4-11/32
Molecule ... B: B Comp.SVol.3/4-5/13, 36/7, 66, 72/3
B: B Comp.SVol.4/4-7/11
Physical properties ... B: B Comp.SVol.3/4-5/13, 36/7, 66, 72/3
B: B Comp.SVol.4/4-7/11
Preparation ... B: B Comp.SVol.3/4-3/5
B: B Comp.SVol.4/4-4/7
Spectra ... B: B Comp.SVol.3/4-7/11, 36/7, 66, 72/3
B: B Comp.SVol.4/4-8/10
Uses ... B: B Comp.SVol.3/4-14/9
B: B Comp.SVol.4/4-11/9
– ... $^{10}BCl_3$ ... B: B Comp.SVol.3/4-4, 11/3
– ... $^{10}B^{35}Cl_3$ ... B: B Comp.SVol.3/4-10
– ... $^{10}B^{37}Cl_3$ ... B: B Comp.SVol.3/4-10
– ... $^{11}BCl_3$ ... B: B Comp.SVol.3/4-4, 10/1, 13
– ... $^{11}B^{35}Cl_3$ ... B: B Comp.SVol.3/4-10
– ... $^{11}B^{37}Cl_3$ ... B: B Comp.SVol.3/4-10
– ... $BCl_3 \cdot H_2O$ ... B: B Comp.SVol.3/4-52
B: B Comp.SVol.4/3b-148
BCl_3^+ ... BCl_3^+ ... B: B Comp.SVol.3/4-6/7
BCl_3^{2+} ... BCl_3^{2+} ... B: B Comp.SVol.3/4-6/7
BCl_3H_2S ... $BCl_3 \cdot H_2S$ ... B: B Comp.SVol.4/4-161
B: B Comp.SVol.4/3b-148
BCl_3H_3N ... $BCl_3 \cdot NH_3$ ... B: B Comp.SVol.3/4-32
B: B Comp.SVol.4/3b-148
B: B Comp.SVol.4/4-32
– ... $BCl_3 \cdot {}^{15}NH_3$ ... B: B Comp.SVol.3/4-32
BCl_3H_3P ... $BCl_3 \cdot PH_3$ ... B: B Comp.SVol.4/3b-148
$BCl_3H_9NSi_3$ ... $(SiH_3)_3N \cdot BCl_3$ ... Si: SVol.B4-105
BCl_3NO ... $ON\text{-}BCl_3 \cdot [C_6(CH_3)_6]$ ... B: B Comp.SVol.4/4-33
BCl_3O_4 ... $Cl_2B(ClO_4)$ ... B: B Comp.SVol.3/4-50/1
BCl_3O_8 ... $ClB(ClO_4)_2$ ... B: B Comp.SVol.3/4-50/1
BCl_3O_{12} ... $B(ClO_4)_3$ ... B: B Comp.SVol.3/4-50
BCl_4 ... BCl_4, radical ... B: B Comp.SVol.3/4-34
BCl_4^- ... $[BCl_4]^-$ ... B: B Comp.SVol.3/4-33
B: B Comp.SVol.4/4-34/5
BCl_4Cs ... $Cs[BCl_4]$ ... B: B Comp.SVol.3/4-33
$BCl_4H_4NO_{16}$ ... $NH_4[B(ClO_4)_4]$ ... B: B Comp.SVol.3/4-62/3

BCl_4NO_{18} .... $NO_2[B(ClO_4)_4]$ .... B: B Comp.SVol.3/4-62/3
$BCl_4O_{16}^-$ .... $[B(ClO_4)_4]^-$ .... B: B Comp.SVol.3/4-62/3
$BCl_5NO_2P_2^+$ .... $[-BCl-O-PCl_2-N=PCl_2-O-]^+$ .... B: B Comp.SVol.4/4-70
$BCl_6H_3N_3O_3P_3$ ... $B[NH(POCl_2)]_3$ .... B: B Comp.SVol.3/3-94
$BCl_6NO_2P_2$ .... $[-BCl-O-PCl_2=N=PCl_2-O-]Cl$ .... B: B Comp.SVol.4/4-70
BCl_6OP .... $Cl_3PO \cdot BCl_3$ .... B: B Comp.SVol.3/4-52
BCl_6Sc_4 .... Sc_4Cl_6B .... Sc: MVol.C11b-471/2
BCl_8NSi_2 .... $Cl_2B-N(SiCl_3)_2$ .... B: B Comp.SVol.4/4-58
BCl_8P .... $[PCl_4][BCl_4]$ .... B: B Comp.SVol.4/4-35
$BCl_9N_2Si_3$ .... $[-N(SiCl_3)-BCl-N(SiCl_3)-]SiCl_2$ .... B: B Comp.SVol.4/4-65
$BCl_{12}FN_2Si_4$ .... $(Cl_3Si)_2N-BF-N(SiCl_3)_2$ .... B: B Comp.SVol.4/3b-245
$BCl_{12}Sc_7$ .... $Sc_7Cl_{12}B$ .... Sc: MVol.C11b-471/2
$BCl_{13}P_2$ .... $[PCl_4]_2[BCl_4][Cl]$ .... B: B Comp.SVol.4/4-35
$BCl_{14}N_3Si_5$ .... $1,3-(Cl_3Si)_2-4-(Cl_3Si)_2N-2,2-Cl_2-1,3,2,4-N_2SiB$
B: B Comp.SVol.4/3a-206
$BCsF_4$ .... $Cs[BF_4]$ .... B: B Comp.SVol.3/3-329/30
B: B Comp.SVol.4/3b-191, 197/8
– .... $Cs[BF_4]$ solid solutions
$Cs[BF_4]-Rb[BF_4]$ .... B: B Comp.SVol.3/3-329, 342
$Cs[BF_4]-[NH_4][BF_4]$ .... B: B Comp.SVol.3/3-342
$BCsF_{20}O_4Te_4$ .... $Cs[B(OTeF_5)_4]$ .... B: B Comp.SVol.3/4-150
$BCuF^+$ .... $[BFCu]^+$ .... B: B Comp.SVol.4/3b-84
– .... $[CuBF]^+$ .... B: B Comp.SVol.4/3b-84
$BCuF_4HO$ .... $CuO \cdot H[BF_4]$ .... B: B Comp.SVol.4/3b-168
$BDyO_3$ .... $DyBO_3$ .... Sc: MVol.C11b-391/6, 399/400, 408, 415/6
– .... $DyBO_3$ solid solutions
$DyBO_3-LaBO_3$ .... Sc: MVol.C11b-417, 418
$DyBO_3-LuBO_3$ .... Sc: MVol.C11b-419
$DyBO_3-YBO_3$ .... Sc: MVol.C11b-416
– .... $DyBO_3$ systems
$DyBO_3-LaBO_3$ .... Sc: MVol.C11b-417, 418
BDy_3O_6 .... Dy_3BO_6 .... Sc: MVol.C11b-389/90
$BErN_2$ .... $ErBN_2$ .... Sc: MVol.C11b-470/1
$BErO_3$ .... $ErBO_3$ .... Sc: MVol.C11b-391/5, 399/400, 404/8, 415/6
– .... $ErBO_3$ solid solutions
$ErBO_3-LaBO_3$ .... Sc: MVol.C11b-417, 418
$ErBO_3-LuBO_3$ .... Sc: MVol.C11b-419
$ErBO_3-YBO_3$ .... Sc: MVol.C11b-416
– .... $ErBO_3$ systems
$ErBO_3-LaBO_3$ .... Sc: MVol.C11b-417, 418
BEr_3O_6 .... Er_3BO_6 .... Sc: MVol.C11b-389/90
$BEuO_3$ .... $EuBO_3$ .... Sc: MVol.C11b-391/408, 415/6
– .... $EuBO_3$ solid solutions
$EuBO_3-GdBO_3$ .... Sc: MVol.C11b-419
$EuBO_3-LaBO_3$ .... Sc: MVol.C11b-417, 418

$BEuO_3$ .... $EuBO_3$ solid solutions
$EuBO_3$-$LuBO_3$ .... Sc: MVol.C11b-419
$EuBO_3$-$ScBO_3$ .... Sc: MVol.C11b-417
$EuBO_3$-YBO_3 .... Sc: MVol.C11b-416
– .... $EuBO_3$ systems
$EuBO_3$-$LaBO_3$ .... Sc: MVol.C11b-417, 418
$EuBO_3$-$ScBO_3$ .... Sc: MVol.C11b-417
BEu_3O_6 .... Eu_3BO_6 .... Sc: MVol.C11b-389/90
– .... Eu_3BO_6 solid solutions
Eu_3BO_6-Gd_3BO_6 .... Sc: MVol.C11b-390/1
Eu_3BO_6-La_3BO_6 .... Sc: MVol.C11b-390/1
Eu_3BO_6-Y_3BO_6 .... Sc: MVol.C11b-390/1
BF .... BF .... B: B Comp.SVol.3/3-231/5
B: B Comp.SVol.4/3b-81/7
– .... ^{11}BF .... B: B Comp.SVol.4/3b-82/4
BF^+ .... $[BF]^+$, radical cation .... B: B Comp.SVol.3/3-231/5
B: B Comp.SVol.4/3b-86
– .... $[^{10}BF]^+$, radical cation .... B: B Comp.SVol.4/3b-84/5
– .... $[^{10}B^{19}F]^+$, radical cation .... B: B Comp.SVol.4/3b-86
– .... $[^{11}BF]^+$, radical cation .... B: B Comp.SVol.4/3b-84/5
BFH^+ .... $[HBF]^+$ .... B: B Comp.SVol.3/3-361
B: B Comp.SVol.4/3b-225
– .... $[H^{11}BF]^+$ .... B: B Comp.SVol.4/3b-225
BFH_2 .... H_2BF .... B: B Comp.SVol.3/3-361
B: B Comp.SVol.4/3b-224
BFH_2O .... FB(H)OH .... B: B Comp.SVol.3/3-365/6
B: B Comp.SVol.4/3b-232
BFH_2O_2 .... $FB(OH)_2$ .... B: B Comp.SVol.3/3-366
B: B Comp.SVol.4/3b-232
BFH_3^- .... $[H_3BF]^-$ .... B: B Comp.SVol.4/3b-225
BFH_3KO_3 .... $KF \cdot B(OH)_3$ .... B: B Comp.SVol.3/3-371
BFH_3KO_6 .... $K[BF(OOH)_3]$ .... B: B Comp.SVol.3/3-372
BFH_3N .... $F(H)B-NH_2$ .... B: B Comp.SVol.4/3b-244
BFH_3NaO_6 .... $Na[BF(OOH)_3]$ .... B: B Comp.SVol.3/3-372
$BFH_3O_3^-$ .... $[FB(OH)_3]^-$ .... B: B Comp.SVol.3/3-371/2
$BFH_3O_6^-$ .... $[FB(O-OH)_3]^-$ .... B: B Comp.SVol.3/3-372
B: B Comp.SVol.4/3b-237
BFH_4Si .... FH_2Si-BH_2 .... B: B Comp.SVol.4/3b-224
– .... $H_3Si-BHF$ .... B: B Comp.SVol.4/3b-224
BFH_7NO_6 .... $NH_4[BF(OOH)_3]$ .... B: B Comp.SVol.3/3-372
BFI_2 .... BFI_2 .... B: B Comp.SVol.3/4-101
B: B Comp.SVol.4/4-115
Adducts with tertiary amines .... B: B Comp.SVol.3/4-65/7, 101
$BFLi^+$ .... $[BFLi]^+$ .... B: B Comp.SVol.4/3b-84/5
– .... $[LiBF]^+$ .... B: B Comp.SVol.4/3b-84/5
BFO .... FBO .... B: B Comp.SVol.3/3-365
B: B Comp.SVol.4/3b-230/1
– .... $F^{10}BO$ .... B: B Comp.SVol.4/3b-230/1

Formula	Compound	Reference
BFO	$F^{10}B^{18}O$	B: B Comp.SVol.4/3b-230
–	$F^{11}BO$	B: B Comp.SVol.4/3b-230/1
–	$F^{11}B^{18}O$	B: B Comp.SVol.4/3b-230
BFS	BSF	B: B Comp.SVol.3/4-136
		B: B Comp.SVol.4/4-154/6
–	FBS	B: B Comp.SVol.3/4-135/6
		B: B Comp.SVol.4/4-154/7, 159/60
BFS^+	$[FBS]^+$	B: B Comp.SVol.4/4-158/60
BF_2	BF_2, radical	B: B Comp.SVol.3/3-235
		B: B Comp.SVol.4/3b-88
$BF_2{}^+$	$[BF_2]^+$	B: B Comp.SVol.3/3-235/6
		B: B Comp.SVol.4/3b-88, 197
BF_2H	HBF_2	B: B Comp.SVol.3/3-359/61
		B: B Comp.SVol.4/3b-223/4
–	$H^{10}BF_2$	B: B Comp.SVol.3/3-359/60
		B: B Comp.SVol.4/3b-224
–	$H^{11}BF_2$	B: B Comp.SVol.3/3-359/60
		B: B Comp.SVol.4/3b-224, 242
–	$D^{10}BF_2$	B: B Comp.SVol.4/3b-224
–	$D^{11}BF_2$	B: B Comp.SVol.4/3b-224
BF_2H^+	$[HBF_2]^+$	B: B Comp.SVol.3/3-361
BF_2HO	F_2B-OH	B: B Comp.SVol.3/3-365
		B: B Comp.SVol.4/3b-231
$BF_2H_2{}^-$	$[H_2BF_2]^-$	B: B Comp.SVol.4/3b-225
$BF_2H_2KO_4$	$K[BF_2(O-OH)_2]$	B: B Comp.SVol.3/3-372
BF_2H_2N	BF_2-NH_2	B: B Comp.SVol.4/3b-241/2
–	$^{10}BF_2-NH_2$	B: B Comp.SVol.4/3b-242
–	$^{11}BF_2-NH_2$	B: B Comp.SVol.4/3b-242
–	$^{11}BF_2-ND_2$	B: B Comp.SVol.4/3b-242
$BF_2H_2NaO_4$	$Na[BF_2(O-OH)_2]$	B: B Comp.SVol.3/3-372
$BF_2H_2O_2{}^-$	$[BF_2(OH)_2]^-$	B: B Comp.SVol.3/3-372
$BF_2H_2O_4{}^-$	$[F_2B(O-OH)_2]^-$	B: B Comp.SVol.3/3-372
		B: B Comp.SVol.4/3b-237
BF_2H_3Si	HF_2Si-BH_2	B: B Comp.SVol.4/3b-224
BF_2H_5O	$[-BH_2-F-H-O(H)-H-F-]$	B: B Comp.SVol.4/3b-224
$BF_2H_6NO_2$	$[NH_4][F_2B(OH)_2]$	B: B Comp.SVol.4/3b-237
$BF_2H_6NO_4$	$NH_4[BF_2(O-OH)_2]$	B: B Comp.SVol.3/3-372
$BF_2H_6NSi_2$	$(SiH_3)_2NBF_2$	Si: SVol.B4-148/9
$BF_2H_{10}NSi_4$	$(SiH_3SiH_2)_2NBF_2 \cdot SiH_3SiH_2F$	Si: SVol.B4-119
BF_2I	BF_2I	B: B Comp.SVol.3/4-101
		B: B Comp.SVol.4/4-115
	Adducts with tertiary amines	B: B Comp.SVol.3/4-65/7, 101
BF_2O	F_2BO, radical	B: B Comp.SVol.4/3b-231
BF_2O^-	$[F_2BO]^-$	B: B Comp.SVol.4/3b-231
BF_3	BF_3	B: B Comp.SVol.3/3-239/67
		B: B Comp.SVol.4/3b-94/123

BF_3 .... BF_3
Adducts with tertiary amines .... B: B Comp.SVol.3/4-65/7
– .... $^{10}BF_3$ .... B: B Comp.SVol.4/3b-94, 98/9
– .... $^{11}BF_3$ .... B: B Comp.SVol.4/3b-94, 97/9
– .... $BF_3 \cdot H_2O$ .... B: B Comp.SVol.3/3-285/6
B: B Comp.SVol.4/3b-141/3, 148
BF_3^+ .... BF_3^+ .... B: B Comp.SVol.3/3-244
BF_3^- .... BF_3^-, radical anion .... B: B Comp.SVol.3/3-245
BF_3^{2+} .... BF_3^{2+} .... B: B Comp.SVol.3/3-244
BF_3H^- .... $[HBF_3]^-$ .... B: B Comp.SVol.4/3b-225
BF_3HKO_2 .... $K[BF_3\text{-}O\text{-}OH]$ .... B: B Comp.SVol.3/3-372
BF_3HN_3 .... $HN_3\text{-}BF_3$ .... B: B Comp.SVol.3/3-293
BF_3HNaO_2 ...... $Na[BF_3\text{-}O\text{-}OH]$ .... B: B Comp.SVol.3/3-372
BF_3HO^- .... $[BF_3\text{-}OH]^-$ .... B: B Comp.SVol.3/3-372
$BF_3HO_2^-$ .... $[F_3B\text{-}O\text{-}OH]^-$ .... B: B Comp.SVol.3/3-372
B: B Comp.SVol.4/3b-235/7
BF_3H_2S .... $BF_3 \cdot H_2S$ .... B: B Comp.SVol.4/3b-148
B: B Comp.SVol.4/4-161
BF_3H_2Si .... $F_3Si\text{-}BH_2$ .... B: B Comp.SVol.4/3b-224
BF_3H_3N .... $BF_3 \cdot NH_3$ .... B: B Comp.SVol.3/3-293
B: B Comp.SVol.4/3b-148
$BF_3H_3O_4P$ .... $BF_3 \cdot H_3PO_4$ .... B: B Comp.SVol.4/3b-145
BF_3H_3P .... $BF_3 \cdot PH_3$ .... B: B Comp.SVol.4/3b-148
$BF_3H_5NO_2$ .... $NH_4[BF_3\text{-}O\text{-}OH]$ .... B: B Comp.SVol.3/3-372
$BF_3H_9NSi_3$ ...... $(SiH_3)_3N \cdot BF_3$ .... Si: SVol.B4-105
$BF_3H_{15}NSi_6$ ...... $(Si_2H_5)_2N\text{-}BF_2 \cdot Si_2H_5\text{-}F$ .... Si: SVol.B4-119
– .... $(Si_2H_5)_3N \cdot BF_3$ .... Si: SVol.B4-119
BF_3NO .... $F_3B(NO)$ .... B: B Comp.SVol.4/3b-150
BF_3N_2 .... $N_2\text{-}BF_3$ .... B: B Comp.SVol.3/3-292/3
B: B Comp.SVol.4/3b-107
BF_3N_2O .... $(N_2O)BF_3$ .... B: B Comp.SVol.3/3-293
BF_3O^{2-} .... $[F_3BO]^{2-}$ .... B: B Comp.SVol.4/3b-107
BF_3O_2S .... $^{10}BF_3 \cdot SO_2$ .... B: B Comp.SVol.3/3-240
– .... $^{11}BF_3 \cdot SO_2$ .... B: B Comp.SVol.3/3-240
BF_4 .... $[BF_4]$, radical .... B: B Comp.SVol.4/3b-156
BF_4^- .... $[BF_4]^-$ .... B: B Comp.SVol.3/3-301/6, 372
B: B Comp.SVol.4/3b-156/9, 225
– .... $[BF_4]^- \cdot H_2O$ .... B: B Comp.SVol.3/3-302
BF_4H .... $H[BF_4]$ .... B: B Comp.SVol.3/3-307/17
B: B Comp.SVol.4/3b-165/8
BF_4HHgO .... $HgO \cdot H[BF_4]$ .... B: B Comp.SVol.3/3-309
B: B Comp.SVol.4/3b-167/8
$BF_4HN_4S_4$ .... $[S_4N_4H][BF_4]$ .... B: B Comp.SVol.3/4-145
$BF_4H_2O^-$ .... $[BF_4]^- \cdot H_2O$ .... B: B Comp.SVol.3/3-302
BF_4H_4N .... $[NH_4][BF_4]$ .... B: B Comp.SVol.3/3-340/3

BF_4H_4N .... $[NH_4][BF_4]$ .... B: B Comp.SVol.4/3b-193, 205/6
– .... $[NH_3D][BF_4]$ .... B: B Comp.SVol.4/3b-205
– .... $[ND_4][BF_4]$ .... B: B Comp.SVol.4/3b-205
– .... $[NH_4][BF_4]$ solid solutions
$[NH_4][BF_4]$-$Cs[BF_4]$ .... B: B Comp.SVol.3/3-342
$[NH_4][BF_4]$-$K[BF_4]$ .... B: B Comp.SVol.3/3-342
– .... $[NH_4][BF_4]$ systems
$[NH_4][BF_4]$-$[NH_4][NO_3]$ .... B: B Comp.SVol.3/3-342
BF_4K .... $K[BF_4]$ .... B: B Comp.SVol.3/3-226/9
B: B Comp.SVol.4/3b-191, 193, 197/8
– .... $K[^{10}BF_4]$ .... B: B Comp.SVol.4/3b-197
– .... $K[^{11}BF_4]$ .... B: B Comp.SVol.4/3b-197
– .... $K[BF_4]$ solid solutions
$K[BF_4]$-$K[ClO_4]$ .... B: B Comp.SVol.3/3-327
$K[BF_4]$-$[NH_4][BF_4]$ .... B: B Comp.SVol.3/3-342
– .... $K[BF_4]$ systems
$K[BF_4]$-$K[BH_4]$-H_2O .... B: B Comp.SVol.3/3-327
BF_4Li .... $Li[BF_4]$ .... B: B Comp.SVol.3/3-318/22
B: B Comp.SVol.4/3b-179/82
– .... $Li[BF_4] \cdot H_2O$ .... B: B Comp.SVol.3/3-318
B: B Comp.SVol.4/3b-179, 180
– .... $Li[BF_4] \cdot 3\,H_2O$ .... B: B Comp.SVol.4/3b-179, 180
– .... $Li[BF_4]$ systems
$Li[BF_4]$ - CH_3O-CH_2CH_2-OCH_3 .... B: B Comp.SVol.4/3b-181
$Li[BF_4]$ - $OC_4H_6(=O\text{-}2)$ - H_2O .... B: B Comp.SVol.4/3b-180
$Li[BF_4]$ - $[4\text{-}CH_3\text{-}1,3\text{-}O_2C_3H_3(=O\text{-}2)]$ - H_2O
B: B Comp.SVol.4/3b-180/1
BF_4NO .... $[NO][BF_4]$ .... B: B Comp.SVol.3/3-350/3
B: B Comp.SVol.4/3b-217
– .... $[^{15}NO][BF_4]$ .... B: B Comp.SVol.3/3-350
B: B Comp.SVol.4/3b-217
BF_4NO_2 .... $[NO_2][BF_4]$ .... B: B Comp.SVol.3/3-353/4
B: B Comp.SVol.4/3b-217/8
– .... $[^{15}NO_2][BF_4]$ .... B: B Comp.SVol.3/3-353
BF_4NS .... $[NS][BF_4]$ .... S: S-N Comp.5-43
$BF_4N_3O_2$ .... $[O_2N\text{-}N{\equiv}N][BF_4]$ .... B: B Comp.SVol.4/3b-218
$BF_4N_5S_5$ .... $[S_5N_5][BF_4]$ .... B: B Comp.SVol.4/4-167
BF_4Na .... $Na[BF_4]$ .... B: B Comp.SVol.3/3-322/6
B: B Comp.SVol.4/3b-190/3, 197
– .... $Na[BF_4] \cdot 2\,H_2O$ .... B: B Comp.SVol.3/3-323
– .... $Na[BF_4]$ systems
$Na[BF_4]$-$Na[BH_4]$-H_2O .... B: B Comp.SVol.3/3-323
BF_4Rb .... $Rb[BF_4]$ .... B: B Comp.SVol.3/3-329

BF_4Rb .... $Rb[BF_4]$ .... B: B Comp.SVol.4/3b-191, 197/8
– .... $Rb[BF_4]$ solid solutions
$Rb[BF_4]-Cs[BF_4]$ .... B: B Comp.SVol.3/3-329, 342

– .... $Rb[BF_4]$ systems
$Rb[BF_4]-RbCl$ .... B: B Comp.SVol.3/3-329
$Rb[BF_4]-RbF$ .... B: B Comp.SVol.3/3-329
$Rb[BF_4]-Rb[NO_3]$ .... B: B Comp.SVol.3/3-329

BF_5H_3N .... $[NH_3F][BF_4]$ .... B: B Comp.SVol.4/3b-108, 210
BF_5OSi .... $F_2B-OSiF_3$ .... B: B Comp.SVol.3/3-367
BF_5Si .... F_3Si-BF_2 .... B: B Comp.SVol.3/3-363
– .... $F_3{}^{28}Si-{}^{10}BF_2$ .... B: B Comp.SVol.3/3-363
– .... $F_3{}^{28}Si-{}^{11}BF_2$ .... B: B Comp.SVol.3/3-363
– .... $F_3{}^{29}Si-{}^{11}BF_2$ .... B: B Comp.SVol.3/3-363
$BF_7O_6P_3^-$ .... $[BF(PO_2F_2)_3]^-$ .... B: B Comp.SVol.3/3-251
BF_8N .... $[NF_4][BF_4]$ .... B: B Comp.SVol.3/3-345
B: B Comp.SVol.4/3b-107
$BF_{15}O_3Te_3$ ...... $B(OTeF_5)_3$ .... B: B Comp.SVol.3/4-150
B: B Comp.SVol.4/4-176
$BF_{20}O_4Te_4^-$ ...... $[B(OTeF_5)_4]^-$ .... B: B Comp.SVol.4/4-176
$BGaH_{12}N_2$ .... $[GaH_2(NH_3)_2][BH_4]$ .... Ga: SVol.D1-210/1
$BGdN_2$ .... $GdBN_2$ .... Sc: MVol.C11b-470/1
$BGdO_3$ .... $GdBO_3$ .... Sc: MVol.C11b-391/6, 399/416
doped with rare earth ions .... Sc: MVol.C11b-408/15
doped with rare earth ions, Bi^{3+}, and Sb^{3+}
Sc: MVol.C11b-412/5

– .... $GdBO_3$ solid solutions
$GdBO_3-EuBO_3$ .... Sc: MVol.C11b-419
$GdBO_3-LaBO_3$ .... Sc: MVol.C11b-417, 418
$GdBO_3-TbBO_3$ .... Sc: MVol.C11b-419
$GdBO_3-YBO_3$, doped with rare earth ions .. Sc: MVol.C11b-413/5
$GdBO_3-YBO_3$, doped with rare earth ions and Bi^{3+}
Sc: MVol.C11b-413/5

– .... $GdBO_3$ systems
$GdBO_3-LaBO_3$ .... Sc: MVol.C11b-417, 418

BGd_3O_6 .... Gd_3BO_6 .... Sc: MVol.C11b-389/90
– .... Gd_3BO_6 solid solutions
$Gd_3BO_6-Eu_3BO_6$ .... Sc: MVol.C11b-390/1
$Gd_3BO_6-Tb_3BO_6$ .... Sc: MVol.C11b-390/1

$BGeNdO_5$ .... $NdGeBO_5$ .... Sc: MVol.C11b-439
$BH_{0.5}N$ .... $BNH_{0.5}$ .... B: B Comp.SVol.3/3-171
BH^+ .... $[HB]^+$ .... B: B Comp.SVol.4/3b-43
BHI_2 .... HBI_2 .... B: B Comp.SVol.4/4-112
$BHLiNdO_4$ .... $LiNd[BO_3(OH)]$ .... Sc: MVol.C11b-463/4
BHN^+ .... HNB^+ .... B: B Comp.SVol.3/3-141
BHOS .... HOBS .... B: B Comp.SVol.3/4-119

BHOS . . . . HOBS . . . . B: B Comp.SVol.4/4-136
BHS. . . . . HBS. . . . . B: B Comp.SVol.3/4-110
B: B Comp.SVol.4/4-126/8
– . . . . $H^{11}BS$ . . . . B: B Comp.SVol.3/4-110
– . . . . DBS. . . . . B: B Comp.SVol.3/4-110
– . . . . $D^{11}BS$ . . . . B: B Comp.SVol.3/4-110
BHS^+ . . . . $[HBS]^+$ . . . . B: B Comp.SVol.3/4-116
B: B Comp.SVol.4/4-127/8
– . . . . $[DBS]^+$ . . . . B: B Comp.SVol.4/4-128
BH_2^+ . . . . $[H_2B]^+$. . . . . B: B Comp.SVol.4/3b-43
BH_2I . . . . H_2BI . . . . B: B Comp.SVol.4/4-112
BH_2N. . . . . $(-BH-NH-)_n$ . . . . B: B Comp.SVol.3/3-102, 133/4
– . . . . HB=NH . . . . B: B Comp.SVol.3/3-134, 141
B: B Comp.SVol.4/3a-210/1
– . . . . $H^{10}B=NH$. . . . . B: B Comp.SVol.4/3a-210
– . . . . $H^{11}B=NH$. . . . . B: B Comp.SVol.4/3a-210
BH_2NO_2. . . . . H_2B-NO_2. . . . . B: B Comp.SVol.3/3-146
BH_2NS. . . . . $S{\equiv}B-NH_2$. . . . . B: B Comp.SVol.3/4-122
B: B Comp.SVol.4/4-139
BH_2NaS_3. . . . . $Na[BH_2S_3]$. . . . . B: B Comp.SVol.3/4-116
B: B Comp.SVol.4/4-129
BH_2S^- . . . . $[SBH_2]^-$. . . . . B: B Comp.SVol.4/4-129
$BH_2S_3^-$ . . . . $[BH_2S_3]^-$. . . . . B: B Comp.SVol.3/4-116
BH_3HoNaO_5 . . . . . $NaHo[BO_2(OH)_2]OH$ . . . . Sc: MVol.C11b-469/70
BH_3I_3P . . . . H_3P-BI_3. . . . . B: B Comp.SVol.3/4-98
BH_3Li_3N . . . . H_3B-NLi_3 . . . . B: B Comp.SVol.4/3b-2/4
BH_3N^+ . . . . $[B-NH_3]^+$ . . . . B: B Comp.SVol.4/3b-43
– . . . . $[HB-NH_2]^+$ . . . . B: B Comp.SVol.4/3b-43
BH_3N^-. . . . . $[H_2BNH]^-$. . . . . B: B Comp.SVol.3/3-146
BH_3NO^+ . . . . $[H_3B-NO]^+$ . . . . B: B Comp.SVol.4/3b-4/5
BH_3N_2. . . . . HB(-NH-NH-). . . . . B: B Comp.SVol.3/3-134
B: B Comp.SVol.4/3a-193
– . . . . $H-N=B-NH_2$ . . . . B: B Comp.SVol.3/3-100
– . . . . $N_2 \cdot BH_3$. . . . . B: B Comp.SVol.3/3-168
BH_3O_3. . . . . H_3BO_3 systems
$H_3BO_3-NH(C_2H_4NH_2)_2-H_2O$ . . . . B: B Comp.SVol.3/3-226
$H_3BO_3-NH(C_2H_4OH)_2-H_2O$. . . . . B: B Comp.SVol.3/3-225
BH_3S. . . . . $HSBH_2$. . . . . B: B Comp.SVol.4/4-128
BH_3S_2 . . . . $HB(SH)_2$ . . . . B: B Comp.SVol.4/4-128
BH_4K. . . . . $K[BH_4]$ systems
$K[BH_4]-K[BF_4]-H_2O$ . . . . B: B Comp.SVol.3/3-327
$K[BH_4]-La[BH_4]_3-C_4H_8O$ . . . . Sc: MVol.C11b-492/3
BH_4Li . . . . $Li[BH_4]$ systems
$Li[BH_4]-Ce[BH_4]_3-C_4H_8O$. . . . . Sc: MVol.C11b-491/2
$Li[BH_4]-Dy[BH_4]_3-C_4H_8O$. . . . . Sc: MVol.C11b-491/2
$Li[BH_4]-Er[BH_4]_3-C_4H_8O$ . . . . Sc: MVol.C11b-491/2

BH_4Li .... $Li[BH_4]$ systems
$Li[BH_4]-Gd[BH_4]_3-C_4H_8O$ .... Sc: MVol.C11b-491/2
$Li[BH_4]-Ho[BH_4]_3-C_4H_8O$ .... Sc: MVol.C11b-491/2
$Li[BH_4]-La[BH_4]_3-C_4H_8O$ .... Sc: MVol.C11b-491/2
$Li[BH_4]-Nd[BH_4]_3-C_4H_8O$ .... Sc: MVol.C11b-491/2
$Li[BH_4]-Tm[BH_4]_3-C_4H_8O$ .... Sc: MVol.C11b-491/2
$Li[BH_4]-Y[BH_4]_3-C_4H_8O$ .... Sc: MVol.C11b-491/2

BH_4Li_2N .... $H_3B-NHLi_2$ .... B: B Comp.SVol.4/3b-2/4
BH_4N .... H_2B-NH_2 .... B: B Comp.SVol.3/3-100/1, 145/7
B: B Comp.SVol.4/3a-219/23
– .... $H_2B-^{15}NH_2$ .... B: B Comp.SVol.4/3a-220
– .... H_2B-ND_2 .... B: B Comp.SVol.4/3a-220
– .... $H_2{}^{10}B-NH_2$ .... B: B Comp.SVol.4/3a-220
– .... $H_2{}^{11}B-NH_2$ .... B: B Comp.SVol.3/3-146
B: B Comp.SVol.4/3b-242
– .... D_2B-NH_2 .... B: B Comp.SVol.4/3a-220
– .... $[-BH_2-NH_2-]_n$ .... B: B Comp.SVol.3/3-146
B: B Comp.SVol.4/3a-203
BH_4N^+ .... $[HB-NH_3]^+$ .... B: B Comp.SVol.4/3b-43/4
– .... $[H_2B-NH_2]^+$ .... B: B Comp.SVol.4/3a-222/3
B: B Comp.SVol.4/3b-43/4
BH_4N^{2+} .... $[HB-NH_3]^{2+}$ .... B: B Comp.SVol.4/3b-42/4
– .... $[H_2B-NH_2]^{2+}$ .... B: B Comp.SVol.4/3a-223
B: B Comp.SVol.4/3b-43/4
BH_4N_3 .... $3-NH_2-N_2BH_2$ .... B: B Comp.SVol.4/3a-193, 194

BH_4Na .... $Na[BH_4]$ systems
$Na[BH_4]-Ho[BH_4]_3-C_4H_8O$ .... Sc: MVol.C11b-492/3
$Na[BH_4]-La[BH_4]_3-C_4H_8O$ .... Sc: MVol.C11b-492/3
$Na[BH_4]-Na[BF_4]-H_2O$ .... B: B Comp.SVol.3/3-323
$Na[BH_4]-Sc[BH_4]_3-C_4H_8O$ .... Sc: MVol.C11b-492/3

$BH_4O_8^-$ .... $[B(O-OH)_4]^-$ .... B: B Comp.SVol.4/3b-237
BH_4S^+ .... $[H_2SBH_2]^+$ .... B: B Comp.SVol.4/4-129
BH_5LiN .... H_3B-NH_2Li .... B: B Comp.SVol.4/3b-2/4
$BH_5LiN_2^+$ .... $[HB(-NH_2-Li-NH_2-)]^+$ .... B: B Comp.SVol.3/3-210
BH_5N .... H_2B-NH_3, radical .... B: B Comp.SVol.4/3b-2
BH_5N_2 .... $NH_2-BH-NH_2$ .... B: B Comp.SVol.3/3-100/2
B: B Comp.SVol.4/3a-165/6
– .... $NH_2-^{10}BH-NH_2$ .... B: B Comp.SVol.3/3-101
– .... $NH_2-^{11}BH-NH_2$ .... B: B Comp.SVol.3/3-101/2
– .... $ND_2-^{10}BH-ND_2$ .... B: B Comp.SVol.3/3-101
– .... $ND_2-^{11}BH-ND_2$ .... B: B Comp.SVol.3/3-101/2
$BH_5O_{16}S_4$ .... $H[B(HSO_4)_4]$ .... B: B Comp.SVol.3/4-121
B: B Comp.SVol.4/4-138
BH_6N .... H_3N-BH_3 .... B: B Comp.SVol.3/3-168/71
B: B Comp.SVol.4/3b-1/4
– .... $H_3N-^{10}BH_3$ .... B: B Comp.SVol.3/3-170
– .... $H_3N-^{11}BH_3$ .... B: B Comp.SVol.3/3-169/70

BH_6N .... $H_3N-^{11}BD_2H$ .... B: B Comp.SVol.3/3-170
– .... $H_3N-^{11}BD_3$ .... B: B Comp.SVol.3/3-170
– .... $H_3{}^{15}N-^{10}BH_3$ .... B: B Comp.SVol.3/3-170
– .... $H_3{}^{15}N-^{11}BH_3$ .... B: B Comp.SVol.3/3-170
– .... $D_2HN-^{11}BH_3$ .... B: B Comp.SVol.3/3-170
– .... $D_3N-^{10}BH_3$ .... B: B Comp.SVol.3/3-170
– .... $D_3N-^{11}BH_3$ .... B: B Comp.SVol.3/3-170
$BH_6N_2^-$ .... $[(H_3N)_2B]^-$ .... B: B Comp.SVol.3/3-211
BH_6N_3 .... $B(NH_2)_3$ .... B: B Comp.SVol.3/3-91
$BH_6S_2^+$ .... $[(H_3S)_2B]^+$ .... B: B Comp.SVol.3/4-117
BH_7N_2 .... $H_3B-NH_2-NH_2$ .... B: B Comp.SVol.3/3-171/2
B: B Comp.SVol.4/3b-4
BH_8NSi_2 .... $(SiH_3)_2NBH_2$ .... Si: SVol.B4-148
BH_8N_3 .... $(H_2N)_2BH-NH_3$ .... B: B Comp.SVol.3/3-172
$BH_{10}N_3^{2+}$ .... $[HB(NH_3)_3]^{2+}$ .... B: B Comp.SVol.3/3-211
BHeN .... HeBN .... B: B Comp.SVol.4/3a-43
BHf .... HfB systems
HfB-ThC .... Th: SVol.C6-133/4
$BHoO_3$ .... $HoBO_3$ .... Sc: MVol.C11b-391/5, 399, 404/8
– .... $HoBO_3$ solid solutions
$HoBO_3-LaBO_3$ .... Sc: MVol.C11b-417, 418
$HoBO_3-LuBO_3$ .... Sc: MVol.C11b-419
$HoBO_3-YBO_3$ .... Sc: MVol.C11b-416
– .... $HoBO_3$ systems
$HoBO_3-LaBO_3$ .... Sc: MVol.C11b-417, 418
BHo_3O_6 .... Ho_3BO_6 .... Sc: MVol.C11b-389/90
BI .... BI .... B: B Comp.SVol.3/4-98
B: B Comp.SVol.4/4-109
BIO .... IBO .... B: B Comp.SVol.3/4-100
BI_2 .... BI_2 .... B: B Comp.SVol.3/4-98
BI_3 .... BI_3 .... B: B Comp.SVol.3/4-72/3, 98/100
B: B Comp.SVol.4/4-109/10, 112, 115
$BI_{12}Sc_7$ .... $Sc_7I_{12}B$ .... Sc: MVol.C11b-472
BKS_2 .... KBS_2 .... B: B Comp.SVol.4/4-123
BLa .... LaB .... Sc: MVol.C11a-164/5
$BLaO_3$ .... $LaBO_3$ .... Sc: MVol.C11b-391/416, 434/6
doped with rare earth ions .... Sc: MVol.C11b-408/15
– .... $LaBO_3$ solid solutions
$LaBO_3-DyBO_3$ .... Sc: MVol.C11b-417, 418
$LaBO_3-ErBO_3$ .... Sc: MVol.C11b-417, 418
$LaBO_3-EuBO_3$ .... Sc: MVol.C11b-417, 418
$LaBO_3-GdBO_3$ .... Sc: MVol.C11b-417, 418
$LaBO_3-HoBO_3$ .... Sc: MVol.C11b-417, 418
$LaBO_3-LuBO_3$ .... Sc: MVol.C11b-417, 418
$LaBO_3-NdBO_3$ .... Sc: MVol.C11b-417, 418
$LaBO_3-SmBO_3$ .... Sc: MVol.C11b-417, 418

$BLaO_3$ .... $LaBO_3$ solid solutions
$LaBO_3$-$TbBO_3$ .... Sc: MVol.C11b-417
$LaBO_3$-$TmBO_3$ .... Sc: MVol.C11b-417, 418
$LaBO_3$-YBO_3 .... Sc: MVol.C11b-417, 418
$LaBO_3$-$YbBO_3$ .... Sc: MVol.C11b-417, 418

– .... $LaBO_3$ systems
$LaBO_3$-$DyBO_3$ .... Sc: MVol.C11b-417, 418
$LaBO_3$-$ErBO_3$ .... Sc: MVol.C11b-417, 418
$LaBO_3$-$EuBO_3$ .... Sc: MVol.C11b-417, 418
$LaBO_3$-$GdBO_3$ .... Sc: MVol.C11b-417, 418
$LaBO_3$-$HoBO_3$ .... Sc: MVol.C11b-417, 418
$LaBO_3$-$LuBO_3$ .... Sc: MVol.C11b-417, 418
$LaBO_3$-$NdBO_3$ .... Sc: MVol.C11b-417, 418
$LaBO_3$-$SmBO_3$ .... Sc: MVol.C11b-417, 418
$LaBO_3$-$TmBO_3$ .... Sc: MVol.C11b-417, 418
$LaBO_3$-YBO_3 .... Sc: MVol.C11b-417, 418
$LaBO_3$-$YbBO_3$ .... Sc: MVol.C11b-417, 418

BLa_2 .... La_2B .... Sc: MVol.C11a-164/5
BLa_3 .... La_3B .... Sc: MVol.C11a-164/5
BLa_3O_6 .... La_3BO_6 .... Sc: MVol.C11b-389/90, 434/5
– .... La_3BO_6 solid solutions
La_3BO_6-Eu_3BO_6 .... Sc: MVol.C11b-390/1
La_3BO_6-Tb_3BO_6 .... Sc: MVol.C11b-390/1

$BLiO_2$ .... $LiBO_2$ glasses
$LiBO_2$-$LiBS_2$ .... B: B Comp.SVol.4/4-123

– .... $LiBO_2$ systems
$(LiBO_2)_2$-La_2O_3 .... Sc: MVol.C11b-455/6
$(LiBO_2)_2$-Nd_2O_3 .... Sc: MVol.C11b-456

BLiPd .... LiPdB .... Pd: SVol.B2-259
$BLiS_2$ .... $LiBS_2$ glasses
$LiBS_2$-$LiBO_2$ .... B: B Comp.SVol.4/4-123

BLi_3N_2 .... Li_3BN_2 .... B: B Comp.SVol.4/3a-149/50
BLi_3O_3 .... Li_3BO_3 systems
Li_3BO_3-YBO_3 .... Sc: MVol.C11b-453/5

$BLuO_3$ .... $LuBO_3$ .... Sc: MVol.C11b-391/416
doped with rare earth ions .... Sc: MVol.C11b-408/11

– .... $LuBO_3$ solid solutions
$LuBO_3$-$DyBO_3$ .... Sc: MVol.C11b-419
$LuBO_3$-$ErBO_3$ .... Sc: MVol.C11b-419
$LuBO_3$-$EuBO_3$ .... Sc: MVol.C11b-419
$LuBO_3$-$HoBO_3$ .... Sc: MVol.C11b-419
$LuBO_3$-$LaBO_3$ .... Sc: MVol.C11b-417, 418
$LuBO_3$-$SmBO_3$ .... Sc: MVol.C11b-419
$LuBO_3$-$TbBO_3$ .... Sc: MVol.C11b-419

– .... $LuBO_3$ systems
$LuBO_3$-$LaBO_3$ .... Sc: MVol.C11b-417, 418

BLu_3O_6 .... Lu_3BO_6 .... Sc: MVol.C11b-389/90

$BMgS_2$ .... $Mg[BS_2]$ .... B: B Comp.SVol.3/4-108

BN............. BN

Analytical chemistry .... B: B Comp.SVol.3/3-33
B: B Comp.SVol.4/3a-55/6

Chemical reactions.................. B: B Comp.SVol.3/3-31/5
B: B Comp.SVol.4/3a-53/8

Crystallographic properties .... B: B Comp.SVol.3/3-3
B: B Comp.SVol.4/3a-6/7, 23/4, 30/1, 49/51

Electronic structure.................. B: B Comp.SVol.3/3-21/4, 28/31
B: B Comp.SVol.4/3a-38/9, 47/9

Formation .... B: B Comp.SVol.3/3-1/12
B: B Comp.SVol.4/3a-2/6, 26/7

General .... B: B Comp.SVol.3/3-1
B: B Comp.SVol.4/3a-1

Molecule .... B: B Comp.SVol.3/3-20/1
B: B Comp.SVol.4/3a-33/8

Physical properties.................. B: B Comp.SVol.3/3-12/31
B: B Comp.SVol.4/3a-23/53

Preparation .... B: B Comp.SVol.3/3-1/12
B: B Comp.SVol.4/3a-2/23

Sorption........................... B: B Comp.SVol.3/3-24/6
B: B Comp.SVol.4/3a-43/7

Spectra .... B: B Comp.SVol.3/3-17, 21/4, 29/30
B: B Comp.SVol.4/3a-39/42, 49/51

Uses.............................. B: B Comp.SVol.4/3a-58/149

– .... BN solid solutions

BN-AlN .... B: B Comp.SVol.3/3-90

– .... BN systems

$BN-Li_3N$ .... B: B Comp.SVol.4/3a-149

BN_2^{3-} .... $[N=B=N]^{3-}$ .... B: B Comp.SVol.4/3a-149/50

BN_2Nd.......... $NdBN_2$.............................. Sc: MVol.C11b-470/1

BN_2Sm .... $SmBN_2$ .... Sc: MVol.C11b-470/1

$BNaS_2$.......... $NaBS_2$.............................. B: B Comp.SVol.4/4-123

$BNdO_3$.......... $NdBO_3$.............................. Sc: MVol.C11b-391/408, 415/6, 439

– .... $NdBO_3$ solid solutions

$NdBO_3-LaBO_3$ .... Sc: MVol.C11b-417, 418

$NdBO_3-SmBO_3$....................... Sc: MVol.C11b-417, 418

– .... $NdBO_3$ systems

$NdBO_3-LaBO_3$ .... Sc: MVol.C11b-417, 418

$NdBO_3-SmBO_3$....................... Sc: MVol.C11b-417, 418

BNd_3O_6......... Nd_3BO_6.............................. Sc: MVol.C11b-389/90, 439

BOS^-........... $[BOS]^-$................................ B: B Comp.SVol.4/4-137

$BOSe^-$ .... $BOSe^-$ .... B: B Comp.SVol.4/4-170
BO_3Pm .... $PmBO_3$ .... Sc: MVol.C11b-391/2, 394/5, 397/8
BO_3Pr .... $PrBO_3$ .... Sc: MVol.C11b-391/401, 405/8, 415/6
– .... $PrBO_3$ solid solutions
$PrBO_3$-YBO_3 .... Sc: MVol.C11b-416
BO_3Sc .... $ScBO_3$ .... Sc: MVol.C11b-391, 394/6, 404/16, 431/3
doped with rare earth ions............. Sc: MVol.C11b-408/15
doped with rare earth ions, Bi^{3+}, and Sb^{3+}
Sc: MVol.C11b-412/5
– .... $ScBO_3$ solid solutions
$ScBO_3$-$EuBO_3$ .... Sc: MVol.C11b-417
– .... $ScBO_3$ systems
$ScBO_3$-$EuBO_3$ .... Sc: MVol.C11b-417
BO_3Sm .... $SmBO_3$ .... Sc: MVol.C11b-391/408, 415/6, 439/40
– .... $SmBO_3$ solid solutions
$SmBO_3$-$LaBO_3$ .... Sc: MVol.C11b-417, 418
$SmBO_3$-$LuBO_3$ .... Sc: MVol.C11b-419
$SmBO_3$-$NdBO_3$ .... Sc: MVol.C11b-417, 418
$SmBO_3$-YBO_3 .... Sc: MVol.C11b-416
– .... $SmBO_3$ systems
$SmBO_3$-$LaBO_3$ .... Sc: MVol.C11b-417, 418
$SmBO_3$-$NdBO_3$ .... Sc: MVol.C11b-417, 418
BO_3Tb .... $TbBO_3$ .... Sc: MVol.C11b-391/6, 399/409, 415/6
doped with rare earth ions............. Sc: MVol.C11b-409
– .... $TbBO_3$ solid solutions
$TbBO_3$-$CeBO_3$ .... Sc: MVol.C11b-417
$TbBO_3$-$GdBO_3$ .... Sc: MVol.C11b-419
$TbBO_3$-$LaBO_3$ .... Sc: MVol.C11b-417
$TbBO_3$-$LuBO_3$ .... Sc: MVol.C11b-419
$TbBO_3$-YBO_3 .... Sc: MVol.C11b-416
BO_3Tm .... $TmBO_3$ .... Sc: MVol.C11b-391/5, 399/407, 415/6
– .... $TmBO_3$ solid solutions
$TmBO_3$-$LaBO_3$ .... Sc: MVol.C11b-417, 418
$TmBO_3$-YBO_3 .... Sc: MVol.C11b-416
– .... $TmBO_3$ systems
$TmBO_3$-$LaBO_3$ .... Sc: MVol.C11b-417, 418
BO_3Y .... YBO_3 .... Sc: MVol.C11b-391/5, 398/400, 404/16, 433/4
doped with rare earth ions............. Sc: MVol.C11b-408/15
doped with rare earth ions, Bi^{3+}, and Sb^{3+}
Sc: MVol.C11b-412/5

BO_3Y . . . . YBO_3 solid solutions
YBO_3-$DyBO_3$ . . . . Sc: MVol.C11b-416
YBO_3-$ErBO_3$ . . . . Sc: MVol.C11b-416
YBO_3-$EuBO_3$ . . . . Sc: MVol.C11b-416
YBO_3-$GdBO_3$, doped with rare earth ions . . Sc: MVol.C11b-413/5
YBO_3-$GdBO_3$, doped with rare earth ions and Bi^{3+}
Sc: MVol.C11b-413/5
YBO_3-$HoBO_3$ . . . . Sc: MVol.C11b-416
YBO_3-$LaBO_3$ . . . . Sc: MVol.C11b-417, 418
YBO_3-$PrBO_3$ . . . . Sc: MVol.C11b-416
YBO_3-$SmBO_3$ . . . . Sc: MVol.C11b-416
YBO_3-$TbBO_3$ . . . . Sc: MVol.C11b-416
YBO_3-$TmBO_3$ . . . . Sc: MVol.C11b-416
YBO_3-$YbBO_3$ . . . . Sc: MVol.C11b-416

– . . . . YBO_3 systems
$(YBO_3)_2$-$Li_2B_4O_7$ . . . . Sc: MVol.C11b-453/5
YBO_3-$LaBO_3$ . . . . Sc: MVol.C11b-417, 418
YBO_3-$LiYO_2$ . . . . Sc: MVol.C11b-453/5
YBO_3-Li_3BO_3 . . . . Sc: MVol.C11b-453/5
YBO_3-$Li_6Y(BO_3)_3$ . . . . Sc: MVol.C11b-453/5

BO_3Yb . . . . $YbBO_3$ . . . . Sc: MVol.C11b-391/408, 415/6
– . . . . $YbBO_3$ solid solutions
$YbBO_3$-$LaBO_3$ . . . . Sc: MVol.C11b-417, 418
$YbBO_3$-YBO_3 . . . . Sc: MVol.C11b-416

– . . . . $YbBO_3$ systems
$YbBO_3$-$LaBO_3$ . . . . Sc: MVol.C11b-417, 418

BO_6Pr_3 . . . . Pr_3BO_6 . . . . Sc: MVol.C11b-389/90
BO_6Sc_3 . . . . Sc_3BO_6 . . . . Sc: MVol.C11b-389
BO_6Sm_3 . . . . Sm_3BO_6 . . . . Sc: MVol.C11b-389/90
BO_6Tb_3 . . . . Tb_3BO_6 . . . . Sc: MVol.C11b-389/90
– . . . . Tb_3BO_6 solid solutions
Tb_3BO_6-Gd_3BO_6 . . . . Sc: MVol.C11b-390/1
Tb_3BO_6-La_3BO_6 . . . . Sc: MVol.C11b-390/1
Tb_3BO_6-Y_3BO_6 . . . . Sc: MVol.C11b-390/1

BO_6Tm_3 . . . . Tm_3BO_6 . . . . Sc: MVol.C11b-389/90
BO_6Y_3 . . . . Y_3BO_6 . . . . Sc: MVol.C11b-389/90
– . . . . Y_3BO_6 solid solutions
Y_3BO_6-Eu_3BO_6 . . . . Sc: MVol.C11b-390/1
Y_3BO_6-Tb_3BO_6 . . . . Sc: MVol.C11b-390/1

BO_6Yb_3 . . . . Yb_3BO_6 . . . . Sc: MVol.C11b-389/90
BPd . . . . PdB . . . . Pd: SVol.B2-258
BPd_2 . . . . Pd_2B . . . . Pd: SVol.B2-256/9
BPd_3 . . . . Pd_3B . . . . Pd: SVol.B2-256/9
BPd_4 . . . . Pd_4B . . . . Pd: SVol.B2-256/7
BPo . . . . BPo . . . . B: B Comp.SVol.3/4-152
Po: SVol.1-308/9
BS . . . . BS, radical . . . . B: B Comp.SVol.3/4-103/6
B: B Comp.SVol.4/4-118/21

BS . . . . . $B^{32}S$, radical . . . . . B: B Comp.SVol.4/4-118/9
– . . . . . $B^{34}S$, radical . . . . . B: B Comp.SVol.4/4-118/9
– . . . . . $^{11}B^{32}S$, radical . . . . . B: B Comp.SVol.4/4-120
BS^- . . . . . $[BS]^-$ . . . . . B: B Comp.SVol.4/4-123/4
BS_2 . . . . . BS_2 . . . . . B: B Comp.SVol.3/4-106
B: B Comp.SVol.4/4-121
– . . . . . $^{11}BS_2$ . . . . . B: B Comp.SVol.3/4-107
– . . . . . $(BS_2)_n$ . . . . . B: B Comp.SVol.3/4-106
B: B Comp.SVol.4/4-121
BS_2^- . . . . . $[BS_2]^-$ . . . . . B: B Comp.SVol.3/4-108
B: B Comp.SVol.4/4-123
BS_3^{3-} . . . . . $[BS_3]^{3-}$ . . . . . B: B Comp.SVol.4/4-123, 125
BS_3Tl_3 . . . . . $Tl_3[BS_3]$ . . . . . B: B Comp.SVol.4/4-123, 125
BSe . . . . . BSe . . . . . B: B Comp.SVol.3/4-147
B: B Comp.SVol.4/4-169
BSe_2 . . . . . BSe_2 . . . . . B: B Comp.SVol.4/4-169/70
– . . . . . $(BSe_2)_n$ . . . . . B: B Comp.SVol.3/4-147
BSe_3Tl_3 . . . . . Tl_3BSe_3 . . . . . B: B Comp.SVol.4/4-169
BTe . . . . . BTe . . . . . B: B Comp.SVol.3/4-150
BTi . . . . . TiB systems
TiB-ThC . . . . . Th: SVol.C6-133/4
$B_{1.33}C_xF_{4.99}$ . . . . . $C_x[BF_4] \cdot 0.33\ BF_3$ (x from 23 to 33) . . . . . B: B Comp.SVol.4/3b-107
$B_{1.5}C_4F_{4.5}H_{12}N_2OS$
$((CH_3)_2N)_2SO \cdot 1.5\ BF_3$ . . . . . S: S-N Comp.8-341
$B_2BrC_4H_8NS_2$ . . . . . $1,3,2\text{-}S_2BC_2H_2\text{-}2\text{-}[BBr\text{-}N(CH_3)_2]$ . . . . . B: B Comp.SVol.4/4-167
$B_2BrC_6H_9N_4$ . . . . . $H_2B[\text{-}1,2\text{-}N_2C_3H_3\text{-}]_2BHBr$ . . . . . B: B Comp.SVol.4/4-100/1
$B_2BrC_6H_{18}NSn$ . . . $(CH_3)_3SnN(B(CH_3)_2)BBrCH_3$ . . . . . Sn: Org.Comp.18-81, 83
$B_2BrC_7H_{19}N_2$ . . . . $CH_3B(\text{-}NCH_3\text{-}CH_2\text{-}CH_2\text{-}NCH_3\text{-}) \cdot (CH_3)_2BBr$ B: B Comp.SVol.3/3-113
$B_2BrC_9H_{15}N_4S_3$ . . $(CH_3S)_2B[\text{-}N_2C_3H_3\text{-}B(Br)SCH_3\text{-}N_2C_3H_3\text{-}]$ . . . . B: B Comp.SVol.3/4-145
$B_2BrC_{10}H_{24}N_3$ . . . $1,3\text{-}(t\text{-}C_4H_9)_2\text{-}2\text{-}[(CH_3)_2N]\text{-}1,3,2,4\text{-}N_2B_2\text{-}4\text{-}Br$ B: B Comp.SVol.4/4-102
$B_2BrC_{11}H_{21}N_4S_3$. $(CH_3S)(Br)B[\text{-}N_2C_3H_3\text{-}B(S(CH_3)_2)_2\text{-}N_2C_3H_3\text{-}]$ B: B Comp.SVol.3/3-200
$B_2BrC_{12}H_{21}N_4$ . . . $(C_2H_5)(Br)B[\text{-}N_2C_3H_3\text{-}B(C_2H_5)_2\text{-}N_2C_3H_3\text{-}]$ . . . . B: B Comp.SVol.3/3-200
$B_2BrC_{12}H_{29}N_2$ . . . $(i\text{-}C_3H_7)_2N\text{-}BBr\text{-}N(C_4H_9\text{-}t)\text{-}B(CH_3)_2$ . . . . . B: B Comp.SVol.4/4-104
$B_2BrC_{14}H_{30}N_2^+$ . . $[CH_3B(\text{-}NC_5H_6(CH_3)_4\text{-}BBr\text{-}N(C_4H_9\text{-}t)\text{-})]^+$ . . . . B: B Comp.SVol.3/4-95
$B_2BrC_{15}H_{33}N_2$ . . . $[(\text{-}C(CH_3)_2\text{-}(CH_2)_3\text{-}C(CH_3)_2\text{-})N(\text{-}BCH_3\text{-}$
$N(t\text{-}C_4H_9)\text{-}BCH_3\text{-})]Br$ . . . . . B: B Comp.SVol.3/3-208
$B_2BrC_{21}H_{41}N_2$ . . . $1\text{-}[(3.3.1)\text{-}9\text{-}BC_8H_{14}\text{-}9\text{-}N(C_4H_9\text{-}t)\text{-}BBr]\text{-}$
$2,2,6,6\text{-}(CH_3)_4\text{-}NC_5H_6$ . . . . . B: B Comp.SVol.4/4-104
B_2BrF_9 . . . . . $[BrF_2][B_2F_7]$ . . . . . Br: SVol.B3-41/5
$B_2Br_2CH_3NS_2$ . . . . $[\text{-}BBr\text{-}N(CH_3)\text{-}BBr\text{-}S\text{-}S\text{-}]$ . . . . . B: B Comp.SVol.3/4-143
B: B Comp.SVol.4/4-167
$B_2Br_2C_2H_6N_2S$ . . . $BrB(\text{-}NCH_3\text{-}NCH_3\text{-}BBr\text{-}S\text{-})$ . . . . . B: B Comp.SVol.3/4-143
$B_2Br_2C_4Cl_2H_{10}N_2$ $ClBrB\text{-}NCH_3\text{-}CH_2\text{-}CH_2\text{-}NCH_3\text{-}BBrCl$ . . . . . B: B Comp.SVol.3/4-95
$B_2Br_2C_4H_9N_3S$ . . . $BrB(\text{-}NCH_3\text{-}BBr\text{-}NCH_3\text{-}CS\text{-}NCH_3\text{-})$ . . . . . B: B Comp.SVol.3/4-93/4
$B_2Br_2C_4H_{12}N_2$ . . . $(CH_3)_2N\text{-}BBr\text{-}BBr\text{-}N(CH_3)_2$ . . . . . B: B Comp.SVol.4/4-106
$B_2Br_2C_5H_{15}NSn$ . . $(CH_3)_3SnN(BBrCH_3)_2$ . . . . . Sn: Org.Comp.18-28, 42

$B_2Br_2C_6Cl_2H_6N_4$ .. $(H)(Br)B(-N_2C_3H_2Cl-BBrH-N_2C_3H_2Cl-)$ ...... B: B Comp.SVol.3/3-194/6, 200

$B_2Br_2C_6F_5NS_2$ ... $SSBBrN(C_6F_5)BBr$ .... F: PFHOrg.SVol.4-33, 58

$B_2Br_2C_6H_5NS_2$ ... $4-C_6H_5-3,5-Br_2-1,2,4,3,5-S_2NB_2$ .... B: B Comp.SVol.3/4-143

B: B Comp.SVol.4/4-140

$B_2Br_2C_6H_8N_4$ .... $BrBH[-1,2-N_2C_3H_3-]_2BHBr$ .... B: B Comp.SVol.3/3-197, 199

B: B Comp.SVol.4/4-100/1

– .... $H_2B[-1,2-N_2C_3H_2Br-]_2BH_2$ .... B: B Comp.SVol.3/3-199

$B_2Br_2C_6H_{15}N$ .... $t-C_4H_9-N(BBr-CH_3)_2$ .... B: B Comp.SVol.4/4-105

$B_2Br_2C_7F_3H_4NS_2$ $BrB[-S-S-BBr-N(C_6H_4-3-CF_3)-]$ .... B: B Comp.SVol.3/4-143

$B_2Br_2C_7H_{10}N_4S_2$. $(-S-CH_2-CH_2-S-)B[-N_2C_2H_3-BBr_2-N_2C_3H_3-]$ B: B Comp.SVol.3/4-145

$B_2Br_2C_7H_{21}N_3Sn$ $(CH_3)_3SnN(BBrN(CH_3)_2)_2$ .... Sn: Org.Comp.18-28, 42

$B_2Br_2C_8H_{12}N_4S_2$. $(CH_3S)(Br)B[-N_2C_3H_3-BBr(SCH_3)-N_2C_3H_3-]$ .. B: B Comp.SVol.3/3-200

$B_2Br_2C_8H_{19}N$ .... $t-C_4H_9-N(BBr-C_2H_5)_2$ .... B: B Comp.SVol.4/4-105

$B_2Br_2C_8H_{20}N_2$ ... $(C_2H_5)_2N-BBr-BBr-N(C_2H_5)_2$ .... B: B Comp.SVol.4/4-106

$B_2Br_2C_8H_{21}N_3$ ... $[-BBr(N(CH_3)_2)-N(CH_3)_2-BBr-N(C_4H_9-t)-]$.... B: B Comp.SVol.4/4-100

$B_2Br_2C_{10}H_{16}N_4$... $BrBH[-1,2-N_2C_3H(CH_3)_2-]_2BHBr$. .... B: B Comp.SVol.4/4-101/2

$B_2Br_2C_{10}H_{18}N_4S_2$ $[(CH_3)_2S]_2B(-N_2C_3H_3-BBr_2-N_2C_3H_3-)$ .... B: B Comp.SVol.3/3-200

$B_2Br_2C_{12}FeH_{14}$... $Fe(C_5H_4BBrCH_3)_2$. .... Fe: Org.Comp.A9-278

$B_2Br_2C_{12}H_8$ ..... $5,10-Br_2-5,10-B_2C_{12}H_8$ .... B: B Comp.SVol.4/4-94

$B_2Br_2C_{13}H_{27}N_2^+$ $[BrB(-NC_5H_6(CH_3)_4-BBr-N(C_4H_9-t)-)]^+$...... B: B Comp.SVol.3/4-95

$B_2Br_2C_{13}H_{33}N_3Si$ $(CH_3)_3Si-N(C_4H_9-t)-BBr-N(C_4H_9-t)-BBr-N(CH_3)_2$

B: B Comp.SVol.4/4-104

$B_2Br_2C_{14}H_{14}N_2$... $2,4-Br_2-8-CH_3-3-(2-CH_3-C_6H_4)-1,3,2,4-N_2B_2C_6H_4$

B: B Comp.SVol.4/4-102

$B_2Br_2C_{14}H_{24}N_4$... $Br_2B[-1,2-N_2C_3H(CH_3)_2-]_2B(C_2H_5)_2$ .... B: B Comp.SVol.4/4-101/2

$B_2Br_2C_{14}H_{30}N_2$... $CH_3B[-NC_5H_6(CH_3)_4-BBr_2-N(C_4H_9-t)-]$ ...... B: B Comp.SVol.3/4-93

$B_2Br_2C_{15}H_{33}N_3$... $2,2,6,6-(CH_3)_4NC_5H_6-1-BBr[-N(CH_3)_2-BBr-N(C_4H_9-t)-]$

B: B Comp.SVol.4/4-106

$B_2Br_2C_{17}H_{36}N_2$... $2,2,6,6-(CH_3)_4NC_5H_6-1-BBr-N(C_4H_9-t)-BBr-C_4H_9-t$

B: B Comp.SVol.4/4-104

$B_2Br_2C_{17}H_{37}N_3$... $2,2,6,6-(CH_3)_4NC_5H_6-1-BBr-N(C_4H_9-t)-BBr-N(C_2H_5)_2$

B: B Comp.SVol.4/4-104

$B_2Br_2C_{17}H_{41}N_3Si$ $(CH_3)_3Si-N(C_4H_9-t)-BBr-N(C_4H_9-t)-BBr-N(C_3H_7-i)_2$

B: B Comp.SVol.4/4-104

$B_2Br_2C_{18}H_{16}N_4$... $Br_2B[-1,2-N_2C_3H_3-]_2B(C_6H_5)_2$ .... B: B Comp.SVol.4/4-100/1

$B_2Br_2C_{19}H_{41}N_3$... $2,2,6,6-(CH_3)_4NC_5H_6-1-BBr-N(C_4H_9-t)-BBr-N(C_3H_7-i)_2$

B: B Comp.SVol.4/4-104

$B_2Br_2C_{46}H_{46}N_2O_4Sn$

$(C_4H_9)_2Sn(OB(C_6H_5)OC_6H_4CHNC_6H_4Br-4)_2$ ... Sn: Org.Comp.15-351

$B_2Br_2C_{54}H_{50}N_2O_4Sn$

$(C_4H_9)_2Sn(OB(C_6H_5)OC_{10}H_6CHNC_6H_4Br-4)_2$ .. Sn: Org.Comp.15-351

$B_2Br_2S_3$ .... $[-BBr-S-BBr-S-S-]$. .... B: B Comp.SVol.3/4-138

B: B Comp.SVol.4/4-157/8

$B_2Br_2S_4$ .... $[-BBr-S-BBr-S-S-S-]$ .... B: B Comp.SVol.3/4-138

$B_2Br_3C_4ClH_{10}N_2$.. $ClBrB-NCH_3-CH_2-CH_2-NCH_3-BBr_2$ .... B: B Comp.SVol.3/4-95

$B_2Br_3C_5H_{13}N_2$ ... $CH_3B(-NCH_3-CH_2-CH_2-NCH_3-) \cdot BBr_3$...... B: B Comp.SVol.3/3-113

$B_2Br_3C_6H_7N_4$ .... $BrBH[-1,2-N_2C_3H_2Br-]_2BH_2$ .... B: B Comp.SVol.3/3-199

$B_2Br_3C_6H_7N_4$ .... $BrBH[-1,2-N_2C_3H_3-]_2BBr_2$ .... B: B Comp.SVol.3/3-197, 200
B: B Comp.SVol.4/4-100/1
$B_2Br_3C_6H_{15}N_2$ ... $[-BBr_2-N(CH_3)_2-BBr-N(C_4H_9-t)-]$........... B: B Comp.SVol.4/4-100
$B_2Br_3C_8H_{18}NS_2$ .. $[(CH_3)_2CH]_2NB(-S-CH_2-CH_2-S(-BBr_3)-)$ ..... B: B Comp.SVol.3/4-144
$B_2Br_3C_{13}H_{27}N_2$... $BrB[-NC_5H_6(CH_3)_4-BBr_2-N(C_4H_9-t)-]$ .... B: B Comp.SVol.3/4-93/4
$B_2Br_3C_{16}GaH_{40}N_8$ $[(-NCH_3-CH_2-CH_2-NCH_3-)_2B]_2 \cdot GaBr_3$ ..... B: B Comp.SVol.3/3-118
$B_2Br_4C_2Cl_4H_6N_2O_2P_2$
$[Cl_2P(O)N(CH_3)BBr_2]_2$ .... B: B Comp.SVol.3/4-92
$B_2Br_4C_3H_9NSn$ ... $(CH_3)_3SnN(BBr_2)_2$........................ Sn: Org.Comp.18-27, 42
$B_2Br_4C_4H_9N$ ..... $t-C_4H_9-N(BBr_2)_2$ .... B: B Comp.SVol.4/4-100
$B_2Br_4C_4H_{10}N_2$ ... $(-NCH_3-CH_2-CH_2-NCH_3-)BBr \cdot BBr_3$ .... B: B Comp.SVol.3/4-92
$B_2Br_4C_6Cl_2H_4N_4$.. $Br_2B[-1,2-N_2C_3H_2(Cl)-]_2BBr_2$.............. B: B Comp.SVol.4/4-102
$B_2Br_4C_6H_4$ ...... $1,3-(BBr_2)_2-C_6H_4$........................ B: B Comp.SVol.4/4-90
– .... $1,4-(BBr_2)_2-C_6H_4$........................ B: B Comp.SVol.4/4-90
$B_2Br_4C_6H_6N_4$ .... $Br_2B[-1,2-N_2C_3H_3-]_2BBr_2$................. B: B Comp.SVol.3/3-200
B: B Comp.SVol.4/4-100/1
$B_2Br_4C_6H_{16}N_2$ ... $[-BBr_2-CH_2-N(CH_3)_2-BBr_2-CH_2-N(CH_3)_2-]$... B: B Comp.SVol.4/4-100
$B_2Br_4C_8H_{10}N_4$ ... $Br_2B[-1,2-N_2C_3H_2(CH_3)-]_2BBr_2$ .... B: B Comp.SVol.4/4-101
$B_2Br_4C_{10}FeH_8$ ... $Fe(C_5H_4BBr_2)_2$......................... Fe: Org.Comp.A9-277, 282
$B_2Br_4C_{10}H_{14}N_4$... $Br_2B[-1,2-N_2C_3H(CH_3)_2-]_2BBr_2$ .... B: B Comp.SVol.4/4-101/2
$B_2Br_4C_{12}H_{36}N_2Si_4$ $[(Si(CH_3)_3)_2N=B=N(Si(CH_3)_3)_2][BBr_4]$ .... B: B Comp.SVol.4/4-81/3
$B_2Br_4C_{13}GaH_{28}N$ $[(C_2H_5)_2B=N=B(-C(CH_3)_2-(CH_2)_3-C(CH_3)_2-)]GaBr_4$
B: B Comp.SVol.3/3-205
$B_2Br_4C_{13}H_{28}N_2$... $[(CH_3)_4C_5H_6N=B=N(C_2H_5)_2][BBr_4]$........... B: B Comp.SVol.3/4-95
$B_2Br_4C_{13}H_{36}N_2Si_3$ $[(CH_3)_3Si-N(C_4H_9-t)=B=N(Si(CH_3)_3)_2][BBr_4]$... B: B Comp.SVol.4/4-81/2
$B_2Br_4C_{14}H_{36}N_2Si_2$ $[(CH_3)_3Si-N(C_4H_9-t)=B=N(C_4H_9-t)-Si(CH_3)_3][BBr_4]$
B: B Comp.SVol.4/4-81/2
– .... $[(t-C_4H_9)_2N=B=N(Si(CH_3)_3)_2][BBr_4]$.......... B: B Comp.SVol.4/4-81/2
$B_2Br_4C_{18}H_{14}N_4$... $Br_2B[-1,2-N_2C_3H_2(C_6H_5)-]_2BBr_2$ .... B: B Comp.SVol.4/4-102
$B_2Br_4C_{18}H_{34}N_2Si$ $[C_6H_5-CH_2-N(C_4H_9-t)=B=N(C_4H_9-t)-Si(CH_3)_3][BBr_4]$
B: B Comp.SVol.4/4-81/2
$B_2Br_4C_{18}H_{36}N_2$... $[(CH_3)_4C_5H_6N=B=NC_5H_6(CH_3)_4][BBr_4]$ .... B: B Comp.SVol.3/3-205
B: B Comp.SVol.3/4-79, 95
$B_2Br_4C_{22}H_{32}N_2$... $[C_6H_5-CH_2-N(C_4H_9-t)=B=N(C_4H_9-t)-CH_2-C_6H_5][BBr_4]$
B: B Comp.SVol.4/4-83
$B_2Br_4C_{24}F_8Ga_2H_{16}N_4$
$[Ga(1,10-N_2C_{12}H_8)_2][BF_4]_2[GaBr_4]$ .... Ga: SVol.D1-316
$B_2Br_4C_{30}Fe_3H_{46}N_2$
$[(CH_3-C_6H_5)Fe(2-CH_3-1-(t-C_4H_9)-1,2-NBC_3H_3)]_2[FeBr_4]$
Fe: Org.Comp.B18-99/102
$B_2Br_6C_6Cl_4N_4$.... $Cl_2B[-1,2-N_2C_3(Br)_3-]_2BCl_2$ .... B: B Comp.SVol.4/4-67/8
$B_2Br_6C_6H_4N_4$ .... $H_2B[-1,2-N_2C_3Br_3-]_2BH_2$................. B: B Comp.SVol.4/3b-33
$B_2Br_6C_{13}H_{28}NTa$ $[(C_2H_5)_2B=N=B(-C(CH_3)_2-(CH_2)_3-C(CH_3)_2-)]TaBr_6$
B: B Comp.SVol.3/3-205
$B_2Br_6C_{16}Ga_2H_{40}N_8$
$[(-NCH_3-CH_2-CH_2-NCH_3-)_2B]_2 \cdot 2\ GaBr_3$.... B: B Comp.SVol.3/3-118
$B_2Br_6C_{20}H_{48}N_4P_2$ $[(C_2H_5)_2N]_2P(CH_2)_4P[N(C_2H_5)_2]_2 \cdot 2\ BBr_3$.... B: B Comp.SVol.3/4-78
$B_2CH_2I_4$......... $I_2B-CH_2-BI_2$.......................... B: B Comp.SVol.4/4-113
B_2CH_9N......... $[-BH_2-H-BH_2-NH(CH_3)-]$ .... B: B Comp.SVol.4/3a-239

Formula	Compound	Reference
$B_2CH_{11}NSi$	$(SiH_3(CH_3)N)B_2H_5$	Si: SVol.B4-317
B_2CTh	ThB_2C	Th: SVol.C6-132/4
$B_2C_2Cl_4H_2$	Cl_2B-CH=CH-BCl_2	B: B Comp.SVol.4/4-44
$B_2C_2Cl_4H_4$	Cl_2B-CH_2CH_2-BCl_2	B: B Comp.SVol.4/4-44
$B_2C_2Cl_6F_6N_2S_2$	CF_3-SCl=N(-BCl_2-$)_2$N=SCl-CF_3	B: B Comp.SVol.4/4-165/6
$B_2C_2Cl_8H_6N_2O_2P_2$	$[Cl_2BN(CH_3)P(O)Cl_2]_2$	B: B Comp.SVol.3/4-55
$B_2C_2H_4$	$C_2B_2H_4$	B: B Comp.SVol.3/4-155
		B: B Comp.SVol.4/4-180
$B_2C_2H_4I_4$	I_2B-CH_2CH_2-BI_2	B: B Comp.SVol.4/4-113
$B_2C_2H_5N_2Na$	$[H_3BCNBH_2CN]Na \cdot 2\ C_4H_8O_2$	B: B Comp.SVol.3/3-212
$B_2C_2H_6N_2$	1,2,3,6-$N_2B_2C_2H_6$	B: B Comp.SVol.3/3-165
–	1,4,2,5-$N_2B_2C_2H_6$	B: B Comp.SVol.3/3-165
$B_2C_2H_6O_2S_3$	CH_3OB(-S-S-$BOCH_3$-S-)	B: B Comp.SVol.3/4-119/20
$B_2C_2H_6S_3$	[-B(CH_3)-S-B(CH_3)-S-S-]	B: B Comp.SVol.3/4-138
		B: B Comp.SVol.4/4-130
–	[-$^{10}B(CH_3)$-S-$^{10}B(CH_3)$-S-S-]	B: B Comp.SVol.3/4-111/2
–	[-$^{11}B(CH_3)$-S-$^{11}B(CH_3)$-S-S-]	B: B Comp.SVol.3/4-111/2
$B_2C_2H_6Se_3$	[-B(CH_3)-Se-B(CH_3)-Se-Se-]	B: B Comp.SVol.4/4-171/2
$B_2C_2H_8N_2$	$(H_2C=N-BH_2)_2$	B: B Comp.SVol.3/3-172
$B_2C_2H_{11}N$	[-BH_2-H-BH_2-N$(CH_3)_2$-]	B: B Comp.SVol.4/3a-239
$B_2C_2H_{12}N_2O_4$	$(H_3N-BH_2COOH)_2$	B: B Comp.SVol.3/3-185
$B_2C_2N_2S_5$	SCNB[-S-S-BNCS-S-]	B: B Comp.SVol.3/4-123
$B_2C_3Cl_4H_9NSi$	$(Cl_2B)_2N$-$Si(CH_3)_3$	B: B Comp.SVol.4/4-59
$B_2C_3Cl_4H_9NSn$	$(Cl_2B)_2N$-$Sn(CH_3)_3$	B: B Comp.SVol.4/4-59
		Sn: Org.Comp.18-27, 42
$B_2C_3F_4H_6O_3$	F_2B[-OCH_3-BF_2-O=CCH_3-O-]	B: B Comp.SVol.3/3-367/9
$B_2C_3H_9NS_2$	CH_3B(-S-S-BCH_3-NCH_3-)	B: B Comp.SVol.3/4-123/4
$B_2C_3H_9N_3O$	CH_3B[-NH-CO-NH-BCH_3-NH-]	B: B Comp.SVol.3/3-118
$B_2C_3H_9N_3S$	CH_3B(-NH-CS-NH-BCH_3-NH-)	B: B Comp.SVol.3/3-119
$B_2C_3H_{11}NS_3$	HB(-S-S-BH-S-) $\cdot$ $N(CH_3)_3$	B: B Comp.SVol.3/4-123
$B_2C_3H_{13}NS$	[-BH_2-S(BH_3)-CH_2CH_2-NH(CH_3)-]	B: B Comp.SVol.4/4-145/6, 148
$B_2C_3Th_3$	$Th_3B_2C_3$	Th: SVol.C6-132/4
$B_2C_4ClH_8NS_2$	1,3,2-$S_2BC_2H_2$-2-[BCl-N$(CH_3)_2$]	B: B Comp.SVol.4/4-167
$B_2C_4Cl_2H_9N_3O$	ClB(-NCH_3-BCl-NCH_3-CO-NCH_3-)	B: B Comp.SVol.3/4-59
$B_2C_4Cl_2H_9N_3S$	ClB(-NCH_3-BCl-NCH_3-CS-NCH_3-)	B: B Comp.SVol.3/4-59
$B_2C_4Cl_2H_{12}N_2$	$(CH_3)_2N$-BCl-BCl-N$(CH_3)_2$	B: B Comp.SVol.4/4-60
$B_2C_4Cl_4H_9N$	$(Cl_2B)_2N$-C_4H_9-t	B: B Comp.SVol.4/4-59
$B_2C_4Cl_7H_{12}N$	$[N(CH_3)_4][B_2Cl_7]$	B: B Comp.SVol.3/4-33
$B_2C_4F_2H_6O_5$	$(CH_3)_2B_2O_5C_2(F)_2$	B: B Comp.SVol.3/3-370/1
$B_2C_4F_4H_4N_4NiO_4$	$[F_2B$(-O-N=CH-CH=N-O-$)_2BF_2]$Ni	B: B Comp.SVol.3/3-370
$B_2C_4F_4H_8O_3$	F_2B[-O(CH_3)-BF_2-O=C(C_2H_5)-O-]	B: B Comp.SVol.3/3-367/8
–	F_2B[-O(C_2H_5)-BF_2-O=C(CH_3)-O-]	B: B Comp.SVol.3/3-367/8
$B_2C_4F_4H_{12}N_2$	$[(CH_3)_2N-BF_2]_2$	B: B Comp.SVol.4/3b-227
$B_2C_4F_6H_4N_2$	1,3-$N_2C_4H_4$ $\cdot$ 2 BF_3	B: B Comp.SVol.4/3b-149
–	1,4-$N_2C_4H_4$ $\cdot$ 2 BF_3	B: B Comp.SVol.4/3b-149
$B_2C_4F_8H_{18}O_4$	$[(CH_3OH)_2H]_2[BF_4]_2$	B: B Comp.SVol.3/3-307
$B_2C_4H_3LiN_4$	$[H_2B(CN)CNBH(CN)_2]Li \cdot 1.5\ C_4H_8O_2$	B: B Comp.SVol.3/3-212/3
$B_2C_4H_6$	1,2,3,5-$C_4B_2H_6$	B: B Comp.SVol.4/4-180
–	2,3,4,5-$C_4B_2H_6$	B: B Comp.SVol.4/4-180

Formula	Compound	Reference
$B_2C_4H_8S_5$	$[1,3,2\text{-}S_2BC_2H_4\text{-}2\text{-}]_2S$	B: B Comp.SVol.4/4-131
$B_2C_4H_{10}N_2$	$2\ H_3B \cdot 1,3\text{-}N_2C_4H_4$	B: B Comp.SVol.4/3b-14
–	$2\ H_3B \cdot 1,4\text{-}N_2C_4H_4$	B: B Comp.SVol.4/3b-14
$B_2C_4H_{10}O_2S_3$	$C_2H_5OB(\text{-}S\text{-}S\text{-}BOC_2H_5\text{-}S\text{-})$	B: B Comp.SVol.3/4-119/20
$B_2C_4H_{10}S_5$	$C_2H_5SB(\text{-}S\text{-}S\text{-}BC_2H_5S\text{-}S\text{-})$	B: B Comp.SVol.3/4-111/2
$B_2C_4H_{10}Se_3$	$[\text{-}B(C_2H_5)\text{-}Se\text{-}B(C_2H_5)\text{-}Se\text{-}Se\text{-}]$	B: B Comp.SVol.4/4-171/2
$B_2C_4H_{12}N_2$	$[\text{-}B(CH_3)\text{-}N(CH_3)\text{-}B(CH_3)\text{-}N(CH_3)\text{-}]$	B: B Comp.SVol.4/3a-195
$B_2C_4H_{12}N_2OS$	$[\text{-}B(CH_3)\text{-}N{=}S(CH_3)_2{=}N\text{-}B(CH_3)\text{-}O\text{-}]$	B: B Comp.SVol.4/4-153
$B_2C_4H_{12}N_2O_4S_2$	$[\text{-}BCH_3\text{-}NSO_2CH_3\text{-}]_2$	B: B Comp.SVol.3/3-135
$B_2C_4H_{12}N_2S$	$[\text{-}B(CH_3)\text{-}N(CH_3)\text{-}N(CH_3)\text{-}B(CH_3)\text{-}S\text{-}]$	B: B Comp.SVol.3/4-124
		B: B Comp.SVol.4/3a-251
		B: B Comp.SVol.4/4-140
$B_2C_4H_{12}N_2S_2$	$2,4\text{-}[(CH_3)_2N]_2\text{-}1,3,2,4\text{-}S_2B_2$	B: B Comp.SVol.4/4-144
–	$[\text{-}B(CH_3)\text{-}S\text{-}B(CH_3)\text{-}N{=}S(CH_3)_2{=}N\text{-}]$	B: B Comp.SVol.4/4-141/2
$B_2C_4H_{12}N_2S_3$	$(CH_3)_2NB[\text{-}S\text{-}S\text{-}BN(CH_3)_2\text{-}S\text{-}]$	B: B Comp.SVol.3/4-123
$B_2C_4H_{12}N_4$	$NC\text{-}BH_2\text{-}NH_2\text{-}CH_2CH_2\text{-}NH_2\text{-}BH_2\text{-}CN$	B: B Comp.SVol.3/3-184
		B: B Comp.SVol.4/3b-23
$B_2C_4H_{13}N_3S$	$[\text{-}B(CH_3)\text{-}NH\text{-}B(CH_3)\text{-}N{=}S(CH_3)_2{=}N\text{-}]$	B: B Comp.SVol.4/4-141/2
$B_2C_4H_{14}N_2$	$(CH_3)_3N\text{-}BH_2\text{-}NC\text{-}BH_3$	B: B Comp.SVol.3/3-184
$B_2C_4H_{14}S_2$	$H_2B(\text{-}SCH_3\text{-}CH_2\text{-}BH_2\text{-}SCH_3\text{-}CH_2\text{-})$	B: B Comp.SVol.3/4-115
$B_2C_4H_{15}NS$	$[\text{-}BH_2\text{-}S(BH_3)\text{-}CH_2CH_2\text{-}NH(C_2H_5)\text{-}]$	B: B Comp.SVol.4/4-148
$B_2C_4H_{16}N_2$	$[H_2B\text{-}N(CH_3)_2]_2$	B: B Comp.SVol.4/3a-224
$B_2C_4H_{18}NP$	$H_3B\text{-}N(CH_3)_2\text{-}P(CH_3)_2\text{-}BH_3$	B: B Comp.SVol.4/3b-15
$B_2C_4H_{20}N_2Si$	$SiH_2[N(CH_3)_2]_2 \cdot 2\ BH_3$	Si: SVol.B4-186
–	$SiH_3(CH_3)NB_2H_5 \cdot N(CH_3)_3$	Si: SVol.B4-318
$B_2C_5Cl_2H_{14}N_2$	$(CH_3)_2N\text{-}BCl\text{-}CH_2\text{-}BCl\text{-}N(CH_3)_2$	B: B Comp.SVol.3/4-55
		B: B Comp.SVol.4/4-61
$B_2C_5Cl_3H_{13}N_2$	$CH_3B(\text{-}NCH_3\text{-}CH_2\text{-}CH_2\text{-}NCH_3\text{-}) \cdot BCl_3$	B: B Comp.SVol.3/3-113
$B_2C_5F_4H_{10}O_3$	$F_2B[\text{-}O(CH_3)\text{-}BF_2\text{-}O{=}C(C_3H_7)\text{-}O\text{-}]$	B: B Comp.SVol.3/3-367/8
–	$F_2B[\text{-}O(C_2H_5)\text{-}BF_2\text{-}O{=}C(C_2H_5)\text{-}O\text{-}]$	B: B Comp.SVol.3/3-367/8
$B_2C_5H_6NaO_5Re$	$Na[(CO)_5Re(BH_3)_2]$	Re: Org.Comp.2-168
$B_2C_5H_6O_5Re^-$	$[(CO)_5Re(BH_3)_2]^-$	Re: Org.Comp.2-168/9
$B_2C_5H_8I_2$	$1,3\text{-}I_2\text{-}4,5\text{-}(CH_3)_2\text{-}1,3\text{-}B_2C_3H_2$	B: B Comp.SVol.4/4-114
$B_2C_5H_8I_4$	$CH_3\text{-}CI{=}C(CH_3)\text{-}BI\text{-}CH_2\text{-}BI_2$	B: B Comp.SVol.4/4-113
$B_2C_5H_{11}NS_3$	$2\text{-}[(CH_3)_2N\text{-}B(SCH_3)]\text{-}1,3,2\text{-}S_2BC_2H_2$	B: B Comp.SVol.4/4-146
$B_2C_5H_{12}N_2$	$2\ H_3B \cdot 1,4\text{-}N_2C_4H_3\text{-}CH_3$	B: B Comp.SVol.4/3b-14
–	$(CH_3)_2B\text{-}N{=}C{=}N\text{-}B(CH_3)_2$	B: B Comp.SVol.4/3a-243
$B_2C_5H_{12}N_2O_2$	$[\text{-}BCH_3\text{-}NCH_3\text{-}CO\text{-}NCH_3\text{-}BCH_3\text{-}O\text{-}]$	B: B Comp.SVol.3/3-165
$B_2C_5H_{13}I_3N_2$	$CH_3B[\text{-}N(CH_3)CH_2CH_2N(CH_3)\text{-}] \cdot BI_3$	B: B Comp.SVol.3/3-113
		B: B Comp.SVol.3/4-99
–	$CH_3B[\text{-}N(CH_3)CH_2CH_2N(CH_3)\text{-}] \cdot CH_3BBr_2$	B: B Comp.SVol.3/3-113
$B_2C_5H_{13}N_3S$	$CH_3B(\text{-}NCH_3\text{-}CS\text{-}NCH_3\text{-}BCH_3\text{-}NH\text{-})$	B: B Comp.SVol.3/3-120
$B_2C_5H_{15}N_3S_2$	$4\text{-}CH_3\text{-}3,5\text{-}[(CH_3)_2N]_2\text{-}1,2,4,3,5\text{-}S_2NB_2$	B: B Comp.SVol.4/4-140
$B_2C_5H_{17}NS$	$[\text{-}BH_2\text{-}S(BH_3)\text{-}CH_2CH_2\text{-}NH(C_3H_7\text{-}i)\text{-}]$	B: B Comp.SVol.4/4-148
$B_2C_5H_{18}N_2$	$[\text{-}N(CH_3)\text{-}CH_2\text{-}N(CH_3)\text{-}CH_2CH_2\text{-}] \cdot 2\ BH_3$	B: B Comp.SVol.4/3b-12
$B_2C_6Cl_2F_2H_6N_4$	$Cl_2B[\text{-}1,2\text{-}N_2C_3H_3\text{-}]_2BF_2$	B: B Comp.SVol.4/3b-248
		B: B Comp.SVol.4/4-68
$B_2C_6Cl_2F_4H_4N_4$	$F_2B[\text{-}1,2\text{-}N_2C_3H_2Cl\text{-}]_2BF_2$	B: B Comp.SVol.4/3b-249/50
$B_2C_6Cl_2H_8N_4$	$H_2B[\text{-}1,2\text{-}N_2C_3H_2Cl\text{-}]_2BH_2$	B: B Comp.SVol.3/3-198
		B: B Comp.SVol.4/3b-33

$B_2C_6Cl_2H_{18}N_2S_2Si_2$
$[-S-N(SiCl(CH_3)_2)-B(CH_3)-S-B(CH_3)-N(SiCl(CH_3)_2)-]$ B: B Comp.SVol.4/4-140/1
$B_2C_6Cl_3H_{15}N_2$ .... $Cl_2B-N(C_4H_9-t)-BCl-N(CH_3)_2$ .... B: B Comp.SVol.4/4-59
$B_2C_6Cl_4H_4$ .... 1,2-$(BCl_2)_2-C_6H_4$ .... B: B Comp.SVol.4/4-45
– .... 1,4-$(BCl_2)_2-C_6H_4$ .... B: B Comp.SVol.4/4-45
$B_2C_6Cl_4H_6N_4$ .... $Cl_2B[-1,2-N_2C_3H_3-]_2BCl_2$ .... B: B Comp.SVol.3/3-194/6, 199
B: B Comp.SVol.4/4-67/8
$B_2C_6Cl_4H_{10}$ .... $Cl_2B-C(C_2H_5)=C(C_2H_5)-BCl_2$ .... B: B Comp.SVol.4/4-44
$B_2C_6Cl_4H_{12}$ .... n-$C_3H_7-C(C_2H_5)(BCl_2)_2$ .... B: B Comp.SVol.4/4-45
– .... n-$C_5H_{11}-CH(BCl_2)_2$ .... B: B Comp.SVol.4/4-44
$B_2C_6Cl_4H_{16}N_2$ .... $[-BCl_2-CH_2-N(CH_3)_2-BCl_2-CH_2-N(CH_3)_2-]$ .... B: B Comp.SVol.4/4-59
$B_2C_6Cl_6H_4N_4$ .... $Cl_2B[-1,2-N_2C_3H_2(Cl)-]_2BCl_2$ .... B: B Comp.SVol.4/4-67/8
$B_2C_6FH_{18}O_6^-$ .... $[B_2F(OCH_3)_6]^-$ .... B: B Comp.SVol.4/3b-237
$B_2C_6F_2H_8N_4$ .... $F_2B[-1,2-N_2C_3H_3-]_2BH_2$ .... B: B Comp.SVol.3/3-199
B: B Comp.SVol.4/3b-248
$B_2C_6F_2H_{10}O_5$ .... $B_2C_2F_2O_5(C_2H_5)_2$ .... B: B Comp.SVol.3/3-370/1
B: B Comp.SVol.4/3b-235
$B_2C_6F_4H_4$ .... 1,4-$(BF_2)_2-C_6H_4$ .... B: B Comp.SVol.4/3b-226
$B_2C_6F_4H_6N_4$ .... $F_2B[-1,2-N_2C_3H_3-]_2BF_2$ .... B: B Comp.SVol.3/3-199
B: B Comp.SVol.4/3b-248/9
$B_2C_6F_4H_{16}N_2$ .... $[(CH_3)_3N-BH_2C{\equiv}N-C_2H_5][BF_4]$ .... B: B Comp.SVol.3/3-186, 335
$B_2C_6F_6H_{16}N_2$ .... $(CH_3)_2N-CH_2CH_2-N(CH_3)_2 \cdot 2\ BF_3$ .... B: B Comp.SVol.4/3b-124
$B_2C_6H_{10}I_2$ .... 1,4-I_2-2,3-$(CH_3)_2$-1,4-$B_2C_4H_4$ .... B: B Comp.SVol.4/4-114
$B_2C_6H_{10}I_2S$ .... 2,5-I_2-3,4-$(C_2H_5)_2$-1,2,5-SB_2C_2 .... B: B Comp.SVol.4/4-163
$B_2C_6H_{10}N_4$ .... $H_2B[-1,2-N_2C_3H_3-]_2BH_2$ .... B: B Comp.SVol.3/3-197/8
B: B Comp.SVol.4/3b-33
$B_2C_6H_{12}S_6$ .... 1,3,2-$S_2BC_2H_4$-2-S-CH_2CH_2-S-2-(1,3,2-$S_2BC_2H_4$)
B: B Comp.SVol.3/4-112
B: B Comp.SVol.4/4-131
$B_2C_6H_{14}N_2$ .... 1,2-$(H_3B-NH_2)_2-C_6H_4$ .... B: B Comp.SVol.4/3b-11
– .... 1,3-$(H_3B-NH_2)_2-C_6H_4$ .... B: B Comp.SVol.4/3b-11
– .... 1,4-$(H_3B-NH_2)_2-C_6H_4$ .... B: B Comp.SVol.4/3b-5, 11
– .... 2 H_3B · 1,4-$N_2C_4H_2$-2,3-$(CH_3)_2$ .... B: B Comp.SVol.4/3b-15
– .... 2 H_3B · 1,4-$N_2C_4H_2$-2,5-$(CH_3)_2$ .... B: B Comp.SVol.4/3b-15
– .... 2 H_3B · 1,4-$N_2C_4H_2$-2,6-$(CH_3)_2$ .... B: B Comp.SVol.4/3b-15
$B_2C_6H_{14}N_2O_2$ .... $B_2C_2H_2N_2O_2(CH_3)_4$ .... B: B Comp.SVol.3/3-123/4
$B_2C_6H_{14}N_2S_2$ .... 1,3,2-$S_2BC_2H_2$-2-$B[N(CH_3)_2]_2$ .... B: B Comp.SVol.4/4-146
$B_2C_6H_{14}O_2S_3$ .... i-$C_3H_7OB[-S-S-BO(C_3H_7-i)-S-]$ .... B: B Comp.SVol.3/4-119/20
$B_2C_6H_{14}O_4S_3$ .... $CH_3OC_2H_4OB[-S-S-B(OCH_2CH_2OCH_3)-S-]$ .... B: B Comp.SVol.3/4-119/20
$B_2C_6H_{15}NS_2$ .... $[-B(CH_3)-S-S-B(CH_3)-N(C_4H_9-t)-]$ .... B: B Comp.SVol.3/4-123/4
B: B Comp.SVol.4/4-140
$B_2C_6H_{15}N_3O$ .... $CH_3B[-NCH_3-CO-NCH_3-BCH_3-NCH_3-]$ .... B: B Comp.SVol.3/3-118
$B_2C_6H_{15}N_3S$ .... $CH_3B(-NCH_3-CS-NCH_3-BCH_3-NCH_3-)$ .... B: B Comp.SVol.3/3-119/20
$B_2C_6H_{16}N_2OS$ .... $[-B(CH_3)-N{=}S(C_2H_5)_2{=}N-B(CH_3)-O-]$ .... B: B Comp.SVol.4/4-153
$B_2C_6H_{16}N_2O_8$ .... $[-B(OH)_2-NH_2-CH(COOH)-CH_2-B(OH)_2-NH_2-$ $CH(COOH)-CH_2-]$ .... B: B Comp.SVol.4/3b-73/4
$B_2C_6H_{16}N_2S_2$ .... 2,3-$[(CH_3)_2N]_2$-1,4,2,3-$S_2B_2C_2H_4$ .... B: B Comp.SVol.4/4-147

$B_2C_6H_{16}N_2S_2$ .... $[-B(CH_3)-S-B(CH_3)-N=S(C_2H_5)_2=N-]$ .... B: B Comp.SVol.4/4-141/2
$B_2C_6H_{17}N_3S$ ..... $[-B(CH_3)-NH-B(CH_3)-N=S(C_2H_5)_2=N-]$ .... B: B Comp.SVol.4/4-141/2
$B_2C_6H_{18}I_2N_2$ ..... $[-BHI-CH_2-N(CH_3)_2-BHI-CH_2-N(CH_3)_2-]$ ..... B: B Comp.SVol.4/4-114
$B_2C_6H_{18}N_2$ ...... $(CH_3)_2B-N(CH_3)-B(CH_3)-N(CH_3)_2$ .... B: B Comp.SVol.3/3-106
– .... $H_3B-N[-CH_2-CH_2-]_3N-BH_3$ .... B: B Comp.SVol.3/3-177
– .... $[(CH_3)_2B]_2N-N(CH_3)_2$ .... B: B Comp.SVol.3/3-159
$B_2C_6H_{18}N_2OSi$ ... $[-Si(CH_3)_2-NCH_3-BCH_3-O-BCH_3-NCH_3-]$ .... B: B Comp.SVol.4/3a-206
$B_2C_6H_{18}N_4$ ...... $[-BCH_3-NCH_3-NCH_3-BCH_3-NCH_3-NCH_3-]$ ... B: B Comp.SVol.4/3a-200
$B_2C_6H_{19}IN_2$ ...... $[-BH_2-CH_2-N(CH_3)_2-BHI-CH_2-N(CH_3)_2-]$ .... B: B Comp.SVol.4/4-114
$B_2C_6H_{20}NP$ ...... $[-N(CH_3)_2-BH_2-CH_2-P(CH_3)_2-CH_2-BH_2-]$ .... B: B Comp.SVol.3/3-187
$B_2C_6H_{20}N_2$ ...... $H_3B-NCH_3(-CH_2-CH_2-)_2NCH_3-BH_3$ .... B: B Comp.SVol.3/3-177
– .... $[-BH_2-CH_2-N(CH_3)_2-BH_2-CH_2-N(CH_3)_2-]$ .... B: B Comp.SVol.3/3-187
B: B Comp.SVol.4/3a-256
– .... $[-NCH_3-CHCH_3-NCH_3-CH_2CH_2-] \cdot 2\ BH_3$ .... B: B Comp.SVol.4/3b-12
$B_2C_6H_{20}N_2S_3$ .... $HB(-S-S-BH-S-) \cdot 2\ N(CH_3)_3$ .... B: B Comp.SVol.3/4-123
$B_2C_6H_{22}NP$ ...... $(CH_3)_3N-BH_2-BH_2-P(CH_3)_3$ .... B: B Comp.SVol.4/3b-25
$B_2C_6H_{22}N_2$ ...... $(CH_3)_3N-BH_2-BH_2-N(CH_3)_3$ .... B: B Comp.SVol.4/3b-25
– .... $H_3B-N(CH_3)_2-CH_2-CH_2-N(CH_3)_2-BH_3$ .... B: B Comp.SVol.3/3-176
$B_2C_6H_{24}N_2Si_2$ .... $[(CH_3)_3Si-NH-BH_2]_2$ .... B: B Comp.SVol.3/3-187
$B_2C_6H_{25}N_3Si$ ..... $SiH(N(CH_3)_2)_3 \cdot 2\ BH_3$ .... Si: SVol.B4-193
$B_2C_7Cl_2H_6O$ ..... 1,3-Cl_2-2,1,3-$OB_2C_7H_6$ .... B: B Comp.SVol.4/4-55/6
$B_2C_7Cl_2H_{12}$ ...... 4,5-$(C_2H_5)_2$-1,3-Cl_2-1,3-$B_2C_3H_2$ .... B: B Comp.SVol.4/4-50
$B_2C_7Cl_2H_{21}N_3Sn$. $[(CH_3)_2N-BCl]_2N-Sn(CH_3)_3$ .... B: B Comp.SVol.4/4-64
Sn: Org.Comp.18-27
$B_2C_7Cl_4H_6$ ...... 2-$Cl_2B-C_6H_4-CH_2-BCl_2$ .... B: B Comp.SVol.4/4-45
$B_2C_7F_6H_{14}N_2O_2$ .. $[F_2B(-O-C(=N(CH_3)_2)-CH_2-C(=N(CH_3)_2)-O-)][BF_4]$
B: B Comp.SVol.3/3-372
$B_2C_7H_{12}I_2$ .... 1,3-I_2-4,5-$(C_2H_5)_2$-1,3-$B_2C_3H_2$ .... B: B Comp.SVol.4/4-114
$B_2C_7H_{12}I_4$ .... $C_2H_5-CI=C(C_2H_5)-BI-CH_2-BI_2$ .... B: B Comp.SVol.4/4-113
$B_2C_7H_{14}N_2$ ...... $H_2B-NH(CH_2)_2$-2-$C_5H_4N-BH_3$ .... B: B Comp.SVol.3/3-147
$B_2C_7H_{14}S_2$ ...... 1,3-$(CH_3S)_2$-4,5-$(CH_3)_2$-1,3-$B_2C_3H_2$ .... B: B Comp.SVol.4/4-133, 163
$B_2C_7H_{16}N_2$ ...... 2 $H_3B \cdot$ 1,4-N_2C_4H-2,3,5-$(CH_3)_3$ .... B: B Comp.SVol.4/3b-15
$B_2C_7H_{16}N_2O_2$ .... $[-BCH_3-NC_2H_5-CO-NC_2H_5-BCH_3-O-]$ .... B: B Comp.SVol.3/3-165
$B_2C_7H_{16}N_2S_2$ .... $CH_3B[-NC_2H_5-CS-NC_2H_5-BCH_3-S-]$ .... B: B Comp.SVol.3/4-128/9
$B_2C_7H_{17}N_3O$ ..... $CH_3B[-NCH_3-CO-N(i-C_3H_7)-BCH_3-NH-]$ ..... B: B Comp.SVol.3/3-119
$B_2C_7H_{20}N_2O_2$ .... $HC(=O)-O-BH[-CH_2-N(CH_3)_2-BH_2-CH_2-N(CH_3)_2-]$
B: B Comp.SVol.4/3b-72
$B_2C_7H_{21}NOSi$ .... $(CH_3)_2B-N[Si(CH_3)_3]-O-B(CH_3)_2$ .... B: B Comp.SVol.4/3b-62
$B_2C_7H_{21}NSn$ ..... $(CH_3)_3SnN(B(CH_3)_2)_2$ .... Sn: Org.Comp.18-23, 26
$B_2C_7H_{21}N_2PS$ .... $(CH_3)_2B-N(CH_3)-P(=S)(CH_3)-N(CH_3)-B(CH_3)_2$ B: B Comp.SVol.4/3a-244
$B_2C_7H_{21}N_3Sn$ .... $(CH_3)_3Sn-N[-BCH_3-NCH_3-NCH_3-BCH_3-]$ ..... B: B Comp.SVol.3/3-137
Sn: Org.Comp.18-84, 96
– .... $(CH_3)_3Sn-N[-NCH_3-BCH_3-NCH_3-BCH_3-]$ ..... Sn: Org.Comp.18-84, 96
$B_2C_7H_{22}N_2$ ...... $[-NCH_3-C(CH_3)_2-NCH_3-CH_2CH_2-] \cdot 2\ BH_3$ ... B: B Comp.SVol.4/3b-12
$B_2C_7H_{23}N_2O_3P$ ... $CH_3O-PH(O)O-BH[-CH_2N(CH_3)_2BH_2CH_2N(CH_3)_2-]$
B: B Comp.SVol.4/3b-72
$B_2C_8ClDyH_{24}O_2$ .. $DyCl[BH_4]_2 \cdot 2\ C_4H_8O$ .... Sc: MVol.C11b-489/90
$B_2C_8ClGdH_{24}O_2$ .. $GdCl[BH_4]_2 \cdot 2\ C_4H_8O$ .... Sc: MVol.C11b-489/90
$B_2C_8ClH_{22}N_2P$ ... $(CH_3)_2N-BCl-B(PH-C_4H_9-t)-N(CH_3)_2$ .... B: B Comp.SVol.4/4-60

Formula	Compound	Reference
$B_2C_8ClH_{24}HoO_2$	$HoCl[BH_4]_2 \cdot 2\ C_4H_8O$	Sc: MVol.C11b-489/90
$B_2C_8ClH_{24}O_2Tb$	$TbCl[BH_4]_2 \cdot 2\ C_4H_8O$	Sc: MVol.C11b-489/90
$B_2C_8ClH_{24}O_2Y$	$YCl[BH_4]_2 \cdot 2\ C_4H_8O$	Sc: MVol.C11b-489/90
$B_2C_8Cl_2H_{16}$	CH_3-BCl-C(C_2H_5)=C(C_2H_5)-BCl-CH_3	B: B Comp.SVol.4/4-50
$B_2C_8Cl_2H_{18}$	(t-C_4H_9-BCl)$_2$	B: B Comp.SVol.3/4-45
–	(t-C_4D_9-BCl)$_2$	B: B Comp.SVol.3/4-45
$B_2C_8Cl_2H_{18}N_2$	[-BCl-N(C_4H_9-t)-BCl-N(C_4H_9-t)-]	B: B Comp.SVol.4/4-67
$B_2C_8Cl_2H_{21}N_3$	[(CH_3)$_2$N-BCl]$_2$N-C_4H_9-t	B: B Comp.SVol.4/4-64
$B_2C_8Cl_2H_{22}N_4Zn$	2 (CH_3)$_3$N-BH_2-CN $\cdot$ $ZnCl_2$	B: B Comp.SVol.4/3b-22
$B_2C_8Cl_4H_{10}N_4$	Cl_2B-[-1,2-$N_2C_3H_2$(CH_3)-]$_2$$BCl_2$	B: B Comp.SVol.4/4-67/8
$B_2C_8F_2H_{14}O_5$	$B_2C_2F_2O_5$(C_3H_7-n)$_2$	B: B Comp.SVol.3/3-370/1
		B: B Comp.SVol.4/3b-235
$B_2C_8F_2H_{20}N_4$	B[-NCH_3-CH_2-CH_2-NCH_3-]$_2$$BF_2$	B: B Comp.SVol.3/3-377
$B_2C_8F_4H_{10}N_4$	F_2B[-1,2-$N_2C_3H_2$(CH_3)-]$_2$$BF_2$	B: B Comp.SVol.4/3b-249/50
$B_2C_8F_8H_{12}MnN_4$	Mn($NCCH_3$)$_4$(BF_4)$_2$	Mn: MVol.D7-9
$B_2C_8F_8H_{15}N_4O_3Re$	[ON-Re(CO)$_2$(1,4,7-$N_3C_6H_{15}$)][BF_4]$_2$	Re: Org.Comp.1-62
$B_2C_8H_{11}NS_2$	4-C_6H_5-3,5-(CH_3)$_2$-1,2,4,3,5-S_2NB_2	B: B Comp.SVol.4/4-140
$B_2C_8H_{12}N_4S_2$	CH_3-B[-$N_2C_3H_3$-]$_2$[-S_2-]B-CH_3	B: B Comp.SVol.4/3b-37
		B: B Comp.SVol.4/4-142
$B_2C_8H_{14}$	2,3,4,5-$C_4B_2H_2$-1,2,4,6-(CH_3)$_4$	B: B Comp.SVol.3/4-155/6
–	2,3,4,5-$C_4B_2H_2$-1,3,4,6-(CH_3)$_4$	B: B Comp.SVol.3/4-155/6
$B_2C_8H_{14}N_4$	H_2B[-1,2-$N_2C_3H_2$(CH_3)-]$_2$$BH_2$	B: B Comp.SVol.4/3b-33
$B_2C_8H_{16}Li_2N_2$	Li_2[1,2-((CH_3)$_2$N)$_2$-1,2-$B_2C_4H_4$]	B: B Comp.SVol.4/3a-246/7
$B_2C_8H_{16}S$	3,4-(C_2H_5)$_2$-2,5-(CH_3)$_2$-1,2,5-SB_2C_2	B: B Comp.SVol.4/4-130
$B_2C_8H_{18}N_2$	1,2-[(CH_3)$_2$N]$_2$-1,2-$B_2C_4H_6$	B: B Comp.SVol.4/3a-246
–	H_3B-N(CH_3)$_2$-C_6H_4-2-(NH_2-BH_3)	B: B Comp.SVol.4/3b-11
–	H_3B-N(CH_3)$_2$-C_6H_4-4-(NH_2-BH_3)	B: B Comp.SVol.4/3b-11
$B_2C_8H_{18}N_2O_2$	$B_2C_2N_2O_2$(CH_3)$_6$	B: B Comp.SVol.3/3-123/4
$B_2C_8H_{18}O_2S_3$	n-C_4H_9OB[-S-S-BO(C_4H_9-n)-S-]	B: B Comp.SVol.3/4-119/20
$B_2C_8H_{18}O_4Sn$	(C_4H_9)$_2$Sn(OBO)$_2$	Sn: Org.Comp.15-350
$B_2C_8H_{18}Se_3$	[-B(C_4H_9-n)-Se-B(C_4H_9-n)-Se-Se-]	B: B Comp.SVol.4/4-171/2
$B_2C_8H_{20}N_2S_3$	t-C_4H_9NHB[-S-S-BNH(C_4H_9-t)-S-]	B: B Comp.SVol.3/4-123
$B_2C_8H_{20}N_4$	NC-BH_2-N(CH_3)$_2$-CH_2CH_2-N(CH_3)$_2$-BH_2-CN	B: B Comp.SVol.3/3-184
		B: B Comp.SVol.4/3b-23
–	[B(-N(CH_3)-CH_2-CH_2-N(CH_3)-)]$_2$	B: B Comp.SVol.3/4-99
$B_2C_8H_{20}N_6S$	1,3-(CH_3)$_2$-1,3,2-$N_2BC_2H_4$-2-N=S=N-2-[1,3,2-$N_2BC_2H_4$(CH_3)$_2$-1,3]	B: B Comp.SVol.3/3-98
		S: S-N Comp.7-179
$B_2C_8H_{22}N_2O_2$	CH_3-C(=O)-O-BH[-CH_2-N(CH_3)$_2$-BH_2-CH_2-N(CH_3)$_2$-]	B: B Comp.SVol.4/3b-72
$B_2C_8H_{22}N_2O_4$	[CH_2N(CH_3)$_2$-BH_2COOH]$_2$	B: B Comp.SVol.3/3-187
$B_2C_8H_{24}N_2O_2$	H_3B-N(CH_3)$_2$-CH_2CH_2-OC(O)-BH_2-N(CH_3)$_3$	B: B Comp.SVol.4/3b-15, 25
$B_2C_8H_{24}N_2Si$	(CH_3)$_2$B-N(CH_3)-Si(CH_3)$_2$-N(CH_3)-B(CH_3)$_2$	B: B Comp.SVol.4/3a-244
$B_2C_8H_{24}N_2Sn_2$	1,3-[(CH_3)$_3$Sn]$_2$-2,4-(CH_3)$_2$-1,3,2,4-N_2B_2	B: B Comp.SVol.4/3a-197
$B_2C_8H_{24}N_4$	[(CH_3)$_2$N]$_2$B-B[N(CH_3)$_2$]$_2$	B: B Comp.SVol.3/3-103
		B: B Comp.SVol.4/3a-166
$B_2C_8H_{24}N_6S$	[(CH_3)$_2$N]$_2$BN=S=NB[N(CH_3)$_2$]$_2$	B: B Comp.SVol.3/3-98
$B_2C_9ClH_{13}$	CH_3-BCl-C_6H_4-2-B(CH_3)$_2$	B: B Comp.SVol.4/4-49
$B_2C_9Cl_2H_{23}NSi$	(i-C_3H_7-BCl)$_2$N-Si(CH_3)$_3$	B: B Comp.SVol.4/4-61

$B_2C_9F_3H_{10}NS_2$ ... $CH_3B[-S-N(C_6H_4-4-CF_3)-BCH_3-S-]$ .... B: B Comp.SVol.3/4-143
$B_2C_9H_{12}N_6$ ...... $H_2B[-N_2C_3H_3-BH(C_3H_3N_2)-N_2C_3H_3-]$ .... B: B Comp.SVol.3/3-199
$B_2C_9H_{17}NS$ ...... $[-BH_2-S(BH_3)-CH_2CH_2-NH(CH_2-C_6H_5)-]$ ..... B: B Comp.SVol.4/4-148
$B_2C_9H_{18}O_3$ ...... $B[-O-(CH_2)_3-]_3B$ .... B: B Comp.SVol.3/3-176
$B_2C_9H_{19}N$ .... $[-BCH_3-NCH_3-BCH_3-CC_2H_5=CC_2H_5-]$ .... B: B Comp.SVol.3/3-164
$B_2C_9H_{20}N_2$ ...... 1,3-$[(CH_3)_2N]_2$-4,5-$(CH_3)_2$-1,3-$B_2C_3H_2$ ...... B: B Comp.SVol.4/3a-246
$B_2C_9H_{20}N_6$ ...... $[-N(CH_3)-CH_2CH_2-N(CH_3)-]B-N=C=N-$
$B[-N(CH_3)-CH_2CH_2-N(CH_3)-]$ .... B: B Comp.SVol.4/3a-157
$B_2C_9H_{21}N_3S$ ..... $CH_3B(-NC_2H_5-CS-NC_2H_5-BCH_3-NC_2H_5-)$ .... B: B Comp.SVol.3/3-119/20
$B_2C_9H_{23}N$ .... 2,2,6,6-$(CH_3)_4$-NC_5H_6[-1-BH_2-H-BH_2-1-] .... B: B Comp.SVol.4/3a-240
$B_2C_9H_{23}N_3S$ ..... 3-CH_3-2,4-$[(C_2H_5)_2N]_2$-1,3,2,4-SNB_2 .... B: B Comp.SVol.4/4-140
$B_2C_9H_{23}N_3S_2$ .... 4-CH_3-3,5-$[(C_2H_5)_2N]_2$-1,2,4,3,5-S_2NB_2 ...... B: B Comp.SVol.4/4-140
$B_2C_9H_{24}N_2SSeSi$ $[-BCH_3-Se-BCH_3-N(C_4H_9-t)-S-]N-Si(CH_3)_3$ .. B: B Comp.SVol.4/4-172
$B_2C_9H_{26}N_2$ ...... $[-N(C_3H_7-i)CH_2N(C_3H_7-i)CH_2CH_2-]$ · 2 BH_3 .. B: B Comp.SVol.4/3b-12
$B_2C_9H_{26}N_3O$ ..... $[-BH_2-CH_2-N(CH_3)_2-BH(OCHN(CH_3)_2)-CH_2-N(CH_3)_2-]$
B: B Comp.SVol.4/4-115
$B_2C_9H_{27}N_3O$ ..... $(CH_3)_2N-CH_2-B(OH)[-CH_2-N(CH_3)_2-BH_2-CH_2$
$-N(CH_3)_2-]$ .... B: B Comp.SVol.3/3-218
B: B Comp.SVol.4/3b-72
$B_2C_9H_{28}IN_3$...... $[-BH_2-CH_2-N(CH_3)_2-BH(N(CH_3)_3)-CH_2-N(CH_3)_2-][I]$
B: B Comp.SVol.4/4-114
$B_2C_9H_{30}N_2P^+$ .... $[(CH_3)_3P-BH_2CH_2N(CH_3)_2BH_2N(CH_3)_3]^+$ ..... B: B Comp.SVol.3/3-210
$B_2C_{10}ClFeH_{11}O_4$. $[(HO)_2B-C_5H_4]Fe[C_5H_3(Cl)-B(OH)_2]$ .... Fe: Org.Comp.A10-299
$B_2C_{10}ClH_{14}N_3O$ .. $ClB(-NCH_3-BC_6H_5-NCH_3-CO-NCH_3-)$ .... B: B Comp.SVol.3/4-59
$B_2C_{10}ClH_{23}N_2$.... $[-N(C_4H_9-t)-BCl-N(C_4H_9-t)-]B-C_2H_5$ .... B: B Comp.SVol.4/4-67
$B_2C_{10}ClH_{24}N_3$.... $[-N(C_4H_9-t)-BCl-N(C_4H_9-t)-]B-N(CH_3)_2$ ..... B: B Comp.SVol.4/4-67
$B_2C_{10}Cl_2H_6O$ .... 2,1,3-$OB_2C_{10}H_6$-1,3-Cl_2 .... B: B Comp.SVol.4/4-45, 55/6
$B_2C_{10}Cl_2H_{16}$ ..... 4,5-$(C_2H_5)_2$-1,3-Cl_2-1,3-B_2C_3-$[=C(CH_3)_2]$-2 .. B: B Comp.SVol.4/4-50
$B_2C_{10}Cl_2H_{16}N_4$... $C_2H_5-BCl[-1,2-N_2C_3H_3-]_2BCl-C_2H_5$......... B: B Comp.SVol.4/4-68/9
$B_2C_{10}Cl_2H_{18}Ti_2$ .. $[(C_5H_5)TiCl(BH_4)]_2$ .... Ti: Org.Comp.5-36/7
$B_2C_{10}Cl_2H_{20}$..... $[Cl(t-C_4H_9)B]CH=CH[B(t-C_4H_9)Cl]$ .... B: B Comp.SVol.3/4-45
$B_2C_{10}Cl_2H_{23}N$.... $(i-C_3H_7-BCl)_2N-C_4H_9-t$................... B: B Comp.SVol.4/4-61
$B_2C_{10}Cl_2H_{28}N_2Si_2$ $Cl_2B-NH-B(C_4H_9-t)-N[Si(CH_3)_3]_2$........... B: B Comp.SVol.4/4-59
$B_2C_{10}Cl_3H_{15}N_2$... $C_6H_5B(-NCH_3-CH_2-CH_2-NCH_3-)$ · BCl_3 ..... B: B Comp.SVol.3/3-113
– .... $[-BCl-NCH_3-(CH_2)_2-N(B(C_6H_5)Cl_2)CH_3-]$..... B: B Comp.SVol.3/4-41
$B_2C_{10}Cl_4FeH_8$.... $Fe(C_5H_4BCl_2)_2$ .... Fe: Org.Comp.A9-277
$B_2C_{10}Cl_4H_6$...... 1,8-$(BCl_2)_2$-$C_{10}H_6$ .... B: B Comp.SVol.4/4-45
$B_2C_{10}Cl_4H_{14}N_4$... $Cl_2B[-1,2-N_2C_3H(CH_3)_2-]_2BCl_2$............. B: B Comp.SVol.4/4-67/8
$B_2C_{10}Cl_4H_{14}O_4$... $[B(-O=CCH_3-CH=CCH_3-O-)_2][BCl_4]$......... B: B Comp.SVol.3/4-52
$B_2C_{10}Cl_4H_{18}$..... $Cl_2B-C(C_4H_9-t)=C(C_4H_9-t)-BCl_2$............ B: B Comp.SVol.3/4-41
$B_2C_{10}Cl_5H_6^-$..... $[C_{10}H_6B_2Cl_5]^-$.......................... B: B Comp.SVol.4/4-45
$B_2C_{10}Co_2H_{12}S_2$ .. $B_2Co_2S_2(H)_2(C_5H_5)_2$..................... B: B Comp.SVol.4/4-130
$B_2C_{10}F_4FeH_8$ .... $Fe(C_5H_4BF_2)_2$.......................... Fe: Org.Comp.A9-277, 282
$B_2C_{10}F_4H_{14}N_4$ ... $F_2B[-1,2-N_2C_3H(CH_3)_2-]_2BF_2$ .... B: B Comp.SVol.4/3b-249/50
$B_2C_{10}FeH_8I_4$..... $Fe(C_5H_4BI_2)_2$ .... Fe: Org.Comp.A9-277, 282
$B_2C_{10}FeH_{12}$ ..... $Fe(C_5H_5BH)_2$............................ Fe: Org.Comp.B17-328, 334
$B_2C_{10}FeH_{12}O_4$ ... $Fe[C_5H_4B(OH)_2]_2$ .... Fe: Org.Comp.A9-279, 284
$B_2C_{10}H_{14}N_4S_4$ ... $[-S-CH_2CH_2-S-]B[-1,2-N_2C_3H_3-]_2B[-S-CH_2CH_2-S-]$
B: B Comp.SVol.3/3-194/6, 200

$B_2C_{10}H_{14}N_4S_4$ ... [-S-CH_2CH_2-S-]B[-1,2-$N_2C_3H_3$-]$_2$B[-S-CH_2CH_2-S-]
B: B Comp.SVol.3/4-129/30
B: B Comp.SVol.4/3b-36
B: B Comp.SVol.4/4-147
$B_2C_{10}H_{15}I_5$ ...... [$(CH_3)_5C_5$-B-I][BI_4] ...... B: B Comp.SVol.4/4-110
$B_2C_{10}H_{15}LiN_4O_3$ .. [H_2B(CN)CNBH$(CN)_2$]Li · 1.5 $C_4H_8O_2$ ...... B: B Comp.SVol.3/3-212/3
$B_2C_{10}H_{15}NS_2$ .... CH_3B[-S-BCH_3-S-N(C_6H_3-1,3-$(CH_3)_2$)-] ..... B: B Comp.SVol.3/4-124
$B_2C_{10}H_{16}N_2S_2$ ... 2-[$(CH_3)_2$N]$_2$B-1,3,2-$S_2BC_6H_4$ ...... B: B Comp.SVol.3/4-127
B: B Comp.SVol.4/4-147
$B_2C_{10}H_{18}MnO_8$ ... [Mn$(C_5H_9BO_4)_2$ · NH_3 · 9 H_2O]$_n$ ...... Mn: MVol.D8-3
$B_2C_{10}H_{18}N_4$ ..... $(CH_3)_2$B[-1,2-$N_2C_3H_3$-]$_2$B$(CH_3)_2$ ...... B: B Comp.SVol.4/3b-34
– ...... $(C_2H_5)_2$B[-1,2-$N_2C_3H_3$-]$_2$$BH_2$ ...... B: B Comp.SVol.3/3-199
– ...... H_2B[-1,2-N_2C_3H$(CH_3)_2$-]$_2$$BH_2$ ...... B: B Comp.SVol.3/3-200
B: B Comp.SVol.4/3b-30
$B_2C_{10}H_{18}N_4S_2$ ... CH_3S(CH_3)B[-$N_2C_3H_3$-B$(SCH_3)CH_3$-$N_2C_3H_3$-] B: B Comp.SVol.3/4-129
$B_2C_{10}H_{18}N_4S_4$ ... $(CH_3S)_2$B[-$N_2C_3H_3$-B$(SCH_3)_2$-$N_2C_3H_3$-] ...... B: B Comp.SVol.3/3-194/6
$B_2C_{10}H_{18}N_6O_5$ ... [1,3,5-$(CH_3)_3$-4,6-(O=)$_2$-1,3,5,2-N_3BC_2-2]$_2$O .. B: B Comp.SVol.4/3b-55
$B_2C_{10}H_{21}MnNO_8$. [Mn$(C_5H_9BO_4)_2$ · NH_3 · 9 H_2O]$_n$ ...... Mn: MVol.D8-3
$B_2C_{10}H_{21}N_2NaO_4$ [H_3BC≡N-BH_2CN]Na · 2 $C_4H_8O_2$ ...... B: B Comp.SVol.3/3-212
$B_2C_{10}H_{22}N_2$ ..... 1,4-[H_3B-N$(CH_3)_2$]$_2$-C_6H_4 ...... B: B Comp.SVol.4/3b-11
$B_2C_{10}H_{22}N_2O_2$ ... $B_2C_2N_2O_2(CH_3)_4(C_2H_5)_2$ ...... B: B Comp.SVol.3/3-123/4
$B_2C_{10}H_{22}N_2S_2$ ... 2,3-[$(C_2H_5)_2$N]$_2$-1,4,2,3-$S_2B_2C_2H_2$ ...... B: B Comp.SVol.4/4-146/7
– ...... 2-[$(C_2H_5)_2$N]$_2$B-1,3,2-$S_2BC_2H_2$ ...... B: B Comp.SVol.4/4-146
$B_2C_{10}H_{24}N_2$ ..... [-B(CH_3)-N(C_4H_9-t)-B(CH_3)-N(C_4H_9-t)-] ..... B: B Comp.SVol.3/3-135
B: B Comp.SVol.4/3a-195
$B_2C_{10}H_{24}N_2S_2$ ... [-S-N(C_4H_9-t)-B(CH_3)-S-B(CH_3)-N(C_4H_9-t)-] B: B Comp.SVol.3/4-124
B: B Comp.SVol.4/4-140/1
$B_2C_{10}H_{28}N_2$ ..... [-N(C_3H_7-i)-CH(CH_3)-N(C_3H_7-i)-CH_2CH_2-] · 2 BH_3
B: B Comp.SVol.4/3b-12
$B_2C_{10}H_{28}N_3O$ .... [-BH_2CH_2N$(CH_3)_2$-BH(OC(CH_3)N$(CH_3)_2$)-CH_2N$(CH_3)_2$-]
B: B Comp.SVol.4/4-115
$B_2C_{10}H_{29}N_3O$ .... $(CH_3)_2$NCH_2-B(OCH_3)[-CH_2N$(CH_3)_2$$BH_2CH_2N(CH_3)_2$-]
B: B Comp.SVol.4/3b-72
$B_2C_{10}H_{30}N_2OSi_2$ .. $(CH_3)_2$B-NCH_3-BCH_3-N[Si$(CH_3)_3$][OSi$(CH_3)_3$] B: B Comp.SVol.3/3-106
$B_2C_{10}H_{30}N_2Si_2$ ... [$(CH_3)_2$N][$(CH_3)_3$Si]B-B[Si$(CH_3)_3$][N$(CH_3)_2$] ... B: B Comp.SVol.3/3-160
$B_2C_{10}H_{30}N_2Sn_2$ .. [$(CH_3)_2$N][$(CH_3)_3$Sn]B-B[Sn$(CH_3)_3$][N$(CH_3)_2$] .. B: B Comp.SVol.3/3-160
$B_2C_{10.4}ClH_{28.8}O_{2.6}Sm$
SmCl[BH_4]$_2$ · 2.6 C_4H_8O ...... Sc: MVol.C11b-489/90
$B_2C_{11}ClH_{30}N_3O_2$. [-B(CH_2-NH$(CH_3)_2$)(O-C(=O)-CH_3)-CH_2-
N$(CH_3)_2$-BH_2-CH_2-N$(CH_3)_2$-]Cl ...... B: B Comp.SVol.4/3b-72
$B_2C_{11}Cl_2H_{25}N$ .... i-C_3H_7-BCl-N(C_4H_9-t)-BCl-C_4H_9-t ...... B: B Comp.SVol.4/4-61
$B_2C_{11}CoH_{15}$ ..... C_5H_5Co[(CH_3)BC_4H_4B(CH_3)] ...... B: B Comp.SVol.3/4-157
$B_2C_{11}F_6H_{30}N_3O_2P$ [-B(CH_2-NH$(CH_3)_2$)(O-C(=O)-CH_3)-CH_2-
N$(CH_3)_2$-BH_2-CH_2-N$(CH_3)_2$-][PF_6] ...... B: B Comp.SVol.4/3b-72
$B_2C_{11}H_{13}N$ ...... (-BCH_3-$NCH_2C_6H_5$-CH_2-CB=CH-) ...... B: B Comp.SVol.3/3-162
$B_2C_{11}H_{17}N_3O$ .... CH_3B[-NC_6H_5-CO-NCH_3-BCH_3-NCH_3-] ..... B: B Comp.SVol.3/3-118/9
$B_2C_{11}H_{20}$ .... 2,3,4,5-$C_4B_2(CH_3)_5$-6-C_2H_5 ...... B: B Comp.SVol.3/4-155/6
$B_2C_{11}H_{20}N_2$ ..... 3-CH_3-BC_4H_7-1-N=C=N-1-BC_4H_7-3-CH_3 .... B: B Comp.SVol.4/3a-244
$B_2C_{11}H_{21}N$ ...... 2,3,4,5-$C_4B_2H_4$-1-CH_3-2-N(C_3H_7-i)$_2$ ...... B: B Comp.SVol.3/4-157
$B_2C_{11}H_{22}N_2$ ..... [-NCH_3-CH(C_6H_5)-NCH_3-CH_2CH_2-] · 2 BH_3 .. B: B Comp.SVol.4/3b-12

$B_2C_{11}H_{24}IN_3$ ..... [-BH_2-CH_2-$N(CH_3)_2$-$BH(NC_5H_5)$-CH_2-$N(CH_3)_2$-][I]
B: B Comp.SVol.4/4-114
$B_2C_{11}H_{24}N_2O_2$ ... [-B(n-C_4H_9)-NCH_3CONCH_3-B(n-C_4H_9)-O-] .. B: B Comp.SVol.3/3-165
$B_2C_{11}H_{24}N_2S_2$ ... CH_3B[-N(C_4H_9-n)-C(S)-N(C_4H_9-n)-BCH_3-S-] B: B Comp.SVol.3/4-128/9
– .... n-C_4H_9-B[-NCH_3-C(S)-NCH_3-B(C_4H_9-n)-S-] B: B Comp.SVol.3/4-128/9
$B_2C_{11}H_{24}N_3O$ .... [-BH_2-CH_2-$N(CH_3)_2$-BH(O-1-NC_5H_5)-CH_2-$N(CH_3)_2$-]
B: B Comp.SVol.4/4-115
$B_2C_{11}H_{26}N_2$ ..... 1,3-(t-C_4H_9)$_2$-2,4-$(CH_3)_2$-1,3,2,4-$N_2B_2CH_2$ ... B: B Comp.SVol.4/3a-176/7
$B_2C_{11}H_{27}N$ ...... $(CH_3)(C_2H_5)$B-N(t-C_4H_9)-$B(C_2H_5)_2$ .... B: B Comp.SVol.3/3-160
$B_2C_{11}H_{29}NSn$ .... $[(C_2H_5)_2B]_2$N-$Sn(CH_3)_3$ .... B: B Comp.SVol.4/3a-240
Sn: Org.Comp.18-23, 27
$B_2C_{11}H_{33}N_3Sn_3$ .. CH_3B[-$NSn(CH_3)_3$-BCH_3-$NSn(CH_3)_3$-$NSn(CH_3)_3$-]
B: B Comp.SVol.3/3-137
$B_2C_{12}ClH_{27}N_2$.... [-N(C_4H_9-t)-BCl-N(C_4H_9-t)-]B-C_4H_9-i ...... B: B Comp.SVol.4/4-67
$B_2C_{12}ClH_{28}N_3$.... [-N(C_4H_9-t)-BCl-N(C_4H_9-t)-]B-$N(C_2H_5)_2$..... B: B Comp.SVol.4/4-67
$B_2C_{12}Cl_2H_8$...... 5,10-Cl_2-5,10-$B_2C_{12}H_8$ .... B: B Comp.SVol.4/4-50
$B_2C_{12}Cl_2H_{24}$..... CH_3-BCl-C(C_4H_9-t)=C(C_4H_9-t)-BCl-CH_3 ..... B: B Comp.SVol.4/4-50
– .... t-C_4H_9-BCl-C(CH_3)=C(CH_3)-BCl-C_4H_9-t..... B: B Comp.SVol.3/4-45
B: B Comp.SVol.4/4-50
$B_2C_{12}Cl_2H_{27}N$.... (t-C_4H_9-BCl)$_2$N-C_4H_9-t................... B: B Comp.SVol.4/4-61
$B_2C_{12}Cl_2H_{29}N_3$... $[(C_2H_5)_2N\text{-}BCl]_2$N-$C_4H_9$-t .... B: B Comp.SVol.4/4-64
$B_2C_{12}Cl_2H_{30}N_2Si_2$ $[(CH_3)_3Si]_2$C=C$[BCl\text{-}N(CH_3)_2]_2$ .... B: B Comp.SVol.3/4-55, 57
$B_2C_{12}Cl_4F_{14}H_4N_4$ Cl_2B[-1,2-$N_2C_3H_2$(C_3F_7-i)-]$_2BCl_2$........... B: B Comp.SVol.4/4-67/8
$B_2C_{12}Cl_4H_{18}N_4$... Cl_2B[-1,2-$N_2C_3(CH_3)_3$-]$_2BCl_2$ .... B: B Comp.SVol.4/4-67/8
– .... Cl_2B-[-1,2-$N_2C_3H_2$(C_3H_7-i)-]$_2BCl_2$ .... B: B Comp.SVol.4/4-67/8
$B_2C_{12}Cl_8H_{10}N_2O_2P_2$
$[Cl_2BN(C_6H_5)P(O)Cl_2]_2$.................... B: B Comp.SVol.3/4-55
$B_2C_{12}CoH_{16}$ ..... $(C_5H_5BCH_3)_2Co$ .... B: B Comp.SVol.4/4-180
$B_2C_{12}CoH_{18}Tl$.... $Tl[C_3B_2H(CH_3)_4]Co(C_5H_5)$ .... B: B Comp.SVol.4/4-180
$B_2C_{12}F_3H_{37}N_4Si_5$ $[(CH_3)_3Si]_2$N-BF-N[-$Si(CH_3)_2$-NH-$Si(CH_3)_2$-
N(BF_2)-$Si(CH_3)_2$-] .... B: B Comp.SVol.4/3b-244
$B_2C_{12}F_6H_{27}OP$ ... $(C_4H_9)_3PO$ · B_2F_6 .... B: B Comp.SVol.3/3-252
$B_2C_{12}F_6H_{32}N_3O_2P$ [-B(CH_2-$N(CH_3)_3$)(O-C(=O)-CH_3)-CH_2-
$N(CH_3)_2$-BH_2-CH_2-$N(CH_3)_2$-][PF_6]......... B: B Comp.SVol.4/3b-72
$B_2C_{12}F_8FeH_{12}$ ... $[(C_6H_6)_2Fe][BF_4]_2$......................... Fe: Org.Comp.B19-355
$B_2C_{12}F_8FeH_{17}NO_2$ $[(C_5H_5)Fe(CO)_2(CH_2{=}CH(CH_2)_3NH_3)][BF_4]_2$.... Fe: Org.Comp.B17-35/6
$B_2C_{12}F_8H_{18}MnN_6$ $[Mn(NCCH_3)_6](BF_4)_2$..................... Mn: MVol.D7-6/7
$B_2C_{12}F_8H_{20}MnN_4S_8$
[Mn((2-S=)C_3H_5NS-3,1)$_4$]$(BF_4)_2$ · 2 H_2O ..... Mn: MVol.D7-61/3
$B_2C_{12}F_8H_{36}MnO_6S_6$
$[Mn(O{=}S(CH_3)_2)_6](BF_4)_2$.................. Mn: MVol.D7-96
$B_2C_{12}F_{14}H_8N_4$ ... H_2B[-1,2-$N_2C_3H_2$(C_3F_7-i)-]$_2BH_2$............ B: B Comp.SVol.4/3b-34
$B_2C_{12}FeH_{16}$ ..... Fe($C_5H_5BCH_3$-1)$_2$......................... Fe: Org.Comp.B17-328, 334
$B_2C_{12}FeH_{16}O_2$ ... Fe($C_5H_5BOCH_3$-1)$_2$ .... Fe: Org.Comp.B17-329, 335
$B_2C_{12}H_8S_4$ ...... 2-(1,3,2-$S_2BC_6H_4$-2)-1,3,2-$S_2BC_6H_4$ .... B: B Comp.SVol.3/4-114
B: B Comp.SVol.4/4-131
$B_2C_{12}H_{10}O_2S_3$ ... C_6H_5OB(-S-S-BOC_6H_5-S-)................ B: B Comp.SVol.3/4-119/20
$B_2C_{12}H_{10}O_4Sn$ ... $(C_6H_5)_2Sn(OBO)_2$ .... Sn: Org.Comp.16-148
$B_2C_{12}H_{10}S_2$ ..... C_6H_5-B[-S-]$_2$B-C_6H_5..................... B: B Comp.SVol.4/4-129/30
$B_2C_{12}H_{10}S_3$ ..... [-B(C_6H_5)-S-B(C_6H_5)-S-S-]............... B: B Comp.SVol.3/4-111/2

$B_2C_{12}H_{10}S_3$ ..... $[-B(C_6H_5)-S-B(C_6H_5)-S-S-]$ ..... B: B Comp.SVol.4/4-130
$B_2C_{12}H_{14}N_8$ ..... $(1,2-N_2C_3H_3-1)_2B[-1,2-N_2C_3H_3-]_2BH_2$ ..... B: B Comp.SVol.3/3-199
B: B Comp.SVol.4/3b-34
– ..... $1,2-N_2C_3H_3-1-BH[-1,2-N_2C_3H_3-]_2BH-1-(1,2-N_2C_3H_3)$
B: B Comp.SVol.3/3-199
$B_2C_{12}H_{16}MnN_8$ ... $Mn[(1,2-N_2C_3H_3-1)_2BH_2]_2$ ..... Mn: MVol.D8-9/10
– ..... $Mn[(1,3-N_2C_3H_3-1)_2BH_2]_2$ ..... Mn: MVol.D8-18/9
$B_2C_{12}H_{16}MoO_4S$ .. $3,4-(C_2H_5)_2-2,5-(CH_3)_2-1,2,5-SB_2C_2-Mo(CO)_4$ Mo: Org.Comp.5-164, 166
$B_2C_{12}H_{18}N_4S_4$ ... $[-S-(CH_2)_3-S-]B[-N_2C_3H_3-B(-S-(CH_2)_3-S-)-N_2C_3H_3-]$
B: B Comp.SVol.3/4-129/30
$B_2C_{12}H_{22}N_4$ ..... $H_2B[-1,2-N_2C_3(CH_3)_3-]_2BH_2$ ..... B: B Comp.SVol.4/3b-33
– ..... $H_2B[-1,2-N_2C_3H_2(C_3H_7-i)-]_2BH_2$ ..... B: B Comp.SVol.4/3b-34
$B_2C_{12}H_{22}N_4O_2$ ... $CH_3O-B(C_2H_5)[-1,2-N_2C_3H_3-]_2B(C_2H_5)-OCH_3$ B: B Comp.SVol.4/3b-59
$B_2C_{12}H_{23}N_2P$ .... $(CH_3)_2B-N(CH_3)-P(C_6H_5)-N(CH_3)-B(CH_3)_2$ ... B: B Comp.SVol.4/3a-244
$B_2C_{12}H_{23}N_2PS$ ... $(CH_3)_2B-N(CH_3)-P(S)(C_6H_5)-N(CH_3)-B(CH_3)_2$ B: B Comp.SVol.4/3a-244
$B_2C_{12}H_{24}N_2$ ..... $[-N(CH_3)-C(CH_3)(C_6H_5)-N(CH_3)-CH_2CH_2-] \cdot 2\,BH_3$
B: B Comp.SVol.4/3b-12
$B_2C_{12}H_{26}N_2O_2$ ... $B_2C_2N_2O_2(CH_3)_4(C_3H_7-n)_2$ ..... B: B Comp.SVol.3/3-123/4
– ..... $B_2C_2N_2O_2(CH_3)_4(C_3H_7-i)_2$ ..... B: B Comp.SVol.3/3-123/4
$B_2C_{12}H_{27}N$ ...... $1,2,3-(t-C_4H_9)_3-NB_2$ ..... B: B Comp.SVol.4/3a-193
$B_2C_{12}H_{28}N_2$ ..... $[-B(C_2H_5)-N(C_4H_9-t)-B(C_2H_5)-N(C_4H_9-t)-]$ ... B: B Comp.SVol.3/3-135/6
B: B Comp.SVol.4/3a-195
– ..... $[-B(C_3H_7-i)-N(C_3H_7-i)-B(C_3H_7-i)-N(C_3H_7-i)-]$ B: B Comp.SVol.4/3a-195
$B_2C_{12}H_{28}N_2S_2$ ... $2,4-[(i-C_3H_7)_2N]_2-1,3,2,4-S_2B_2$ ..... B: B Comp.SVol.4/4-144
$B_2C_{12}H_{28}N_4$ ..... $N_3-B(C_3H_7-n)-N(C_3H_7-n)-B(C_3H_7-n)_2$ ..... B: B Comp.SVol.3/3-106
– ..... $N_3-B(C_3H_7-i)-N(C_3H_7-i)-B(C_3H_7-i)_2$ ..... B: B Comp.SVol.3/3-106
$B_2C_{12}H_{28}N_6S$ .... $(-C_4H_8-)S[=NB(-NCH_3-CH_2-CH_2-NCH_3-)]_2$ ... B: B Comp.SVol.3/4-130
$B_2C_{12}H_{29}N$ ...... $(C_2H_5)_2B-N(C_4H_9-t)-B(C_2H_5)_2$ ..... B: B Comp.SVol.3/3-160
– ..... $(C_2H_5)_2B-NH-B(C_4H_9-t)_2$ ..... B: B Comp.SVol.4/3a-240
– ..... $(i-C_3H_7)_2B-NH-B(C_3H_7-i)_2$ ..... B: B Comp.SVol.4/3a-240
– ..... $n-C_3H_7-B(C_2H_5)-N(C_3H_7-n)-B(C_2H_5)_2$ ..... B: B Comp.SVol.3/3-160
$B_2C_{12}H_{29}N_3$ ..... $[-N(CH_3)-CH_2CH_2-N(CH_3)-]B-NH-B(C_4H_9-t)_2$ B: B Comp.SVol.4/3a-157
$B_2C_{12}H_{30}N_2$ ..... $[(CH_3)_2N][t-C_4H_9]B-B[t-C_4H_9][N(CH_3)_2]$ ..... B: B Comp.SVol.3/3-160
$B_2C_{12}H_{30}N_4$ ..... $1,3-(C_4H_9-t)_2-2,4-[(CH_3)_2N]_2-1,3,2,4-N_2B_2$ ... B: B Comp.SVol.4/3a-198
$B_2C_{12}H_{30}N_6$ ..... $[(CH_3)_2N_2C_2H_4B]N(CH_3)C_2H_4N(CH_3)[BC_2H_4N_2(CH_3)_2]$
B: B Comp.SVol.3/3-98
$B_2C_{12}H_{30}N_{10}$ .... $1-C_2H_5-N_4C-5-BH_2-N(CH_3)_2-CH_2CH_2-$
$N(CH_3)_2-BH_2-5-N_4C-1-C_2H_5$ ..... B: B Comp.SVol.4/3b-23, 25
$B_2C_{12}H_{31}N$ ...... $(C_2H_7)(i-C_3H_7)B-N(i-C_3H_7)-B(C_2H_5)_2$ ..... B: B Comp.SVol.3/3-160
$B_2C_{12}H_{32}IN_3O_2$ ... $[-B(CH_2-N(CH_3)_3)(O-C(=O)-CH_3)-CH_2-$
$N(CH_3)_2-BH_2-CH_2-N(CH_3)_2-]I$ ..... B: B Comp.SVol.4/3b-72
$B_2C_{12}H_{32}N_4O_2$ ... $C_2H_5-NH-C(=O)-BH_2-N(CH_3)_2-CH_2CH_2-$
$N(CH_3)_2-BH_2-C(=O)-NH-C_2H_5$ ..... B: B Comp.SVol.3/3-187
B: B Comp.SVol.4/3b-25
$B_2C_{12}H_{32}N_4S_2$ ... $[C_2H_5-NH-C(S)-BH_2-N(CH_3)_2-CH_2-]_2$ ..... B: B Comp.SVol.4/3b-23
$B_2C_{12}H_{34}N_6$ ..... $[C_2H_5-N=C(NH_2)-BH_2-N(CH_3)_2-CH_2-]_2$ ..... B: B Comp.SVol.4/3b-23
$B_2C_{13}ClH_{27}Sn$ .... $t-C_4H_9B[-C(Sn(CH_3)_3)=C(B(C_4H_9-t)Cl)-]$ ..... B: B Comp.SVol.3/4-45
$B_2C_{13}ClH_{29}N_2$ .... $1-(i-C_3H_7)-5,5-(CH_3)_2-4-[(C_3H_7-i)_2N]-2-Cl-$
$1,2,4-NB_2C_2H_2$ ..... B: B Comp.SVol.4/4-62
$B_2C_{13}Cl_2H_{28}Si$ ... $[Cl(t-C_4H_9)B]CH=C[Si(CH_3)_3][B(t-C_4H_9)Cl]$ .... B: B Comp.SVol.3/4-45

$B_2C_{13}Cl_2H_{30}N_2$ ... $(i\text{-}C_3H_7)_2N\text{-}BCl\text{-}CH_2\text{-}BCl\text{-}N(C_3H_7\text{-}i)_2$ .... B: B Comp.SVol.4/4-61
$B_2C_{13}Cl_2H_{32}N_2Si$ $Si(CH_3)_3\text{-}N(C_4H_9\text{-}t)\text{-}BCl\text{-}N(C_4H_9\text{-}t)\text{-}BCl\text{-}C_2H_5$ B: B Comp.SVol.4/4-64
$B_2C_{13}Cl_3H_{27}N_2$ ... $ClB[\text{-}(1\text{-}(NC_5H_6(CH_3)_4\text{-}2,2,6,6))\text{-}BCl_2\text{-}N(C_4H_9\text{-}t)\text{-}]$
B: B Comp.SVol.3/4-61
$B_2C_{13}Cl_4GaH_{28}N$ $[(C_2H_5)_2B{=}N{=}B(\text{-}C(CH_3)_2\text{-}(CH_2)_3\text{-}C(CH_3)_2\text{-})]GaCl_4$
B: B Comp.SVol.3/3-205
$B_2C_{13}Cl_4H_{28}N_2$ ... $[(CH_3)_4C_5H_6N{=}B{=}N(C_2H_5)_2][BCl_4]$ .... B: B Comp.SVol.3/4-62
$B_2C_{13}Cl_6H_{28}NTa$ .. $[(C_2H_5)_2B{=}N{=}B(\text{-}C(CH_3)_2\text{-}(CH_2)_3\text{-}C(CH_3)_2\text{-})]TaCl_6$
B: B Comp.SVol.3/3-205
$B_2C_{13}CoH_{21}S$ .... $(C_5H_5)Co[C_2B_2S(C_2H_5)_2(CH_3)_2]$ .... B: B Comp.SVol.4/4-180
$B_2C_{13}F_3H_{27}N_2$ ... $[\text{-}C(CH_3)_2CH_2CH_2CH_2C(CH_3)_2\text{-}]N[\text{-}BF\text{-}N(C_4H_9\text{-}t)\text{-}BF_2\text{-}]$
B: B Comp.SVol.4/3b-247
$B_2C_{13}F_6H_{19}N_6P$ .. $[C_2H_5\text{-}B(\text{-}N_2C_3H_3\text{-})_3B\text{-}C_2H_5][PF_6]$ .... B: B Comp.SVol.4/3b-38
$B_2C_{13}F_8FeH_{19}NO_2$ $[(C_5H_5)Fe(CO)_2(CH_2{=}CH(CH_2)_4NH_3)][BF_4]_2$.... Fe: Org.Comp.B17-36
$B_2C_{13}GaH_{28}I_4N$ .. $[(C_2H_5)_2B{=}N{=}B(\text{-}C(CH_3)_2\text{-}(CH_2)_3\text{-}C(CH_3)_2\text{-})]GaI_4$
B: B Comp.SVol.3/3-205
$B_2C_{13}H_{21}N_7$ ..... $[\text{-}B(C_2H_5)(N_2C_3H_3)\text{-}NH_2\text{-}B(C_2H_5)(N_2C_3H_3)\text{-}N_2C_3H_3\text{-}]$
B: B Comp.SVol.4/3b-36/7
$B_2C_{13}H_{21}NiS$..... $(C_5H_5)Ni[C_2B_2S(C_2H_5)_2(CH_3)_2]$ .... B: B Comp.SVol.4/4-180
$B_2C_{13}H_{24}N_4$ ..... $2\text{-}C_6H_5\text{-}1,3,2\text{-}N_2BC_3H_7\text{-}1\text{-}B[N(CH_3)_2]_2$ ...... B: B Comp.SVol.4/3a-158
$B_2C_{13}H_{26}NO_5Re$ .. $[(C_2H_5)_4N][(CO)_5Re(BH_3)_2]$ .... Re: Org.Comp.2-168
$B_2C_{13}H_{28}HgI_3N$... $[(C_2H_5)_2B{=}N{=}B(\text{-}C(CH_3)_2\text{-}(CH_2)_3\text{-}C(CH_3)_2\text{-})]HgI_3$
B: B Comp.SVol.3/3-205
$B_2C_{13}H_{28}I_4N_2$ .... $[(CH_3)_4C_5H_6N{=}B{=}N(C_2H_5)_2][BI_4]$ .... B: B Comp.SVol.3/4-101
$B_2C_{13}H_{28}N_2S_2$ ... $n\text{-}C_4H_9B[\text{-}NC_2H_5\text{-}CS\text{-}NC_2H_5\text{-}B(C_4H_9\text{-}n)\text{-}S\text{-}]$ B: B Comp.SVol.3/4-128/9
$B_2C_{13}H_{29}NSi$..... $1\text{-}(C_2H_5)_2B\text{-}4,5\text{-}(C_2H_5)_2\text{-}2,2,3\text{-}(CH_3)_3\text{-}1,2,5\text{-}NSiBC_2$
B: B Comp.SVol.4/3a-250/1
$B_2C_{13}H_{29}NSn$ .... $(CH_3)_3SnN(B(CH_2CH(CH_3)CH_2CH_2))_2$ .... Sn: Org.Comp.18-23, 27
$B_2C_{13}H_{30}N_2$ ..... $(i\text{-}C_3H_7)_2N\text{-}B[\text{-}CH_2\text{-}BH\text{-}N(C_3H_7\text{-}i)\text{-}C(CH_3)_2\text{-}]$ B: B Comp.SVol.4/3a-250
– .... $[\text{-}C(CH_3)_2\text{-}CH_2CH_2CH_2\text{-}C(CH_3)_2\text{-}]N[\text{-}BH$
$\text{-}N(C_4H_9\text{-}t)\text{-}BH_2\text{-}]$ .... B: B Comp.SVol.4/3a-202
$B_2C_{13}H_{31}N$ ...... $(C_2H_5)_2B\text{-}N(CH_3)\text{-}B(C_4H_9\text{-}t)_2$ .... B: B Comp.SVol.4/3a-241
– .... $(i\text{-}C_3H_7)_2B\text{-}N(CH_3)\text{-}B(C_3H_7\text{-}i)_2$ .... B: B Comp.SVol.4/3a-241
– .... $n\text{-}C_3H_7\text{-}B(C_2H_5)\text{-}N(C_4H_9\text{-}t)\text{-}B(C_2H_5)_2$ .... B: B Comp.SVol.3/3-160
– .... $i\text{-}C_3H_7\text{-}B(C_2H_5)\text{-}N(C_4H_9\text{-}t)\text{-}B(C_2H_5)_2$ .... B: B Comp.SVol.3/3-160
$B_2C_{13}H_{31}NO$..... $(t\text{-}C_4H_9)_2B\text{-}NH\text{-}B(C_4H_9\text{-}t)\text{-}OCH_3$ .... B: B Comp.SVol.4/3b-62
$B_2C_{13}H_{32}N_2O_2$ ... $2\text{-}CH_3C_5H_8\text{-}BH_2\text{-}N(CH_3)_2CH_2CH_2N(CH_3)_2\text{-}BH_2\text{-}COOH$
B: B Comp.SVol.4/3b-25
$B_2C_{13}H_{33}NSi$..... $t\text{-}C_4H_9\text{-}B(C_2H_5)\text{-}N[Si(CH_3)_3]\text{-}B(C_2H_5)_2$ ...... B: B Comp.SVol.4/3a-243
$B_2C_{13}H_{33}N_2OP$... $(CH_3)_2B\text{-}N(C_4H_9\text{-}t)\text{-}P({=}O)(CH_3)\text{-}N(C_4H_9\text{-}t)\text{-}B(CH_3)_2$
B: B Comp.SVol.4/3a-244
$B_2C_{13}H_{33}N_2P$ .... $(CH_3)_2B\text{-}N(C_4H_9\text{-}t)\text{-}P(CH_3)\text{-}N(C_4H_9\text{-}t)\text{-}B(CH_3)_2$
B: B Comp.SVol.4/3a-244
$B_2C_{13}H_{33}N_2PS$ ... $(CH_3)_2B\text{-}N(C_4H_9\text{-}t)\text{-}P({=}S)(CH_3)\text{-}N(C_4H_9\text{-}t)\text{-}B(CH_3)_2$
B: B Comp.SVol.4/3a-244
$B_2C_{14}ClH_{27}$...... $t\text{-}C_4H_9B[\text{-}C(C_4H_9\text{-}t){=}C(B(C_4H_9\text{-}t)Cl)\text{-}]$ .... B: B Comp.SVol.3/4-45
$B_2C_{14}Cl_2H_{12}$..... $5,11\text{-}Cl_2\text{-}5,11\text{-}B_2C_{14}H_{12}$ .... B: B Comp.SVol.4/4-50
$B_2C_{14}Cl_2H_{22}In_2N_8$ $[CH_3\text{-}In(Cl)(\text{-}N_2C_3H_3\text{-}BH_2\text{-}N_2C_3H_3\text{-})]_2$....... In: Org.Comp.1-296/7
$B_2C_{14}Cl_2H_{26}N_6$... $(C_2H_5)_2N\text{-}BCl[\text{-}1,2\text{-}N_2C_3H_3\text{-}]_2BCl\text{-}N(C_2H_5)_2$ .. B: B Comp.SVol.4/4-68
$B_2C_{14}Cl_2H_{28}$..... $t\text{-}C_4H_9\text{-}BCl\text{-}C(C_2H_5){=}C(C_2H_5)\text{-}BCl\text{-}C_4H_9\text{-}t$ ... B: B Comp.SVol.3/4-45

Formula	Compound	Reference
$B_2C_{14}Cl_2H_{28}$	$t\text{-}C_4H_9\text{-}BCl\text{-}C(C_2H_5)\text{=}C(C_2H_5)\text{-}BCl\text{-}C_4H_9\text{-}t$	B: B Comp.SVol.4/4-50
$B_2C_{14}Cl_2H_{30}N_2$	$(CH_3)_2N\text{-}BClC(t\text{-}C_4H_9)\text{=}C(t\text{-}C_4H_9)BCl\text{-}N(CH_3)_2$	B: B Comp.SVol.3/4-57
$B_2C_{14}Cl_2H_{30}O_2$	$CH_3CH[O\text{-}BCl\text{-}C(CH_3)_2C_3H_7\text{-}i]_2$	B: B Comp.SVol.4/4-55
$B_2C_{14}Cl_2H_{30}Si$	$CH_3[(CH_3)_3Si]C\text{=}C[B(t\text{-}C_4H_9)Cl]_2$	B: B Comp.SVol.3/4-45
$B_2C_{14}Cl_3GaH_{26}N_4$	$[(C_2H_5)_2B(\text{-}N_2C_3H_3\text{-})_2B\text{-}C_2H_5][C_2H_5\text{-}GaCl_3]$	B: B Comp.SVol.4/3b-39
$B_2C_{14}Cl_4H_{22}N_4$	$Cl_2B\text{-}[\text{-}1,2\text{-}N_2C_3H_2(C_4H_9\text{-}t)\text{-}]_2BCl_2$	B: B Comp.SVol.4/4-67/8
$B_2C_{14}CoH_{23}$	$1\text{-}C_5H_5\text{-}2,3\text{-}(C_2H_5)_2\text{-}4,6\text{-}(CH_3)_2\text{-}5\text{-}H\text{-}1,2,3,5\text{-}C_3CoB_2H$	B: B Comp.SVol.3/4-157
$B_2C_{14}F_2FeH_{20}S$	$(1,4\text{-}F_2C_6H_4)Fe[1,2,5\text{-}SB_2C_2\text{-}3,4\text{-}(C_2H_5)_2\text{-}2,5\text{-}(CH_3)_2]$	Fe: Org.Comp.B18-81, 84, 92
$B_2C_{14}F_2FeH_{20}S^-$	$[(1,4\text{-}F_2C_6H_4)Fe(1,2,5\text{-}SB_2C_2\text{-}3,4\text{-}(C_2H_5)_2\text{-}2,5\text{-}(CH_3)_2)]^-$	Fe: Org.Comp.B18-92
$B_2C_{14}F_3FeH_{19}S$	$(1,3,5\text{-}F_3C_6H_3)Fe[1,2,5\text{-}SB_2C_2\text{-}3,4\text{-}(C_2H_5)_2\text{-}2,5\text{-}(CH_3)_2]$	Fe: Org.Comp.B18-81, 83, 92
$B_2C_{14}F_3FeH_{19}S^-$	$[(1,3,5\text{-}F_3C_6H_3)Fe(1,2,5\text{-}SB_2C_2\text{-}3,4\text{-}(C_2H_5)_2\text{-}2,5\text{-}(CH_3)_2)]^-$	Fe: Org.Comp.B18-92
$B_2C_{14}F_3H_{19}N_6O_3S$	$[C_2H_5\text{-}B(\text{-}N_2C_3H_3\text{-})_3B\text{-}C_2H_5][CF_3\text{-}SO_3]$	B: B Comp.SVol.4/3b-38
$B_2C_{14}F_5H_{20}N_3S_2$	$(C_2H_5)_2NB[\text{-}S\text{-}S\text{-}BN(C_2H_5)_2\text{-}NC_6F_5\text{-}]$	B: B Comp.SVol.3/4-143
$B_2C_{14}F_6H_{16}N_4O_4$	$CF_3\text{-}COO\text{-}B(C_2H_5)[\text{-}1,2\text{-}N_2C_3H_3\text{-}]_2B(C_2H_5)\text{-}OC(O)\text{-}CF_3$	B: B Comp.SVol.4/3b-59
$B_2C_{14}F_6H_{21}N_6P$	$[HB(\text{-}N_2C_3H_2(CH_3)\text{-})_3B\text{-}C_2H_5][PF_6]$	B: B Comp.SVol.4/3b-38
$B_2C_{14}F_8H_{21}MoN_7$	$[Mo(CN\text{-}CH_3)_7][BF_4]_2$	Mo:Org.Comp.5-83, 84, 86/7
$B_2C_{14}FeH_{18}O$	$1\text{-}H_3CBC_5H_5FeC_5H_4(COCH_3\text{-}2)BCH_3\text{-}1$	Fe: Org.Comp.B17-329
$B_2C_{14}FeH_{20}$	$Fe[C_5H_4B(CH_3)_2]_2$	Fe: Org.Comp.A9-277, 281
$B_2C_{14}FeH_{20}{}^+$	$[Fe(C_5H_4B(CH_3)_2)_2]^+$	Fe: Org.Comp.A9-281
$B_2C_{14}FeH_{20}I$	$[Fe(C_5H_4B(CH_3)_2)_2]I$	Fe: Org.Comp.A9-281
$B_2C_{14}FeH_{20}IS_4$	$[Fe(C_5H_4B(SCH_3)_2)_2]I$	Fe: Org.Comp.A9-284
$B_2C_{14}FeH_{20}I_4S_2$	$Fe(C_5H_4BI_2)_2 \cdot 2\ S(CH_3)_2$	Fe: Org.Comp.A9-282
$B_2C_{14}FeH_{20}S_4$	$Fe[C_5H_4B(SCH_3)_2]_2$	Fe: Org.Comp.A9-279, 284
$B_2C_{14}FeH_{22}S$	$(C_6H_6)Fe[1,2,5\text{-}SB_2C_2\text{-}3,4\text{-}(C_2H_5)_2\text{-}2,5\text{-}(CH_3)_2]$	Fe: Org.Comp.B18-81, 84, 92
$B_2C_{14}FeH_{22}S^-$	$[(C_6H_6)Fe(1,2,5\text{-}SB_2C_2\text{-}3,4\text{-}(C_2H_5)_2\text{-}2,5\text{-}(CH_3)_2)]^-$	Fe: Org.Comp.B18-92
$B_2C_{14}H_{16}N_2$	$[\text{-}BCH_3\text{-}NC_6H_5\text{-}]_2$	B: B Comp.SVol.3/3-135
$B_2C_{14}H_{20}N_2O_2$	$[\text{-}O\text{-}BH_2\text{-}1\text{-}NC_5H_4\text{-}2\text{-}CH_2CH_2\text{-}O\text{-}BH_2\text{-}1\text{-}NC_5H_4\text{-}2\text{-}CH_2CH_2\text{-}]$	B: B Comp.SVol.4/3b-72
$B_2C_{14}H_{22}N_4O_4$	$CH_3\text{-}COO\text{-}B(C_2H_5)[\text{-}1,2\text{-}N_2C_3H_3\text{-}]_2B(C_2H_5)\text{-}OC(O)\text{-}CH_3$	B: B Comp.SVol.4/3b-59
$B_2C_{14}H_{23}N_7$	$[\text{-}B(C_2H_5)(N_2C_3H_3)\text{-}NH(CH_3)\text{-}B(C_2H_5)(N_2C_3H_3)\text{-}N_2C_3H_3\text{-}]$	B: B Comp.SVol.4/3b-36/7
$B_2C_{14}H_{25}N_3S$	$3\text{-}C_6H_5\text{-}2,4\text{-}[(C_2H_5)_2N]_2\text{-}1,3,2,4\text{-}SNB_2$	B: B Comp.SVol.4/4-140
$B_2C_{14}H_{25}N_3S_2$	$4\text{-}C_6H_5\text{-}3,5\text{-}[(C_2H_5)_2N]_2\text{-}1,2,4,3,5\text{-}S_2NB_2$	B: B Comp.SVol.4/4-140
$B_2C_{14}H_{26}$	$2,3,4,5\text{-}C_4B_2\text{-}1,3,4,6\text{-}(C_2H_5)_4\text{-}2,5\text{-}(CH_3)_2$	B: B Comp.SVol.3/4-155/6
–	$2,3,4,5\text{-}C_4B_2\text{-}1,3,5,6\text{-}(C_2H_5)_4\text{-}2,4\text{-}(CH_3)_2$	B: B Comp.SVol.3/4-155/6
$B_2C_{14}H_{26}N_4$	$(CH_3)_2B[\text{-}1,2\text{-}N_2C_3H(CH_3)_2\text{-}]_2B(CH_3)_2$	B: B Comp.SVol.4/3b-34
–	$(C_2H_5)_2B[\text{-}1,2\text{-}N_2C_3H_3\text{-}]_2B(C_2H_5)_2$	B: B Comp.SVol.3/3-199 B: B Comp.SVol.4/3b-34
–	$(n\text{-}C_4H_9)_2B[\text{-}1,2\text{-}N_2C_3H_3\text{-}]_2BH_2$	B: B Comp.SVol.3/3-199
–	$H_2B[\text{-}1,2\text{-}N_2C_3H(CH_3)_2\text{-}]_2B(C_2H_5)_2$	B: B Comp.SVol.3/3-200 B: B Comp.SVol.4/3b-34

$B_2C_{14}H_{26}N_4$ ..... $H_2B[-1,2-N_2C_3H_2(C_4H_9-t)-]_2BH_2$ .... B: B Comp.SVol.4/3b-34
$B_2C_{14}H_{26}N_4O_2$ ... $C_2H_5-O-B(C_2H_5)[-1,2-N_2C_3H_3-]_2B(C_2H_5)-O-C_2H_5$
B: B Comp.SVol.4/3b-59
$B_2C_{14}H_{26}N_4S_2$ ... $CH_3S-CH_2CH_2CH_2-BH[-1,2-N_2C_3H_3-]_2BH$
$-CH_2CH_2CH_2-SCH_3$ .... B: B Comp.SVol.4/3b-35
$B_2C_{14}H_{26}N_4S_4$ ... $(C_2H_5-S)_2B[-1,2-N_2C_3H_3-]_2B(S-C_2H_5)_2$ ...... B: B Comp.SVol.3/4-129
B: B Comp.SVol.4/3b-36
$B_2C_{14}H_{26}N_8$ ..... $[-N(CH_3)-CH_2CH_2-N(CH_3)-]B[-1,2-N_2$
$C_3H_3-]_2B[-N(CH_3)-CH_2CH_2-N(CH_3)-]$ ...... B: B Comp.SVol.4/3b-36
$B_2C_{14}H_{29}N$ ...... 9-$[(i-C_3H_7)_2B-NH]$-[3.3.1]-9-BC_8H_{14} .... B: B Comp.SVol.4/3a-241
$B_2C_{14}H_{30}N_2$ ..... $1,2-[(CH_3)_2N]_2-1,2-B_2C_2-3,4-(C_4H_9-t)_2$ ...... B: B Comp.SVol.4/3a-244
– .... $1,2-[(i-C_3H_7)_2N]_2-1,2-B_2C_2H_2$ .... B: B Comp.SVol.4/3a-244/5
– .... $1,3-[(CH_3)_2N]_2-1,3-B_2C_2-2,4-(C_4H_9-t)_2$ ...... B: B Comp.SVol.3/3-161
B: B Comp.SVol.4/3a-244/5
– .... $1,3-[(i-C_3H_7)_2N]_2-1,3-B_2C_2H_2$ .... B: B Comp.SVol.4/3a-244/5
– .... $t-C_4H_9-NH-B(C_2H_5)-N(C_4H_9-t)-B(C_2H_5)-CCH$ B: B Comp.SVol.4/3a-172
$B_2C_{14}H_{30}N_2^-$..... $[1,3-((CH_3)_2N)2-1,3-B_2C_2-2,4-(C_4H_9-t)_2]^-$,
radical anion .... B: B Comp.SVol.4/3a-245
$B_2C_{14}H_{30}N_2O_2$ ... $B_2C_2N_2O_2(CH_3)_4(C_4H_9-n)_2$ .... B: B Comp.SVol.3/3-123/4
$B_2C_{14}H_{30}N_4S_4$ ... $[(CH_3)_2S]_2B[-N_2C_3H_3-B(S(CH_3)_2)_2-N_2C_3H_3-]$.. B: B Comp.SVol.3/3-200
$B_2C_{14}H_{30}N_6$ ..... $[C_2H_5-N=C(CN)-BH_2-N(CH_3)_2-CH_2-]_2$ .... B: B Comp.SVol.4/3b-23
$B_2C_{14}H_{30}N_8$ ..... $[(CH_3)_2N]_2B[-N_2C_3H_3-B(N(CH_3)_2)_2-N_2C_3H_3-]$ B: B Comp.SVol.3/3-199
$B_2C_{14}H_{32}N_2$ ..... $(i-C_3H_7)_2N-BH-CH=CH-BH-N(C_3H_7-i)_2$ ...... B: B Comp.SVol.4/3a-244
– .... $(i-C_3H_7)_2N-B[-CH_2BCH_3-N(C_3H_7-i)-C(CH_3)_2-]$
B: B Comp.SVol.4/3a-250
– .... $[-B(C_3H_7-i)-N(C_4H_9-t)-B(C_3H_7-i)-N(C_4H_9-t)-]$ B: B Comp.SVol.3/3-136
B: B Comp.SVol.4/3a-196
– .... $[-B(C_3H_7-n)-N(C_4H_9-t)-B(C_3H_7-n)-N(C_4H_9-t)-]$
B: B Comp.SVol.3/3-135/6
B: B Comp.SVol.4/3a-196
– .... $[-B(C_4H_9-t)-N(C_3H_7-i)-B(C_4H_9-t)-N(C_3H_7-i)-]$ B: B Comp.SVol.4/3a-197
$B_2C_{14}H_{32}N_4$ ..... $(n-C_3H_7)_2B-N(C_4H_9-t)-B(N_3)-C_4H_9-n$ .... B: B Comp.SVol.4/3a-171
– .... $(i-C_3H_7)_2B-N(C_4H_9-i)-B(N_3)-C_4H_9-i$ .... B: B Comp.SVol.3/3-106
– .... $(n-C_4H_9)_2B-N(C_3H_7-i)-B(N_3)-C_3H_7-i$ .... B: B Comp.SVol.4/3a-171
– .... $(i-C_4H_9)_2B-N(C_3H_7-i)-B(N_3)-C_3H_7-i$ .... B: B Comp.SVol.3/3-106
$B_2C_{14}H_{33}N$ ...... $(t-C_4H_9)_2B-NH-B(C_3H_7-i)_2$ .... B: B Comp.SVol.4/3a-240
– .... $n-C_4H_9-B(CH_3)-N(CH_3)-B(C_4H_9-n)_2$ .... B: B Comp.SVol.4/3a-242
– .... $n-C_4H_9-B(C_2H_5)-N(C_4H_9-n)-B(C_2H_5)_2$ .... B: B Comp.SVol.3/3-160
– .... $n-C_4H_9-B(C_2H_5)-N(C_4H_9-t)-B(C_2H_5)_2$ .... B: B Comp.SVol.3/3-160
– .... $i-C_4H_9-B(C_2H_5)-N(C_4H_9-i)-B(C_2H_5)_2$ .... B: B Comp.SVol.3/3-160
– .... $s-C_4H_9-B(C_2H_5)-N(C_4H_9-s)-B(C_2H_5)_2$ .... B: B Comp.SVol.3/3-160
– .... $s-C_4H_9-B(C_2H_5)-N(C_4H_9-t)-B(C_2H_5)_2$ .... B: B Comp.SVol.4/3a-243
– .... $t-C_4H_9-B(C_2H_5)-N(C_4H_9-t)-B(C_2H_5)_2$ .... B: B Comp.SVol.3/3-160
$B_2C_{14}H_{34}N_2O_2$ ... $2-CH_3C_6H_{10}-BH_2-N(CH_3)_2CH_2CH_2N(CH_3)_2-BH_2-COOH$
B: B Comp.SVol.4/3b-25
$B_2C_{14}H_{34}N_4S$ .... $(i-C_3H_7)_2NB[-NCH_3-NCH_3-BN(C_3H_7-i)_2-S-]$ .. B: B Comp.SVol.3/4-124
$B_2C_{14}H_{36}N_2O_3Si$.. $(CH_3)_3Si-N(C_4H_9-t)-B(OCH_3)-N(C_4H_9-t)-B(OCH_3)_2$
B: B Comp.SVol.4/3b-53
$B_2C_{14}H_{36}N_2Si$.... $(CH_3)_2B-N(C_4H_9-t)-Si(CH_3)_2-N(C_4H_9-t)-B(CH_3)_2$
B: B Comp.SVol.4/3a-244

$B_2C_{14}H_{36}N_2Si_2$ . . . [-B(C_4H_9-t)-N(Si(CH_3)$_3$)-B(C_4H_9-t)-N(Si(CH_3)$_3$)-]
B: B Comp.SVol.4/3a-197
B: B Comp.SVol.4/3b-247

$B_2C_{14}H_{38}N_2OSi_2$. . (C_2H_5)$_2$B-NC_2H_5-BC_2H_5-N[Si(CH_3)$_3$][OSi(CH_3)$_3$]
B: B Comp.SVol.3/3-106

$B_2C_{14}H_{38}N_4Si_2$ . . . 1,3-[(CH_3)$_3$Si]$_2$-2,4-[(C_2H_5)$_2$N]$_2$-1,3,2,4-N_2B_2 B: B Comp.SVol.4/3a-198

$B_2C_{14}H_{40}N_4Si_2$ . . . [(CH_3)$_2$N]$_2$B-NH-B(C_4H_9-t)-N[Si(CH_3)$_3$]$_2$ . . . . . B: B Comp.SVol.4/3a-157

$B_2C_{15}ClH_{17}N_2$. . . . 1-(C_6H_5-BCl)-2-C_6H_5-1,3,2-$N_2BC_3H_7$ . . . . B: B Comp.SVol.4/4-64

$B_2C_{15}Cl_2H_{26}N_2Si$ [(CH_3)$_3$Si]C_6H_5C=C[BCl(N(CH_3)$_2$)]$_2$ . . . . B: B Comp.SVol.3/4-55, 57

$B_2C_{15}Cl_2H_{33}N_3$ . . . 2,2,6,6-(CH_3)$_4$-NC_5H_6-1-BCl[-N(CH_3)$_2$-BCl-N(C_4H_9-t)-]
B: B Comp.SVol.4/4-67

$B_2C_{15}Cl_2H_{36}N_2Si$ Si(CH_3)$_3$-N(C_4H_9-t)-BCl-N(C_4H_9-t)-BCl-C_4H_9-i
B: B Comp.SVol.4/4-64

– . . . . Si(CH_3)$_3$-N(C_4H_9-t)-BCl-N(C_4H_9-t)-BCl-C_4H_9-t
B: B Comp.SVol.4/4-64

$B_2C_{15}Cl_2H_{37}N_3Si$ Si(CH_3)$_3$-N(C_4H_9-t)-BCl-N(C_4H_9-t)-BCl-N(C_2H_5)$_2$
B: B Comp.SVol.4/4-64

$B_2C_{15}Cl_6H_{23}N_6Nb$ [HB((CH_3)$_2C_3HN_2$)$_3$BH]$NbCl_6$. . . . . B: B Comp.SVol.3/3-201

$B_2C_{15}Cl_6H_{23}N_6Ta$ [HB((CH_3)$_2C_3HN_2$)$_3$BH]$TaCl_6$ . . . . B: B Comp.SVol.3/3-201/2

$B_2C_{15}CoH_{23}O_3$ . . . (CO)$_3$Co[C_3B_2(C_2H_5)$_4$(CH_3)] . . . . B: B Comp.SVol.4/4-180

$B_2C_{15}F_4FeH_{13}N$ . . Fe($C_5H_4BF_2$)$_2$ · NC_5H_5 . . . . Fe: Org.Comp.A9-282

$B_2C_{15}F_8H_{45}MnO_5P_5$
[Mn((CH_3)$_3$P=O)$_5$][BF_4]$_2$. Mn:MVol.D8-84

$B_2C_{15}FeH_{15}NO_4$ . . [(HO)$_2$B-C_5H_4]Fe[C_5H_3(B(OH)$_2$)-2-C_5H_4N]. . . . Fe: Org.Comp.A10-299

$B_2C_{15}FeH_{24}S$ . . . . ($CH_3C_6H_5$)Fe[1,2,5-SB_2C_2-3,4-(C_2H_5)$_2$-2,5-(CH_3)$_2$]
Fe: Org.Comp.B18-81, 84, 92

$B_2C_{15}FeH_{24}S^-$ . . . [($CH_3C_6H_5$)Fe(1,2,5-SB_2C_2-3,4-(C_2H_5)$_2$-2,5-(CH_3)$_2$)]$^-$
Fe: Org.Comp.B18-92

$B_2C_{15}H_{16}N_2O_2$ . . . [-BC_6H_5-NCH_3-CO-NCH_3-BC_6H_5-O-]. . . . . . . B: B Comp.SVol.3/3-165

$B_2C_{15}H_{17}NO_4Si$ . . [1,3,2-$O_2BC_6H_4$-2]$_2$N-Si(CH_3)$_3$ . . . . B: B Comp.SVol.4/3b-63

$B_2C_{15}H_{17}N_3O$ . . . . C_6H_5B[-NCH_3-CO-NCH_3-BC_6H_5-NH-]. . . . . . . B: B Comp.SVol.3/3-119

$B_2C_{15}H_{17}N_3S$ . . . . CH_3B[-S-C(=NC_6H_5)-NC_6H_5-BCH_3-NH-]. . . . . B: B Comp.SVol.3/4-128

$B_2C_{15}H_{18}N_2$ . . . . . CH_3B[-(1,2-C_6H_4)-NH-BCH_3-NC_6H_4(2-CH_3)-] B: B Comp.SVol.3/3-121

$B_2C_{15}H_{23}N_6^+$ . . . . [HB((CH_3)C_3HN_2)$_3$BH]$^+$ . . . . B: B Comp.SVol.3/3-201/2

$B_2C_{15}H_{25}N_3S$ . . . . 2,6-(CH_3)$_2C_6H_3$-N[-BCH_3-NC_2H_5-BCH_3-NC_2H_5-CS-]
B: B Comp.SVol.3/3-120

$B_2C_{15}H_{28}N_2S_2$ . . . CH_3B[-N(C_6H_{11}-c)-CS-N(C_6H_{11}-c)-BCH_3-S-] B: B Comp.SVol.3/4-128/9

$B_2C_{15}H_{29}N$ . . . . . . CH_2=$CHCH_2$-B(C_3H_7-i)-N(C_3H_7-i)-B(CH_2CH=CH_2)$_2$
B: B Comp.SVol.4/3a-242

– . . . . CH_2=$CHCH_2$-B[-CH_2CH(CH_2CH=CH_2)CH_2-
B(C_3H_7-i)-N(C_3H_7-i)-]. . . . . B: B Comp.SVol.4/3a-252

$B_2C_{15}H_{30}N_2$ . . . . . [-N(C_3H_7-i)-CH(C_6H_5)-N(C_3H_7-i)-CH_2CH_2-] · 2 BH_3
B: B Comp.SVol.4/3b-12

$B_2C_{15}H_{31}N$ . . . . . . 9-[(i-C_3H_7)$_2$B-N(CH_3)]-[3.3.1]-9-BC_8H_{14}. . . . . B: B Comp.SVol.4/3a-241

$B_2C_{15}H_{33}N_2^+$ . . . . [-C(CH_3)$_2$-(CH_2)$_3$-C(CH_3)$_2$-]N[-BCH_3-
N(t-C_4H_9)-BCH_3-]$^+$ . . . . B: B Comp.SVol.3/3-208

$B_2C_{15}H_{33}N_3S$ . . . . CH_3B[-N(n-C_4H_9)-CS-N(n-C_4H_9)-BCH_3-N(n-C_4H_9)-]
B: B Comp.SVol.3/3-119/20

$B_2C_{15}H_{35}N_3$ . . . . . (i-C_3H_7)$_2$N-B[-CH_2-B(N(CH_3)$_2$)-N(C_3H_7-i)-C(CH_3)$_2$-]
B: B Comp.SVol.4/3a-250

$B_2C_{15}H_{36}N_6$ ..... $[(CH_3)_2N_2C_3H_6B]N(CH_3)C_3H_6N(CH_3)[BC_3H_6N_2(CH_3)_2]$
B: B Comp.SVol.3/3-98

$B_2C_{15}H_{38}N_4Si$.... 1-$(CH_3)_3Si$-3-$(C_4H_9$-t)-2,4-$[(C_2H_5)_2N]_2$-1,3,2,4-N_2B_2
B: B Comp.SVol.4/3a-198

$B_2C_{16}ClH_{33}N_2Si_2$ C_6H_5-BCl-NH-B$(C_4H_9$-t)-N$[Si(CH_3)_3]_2$....... B: B Comp.SVol.4/4-61

$B_2C_{16}ClH_{42}N_3O_2Si$
$[(CH_3)_2N]ClB$-Si(O-t-$C_4H_9)_2$-B$[N(CH_3)_3]_2$ .... B: B Comp.SVol.3/4-57

$B_2C_{16}Cl_2H_{12}$..... 5,6-Cl_2-5,6-$B_2C_{16}H_{12}$ .... B: B Comp.SVol.4/4-50/1

$B_2C_{16}Cl_2H_{36}Si_2$ .. $[(CH_3)_3Si]_2C=C[BCl-C_4H_9-t]_2$ .... B: B Comp.SVol.3/4-45
B: B Comp.SVol.4/4-50

$B_2C_{16}Cl_3GaH_{40}N_8$ $[(-NCH_3-CH_2-CH_2-NCH_3-)_2B]_2 \cdot GaCl_3$ ..... B: B Comp.SVol.3/3-118

$B_2C_{16}Cl_4F_{16}N_2O_2$ $[CF_3-C(=O)N(C_6F_5)BCl_2]_2$ .... F: PFHOrg.SVol.6-83, 93

$B_2C_{16}Cl_6Ga_2H_{40}N_8$
$[(-NCH_3-CH_2-CH_2-NCH_3-)_2B]_2 \cdot 2\ GaCl_3$.... B: B Comp.SVol.3/3-118

$B_2C_{16}Cl_{10}F_{10}N_2O_2$
$[CCl_3-C(=O)N(C_6F_5)BCl_2]_2$................. F: PFHOrg.SVol.6-83, 93

$B_2C_{16}CrH_{22}O_4$ ... $B_2C_3Cr(CO)_4(C_2H_5)_2(CH_3)_2[=C(CH_3)_2]$ .... B: B Comp.SVol.4/4-180

$B_2C_{16}CrH_{28}N_2O_4$ $[-B(C_2H_5)-N(t-C_4H_9)-]_2 \cdot Cr(CO)_4$.......... B: B Comp.SVol.3/3-136

$B_2C_{16}F_3H_{23}N_6O_3S$ $[HB(-N_2C_3H(CH_3)_2-)_3BH][CF_3-SO_3]$ .... B: B Comp.SVol.4/3b-38

$B_2C_{16}F_6H_{14}MnN_2O_2$
$[Mn(-O-C_6H_4$-2-CH=NCH_2CH_2N=CH-C_6H_4-2-O-$)(BF_3)_2]$
Mn:MVol.D6-98, 101

$B_2C_{16}F_8FeH_{26}N_2$. $[Fe(C_5H_4-N(CH_3)_3)_2][BF_4]_2$ .... Fe: Org.Comp.A9-20, 22

$B_2C_{16}F_{12}H_{28}N_2O_{12}S_4$
$[(i-C_3H_7)_2N=B=N(C_3H_7-i)_2][B(O-SO_2-CF_3)_4]$ .. B: B Comp.SVol.4/4-165
B: B Comp.SVol.4/3b-43

$B_2C_{16}FeH_{21}MnO_3S$
$(C_5H_5)Fe(CC_2H_5BCH_3)_2S(Mn(CO)_3)$.......... Fe: Org.Comp.B16b-12/3

$B_2C_{16}FeH_{26}S$ .... $[1,4-(CH_3)_2C_6H_4]Fe[1,2,5-SB_2C_2-3,4-(C_2H_5)_2$
$-2,5-(CH_3)_2]$......................... Fe: Org.Comp.B18-81, 84, 92

$B_2C_{16}FeH_{26}S^-$ ... $[(1,4-(CH_3)_2C_6H_4)Fe(1,2,5-SB_2C_2-3,4-(C_2H_5)_2$
$-2,5-(CH_3)_2)]^-$......................... Fe: Org.Comp.B18-92

$B_2C_{16}GaH_{40}I_3N_8$.. $[(-NCH_3-CH_2-CH_2-NCH_3-)_2B]_2 \cdot GaI_3$ ...... B: B Comp.SVol.3/3-118

$B_2C_{16}Ga_2H_{40}I_6N_8$ $[(-NCH_3-CH_2-CH_2-NCH_3-)_2B]_2 \cdot 2\ GaI_3$..... B: B Comp.SVol.3/3-118

$B_2C_{16}H_{13}NS_4$ .... $HN[B(-S-CH=CC_6H_5-S-)]_2$ .... B: B Comp.SVol.3/4-127

$B_2C_{16}H_{18}O_2S_3$ ... 2,6-$(CH_3)_2C_6H_3OB[-S-S-B(OC_6H_3-2,6-(CH_3)_2)-S-]$
B: B Comp.SVol.3/4-119/20

$B_2C_{16}H_{19}N_3S$ .... $C_6H_5B(-NCH_3-CS-NCH_3-BC_6H_5-NCH_3-)$..... B: B Comp.SVol.3/3-119/20

$B_2C_{16}H_{22}MoO_4$... $B_2C_3Mo(CO)_4(C_2H_5)_2(CH_3)_2[=C(CH_3)_2]$....... B: B Comp.SVol.4/4-180

$B_2C_{16}H_{22}N_2$ ..... $[(CH_3)_2B][(C_6H_5)_2B]N-N(CH_3)_2$ .... B: B Comp.SVol.3/3-159/60

$B_2C_{16}H_{22}N_2O$ .... $(CH_3)_2N-B(C_6H_5)-O-B(C_6H_5)-N(CH_3)_2$....... B: B Comp.SVol.4/3b-62

$B_2C_{16}H_{22}N_2O_2Si$.. $[-O-N(Si(CH_3)_3)-B(C_6H_5)-N(OCH_3)-B(C_6H_5)-]$ B: B Comp.SVol.4/3b-64

$B_2C_{16}H_{22}N_4$ ..... $[-B(C_6H_5)-N(CH_3)-N(CH_3)-B(C_6H_5)-N(CH_3)-N(CH_3)-]$
B: B Comp.SVol.4/3a-200

$B_2C_{16}H_{22}N_8$ ..... $(1,2-N_2C_3H_3-1)_2B[-1,2-N_2C_3H_3-]_2B(C_2H_5)_2$... B: B Comp.SVol.3/3-199
B: B Comp.SVol.4/3b-35

– .... 1,2-$N_2C_3H_3$-1-B$(C_2H_5)[-1,2-N_2C_3H_3-]_2$
$B(C_2H_5)-1-(1,2-N_2C_3H_3)$ .... B: B Comp.SVol.4/3b-35

$B_2C_{16}H_{23}N_3$ ..... $(CH_3)_2N-B(C_6H_5)-NH-B(C_6H_5)-N(CH_3)_2$...... B: B Comp.SVol.4/3a-168

$B_2C_{16}H_{24}N_{10}$ .... $[(CH_3)_2N](C_3H_3N_2)B[-N_2C_3H_3-B(C_3H_3N_2)(N(CH_3)_2)-N_2C_3H_3-]$ .... B: B Comp.SVol.3/3-199
$B_2C_{16}H_{25}N_3S_2$ ... $[-S-C(C_6H_5)=CH-S-]B-N(C_4H_9-t)-B[-N(CH_3)-CH_2-CH_2-N(CH_3)-]$ .... B: B Comp.SVol.3/4-127
$B_2C_{16}H_{25}Rh$ ..... $[(CH_3)_5C_5]Rh[C_4H_4B_2(CH_3)_2]$ .... B: B Comp.SVol.3/4-157
$B_2C_{16}H_{28}LiN$ ..... $[3.3.1]-9-BC_8H_{14}-9-NLi-9-[3.3.1]-9-BC_8H_{14}$ .. B: B Comp.SVol.4/3a-243
$B_2C_{16}H_{28}S$ ...... $[3.3.1]-9-BC_8H_{14}-9-S-9-[3.3.1]-9-BC_8H_{14}$ ... B: B Comp.SVol.4/4-132
$B_2C_{16}H_{28}S_2$ ..... $[3.3.1]-9-BC_8H_{14}-9-SS-9-[3.3.1]-9-BC_8H_{14}$ .. B: B Comp.SVol.4/4-132
$B_2C_{16}H_{28}Se$ ..... $[3.3.1]-9-BC_8H_{14}-9-Se-9-[3.3.1]-9-BC_8H_{14}$ .. B: B Comp.SVol.4/4-170
$B_2C_{16}H_{28}Se_2$ .... $[3.3.1]-9-BC_8H_{14}-9-SeSe-9-[3.3.1]-9-BC_8H_{14}$ B: B Comp.SVol.4/4-170/1
$B_2C_{16}H_{29}N$ ...... $C_6H_5-B(C_2H_5)-N(C_4H_9-t)-B(C_2H_5)_2$ .... B: B Comp.SVol.4/3a-243
– .... $[3.3.1]-9-BC_8H_{14}-9-NH-9-[3.3.1]-9-BC_8H_{14}$ .. B: B Comp.SVol.4/3a-243
$B_2C_{16}H_{30}$ .... $2,3,4,5-C_4B_2-1,2,3,4,5,6-(C_2H_5)_6$ .... B: B Comp.SVol.3/4-155/6
– .... $2,3,4,5-C_4B_2-1,2,4-(CH_3)_3-3,5,6-(C_3H_7-i)_3$ .... B: B Comp.SVol.3/4-155/6
– .... $2,3,4,5-C_4B_2-1,2,5-(CH_3)_3-3,4,6-(C_3H_7-i)_3$ .... B: B Comp.SVol.3/4-155/6
$B_2C_{16}H_{30}N_2O_4$ ... $1,3-(t-C_4H_9)_2-2,4-(CH_3)_2-1,3,2,4-N_2B_2C_2-5,6-(COOCH_3)_2$ .... B: B Comp.SVol.4/3a-177
$B_2C_{16}H_{30}N_4$ ..... $(C_2H_5)_2B[-N_2C_3H_2(CH_3)-B(C_2H_5)_2-N_2C_3H_2(CH_3)-]$ B: B Comp.SVol.3/3-200
$B_2C_{16}H_{32}N_2$ ..... $[(3.3.1)-9-BC_8H_{14}-9-NH_2]_2$ .... B: B Comp.SVol.4/3a-228
– .... $[(i-C_3H_7)_2N]_2B_2C_4H_4$ .... B: B Comp.SVol.3/4-157
$B_2C_{16}H_{32}N_2O_2$ ... $t-C_4H_9-NH-B(C_2H_5)-N(C_4H_9-t)-B(C_2H_5)-CC-COO-CH_3$ B: B Comp.SVol.4/3a-172
$B_2C_{16}H_{33}N$ ...... $9-[(t-C_4H_9)_2B-NH]-[3.3.1]-9-BC_8H_{14}$ .... B: B Comp.SVol.4/3a-241
$B_2C_{16}H_{34}N_2$ ..... $BC_5H_{11}-N(-CH_2-CH_2-)_3N-BC_5H_{11}$ .... B: B Comp.SVol.3/3-190
– .... $t-C_4H_9-NH-BCH_3-N(C_4H_9-t)-BCH_3-CC-C_4H_9-t$ B: B Comp.SVol.4/3a-172
$B_2C_{16}H_{34}N_2^-$ ..... $[(C_2H_5)_3B-1,4-N_2C_4H_4-B(C_2H_5)_3]^-$, radical anion B: B Comp.SVol.3/3-215
$B_2C_{16}H_{34}N_2O_2$ ... $(i-C_3H_7)_2N-B(OCH_3)-C{\equiv}C-B(OCH_3)-N(C_3H_7-i)_2$ B: B Comp.SVol.4/3b-62
$B_2C_{16}H_{34}N_4O_4$ ... $1,5-(t-C_4H_9)_2-2,3-(C_2H_5-COO)_2-4,6-(CH_3)_2-1,2,3,5,4,6-N_4B_2$ .... B: B Comp.SVol.4/3a-200/1
$B_2C_{16}H_{34}N_8$ ..... $(n-C_3H_7)_2B[-N_3C_2H(NH_2)-]_2B(C_3H_7-n)_2$ ...... B: B Comp.SVol.4/3b-37/8
$B_2C_{16}H_{36}N$ ...... $(t-C_4H_9)B[-(t-C_4H_9)-B(t-C_4H_9)-N(t-C_4H_9)-]$ .. B: B Comp.SVol.3/3-142
$B_2C_{16}H_{36}N_2$ ..... $1-[(CH_3)_2B-N(C_4H_9-t)-B(CH_3)]-2,2,6,6-(CH_3)_4-NC_5H_6$ B: B Comp.SVol.4/3a-171
– .... $[-B(C_4H_9-n)-N(C_4H_9-t)-B(C_4H_9-n)-N(C_4H_9-t)-]$ B: B Comp.SVol.3/3-135/6
B: B Comp.SVol.4/3a-196
– .... $[-B(C_4H_9-i)-N(C_4H_9-t)-B(C_4H_9-i)-N(C_4H_9-t)-]$ B: B Comp.SVol.4/3a-196
– .... $[-B(C_4H_9-s)-N(C_4H_9-t)-B(C_4H_9-s)-N(C_4H_9-t)-]$ B: B Comp.SVol.4/3a-196
– .... $[-B(C_4H_9-s)-N(C_4H_9-s)-B(C_4H_9-s)-N(C_4H_9-s)-]$ B: B Comp.SVol.3/3-136, 142
– .... $[-B(C_4H_9-t)-N(C_4H_9-t)-B(C_4H_9-t)-N(C_4H_9-t)-]$ B: B Comp.SVol.3/3-136
B: B Comp.SVol.4/3a-196
B: B Comp.SVol.4/3b-247

$B_2C_{16}H_{36}N_2O_3$ ... 2,2,6,6-$(CH_3)_4$-NC_5H_6-1-$B(OCH_3)$-$N(C_4H_9$-t)-$B(OCH_3)_2$
B: B Comp.SVol.4/3b-53
$B_2C_{16}H_{36}N_4$ ..... (n-$C_4H_9)_2B$-$N(C_4H_9$-n)-$B(N_3)$-C_4H_9-n .... B: B Comp.SVol.3/3-106
– ..... (n-$C_4H_9)_2B$-$N(C_4H_9$-i)-$B(N_3)$-C_4H_9-i .... B: B Comp.SVol.4/3a-171
– ..... (i-$C_4H_9)_2B$-$N(C_4H_9$-i)-$B(N_3)$-C_4H_9-i .... B: B Comp.SVol.3/3-106
– ..... (s-$C_4H_9)_2B$-$N(C_4H_9$-s)-$B(N_3)$-C_4H_9-s .... B: B Comp.SVol.3/3-106
$B_2C_{16}H_{37}N$ ...... (t-$C_4H_9)_2B$-NH-$B(C_4H_9$-t$)_2$ .... B: B Comp.SVol.4/3a-240
– ..... n-C_4H_9-$B(C_2H_5)$-$N(C_2H_5)$-$B(C_4H_9$-n$)_2$ .... B: B Comp.SVol.3/3-160
– ..... n-C_5H_{11}-$B(C_2H_5)$-$N(C_5H_{11}$-n)-$B(C_2H_5)_2$ ..... B: B Comp.SVol.3/3-160
$B_2C_{16}H_{37}N_3$ ..... 1-(t-C_4H_9)-2,3-[(i-$C_3H_7)_2N]_2$-NB_2 .... B: B Comp.SVol.4/3a-194
$B_2C_{16}H_{38}I_2Si_2$.... $[(CH_3)_3Si]_2CH$-CH(BI-C_4H_9-t$)_2$ .... B: B Comp.SVol.4/4-114
$B_2C_{16}H_{38}N_2$ ..... (i-$C_3H_7)_2N$-$B(C_2H_5)$-$N(C_4H_9$-t)-$B(C_2H_5)_2$..... B: B Comp.SVol.4/3a-171
$B_2C_{16}H_{38}N_2O$ .... t-C_4H_9-NH-$B(C_2H_5)$-$N(C_4H_9$-t)-$B(C_2H_5)$-O-C_4H_9-t
B: B Comp.SVol.4/3b-53
$B_2C_{16}H_{38}N_2O_4$ ... $[H_3B$-$N(CH_3)_2$-CH_2CH_2-OC(O)-$CH_2CH_2CH_2]_2$ B: B Comp.SVol.4/3b-15
$B_2C_{16}H_{38}N_4O_2Si$.. [(-NCH_3-CH_2-CH_2-NCH_3-)B$]_2Si(OC_4H_9$-t$)_2$... B: B Comp.SVol.3/3-118
$B_2C_{16}H_{39}N_3$ ..... t-C_4H_9-NH-$B(C_2H_5)$-$N(C_4H_9$-t)-$B(C_2H_5)$-NH-C_4H_9-t
B: B Comp.SVol.4/3a-172
$B_2C_{16}H_{40}N_2$ ..... [-BH(CH_2-C_4H_9-t)-CH_2-$N(CH_3)_2$-BH
(CH_2-C_4H_9-t)-CH_2-$N(CH_3)_2$-] .... B: B Comp.SVol.4/3a-256
$B_2C_{16}H_{40}N_4O_2$ ... $[C_2H_5$-N=C(OC_2H_5)-BH_2-$N(CH_3)_2$-CH_2-$]_2$ .... B: B Comp.SVol.4/3b-23
$B_2C_{16}H_{40}N_8$ ..... [(-NCH_3-CH_2-CH_2-NCH_3-$)_2B]_2$............ B: B Comp.SVol.3/3-118
$B_2C_{16}H_{42}N_2OSi_2$.. (n-$C_3H_7)(C_2H_5)B$-$N(C_2H_5)$-B(n-C_3H_7)
-N$[Si(CH_3)_3]$-O-$Si(CH_3)_3$................ B: B Comp.SVol.3/3-106
$B_2C_{16}H_{42}N_4O_2Si$.. $[((CH_3)_2N)_2B]_2Si$(O-t-$C_4H_9)_2$ .... B: B Comp.SVol.3/3-106
$B_2C_{17}ClH_{37}N_2$.... 1-$[(C_2H_5)_2B$-$N(C_4H_9$-t)-BCl]-2,2,6,6-$(CH_3)_4$-NC_5H_6
B: B Comp.SVol.4/4-70
– ..... [(-$C(CH_3)_2$-$CH_2CH_2CH_2$-$C(CH_3)_2$-)N(-$B(C_2H_5)$
-$N(C_4H_9$-t)-$B(C_2H_5)$-)]Cl................ B: B Comp.SVol.4/3b-46
B: B Comp.SVol.4/4-70
$B_2C_{17}ClH_{37}N_2O_2$. 1-$[(C_2H_5$-O$)_2B$-$N(C_4H_9$-t)-BCl]-2,2,6,6-$(CH_3)_4$-NC_5H_6
B: B Comp.SVol.4/4-55
– ..... 1-$[C_2H_5$-O-BCl-$N(C_4H_9$-t)-B(O-C_2H_5)]-
2,2,6,6-$(CH_3)_4$-NC_5H_6 .... B: B Comp.SVol.4/4-55
$B_2C_{17}ClH_{39}N_4$.... 2,2,6,6-$(CH_3)_4$-NC_5H_6-1-BCl-$N(C_4H_9$-t)-$B[N(CH_3)_2]_2$
B: B Comp.SVol.4/4-64
$B_2C_{17}Cl_2H_{37}N_3$... $(C_2H_5)_2N$-BCl-$N(C_4H_9$-t)-BCl-1-NC_5H_6-2,2,6,6-$(CH_3)_4$
B: B Comp.SVol.4/4-64
$B_2C_{17}CoH_{27}Ni$ ... $(C_3H_5)Ni[C_3B_2H(C_2H_5)_2(CH_3)_2]Co(C_5H_5)$...... B: B Comp.SVol.4/4-180
$B_2C_{17}FH_{37}N_2$ .... 2,2,6,6-$(CH_3)_4$-NC_5H_6-1-[BF-NH-$B(C_4H_9$-t$)_2$] B: B Comp.SVol.4/3b-247
$B_2C_{17}F_4FeH_{16}O$ .. $[C_6H_7FeC_4H_4BC_6H_5CO][BF_4]$.............. Fe: Org.Comp.B17-245
$B_2C_{17}FeH_{22}MnO_3S^+$
$[(C_6H_6)$Fe(1,2,5-SB_2C_2-3,4-$(C_2H_5)_2$-
2,5-$(CH_3)_2)Mn(CO)_3]^+$ .... Fe: Org.Comp.B18-84
$B_2C_{17}FeH_{26}$ ..... $(CH_3C_6H_5)$Fe[1,4-$(CH_3)_2$-2,3-$(C_2H_5)_2$-1,4-$B_2C_4H_2$]
Fe: Org.Comp.B18-114, 135
$B_2C_{17}H_{20}N_2O_2$ ... [-BC_6H_5-NC_2H_5-CO-NC_2H_5-BC_6H_5-O-] ..... B: B Comp.SVol.3/3-165
$B_2C_{17}H_{20}N_2S_2$ ... C_6H_5B[-NC_2H_5-CS-NC_2H_5-BC_6H_5-S-] ...... B: B Comp.SVol.3/4-128/9
$B_2C_{17}H_{20}N_6$ ..... 1,3-$(CH_3)_2$-1,3,2-$N_2BC_6H_4$-2-N=C=N-2-
[1,3,2-$N_2BC_6H_4$-1,3-$(CH_3)_2$] .... B: B Comp.SVol.4/3a-158

$B_2C_{17}H_{21}N_3S$ .... $C_6H_5B(-NC_2H_5-CS-NC_2H_5-BC_6H_5-NH-)$ ..... B: B Comp.SVol.3/3-120/1

$B_2C_{17}H_{21}N_{11}$ .... $[(CH_3)_2N](C_3H_3N_2)B[-N_2C_3H_3-B(C_3H_3N_2)_2-N_2C_3H_3-]$ B: B Comp.SVol.3/3-199

$B_2C_{17}H_{23}N_3$ ..... $1-[(CH_3)_2N-B(C_6H_5)]-2-C_6H_5-1,3,2-N_2BC_3H_7$ B: B Comp.SVol.4/3a-177

$B_2C_{17}H_{25}N_3$ ..... $(CH_3)_2N-B(C_6H_5)-N(CH_3)-B(C_6H_5)-N(CH_3)_2$... B: B Comp.SVol.4/3a-168

$B_2C_{17}H_{26}N_2$ ..... $[-N(CH_2-C_6H_5)-CH_2-N(CH_2-C_6H_5)-CH_2CH_2-] \cdot 2\ BH_3$ B: B Comp.SVol.4/3b-12

$B_2C_{17}H_{29}Rh$ ..... $(C_5H_5)Rh[C_3B_2H(C_2H_5)_4(CH_3)]$.............. B: B Comp.SVol.4/4-180

$B_2C_{17}H_{32}N_2$ ..... $1-[(3.3.1)-9-BC_8H_{14}-9-N{=}B]-2,2,6,6-(CH_3)_4-NC_5H_6$ B: B Comp.SVol.4/3a-160/5

$B_2C_{17}H_{32}S$ ...... $[-BC_8H_{14}-S(CH_3)-BC_8H_{14}-H-]$ .... B: B Comp.SVol.4/4-132

$B_2C_{17}H_{33}N$ ...... $CH_2{=}CHCH_2-B(C_4H_9-n)-N(C_4H_9-t)-B(CH_2CH{=}CH_2)_2$ B: B Comp.SVol.4/3a-243

– .... $CH_2{=}CHCH_2-B[-CH_2CH(CH_2CH{=}CH_2)CH_2$ $-B(C_4H_9-n)-N(C_4H_9-t)-]$ .... B: B Comp.SVol.4/3a-252

$B_2C_{17}H_{33}NSi$..... $1-[(3.3.1)-9-BC_8H_{14}-9]-4,5-(C_2H_5)_2-$ $2,2,3-(CH_3)_3-1,2,5-NSiBC_2$ .... B: B Comp.SVol.4/3a-250/1

$B_2C_{17}H_{34}N_4Si$.... $1-(CH_3)_3Si-3-C_6H_5-2,4-[(C_2H_5)_2N]_2-1,3,2,4-N_2B_2$ B: B Comp.SVol.4/3a-198

$B_2C_{17}H_{35}N$ ...... $9-[(t-C_4H_9)_2B-N(CH_3)]-[3.3.1]-9-BC_8H_{14}$..... B: B Comp.SVol.4/3a-242

$B_2C_{17}H_{37}IN_2$..... $[(-C(CH_3)_2-(CH_2)_3-C(CH_3)_2-)N(-B(C_2H_5)$ $-N(C_4H_9-t)-B(C_2H_5)-)]I$ .... B: B Comp.SVol.4/3b-46

$B_2C_{17}H_{37}NSn_2$ ... $(CH_3)_3Sn-CC-B[N(C_2H_5)_2]-C[Sn(CH_3)_3]{=}C(CH_3)-B(CH_3)_2$ B: B Comp.SVol.4/3a-244

$B_2C_{17}H_{37}N_2^+$ .... $[(-C(CH_3)_2-(CH_2)_3-C(CH_3)_2-)N(-B(C_2H_5)$ $-N(C_4H_9-t)-B(C_2H_5)-)]^+$ .... B: B Comp.SVol.4/4-70

$B_2C_{17}H_{37}N_3O_2$ ... $(i-C_3H_7)_2B-NCH_3-CO-NCH_3-C[OB(i-C_3H_7)_2]{=}NCH_3$ B: B Comp.SVol.3/3-161

$B_2C_{17}H_{37}N_3S$ .... $n-C_4H_9B[-S-C({=}NC_4H_9-n)-N(C_4H_9-n)-B(C_4H_9-n)-NH-]$ B: B Comp.SVol.3/4-128

$B_2C_{17}H_{41}N_2OP$ ... $(C_2H_5)_2B-N(C_4H_9-t)-P({=}O)(CH_3)-N(C_4H_9-t)-B(C_2H_5)_2$ B: B Comp.SVol.4/3a-244

$B_2C_{17}H_{41}N_2PS$ ... $(C_2H_5)_2B-N(C_4H_9-t)-P({=}S)(CH_3)-N(C_4H_9-t)-B(C_2H_5)_2$ B: B Comp.SVol.4/3a-244

$B_2C_{17}H_{41}N_3Si$.... $(CH_3)_3Si-N(C_4H_9-t)-B[-N(C_4H_9-t)-B(C_3H_7-i)$ $-N(C_3H_7-i)-]$.......................... B: B Comp.SVol.4/3a-197

$B_2C_{17}H_{42}N_2OSi_2$.. $3-CH_3-BC_4H_7-1-N(C_3H_7-i)-B(C_3H_7-i)$ $-N[Si(CH_3)_3]-O-Si(CH_3)_3$................ B: B Comp.SVol.4/3a-172

$B_2C_{17}H_{42}N_2Si$.... $(CH_3)_3Si-N(C_4H_9-t)-B(C_2H_5)-N(C_4H_9-t)-B(C_2H_5)_2$ B: B Comp.SVol.4/3a-171

$B_2C_{17}H_{43}LiN_4$.... $[-N(CH_3)_2-CH_2CH_2-N(CH_3)_2-]Li-N(C_4H_9-t)[$ $-B(CH_3)_2-N(C_4H_9-t)-B(CH_3)-]$ .... B: B Comp.SVol.4/3a-201/2

$B_2C_{18}ClH_{25}N_2Si$.. $1-(C_6H_5-BCl)-2-C_6H_5-3-Si(CH_3)_3-1,3,2-N_2BC_3H_6$ B: B Comp.SVol.4/4-64

$B_2C_{18}Cl_2H_{16}N_4$... $C_6H_5-BCl[-1,2-N_2C_3H_3-]_2BCl-C_6H_5$......... B: B Comp.SVol.4/4-68

– .... $Cl_2B[-1,2-N_2C_3H_3-]_2B(C_6H_5)_2$.............. B: B Comp.SVol.4/4-68

$B_2C_{18}Cl_2H_{38}Si$ ... $t-C_4H_9-CH_2[(CH_3)_3Si]C{=}C[B(t-C_4H_9)Cl]_2$ ..... B: B Comp.SVol.3/4-45

$B_2C_{18}Cl_4H_{14}N_4$... $Cl_2B-[-1,2-N_2C_3H_2(C_6H_5)-]_2BCl_2$ .... B: B Comp.SVol.4/4-67/8

$B_2C_{18}Cl_4H_{30}N_4$... $Cl_2B[-1,2-N_2C_3(CH_3)_2(C_4H_9)-]_2BCl_2$......... B: B Comp.SVol.4/4-67/8

$B_2C_{18}Cl_4H_{32}Mn_2N_{12}O_{22}$
$[Mn((1,2\text{-}N_2C_3H_3\text{-}1)_3BH)_2][Mn(H_2O)_6][ClO_4]_4$ .. Mn: MVol.D8-14/5
$B_2C_{18}Cl_4H_{36}N_2$ ... $[(CH_3)_4C_5H_6N{=}B{=}NC_5H_6(CH_3)_4][BCl_4]$ .... B: B Comp.SVol.3/4-62
$B_2C_{18}Cl_6H_{14}MnN_{12}$
$Mn[(4\text{-}Cl\text{-}1,2\text{-}N_2C_3H_2\text{-}1)_3BH]_2$ .... Mn: MVol.D8-9/10
$B_2C_{18}CoH_{31}$ ..... $(C_5H_5)Co[C_3B_2(CH_3)_2(C_2H_5)_4]$ .... B: B Comp.SVol.4/4-180
$B_2C_{18}CrH_{32}N_2O_4$ $[\text{-}B(n\text{-}C_3H_7)\text{-}N(t\text{-}C_4H_9)\text{-}]_2 \cdot Cr(CO)_4$ .... B: B Comp.SVol.3/3-136
$B_2C_{18}F_2H_{16}N_4$ ... $C_6H_5\text{-}BF[\text{-}1,2\text{-}N_2C_3H_3\text{-}]_2BF\text{-}C_6H_5$ .... B: B Comp.SVol.4/3b-248/9
– .... $F_2B[\text{-}1,2\text{-}N_2C_3H_3\text{-}]_2B(C_6H_5)_2$ .... B: B Comp.SVol.4/3b-248/9
$B_2C_{18}F_2H_{20}N_2O_2$ $B_2C_2N_2O_2(CH_3)_4(C_6H_4\text{-}2\text{-}F)_2$ .... B: B Comp.SVol.3/3-123/4
– .... $B_2C_2N_2O_2(CH_3)_4(C_6H_4\text{-}4\text{-}F)_2$ .... B: B Comp.SVol.3/3-123/4
– .... $(CH_3)_2B\text{-}N(C_6H_4\text{-}2\text{-}F)\text{-}C(O)\text{-}C(O)\text{-}N(C_6H_4\text{-}2\text{-}F)\text{-}B(CH_3)_2$
B: B Comp.SVol.3/3-161
– .... $(CH_3)_2B\text{-}N(C_6H_4\text{-}4\text{-}F)\text{-}C(O)\text{-}C(O)\text{-}N(C_6H_4\text{-}4\text{-}F)\text{-}B(CH_3)_2$
B: B Comp.SVol.3/3-161
$B_2C_{18}F_3H_{37}N_2O_3S$ $[(\text{-}C(CH_3)_2\text{-}CH_2CH_2CH_2\text{-}C(CH_3)_2\text{-})N(\text{-}B(C_2H_5)$
$\text{-}N(C_4H_9\text{-}t)\text{-}B(C_2H_5)\text{-})][CF_3\text{-}SO_3]$ .... B: B Comp.SVol.4/3b-46
$B_2C_{18}F_4FeH_{16}O_4$ $Fe(C_5H_4\text{-}F_2BOC{=}CHCOCH_3)_2$ .... Fe: Org.Comp.A9-280
$B_2C_{18}F_4H_{36}N_2$ ... $[(\text{-}C(CH_3)_2\text{-}(CH_2)_3\text{-}C(CH_3)_2\text{-})N{=}B{=}$
$N(\text{-}C(CH_3)_2\text{-}(CH_2)_3\text{-}C(CH_3)_2\text{-})][BF_4]$ .... B: B Comp.SVol.3/3-205
$B_2C_{18}F_6H_{18}MnN_2O_2$
$[Mn(\text{-}O\text{-}C_6H_4\text{-}2\text{-}C(CH_3){=}N\text{-}CH_2CH_2\text{-}N{=}$
$C(CH_3)\text{-}C_6H_4\text{-}2\text{-}O\text{-})(BF_3)_2]$ .... Mn: MVol.D6-196, 197
$B_2C_{18}F_8FeH_{24}$ ... $[(1,3,5\text{-}(CH_3)_3\text{-}C_6H_3)_2Fe][BF_4]_2$ .... Fe: Org.Comp.B19-347, 364
– .... $[((CH_3)_6C_6)Fe(C_6H_6)][BF_4]_2$ .... Fe: Org.Comp.B19-405
$B_2C_{18}F_{10}H_{15}N$ ... $(C_6F_5)_2B\text{-}N(C_2H_5)\text{-}B(C_2H_5)_2$ .... B: B Comp.SVol.3/3-161
– .... $C_6F_5\text{-}B(C_2H_5)\text{-}N(C_6F_5)\text{-}B(C_2H_5)_2$ .... B: B Comp.SVol.3/3-160
$B_2C_{18}F_{12}H_{14}N_4O_8$ $(CF_3\text{-}COO)_2B[\text{-}1,2\text{-}N_2C_3H(CH_3)_2\text{-}]_2B[O\text{-}C(O)\text{-}CF_3]_2$
B: B Comp.SVol.4/3b-59
$B_2C_{18}FeH_{28}$ ..... $Fe(t\text{-}C_4H_9\text{-}BC_5H_5)_2$ .... Fe: Org.Comp.B17-328/9
– .... $Fe(n\text{-}C_4H_9\text{-}BC_5H_5)_2$ .... Fe: Org.Comp.B17-328
$B_2C_{18}FeH_{28}O_4$ ... $Fe[C_5H_4B(OC_2H_5)_2]_2$ .... Fe: Org.Comp.A9-279
$B_2C_{18}H_{16}N_4O_3$ ... $[2\text{-}(CH_3\text{-}C(O))\text{-}2,3,1\text{-}N_2BC_7H_5\text{-}1]_2O$ .... B: B Comp.SVol.4/3b-67
$B_2C_{18}H_{18}N_4$ ..... $C_6H_5\text{-}BH[\text{-}1,2\text{-}N_2C_3H_3\text{-}]_2BH\text{-}C_6H_5$ .... B: B Comp.SVol.3/3-199
– .... $H_2B[\text{-}1,2\text{-}N_2C_3H_2(C_6H_5)\text{-}]_2BH_2$ .... B: B Comp.SVol.4/3b-34
– .... $H_2B[\text{-}1,2\text{-}N_2C_3H_3\text{-}]_2B(C_6H_5)_2$ .... B: B Comp.SVol.4/3b-34
$B_2C_{18}H_{18}N_{12}$ .... $(1,2\text{-}N_2C_3H_3\text{-}1)_2B[\text{-}1,2\text{-}N_2C_3H_3\text{-}]_2B[\text{-}1\text{-}(1,2\text{-}N_2C_3H_3)]_2$
B: B Comp.SVol.3/3-199
B: B Comp.SVol.4/3b-34
$B_2C_{18}H_{20}MnN_{12}$ .. $Mn[(1,2\text{-}N_2C_3H_3\text{-}1)_3BH]_2$ .... Mn: MVol.D8-9/10
– .... $Mn[(1,3\text{-}N_2C_3H_3\text{-}1)_3BH]_2$ .... Mn: MVol.D8-18/9
$B_2C_{18}H_{20}N_{12}$ .... $(C_3H_4N_2)(C_3H_3N_2)_2B\text{-}B(N_2C_3H_3)_2(N_2C_3H_4)$ .... B: B Comp.SVol.4/3b-36
$B_2C_{18}H_{22}N_2O_2$ ... $B_2C_2N_2O_2(CH_3)_4(C_6H_5)_2$ .... B: B Comp.SVol.3/3-123/4
$B_2C_{18}H_{24}N_2$ ..... $[(CH_3)_2B\text{-}N{=}CH(C_6H_5)]_2$ .... B: B Comp.SVol.3/3-148/9
$B_2C_{18}H_{24}N_4$ ..... $(C_6H_5)_2B\text{-}N(C_3H_7\text{-}i)\text{-}B(N_3)\text{-}C_3H_7\text{-}i$ .... B: B Comp.SVol.4/3a-171
$B_2C_{18}H_{25}N$ ...... $(C_6H_5)_2B\text{-}N(C_2H_5)\text{-}B(C_2H_5)_2$ .... B: B Comp.SVol.3/3-161
– .... $C_6H_5\text{-}B(C_2H_5)\text{-}N(C_6H_5)\text{-}B(C_2H_5)_2$ .... B: B Comp.SVol.3/3-160
$B_2C_{18}H_{28}N_2$ ..... $[\text{-}N(CH_2C_6H_5)\text{-}CH(CH_3)\text{-}N(CH_2C_6H_5)\text{-}CH_2CH_2\text{-}] \cdot 2\ BH_3$
B: B Comp.SVol.4/3b-12
$B_2C_{18}H_{34}N_2$ ..... $1,4\text{-}[(i\text{-}C_3H_7)_2N]_2\text{-}1,4\text{-}B_2C_6H_6$ .... B: B Comp.SVol.4/3a-247

$B_2C_{18}H_{34}N_2O_4$ . . . 1,3-(t-C_4H_9)$_2$-2,4-(C_2H_5)$_2$-1,3,2,4-$N_2B_2C_2$ -5,6-($COOCH_3$)$_2$ . . . . B: B Comp.SVol.4/3a-177

$B_2C_{18}H_{34}N_4$ . . . . . (C_2H_5)$_2$B[-1,2-$N_2C_3H(CH_3)_2$-]$_2$B(C_2H_5)$_2$ . . . . . . B: B Comp.SVol.3/3-200

– . . . . (n-C_3H_7)$_2$B[-1,2-$N_2C_3H_3$-]$_2$B(C_3H_7-n)$_2$ . . . . . . B: B Comp.SVol.3/3-199

$B_2C_{18}H_{34}N_4S_2$ . . . CH_3S-$CH_2CH_2CH_2$-BH[-1,2-$N_2C_3H(CH_3)_2$-]$_2$ BH-$CH_2CH_2CH_2$-SCH_3 . . . . B: B Comp.SVol.4/3b-35

$B_2C_{18}H_{34}S$ . . . . . . [-BC_8H_{14}-S(C_2H_5)-BC_8H_{14}-H-] . . . . B: B Comp.SVol.4/4-132

$B_2C_{18}H_{34}S_2$ . . . . . [-S(CH_3)-BC_8H_{14}-S(CH_3)-BC_8H_{14}-] . . . . B: B Comp.SVol.4/4-132/3

$B_2C_{18}H_{35}N$ . . . . . . [-(BC_8H_{14})-H-(BC_8H_{14})-N(CH_3)$_2$-] . . . . B: B Comp.SVol.3/3-161, 162

$B_2C_{18}H_{36}N_2O_2$ . . . 2,4-[2,2,6,6-(CH_3)$_4$-NC_5H_6-1]$_2$-1,3,2,4-O_2B_2 . . B: B Comp.SVol.4/3b-62

$B_2C_{18}H_{36}N_2S_2$ . . . 2,4-[2,2,6,6-(CH_3)$_4$-NC_5H_6-1]$_2$-1,3,2,4-S_2B_2 . . B: B Comp.SVol.4/4-144/5

$B_2C_{18}H_{36}N_2Se_2$ . . 2,4-[2,2,6,6-(CH_3)$_4$-NC_5H_6-1]$_2$-1,3,2,4-Se_2B_2 B: B Comp.SVol.4/4-171

$B_2C_{18}H_{38}N_2$ . . . . . t-C_4H_9-NH-B(C_2H_5)-N(C_4H_9-t)-B(C_2H_5)-CC-C_4H_9-t B: B Comp.SVol.4/3a-172

– . . . . t-C_4H_9-NH-B(C_4H_9-t)-CC-B(C_4H_9-t)-NH-C_4H_9-t B: B Comp.SVol.4/3a-244

– . . . . HCC-B(C_4H_9-t)[-N(C_4H_9-t)-B(C_4H_9-t)-NH(C_4H_9-t)-] B: B Comp.SVol.4/3a-201

$B_2C_{18}H_{38}N_4O_4$ . . . 1,5-(t-C_4H_9)$_2$-2,3-[C_2H_5-OC(O)]$_2$-4,6-(C_2H_5)$_2$-1,2,3,5,4,6-N_4B_2 . . . . B: B Comp.SVol.4/3a-200/1

$B_2C_{18}H_{38}NiSe_2Si_2$ $B_2C_4NiSe_2Si_2(C_2H_5)_4(CH_3)_6$ . . . . B: B Comp.SVol.4/4-180

$B_2C_{18}H_{40}N_2$ . . . . . [2,2,6,6-(CH_3)$_4$-NC_5H_6-1-BH_2]$_2$ = [-BH_2-$C_9H_{18}N$-BH_2-$C_9H_{18}N$-] . . . . B: B Comp.SVol.4/3a-254

$B_2C_{18}H_{40}N_2O$ . . . . [-$CHCH_3$-N(C_4H_9-t)-B(C_4H_9-n)-N(C_4H_9-t) -B(C_4H_9-n)-O-] . . . . B: B Comp.SVol.4/3b-54

$B_2C_{18}H_{40}N_4$ . . . . . [(C_2H_5)$_2$N]$_2$BC≡CB[N(C_2H_5)$_2$]$_2$ . . . . B: B Comp.SVol.3/3-106

$B_2C_{18}H_{43}N_3$ . . . . . [(i-C_3H_7)$_2$N]$_2$B-NC_2H_5-B(C_2H_5)$_2$ . . . . B: B Comp.SVol.3/3-98

$B_2C_{18}H_{45}N_3O_3Si$ . . [(CH_3)$_2$N]$_2$B-B[N(CH_3)$_2$]Si(O-t-C_4H_9)$_3$ . . . . B: B Comp.SVol.3/3-106

$B_2C_{18}H_{46}N_2OSi_2$ . . (CH_3)$_3$Si-O-N[Si(CH_3)$_3$]-B(C_3H_7-n) -N(C_3H_7-n)-B(C_3H_7-n)$_2$ . . . . B: B Comp.SVol.3/3-106

– . . . . (CH_3)$_3$Si-O-N[Si(CH_3)$_3$]-B(C_4H_9-n)-N(C_4H_9-t)-B(C_2H_5)$_2$ B: B Comp.SVol.4/3a-172

$B_2C_{18}H_{56}N_4OSi_6$ . . [(CH_3)$_3$Si]$_2$N-B[NH-Si(CH_3)$_3$]-O-B[NH -Si(CH_3)$_3$]-N[Si(CH_3)$_3$]$_2$ . . . . B: B Comp.SVol.4/3b-53

$B_2C_{18}H_{57}N_5Si_6$ . . . [(CH_3)$_3$Si]$_2$N-B[NH-Si(CH_3)$_3$]-NH-B[NH -Si(CH_3)$_3$]-N[Si(CH_3)$_3$]$_2$ . . . . B: B Comp.SVol.4/3a-157

$B_2C_{19}Cl_2H_{32}N_2$ . . . [-C(CH_3)$_2$-$CH_2CH_2CH_2$-C(CH_3)$_2$-]N[-BCl(C_6H_5)-N(C_4H_9-t)-BCl-] . . . . B: B Comp.SVol.3/4-61

– . . . . [-C(CH_3)$_2$-$CH_2CH_2CH_2$-C(CH_3)$_2$-]N[-BCl_2-N(C_4H_9-t)-B(C_6H_5)-] . . . . B: B Comp.SVol.3/4-61

$B_2C_{19}Cl_2H_{32}Si$ . . . C_6H_5[(CH_3)$_3$Si]C=C[B(t-C_4H_9)Cl]$_2$ . . . . B: B Comp.SVol.3/4-45

$B_2C_{19}Cl_2H_{41}N_3$ . . . (i-C_3H_7)$_2$N-BCl-N(C_4H_9-t)-BCl-1-NC_5H_6-2,2,6,6-(CH_3)$_4$ B: B Comp.SVol.4/4-64

$B_2C_{19}CoFeH_{27}$ . . . C_5H_5Fe(1,3-(CH_3)$_2$-4,5-(C_2H_5)$_2$$C_3HB_2$-1,3)Co$C_5H_5$ Fe: Org.Comp.B17-175/6

$B_2C_{19}CoFeH_{27}^+$ . . [C_5H_5Fe(1,3-(CH_3)$_2$-4,5-(C_2H_5)$_2$$C_3HB_2$-1,3)Co$C_5H_5$]$^+$ Fe: Org.Comp.B17-175/6

$B_2C_{19}CoFeH_{27}^-$ .. $[C_5H_5Fe(1,3\text{-}(CH_3)_2\text{-}4,5\text{-}(C_2H_5)_2C_3HB_2\text{-}1,3)CoC_5H_5]^-$ Fe: Org.Comp.B17-175/6

$B_2C_{19}CoH_{31}$ ..... $B_2C_4Co(C_5H_5)(CH_3)_2(C_2H_5)_4$ .... B: B Comp.SVol.4/4-180

$B_2C_{19}F_6H_{36}N_2O$ .. $[\text{-}C(CF_3)_2\text{-}N(C_4H_9\text{-}t)\text{-}B(C_4H_9\text{-}n)\text{-}N(C_4H_9\text{-}t)$ $\text{-}B(C_4H_9\text{-}n)\text{-}O\text{-}]$ .... B: B Comp.SVol.4/3b-54

$B_2C_{19}FeH_{32}$ ..... $(CH_3C_6H_5)Fe[1,3\text{-}B_2C_3H\text{-}(CH_3)_2\text{-}1,3\text{-}$ $(C_2H_5)_2\text{-}4,5\text{-}(C_3H_7\text{-}i)\text{-}2]$ .... Fe: Org.Comp.B18-98

– .... $(CH_3C_6H_5)Fe[1,3\text{-}B_2C_3H\text{-}CH_3\text{-}2\text{-}(C_2H_5)_4\text{-}1,3,4,5]$ Fe: Org.Comp.B18-97/8

$B_2C_{19}H_{25}N_3O$ .... $C_6H_5B[\text{-}NC_2H_5\text{-}CO\text{-}NC_2H_5\text{-}BC_6H_5\text{-}NC_2H_5\text{-}]$ .. B: B Comp.SVol.3/3-118

$B_2C_{19}H_{25}N_3S$ .... $C_6H_5B(\text{-}NC_2H_5\text{-}CS\text{-}NC_2H_5\text{-}BC_6H_5\text{-}NC_2H_5\text{-})$ .. B: B Comp.SVol.3/3-119/20

$B_2C_{19}H_{31}N_2Se$ ... $[\text{-}BC_8H_{14}\text{-}C_3H_3N_2\text{-}BC_8H_{14}\text{-}Se\text{-}]$ .... B: B Comp.SVol.4/4-170

$B_2C_{19}H_{32}N_2O_2$ ... $1\text{-}[1,3,2\text{-}O_2BC_6H_4\text{-}2\text{-}N(C_4H_9\text{-}t)\text{-}BH]\text{-}$ $2,2,6,6\text{-}(CH_3)_4\text{-}NC_5H_6$ .... B: B Comp.SVol.4/3a-168

$B_2C_{19}H_{32}N_2S$ .... $[3.3.1]\text{-}9\text{-}BC_8H_{14}\text{-}9\text{-}S\text{-}9\text{-}[3.3.1]\text{-}9\text{-}BC_8H_{14} \cdot 1,2\text{-}N_2C_3H_4$ B: B Comp.SVol.4/4-132

– .... $[\text{-}SH\text{-}BC_8H_{14}\text{-}N_2C_3H_3\text{-}BC_8H_{14}\text{-}]$ .... B: B Comp.SVol.4/4-142/3

$B_2C_{19}H_{35}N$ ...... $t\text{-}C_4H_9\text{-}B(C_2H_5)\text{-}N[B(C_2H_5)_2]\text{-}C_6H_2\text{-}2,4,6\text{-}(CH_3)_3$ B: B Comp.SVol.4/3a-243

$B_2C_{19}H_{37}PSe$ .... $[3.3.1]\text{-}9\text{-}BC_8H_{14}\text{-}9\text{-}Se\text{-}9\text{-}[3.3.1]\text{-}9\text{-}BC_8H_{14} \cdot P(CH_3)_3$ B: B Comp.SVol.4/4-170

$B_2C_{19}H_{41}N_3$ ..... $1\text{-}(C_4H_9\text{-}t)\text{-}2\text{-}(i\text{-}C_3H_7)_2N\text{-}$ $3\text{-}[2,2,6,6\text{-}(CH_3)_4\text{-}NC_5H_6\text{-}1]\text{-}NB_2$ .... B: B Comp.SVol.4/3a-194

$B_2C_{19}H_{42}N_2$ ..... $1\text{-}[i\text{-}C_3H_7\text{-}C(CH_3)_2\text{-}BH\text{-}N(C_4H_9\text{-}t)\text{-}BH]\text{-}$ $2,2,6,6\text{-}(CH_3)_4\text{-}NC_5H_6$ .... B: B Comp.SVol.4/3a-171

$B_2C_{19}H_{42}N_2O$ .... $t\text{-}C_4H_9\text{-}NH\text{-}B(C_4H_9\text{-}n)\text{-}N(C_4H_9\text{-}t)\text{-}B(C_4H_9\text{-}n)$ $\text{-}O\text{-}C(CH_3)\text{=}CH_2$ .... B: B Comp.SVol.4/3b-53

$B_2C_{19}H_{43}N_3$ ..... $1\text{-}[(i\text{-}C_3H_7)_2N\text{-}BH\text{-}N(C_4H_9\text{-}t)\text{-}BH]\text{-}2,2,6,6\text{-}(CH_3)_4\text{-}NC_5H_6$ B: B Comp.SVol.4/3a-172

$B_2C_{19}H_{44}N_2Sn$ ... $[(C_2H_5)_2N]_2B\text{-}C[Sn(CH_3)_3]\text{=}C(C_2H_5)\text{-}B(C_2H_5)_2$ B: B Comp.SVol.4/3a-168

$B_2C_{19}H_{45}NSn$ .... $(n\text{-}C_4H_9)_2B\text{-}N[Sn(CH_3)_3]\text{-}B(C_4H_9\text{-}n)_2$ .... B: B Comp.SVol.4/3a-242 Sn: Org.Comp.18-23, 27

$B_2C_{19}H_{45}N_5Si$.... $(CH_3)_3Si\text{-}N(C_4H_9\text{-}t)\text{-}B(N_3)\text{-}N(C_4H_9\text{-}t)\text{-}B(C_4H_9\text{-}n)_2$ B: B Comp.SVol.4/3a-157

$B_2C_{19}H_{46}N_2OSi_2$.. $3\text{-}CH_3\text{-}BC_4H_7\text{-}1\text{-}N(C_4H_9\text{-}t)\text{-}B(C_4H_9\text{-}n)$ $\text{-}N[Si(CH_3)_3]\text{-}O\text{-}Si(CH_3)_3$................. B: B Comp.SVol.4/3a-172

$B_2C_{19}H_{47}LiN_4$.... $[\text{-}N(CH_3)_2\text{-}CH_2CH_2\text{-}N(CH_3)_2\text{-}]Li\text{-}N(C_4H_9\text{-}t)[$ $\text{-}B(CH_3)(C_2H_5)\text{-}N(C_4H_9\text{-}t)\text{-}B(C_2H_5)\text{-}]$ .... B: B Comp.SVol.4/3a-201/2

$B_2C_{20}Cl_2H_{45}N_3$... $ClB[N(n\text{-}C_4H_9)_2]\text{-}N(n\text{-}C_4H_9)\text{-}BCl[N(n\text{-}C_4H_9)_2]$ B: B Comp.SVol.3/4-59

$B_2C_{20}Cl_2H_{46}N_2Si_2$ $[(i\text{-}C_3H_7)_2N\text{-}BCl]_2C\text{=}C[Si(CH_3)_3]_2$ .... B: B Comp.SVol.4/4-61

$B_2C_{20}Cl_2H_{46}N_2Si_5$ $[1,3\text{-}((CH_3)_3Si)_2\text{-}2\text{-}CH_3\text{-}1,2\text{-}NBC_3H_2\text{-}3]_2SiCl_2$ B: B Comp.SVol.4/3a-253

$B_2C_{20}FH_{36}N_3S$ ... $[((CH_3)_2N)_3S][B_2C_{10}FH_6(CH_3)_4]$ .... B: B Comp.SVol.4/3b-228

$B_2C_{20}F_6H_{20}N_2O_2$ $(CH_3)_2B\text{-}N(C_6H_4\text{-}2\text{-}CF_3)\text{-}C(O)\text{-}C(O)$ $\text{-}N(C_6H_4\text{-}2\text{-}CF_3)\text{-}B(CH_3)_2$ .... B: B Comp.SVol.3/3-161

$B_2C_{20}F_8FeH_{28}$ ... $[(1,2,4,5\text{-}(CH_3)_4\text{-}C_6H_2)_2Fe][BF_4]_2$ .... Fe: Org.Comp.B19-347, 367

$B_2C_{20}F_8FeH_{32}N_4$. $[Fe(C_5H_4\text{-}C(N(CH_3)_2)_2)_2][BF_4]_2$ .... Fe: Org.Comp.A9-15

$B_2C_{20}FeH_{18}I_4N_2$.. $Fe(C_5H_4BI_2)_2 \cdot 2\ NC_5H_5$ .... Fe: Org.Comp.A9-282

$B_2C_{20}H_{22}N_4$ ..... $(CH_3)_2B[\text{-}1,2\text{-}N_2C_3H_3\text{-}]_2B(C_6H_5)_2$ .... B: B Comp.SVol.4/3b-35

– .... $C_6H_5\text{-}B(CH_3)[\text{-}1,2\text{-}N_2C_3H_3\text{-}]_2B(CH_3)\text{-}C_6H_5$ ... B: B Comp.SVol.4/3b-35

$B_2C_{20}H_{22}N_4O_2$ ... $(CH_3O)_2B[\text{-}1,2\text{-}N_2C_3H_3\text{-}]_2B(C_6H_5)_2$.......... B: B Comp.SVol.4/3b-59

$B_2C_{20}H_{22}N_4O_2$ ... $CH_3O-B(C_6H_5)[-1,2-N_2C_3H_3-]_2B(C_6H_5)-OCH_3$ B: B Comp.SVol.4/3b-59
$B_2C_{20}H_{25}I_5N_2$ .... $[(CH_3)_5C_5-B(NC_5H_5)_2-I][BI_4]$ .... B: B Comp.SVol.4/4-110
$B_2C_{20}H_{26}N_2O_2$ ... $(CH_3)_2B-N(CH_2C_6H_5)-C(O)-C(O)-N(CH_2C_6H_5)-B(CH_3)_2$
B: B Comp.SVol.3/3-161
$B_2C_{20}H_{28}N_2$ ..... $[-B(C_6H_5)-N(C_4H_9-t)-B(C_6H_5)-N(C_4H_9-t)-]$ ... B: B Comp.SVol.4/3a-196
$B_2C_{20}H_{29}N$ ...... $2-CH_3C_6H_4-B(C_2H_5)-N(C_6H_4CH_3-2)-B(C_2H_5)_2$ B: B Comp.SVol.3/3-161
– .... $(C_6H_5)_2B-NH-B(C_4H_9-t)_2$................ B: B Comp.SVol.4/3a-241
$B_2C_{20}H_{30}N_8$ ..... $[(CH_3)_2C_3HN_2]BH[-(CH_3)_2C_3HN_2-$
$BH(C_3HN_2(CH_3)_2)-(CH_3)_2C_3HN_2-]$.......... B: B Comp.SVol.3/3-200
$B_2C_{20}H_{31}N_3Si$.... $1-[(CH_3)_2N-B(C_6H_5)]-2-C_6H_5-3-(CH_3)_3Si-$
$1,3,2-N_2BC_3H_6$ .... B: B Comp.SVol.4/3a-177
$B_2C_{20}H_{32}MnN_8$... $Mn[(3,5-(CH_3)_2-1,2-N_2C_3H-1)_2BH_2]_2$ Mn:MVol.D8-9/10
$B_2C_{20}H_{32}N_{10}$ .... $(1,2-N_2C_3H_3-1)_2B[-1,2-N_2C_3H_3-]_2B[N(C_2H_5)_2]_2$
B: B Comp.SVol.4/3b-35
$B_2C_{20}H_{33}N_2Se$ ... $[-BC_8H_{14}-C_4H_5N_2-BC_8H_{14}-Se-]$............ B: B Comp.SVol.4/4-170
$B_2C_{20}H_{34}N_2$ ..... $t-C_4H_9-NH-B(C_2H_5)-N(C_4H_9-t)-B(C_2H_5)-CC-C_6H_5$
B: B Comp.SVol.4/3a-172
$B_2C_{20}H_{34}N_2O_3$ ... $2,2,6,6-(CH_3)_4-NC_5H_6-1-B(OCH_3)$
$-N(C_4H_9-t)-2-(1,3,2-O_2BC_6H_4)$ .... B: B Comp.SVol.4/3b-53
$B_2C_{20}H_{34}N_2Ru$ ... $[C_6(CH_3)_6]Ru[C_4H_4B_2(N(CH_3)_2)_2]$............ B: B Comp.SVol.4/4-180
$B_2C_{20}H_{34}N_2S$ .... $[3.3.1]-9-BC_8H_{14}-9-S-9-[3.3.1]-9-BC_8H_{14}$
$\cdot$ $1,2-N_2C_3H_3-3-CH_3$ .... B: B Comp.SVol.4/4-132
– .... $[-S-BC_8H_{14}-N_2C_3H_2(CH_3)-B(C_8H_{15}-c)-]$ ..... B: B Comp.SVol.4/4-143/4
$B_2C_{20}H_{36}N_2$ ..... $1,8-[(i-C_3H_7)_2N]_2-1,8-B_2C_8H_8$............. B: B Comp.SVol.4/3a-247
$B_2C_{20}H_{38}S_2$ ..... $[-S(C_2H_5)-BC_8H_{14}-S(C_2H_5)-BC_8H_{14}-]$ .... B: B Comp.SVol.4/4-132/3
$B_2C_{20}H_{40}N_2O_2$ ... $CH_3-OC(=O)-CC-B(C_4H_9-t)[-N(C_4H_9-t)$
$-B(C_4H_9-t)-NH(C_4H_9-t)-]$ .... B: B Comp.SVol.4/3a-201
$B_2C_{20}H_{42}N_2O$ .... $[-CH(C(CH_3)=CH_2)-N(C_4H_9-t)-B(C_4H_9-n)$
$-N(C_4H_9-t)-B(C_4H_9-n)-O-]$.............. B: B Comp.SVol.4/3b-54
$B_2C_{20}H_{42}N_2Si_2$ ... $4,5-(C_2H_5)_2-2,2,3-(CH_3)_3-1,2,5-NSiBC_2-1-$
$CH_2CH_2-1-[1,2,5-NSiBC_2-2,2,3-(CH_3)_3-4,5-(C_2H_5)_2]$
B: B Comp.SVol.4/3a-253
$B_2C_{20}H_{42}N_8$ ..... $(n-C_4H_9)_2B[-N_3C_2H(NH_2)-]_2B(C_4H_9-n)_2$...... B: B Comp.SVol.4/3b-37/8
$B_2C_{20}H_{43}NSn_2$ ... $1-(C_2H_5)_2N-2,5-[(CH_3)_3Sn]_2-3-(C_2H_5)_2B-4-C_2H_5-BC_4$
B: B Comp.SVol.4/3a-246
– .... $1-(C_2H_5)_2N-3,5-[(CH_3)_3Sn]_2-2-(C_2H_5)_2B-4-C_2H_5-BC_4$
B: B Comp.SVol.4/3a-246
– .... $(CH_3)_3Sn-CC-B[N(C_2H_5)_2]-C[Sn(CH_3)_3]=$
$C(C_2H_5)-B(C_2H_5)_2$ .... B: B Comp.SVol.4/3a-244
$B_2C_{20}H_{43}N_3O_2$ ... $2,2,6,6-(CH_3)_4-NC_5H_6-1-O-B(CH_3)-NH$
$-B(CH_3)-O-1-NC_5H_6-2,2,6,6-(CH_3)_4$........ B: B Comp.SVol.4/3b-62
$B_2C_{20}H_{44}N_2Si_3$ ... $4,5-(C_2H_5)_2-2,2,3-(CH_3)_3-1,2,5-NSiBC_2-1-$
$Si(CH_3)_2-1-[1,2,5-NSiBC_2-2,2,3-(CH_3)_3-4,5-(C_2H_5)_2]$
B: B Comp.SVol.4/3a-253
$B_2C_{20}H_{44}N_4$ ..... $N_3-B(n-C_5H_{11})-N(n-C_5H_{11})-B(n-C_5H_{11})_2$ .... B: B Comp.SVol.3/3-106
$B_2C_{20}H_{45}N$ ...... $n-C_5H_{11}-B(C_4H_9-t)-N(CH_3)-B(C_5H_{11}-n)_2$..... B: B Comp.SVol.4/3a-243
– .... $[(n-C_4H_9)_2B]_2N-C_4H_9-t$ .... B: B Comp.SVol.3/3-160
$B_2C_{20}H_{46}K_2N_2Si_2$ $K_2[1,3-((i-C_3H_7)_2N)_2-1,3-B_2C_2-2,4-(Si(CH_3)_3)_2]$
B: B Comp.SVol.4/3a-245
$B_2C_{20}H_{46}N_2Si_2$ ... $1,2-[(i-C_3H_7)_2N]_2-1,2-B_2C_2-3,4-[Si(CH_3)_3]_2$ .. B: B Comp.SVol.4/3a-244

$B_2C_{20}H_{46}N_2Si_2$ . . . 1,3-[(i-$C_3H_7)_2N]_2$-1,3-B_2C_2-2,4-[Si$(CH_3)_3]_2$ . . B: B Comp.SVol.4/3a-244/5
$B_2C_{20}H_{46}N_2Si_2^{2-}$ [1,3-((i-$C_3H_7)_2N)_2$-1,3-B_2C_2-2,4-(Si$(CH_3)_3)_2]^{2-}$
B: B Comp.SVol.4/3a-245
$B_2C_{20}H_{46}N_4$ . . . . . 1,3-(C_4H_9-t$)_2$-2,4-[(i-$C_3H_7)_2N]_2$-1,3,2,4-N_2B_2 B: B Comp.SVol.4/3a-198
$B_2C_{20}H_{48}Li_2N_6$ . . . [Li(($CH_3)_2NCH_2CH_2N(CH_3)_2$)]$_2$
[1,2-(($CH_3)_2N)_2$-1,2-$B_2C_4H_4$]. . . . . B: B Comp.SVol.4/3b-49
$B_2C_{20}H_{48}N_2Si_2$ . . . 1,3-[(i-$C_3H_7)_2N]_2$-1,3-$B_2C_2H_2$-2,4-[Si$(CH_3)_3]_2$ B: B Comp.SVol.4/3a-245
$B_2C_{20}H_{50}N_2OSi_2$. . ($CH_3)_3$Si-O-N[Si$(CH_3)_3$]-B(C_3H_7-i)-N(C_3H_7-i)
-B(C_4H_9-n$)_2$ . . . . B: B Comp.SVol.4/3a-172
$B_2C_{20}H_{50}N_4Si_2$ . . . 1,3-(C_3H_7-i$)_2$-2,4-[($CH_3)_3$Si-N(C_4H_9-t)]$_2$-1,3,2,4-N_2B_2
B: B Comp.SVol.4/3a-198
$B_2C_{20}H_{52}N_2O_6$ . . . 2 H_3B-NH_3 · 2,2,3,3,11,11,12,12-$(CH_3)_8$-
1,4,7,10,13,16-$O_6C_{12}H_{16}$ . . . . B: B Comp.SVol.4/3b-2
$B_2C_{20}H_{54}N_4Si_4$ . . . 1,3-(t-$C_4H_9)_2$-2,4-[(($CH_3)_3Si)_2N]_2$-1,3,2,4-N_2B_2
B: B Comp.SVol.3/3-136
– . . . . 1,3-[($CH_3)_3Si]_2$-2,4-[($CH_3)_3$Si-N(C_4H_9-t)]$_2$-1,3,2,4-N_2B_2
B: B Comp.SVol.4/3a-198
$B_2C_{21}ClH_{41}N_2$. . . . 9-[2,2,6,6-$(CH_3)_4$-NC_5H_6-1-BCl-N(C_4H_9-t)]
-[3.3.1]-9-BC_8H_{14} . . . . B: B Comp.SVol.4/4-64
$B_2C_{21}F_3H_{35}N_2$ . . . 2,6-(i-$C_3H_7)_2$-C_6H_3-N(BF_2)-BF-1-NC_5H_6-2,2,6,6-$(CH_3)_4$
B: B Comp.SVol.4/3b-244
$B_2C_{21}F_8H_{35}MoN_7$ [Mo(CN-$C_2H_5)_7$][$BF_4]_2$. Mo:Org.Comp.5-84, 87
$B_2C_{21}H_{21}N_3S$ . . . . CH_3B[-S-C(=NC_6H_5)-NC_6H_5-BCH_3-NC_6H_5-]. . B: B Comp.SVol.3/4-128
$B_2C_{21}H_{21}N_7$ . . . . . [-B(C_6H_5)($N_2C_3H_3$)-NH_2-B(C_6H_5)($N_2C_3H_3$)-$N_2C_3H_3$-]
B: B Comp.SVol.4/3b-36/7
$B_2C_{21}H_{29}N$ . . . . . . 9-[(t-$C_4H_9)_2$B-N(CH_3)]-9-$BC_{12}H_8$. . . . . B: B Comp.SVol.4/3a-242
$B_2C_{21}H_{36}N_2S$ . . . . [3.3.1]-9-BC_8H_{14}-9-S-9-[3.3.1]-9-BC_8H_{14}
· 1,2-$N_2C_3H_2$-3,5-$(CH_3)_2$. . . . . B: B Comp.SVol.4/4-132
$B_2C_{21}H_{42}N_2$ . . . . . 1-[(3.3.1)-9-BC_8H_{14}-9-N(C_4H_9-t)-BH]-
2,2,6,6-$(CH_3)_4$-NC_5H_6 . . . . B: B Comp.SVol.4/3a-171
$B_2C_{21}H_{42}O_5PRe$ . . [(n-$C_4H_9)_4$P][(CO)$_5$Re($BH_3)_2$] . . . . Re: Org.Comp.2-169
$B_2C_{21}H_{45}N_3O_2$ . . . (n-$C_4H_9)_2$B-NCH_3-CO-NCH_3-C[OB(n-$C_4H_9)_2$]=NCH_3
B: B Comp.SVol.3/3-161
$B_2C_{21}H_{46}N_2O$ . . . . [-CH(C_4H_9-t)-N(C_4H_9-t)-B(C_4H_9-n)-N(C_4H_9-t)
-B(C_4H_9-n)-O-]. . . . . B: B Comp.SVol.4/3b-54
$B_2C_{22}Cl_2H_{45}N_3$. . . [2,2,6,6-$(CH_3)_4$-NC_5H_6-1-BCl]$_2$N-C_4H_9-t . . . . . B: B Comp.SVol.3/4-59
B: B Comp.SVol.4/4-64
$B_2C_{22}Cl_8FeH_{38}Li_2N_6P_6$
Li_2[Fe(C_5H_4-$P_3N_3(Cl)_4$-B$(C_2H_5)_3)_2$]. . . . . Fe: Org.Comp.A10-84
$B_2C_{22}Cl_8FeH_{38}N_6P_6^{2-}$
[Fe(C_5H_4-$P_3N_3(Cl)_4$-B$(C_2H_5)_3)_2]^{2-}$ . . . . Fe: Org.Comp.A10-84
$B_2C_{22}CoFeH_{33}$ . . . C_5H_5Fe(1,3,4,5-$(C_2H_5)_4$-2-(CH_3)C_3B_2-1,3)CoC_5H_5
Fe: Org.Comp.B17-177
$B_2C_{22}CoFeH_{33}^+$ . . [C_5H_5Fe(1,3,4,5-$(C_2H_5)_4$-2-(CH_3)C_3B_2-1,3)Co$C_5H_5]^+$
Fe: Org.Comp.B17-177
$B_2C_{22}CoFeH_{33}^-$ . . [C_5H_5Fe(1,3,4,5-$(C_2H_5)_4$-2-(CH_3)C_3B_2-1,3)Co$C_5H_5]^-$
Fe: Org.Comp.B17-177
$B_2C_{22}F_3H_{41}N_2O_3S$ 2,2,6,6-$(CH_3)_4$-NC_5H_6-1-B[O-S(O)$_2$-CF_3]
-N(C_4H_9-t)-9-[(3.3.1)-9-BC_8H_{14}] . . . . B: B Comp.SVol.4/3b-53
$B_2C_{22}F_6H_{16}N_4O_4$ (CF_3-COO)$_2$B[-1,2-$N_2C_3H_3$-]$_2$B($C_6H_5)_2$ . . . . . . B: B Comp.SVol.4/3b-59

$B_2C_{22}F_8FeH_{23}NO_6$ $[((HO)(CH_3O)_2C_{16}H_{13}N-CH_3)Fe(CO)_3][BF_4]_2$ .. Fe: Org.Comp.B15-133, 184
$B_2C_{22}F_{12}H_{36}N_2O_{12}S_4$
$[(-C(CH_3)_2-(CH_2)_3-C(CH_3)_2-)N=B=N(-C(CH_3)_2$
$-(CH_2)_3-C(CH_3)_2-)][B(OSO_2CF_3)_4]$ .... B: B Comp.SVol.3/3-205
$B_2C_{22}FeH_{20}$ ..... $Fe(C_5H_5BC_6H_5-1)_2$ .... Fe: Org.Comp.B17-329, 334
$B_2C_{22}FeH_{20}N_4$ ... $Fe(C_5H_4BNHC_6H_4NH)_2$ .... Fe: Org.Comp.A9-280
$B_2C_{22}H_{22}N_8$ ..... $(C_4H_4N)_2B[-N_2C_3H_3-B(C_4H_4N)_2-N_2C_3H_3-]$.... B: B Comp.SVol.3/3-199
$B_2C_{22}H_{23}N_5O_9$ ... $(CH_3-COO)_4B_2C_2HN_5O(C_6H_5)_2$ .... B: B Comp.SVol.4/3b-57
$B_2C_{22}H_{23}N_7$ ..... $[-B(C_6H_5)(N_2C_3H_3)-NH(CH_3)-B(C_6H_5)(N_2C_3H_3)$
$-N_2C_3H_3-]$ .... B: B Comp.SVol.4/3b-36/7
$B_2C_{22}H_{24}Mn_2N_{12}O_9$
$[Mn_2O((HCOO)_2(1,2-N_2C_3H_3-1)_3BH)_2]$ .... Mn: MVol.D8-10/2
$B_2C_{22}H_{26}N_4$ ..... $(C_2H_5)_2B[-1,2-N_2C_3H_3-]_2B(C_6H_5)_2$ .... B: B Comp.SVol.4/3b-35
– .... $C_6H_5-B(C_2H_5)[-1,2-N_2C_3H_3-]_2B(C_2H_5)-C_6H_5$.. B: B Comp.SVol.4/3b-35
$B_2C_{22}H_{26}N_4S_2$ ... $C_2H_5-S-B(C_6H_5)[-1,2-N_2C_3H_3-]_2B(C_6H_5)-S-C_2H_5$
B: B Comp.SVol.3/4-111
B: B Comp.SVol.4/3b-36
$B_2C_{22}H_{26}N_4S_4$ ... $C_2H_5S(C_6H_5S)B[-N_2C_3H_3-B(SC_2H_5)SC_6H_5-N_2C_3H_3-]$
B: B Comp.SVol.3/4-129
$B_2C_{22}H_{26}N_{12}$ .... $(1,2-N_2C_3H_3-1)_2B[-1,2-N_2C_3H(CH_3)_2-]_2$
$B[-1-(1,2-N_2C_3H_3)]_2$ .... B: B Comp.SVol.4/3b-34
$B_2C_{22}H_{28}N_4S_2$ ... $(C_6H_5)(C_2H_5S)B[-N_2C_3H_3-B(C_6H_5)(SC_2H_5)-N_2C_3H_5-]$
B: B Comp.SVol.3/3-200
$B_2C_{22}H_{28}N_6$ ..... $[(CH_3)_2N](C_6H_5)B[-N_2C_3H_3-B(C_6H_5)(N(CH_3)_2)-N_2C_3H_3-]$
B: B Comp.SVol.3/3-199
$B_2C_{22}H_{32}N_2O_2$ ... $(CH_3)_2B-N[C_6H_5-2,6-(CH_3)_2]-C(O)-C(O)$
$-N[C_6H_3-2,6-(CH_3)_2]-B(CH_3)_2$............. B: B Comp.SVol.3/3-161
$B_2C_{22}H_{34}N_2O_4$ ... $[-O-(1,2-C_6H_4)-O-]B[N(CH_3)_3]CH_2CH_2CH_2$
$CH_2B[N(CH_3)_3][-O-(1,2-C_6H_4)-O-]$......... B: B Comp.SVol.3/3-221
$B_2C_{22}H_{36}N_2$ ..... $2,10-[(i-C_3H_7)_2N]_2-2,10-B_2C_{10}H_8$ .... B: B Comp.SVol.4/3a-248
$B_2C_{22}H_{38}N_2^+$ .... $[(C_2H_5)_3B-NC_5H_4-C_5H_4N-B(C_2H_5)_3]^+$, radical cation
B: B Comp.SVol.3/3-192
$B_2C_{22}H_{38}N_2^-$..... $[(C_2H_5)_3B-NC_5H_4-C_5H_4N-B(C_2H_5)_3]^-$, radical anion
B: B Comp.SVol.3/3-215
$B_2C_{22}H_{38}N_2O_2Se_2$ $[2-CH_3-4-(O=)-5,6-(C_2H_5)_2-$
$1,3,2-SeNBC_3-3-CH_2CH_2CH_2-]_2$........... B: B Comp.SVol.4/4-172/3
$B_2C_{22}H_{42}N_4$ ..... $(n-C_4H_9)_2B[-N_2C_3H_3-B(n-C_4H_9)_2-N_2C_3H_3-]$ .. B: B Comp.SVol.3/3-199
$B_2C_{22}H_{42}NiSe_2Si_2$ $B_2C_4NiSe_2Si_2(C_2H_5)_4(CH_3)_4[C(CH_3)=CH_2]_2$.... B: B Comp.SVol.4/4-180
$B_2C_{22}H_{45}N_3$ ..... $1-(C_4H_9-t)-2,3-[2,2,6,6-(CH_3)_4-NC_5H_6-1]_2-NB_2$
B: B Comp.SVol.3/3-134/5
B: B Comp.SVol.4/3a-194
$B_2C_{22}H_{46}N_2$ ..... $t-C_4H_9-CC-B(C_4H_9-t)[-N(C_4H_9-t)-B(C_4H_9-t)$
$-NH(C_4H_9-t)-]$ .... B: B Comp.SVol.4/3a-201
– .... $[BC_8H_{15}]-N(CH_3)_2-CH_2CH_2-N(CH_3)_2-[BC_8H_{15}]$
B: B Comp.SVol.3/3-190
$B_2C_{22}H_{46}N_4O_4$ ... $1,4,5,6-(t-C_4H_9)_4-2,3-[C_2H_5-OC(O)]_2-$
$[2.2.0]-1,2,3,5,4,6-N_4B_2$ .... B: B Comp.SVol.4/3a-200
$B_2C_{22}H_{46}N_4S_4$ ... $[(C_2H_5)_2S]_2B[-N_2C_3H_3-B(S(C_2H_5)_2)_2-N_2C_3H_3-]$
B: B Comp.SVol.3/3-200

$B_2C_{22}H_{48}N_2$ ..... 1-[(i-C_3H_7)$_2$B-N(C_4H_9-t)-B(C_3H_7-i)]-2,2,6,6-(CH_3)$_4$-NC_5H_6 .... B: B Comp.SVol.4/3a-172

$B_2C_{22}H_{48}N_2O$ .... t-C_4H_9-NH-B(C_4H_9-n)-N(C_4H_9-t)-B(C_4H_9-n)-O-C(C_4H_9-t)=CH_2..................... B: B Comp.SVol.4/3b-53

$B_2C_{22}H_{54}N_4Si_2$ ... 1,3-(C_4H_9-n)$_2$-2,4-[(CH_3)$_3$Si-N(C_4H_9-t)]$_2$-1,3,2,4-N_2B_2 B: B Comp.SVol.4/3a-198

– .... 1,3-(C_4H_9-t)$_2$-2,4-[(CH_3)$_3$Si-N(C_4H_9-t)]$_2$-1,3,2,4-N_2B_2 B: B Comp.SVol.4/3a-198

$B_2C_{23}GeH_{24}O_4$ ... Ge(C_6H_5)$_3$CH(B(-OCH_2CH_2O-))$_2$............ Ge: Org.Comp.3-68

$B_2C_{23}H_{25}N_5O_9$ ... (CH_3-COO)$_4B_2C_2N_5$O(CH_3)(C_6H_5)$_2$ .... B: B Comp.SVol.4/3b-57

$B_2C_{23}H_{25}N_7$ ..... [(CH_3)$_2$N](C_6H_5)B[-$N_2C_3H_3$-B($C_3H_3N_2$)(C_6H_5)-$N_2C_3H_3$-] B: B Comp.SVol.3/3-199

$B_2C_{23}H_{30}N_2$ ..... [-N($CH_2C_6H_5$)-CH(C_6H_5)-N($CH_2C_6H_5$)-CH_2CH_2-] · 2 BH_3..................... B: B Comp.SVol.4/3b-12

$B_2C_{23}H_{33}N_2Se$ ... [-BC_8H_{14}-$C_7H_5N_2$-BC_8H_{14}-Se-]............ B: B Comp.SVol.4/4-170

$B_2C_{23}H_{34}N_2S$ .... [3.3.1]-9-BC_8H_{14}-9-S-9-[3.3.1]-9-BC_8H_{14} · 1,2-$N_2C_7H_6$ B: B Comp.SVol.4/4-132

$B_2C_{23}H_{38}Pt_2$ ..... (C_8H_{12})Pt[C_3B_2H(CH_3)$_4$]Pt(C_8H_{13})........... B: B Comp.SVol.4/4-180

$B_2C_{23}H_{40}N_2O_3$ ... 2,2,6,6-(CH_3)$_4$-NC_5H_6-1-B(O-C_4H_9-n)-N(C_4H_9-t)-2-(1,3,2-$O_2BC_6H_4$)............ B: B Comp.SVol.4/3b-53

$B_2C_{23}H_{42}N_2O$ .... [-CH(C_6H_5)-N(C_4H_9-t)-B(C_4H_9-n)-N(C_4H_9-t)-B(C_4H_9-n)-O-]...................... B: B Comp.SVol.4/3b-54

$B_2C_{23}H_{50}N_2$ ..... 2,2,6,6-(CH_3)$_4$-NC_5H_6-1-BH-N(C_4H_9-t)-B[CH(CH_3)-C_3H_7-i]$_2$ .... B: B Comp.SVol.4/3a-171

$B_2C_{23}H_{55}LiN_4$.... [-N(CH_3)$_2$-CH_2CH_2-N(CH_3)$_2$-]Li-N(C_4H_9-t)[-B(CH_3)(C_4H_9-n)-N(C_4H_9-t)-B(C_4H_9-n)-]... B: B Comp.SVol.4/3a-201/2

$B_2C_{24}Cl_4F_8Ga_2H_{16}N_4$
[Ga(1,10-$N_2C_{12}H_8$)$_2$][BF_4]$_2$[$GaCl_4$] .... Ga: SVol.D1-316

$B_2C_{24}F_6H_{18}MnN_2O_2$
[Mn(-O-2-$C_{10}H_6$-1-CH=N-C_2H_4-N=CH-1-$C_{10}H_6$-2-O-)(BF_3)$_2$].................... Mn: MVol.D6-185, 186

$B_2C_{24}F_6H_{44}N_4$ ... [CH_2=C(NC_5H_{10})-NC_5H_{10}-BF_2-NC_5H_{10}-C(NC_5H_{10})=CH_2][BF_4] .... B: B Comp.SVol.3/3-379

$B_2C_{24}F_8FeH_{25}NO_7$ [((CH_3COO)(CH_3O)$_2C_{16}H_{13}$N-CH_3)Fe(CO)$_3$][BF_4]$_2$ Fe: Org.Comp.B15-133, 184

$B_2C_{24}F_8H_{42}MnN_6O_6S_6$
[Mn(1,4-SNC_4H_7(=O-3))$_6$][BF_4]$_2$ .... Mn: MVol.D7-240/1

$B_2C_{24}F_8H_{48}MnO_6S_{12}$
[Mn(1,3-$S_2C_4H_8$(=O-1))$_6$][BF_4]$_2$............. Mn: MVol.D7-104

– .... [Mn(1,4-$S_2C_4H_8$(=O-1))$_6$][BF_4]$_2$............. Mn: MVol.D7-104/5

$B_2C_{24}F_8H_{72}MnN_{12}O_4P_4$
[Mn(O=P(N(CH_3)$_2$)$_3$)$_4$][BF_4]$_2$ .... Mn: MVol.D8-172/4

$B_2C_{24}F_8H_{72}MnN_{14}O_6P_6$
[Mn((O=P(N(CH_3)$_2$)$_2$NCH_3)$_2$P(N(CH_3)$_2$)=O)$_2$][BF_4]$_2$ Mn: MVol.D8-181

$B_2C_{24}F_{20}N_4$...... (C_6F_5)$_2$B-N[-B(C_6F_5)-N(C_6F_5)-N=N-]........ B: B Comp.SVol.3/3-137/8
F: PFHOrg.SVol.4-46, 71, 78

$B_2C_{24}FeH_{24}$ ..... Fe[BC_5H_4(CH_3)(C_6H_5)]$_2$ .... Fe: Org.Comp.B17-329/30

$B_2C_{24}H_{21}N$ ...... (C_6H_5)$_2$B-NH-B(C_6H_5)$_2$ .... B: B Comp.SVol.4/3a-241

$B_2C_{24}H_{22}N_8$ ..... 1,2-$N_2C_3H_3$-1-$B(C_6H_5)[$-1,2-$N_2C_3H_3$-$]_2$
$B(C_6H_5)$-1-(1,2-$N_2C_3H_3$) .... B: B Comp.SVol.3/3-194/7, 199
B: B Comp.SVol.4/3b-35

$B_2C_{24}H_{24}MnN_{16}$ .. $Mn[(1,2\text{-}N_2C_3H_3\text{-}1)_4B]_2$ Mn:MVol.D8-9/10

– .... $Mn[(1,3\text{-}N_2C_3H_3\text{-}1)_4B]_2$ Mn:MVol.D8-18/9

$B_2C_{24}H_{24}MnO_{28}$ .. $Mn[B(C_6H_6O_7)_2]_2 \cdot 8\,H_2O$ =
$[Mn(H_2O)_6][(B(C_6H_6O_7)_2)(H_2O)]_2$ Mn:MVol.D8-3/4

$B_2C_{24}H_{26}MnN_6O_{12}$
$Mn[(2,5\text{-}(O{=})_2\text{-}NC_4H_4\text{-}1)_3BH]_2$ Mn:MVol.D8-7/8

$B_2C_{24}H_{30}N_{12}$ .... $(3\text{-}CH_3\text{-}1,2\text{-}N_2C_3H_2\text{-}1)_2B[\text{-}1,2\text{-}N_2C_3H_2(CH_3)$
$\text{-}]_2B[\text{-}1\text{-}(1,2\text{-}N_2C_3H_2\text{-}3\text{-}CH_3)]_2$ .... B: B Comp.SVol.4/3b-35

$B_2C_{24}H_{32}MnN_{12}$ .. $Mn[(2\text{-}CH_3\text{-}1,3\text{-}N_2C_3H_2\text{-}1)_3BH]_2$ Mn:MVol.D8-18/9

$B_2C_{24}H_{34}N_2O_2$ ... $(CH_3)_2B\text{-}N[C_6H_2\text{-}2,4,6\text{-}(CH_3)_3]\text{-}C(O)\text{-}C(O)$
$\text{-}N[C_6H_2\text{-}2,4,6\text{-}(CH_3)_3]\text{-}B(CH_3)_2$ .... B: B Comp.SVol.3/3-161

$B_2C_{24}H_{37}N$ ...... $[C_2H_5][2,4,6\text{-}(CH_3)_3C_6H_2]B\text{-}N[C_6H_2$
$\text{-}2,4,6\text{-}(CH_3)_3]\text{-}B(C_2H_5)_2$ .... B: B Comp.SVol.3/3-161

$B_2C_{24}H_{38}N_8S_2$ ... $1,2\text{-}N_2C_3H_3\text{-}1\text{-}B(C_3H_6\text{-}SCH_3)$
$[\text{-}1,2\text{-}N_2C_3H(CH_3)_2\text{-}]_2B(C_3H_6\text{-}SCH_3)\text{-}1\text{-}(1,2\text{-}N_2C_3H_3)$
B: B Comp.SVol.4/3b-36

$B_2C_{24}H_{40}MnO_{36}$ .. $[Mn(H_2O)_6][(B(C_6H_6O_7)_2)(H_2O)]_2$
= $Mn[B(C_6H_6O_7)_2]_2 \cdot 8\,H_2O$ Mn:MVol.D8-3/4

$B_2C_{24}H_{40}N_2Rh_2$ .. $(C_8H_{12})_2Rh_2[C_4H_4B_2(N(CH_3)_2)_2]$ .... B: B Comp.SVol.4/4-180

$B_2C_{24}H_{42}N_2$ ..... $C_6H_5\text{-}CC\text{-}B(C_4H_9\text{-}t)[\text{-}N(C_4H_9\text{-}t)\text{-}B(C_4H_9\text{-}t)$
$\text{-}NH(C_4H_9\text{-}t)\text{-}]$ .... B: B Comp.SVol.4/3a-201

$B_2C_{24}H_{42}N_2Si_2$ ... $1,4\text{-}[4,5\text{-}(C_2H_5)_2\text{-}2,2,3\text{-}(CH_3)_3\text{-}1,2,5\text{-}NSiBC_2\text{-}1]_2\text{-}C_6H_4$
B: B Comp.SVol.4/3a-253

$B_2C_{24}H_{44}N_2O$ .... $t\text{-}C_4H_9\text{-}NH\text{-}B(C_4H_9\text{-}n)\text{-}N(C_4H_9\text{-}t)\text{-}B(C_4H_9\text{-}n)$
$\text{-}O\text{-}C(C_6H_5){=}CH_2$ .... B: B Comp.SVol.4/3b-53

– .... $[\text{-}C(CH_3)(C_6H_5)\text{-}N(C_4H_9\text{-}t)\text{-}B(C_4H_9\text{-}n)$
$\text{-}N(C_4H_9\text{-}t)\text{-}B(C_4H_9\text{-}n)\text{-}O\text{-}]$ .... B: B Comp.SVol.4/3b-54

$B_2C_{24}H_{47}NSi_2$.... $1\text{-}(t\text{-}C_4H_9)\text{-}2\text{-}[2,4,6\text{-}(CH_3)_3\text{-}NC_5H_2\text{-}1\text{-}$
$B(C_4H_9\text{-}t){=}]\text{-}3,3\text{-}[(CH_3)_3Si]_2\text{-}BC_2$ .... B: B Comp.SVol.4/3a-217/8

$B_2C_{24}H_{50}N_2Si_2$ ... $4,5\text{-}(C_2H_5)_2\text{-}2,2,3\text{-}(CH_3)_3\text{-}1,2,5\text{-}NSiBC_2\text{-}1\text{-}$
$(CH_2)_6\text{-}1\text{-}[1,2,5\text{-}NSiBC_2\text{-}2,2,3\text{-}(CH_3)_3\text{-}4,5\text{-}(C_2H_5)_2]$
B: B Comp.SVol.4/3a-253

$B_2C_{24}H_{54}N_4$ ..... $1,3\text{-}(C_4H_9\text{-}t)_2\text{-}2,4\text{-}[(t\text{-}C_4H_9)_2N]_2\text{-}1,3,2,4\text{-}N_2B_2$ B: B Comp.SVol.4/3a-198

$B_2C_{24}H_{54}N_4Si_2$ ... $1,3\text{-}[(CH_3)_3Si]_2\text{-}$
$2,4\text{-}[2,2,6,6\text{-}(CH_3)_4\text{-}NC_5H_6\text{-}1]_2\text{-}1,3,2,4\text{-}N_2B_2$. B: B Comp.SVol.4/3a-198

$B_2C_{24}H_{56}N_4O_4Si_2$ $[(\text{-}NCH_3\text{-}CH_2CH_2\text{-}NCH_3\text{-})B]_2[Si(OC_4H_9\text{-}t)_2]_2$ .. B: B Comp.SVol.3/3-118

$B_2C_{24}H_{56}N_6$ ..... $[(i\text{-}C_3H_7)_2N]_2B\text{-}N[C(CH_3){=}CH_2]\text{-}N(i\text{-}C_3H_7)$
$\text{-}B(NH_2)\text{-}N(i\text{-}C_3H_7)_2$ .... B: B Comp.SVol.3/3-98

$B_2C_{24}H_{56}N_7$ ..... $[(i\text{-}C_3H_7)_2N]_2B\text{-}N(i\text{-}C_3H_7)\text{-}N(i\text{-}C_3H_7)\text{-}B(N_3)\text{-}(i\text{-}C_3H_7)_2$
B: B Comp.SVol.3/3-98

$B_2C_{25}ClH_{35}N_2$.... $BC_{12}H_8[\text{-}NC_5H_6(CH_3)_4\text{-}BCl\text{-}N(C_4H_9\text{-}t)\text{-}]$ ..... B: B Comp.SVol.4/4-67

$B_2C_{25}CoFeH_{23}$ ... $C_5H_5FeC_4H_4BC_6H_5CoC_4H_4BC_6H_5$ .... Fe: Org.Comp.B17-225

$B_2C_{25}CoFeH_{33}NiO_3$
$(C_5H_5)(CO)Fe(CO)_2Ni(C_3B_2(C_2H_5)_4CH_3)Co(C_5H_5)$
Fe: Org.Comp.B16b-146/7

$B_2C_{25}F_8FeH_{27}NO_7$ $[((C_2H_5COO)(CH_3O)_2C_{16}H_{13}N\text{-}CH_3)Fe(CO)_3][BF_4]_2$ Fe: Org.Comp.B15-133, 184

$B_2C_{25}F_8H_{33}MoN_7$ $[Mo(CN\text{-}C_2H_5)_5(NC_5H_4\text{-}2\text{-}(2\text{-}C_5H_4N))][BF_4]_2$ . . Mo:Org.Comp.5-55, 64

$B_2C_{25}FeH_{32}NP$ . . . $(C_5H_5)Fe[C_5H_3(P(C_6H_5)_2 \cdot BH_3)\text{-}CH_2\text{-}N(CH_3)_2 \cdot BH_3]$ Fe: Org.Comp.A10-67, 68, 71

$B_2C_{25}GeH_{28}O_4$ . . . $Ge(C_6H_5)_3CH(B(OCH_2)_2CH_2)_2$ . . . . Ge: Org.Comp.3-68

$B_2C_{25}H_{20}N_2$ . . . . . $(C_6H_5)_2B\text{-}N{=}C{=}N\text{-}B(C_6H_5)_2$ . . . . B: B Comp.SVol.4/3a-243

$B_2C_{25}H_{23}N$ . . . . . . $(C_6H_5)_2B\text{-}N(CH_3)\text{-}B(C_6H_5)_2$ . . . . B: B Comp.SVol.4/3a-242

$B_2C_{25}H_{33}N_3S$ . . . . $C_6H_5B[\text{-}N(c\text{-}C_6H_{11})\text{-}CS\text{-}N(c\text{-}C_6H_{11})\text{-}BC_6H_5\text{-}NH\text{-}]$ B: B Comp.SVol.3/3-120/1

$B_2C_{25}H_{37}N_3S$ . . . . $C_6H_5B[\text{-}N(n\text{-}C_4H_9)\text{-}CS\text{-}N(n\text{-}C_4H_9)\text{-}BC_6H_5\text{-}N(n\text{-}C_4H_9)\text{-}]$ B: B Comp.SVol.3/3-119/20

$B_2C_{25}H_{41}NO$ . . . . . $1\text{-}C_6H_5\text{-}2\text{-}[(n\text{-}C_3H_7)_2B\text{-}O]\text{-}3,3\text{-}(n\text{-}C_3H_7)_2\text{-}1,2\text{-}NBC_7H_8$ B: B Comp.SVol.4/3b-65

$B_2C_{25}H_{50}N_2$ . . . . . $1\text{-}[(c\text{-}C_6H_{11})_2B\text{-}N(C_4H_9\text{-}t)\text{-}BH]\text{-}2,2,6,6\text{-}(CH_3)_4\text{-}NC_5H_6$ B: B Comp.SVol.4/3a-171

$B_2C_{25}H_{54}N_2$ . . . . . $1\text{-}[(n\text{-}C_4H_9)_2B\text{-}N(C_4H_9\text{-}t)\text{-}B(C_4H_9\text{-}n)]\text{-}$ $2,2,6,6\text{-}(CH_3)_4\text{-}NC_5H_6$ . . . . B: B Comp.SVol.4/3a-172

$B_2C_{26}Cl_2H_{40}Si_2$ . . $[(CH_3)_3Si]_2C{=}C[BCl\text{-}C_6H_2\text{-}2,4,6\text{-}(CH_3)_3]_2$ . . . . . B: B Comp.SVol.4/4-50

$B_2C_{26}Cl_4H_{54}N_4Pd_2$ $[\text{-}N(C_4H_9\text{-}t)\text{-}BCl\text{-}NC_9H_{18}\text{-}]PdCl_2Pd[$ $\text{-}N(C_4H_9\text{-}t)\text{-}BCl\text{-}NC_9H_{18}\text{-}]$. . . . . B: B Comp.SVol.4/4-65

$B_2C_{26}F_4H_{54}N_4Si$ . . $[2,2,6,6\text{-}(CH_3)_4\text{-}NC_5H_6\text{-}1\text{-}BF\text{-}N(C_4H_9\text{-}t)]_2SiF_2$ B: B Comp.SVol.4/3b-246

$B_2C_{26}FeH_{28}$ . . . . . $Fe(C_5H_3(CH_3)_2\text{-}3,4BC_6H_5\text{-}1)_2$ . . . . Fe: Org.Comp.B17-330

$B_2C_{26}FeH_{44}O_4$ . . . $Fe[C_5H_4B(OC_4H_9\text{-}n)_2]_2$ . . . . Fe: Org.Comp.A9-279, 284

$B_2C_{26}FeH_{48}N_4$ . . . $Fe(C_5H_4B(N(C_2H_5)_2)_2)_2$ . . . . Fe: Org.Comp.A9-279

$B_2C_{26}H_{20}N_2O_4$ . . . $[C_6H_5B(\text{-}O\text{-}(1,2\text{-}C_6H_4))\text{-}CH{=}N(\text{-}O\text{-})\text{-}]_2$ . . . . B: B Comp.SVol.3/3-220/1

$B_2C_{26}H_{32}Mn_2N_{12}O_9$ $[Mn_2O((CH_3COO)_2(1,2\text{-}N_2C_3H_3\text{-}1)_3BH)_2]$ Mn:MVol.D8-10/3

$B_2C_{26}H_{36}MnN_{12}$ . . $Mn[(1,2\text{-}N_2C_3H_3\text{-}1)_3B\text{-}C_4H_9]_2$ Mn:MVol.D8-9/10

$B_2C_{26}H_{38}N_2O$ . . . . $[\text{-}C({=}C(C_6H_5)_2)\text{-}N(C_4H_9\text{-}t)\text{-}B(C_2H_5)\text{-}N(C_4H_9\text{-}t)$ $\text{-}B(C_2H_5)\text{-}O\text{-}]$. . . . . B: B Comp.SVol.4/3b-54

$B_2C_{26}H_{40}N_2$ . . . . . $1,3\text{-}[2,4,6\text{-}(CH_3)_3\text{-}C_6H_2]_2\text{-}2,4\text{-}(t\text{-}C_4H_9)_2\text{-}1,3,2,4\text{-}N_2B_2$ B: B Comp.SVol.4/3a-197

– . . . . $(i\text{-}C_3H_7)_2N\text{-}B[\text{-}(1,2\text{-}C_6H_4)\text{-}CH_2\text{-}B(N(C_3H_7\text{-}i)_2)$ $\text{-}(1,2\text{-}C_6H_4)\text{-}CH_2\text{-}]$. . . . . B: B Comp.SVol.4/3a-247

$B_2C_{26}H_{40}Rh_2^{2+}$ . . $[((CH_3)_5C_5)Rh]_2[C_4H_4B_2(CH_3)_2]^{2+}$ . . . . B: B Comp.SVol.3/4-157

$B_2C_{26}H_{48}N_2$ . . . . . $(C_{10}H_{16})\text{-}BH_2\text{-}N(\text{-}CH_2\text{-}CH_2\text{-})_3N\text{-}BH_2\text{-}(C_{10}H_{16})$ B: B Comp.SVol.3/3-188

$B_2C_{26}H_{51}N_3OSi$ . . $(CH_3)_3Si\text{-}N(C_4H_9\text{-}t)\text{-}B[\text{-}N(C_4H_9\text{-}t)\text{-}B(C_4H_9\text{-}n)$ $\text{-}N(C_4H_9\text{-}t)\text{-}CH(C_6H_5)\text{-}O\text{-}]$. . . . . B: B Comp.SVol.4/3b-54

– . . . . $(CH_3)_3Si\text{-}N(C_4H_9\text{-}t)\text{-}B[\text{-}N(C_4H_9\text{-}t)\text{-}B(C_4H_9\text{-}n)$ $\text{-}O\text{-}CH(C_6H_5)\text{-}N(C_4H_9\text{-}t)\text{-}]$. . . . . B: B Comp.SVol.4/3b-54

$B_2C_{26}H_{52}N_2$ . . . . . $(C_{10}H_{16})\text{-}BH_2\text{-}N(CH_3)_2CH_2CH_2N(CH_3)_2\text{-}BH_2\text{-}(C_{10}H_{16})$ B: B Comp.SVol.3/3-188

$B_2C_{26}H_{54}N_4$ . . . . . $1,3\text{-}(C_4H_9\text{-}t)_2\text{-}2,4\text{-}[2,2,6,6\text{-}(CH_3)_4\text{-}NC_5H_6\text{-}1]_2\text{-}$ $1,3,2,4\text{-}N_2B_2$. . . . . B: B Comp.SVol.4/3a-198

$B_2C_{27}F_8H_{81}MnN_{15}O_6P_6$ $[Mn((O{=}P(N(CH_3)_2)_2)_2NCH_3)_3][BF_4]_2$. Mn:MVol.D8-179/80

$B_2C_{27}Fe_2H_{24}O_3Os$ $[(C_5H_5)_2FeBC_4H_4BFe(C_5H_5)_2]Os(CO)_3$ . . . . B: B Comp.SVol.3/4-157

$B_2C_{27}Fe_2H_{24}O_3Ru$ $[(C_5H_5)_2FeBC_4H_4BFe(C_5H_5)_2]Ru(CO)_3$ .... B: B Comp.SVol.3/4-157
$B_2C_{27}Fe_3H_{24}O_3$ .. $[(C_5H_5)_2FeBC_4H_4BFe(C_5H_5)_2]Fe(CO)_3$ .... B: B Comp.SVol.3/4-157
$B_2C_{27}H_{43}N_3$ ..... 2,2,6,6-$(CH_3)_4$-NC_5H_6-1-B[-N(C_4H_9-t)
-$B(C_6H_5)_2$-$NH(C_2H_5)$-]. .... B: B Comp.SVol.4/3a-202
$B_2C_{27}H_{48}N_2S$ .... [3.3.1]-9-BC_8H_{14}-9-S-9-[3.3.1]-9-BC_8H_{14}
· 1,2-$N_2C_3H_2$-3,5-$(C_4H_9$-t$)_2$ .... B: B Comp.SVol.4/4-132
$B_2C_{28}Cl_4Fe_3H_{42}N_2$
$[(C_6H_6)Fe(2$-CH_3-1-(t-C_4H_9)-1,2-$NBC_3H_3))]_2[FeCl_4]$
Fe: Org.Comp.B18-99, 101
$B_2C_{28}CoFe_2H_{24}O_4$ $[(C_5H_5)_2FeBC_4H_4BFe(C_5H_5)_2]Co(CO)_4$ .... B: B Comp.SVol.3/4-157
$B_2C_{28}CrFe_2H_{24}O_4$ $[(C_5H_5)_2FeBC_4H_4BFe(C_5H_5)_2]Cr(CO)_4$ .... B: B Comp.SVol.3/4-157
$B_2C_{28}Fe_2H_{24}O_4W$ $[(C_5H_5)_2FeBC_4H_4BFe(C_5H_5)_2]W(CO)_4$ .... B: B Comp.SVol.3/4-157
$B_2C_{28}H_{18}N_2OS_4$ .. [4-C_6H_5-1,6,4,5-$S_2NBC_8H_4$-5$]_2$O .... B: B Comp.SVol.4/3b-67
$B_2C_{28}H_{20}MnN_4S_8$ Mn[(2-(S=)-1,3-SNC_7H_4-3$)_2BH_2]_2$. .... Mn: MVol.D8-21
$B_2C_{28}H_{20}MnN_{12}O_8$
Mn[(5-NO_2-1,2-$N_2C_7H_4$-1$)_2BH_2]_2$. .... Mn: MVol.D8-16
$B_2C_{28}H_{20}N_{12}O_8Pd$ Pd(H_2B(N(-N=$CHC_6H_3(NO_2)$)-$)_2)_2$ .... Pd: SVol.B2-260/1
$B_2C_{28}H_{24}N_2O_4$ ... [C_6H_5B(-O-(1,2-C_6H_4))-CCH_3=N(-O-)-$]_2$. .... B: B Comp.SVol.3/3-220/1
$B_2C_{28}H_{35}Mn_2N_{13}O_9$
$[Mn_2O((CH_3COO)_2(1,2$-$N_2C_3H_3$-1$)_3BH)_2]$ · CH_3CN
Mn: MVol.D8-12/3
$B_2C_{28}H_{42}N_2Se$ ... [3.3.1]-9-BC_8H_{14}-9-Se-9-[3.3.1]-9-BC_8H_{14}
· 2 NC_5H_4-3-CH_3. .... B: B Comp.SVol.4/4-170
$B_2C_{28}H_{48}N_2O_8$ ... (t-$C_4H_9)_4$-$B_2C_4N_2$-(COO-$CH_3)_4$ .... B: B Comp.SVol.4/3a-201
$B_2C_{28}H_{52}N_4$ ..... [1,3-(c-$C_6H_{11})_2$-1,3,2-$N_2BC_2H_4$-2$]_2$ .... B: B Comp.SVol.4/3a-177
$B_2C_{28}H_{56}N_4$ ..... 1-[2,2,6,6-$(CH_3)_4$-NC_5H_6-1-B(NH-C_4H_9-t)
-CC-B(NH-C_4H_9-t)]-2,2,6,6-$(CH_3)_4$-NC_5H_6 .. B: B Comp.SVol.4/3a-168
$B_2C_{28}H_{56}N_6$ ..... $[(CH_3)_2C_5H_8N]B[-(NC_5H_7(CH_3)_2)$-NH
-$B(NC_5H_8(CH_3)_2)$-$N(NC_5H_8(CH_3)_2)$-] .... B: B Comp.SVol.3/3-98
$B_2C_{28}H_{56}N_8$ ..... $[(CH_3)_2C_5H_8N]_2BN[NC_5H_8(CH_3)_2]B(N_3)[NC_5H_8(CH_3)_2]$
B: B Comp.SVol.3/3-98
$B_2C_{28}H_{59}N_5O_2$ ... 2,2,6,6-$(CH_3)_4$-NC_5H_6-1-B(NH-C_4H_9-t)-OC(O)
-CH_2-NH-B(NH-C_4H_9-t)-1-NC_5H_6-2,2,6,6-$(CH_3)_4$
B: B Comp.SVol.4/3b-53
$B_2C_{28}H_{62}K_2NO_2Si_2$
$[K(OC_4H_8)]_2$[1-(i-$C_3H_7)_2$N-3,3-(i-$C_3H_7)_2$-
2,4-$((CH_3)_3Si)_2$-1,3-B_2C_2] .... B: B Comp.SVol.4/3b-49
$B_2C_{28}H_{68}LiN_5$.... $[Li((CH_3)_2NCH_2CH_2N(CH_3)_2)_2]$[(t-$C_4H_9)_2$B=N=B($C_4H_9$-t$)_2$]
B: B Comp.SVol.4/3b-46/7
$B_2C_{29}F_8H_{31}MoN_7$ [Mo(CN-$C_2H_5)_3(NC_5H_4$-2-(2-$C_5H_4N))_2][BF_4]_2$ Mo: Org.Comp.5-31/3
$B_2C_{29}H_{29}N_5O_9$ ... (CH_3-COO$)_4B_2C_2N_5O(C_6H_4$-4-$CH_3)(C_6H_5)_2$ ... B: B Comp.SVol.4/3b-57
$B_2C_{29}H_{45}N_3$ ..... 2,2,6,6-$(CH_3)_4$-NC_5H_6-1-B[N$(C_2H_5)_2$]
-N(C_4H_9-t)-9-(9-$BC_{12}H_8$) .... B: B Comp.SVol.4/3a-158
$B_2C_{29}H_{61}N_5O_2$ ... 2,2,6,6-$(CH_3)_4$-NC_5H_6-1-B(NH-C_4H_9-t)-OC(O)
-CH(CH_3)-NH-B(NH-C_4H_9-t)-1-NC_5H_6-2,2,6,6-$(CH_3)_4$
B: B Comp.SVol.4/3b-53
$B_2C_{30}ClF_{10}H_{23}N_2$ ClB(C_6F_5)-N[2,4,6-$(CH_3)_3C_6H_2$]-B(C_6F_5)
-NH[2,4,6-$(CH_3)_3C_6H_2$] .... B: B Comp.SVol.3/4-59
$B_2C_{30}Cl_2H_{44}MnN_{12}O_8$
$[Mn((3,5$-$(CH_3)_2$-1,2-N_2C_3H-1$)_3BH)_2][ClO_4]_2$ .. Mn: MVol.D8-14/5

$B_2C_{30}F_8H_{54}MoN_6$ $[Mo(CN\text{-}C_4H_9\text{-}t)_6][BF_4]_2$ Mo:Org.Comp.5-71
$B_2C_{30}FeH_{24}$ ..... $Fe(C_9H_7BC_6H_5)_2$ .... Fe: Org.Comp.B17-330
$B_2C_{30}FeH_{42}N_2$ ... $Fe(C_9H_7BN(C_3H_7\text{-}i)_2)_2$ .... Fe: Org.Comp.B17-330
$B_2C_{30}H_{26}N_4$ ..... $(C_6H_5)_2B[\text{-}1,2\text{-}N_2C_3H_3\text{-}]_2B(C_6H_5)_2$ .... B: B Comp.SVol.3/3-199
B: B Comp.SVol.4/3b-34
– .... $H_2B[\text{-}1,2\text{-}N_2C_3H(C_6H_5)_2\text{-}]_2BH_2$ .... B: B Comp.SVol.3/3-200
$B_2C_{30}H_{26}N_4O_4$ ... $(C_6H_5O)_2B[\text{-}N_2C_3H_3\text{-}B(OC_6H_5)_2\text{-}N_2C_3H_3]$ .... B: B Comp.SVol.3/3-199
$B_2C_{30}H_{28}MnN_{12}$ .. $Mn[(1,2\text{-}N_2C_3H_3\text{-}1)_3B\text{-}C_6H_5]_2$ Mn:MVol.D8-9/10
$B_2C_{30}H_{40}Mn_2N_{12}O_9$
$[Mn_2O((C_2H_5COO)_2(1,2\text{-}N_2C_3H_3\text{-}1)_3BH)_2]$ Mn:MVol.D8-10/2
$B_2C_{30}H_{42}N_{12}$ .... $[(CH_3)_2C_3HN_2]_2B[\text{-}(CH_3)_2C_3HN_2\text{-}$
$B(C_3HN_2(CH_3)_2)_2\text{-}(CH_3)_2C_3HN_2\text{-}]$ .... B: B Comp.SVol.3/3-200
$B_2C_{30}H_{44}MnN_{12}$ .. $Mn[(3,5\text{-}(CH_3)_2\text{-}1,2\text{-}N_2C_3H\text{-}1)_3BH]_2$ Mn:MVol.D8-9/10
$B_2C_{30}H_{48}NiP_2Si_2$ $B_2C_4NiP_2Si_2(C_2H_5)_4(CH_3)_6(C_6H_5)_2$ .... B: B Comp.SVol.4/4-180
$B_2C_{31}H_{25}N_3S$ .... $C_6H_5B[\text{-}S\text{-}C(=NC_6H_5)\text{-}NC_6H_5\text{-}BC_6H_5\text{-}NC_6H_5\text{-}]$
B: B Comp.SVol.3/4-128
$B_2C_{31}H_{40}N_2S$ .... $[3.3.1]\text{-}9\text{-}BC_8H_{14}\text{-}9\text{-}S\text{-}9\text{-}[3.3.1]\text{-}9\text{-}BC_8H_{14}$
$\cdot$ $1,2\text{-}N_2C_3H_2\text{-}3,5\text{-}(C_6H_5)_2$ .... B: B Comp.SVol.4/4-132
$B_2C_{31}H_{42}N_2$ ..... $1\text{-}[(C_6H_5)_2B\text{-}N(C_4H_9\text{-}t)\text{-}B(C_6H_5)]\text{-}2,2,6,6\text{-}(CH_3)_4\text{-}NC_5H_6$
B: B Comp.SVol.4/3a-172
$B_2C_{31}H_{65}N_5O_2$ ... $2,2,6,6\text{-}(CH_3)_4\text{-}NC_5H_6\text{-}1\text{-}B(NH\text{-}C_4H_9\text{-}t)\text{-}OC(O)$
$\text{-}CH(C_3H_7\text{-}i)\text{-}NH\text{-}B(NH\text{-}C_4H_9\text{-}t)\text{-}1\text{-}NC_5H_6$
$\text{-}2,2,6,6\text{-}(CH_3)_4$ .... B: B Comp.SVol.4/3b-53
$B_2C_{32}H_{24}MnO_{12}$ .. $Mn[B[\text{-}O\text{-}C(=O)\text{-}CH(C_6H_5)\text{-}O\text{-}]_2]_2$ Mn:MVol.D8-3/4
$B_2C_{32}H_{32}MnN_8O_{16}$
$Mn[(2,5\text{-}(O=)_2\text{-}NC_4H_4\text{-}1)_4B]_2$ Mn:MVol.D8-7/8
$B_2C_{32}H_{40}MnN_{16}$ .. $Mn[(2\text{-}CH_3\text{-}1,3\text{-}N_2C_3H_2\text{-}1)_4B]_2$ Mn:MVol.D8-18/9
$B_2C_{32}H_{40}N_2$ ..... $[BC_9H_{15}\text{-}CC_6H_5=N]_2$ .... B: B Comp.SVol.3/3-166
$B_2C_{32}H_{48}MoO_2P_2Si_2$
$B_2C_4MoP_2Si_2(CO)_2(C_2H_5)_4(C_6H_5)_2(CH_3)_6$ ..... B: B Comp.SVol.4/4-180
$B_2C_{32}H_{52}N_2$ ..... $1,3\text{-}[2,6\text{-}(i\text{-}C_3H_7)_2\text{-}C_6H_3]_2\text{-}2,4\text{-}(t\text{-}C_4H_9)_2\text{-}1,3,2,4\text{-}N_2B_2$
B: B Comp.SVol.4/3a-197
$B_2C_{32}H_{65}N_5O_4$ ... $2,2,6,6\text{-}(CH_3)_4\text{-}NC_5H_6\text{-}1\text{-}B(NH\text{-}C_4H_9\text{-}t)\text{-}OC(O)$
$\text{-}CH[N(CH_3)_2]\text{-}CH_2\text{-}C(O)O\text{-}B(NH\text{-}C_4H_9\text{-}t)$
$\text{-}1\text{-}NC_5H_6\text{-}2,2,6,6\text{-}(CH_3)_4$ .... B: B Comp.SVol.4/3b-53
$B_2C_{32}H_{67}N_5O_2$ ... $2,2,6,6\text{-}(CH_3)_4\text{-}NC_5H_6\text{-}1\text{-}B(NH\text{-}C_4H_9\text{-}t)\text{-}OC(O)$
$\text{-}CH(C_4H_9\text{-}s)\text{-}NH\text{-}B(NH\text{-}C_4H_9\text{-}t)\text{-}1\text{-}NC_5H_6$
$\text{-}2,2,6,6\text{-}(CH_3)_4$ .... B: B Comp.SVol.4/3b-53
– .... $2,2,6,6\text{-}(CH_3)_4\text{-}NC_5H_6\text{-}1\text{-}B(NH\text{-}C_4H_9\text{-}t)\text{-}OC(O)$
$\text{-}CH(C_4H_9\text{-}i)\text{-}NH\text{-}B(NH\text{-}C_4H_9\text{-}t)\text{-}1\text{-}NC_5H_6$
$\text{-}2,2,6,6\text{-}(CH_3)_4$ .... B: B Comp.SVol.4/3b-53
$B_2C_{32}H_{70}O_6Sn$ ... $(C_4H_9)_2Sn(OB(OH)OC_{12}H_{25})_2$ .... Sn: Org.Comp.15-351
$B_2C_{34}H_{34}N_4$ ..... $(C_6H_5)_2B[\text{-}1,2\text{-}N_2C_3H(CH_3)_2\text{-}]_2B(C_6H_5)_2$ ...... B: B Comp.SVol.3/3-200
B: B Comp.SVol.4/3b-34
$B_2C_{34}H_{44}Mn_2N_{16}O_9$
$[Mn_2O((CH_3COO)_2(1,2\text{-}N_2C_3H_3\text{-}1)_3BH)_2]$ $\cdot$ 4 CH_3CN
Mn:MVol.D8-12/3
$B_2C_{34}H_{44}N_2$ ..... $[BC_9H_{15}\text{-}C(C_6H_4\text{-}4\text{-}CH_3)=N]_2$ .... B: B Comp.SVol.3/3-166
$B_2C_{34}H_{56}N_2S_2$ ... $C_6H_5CH_2\text{-}C(S\text{-}C_3H_7\text{-}n)=N[\text{-}B(C_3H_7\text{-}n)_2\text{-}]_2N=$
$C(S\text{-}C_3H_7\text{-}n)\text{-}CH_2C_6H_5$ .... B: B Comp.SVol.4/3a-254

$B_2C_{34}H_{56}N_2S_2$ ... $C_6H_5CH_2-C(S-C_3H_7-n)=N[-B(C_3H_7-n)_2-]_2N=C(S-C_3H_7-n)-CH_2C_6H_5$ ... B: B Comp.SVol.4/4-149
$B_2C_{34}H_{62}N_2O$ .... $[1-(i-C_4H_9)-3,3-(n-C_3H_7)_2-1,2-NBC_7H_8-2]_2O$ B: B Comp.SVol.4/3b-66
$B_2C_{35}H_{53}N_3$ ..... $1-(t-C_4H_9)-2,4-[2,2,6,6-(CH_3)_4-NC_5H_6-1-]_2-1,2,4-NB_2C_{13}H_8$ ... B: B Comp.SVol.4/3a-207
– ... $2,2,6,6-(CH_3)_4-NC_5H_6-1-B[-N(C_4H_9-t)-B(1-NC_5H_6-2,2,6,6-(CH_3)_4)-9-C_{13}H_8-9-]$ ... B: B Comp.SVol.4/3a-248
$B_2C_{35}H_{65}N_5O_2$ ... $2,2,6,6-(CH_3)_4-NC_5H_6-1-B(NH-C_4H_9-t)-OC(O)-CH(CH_2-C_6H_5)-NH-B(NH-C_4H_9-t)-1-NC_5H_6-2,2,6,6-(CH_3)_4$ ... B: B Comp.SVol.4/3b-53
$B_2C_{36}F_2H_{53}N_3$ ... $2,6-(i-C_3H_7)_2-C_6H_3-NH-BF-N[C_6H_3-2,6-(C_3H_7-i)_2]-BF-NH-C_6H_3-2,6-(C_3H_7-i)_2$. B: B Comp.SVol.4/3b-247
$B_2C_{36}H_{26}MnN_{18}$ .. $Mn[(1,2,3-N_3C_6H_4-1)_3BH]_2$ ... Mn:MVol.D8-20
– ... $Mn[(1,2,3-N_3C_6H_4-2)_3BH]_2$ ... Mn:MVol.D8-20
$B_2C_{36}H_{44}N_2$ ..... $[-B(C_6H_2-2,4,6-(CH_3)_3)-N(C_6H_2-2,4,6-(CH_3)_3)-]_2$ B: B Comp.SVol.3/3-136
$B_2C_{36}H_{44}N_4$ ..... $2,4,6-(CH_3)_3C_6H_2-B[-N(C_6H_2(CH_3)_3-2,4,6)-N=N-N(B(C_6H_2(CH_3)_3-2,4,6)_2)-]$ ... B: B Comp.SVol.3/3-137/8
$B_2C_{36}H_{45}N$ ...... $[2,4,6-(CH_3)_3-C_6H_2]_2B-NH-B[C_6H_2-2,4,6-(CH_3)_3]_2$ B: B Comp.SVol.4/3a-241; B: B Comp.SVol.4/3b-228
$B_2C_{36}H_{56}MnN_{12}$ .. $Mn[(3,4,5-(CH_3)_3-1,2-N_2C_3-1)_3BH]_2$ ... Mn:MVol.D8-9/10
$B_2C_{36}H_{60}N_2S_2$ ... $C_6H_5CH_2-C(S-C_4H_9-n)=N[-B(C_3H_7-n)_2-]_2N=C(S-C_4H_9-n)CH_2C_6H_5$ ... B: B Comp.SVol.4/3a-254; B: B Comp.SVol.4/4-149
$B_2C_{38}H_{52}N_2Si_4$ ... $[(CH_3)_3Si]_2N-B(-9-C_{13}H_8-9-)_2B-N[Si(CH_3)_3]_2$ B: B Comp.SVol.4/3a-246
$B_2C_{38}H_{54}N_2O$ .... $[1-C_6H_5-3,3-(n-C_3H_7)_2-1,2-NBC_7H_8-2]_2O$ ... B: B Comp.SVol.4/3b-66
$B_2C_{38}H_{62}N_2O$ .... $[-N(C_4H_9-i)-(1,2-C_6H_4)-C(C_4H_9-n)_2-]B-O-B[-N(C_4H_9-i)-(1,2-C_6H_4)-C(C_4H_9-n)_2-]$ ... B: B Comp.SVol.3/3-164
$B_2C_{40}F_8H_{68}MnN_8O_4P_4$ $[Mn((C_6H_5)P(=O)(N(CH_3)_2)_2)_4][BF_4]_2$ ... Mn:MVol.D8-170
$B_2C_{40}H_{60}N_2O_{10}$ .. $2\ H_3B-NH_3 \cdot 2,3,11,12-(4-CH_3O-C_6H_4)_4-1,4,7,10,13,16-O_6C_{12}H_{20}$ ... B: B Comp.SVol.4/3b-2
$B_2C_{40}H_{86}O_6Sn$ ... $(C_4H_9)_2Sn(OB(OCH_2CH(C_2H_5)C_4H_9)_2)_2$ ... Sn: Org.Comp.15-351
$B_2C_{42}H_{26}MnN_6S_{12}$ $Mn[(2-(S=)-1,3-SNC_7H_4-3)_3BH]_2$ ... Mn:MVol.D8-21
$B_2C_{42}H_{26}N_{18}O_{12}Pd$ $Pd(HB(N(-N=CHC_6H_3(NO_2))-)_3)_2$ ... Pd: SVol.B2-260/1
$B_2C_{42}H_{62}N_2O$ .... $[1-C_6H_5-3,3-(n-C_4H_9)_2-1,2-NBC_7H_8-2]_2O$ ... B: B Comp.SVol.4/3b-66
$B_2C_{43}GeH_{42}O_4Pb$ $Ge(C_6H_5)_3C(B(OCH_2)_2CH_2)_2Pb(C_6H_5)_3$ ... Ge:Org.Comp.3-69
$B_2C_{43}GeH_{42}O_4Sn$ $Ge(C_6H_5)_3C(B(OCH_2)_2CH_2)_2Sn(C_6H_5)_3$ ... Ge:Org.Comp.3-68/9, 77
$B_2C_{43}H_{61}N_3$ ..... $2,2,6,6-(CH_3)_4-NC_5H_6-1-B[-N(C_6H_3-2,6-(C_3H_7-i)_2)-B(1-NC_5H_6-2,2,6,6-(CH_3)_4)-9-C_{13}H_8-9-]$ ... B: B Comp.SVol.4/3a-248
$B_2C_{44}F_8H_{37}MoN_7$ $[Mo(CN-C_6H_4-CH_3-4)_3(NC_5H_4-2-(2-C_5H_4N))_2][BF_4]_2$ Mo:Org.Comp.5-31/3
$B_2C_{44}F_8H_{41}MoNO_2P_2$ $[C_5H_5Mo(NO)(P(C_6H_5)_3)_2(O=C(CH_3)_2)][BF_4]_2$ .. Mo:Org.Comp.6-41
$B_2C_{44}H_{66}Li_2N_2O_2$ $[(2,4,6-(CH_3)_3-C_6H_2)_2B-NH-Li(O(C_2H_5)_2)]_2$ ... B: B Comp.SVol.4/3a-231

$B_2C_{45}F_8H_{45}MoNO_2P_2$
$[C_5H_5Mo(NO)(P(C_6H_5)_3)_2O(C_2H_5)_2][BF_4]_2$ Mo:Org.Comp.6-41
$B_2C_{46}Cl_2H_{46}N_2O_4Sn$
$(C_4H_9)_2Sn(OB(C_6H_5)OC_6H_4CHNC_6H_4Cl\text{-}4)_2$ . . . Sn: Org.Comp.15-351
$B_2C_{46}Cl_5H_{36}NP_2$ . . $[(C_6H_5)_3P{=}N{=}P(C_6H_5)_3][C_{10}H_6B_2Cl_5]$ . . . . B: B Comp.SVol.4/4-45
$B_2C_{46}F_2H_{46}N_2O_4Sn$
$(C_4H_9)_2Sn(OB(C_6H_5)OC_6H_4CHNC_6H_4F\text{-}4)_2$ . . . . Sn: Org.Comp.15-351
$B_2C_{46}F_8FeH_{54}P_2$ . . $[Fe(C_5H_2(C_6H_5)_2\text{-}2,6(C_4H_9\text{-}t)\text{-}4P(CH_3)_2\text{-}1,1)_2][BF_4]_2$
Fe: Org.Comp.B17-333
$B_2C_{46}H_{46}N_4O_8Sn$ $(C_4H_9)_2Sn(OB(C_6H_5)OC_6H_4CHNC_6H_4NO_2\text{-}4)_2$ . . Sn: Org.Comp.15-351
$B_2C_{48}F_8FeH_{42}P_2$ . . $[Fe(C_5H_3P(C_6H_5)_3\text{-}2,4,6(CH_3\text{-}1))_2][BF_4]_2$ . . . . . Fe: Org.Comp.B17-334
$B_2C_{48}F_8H_{50}N_4P_2Re$
$[Re(CNCH(CH_3)_2)_2(P(C_6H_5)_3)_2(NCCH_3)_2][BF_4]_2$ Re: Org.Comp.2-257
$B_2C_{48}F_8H_{96}MnN_{12}O_{16}P_4$
$[Mn(O{=}P(N(CH_2CH_2)_2O)_3)_4][BF_4]_2$ Mn:MVol.D8-176
$B_2C_{48}H_{26}MnN_6O_{12}$
$Mn[(1,3\text{-}(O{=})_2\text{-}2\text{-}NC_8H_4\text{-}2)_3BH]_2$ Mn:MVol.D8-7/8
$B_2C_{48}H_{44}N_4P_2$ . . . $[\text{-}BH_2\text{-}N(C_6H_5)\text{-}P(C_6H_5)_2\text{-}N(C_6H_5)\text{-}BH_2\text{-}$
$N(C_6H_5)\text{-}P(C_6H_5)_2\text{-}N(C_6H_5)\text{-}]$ . . . . B: B Comp.SVol.4/3a-260
$B_2C_{48}H_{74}LiNO_3$ . . $[Li(C_2H_5\text{-}O\text{-}C_2H_5)_3][(2,4,6\text{-}(CH_3)_3\text{-}C_6H_2)_2B{=}$
$N{=}B(C_6H_2\text{-}2,4,6\text{-}(CH_3)_3)_2]$ . . . . B: B Comp.SVol.4/3b-47
$B_2C_{49}F_8H_{57}N_4P_2Re$
$[ReH(CNC_4H_9\text{-}t)(P(C_6H_5)_3)_2(NCCH_3)_2NH_2C_4H_9\text{-}t][BF_4]_2$
Re: Org.Comp.2-230, 238
$B_2C_{50}F_8FeH_{46}P_2$ . . $[Fe(C_5H_2P(C_6H_5)_3\text{-}2,4,6(CH_3)_2\text{-}1,1)_2][BF_4]_2$ . . . Fe: Org.Comp.B17-333, 335/6
$B_2C_{50}F_8H_{43}MoN_7$ $[Mo(CN\text{-}C_6H_4\text{-}CH_3\text{-}4)_5(NC_5H_4\text{-}2\text{-}(2\text{-}C_5H_4N))][BF_4]_2$
Mo:Org.Comp.5-62, 67
$B_2C_{50}F_8H_{54}N_4P_2Re$
$[Re(CNC_4H_9\text{-}t)_2(P(C_6H_5)_3)_2(NCCH_3)_2][BF_4]_2$. . . Re: Org.Comp.2-258, 263
$B_2C_{50}H_{46}MnN_6$ . . . $Mn(NC\text{-}CH_2CH_2CH_2CH_2\text{-}CN)_2[NCB(C_6H_5)_3]_2$ . . Mn:MVol.D7-14/9
– . . . . $Mn[NC\text{-}CH(CH_3)\text{-}CH_2CH_2\text{-}CN]_2[NCB(C_6H_5)_3]_2$ Mn:MVol.D7-14/9
$B_2C_{51}F_8H_{52}N_4P_2Re$
$[Re(CNCH(CH_3)_2)_2(P(C_6H_5)_3)_2(NCCH_3)C_5H_5N][BF_4]_2$
Re: Org.Comp.2-257
$B_2C_{51}F_8H_{59}N_4P_2Re$
$[ReH(CNC_4H_9\text{-}t)(P(C_6H_5)_3)_2(NCCH_3)_2NH_2C_6H_{11}][BF_4]_2$
Re: Org.Comp.2-230, 238
$B_2C_{53}F_8H_{56}N_4P_2Re$
$[Re(CNC_4H_9\text{-}t)_2(P(C_6H_5)_3)_2(NCCH_3)C_5H_5N][BF_4]_2$
Re: Org.Comp.2-258
$B_2C_{54}Cl_2H_{50}N_2O_4Sn$
$(C_4H_9)_2Sn(OB(C_6H_5)OC_{10}H_6CHNC_6H_4Cl\text{-}4)_2$ . . Sn: Org.Comp.15-351
$B_2C_{54}F_2H_{50}N_2O_4Sn$
$(C_4H_9)_2Sn(OB(C_6H_5)OC_{10}H_6CHNC_6H_4F\text{-}4)_2$ . . . Sn: Org.Comp.15-351
$B_2C_{54}H_{50}N_4O_8Sn$ $(C_4H_9)_2Sn(OB(C_6H_5)OC_{10}H_6CHNC_6H_4NO_2\text{-}4)_2$ Sn: Org.Comp.15-351
$B_2C_{54}H_{66}MnN_2$ . . . $Mn[N(C_6H_2\text{-}2,4,6\text{-}(CH_3)_3)\text{-}B(C_6H_2$
$\text{-}2,4,6\text{-}(CH_3)_3)_2]_2 \cdot 3\ C_6H_5\text{-}CH_3$ . . . . B: B Comp.SVol.4/3a-237
Mn:MVol.D8-5/6

$B_2C_{56}F_8H_{54}MoN_2P_4$
$[Mo(CN-CH_3)_2(P(C_6H_5)_2CH_2CH_2P(C_6H_5)_2)_2][BF_4]_2$ Mo:Org.Comp.5-13, 19

$B_2C_{56}F_8H_{56}MoN_2P_4$
$[(CH_3NH-C)_2Mo(P(C_6H_5)_2CH_2CH_2P(C_6H_5)_2)_2][BF_4]_2$ Mo:Org.Comp.5-119/20

– . . . . $[CH_3NH-CMo(H)(CN-CH_3)(P(C_6H_5)_2CH_2CH_2P(C_6H_5)_2)_2][BF_4]_2$ Mo:Org.Comp.5-115/8

$B_2C_{56}F_8H_{64}MnN_4O_4P_4$
$[Mn((C_6H_5)_2P(=O)N(CH_3)_2)_4][BF_4]_2$ Mn:MVol.D8-170

$B_2C_{56}H_{32}MnN_8S_{16}$
$Mn[(2-(S=)-1,3-SNC_7H_4-3)_4B]_2$ Mn:MVol.D8-21

$B_2C_{56}H_{32}N_{24}O_{16}Pd$
$Pd(B(N(-N=CHC_6H_3(NO_2))-)_4)_2$ . . . . Pd: SVol.B2-260/1

$B_2C_{56}H_{52}N_8Ti_2$ . . . $[((C_5H_5)_2Ti)_2((C_3H_3N_2)_4B)][B(C_6H_5)_4]$ . . . . Ti: Org.Comp.5-176

$B_2C_{58}F_8H_{38}OS_{16}$. $[(C_{14}H_8S_4)_4(BF_4)_2(C_2H_5OH)]$ . . . . B: B Comp.SVol.4/4-163

$B_2C_{60}ClCoH_{82}LiN_2O_3$
$[(C_2H_5)_2O]_2Li(OC_4H_8)-Cl-Co[N(C_6H_5)-B(C_6H_2-2,4,6-(CH_3)_3)_2]_2$ . . . . B: B Comp.SVol.4/3a-237

$B_2C_{60}F_8H_{72}MnO_6S_{12}$
$[Mn((2-C_6H_5)C_4H_7S_2-1,3-(=O-1))_6](BF_4)_2$ Mn:MVol.D7-104

$B_2C_{64}FeH_{60}$ . . . . . $[(1,3-(CH_3)_2-C_6H_4)_2Fe][B(C_6H_5)_4]_2$ . . . . Fe: Org.Comp.B19-347, 359

– . . . . $[(C_2H_5-C_6H_5)_2Fe][B(C_6H_5)_4]_2$ . . . . Fe: Org.Comp.B19-347, 357

$B_2C_{64}H_{32}MnN_8O_{16}$
$Mn[(1,3-(O=)_2-2-NC_8H_4-2)_4B]_2$ Mn:MVol.D8-7/8

$B_2C_{65}H_{71}MnN_7$ . . . $[Mn(2,6-(NH_2-C_2H_4-NH-C_2H_4-N=C(CH_3))_2-C_5H_3N)][B(C_6H_5)_4]_2$ Mn:MVol.D6-221

$B_2C_{66}FeH_{64}$ . . . . . $[(1,2,4-(CH_3)_3-C_6H_3)_2Fe][B(C_6H_5)_4]_2$ . . . . Fe: Org.Comp.B19-347, 361

– . . . . $[(1,3,5-(CH_3)_3-C_6H_3)_2Fe][B(C_6H_5)_4]_2$ . . . . Fe: Org.Comp.B19-347, 364, 378

$B_2C_{67}Cl_3H_{84}MoN_7Sn$
$[Cl_3Sn-Mo(CN-C_4H_9-t)_6][(C_6H_5)_3B-CN-B(C_6H_5)_3]$ Mo:Org.Comp.5-76, 81/2

$B_2C_{68}F_8H_{62}MoN_2P_4$
$[Mo(CN-C_6H_4CH_3-4)_2(P(C_6H_5)_2CH_2CH_2P(C_6H_5)_2)_2][BF_4]_2$ Mo:Org.Comp.5-16, 20

$B_2C_{68}FeH_{64}$ . . . . . $[(C_{10}H_{12})_2Fe][B(C_6H_5)_4]_2$ . . . . Fe: Org.Comp.B19-400

$B_2C_{68}H_{72}O_6Ti_2$ . . . $[(C_5H_5)_2Ti(H_2O)_3]_2[B(C_6H_5)_4]_2$ . . . . Ti: Org.Comp.5-164/5

$B_2C_{68}H_{126}O_6Sn$ . . $(4-i-C_3H_7C_6H_4CH_2)_2Sn(OB(OC_{12}H_{25})_2)_2$ . . . . . . Sn: Org.Comp.16-88

$B_2C_{72}F_8H_{60}MnO_4P_4$
$[Mn((C_6H_5)_3P=O)_4(BF_4)][BF_4]$ Mn:MVol.D8-89

$B_2C_{72}F_8H_{60}MnO_6S_6$
$[Mn(O=S(C_6H_5)_2)_6](BF_4)_2$ Mn:MVol.D7-102

$B_2C_{72}FeH_{76}$ . . . . . $[((CH_3)_6C_6)_2Fe][B(C_6H_5)_4]_2$ . . . . Fe: Org.Comp.B19-371

$B_2C_{72}H_{64}MnN_{12}$ . . $[Mn(NC_5H_4-2-NHN=CHCH=NNH-2-NC_5H_4)_2][B(C_6H_5)_4]_2$ Mn:MVol.D6-264, 266

– . . . . $[Mn(NC_5H_4-2-NHN=CHCH=NNH-2-NC_5H_4)_2][B(C_6H_5)_4]_2 \cdot n\ H_2O$ Mn:MVol.D6-264, 266

$B_2C_{75}H_{90}MnN_2$ . . . $Mn[N(C_6H_2-2,4,6-(CH_3)_3)-B(C_6H_2-2,4,6-(CH_3)_3)_2]_2 \cdot 3\ C_6H_5-CH_3$ . . . . B: B Comp.SVol.4/3a-237

$B_2C_{75}H_{90}MnN_2$ ... $Mn[N(C_6H_2\text{-}2,4,6\text{-}(CH_3)_3)\text{-}B(C_6H_2\text{-}2,4,6\text{-}(CH_3)_3)_2]_2 \cdot 3\ C_6H_5\text{-}CH_3$ ... Mn:MVol.D8-5/6

$B_2C_{76}H_{68}MnN_8$ ... $[Mn(NC_5H_4\text{-}2\text{-}CH{=}N\text{-}C_2H_4\text{-}N{=}CH\text{-}2\text{-}C_5H_4N)_2][B(C_6H_5)_4]_2$ ... Mn:MVol.D6-189/90

$B_2C_{80}H_{76}MnN_{12}$ .. $[Mn(1,2\text{-}(NC_5H_4\text{-}2\text{-}NH\text{-}N{=})_2\text{-}C_6H_8)_2][B(C_6H_5)_4]_2$ Mn:MVol.D6-264, 266

– ... $[Mn(1,2\text{-}(NC_5H_4\text{-}2\text{-}NH\text{-}N{=})_2\text{-}C_6H_8)_2][B(C_6H_5)_4]_2 \cdot n\ H_2O$ ... Mn:MVol.D6-264, 266

$B_2C_{83}H_{103}MoN_7$ .. $[Mo(CN\text{-}C_4H_9\text{-}t)_7][B(C_6H_5)_4]_2$ ... Mo:Org.Comp.5-85, 88

$B_2C_{83}H_{105}MoN_7$ .. $[(t\text{-}C_4H_9\text{-}NH\text{-}CC\text{-}NH\text{-}C_4H_9\text{-}t)Mo(CN\text{-}C_4H_9\text{-}t)_5][B(C_6H_5)_4]_2 \cdot CH_3\text{-}OH$ ... Mo:Org.Comp.5-179

$B_2C_{84}H_{109}MoN_7O$ $[(t\text{-}C_4H_9\text{-}NH\text{-}CC\text{-}NH\text{-}C_4H_9\text{-}t)Mo(CN\text{-}C_4H_9\text{-}t)_5][B(C_6H_5)_4]_2 \cdot CH_3\text{-}OH$ ... Mo:Org.Comp.5-179

B_2Ce ... CeB_2 ... Sc: MVol.C11a-2

B_2CeClH_8 ... $CeCl[BH_4]_2 \cdot n\ C_4H_8O$ ... Sc: MVol.C11b-489

B_2ClDyH_8 ... $DyCl[BH_4]_2 \cdot n\ C_4H_8O$ ... Sc: MVol.C11b-489/90

B_2ClErH_8 ... $ErCl[BH_4]_2 \cdot n\ C_4H_8O$ ... Sc: MVol.C11b-489/90

B_2ClEuH_8 ... $EuCl[BH_4]_2 \cdot n\ C_4H_8O$ ... Sc: MVol.C11b-489

B_2ClGdH_6 ... $GdCl[B_2H_6]$ ... Sc: MVol.C11b-491

B_2ClGdH_8 ... $GdCl[BH_4]_2 \cdot n\ C_4H_8O$ ... Sc: MVol.C11b-489/90

B_2ClH_6Sm ... $SmCl[B_2H_6]$ ... Sc: MVol.C11b-491

B_2ClH_6Tb ... $TbCl[B_2H_6]$ ... Sc: MVol.C11b-491

B_2ClH_8Ho ... $HoCl[BH_4]_2 \cdot n\ C_4H_8O$ ... Sc: MVol.C11b-489/90

B_2ClH_8La ... $LaCl[BH_4]_2 \cdot n\ C_4H_8O$ ... Sc: MVol.C11b-489

B_2ClH_8Nd ... $NdCl[BH_4]_2 \cdot n\ C_4H_8O$ ... Sc: MVol.C11b-489

B_2ClH_8Pr ... $PrCl[BH_4]_2 \cdot n\ C_4H_8O$ ... Sc: MVol.C11b-489

B_2ClH_8Sm ... $SmCl[BH_4]_2 \cdot n\ C_4H_8O$ ... Sc: MVol.C11b-489/90

B_2ClH_8Tb ... $TbCl[BH_4]_2 \cdot n\ C_4H_8O$ ... Sc: MVol.C11b-489/90

B_2ClH_8Y ... $YCl[BH_4]_2 \cdot n\ C_4H_8O$ ... Sc: MVol.C11b-489/90

B_2ClH_8Yb ... $YbCl[BH_4]_2 \cdot n\ C_4H_8O$ ... Sc: MVol.C11b-489/90

B_2Cl_2 ... B_2Cl_2 ... B: B Comp.SVol.3/4-34

$B_2Cl_2H_6Sm_2$ ... $Sm_2Cl_2[B_2H_6]$ ... Sc: MVol.C11b-491

B_2Cl_4 ... B_2Cl_4 ... B: B Comp.SVol.3/4-34
B: B Comp.SVol.4/4-35

$B_2Cl_7^-$ ... $[B_2Cl_7]^-$ ... B: B Comp.SVol.3/4-33

$B_2Cs_2F_8$ ... $Cs_2[BF_4]_2$ ... B: B Comp.SVol.4/3b-197

B_2Dy ... DyB_2 ... Sc: MVol.C11a-2/5
Sc: MVol.C11b-316/7

B_2Er ... ErB_2 ... Sc: MVol.C11a-2/5
Sc: MVol.C11b-330/1

B_2EuO_4 ... EuB_2O_4 ... Sc: MVol.C11b-443/7

$B_2Eu_2O_5$ ... $Eu_2B_2O_5$ ... Sc: MVol.C11b-442/3

$B_2Eu_3O_6$ ... $Eu_3(BO_3)_2$ ... Sc: MVol.C11b-441/2

$B_2FH_6^-$ ... $[FB_2H_6]^-$ ... B: B Comp.SVol.4/3b-224

– ... $[(H_3B)_2F]^-$ ... B: B Comp.SVol.3/3-362

$B_2F_2H_2O_8^{2-}$ ... $[B_2(O_2)_2F_2(O\text{-}OH)_2]^{2-}$ ... B: B Comp.SVol.4/3b-237

$B_2F_2H_8N_2O_6$ ... $[NH_4]_2[B_2(O_2)_3F_2]$ ... B: B Comp.SVol.4/3b-237

$B_2F_2O_5Sc_2$ ... $Sc_2B_2O_5F_2$ ... Sc: MVol.C11b-449

$B_2F_2S_2$ ... $F_2B_2S_2$ ... B: B Comp.SVol.3/4-136

$B_2F_4O_4^{2-}$ ... $[F_2B(O_2)_2BF_2]^{2-}$ ... B: B Comp.SVol.4/3b-237

B_2F_6 ... $(BF_3)_2$ ... B: B Comp.SVol.4/3b-97/8
– ... $(^{10}BF_3)(^{11}BF_3)$ ... B: B Comp.SVol.4/3b-98
– ... $(^{11}BF_3)_2$ ... B: B Comp.SVol.4/3b-98
$B_2F_6O_4S_2$ ... $(F_3B-SO_2)_2$ ... B: B Comp.SVol.4/3b-108
$B_2F_7^-$ ... $[B_2F_7]^-$ ... B: B Comp.SVol.3/3-251
– ... $[(F_3B)_2F]^-$ ... B: B Comp.SVol.3/3-362
$B_2F_8K_2$ ... $K_2[BF_4]_2$ ... B: B Comp.SVol.4/3b-197
$B_2F_8Rb_2$ ... $Rb_2[BF_4]_2$ ... B: B Comp.SVol.4/3b-197
B_2Gd ... GdB_2 ... Sc: MVol.C11a-2/5
Sc: MVol.C11b-264/6
$B_2H_2S_3$ ... (-BH-S-S-BH-S-) ... B: B Comp.SVol.3/4-111
B_2H_3N ... HB(-NH-BH-) ... B: B Comp.SVol.3/3-134
B: B Comp.SVol.4/3a-193
– ... $H_2B-N{=}BH$ ... B: B Comp.SVol.3/3-159
– ... $H_2N-B{=}BH$ ... B: B Comp.SVol.4/3a-239
$B_2H_4N_2$ ... $H_2N-B{=}B-NH_2$ ... B: B Comp.SVol.4/3a-239
$B_2H_4O_{12}^-$ ... $[B_2(O_2)_2(O-OH)_4]^-$ ... B: B Comp.SVol.4/3b-237
B_2H_5N ... $(H_2B)_2NH$ ... B: B Comp.SVol.3/3-159
$B_2H_5N_3$ ... $2,3-(NH_2)_2-NB_2H$ ... B: B Comp.SVol.4/3a-193, 194
– ... (-BH-NH-BH-NH-NH-) ... B: B Comp.SVol.3/3-137
$B_2H_6N^{3-}$ ... $[(H_3B)_2N]^{3-}$ ... B: B Comp.SVol.3/3-211
B_2H_6S ... $[-BH_2-H-BH_2-SH-]$ ... B: B Comp.SVol.4/4-128
$B_2H_6S_2$ ... [-BH(SH)-H-BH(SH)-H-] ... B: B Comp.SVol.4/4-128
– ... $[-BH_2-SH-BH_2-SH-]$ ... B: B Comp.SVol.4/4-128
B_2H_7N ... $[-BH_2-H-BH_2-NH_2-]$ ... B: B Comp.SVol.4/3a-239
$B_2H_{10}N_2$ ... $H_3B-NH_2-NH_2-BH_3$ ... B: B Comp.SVol.3/3-172
$B_2H_{11}NSi_2$ ... $(SiH_3)_2NB_2H_5$ ... Si: SVol.B4-148
B_2Ho ... HoB_2 ... Sc: MVol.C11a-2/5
Sc: MVol.C11b-323/4
$B_2I_2S_3$ ... [-B(I)-S-B(I)-S-S-] ... B: B Comp.SVol.3/4-138
$B_2I_2Se_3$ ... [-B(I)-Se-B(I)-Se-Se-] ... B: B Comp.SVol.4/4-171/2
B_2La ... LaB_2 ... Sc: MVol.C11a-2
$B_2LaNa_3O_6$ ... $Na_3La(BO_3)_2$ ... Sc: MVol.C11b-467/8
doped with Nd^{3+} ... Sc: MVol.C11b-469
B_2Lu ... LuB_2 ... Sc: MVol.C11a-2/5
Sc: MVol.C11b-377/8
$B_2Na_3NdO_6$ ... $Na_3Nd(BO_3)_2$ ... Sc: MVol.C11b-466, 467/8
B_2Nd ... NdB_2 ... Sc: MVol.C11b-115
B_2O_3 ... B_2O_3 glasses
$B_2O_3-CdO-La_2O_3$ ... Sc: MVol.C11b-435
$B_2O_3-CeO_2-Na_2O$ ... Sc: MVol.C11b-466
$B_2O_3-Ce_2O_3$ ... Sc: MVol.C11b-386
$B_2O_3-Dy_2O_3$ ... Sc: MVol.C11b-385/6
$B_2O_3-Dy_2O_3-La_2O_3$ ... Sc: MVol.C11b-388
$B_2O_3-Dy_2O_3-Pr_2O_3$ ... Sc: MVol.C11b-388
$B_2O_3-Er_2O_3$ ... Sc: MVol.C11b-385/6
$B_2O_3-Er_2O_3-La_2O_3$ ... Sc: MVol.C11b-388
$B_2O_3-Er_2O_3-Pr_2O_3$ ... Sc: MVol.C11b-388
$B_2O_3-Eu_2O_3$ ... Sc: MVol.C11b-385/6

B_2O_3 .... B_2O_3 glasses
B_2O_3-Eu_2O_3-La_2O_3 .... Sc: MVol.C11b-388
B_2O_3-Eu_2O_3-Na_2O .... Sc: MVol.C11b-466
B_2O_3-Eu_2O_3-Pr_2O_3 .... Sc: MVol.C11b-388
B_2O_3-Gd_2O_3 .... Sc: MVol.C11b-385/6
B_2O_3-Gd_2O_3-La_2O_3 .... Sc: MVol.C11b-388
B_2O_3-Gd_2O_3-Pr_2O_3 .... Sc: MVol.C11b-388
B_2O_3-Ho_2O_3 .... Sc: MVol.C11b-385/6
B_2O_3-Ho_2O_3-La_2O_3 .... Sc: MVol.C11b-388
B_2O_3-Ho_2O_3-Pr_2O_3 .... Sc: MVol.C11b-388
B_2O_3-La_2O_3 .... Sc: MVol.C11b-385/6, 388, 435
B_2O_3-La_2O_3-Li_2O .... Sc: MVol.C11b-455
B_2O_3-La_2O_3-Nd_2O_3 .... Sc: MVol.C11b-388
B_2O_3-La_2O_3-Pr_2O_3 .... Sc: MVol.C11b-388
B_2O_3-La_2O_3-Sm_2O_3 .... Sc: MVol.C11b-388
B_2O_3-La_2O_3-Tb_2O_3 .... Sc: MVol.C11b-388
B_2O_3-La_2O_3-Tm_2O_3 .... Sc: MVol.C11b-388
B_2O_3-La_2O_3-Y_2O_3 .... Sc: MVol.C11b-388, 435
B_2O_3-La_2O_3-Yb_2O_3 .... Sc: MVol.C11b-388
B_2O_3-LiCl-Li_2O .... B: B Comp.SVol.3/4-33
B_2O_3-Li_2O-TeO_2 .... B: B Comp.SVol.4/4-176
B_2O_3-Lu_2O_3 .... Sc: MVol.C11b-385/6
B_2O_3-Na_2O-Nd_2O_3 .... Sc: MVol.C11b-466
B_2O_3-Nd_2O_3 .... Sc: MVol.C11b-385/6
B_2O_3-Nd_2O_3-Pr_2O_3 .... Sc: MVol.C11b-388
B_2O_3-Pr_2O_3 .... Sc: MVol.C11b-386, 388
B_2O_3-Pr_2O_3-Sm_2O_3 .... Sc: MVol.C11b-388
B_2O_3-Pr_2O_3-Tb_2O_3 .... Sc: MVol.C11b-388
B_2O_3-Pr_2O_3-Tm_2O_3 .... Sc: MVol.C11b-388
B_2O_3-Pr_2O_3-Yb_2O_3 .... Sc: MVol.C11b-388
B_2O_3-Sc_2O_3 .... Sc: MVol.C11b-385/6
B_2O_3-Sm_2O_3 .... Sc: MVol.C11b-385/6
B_2O_3-TeO_2 .... B: B Comp.SVol.4/4-176
B_2O_3-Tm_2O_3 .... Sc: MVol.C11b-385/6
B_2O_3-Y_2O_3 .... Sc: MVol.C11b-385/6
B_2O_3-Yb_2O_3 .... Sc: MVol.C11b-385/6
– .... B_2O_3 melts
B_2O_3-Ce_2O_3 .... Sc: MVol.C11b-386/7
B_2O_3-La_2O_3 .... Sc: MVol.C11b-386/7
B_2O_3-Nd_2O_3 .... Sc: MVol.C11b-386/7
B_2O_3-Sm_2O_3 .... Sc: MVol.C11b-386/7
B_2O_3-Y_2O_3 .... Sc: MVol.C11b-386/7
– .... B_2O_3 systems
B_2O_3-CeO_2-Na_2O .... Sc: MVol.C11b-465/6
B_2O_3-Ce_2O_3 .... Sc: MVol.C11b-386
B_2O_3-CoO-Sm_2O_3 .... Sc: MVol.C11b-440
B_2O_3-Dy_2O_3 .... Sc: MVol.C11b-385/6
B_2O_3-Dy_2O_3-Na_2O .... Sc: MVol.C11b-465/6

B_2O_3 . . . . B_2O_3 systems
B_2O_3-Er_2O_3 . . . . Sc: MVol.C11b-385/6
B_2O_3-Er_2O_3-Na_2O . . . . Sc: MVol.C11b-465/6
B_2O_3-Eu_2O_3 . . . . Sc: MVol.C11b-385/6
B_2O_3-Gd_2O_3 . . . . Sc: MVol.C11b-385/6
B_2O_3-Gd_2O_3-Na_2O . . . . Sc: MVol.C11b-465/6
B_2O_3-GeO_2-Nd_2O_3 . . . . Sc: MVol.C11b-439
B_2O_3-Ho_2O_3 . . . . Sc: MVol.C11b-385/6
B_2O_3-La_2O_3 . . . . Sc: MVol.C11b-385/6, 434/5
B_2O_3-La_2O_3-H_2O . . . . Sc: MVol.C11b-437
B_2O_3-La_2O_3-Li_2O . . . . Sc: MVol.C11b-455/6
B_2O_3-Li_2O-Nd_2O_3 . . . . Sc: MVol.C11b-456
B_2O_3-Li_2O-Y_2O_3 . . . . Sc: MVol.C11b-453/5
B_2O_3-Lu_2O_3 . . . . Sc: MVol.C11b-385/6
B_2O_3-Lu_2O_3-Na_2O . . . . Sc: MVol.C11b-465/6
B_2O_3-Na_2O-Nd_2O_3 . . . . Sc: MVol.C11b-465/6
B_2O_3-Na_2O-Pr_2O_3 . . . . Sc: MVol.C11b-465/6
B_2O_3-Na_2O-Sm_2O_3 . . . . Sc: MVol.C11b-465/6
B_2O_3-Na_2O-Tb_2O_3 . . . . Sc: MVol.C11b-465/6
B_2O_3-Na_2O-Yb_2O_3 . . . . Sc: MVol.C11b-465/6
B_2O_3-Nd_2O_3 . . . . Sc: MVol.C11b-385/6
B_2O_3-Pr_2O_3 . . . . Sc: MVol.C11b-386
B_2O_3-Sc_2O_3 . . . . Sc: MVol.C11b-385/6, 431
B_2O_3-Sm_2O_3 . . . . Sc: MVol.C11b-385/6
B_2O_3-Tb_2O_3 . . . . Sc: MVol.C11b-386
B_2O_3-Tm_2O_3 . . . . Sc: MVol.C11b-385/6
B_2O_3-Y_2O_3 . . . . Sc: MVol.C11b-385/6
B_2O_3-Yb_2O_3 . . . . Sc: MVol.C11b-385/6

B_2O_4Pd . . . . PdB_2O_4 . . . . Pd: SVol.B2-260/1
B_2Pd . . . . PdB_2 . . . . Pd: SVol.B2-258
B_2Pd_3 . . . . Pd_3B_2 . . . . Pd: SVol.B2-258
B_2Pd_5 . . . . Pd_5B_2 . . . . Pd: SVol.B2-256/8
B_2S_2Se . . . . B_2S_2Se . . . . B: B Comp.SVol.4/4-169/70
B_2S_3 . . . . B_2S_3 . . . . B: B Comp.SVol.3/4-107
B: B Comp.SVol.4/4-121/2
– . . . . $^{10}B_2{}^{32}S_3$ . . . . B: B Comp.SVol.4/4-122
– . . . . $^{10}B^{11}B^{32}S_3$ . . . . B: B Comp.SVol.4/4-122
– . . . . $^{11}B_2{}^{32}S_3$ . . . . B: B Comp.SVol.4/4-122
– . . . . B_2S_3 glasses
B_2S_3-B_2Se_3 . . . . B: B Comp.SVol.4/4-170
B_2S_3-LiBr-Li_2S . . . . B: B Comp.SVol.4/4-122
B_2S_3-LiCl-Li_2S . . . . B: B Comp.SVol.4/4-122
B_2S_3-LiI-Li_2S . . . . B: B Comp.SVol.4/4-121/2
B_2S_3-LiI-Li_2S-P_2S_5-SiS_2 . . . . B: B Comp.SVol.4/4-122
B_2S_3-Li_2O . . . . B: B Comp.SVol.4/4-121
B_2S_3-Li_2S . . . . B: B Comp.SVol.4/4-121/2
B_2S_3-Li_2S-P_2S_5 . . . . B: B Comp.SVol.4/4-122
B_2S_3-Na_2S . . . . B: B Comp.SVol.4/4-122

B_2S_4 . . . . B_2S_4 . . . . B: B Comp.SVol.3/4-107

Formula	Compound	Reference
B_2S_5	B_2S_5	B: B Comp.SVol.4/4-121
B_2Sc	ScB_2	Sc: MVol.C11a-2/5, 119/25
B_2Se	B_2Se	B: B Comp.SVol.4/4-169
B_2Se_3	B_2Se_3 glasses	
	B_2Se_3-B_2S_3	B: B Comp.SVol.4/4-170
B_2Sm	SmB_2	Sc: MVol.C11a-2/5
		Sc: MVol.C11b-137/8
B_2Tb	TbB_2	Sc: MVol.C11a-2/5
		Sc: MVol.C11b-307/9
B_2Tm	TmB_2	Sc: MVol.C11a-2/5
		Sc: MVol.C11b-340/2
B_2Y	YB_2	Sc: MVol.C11a-2/5, 129/32
B_2Yb	YbB_2	Sc: MVol.C11a-2/5
		Sc: MVol.C11b-346/7
$B_3BrC_6F_5NS_2$	$BB(-S-S-BBr-NC_6F_5-)$	B: B Comp.SVol.3/4-143
$B_3BrC_{26}FeH_{40}$	$[(CH_3)_6C_6]Fe[2-(BrCH_2-4-C_6H_4-CH_2)-$	
	$4,5-(C_2H_5)_2-1,2,3-B_3C_2H_4]$	Fe: Org.Comp.B18-81, 86
$B_3Br_2C_5H_{13}N_2S_4$	$S_4B_2Br_2-CH_3B(-NCH_3-C_2H_4-NCH_3-)$	B: B Comp.SVol.3/4-144
$B_3Br_2C_{17}H_{39}N_4$	$2,2,6,6-(CH_3)_4-NC_5H_6-1-BBr-N(C_4H_9-t)$	
	$-B[N(CH_3)_2]-BBr-N(CH_3)_2$	B: B Comp.SVol.4/4-106
$B_3Br_2HS_3$	$(BrBS)_2BSH$	B: B Comp.SVol.4/4-161
$B_3Br_3C_3H_9N$	$(CH_3-BBr)_3N$	B: B Comp.SVol.4/4-99/100
$B_3Br_3C_3H_9NS_2$	$CH_3B[-S-S-(-BBr_3)-BCH_3-NCH_3-]$	B: B Comp.SVol.3/4-144
$B_3Br_3C_3H_9N_3$	$(-BBr-NCH_3-)_3$	B: B Comp.SVol.3/4-92
$B_3Br_3C_9H_{27}N_3Si_3$	$[-BBr-N(Si(CH_3)_3)-BBr-N(Si(CH_3)_3)-BBr-N(Si(CH_3)_3)-]$	
		B: B Comp.SVol.4/4-106
$B_3Br_3C_{16}H_{40}N_8$	$[(-NCH_3-CH_2-CH_2-NCH_3-)_2B]_2 \cdot BBr_3$	B: B Comp.SVol.3/3-118
$B_3Br_3C_{18}F_{15}N_3$	$N(C_6F_5)BBrN(C_6F_5)BBrN(C_6F_5)BBr$	F: PFHOrg.SVol.4-233/4, 246
$B_3Br_3C_{18}H_{15}N_3$	$[-BH-N(C_6H_4-4-Br)-]_3$	B: B Comp.SVol.3/3-129
$B_3Br_3S_3$	$[-BBr-S-BBr-S-BBr-S-]$	B: B Comp.SVol.4/4-157
B_3Br_6N	$N(BBr_2)_3$	B: B Comp.SVol.4/4-99
$B_3CH_4^{3-}$	$[CB_3H_4]^{3-}$	B: B Comp.SVol.3/4-158
$B_3CH_7NS^-$	$[B_3H_7(NCS)]^-$	B: B Comp.SVol.4/4-151
$B_3CH_8N_3$	$HB(-NH-BH-NH-BH-NCH_3-)$	B: B Comp.SVol.3/3-133
$B_3C_2Cl_4H_6N$	$Cl_2B-N[BCl-CH_3]_2$	B: B Comp.SVol.4/4-59
$B_3C_2H_5$	$1,2-C_2B_3H_5$	B: B Comp.SVol.4/4-181
–	$1,5-C_2B_3H_5$	B: B Comp.SVol.3/4-158
		B: B Comp.SVol.4/4-181/2
–	$2,3-C_2B_3H_5$	B: B Comp.SVol.4/4-181
$B_3C_2H_7$	$1,2-C_2B_3H_7$	B: B Comp.SVol.4/4-182
$B_3C_2H_8N_3$	$CH_2{=}CH-B[-NH-BH-NH-BH-NH-]$	B: B Comp.SVol.4/3a-189
$B_3C_2H_9OS$	$[-B(H)_2-H-B(H)-OC(CH_3)S-B(H)-H-]$	B: B Comp.SVol.4/4-137
$B_3C_2H_{14}In$	$(CH_3)_2In-B_3H_8$	In: Org.Comp.1-326
$B_3C_3Cl_3H_9N_3$	$(-BCl-NCH_3-)_3$	B: B Comp.SVol.3/4-60
$B_3C_3Cl_3H_9N_3O_3$	$[-BCl-N(OCH_3)-BCl-N(OCH_3)-BCl-N(OCH_3)-]$	B: B Comp.SVol.4/4-66
$B_3C_3F_3H_9N_3$	$[-BF-N(CH_3)-]_3$	B: B Comp.SVol.3/3-377
		B: B Comp.SVol.4/3b-247

$B_3C_3F_9H_6N_6S_3$ . . . [-B(NH-SCF_3)-NH-B(NH-SCF_3)-NH-B(NH-SCF_3)-NH-] B: B Comp.SVol.4/4-148, 166
B: B Comp.SVol.4/3a-186
$B_3C_3H_9I_3N_3$ . . . . . . (-BI-NCH_3-)$_3$ . . . . . . B: B Comp.SVol.3/4-101
$B_3C_3H_{10}NO_2$ . . . . . [-B(CH_3)-NH-B(CH_3)-O-B(CH_3)-O-] . . . . . . B: B Comp.SVol.4/3a-203/5
$B_3C_3H_{10}N_3$ . . . . . . CH_2=C(CH_3)-B[-NH-BH-NH-BH-NH-] . . . . . . B: B Comp.SVol.4/3a-189/90
– . . . . . . CH_3-CH=CH-B[-NH-BH-NH-BH-NH-] . . . . . . B: B Comp.SVol.4/3a-189/90
$B_3C_3H_{11}N_2O$ . . . . . [-B(CH_3)-NH-B(CH_3)-NH-B(CH_3)-O-] . . . . . . B: B Comp.SVol.4/3a-203/5
$B_3C_3H_{12}N_3$ . . . . . . [-BH-N(CH_3)-BH-N(CH_3)-BH-N(CH_3)-] . . . . . . B: B Comp.SVol.3/3-129
– . . . . . . [-NH-B(CH_3)-NH-B(CH_3)-NH-B(CH_3)-] . . . . . . B: B Comp.SVol.3/3-130
$B_3C_3H_{14}N_3Si$ . . . . . HB[-NH-BH-NH-BH-NSi(CH_3)$_3$-] . . . . . . B: B Comp.SVol.3/3-133
$B_3C_3H_{15}N_6$ . . . . . . [-B(NH_2)-NCH_3-B(NH_2)-NCH_3-B(NH_2)-NCH_3-] B: B Comp.SVol.3/3-130
B: B Comp.SVol.4/3a-186
$B_3C_3H_{18}N_3$ . . . . . . [-BH_2-$NHCH_3$-BH_2-$NHCH_3$-BH_2-$NHCH_3$-] . . . B: B Comp.SVol.4/3a-203
$B_3C_4CeH_{20}O$ . . . . . Ce[BH_4]$_3$ · C_4H_8O . . . . . . Sc: MVol.C11b-479, 491
$B_3C_4Cl_2H_{12}N_3$ . . . . [-B(CH_3)-N(CH_3)-BCl-N(CH_3)-BCl-N(CH_3)-] . . B: B Comp.SVol.4/4-66
$B_3C_4H_7$ . . . . 1,5-$C_2B_3H_4$-2-CH=CH_2 . . . . . . B: B Comp.SVol.3/4-159
$B_3C_4H_8O_4Re$ . . . . . (OC)$_4$Re(B_3H_8) . . . . . . Re: Org.Comp.1-350/1
$B_3C_4H_{12}NO_2$ . . . . . [-B(CH_3)-N(CH_3)-B(CH_3)-O-B(CH_3)-O-] . . . . . B: B Comp.SVol.4/3a-203/5
$B_3C_4H_{16}NO_9$ . . . . . [($HOCH_2CH_2$)$_2NH_2$][$H_4B_3O_7$ · 2 H_2O] . . . . . . B: B Comp.SVol.3/3-225/6
$B_3C_4H_{20}LaO$ . . . . . La[BH_4]$_3$ · C_4H_8O . . . . . . Sc: MVol.C11b-479, 491
$B_3C_4H_{20}OSc$ . . . . . Sc[BH_4]$_3$ · C_4H_8O . . . . . . Sc: MVol.C11b-477/8, 484, 485
$B_3C_5ClH_{15}N_3$ . . . . [-BCH_3-NCH_3-BCl-NCH_3-BCH_3-NCH_3-] . . . . . B: B Comp.SVol.4/4-66
$B_3C_5FeH_5O_3$ . . . . . 1,1,1-(CO)$_3$-1,2,6-$FeC_2B_3H_5$ . . . . . . B: B Comp.SVol.3/4-159
$B_3C_5FeH_7O_3$ . . . . . $C_2B_3H_7Fe(CO)_3$ . . . . . . B: B Comp.SVol.3/4-159
$B_3C_5H_{14}NO_2$ . . . . . [-B(CH_3)-N(C_2H_5)-B(CH_3)-O-B(CH_3)-O-] . . . . B: B Comp.SVol.4/3a-203/5
$B_3C_5H_{15}N_2O$ . . . . . [-B(CH_3)-N(CH_3)-B(CH_3)-N(CH_3)-B(CH_3)-O-] B: B Comp.SVol.4/3a-203/5
$B_3C_5H_{15}N_2O_5S_2$ . . CH_3B[-N(SO_2CH_3)-BCH_3-N(SO_2CH_3)-BCH_3-O-] B: B Comp.SVol.3/4-135
$B_3C_5H_{15}N_5$ . . . . . . [-N_3B_3(CH_3)$_3$(NCH_3-)-NCH_3-]$_n$ . . . . . . B: B Comp.SVol.4/3a-190
$B_3C_6Cl_3GaH_{18}N_3$ (-BCH_3-NCH_3-)$_3$ · $GaCl_3$ . . . . . . B: B Comp.SVol.3/3-130
$B_3C_6F_3H_{15}N_3$ . . . . [-BF-N(C_2H_5)-]$_3$ . . . . . . B: B Comp.SVol.4/3b-247
$B_3C_6F_5H_{10}O_5$ . . . . $B_3C_2F_5O_5(C_2H_5)_2$ . . . . . . B: B Comp.SVol.4/3b-235
$B_3C_6GdH_{24}O_{1.5}$ . . Gd[BH_4]$_3$ · 1.51 C_4H_8O . . . . . . Sc: MVol.C11b-478/80
$B_3C_6H_9N_6O_3$ . . . . . [-B(NCO)-N(CH_3)-B(NCO)-N(CH_3)-B(NCO)-N(CH_3)-] B: B Comp.SVol.4/3a-187
$B_3C_6H_9N_6S_3$ . . . . . (-BNCS-NCH_3-)$_3$ . . . . . . B: B Comp.SVol.3/3-130
$B_3C_6H_{11}$ . . . . 1,5-$C_2B_3H_4$-2-[C(=CH_2)-C_2H_5] . . . . . . B: B Comp.SVol.3/4-159
– . . . . . . 1,5-$C_2B_3H_4$-2-[C(CH_3)=$CHCH_3$] . . . . . . B: B Comp.SVol.3/4-159
– . . . . . . 1,5-$C_2B_3H_4$-2-[CH=CH-C_2H_5] . . . . . . B: B Comp.SVol.3/4-159
$B_3C_6H_{12}NS_6$ . . . . . [1,3,2-$S_2BC_2H_4$-2-]$_3$N . . . . . . B: B Comp.SVol.3/4-122/3
B: B Comp.SVol.4/4-147
$B_3C_6H_{16}NO_2$ . . . . . [-B(CH_3)-N(C_3H_7-i)-B(CH_3)-O-B(CH_3)-O-] . . . B: B Comp.SVol.4/3a-203/5
$B_3C_6H_{18}I_2N_3$ . . . . . (-BCH_3-NCH_3-)$_3$ · I_2 . . . . . . B: B Comp.SVol.3/3-130
$B_3C_6H_{18}N$ . . . . [(CH_3)$_2$B]$_3$N . . . . . . B: B Comp.SVol.3/3-159
$B_3C_6H_{18}NO_2Si$ . . . [-B(CH_3)-N(Si(CH_3)$_3$)-B(CH_3)-O-B(CH_3)-O-] B: B Comp.SVol.4/3a-203/5
$B_3C_6H_{18}NS_6$ . . . . . N[B(SCH_3)$_2$]$_3$ . . . . . . B: B Comp.SVol.3/4-122/3

$B_3C_6H_{18}N_3$ ...... $[-BCH_3-NCH_3-BCH_3-NCH_3-BCH_3-NCH_3-]$ ... B: B Comp.SVol.3/3-130
B: B Comp.SVol.4/3a-187
– .... $[-B(C_2H_5)-NH-B(C_2H_5)-NH-B(C_2H_5)-NH]$ .... B: B Comp.SVol.3/3-130
B: B Comp.SVol.4/3a-185/6
$B_3C_6H_{18}N_3O_3$ .... $[-B(N(CH_3)_2)-O-]_3$ .... B: B Comp.SVol.4/3b-67
– .... $[-B(OCH_3)-N(CH_3)-]_3$ .... B: B Comp.SVol.3/3-130
$B_3C_6H_{18}N_3O_6$ .... $[-B(OCH_3)-N(OCH_3)-B(OCH_3)-N(OCH_3)$
$-B(OCH_3)-N(OCH_3)-]$ .... B: B Comp.SVol.4/3a-189
$B_3C_6H_{18}N_3S_3$ .... $(-BSCH_3-NCH_3-)_3$ .... B: B Comp.SVol.3/4-129
$B_3C_6H_{19}N_2O_3$ .... $[-B(C_2H_5)-O-]_3 \cdot N_2H_4$ .... B: B Comp.SVol.4/3b-73
$B_3C_6H_{22}N_3Si_2$ .... $HB[-NH-BH-NSi(CH_3)_3-BH-NSi(CH_3)_3-]$ ..... B: B Comp.SVol.3/3-133
$B_3C_6H_{23}N_4O_3$ .... $[-B(C_2H_5)-O-]_3 \cdot 2\ N_2H_4$ .... B: B Comp.SVol.4/3b-73
$B_3C_7H_{13}MoO_2$ ... $(C_5H_5)Mo(CO)_2(B_3H_8)$ Mo:Org.Comp.7-43
$B_3C_7H_{18}NO_2$ ..... $[-BCH_3-N(C_4H_9-i)-BCH_3-O-BCH_3-O-]$ ...... B: B Comp.SVol.4/3a-203/5
– .... $[-BCH_3-N(C_4H_9-t)-BCH_3-O-BCH_3-O-]$ ...... B: B Comp.SVol.3/3-139
$B_3C_7H_{19}N_2O$ ..... $[-BCH_3-N(C_2H_5)-BCH_3-N(C_2H_5)-BCH_3-O-]$ .. B: B Comp.SVol.4/3a-203/5
$B_3C_8CeH_{28}O_2$ .... $Ce[BH_4]_3 \cdot 2\ C_4H_8O$ .... Sc: MVol.C11b-479, 488, 491
$B_3C_8DyH_{28}O_2$ .... $Dy[BH_4]_3 \cdot 2\ C_4H_8O$ .... Sc: MVol.C11b-478/80, 486
$B_3C_8ErH_{28}O_2$ .... $Er[BH_4]_3 \cdot 2\ C_4H_8O$ .... Sc: MVol.C11b-478/80, 486
$B_3C_8EuH_{28}O_2$ .... $Eu[BH_4]_3 \cdot 2\ C_4H_8O$ .... Sc: MVol.C11b-478
$B_3C_8F_5H_{14}O_5$ .... $B_3C_2F_5O_5(C_3H_7-n)_2$ .... B: B Comp.SVol.4/3b-235
$B_3C_8GdH_{28}O_2$ .... $Gd[BH_4]_3 \cdot 2\ C_4H_8O$ .... Sc: MVol.C11b-478/80, 486, 488
$B_3C_8GdH_{44}N_8$ .... $Gd[BH_4]_3 \cdot 4\ (NH_2-CH_2CH_2-NH_2)$ .... Sc: MVol.C11b-488
$B_3C_8H_{23}N_2O_3$ .... $[-B(C_2H_5)-O-]_3 \cdot (CH_3)_2N-NH_2$ .... B: B Comp.SVol.4/3b-73
– .... $[-B(C_2H_5)-O-]_3 \cdot CH_3-NHNH-CH_3$ .... B: B Comp.SVol.4/3b-73
$B_3C_8H_{28}HoO_2$ .... $Ho[BH_4]_3 \cdot 2\ C_4H_8O$ .... Sc: MVol.C11b-486
$B_3C_8H_{28}LaO_2$ .... $La[BH_4]_3 \cdot 2\ C_4H_8O$ .... Sc: MVol.C11b-477/8, 486, 488, 492
$B_3C_8H_{28}LuO_2$ .... $Lu[BH_4]_3 \cdot 2\ C_4H_8O$ .... Sc: MVol.C11b-478/80, 486
$B_3C_8H_{28}NdO_2$ .... $Nd[BH_4]_3 \cdot 2\ C_4H_8O$ .... Sc: MVol.C11b-478/80, 486, 491
$B_3C_8H_{28}O_2Sc$ .... $Sc[BH_4]_3 \cdot 2\ C_4H_8O$ .... Sc: MVol.C11b-475, 477/81, 484/8, 492
$B_3C_8H_{28}O_2Sm$ ... $Sm[BH_4]_3 \cdot 2\ C_4H_8O$ .... Sc: MVol.C11b-478
$B_3C_8H_{28}O_2Tm$ ... $Tm[BH_4]_3 \cdot 2\ C_4H_8O$ .... Sc: MVol.C11b-478/80
$B_3C_8H_{28}O_2Y$ ..... $Y[BH_4]_3 \cdot 2\ C_4H_8O$ .... Sc: MVol.C11b-475, 478, 481/2, 484/7
$B_3C_8H_{28}O_2Yb$ .... $Yb[BH_4]_3 \cdot 2\ C_4H_8O$ .... Sc: MVol.C11b-478/80
$B_3C_8H_{44}LaN_8$ .... $La[BH_4]_3 \cdot 4\ (NH_2-CH_2CH_2-NH_2)$ .... Sc: MVol.C11b-488
$B_3C_8H_{44}LuN_8$ .... $Lu[BH_4]_3 \cdot 4\ (NH_2-CH_2CH_2-NH_2)$ .... Sc: MVol.C11b-488
$B_3C_9ClCoH_{15}Hg$ .. $[(C_5H_5)Co(CH_3)_2C_2B_3H_4]HgCl$ .... B: B Comp.SVol.3/4-159
$B_3C_9Cl_2CoH_{15}Hg^-$
$[(C_5H_5)Co(CH_3)_2C_2B_3H_4 \cdot HgCl_2]^-$ .... B: B Comp.SVol.3/4-159
$B_3C_9Cl_3H_{27}N_3Si_3$ $[-BCl-N(Si(CH_3)_3)-BCl-N(Si(CH_3)_3)-BCl-N(Si(CH_3)_3)-]$
B: B Comp.SVol.3/4-60
B: B Comp.SVol.4/4-66
$B_3C_9CrH_{18}N_3O_3$ .. $(-BCH_3-NCH_3-)_3 \cdot Cr(CO)_3$ .... B: B Comp.SVol.3/3-130
$B_3C_9FH_{13}NO_2$ .... $CH_3B(-NC_6H_4F-BCH_3-O-BCH_3-O-)$ .... B: B Comp.SVol.3/3-139
$B_3C_9F_3H_{27}N_3Si_3$ .. $[-BF-N(Si(CH_3)_3)-]_3$ .... B: B Comp.SVol.4/3b-247

$B_3C_9H_{14}NO_2$ ..... $[-B(CH_3)-N(C_6H_5)-B(CH_3)-O-B(CH_3)-O-]$ .... B: B Comp.SVol.3/3-139
B: B Comp.SVol.4/3a-203/5

$B_3C_9H_{18}N_3O_3$ .... $(BC_3H_6NO)_3$ .... B: B Comp.SVol.4/3a-187

– .... $[(BC_3H_6NO)(BC_2H_3NO(CH_3))_2]$ .... B: B Comp.SVol.4/3a-187

$B_3C_9H_{19}N_2O_3$ .... $HO-B(C_2H_5)-O-B(C_2H_5)-O-B(C_2H_5)-N_2C_3H_3$. B: B Comp.SVol.4/3b-58/9

– .... $[-B(C_2H_5)-O-B(C_2H_5)-O-B(C_2H_5)-O-] \cdot 1,2-N_2C_3H_4$
B: B Comp.SVol.4/3b-58/9

$B_3C_9H_{20}NO_2$ ..... $CH_3B[-N(c-C_6H_{11})-BCH_3-O-BCH_3-O-]$ ...... B: B Comp.SVol.3/3-139

$B_3C_9H_{21}N_6$ ...... $[BC_2H_4N_2(CH_3)]_3$ .... B: B Comp.SVol.4/3a-187

$B_3C_9H_{22}N_5$ ...... $NH_2-B(C_2H_5)-NH-B(C_2H_5)-NH-B(C_2H_5)-N_2C_3H_3$
B: B Comp.SVol.4/3b-37

$B_3C_9H_{23}N_2O$ ..... $[-B(CH_3)-N(C_3H_7-i)-B(CH_3)-N(C_3H_7-i)-B(CH_3)-O-]$
B: B Comp.SVol.4/3a-203/5

$B_3C_9H_{24}N_3$ ...... $[-B(C_2H_5)-N(CH_3)-B(C_2H_5)-N(CH_3)-B(C_2H_5)-N(CH_3)-]$
B: B Comp.SVol.4/3a-188

$B_3C_9H_{27}N_2OSi_2$ .. $[-B(CH_3)-N(Si(CH_3)_3)-B(CH_3)-N(Si(CH_3)_3)-B(CH_3)-O-]$
B: B Comp.SVol.4/3a-203/5

$B_3C_9H_{27}N_6$ ...... $[-BN(CH_3)_2-NCH_3-]_3$ .... B: B Comp.SVol.3/3-130

$B_3C_9H_{27}N_6O_3$ .... $[-B(OCH_3)-N(N(CH_3)_2)-B(OCH_3)-N(N(CH_3)_2)$
$-B(OCH_3)-N(N(CH_3)_2)-]$ .... B: B Comp.SVol.4/3a-189

$B_3C_9H_{30}N_3O$ ..... $BH_3-N(CH_3)_2-CH_2-B(OH)[-CH_2-N(CH_3)_2-BH_2$
$-CH_2-N(CH_3)_2-]$ .... B: B Comp.SVol.4/3b-72

$B_3C_9H_{30}N_3Si_3$ .... $[-BH-NSi(CH_3)_3-]_3$ .... B: B Comp.SVol.3/3-129

$B_3C_9H_{33}N_6Si_3$ .... $[-B(NH_2)-N(Si(CH_3)_3)-B(NH_2)-N(Si(CH_3)_3)$
$-B(NH_2)-N(Si(CH_3)_3)-]$ .... B: B Comp.SVol.4/3a-186

– .... $[-B(NH-Si(CH_3)_3)-NH-B(NH-Si(CH_3)_3)-NH$
$-B(NH-Si(CH_3)_3)-NH-]$ .... B: B Comp.SVol.4/3a-186

$B_3C_9H_{36}N_3Si_3$ .... $[H_2B-NH(Si(CH_3)_3)]_3$ .... B: B Comp.SVol.3/3-146

$B_3C_{10}CeH_{22}N_2$ ... $Ce[BH_4]_3 \cdot 2\ C_5H_5N$ .... Sc: MVol.C11b-488/9

$B_3C_{10}H_{21}$ .... $1,5-C_2B_3-1,5-(CH_3)_2-2,3,4-(C_2H_5)_3$ .... B: B Comp.SVol.3/4-158
B: B Comp.SVol.4/4-182

$B_3C_{10}H_{21}{}^{2-}$ ...... $[1,5-C_2B_3-1,5-(CH_3)_2-2,3,4-(C_2H_5)_3]^{2-}$ .... B: B Comp.SVol.4/4-182

$B_3C_{10}H_{22}LaN_2$ ... $La[BH_4]_3 \cdot 2\ C_5H_5N$ .... Sc: MVol.C11b-488/9

$B_3C_{10}H_{22}N_2Nd$ ... $Nd[BH_4]_3 \cdot 2\ C_5H_5N$ .... Sc: MVol.C11b-488/9

$B_3C_{10}H_{25}N_4S_2$ ... $6-(t-C_4H_9)-3,5-(CH_3)_2-2-[1,3-(CH_3)_2-$
$1,3,2-N_2BC_2H_4-2]-1,4,2,6,3,5-S_2N_2B_2$ B: B Comp.SVol.4/4-140/1

$B_3C_{10}H_{32}LaO_{2.5}$ .. $La[BH_4]_3 \cdot 2.5\ C_4H_8O$ .... Sc: MVol.C11b-479, 491

$B_3C_{11}FeH_{17}$ ..... $1,2,3-FeC_2B_3H_6-2-(CH_2CH_2CH_2-C_6H_5)$ ...... B: B Comp.SVol.4/4-197/8
Fe: Org.Comp.B19-409

$B_3C_{11}H_{20}NO_2$ .... $CH_3B[-N(c-C_6H_{10}-2-C(=CH))-BCH_3-O-BCH_3-O-]$
B: B Comp.SVol.3/3-139

$B_3C_{11}H_{23}N_2O_3$ ... $[-B(C_2H_5)-O-]_3 \cdot 1,2-N_2C_3H_2-3,5-(CH_3)_2$ .... B: B Comp.SVol.4/3b-73

$B_3C_{11}H_{27}N_2O$ .... $[-BCH_3-N(C_4H_9-i)-BCH_3-N(C_4H_9-i)-BCH_3-O-]$
B: B Comp.SVol.4/3a-203/5

– .... $[-BCH_3-N(C_4H_9-t)-BCH_3-N(C_4H_9-t)-BCH_3-O-]$
B: B Comp.SVol.3/3-139

$B_3C_{12}CeH_{36}O_3$ ... $Ce[BH_4]_3 \cdot 3\ OC_4H_8$ .... Sc: MVol.C11b-487/8

$B_3C_{12}DyH_{36}O_3$ ... $Dy[BH_4]_3 \cdot 2.95\ OC_4H_8$ .... Sc: MVol.C11b-478/80

– .... $Dy[BH_4]_3 \cdot 3\ OC_4H_8$ .... Sc: MVol.C11b-475, 478/80, 491

$B_3C_{12}ErH_{36}O_3$ ... $Er[BH_4]_3 \cdot 3\ OC_4H_8$ ... Sc: MVol.C11b-475, 477/80, 483, 485/6, 491

– ... $Er[BH_4]_3 \cdot 3\ OC_4H_8$ solid solutions
$Er[BH_4]_3 \cdot 3\ OC_4H_8$-$Gd[BH_4]_3 \cdot 3\ OC_4H_8$ ... Sc: MVol.C11b-486/7
$Er[BH_4]_3 \cdot 3\ OC_4H_8$-$La[BH_4]_3 \cdot 3\ OC_4H_8$ ... Sc: MVol.C11b-486/7
$Er[BH_4]_3 \cdot 3\ OC_4H_8$-$Y[BH_4]_3 \cdot 3\ OC_4H_8$ ... Sc: MVol.C11b-486/7

$B_3C_{12}F_3H_{27}N_3$ ... $[-BF-N(C_4H_9-t)-]_3$ ... B: B Comp.SVol.4/3b-247
$B_3C_{12}F_6H_{16}N_8P$ ... $[H_2B(-N_2C_3H_3-)_2B(-N_2C_3H_3-)_2BH_2][PF_6]$ ... B: B Comp.SVol.4/3b-39
$B_3C_{12}FeH_{21}$ ... $(C_6H_6)FeC_2B_3H_5(C_2H_5)_2$ ... B: B Comp.SVol.4/4-182
Fe: Org.Comp.B18-81, 84/5

$B_3C_{12}GdH_{36}O_3$ ... $Gd[BH_4]_3 \cdot 3\ OC_4H_8$ ... Sc: MVol.C11b-475, 477/80, 483/4, 486/8, 491

– ... $Gd[BH_4]_3 \cdot 3\ OC_4H_8$ solid solutions
$Gd[BH_4]_3 \cdot 3\ OC_4H_8$-$Er[BH_4]_3 \cdot 3\ OC_4H_8$ ... Sc: MVol.C11b-486/7

$B_3C_{12}H_{21}N_4O_2$ ... $C_2H_5-B[-N_2C_3H_3-]_2[-O-B(C_2H_5)-O-]B-C_2H_5$ B: B Comp.SVol.4/3b-58
$B_3C_{12}H_{25}$ ... $1,5-C_2B_3(C_2H_5)_5$ ... B: B Comp.SVol.3/4-158
$B_3C_{12}H_{27}N_6$ ... $[BC_2H_4N_2(C_2H_5)]_3$ ... B: B Comp.SVol.4/3a-187
– ... $[BC_3H_6N_2(CH_3)]_3$ ... B: B Comp.SVol.4/3a-187
$B_3C_{12}H_{28}N_5$ ... $NH(CH_3)-B(C_2H_5)-N(CH_3)-B(C_2H_5)-N(CH_3)$
$-B(C_2H_5)-N_2C_3H_3$ ... B: B Comp.SVol.4/3b-37
$B_3C_{12}H_{30}N_3$ ... $[-B(C_2H_5)-N(C_2H_5)-B(C_2H_5)-N(C_2H_5)-B(C_2H_5)$
$-N(C_2H_5)-]$ ... B: B Comp.SVol.3/3-130
– ... $[-B(C_4H_9-t)-NH-B(C_4H_9-t)-NH-B(C_4H_9-t)-NH-]$
B: B Comp.SVol.4/3a-186
– ... $[H_2B(-N(CH_3)_2-(1,2-C_6H_4)-N(CH_3)_2-)][(H_3B)_2N(CH_3)_2]$
B: B Comp.SVol.3/3-211

$B_3C_{12}H_{36}HoO_3$ ... $Ho[BH_4]_3 \cdot 3\ OC_4H_8$ ... Sc: MVol.C11b-475, 478/80, 491
$B_3C_{12}H_{36}LaO_3$ ... $La[BH_4]_3 \cdot 3\ OC_4H_8$ ... Sc: MVol.C11b-477/8, 483/8, 492

– ... $La[BH_4]_3 \cdot 3\ OC_4H_8$ solid solutions
$La[BH_4]_3 \cdot 3\ OC_4H_8$-$Er[BH_4]_3 \cdot 3\ OC_4H_8$ ... Sc: MVol.C11b-486/7

$B_3C_{12}H_{36}LuO_3$ ... $Lu[BH_4]_3 \cdot 3\ OC_4H_8$ ... Sc: MVol.C11b-475, 478/80, 483, 486, 488
$B_3C_{12}H_{36}NdO_3$ ... $Nd[BH_4]_3 \cdot 3\ OC_4H_8$ ... Sc: MVol.C11b-478/80, 483, 491
$B_3C_{12}H_{36}O_3Pr$ ... $Pr[BH_4]_3 \cdot 3\ OC_4H_8$ ... Sc: MVol.C11b-478/80, 486
$B_3C_{12}H_{36}O_3Sm$ ... $Sm[BH_4]_3 \cdot 3\ OC_4H_8$ ... Sc: MVol.C11b-478/80
$B_3C_{12}H_{36}O_3Tm$ ... $Tm[BH_4]_3 \cdot 3\ OC_4H_8$ ... Sc: MVol.C11b-475, 478/80, 491
$B_3C_{12}H_{36}O_3Y$ ... $Y[BH_4]_3 \cdot 3\ OC_4H_8$ ... Sc: MVol.C11b-475, 477/80, 483/4, 486, 491

– ... $Y[BH_4]_3 \cdot 3\ OC_4H_8$ solid solutions
$Y[BH_4]_3 \cdot 3\ OC_4H_8$-$Er[BH_4]_3 \cdot 3\ OC_4H_8$ ... Sc: MVol.C11b-486/7

$B_3C_{12}H_{36}O_3Yb$ ... $Yb[BH_4]_3 \cdot 3\ OC_4H_8$ ... Sc: MVol.C11b-475, 478/80, 486
$B_3C_{12}H_{40}N_2P_2^+$ ... $[((CH_3)_3P-BH_2CH_2N(CH_3)_2)_2BH_2]^+$ ... B: B Comp.SVol.3/3-210
$B_3C_{12}H_{40}N_3P^+$ ... $[(CH_3)_3P-BH_2CH_2N(CH_3)_2BH_2N(CH_3)_2CH_2BH_2N(CH_3)_3]^+$
B: B Comp.SVol.3/3-210

$B_3C_{14}CeH_{40}O_{3.5}$ . . $Ce[BH_4]_3 \cdot 3.5\ C_4H_8O$ . . . Sc: MVol.C11b-478
$B_3C_{14}ClCoH_{25}Hg$ $[(C_5(CH_3)_5)Co(CH_3)_2C_2B_3H_4]HgCl$ . . . B: B Comp.SVol.3/4-159
$B_3C_{14}Cl_2CoH_{25}Hg^-$
$[(C_5(CH_3)_5)Co(CH_3)_2C_2B_3H_4 \cdot HgCl_2]^-$ . . . B: B Comp.SVol.3/4-159
$B_3C_{14}CoFeH_{20}$ . . . $(C_5H_5)FeH(2,4,5\text{-}CoC_2B_3H_3(C_5H_5\text{-}2)(CH_3)_2\text{-}4,5)$
Fe: Org.Comp.B16b-53, 60
$B_3C_{14}CoH_{26}$ . . . $1,2,3\text{-}[C_5(CH_3)_5]Co(CH_3)_2C_2B_3H_5$ . . . B: B Comp.SVol.3/4-159
$B_3C_{14}FeH_{25}$ . . . $(C_8H_{10})Fe[1,2,3\text{-}B_3C_2H_5\text{-}4,5\text{-}(C_2H_5)_2]$ . . . Fe: Org.Comp.B18-81, 85
$B_3C_{14}H_{23}$ . . . $1,5\text{-}C_2B_3H_2\text{-}2,3,4\text{-}(CH_3C{=}CHCH_3)_3$ . . . B: B Comp.SVol.3/4-159
$B_3C_{14}H_{40}LaO_{3.5}$ . . $La[BH_4]_3 \cdot 3.5\ C_4H_8O$ . . . Sc: MVol.C11b-478
$B_3C_{14}H_{40}O_{3.5}Yb$ . . $Yb[BH_4]_3 \cdot 3.53\ C_4H_8O$ . . . Sc: MVol.C11b-478/80
$B_3C_{15}CrH_{30}N_3O_3$ $(\text{-}BC_2H_5\text{-}NC_2H_5\text{-})_3 \cdot Cr(CO)_3$ . . . B: B Comp.SVol.3/3-130
$B_3C_{15}FeH_{23}$ . . . $(CH_3\text{-}C_6H_5)Fe[(CH_3)_4\text{-}C_4B_3H_3]$ . . . Fe: Org.Comp.B18-81, 87
$B_3C_{15}H_{19}N_2O$ . . . $[\text{-}BCH_3\text{-}N(C_6H_5)\text{-}BCH_3\text{-}N(C_6H_5)\text{-}BCH_3\text{-}O\text{-}]$ . . B: B Comp.SVol.3/3-139
B: B Comp.SVol.4/3a-203/5
$B_3C_{15}H_{21}Rh_2$ . . . $(C_4H_4BCH_3)Rh(C_4H_4BCH_3)Rh(C_4H_4BCH_3)$ . . . B: B Comp.SVol.4/4-182
$B_3C_{15}H_{27}N_6$ . . . $2\text{-}C_6H_5\text{-}1,3\text{-}(1,3,2\text{-}N_2BC_3H_8)_2\text{-}1,3,2\text{-}N_2BC_3H_6$ B: B Comp.SVol.4/3a-158
$B_3C_{15}H_{31}$ . . . $1,5\text{-}C_2B_3(C_3H_7\text{-}n)_3\text{-}1,5\text{-}(C_2H_5)_2$ . . . B: B Comp.SVol.3/4-158
$B_3C_{15}H_{33}N_6$ . . . $[BC_3H_6N_2(C_2H_5)]_3$ . . . B: B Comp.SVol.4/3a-187
$B_3C_{15}H_{36}N_3$ . . . $[\text{-}BCH_3\text{-}N(C_4H_9\text{-}t)\text{-}BCH_3\text{-}N(C_4H_9\text{-}t)\text{-}BCH_3\text{-}N(C_4H_9\text{-}t)\text{-}]$
B: B Comp.SVol.3/3-131
B: B Comp.SVol.4/3a-188
– . . . $[\text{-}B(C_4H_9\text{-}t)\text{-}NCH_3\text{-}B(C_4H_9\text{-}t)\text{-}NCH_3\text{-}B(C_4H_9\text{-}t)\text{-}NCH_3\text{-}]$
B: B Comp.SVol.4/3a-188
B: B Comp.SVol.4/3b-247
$B_3C_{15}H_{36}N_3O_3$ . . . $[\text{-}B(C_4H_9\text{-}t)\text{-}N(OCH_3)\text{-}B(C_4H_9\text{-}t)\text{-}N(OCH_3)$
$\text{-}B(C_4H_9\text{-}t)\text{-}N(OCH_3)\text{-}]$ . . . B: B Comp.SVol.4/3a-189
$B_3C_{15}H_{36}N_3O_6$ . . . $[\text{-}B(O\text{-}C_4H_9\text{-}t)\text{-}N(OCH_3)\text{-}B(O\text{-}C_4H_9\text{-}t)$
$\text{-}N(OCH_3)\text{-}B(O\text{-}C_4H_9\text{-}t)\text{-}N(OCH_3)\text{-}]$ . . . B: B Comp.SVol.4/3a-189
$B_3C_{16}CeH_{44}O_4$ . . . $Ce[BH_4]_3 \cdot 4\ OC_4H_8$ . . . Sc: MVol.C11b-475, 479, 486, 491
$B_3C_{16}Cl_3H_{40}N_8$ . . . $[(\text{-}NCH_3\text{-}CH_2\text{-}CH_2\text{-}NCH_3\text{-})_2B]_2 \cdot BCl_3$ . . . B: B Comp.SVol.3/3-118
$B_3C_{16}DyH_{44}O_4$ . . . $Dy[BH_4]_3 \cdot 4\ OC_4H_8$ . . . Sc: MVol.C11b-475
$B_3C_{16}ErH_{44}O_4$ . . . $Er[BH_4]_3 \cdot 4\ OC_4H_8$ . . . Sc: MVol.C11b-475
– . . . $Er[BH_4]_3 \cdot 4.24\ OC_4H_8$ . . . Sc: MVol.C11b-478/80
$B_3C_{16}EuH_{44}O_4$ . . . $Eu[BH_4]_3 \cdot 4\ OC_4H_8$ . . . Sc: MVol.C11b-475
$B_3C_{16}F_6H_{24}N_8P$ . . $[H_2B(\text{-}N_2C_3H_3\text{-})_2B(\text{-}N_2C_3H_3\text{-})_2B(C_2H_5)_2][PF_6]$ B: B Comp.SVol.4/3b-39
$B_3C_{16}H_{22}N_3$ . . . $C_6H_5B(\text{-}NCH_3\text{-}BC_6H_5\text{-}NCH_3\text{-}BCH_3\text{-}NCH_3\text{-})$ . . . B: B Comp.SVol.3/3-133
$B_3C_{16}H_{29}N_4$ . . . $HB[NH\text{-}C_6H_4\text{-}2\text{-}(N(CH_3)_2\text{-}BH_3)]_2$ . . . B: B Comp.SVol.4/3b-11
– . . . $HB[NH\text{-}C_6H_4\text{-}4\text{-}(N(CH_3)_2\text{-}BH_3)]_2$ . . . B: B Comp.SVol.4/3b-11
$B_3C_{16}H_{31}N_6$ . . . $1\text{-}[(CH_3)_2N]_2B\text{-}2\text{-}C_6H_5\text{-}3\text{-}(1,3,2\text{-}N_2BC_3H_8)\text{-}$
$1,3,2\text{-}N_2BC_3H_6$ . . . B: B Comp.SVol.4/3a-158
$B_3C_{16}H_{33}S$ . . . $(CH_3)_2S\text{-}B(C_{14}H_{27})B_2$ . . . B: B Comp.SVol.3/4-137
$B_3C_{16}H_{40}I_3N_8$ . . . $[(\text{-}NCH_3\text{-}CH_2\text{-}CH_2\text{-}NCH_3\text{-})_2B]_2 \cdot BI_3$ . . . B: B Comp.SVol.3/3-118
$B_3C_{16}H_{44}HoO_4$ . . . $Ho[BH_4]_3 \cdot 4\ OC_4H_8$ . . . Sc: MVol.C11b-475
$B_3C_{16}H_{44}LaO_4$ . . . $La[BH_4]_3 \cdot 4\ OC_4H_8$ . . . Sc: MVol.C11b-475, 479, 486, 491/2
$B_3C_{16}H_{44}NdO_4$ . . . $Nd[BH_4]_3 \cdot 4\ OC_4H_8$ . . . Sc: MVol.C11b-475, 479, 491
$B_3C_{16}H_{44}O_4Pr$ . . . $Pr[BH_4]_3 \cdot 4\ OC_4H_8$ . . . Sc: MVol.C11b-475
$B_3C_{16}H_{44}O_4Sc$ . . . $Sc[BH_4]_3 \cdot 4\ OC_4H_8$ . . . Sc: MVol.C11b-475

$B_3C_{16}H_{44}O_4Sm$ ... $Sm[BH_4]_3 \cdot 4\ OC_4H_8$ ... Sc: MVol.C11b-475
$B_3C_{16}H_{44}O_4Tm$ ... $Tm[BH_4]_3 \cdot 4\ OC_4H_8$ ... Sc: MVol.C11b-475
$B_3C_{17}Cl_2H_{39}N_4$ ... 2,2,6,6-$(CH_3)_4$-NC_5H_6-1-BCl-N(C_4H_9-t)
-B[N$(CH_3)_2$]-BCl-N$(CH_3)_2$ ... B: B Comp.SVol.4/4-64
$B_3C_{17}H_{35}N_6$ ... 1,3-[((CH_3)_2N)_2B]_2-2-C_6H_5-1,3,2-$N_2BC_3H_6$ ... B: B Comp.SVol.4/3a-158
$B_3C_{18}Cl_3F_{15}N_3$ ... N(C_6F_5)BClN(C_6F_5)BClN(C_6F_5)BCl ... F: PFHOrg.SVol.4-233/4, 246
$B_3C_{18}Cl_3H_{15}N_3$ ... [-BCl-N(C_6H_5)-BCl-N(C_6H_5)-BCl-N(C_6H_5)-] ... B: B Comp.SVol.4/4-66
$B_3C_{18}F_3H_{15}N_3$ ... [-BF-N(C_6H_5)-]$_3$ ... B: B Comp.SVol.4/3b-247
$B_3C_{18}F_{18}N_3$ ... N(C_6F_5)BFN(C_6F_5)BFN(C_6F_5)BF ... F: PFHOrg.SVol.4-233/4, 246
$B_3C_{18}FeH_{32}^-$ ... [((CH_3)$_6C_6$)Fe(1,2,3-$B_3C_2H_4$-4,5-(C_2H_5)$_2$)]$^-$ ... Fe: Org.Comp.B18-81/2
$B_3C_{18}FeH_{33}$ ... 1,2,3-[$C_6(CH_3)_6$]Fe(C_2H_5)$_2C_2B_3H_5$ ... B: B Comp.SVol.3/4-159
Fe: Org.Comp.B18-81, 85
$B_3C_{18}H_{15}N_6$ ... ($BC_6H_5N_2$)$_3$ ... B: B Comp.SVol.4/3a-187
$B_3C_{18}H_{15}S_3$ ... [-B(C_6H_5)-S-B(C_6H_5)-S-B(C_6H_5)-S-] ... B: B Comp.SVol.4/4-129/30
$B_3C_{18}H_{18}N_3$ ... [-B(C_6H_5)-NH-B(C_6H_5)-NH-B(C_6H_5)-NH-] ... B: B Comp.SVol.3/3-130
B: B Comp.SVol.4/3a-186
– ... [-BH-N(C_6H_5)-BH-N(C_6H_5)-BH-N(C_6H_5)-] ... B: B Comp.SVol.3/3-129
$B_3C_{18}H_{19}N_2O_3$ ... [-B(C_6H_5)-O-]$_3 \cdot N_2H_4$ ... B: B Comp.SVol.4/3b-73
$B_3C_{18}H_{21}N_6$ ... [-B(NH_2)-N(C_6H_5)-B(NH_2)-N(C_6H_5)-B(NH_2)-N(C_6H_5)-]
B: B Comp.SVol.3/3-130
B: B Comp.SVol.4/3a-186
– ... [-B(NH-C_6H_5)-NH-B(NH-C_6H_5)-NH-B(NH-C_6H_5)-NH-]
B: B Comp.SVol.3/3-130
B: B Comp.SVol.4/3a-186
$B_3C_{18}H_{23}N_4O_3$ ... [-B(C_6H_5)-O-]$_3 \cdot 2\ N_2H_4$ ... B: B Comp.SVol.4/3b-73
$B_3C_{18}H_{32}Ni^+$ ... [(1,4,6-(CH_3)$_3$-2,3-(C_2H_5)$_2$-2,3,5-C_3B_3H)Ni(C_8H_{12})]$^+$
B: B Comp.SVol.4/4-182
$B_3C_{18}H_{42}N_3$ ... [-B(C_2H_5)-N(C_4H_9-t)-B(C_2H_5)-N(C_4H_9-t)
-B(C_2H_5)-N(C_4H_9-t)-] ... B: B Comp.SVol.3/3-131
B: B Comp.SVol.4/3a-188
– ... [-B(C_3H_7-n)-N(C_3H_7-n)-B(C_3H_7-n)
-N(C_3H_7-n)-B(C_3H_7-n)-N(C_3H_7-n)-] ... B: B Comp.SVol.3/3-131
– ... [-B(C_3H_7-i)-N(C_3H_7-i)-B(C_3H_7-i)-N(C_3H_7-i)
-B(C_3H_7-i)-N(C_3H_7-i)-] ... B: B Comp.SVol.3/3-131
– ... [-B(C_4H_9-t)-N(C_2H_5)-B(C_4H_9-t)-N(C_2H_5)
-B(C_4H_9-t)-N(C_2H_5)-] ... B: B Comp.SVol.4/3a-188
B: B Comp.SVol.4/3b-247
$B_3C_{18}H_{42}N_3P_2$ ... P[-B(N(C_3H_7-i)$_2$)-]$_3$P ... B: B Comp.SVol.4/3a-248
$B_3C_{18}H_{51}N_6Si_3$ ... [-B(NH-Si(C_2H_5)$_3$)-NH-B(NH-Si(C_2H_5)$_3$)-NH
-B(NH-Si(C_2H_5)$_3$)-NH-] ... B: B Comp.SVol.4/3a-186
$B_3C_{19}CoF_4FeH_{27}$ [C_5H_5Fe(1,3-(CH_3)$_2$-4,5-(C_2H_5)$_2C_3HB_2$-1,3)
CoC_5H_5][BF_4] ... Fe: Org.Comp.B17-175/6
$B_3C_{19}FeH_{25}$ ... ($C_{13}H_{10}$)Fe[1,2,3-$B_3C_2H_5$-4,5-(C_2H_5)$_2$] ... Fe: Org.Comp.B18-81, 86/7
$B_3C_{19}FeH_{34}I$ ... [(CH_3)$_6C_6$]Fe[1-I-2-CH_3-4,5-(C_2H_5)$_2$-1,2,3-$B_3C_2H_3$]
Fe: Org.Comp.B18-81, 85
$B_3C_{19}FeH_{35}$ ... [(CH_3)$_6C_6$]Fe[2-CH_3-4,5-(C_2H_5)$_2$-1,2,3-$B_3C_2H_4$]
Fe: Org.Comp.B18-81, 85
$B_3C_{19}H_{31}N_2O$ ... CH_3B[-O-BCH_3-N(c-C_6H_{10}-2-C≡CH)
-BCH_3-N(c-C_6H_{10}-2-C≡CH)-] ... B: B Comp.SVol.3/3-139

$B_3C_{20}F_6H_{32}N_8P$ .. $[(C_2H_5)_2B(-N_2C_3H_3-)_2B(-N_2C_3H_3-)_2B(C_2H_5)_2][PF_6]$ B: B Comp.SVol.4/3b-39

$B_3C_{20}FeH_{33}$ ..... $[(CH_3)_6C_6]Fe[(CH_3)_4C_4B_3H_3]$ .... Fe: Org.Comp.B18-81, 87, 93

$B_3C_{20}FeH_{37}$ ..... $[(CH_3)_6C_6]Fe[1,2-(CH_3)_2-4,5-(C_2H_5)_2-1,2,3-B_3C_2H_3]$ Fe: Org.Comp.B18-81, 86

$B_3C_{20}H_{23}N_2O_3$ ... $[-B(C_6H_5)-O-]_3 \cdot (CH_3)_2N-NH_2$ .... B: B Comp.SVol.4/3b-73

– .... $[-B(C_6H_5)-O-]_3 \cdot CH_3-NHNH-CH_3$.......... B: B Comp.SVol.4/3b-73

$B_3C_{21}Cl_3H_{21}N_3$... $[-BCl-N(C_6H_4-2-CH_3)-BCl-N(C_6H_4-2-CH_3)$ $-BCl-N(C_6H_4-2-CH_3)-]$ .... B: B Comp.SVol.4/4-66

$B_3C_{21}F_{15}H_9N_3$ ... $(-BC_6F_5-NCH_3-)_3$........................ B: B Comp.SVol.3/3-133

$B_3C_{21}F_{15}N_6O_3$ ... $[N(C_6F_5)B(NCO)]_3$........................ F: PFHOrg.SVol.4-233/4, 246

$B_3C_{21}F_{15}N_6S_3$.... $[N(C_6F_5)B(NCS)]_3$........................ F: PFHOrg.SVol.4-233/4, 247

$B_3C_{21}FeH_{39}$ ..... $[(CH_3)_6C_6]Fe[1,2,3-(CH_3)_3-4,5-(C_2H_5)_2-1,2,3-B_3C_2H_2]$ Fe: Org.Comp.B18-81, 86

$B_3C_{21}H_{18}NS_6$ .... $N[B(-S-(1,2-C_6H_3-4-CH_3)-S-)]_3$............ B: B Comp.SVol.3/4-122/3

$B_3C_{21}H_{19}N_2O_3$ ... $HO-B(C_6H_5)-O-B(C_6H_5)-O-B(C_6H_5)-N_2C_3H_3$. B: B Comp.SVol.4/3b-58/9

– .... $[-B(C_6H_5)-O-B(C_6H_5)-O-B(C_6H_5)-O-] \cdot 1,2-N_2C_3H_4$ B: B Comp.SVol.4/3b-58/9, 73

– .... $[-B(C_6H_5)-O-B(C_6H_5)-O-B(C_6H_5)-O-] \cdot 1,3-N_2C_3H_4$ B: B Comp.SVol.4/3b-73

$B_3C_{21}H_{22}N_5$ ..... $NH_2-B(C_6H_5)-NH-B(C_6H_5)-NH-B(C_6H_5)-N_2C_3H_3$ B: B Comp.SVol.4/3b-37

$B_3C_{21}H_{24}N_3$ ..... $[-B(C_6H_5)-N(CH_3)-]_3$ .... B: B Comp.SVol.3/3-132
B: B Comp.SVol.4/3a-188

– .... $[-BH-N(C_6H_4-4-CH_3)-]_3$ .... B: B Comp.SVol.3/3-129

$B_3C_{21}H_{24}N_3O_3$ ... $[-BH-N(C_6H_4-4-OCH_3)-]_3$................ B: B Comp.SVol.3/3-129

$B_3C_{21}H_{48}N_3$ ..... $[-B(C_3H_7-i)-N(C_4H_9-t)-]_3$ .... B: B Comp.SVol.3/3-131/2

– .... $[-B(C_3H_7-n)-N(C_4H_9-t)-]_3$................ B: B Comp.SVol.3/3-131

$B_3C_{22}H_{25}N_2O_4$ ... $(C_6H_5)_3C_2B_3H_4N_2O_4(CH_3)_2$ .... B: B Comp.SVol.4/3b-73

$B_3C_{22}H_{26}N_3O_2$ ... $1,3,5-(2-CH_3-C_6H_4)_3-2,4-(HO)_2-6-CH_3-1,3,5,2,4,6-N_3B_3$ B: B Comp.SVol.4/3a-190

$B_3C_{22}H_{31}N_4O_3$ ... $[-B(C_6H_5)-O-]_3 \cdot 2\ (CH_3)_2N-NH_2$........... B: B Comp.SVol.4/3b-73

$B_3C_{23}H_{23}N_2O_3$ ... $[-B(C_6H_5)-O-]_3 \cdot 1,2-N_2C_3H_2-3,5-(CH_3)_2$ .... B: B Comp.SVol.4/3b-73

$B_3C_{23}H_{28}N_3O$ .... $1,3,5-(2-CH_3-C_6H_4)_3-2-HO-4,6-(CH_3)_2-1,3,5,2,4,6-N_3B_3$ B: B Comp.SVol.4/3a-190

$B_3C_{24}Co_2H_{39}$ .... $[C_5(CH_3)_5]_2(CH_3)_2-1,7,2,3,-Co_2C_2B_3H_3$....... B: B Comp.SVol.3/4-159

$B_3C_{24}EuH_{60}O_6$ ... $Eu[BH_4]_3 \cdot 6\ C_4H_8O$ .... Sc: MVol.C11b-475

$B_3C_{24}F_3H_{72}N_6Si_9$ $1,3,5-[((CH_3)_3Si)_2N-BF]_3-2,2,4,4,6,6-(CH_3)_6-$ $1,3,5,2,4,6-N_3Si_3$........................ B: B Comp.SVol.4/3b-247

$B_3C_{24}H_{21}N_4O_2$ ... $C_6H_5-B[-N_2C_3H_3-]_2[-O-B(C_6H_5)-O-]B-C_6H_5$ B: B Comp.SVol.4/3b-58

$B_3C_{24}H_{27}N_6$ ..... $[BC_2H_4N_2(C_6H_5)]_3$ .... B: B Comp.SVol.4/3a-187

$B_3C_{24}H_{28}N_5$ ..... $NH(CH_3)-B(C_6H_5)-N(CH_3)-B(C_6H_5)-N(CH_3)$ $-B(C_6H_5)-N_2C_3H_3$ .... B: B Comp.SVol.4/3b-37

$B_3C_{24}H_{30}N_3$ ..... $[-B(CH_3)-N(C_6H_4-2-CH_3)-]_3$.............. B: B Comp.SVol.3/3-132
B: B Comp.SVol.4/3a-188

$B_3C_{24}H_{48}N_3$ ..... $2,8,9-[(i-C_3H_7)_2N]_3-[3.3.1]-2,8,9-B_3C_6H_6$ ..... B: B Comp.SVol.4/3a-248

$B_3C_{24}H_{54}N_3$ ..... $[-B(C_4H_9-n)-N(C_4H_9-n)-]_3$ .... B: B Comp.SVol.3/3-131

– .... $[-B(C_4H_9-n)-N(C_4H_9-t)-]_3$................ B: B Comp.SVol.3/3-131

– .... $[-B(C_4H_9-i)-N(C_4H_9-i)-]_3$ .... B: B Comp.SVol.3/3-131

$B_3C_{24}H_{54}N_3$ ..... $[-B(C_4H_9-s)-N(C_4H_9-s)-]_3$ .... B: B Comp.SVol.3/3-131
$B_3C_{24}H_{60}NdO_6$ ... $Nd[BH_4]_3 \cdot 6\ C_4H_8O$ .... Sc: MVol.C11b-475
$B_3C_{25}FeH_{39}$ ..... $[(CH_3)_6C_6]Fe[2-(C_6H_5-CH_2)-4,5-(C_2H_5)_2-1,2,3-B_3C_2H_4]$ Fe: Org.Comp.B18-81, 86, 92/3
$B_3C_{25}H_{33}N_4$ ..... $1,3-[(CH_3)_2N-B(C_6H_5)]_2-2-C_6H_5-1,3,2-N_2BC_3H_6$ B: B Comp.SVol.4/3a-177
$B_3C_{27}H_{33}N_6$ ..... $[BC_6H_2N_2(CH_3)_3]_3$ .... B: B Comp.SVol.4/3a-187
$B_3C_{27}H_{33}O_3S_3$ ... $2,4,6-(C_2H_5SCH_2-2-C_6H_4)_3-1,3,5,2,4,6-O_3B_3$ .. B: B Comp.SVol.4/4-137
$B_3C_{27}H_{36}N_3$ ..... $[-B(C_2H_5)-N(C_6H_4-2-CH_3)-B(C_2H_5)$ $-N(C_6H_4-2-CH_3)-B(C_2H_5)-N(C_6H_4-2-CH_3)-]$ B: B Comp.SVol.4/3a-188
$B_3C_{27}H_{39}N_8O_3S$ .. $[(C_2H_5)_2B(-N_2C_3H_3-)_2B(-N_2C_3H_3-)_2B(C_2H_5)_2]$ $[O_3S-C_6H_4-4-CH_3]$ .... B: B Comp.SVol.4/3b-39
$B_3C_{27}H_{54}N_3O_3$ ... $[-B(1-NC_5H_6-2,2,6,6-(CH_3)_4)-O-]_3$ .... B: B Comp.SVol.4/3b-67
$B_3C_{27}H_{60}N_3$ ..... $[-B(C_5H_{11}-n)-N(C_4H_9-t)-B(C_5H_{11}-n)$ $-N(C_4H_9-t)-B(C_5H_{11}-n)-N(C_4H_9-t)-]$ .... B: B Comp.SVol.4/3a-188
$B_3C_{28}GeH_{33}O_6$ ... $Ge(C_6H_5)_3C(B(OCH_2)_2CH_2)_3$ .... Ge: Org.Comp.3-68
$B_3C_{28}H_{60}N_3Si_4$ ... $[4,5-(C_2H_5)_2-2,2,3-(CH_3)_3-1,2,5-NSiBC_2-1]_3Si-CH_3$ B: B Comp.SVol.4/3a-253
$B_3C_{28}H_{68}O_7Y$ .... $Y[BH_4]_3 \cdot 7\ C_4H_8O$ .... Sc: MVol.C11b-479, 491
$B_3C_{30}H_{27}Rh_2$ .... $[(C_4H_4BC_6H_5)Rh]_2(C_4H_4BC_6H_5)$ .... B: B Comp.SVol.3/4-159
$B_3C_{30}H_{42}N_3$ ..... $[-B(C_4H_9-t)-N(C_6H_5)-B(C_4H_9-t)-N(C_6H_5)$ $-B(C_4H_9-t)-N(C_6H_5)-]$ .... B: B Comp.SVol.4/3b-247
$B_3C_{30}H_{66}N_3$ ..... $[-B(n-C_5H_{11})-N(n-C_5H_{11})-]_3$ .... B: B Comp.SVol.3/3-131
$B_3C_{32}CeH_{76}O_8$ ... $Ce[BH_4]_3 \cdot 8\ OC_4H_8$ .... Sc: MVol.C11b-475
$B_3C_{32}DyH_{76}O_8$ ... $Dy[BH_4]_3 \cdot 8\ OC_4H_8$ .... Sc: MVol.C11b-475
$B_3C_{32}ErH_{76}O_8$ ... $Er[BH_4]_3 \cdot 8\ OC_4H_8$ .... Sc: MVol.C11b-475
$B_3C_{32}EuH_{76}O_8$ ... $Eu[BH_4]_3 \cdot 8\ OC_4H_8$ .... Sc: MVol.C11b-475
$B_3C_{32}GdH_{76}O_8$ ... $Gd[BH_4]_3 \cdot 8\ OC_4H_8$ .... Sc: MVol.C11b-475
$B_3C_{32}H_{76}HoO_8$ ... $Ho[BH_4]_3 \cdot 8\ OC_4H_8$ .... Sc: MVol.C11b-475
$B_3C_{32}H_{76}LaO_8$ ... $La[BH_4]_3 \cdot 8\ OC_4H_8$ .... Sc: MVol.C11b-475
$B_3C_{32}H_{76}LuO_8$ ... $Lu[BH_4]_3 \cdot 8\ OC_4H_8$ .... Sc: MVol.C11b-475
$B_3C_{32}H_{76}NdO_8$ ... $Nd[BH_4]_3 \cdot 8\ OC_4H_8$ .... Sc: MVol.C11b-475
$B_3C_{32}H_{76}O_8Pr$ ... $Pr[BH_4]_3 \cdot 8\ OC_4H_8$ .... Sc: MVol.C11b-475
$B_3C_{32}H_{76}O_8Sc$ ... $Sc[BH_4]_3 \cdot 8\ OC_4H_8$ .... Sc: MVol.C11b-475
$B_3C_{32}H_{76}O_8Sm$ ... $Sm[BH_4]_3 \cdot 8\ OC_4H_8$ .... Sc: MVol.C11b-475
$B_3C_{32}H_{76}O_8Tb$ ... $Tb[BH_4]_3 \cdot 8\ OC_4H_8$ .... Sc: MVol.C11b-475
$B_3C_{32}H_{76}O_8Tm$ ... $Tm[BH_4]_3 \cdot 8\ OC_4H_8$ .... Sc: MVol.C11b-475
$B_3C_{32}H_{76}O_8Yb$ ... $Yb[BH_4]_3 \cdot 8\ OC_4H_8$ .... Sc: MVol.C11b-475
$B_3C_{36}F_{30}H_3N_6$ ... $[N(C_6F_5)B(NHC_6F_5)]_3$ .... F: PFHOrg.SVol.4-233/4
$B_3C_{36}F_{30}N_3$ ...... $[N(C_6F_5)B(C_6F_5)]_3$ .... F: PFHOrg.SVol.4-233/4, 247
$B_3C_{36}H_{30}N_3$ ..... $(-BC_6H_5-NC_6H_5-)_3$ .... B: B Comp.SVol.3/3-132
$B_3C_{36}H_{87}N_6Si_3$ ... $[-B(NH-Si(C_4H_9-n)_3)-NH-B(NH-Si(C_4H_9-n)_3)$ $-NH-B(NH-Si(C_4H_9-n)_3)-NH-]$ .... B: B Comp.SVol.4/3a-186
$B_3C_{42}H_{42}N_3$ ..... $[-B(C_6H_4-2-CH_3)-N(C_6H_4-2-CH_3)-]_3$ .... B: B Comp.SVol.3/3-132
$B_3C_{86}H_{99}MoN_8$ ... $[Mo(CN-C_4H_9-t)_6(NC-B(C_6H_5)_3)][(C_6H_5)_3B$ $-CN-B(C_6H_5)_3]$ .... Mo: Org.Comp.5-76, 81/2
B_3CeH_{12} .... $Ce[BH_4]_3$ .... Sc: MVol.C11b-473/4
– .... $Ce[BH_4]_3$ systems
$Ce[BH_4]_3-Li[BH_4]-OC_4H_8$ .... Sc: MVol.C11b-491/2

B_3CeH_{12} .... $Ce[BH_4]_3$ systems
$Ce[BH_4]_3$-OC_4H_8 .... Sc: MVol.C11b-475/6
– .... $Ce[BH_4]_3 \cdot n\ OC_4H_8$ .... Sc: MVol.C11b-475/88, 491
B_3CeO_6 .... CeB_3O_6 .... Sc: MVol.C11b-420/2, 425/6, 430
– .... $CeB_3O_6 \cdot 4\ H_2O$ .... Sc: MVol.C11b-430/1
– .... $CeB_3O_6 \cdot 6\ H_2O$ .... Sc: MVol.C11b-430/1
$B_3ClH_2O_3$ .... [-BH-O-BH-O-BCl-O-] .... B: B Comp.SVol.4/4-54
$B_3Cl_2HO_3$ .... [-BH-O-BCl-O-BCl-O-] .... B: B Comp.SVol.4/4-54
$B_3Cl_3H_3N_3$ .... $(-BCl-NH-)_3$ .... B: B Comp.SVol.3/4-60
B: B Comp.SVol.4/4-66
– .... $(-BCl-^{15}NH-)_3$ .... B: B Comp.SVol.3/4-60
– .... $(-BCl-^{15}ND-)_3$ .... B: B Comp.SVol.3/4-60
– .... $(-BCl-ND-)_3$ .... B: B Comp.SVol.3/4-60
– .... $(-^{10}BCl-NH-)_3$ .... B: B Comp.SVol.3/4-60
– .... $(-^{10}BCl-^{15}NH-)_3$ .... B: B Comp.SVol.3/4-60
– .... $(-^{10}BCl-^{15}ND-)_3$ .... B: B Comp.SVol.3/4-60
– .... $(-^{10}BCl-ND-)_3$ .... B: B Comp.SVol.3/4-60
$B_3Cl_3O_3$ .... [-BCl-O-BCl-O-BCl-O-] .... B: B Comp.SVol.3/4-49
B: B Comp.SVol.4/4-54
$B_3Cl_3S_3$ .... [-BCl-S-BCl-S-BCl-S-] .... B: B Comp.SVol.4/4-155
B_3Cl_6N .... $N(BCl_2)_3$ .... B: B Comp.SVol.3/4-54
B: B Comp.SVol.4/4-57/8
$B_3Cl_6N_3$ .... $(-BCl-NCl-)_3$ .... B: B Comp.SVol.3/4-60
B_3DyH_{12} .... $Dy[BH_4]_3$ systems
$Dy[BH_4]_3$-$Li[BH_4]$-OC_4H_8 .... Sc: MVol.C11b-491/2
$Dy[BH_4]_3$-OC_4H_8 .... Sc: MVol.C11b-475/6
– .... $Dy[BH_4]_3 \cdot n\ OC_4H_8$ .... Sc: MVol.C11b-475/88, 491
$B_3DyLi_6O_9$ .... $Li_6Dy(BO_3)_3$ .... Sc: MVol.C11b-460/3
B_3DyO_6 .... DyB_3O_6 .... Sc: MVol.C11b-420/1, 424/5, 430
– .... $DyB_3O_6 \cdot 4\ H_2O$ .... Sc: MVol.C11b-430/1
– .... $DyB_3O_6 \cdot 6\ H_2O$ .... Sc: MVol.C11b-430/1
B_3ErH_{12} .... $Er[BH_4]_3$ solid solutions
$Er[BH_4]_3 \cdot 3\ OC_4H_8$-$Gd[BH_4]_3 \cdot 3\ OC_4H_8$ .... Sc: MVol.C11b-486/7
$Er[BH_4]_3 \cdot 3\ OC_4H_8$-$La[BH_4]_3 \cdot 3\ OC_4H_8$ .... Sc: MVol.C11b-486/7
$Er[BH_4]_3 \cdot 3\ OC_4H_8$-$Y[BH_4]_3 \cdot 3\ OC_4H_8$ .... Sc: MVol.C11b-486/7
– .... $Er[BH_4]_3$ systems
$Er[BH_4]_3$-$Li[BH_4]$-OC_4H_8 .... Sc: MVol.C11b-491/2
$Er[BH_4]_3$-OC_4H_8 .... Sc: MVol.C11b-475/6
– .... $Er[BH_4]_3 \cdot n\ OC_4H_8$ .... Sc: MVol.C11b-475/88, 491
B_3ErO_6 .... ErB_3O_6 .... Sc: MVol.C11b-420/1, 424/5, 430
– .... $ErB_3O_6 \cdot 4\ H_2O$ .... Sc: MVol.C11b-430/1
– .... $ErB_3O_6 \cdot 6\ H_2O$ .... Sc: MVol.C11b-430/1
B_3EuH_{12} .... $Eu[BH_4]_3$ systems
$Eu[BH_4]_3$-OC_4H_8 .... Sc: MVol.C11b-475/6
– .... $Eu[BH_4]_3 \cdot n\ OC_4H_8$ .... Sc: MVol.C11b-475/88
$B_3EuLi_6O_9$ .... $Li_6Eu(BO_3)_3$ .... Sc: MVol.C11b-460/3

B_3EuO_6 . . . . EuB_3O_6 . . . . Sc: MVol.C11b-420/2, 425/6
$B_3Eu_2Li_3O_9$ . . . . . . $Li_3Eu_2(BO_3)_3$ . . . . Sc: MVol.C11b-456/9
$B_3FH_2O_3$ . . . . [-BF-O-BH-O-BH-O-] . . . . B: B Comp.SVol.4/3b-232
$B_3F_2HO_3$ . . . . [-BF-O-BF-O-BH-O-] . . . . B: B Comp.SVol.4/3b-232
$B_3F_3H_3N_3$ . . . . $(-BF-NH-)_3$ . . . . B: B Comp.SVol.3/3-376/7
– . . . . $(-BF-ND-)_3$ . . . . B: B Comp.SVol.3/3-376/7
– . . . . $(-^{10}BF-NH-)_3$ . . . . B: B Comp.SVol.3/3-376/7
$B_3F_3H_9N_3Si_3$ . . . . . $(SiH_3NBF)_3$ . . . . Si: SVol.B4-149
$B_3F_3O_3$ . . . . $(-BF-O-)_3$ . . . . B: B Comp.SVol.3/3-365
B: B Comp.SVol.4/3b-231
$B_3F_3S_3$ . . . . $F_3B_3S_3$ . . . . B: B Comp.SVol.3/4-136
B: B Comp.SVol.4/4-155
$B_3F_4HO_{21}S_6$ . . . . . [H][-B(O-SO_2-O-SO_2-F)-O-B(O-SO_2-O
-SO_2-F)-O-BF(O-SO_2-O-SO_2-F)-O-] . . . . . . B: B Comp.SVol.4/4-165
$B_3F_6K_3O_3$ . . . . $K_3[B_3O_3F_6]$ . . . . B: B Comp.SVol.4/3b-197
B_3F_6N . . . . $(F_2B)_3N$ . . . . B: B Comp.SVol.3/3-375
B: B Comp.SVol.4/3b-241
B_3Gd . . . . GdB_{3+x} . . . . Sc: MVol.C11a-6
B_3GdH_{12} . . . . $Gd[BH_4]_3$ solid solutions
$Gd[BH_4]_3 \cdot 3\ OC_4H_8$-$Er[BH_4]_3 \cdot 3\ OC_4H_8$ . . Sc: MVol.C11b-486/7
– . . . . $Gd[BH_4]_3$ systems
$Gd[BH_4]_3$-$Li[BH_4]$-OC_4H_8 . . . . Sc: MVol.C11b-491/2
$Gd[BH_4]_3$-OC_4H_8 . . . . Sc: MVol.C11b-475/6
– . . . . $Gd[BH_4]_3 \cdot n\ OC_4H_8$ . . . . Sc: MVol.C11b-475/88, 491
$B_3GdH_{28}N_8$ . . . . . . $Gd[BH_4]_3 \cdot 4\ N_2H_4$ . . . . Sc: MVol.C11b-488
$B_3GdLi_6O_9$ . . . . . . $Li_6Gd(BO_3)_3$ . . . . Sc: MVol.C11b-460/3
doped with Eu^{3+} . . . . Sc: MVol.C11b-462
doped with Nd^{3+} . . . . Sc: MVol.C11b-462
B_3GdO_6 . . . . GdB_3O_6 . . . . Sc: MVol.C11b-420/30
doped with Bi^{3+} . . . . Sc: MVol.C11b-428
doped with Bi^{3+} and Sb^{3+} . . . . Sc: MVol.C11b-428
doped with rare earth ions . . . . Sc: MVol.C11b-426/9
doped with rare earth ions, Bi^{3+}, and Sb^{3+}
Sc: MVol.C11b-428/9
– . . . . $GdB_3O_6 \cdot 4\ H_2O$ . . . . Sc: MVol.C11b-430/1
– . . . . $GdB_3O_6 \cdot 6\ H_2O$ . . . . Sc: MVol.C11b-430/1
– . . . . GdB_3O_6 solid solutions
GdB_3O_6-LaB_3O_6
doped with rare earth ions . . . . Sc: MVol.C11b-429
doped with rare earth ions and Bi^{3+} . . . . . Sc: MVol.C11b-429
GdB_3O_6-YB_3O_6, doped with rare earth ions Sc: MVol.C11b-429
$B_3H_3S_6$ . . . . $(-BSH-S-)_3$ . . . . B: B Comp.SVol.3/4-111
$B_3H_5N_2$ . . . . HB(-BH-NH-BH-NH-) . . . . B: B Comp.SVol.3/3-138
$B_3H_5N_3^+$ . . . . $[B_3N_3H_5]^+$ . . . . B: B Comp.SVol.4/3a-182
$B_3H_5N_5$ . . . . $[-N_3B_3H_3(NH-)-NH-]_n$ . . . . B: B Comp.SVol.4/3a-190
B_3H_6N . . . . $(H_2B)_3N$ . . . . B: B Comp.SVol.3/3-159
$B_3H_6N_3$ . . . . [-BH-BH-NH-B(NH_2)-NH-] . . . . B: B Comp.SVol.3/3-138/9
– . . . . [-BH-NH-BH-NH-BH-NH-] . . . . B: B Comp.SVol.3/3-126/9

$B_3H_6N_3$ . . . . [-BH-NH-BH-NH-BH-NH-]. . . . . B: B Comp.SVol.4/3a-182/5
– . . . . [-BH-^{14}NH-BH-^{14}NH-BH-^{14}NH-]. . . . . B: B Comp.SVol.4/3a-183
– . . . . [-BH-^{15}NH-BH-^{15}NH-BH-^{15}NH-]. . . . . B: B Comp.SVol.4/3a-183
– . . . . [-BH-NH-NH-BH-N(BH_2)-]. . . . . B: B Comp.SVol.3/3-138/9
$B_3H_6N_3^+$ . . . . [-BH-NH-BH-NH-BH-NH-]$^+$, radical cation . . B: B Comp.SVol.3/3-126, 128/9
B: B Comp.SVol.4/3a-182
$B_3H_9N^{3-}$ . . . . [$(H_3B)_3N$]$^{3-}$ . . . . B: B Comp.SVol.3/3-214
$B_3H_9N_6$ . . . . [-B(NH_2)-NH-B(NH_2)-NH-B(NH_2)-NH-] . . . . . . B: B Comp.SVol.4/3a-186
$B_3H_{12}Ho$ . . . . Ho[BH_4]$_3$ systems
Ho[BH_4]$_3$-Li[BH_4]-OC_4H_8. . . . . Sc: MVol.C11b-491/2
Ho[BH_4]$_3$-Na[BH_4]-OC_4H_8 . . . . Sc: MVol.C11b-492/3
Ho[BH_4]$_3$-OC_4H_8. . . . . Sc: MVol.C11b-475/6
– . . . . Ho[BH_4]$_3$ · n OC_4H_8 . . . . Sc: MVol.C11b-475/88, 491
$B_3H_{12}La$ . . . . La[BH_4]$_3$ . . . . Sc: MVol.C11b-473/4
– . . . . La[BH_4]$_3$ solid solutions
La[BH_4]$_3$ · 3 OC_4H_8-Er[BH_4]$_3$ · 3 OC_4H_8 . . Sc: MVol.C11b-486/7
– . . . . La[BH_4]$_3$ systems
La[BH_4]$_3$-K[BH_4]-OC_4H_8 . . . . Sc: MVol.C11b-492/3
La[BH_4]$_3$-Li[BH_4]-OC_4H_8. . . . . Sc: MVol.C11b-491/2
La[BH_4]$_3$-Na[BH_4]-OC_4H_8 . . . . Sc: MVol.C11b-492/3
La[BH_4]$_3$-OC_4H_8. . . . . Sc: MVol.C11b-475/6
– . . . . La[BH_4]$_3$ · n OC_4H_8 . . . . Sc: MVol.C11b-475/88, 491/2
$B_3H_{12}Lu$ . . . . Lu[BH_4]$_3$ systems
Lu[BH_4]$_3$-OC_4H_8. . . . . Sc: MVol.C11b-475/6
– . . . . Lu[BH_4]$_3$ · n OC_4H_8 . . . . Sc: MVol.C11b-475/88
$B_3H_{12}Nd$ . . . . Nd[BH_4]$_3$. . . . . Sc: MVol.C11b-473/4
– . . . . Nd[BH_4]$_3$ systems
Nd[BH_4]$_3$-Li[BH_4]-OC_4H_8. . . . . Sc: MVol.C11b-491/2
Nd[BH_4]$_3$-OC_4H_8. . . . . Sc: MVol.C11b-475/6
– . . . . Nd[BH_4]$_3$ · n OC_4H_8 . . . . Sc: MVol.C11b-475/88, 491
$B_3H_{12}Pr$ . . . . Pr[BH_4]$_3$ systems
Pr[BH_4]$_3$-OC_4H_8 . . . . Sc: MVol.C11b-475/6
– . . . . Pr[BH_4]$_3$ · n OC_4H_8. Sc: MVol.C11b-475/88
$B_3H_{12}Sc$ . . . . Sc[BH_4]$_3$ . . . . Sc: MVol.C11b-472
– . . . . Sc[BH_4]$_3$ systems
Sc[BH_4]$_3$-Na[BH_4]-OC_4H_8 . . . . Sc: MVol.C11b-492/3
Sc[BH_4]$_3$-OC_4H_8 . . . . Sc: MVol.C11b-475/6
– . . . . Sc[BH_4]$_3$ · n OC_4H_8 . . . . Sc: MVol.C11b-475/88, 492
$B_3H_{12}Sm$. . . . . Sm[BH_4]$_3$ systems
Sm[BH_4]$_3$-OC_4H_8 . . . . Sc: MVol.C11b-475/6
– . . . . Sm[BH_4]$_3$ · n OC_4H_8. Sc: MVol.C11b-475/88
$B_3H_{12}Tb$ . . . . Tb[BH_4]$_3$ systems
Tb[BH_4]$_3$-OC_4H_8. . . . . Sc: MVol.C11b-475/6
– . . . . Tb[BH_4]$_3$ · n OC_4H_8 . . . . Sc: MVol.C11b-475/88
$B_3H_{12}Tm$. . . . . Tm[BH_4]$_3$ systems
Tm[BH_4]$_3$-Li[BH_4]-OC_4H_8 . . . . Sc: MVol.C11b-491/2

$B_3H_{12}Tm$ .... $Tm[BH_4]_3$ systems
  $Tm[BH_4]_3$-OC_4H_8 .... Sc: MVol.C11b-475/6

– .... $Tm[BH_4]_3 \cdot n\ OC_4H_8$ .... Sc: MVol.C11b-475/88, 491

$B_3H_{12}Y$ .... $Y[BH_4]_3$ .... Sc: MVol.C11b-472

– .... $Y[BH_4]_3$ solid solutions
  $Y[BH_4]_3 \cdot 3\ OC_4H_8$-$Er[BH_4]_3 \cdot 3\ OC_4H_8$ ... Sc: MVol.C11b-486/7

– .... $Y[BH_4]_3$ systems
  $Y[BH_4]_3$-$Li[BH_4]$-OC_4H_8 .... Sc: MVol.C11b-491/2
  $Y[BH_4]_3$-OC_4H_8 .... Sc: MVol.C11b-475/6

– .... $Y[BH_4]_3 \cdot n\ OC_4H_8$ .... Sc: MVol.C11b-475/88, 491

$B_3H_{12}Yb$ .... $Yb[BH_4]_3$ systems
  $Yb[BH_4]_3$-OC_4H_8 .... Sc: MVol.C11b-475/6

– .... $Yb[BH_4]_3 \cdot n\ OC_4H_8$ .... Sc: MVol.C11b-475/88

$B_3H_{24}LaN_4$ ...... $La[BH_4]_3 \cdot 4\ NH_3$ .... Sc: MVol.C11b-474/5

$B_3H_{28}LaN_8$ ...... $La[BH_4]_3 \cdot 4\ N_2H_4$ .... Sc: MVol.C11b-473, 488

$B_3H_{28}LuN_8$ ...... $Lu[BH_4]_3 \cdot 4\ N_2H_4$ .... Sc: MVol.C11b-488

$B_3H_{30}LaN_6$ ...... $La[BH_4]_3 \cdot 6\ NH_3$ .... Sc: MVol.C11b-474/5

$B_3H_{30}N_6Sc$ ...... $Sc[BH_4]_3 \cdot 6\ NH_3$ .... Sc: MVol.C11b-474/5

$B_3H_{30}N_6Y$ .... $Y[BH_4]_3 \cdot 6\ NH_3$ .... Sc: MVol.C11b-474/5

$B_3HoLi_6O_9$ ...... $Li_6Ho(BO_3)_3$ .... Sc: MVol.C11b-460/3

B_3HoO_6 .... HoB_3O_6 .... Sc: MVol.C11b-420/1, 424/5, 430

– .... $HoB_3O_6 \cdot 4\ H_2O$ .... Sc: MVol.C11b-430/1

– .... $HoB_3O_6 \cdot 6\ H_2O$ .... Sc: MVol.C11b-430/1

B_3La .... LaB_{3+x} .... Sc: MVol.C11a-6

B_3LaO_6 .... LaB_3O_6 .... Sc: MVol.C11b-420/30, 434/5, 436/7
    doped with Bi^{3+} .... Sc: MVol.C11b-428
    doped with Bi^{3+} and Sb^{3+} .... Sc: MVol.C11b-428
    doped with rare earth ions .... Sc: MVol.C11b-426/9, 436/7

– .... $LaB_3O_6 \cdot 4\ H_2O = La_2O_3 \cdot 3\ B_2O_3 \cdot 8\ H_2O$ Sc: MVol.C11b-430/1, 437

– .... $LaB_3O_6 \cdot 6\ H_2O$ .... Sc: MVol.C11b-430/1

– .... LaB_3O_6 solid solutions
  LaB_3O_6-GdB_3O_6
    doped with rare earth ions .... Sc: MVol.C11b-429
    doped with rare earth ions and Bi^{3+} ..... Sc: MVol.C11b-429

$B_3La_2Li_3O_9$ ...... $Li_3La_2(BO_3)_3$ .... Sc: MVol.C11b-455, 456/9

$B_3La_2Na_3O_9$ ..... $Na_3La_2(BO_3)_3$ .... Sc: MVol.C11b-467
    doped with Nd^{3+} .... Sc: MVol.C11b-469

$B_3Li_3Nd_2O_9$ ...... $Li_3Nd_2(BO_3)_3$ .... Sc: MVol.C11b-456/9

$B_3Li_3O_9Pr_2$ ...... $Li_3Pr_2(BO_3)_3$ .... Sc: MVol.C11b-456/9

$B_3Li_3O_9Y_2$ .... $Li_3Y_2(BO_3)_3$ .... Sc: MVol.C11b-453/5, 456/9

$B_3Li_6NdO_9$ ...... $Li_6Nd(BO_3)_3$ .... Sc: MVol.C11b-460/3

$B_3Li_6O_9Sm$ ...... $Li_6Sm(BO_3)_3$ .... Sc: MVol.C11b-460/3

$B_3Li_6O_9Y$ .... $Li_6Y(BO_3)_3$ .... Sc: MVol.C11b-453/5, 460/3

– .... $Li_6Y(BO_3)_3$ systems
  $Li_6Y(BO_3)_3$-$Li_2YB_5O_{10}$ .... Sc: MVol.C11b-453/5

$B_3Li_6O_9Y$ .... $Li_6Y(BO_3)_3$ systems
$Li_6Y(BO_3)_3$-YBO_3 .... Sc: MVol.C11b-453/5
$B_3Li_6O_9Yb$ ...... $Li_6Yb(BO_3)_3$ .... Sc: MVol.C11b-460/3
B_3LuO_6 .... LuB_3O_6 .... Sc: MVol.C11b-420/1, 424/5
$B_3Na_3Nd_2O_9$ ..... $Na_3Nd_2(BO_3)_3$ .... Sc: MVol.C11b-466/7
B_3NdO_6 .... NdB_3O_6 .... Sc: MVol.C11b-420/26, 430, 439
– .... $NdB_3O_6 \cdot 4\ H_2O$ .... Sc: MVol.C11b-430/1
– .... $NdB_3O_6 \cdot 6\ H_2O$ .... Sc: MVol.C11b-430/1
B_3O_6Pr .... PrB_3O_6 .... Sc: MVol.C11b-420/26, 430
– .... $PrB_3O_6 \cdot 4\ H_2O$ .... Sc: MVol.C11b-430/1
– .... $PrB_3O_6 \cdot 6\ H_2O$ .... Sc: MVol.C11b-430/1
B_3O_6Sm .... SmB_3O_6 .... Sc: MVol.C11b-420/26, 430, 440
– .... $SmB_3O_6 \cdot 4\ H_2O$ .... Sc: MVol.C11b-430/1
– .... $SmB_3O_6 \cdot 6\ H_2O$ .... Sc: MVol.C11b-430/1
B_3O_6Tb .... TbB_3O_6 .... Sc: MVol.C11b-420/2, 425/7
doped with rare earth ions .... Sc: MVol.C11b-427
B_3O_6Tm .... TmB_3O_6 .... Sc: MVol.C11b-420/1, 424/5
B_3O_6Y .... YB_3O_6 .... Sc: MVol.C11b-420, 424/5, 427, 430
doped with rare earth ions .... Sc: MVol.C11b-427
– .... $YB_3O_6 \cdot 4\ H_2O$ .... Sc: MVol.C11b-430/1
– .... $YB_3O_6 \cdot 6\ H_2O$ .... Sc: MVol.C11b-430/1
– .... YB_3O_6 solid solutions
YB_3O_6-GdB_3O_6, doped with rare earth ions Sc: MVol.C11b-429
B_3O_6Yb .... YbB_3O_6 .... Sc: MVol.C11b-420/1, 424/5, 430
– .... $YbB_3O_6 \cdot 4\ H_2O$ .... Sc: MVol.C11b-430/1
– .... $YbB_3O_6 \cdot 6\ H_2O$ .... Sc: MVol.C11b-430/1
B_3Pd_{16} .... $Pd_{16}B_3$ .... Pd: SVol.B2-257
B_3Pr .... PrB_{3+x} .... Sc: MVol.C11a-6
B_3Sm .... SmB_{3+x} .... Sc: MVol.C11a-6
B_3Y .... YB_3 .... Sc: MVol.C11a-133
– .... YB_{3+x} .... Sc: MVol.C11a-6
B_3Yb .... YbB_{3+x} .... Sc: MVol.C11a-6
$B_{3.7}FN_{3.7}O_3S$ .... $(BN)_{3.7}SO_3F$ .... B: B Comp.SVol.3/3-32
$B_4BrC_2ClH_4$ ..... 2-Cl-4-Br-1,6-$C_2B_4H_4$ .... B: B Comp.SVol.4/4-184
$B_4BrC_2H_5$ .... 2-Br-1,6-$C_2B_4H_5$ .... B: B Comp.SVol.4/4-184
$B_4BrC_{42}H_{43}NiP_2$ .. 1-Br-1,5-$[(C_6H_5)_3P]_2$-2,3-$(C_2H_5)_2$-1,2,3-$NiC_2B_4H_3$
B: B Comp.SVol.4/4-196
$B_4Br_2C_2H_4$ ...... 2,4-Br_2-1,6-$C_2B_4H_4$ .... B: B Comp.SVol.4/4-184
$B_4Br_2C_8H_{24}N_4$ ... $(CH_3)_2N$-BBr-B$[N(CH_3)_2]$-B$[N(CH_3)_2]$-BBr-$N(CH_3)_2$
B: B Comp.SVol.4/4-106
$B_4Br_4C_4H_{12}N_4S_2$. $[BrB(-NCH_3-NCH_3-BBr-S-)]_2$ .... B: B Comp.SVol.3/4-143
$B_4Br_4C_{12}H_{36}N_4Si_4$ [-BBr-$N(Si(CH_3)_3)$-BBr-$N(Si(CH_3)_3)$-BBr-
$N(Si(CH_3)_3)$-BBr-$N(Si(CH_3)_3)$-] .... B: B Comp.SVol.4/4-83, 106
$B_4Br_6C_{16}H_{40}N_8$... $[(-NCH_3-CH_2-CH_2-NCH_3-)_2B]_2 \cdot 2\ BBr_3$ ..... B: B Comp.SVol.3/3-118

$B_4CH_5^{3-}$	$[CB_4H_5]^{3-}$	B: B Comp.SVol.3/4-160
$B_4C_2ClH_5$	2-Cl-1,6-$C_2B_4H_5$	B: B Comp.SVol.4/4-184
$B_4C_2Cl_2H_4$	2,4-Cl_2-1,6-$C_2B_4H_4$	B: B Comp.SVol.4/4-184
$B_4C_2Cl_4H_2$	$C_2B_4Cl_4H_2$	B: B Comp.SVol.4/4-184
$B_4C_2H_6$	1,2-$C_2B_4H_6$	B: B Comp.SVol.3/4-160/1
		B: B Comp.SVol.4/4-183
–	1,6-$C_2B_4H_6$	B: B Comp.SVol.3/4-160/1
		B: B Comp.SVol.4/4-183
$B_4C_2H_6^{2-}$	$[C_2B_4H_6]^{2-}$	B: B Comp.SVol.3/4-162
$B_4C_2H_6Sn$	1,2,3-$SnC_2B_4H_6$	B: B Comp.SVol.4/4-195
$B_4C_2H_7^+$	$[1,2\text{-}C_2B_4H_7]^+$	B: B Comp.SVol.4/4-183
–	$[1,6\text{-}C_2B_4H_7]^+$	B: B Comp.SVol.4/4-183/4
$B_4C_2H_7^-$	$[2,4\text{-}C_2B_4H_7]^-$	B: B Comp.SVol.4/4-193
$B_4C_2H_8$	2,3-$C_2B_4H_8$	B: B Comp.SVol.4/4-185
$B_4C_2H_{12}P_2Pt$	$(H_3P)_2PtC_2B_4H_6$	B: B Comp.SVol.3/4-163
$B_4C_2H_{12}S_4$	$[CH_2(SBH_2)_2]_2$	B: B Comp.SVol.3/4-115/6
$B_4C_3H_9In$	CH_3-$InC_2B_4H_6$	In: Org.Comp.1-327/8
$B_4C_4H_8$	$C_4B_4H_8$	B: B Comp.SVol.3/4-161
$B_4C_4H_{11}^-$	$[2,3\text{-}C_2B_4H_5\text{-}2,3\text{-}(CH_3)_2]^-$	B: B Comp.SVol.4/4-185
$B_4C_4H_{12}$	2,3-$C_2B_4H_6$-2,3-$(CH_3)_2$	B: B Comp.SVol.4/4-185
$B_4C_4H_{12}N_2S_2$	1,3,5,7-$(CH_3)_4$-$B_4N_2S_2$	B: B Comp.SVol.4/4-140
$B_4C_5GeH_{14}Si$	1,2,3-$GeC_2B_4H_5$-2-$Si(CH_3)_3$	B: B Comp.SVol.4/4-195
$B_4C_5H_{14}SiSn$	1,2,3-$SnC_2B_4H_5$-2-$Si(CH_3)_3$	B: B Comp.SVol.3/4-162
		B: B Comp.SVol.4/4-195, 196, 235
$B_4C_5H_{16}Si$	2,3-$C_2B_4H_7$-2-$Si(CH_3)_3$	B: B Comp.SVol.4/4-185/6
$B_4C_6ClH_{17}Si_2$	$[2\text{-}(CH_3)_3Si\text{-}3\text{-}CH_3\text{-}2,3\text{-}C_2B_4H_4]SiH(Cl)$	B: B Comp.SVol.4/4-194
$B_4C_6GeH_{16}Si$	1,2,3-$GeC_2B_4H_4$-2-$Si(CH_3)_3$-3-CH_3	B: B Comp.SVol.4/4-195
$B_4C_6H_{12}$	1,6-$C_2B_4H_5$-2-$CH_3C{=}CHCH_3$	B: B Comp.SVol.3/4-161
$B_4C_6H_{14}$	2,3-$C_2B_4H_4$-2,3-$(C_2H_5)_2$	B: B Comp.SVol.3/4-163
–	2,3-$C_2B_4H_7$-1-$CCH_3{=}CHCH_3$	B: B Comp.SVol.3/4-162
–	2,3-$C_2B_4H_7$-4-$CCH_3{=}CHCH_3$	B: B Comp.SVol.3/4-162
–	2,3-$C_2B_4H_7$-5-$CCH_3{=}CHCH_3$	B: B Comp.SVol.3/4-162
$B_4C_6H_{15}^-$	$[2,3\text{-}C_2B_4H_5\text{-}2,3\text{-}(C_2H_5)_2]^-$	B: B Comp.SVol.3/4-162
		B: B Comp.SVol.4/4-185, 189
$B_4C_6H_{15}Na$	$Na[2,3\text{-}C_2B_4H_5\text{-}2,3\text{-}(C_2H_5)_2]$	B: B Comp.SVol.3/4-161, 243
$B_4C_6H_{16}$	2,3-$C_2B_4H_6$-2,3-$(C_2H_5)_2$	B: B Comp.SVol.3/4-162
		B: B Comp.SVol.4/4-185
$B_4C_6H_{16}SiSn$	1,2,3-$SnC_2B_4H_4$-2-$Si(CH_3)_3$-3-CH_3	B: B Comp.SVol.3/4-162
		B: B Comp.SVol.4/4-195, 196, 235
$B_4C_6H_{18}Si$	2,3-$C_2B_4H_6$-2-$Si(CH_3)_3$-3-CH_3	B: B Comp.SVol.4/4-185/6
$B_4C_6H_{18}Si_2$	$[2\text{-}(CH_3)_3Si\text{-}3\text{-}CH_3\text{-}2,3\text{-}C_2B_4H_4]SiH_2$	B: B Comp.SVol.4/4-194
$B_4C_7CoH_{11}$	1-C_5H_5-1,2,3-$CoC_2B_4H_6$	B: B Comp.SVol.3/4-163
–	1-C_5H_5-1,2,4-$CoC_2B_4H_6$	B: B Comp.SVol.3/4-163
$B_4C_7FeH_{11}$	$(C_5H_5)Fe(2,3\text{-}C_2B_4H_6)$	Fe: Org.Comp.B16b-53, 60/1
–	$(C_5H_5)Fe(2,4\text{-}C_2B_4H_6)$	Fe: Org.Comp.B16b-54
$B_4C_7FeH_{12}$	$(C_5H_5)FeH(2,3\text{-}C_2B_4H_6)$	Fe: Org.Comp.B16b-53, 61

$B_4C_7FeH_{12}$ ...... $(C_5H_5)FeH(2,4\text{-}C_2B_4H_6)$ ...... Fe: Org.Comp.B16b-54

$B_4C_8Cl_2H_{24}N_4$ .... $(CH_3)_2N\text{-}BCl\text{-}B[N(CH_3)_2]\text{-}B[N(CH_3)_2]\text{-}BCl\text{-}N(CH_3)_2$ B: B Comp.SVol.4/4-60

$B_4C_8Cl_3Ge_2H_{21}Si_2$ $2,3\text{-}[(CH_3)_3Si]_2\text{-}5\text{-}GeCl_3\text{-}1,2,3\text{-}GeC_2B_4H_3$ .... B: B Comp.SVol.4/4-196

$B_4C_8GeH_{22}Si_2$ ... $1,2,3\text{-}GeC_2B_4H_4\text{-}2,3\text{-}[Si(CH_3)_3]_2$ ...... B: B Comp.SVol.4/4-195

$B_4C_8H_{20}$ .... $2,3\text{-}C_2B_4H_5\text{-}2,3,4\text{-}(C_2H_5)_3$ ...... B: B Comp.SVol.4/4-189/90

$B_4C_8H_{22}Si_2Sn$.... $1,2,3\text{-}SnC_2B_4H_4\text{-}2,3\text{-}[Si(CH_3)_3]_2$ ...... B: B Comp.SVol.3/4-162
B: B Comp.SVol.4/4-195, 196, 235

$B_4C_8H_{22}Si_3$...... $1,2,3\text{-}SiC_2B_4H_4\text{-}2,3\text{-}[Si(CH_3)_3]_2$ ...... B: B Comp.SVol.4/4-194

$B_4C_8H_{24}Si_2$...... $2,3\text{-}C_2B_4H_6\text{-}2,3\text{-}[Si(CH_3)_3]_2$...... B: B Comp.SVol.4/4-185/6

$B_4C_9FeH_{15}$ ...... $(C_5H_5)Fe(2,3\text{-}C_2B_4H_4(CH_3)_2\text{-}2,3)$...... Fe: Org.Comp.B16b-54

$B_4C_{10}H_{16}$ .... $2,3\text{-}C_2B_4H_7\text{-}2\text{-}(CH_2CH_2\text{-}C_6H_5)$ ...... B: B Comp.SVol.4/4-185

$B_4C_{10}H_{18}$ .... $1,6\text{-}C_2B_4H_4\text{-}2,3\text{-}(CH_3C{=}CHCH_3)_2$ ...... B: B Comp.SVol.3/4-161

– .... $1,6\text{-}C_2B_4H_4\text{-}2,4\text{-}(CH_3C{=}CHCH_3)_2$ ...... B: B Comp.SVol.3/4-161

$B_4C_{10}H_{20}$ .... $4,5,7,8\text{-}C_4B_4H_4\text{-}4,5\text{-}(CH_3)_2\text{-}7,8\text{-}(C_2H_5)_2$ ...... B: B Comp.SVol.4/4-194

$B_4C_{10}H_{24}$ .... $2,3\text{-}C_2B_4H_6\text{-}2,3\text{-}(C_4H_9\text{-}n)_2$ ...... B: B Comp.SVol.4/4-185, 187

$B_4C_{10}H_{28}N_4$ ..... $1,2,4,5\text{-}[(CH_3)_2N]_4\text{-}1,2,4,5\text{-}B_4C_2H_4$ ...... B: B Comp.SVol.3/3-162
B: B Comp.SVol.4/3a-246/7

$B_4C_{11}FeH_{16}$ ..... $[1\text{-}(C_6H_5\text{-}CH_2CH_2CH_2)\text{-}1,2\text{-}C_2B_4H_5]Fe$...... Fe: Org.Comp.B19-408, 410/1

$B_4C_{11}FeH_{19}O$ .... $(C_5H_5)Fe(2,3\text{-}C_2B_4H_3(OC_2H_5)(CH_3)_2\text{-}2,3)$ ..... Fe: Org.Comp.B16b-54

$B_4C_{11}H_{18}$ .... $2,3\text{-}C_2B_4H_7\text{-}2\text{-}(CH_2CH_2CH_2\text{-}C_6H_5)$...... B: B Comp.SVol.4/4-185

$B_4C_{11}H_{22}O_3OsSi_2$ $1\text{-}Os(CO)_3\text{-}2,3\text{-}[(CH_3)_3Si]_2\text{-}2,3\text{-}C_2B_4H_4$ ...... B: B Comp.SVol.3/4-163

$B_4C_{11}H_{29}InSi_2$ ... $i\text{-}C_3H_7\text{-}InC_2B_4H_4\text{-}[Si(CH_3)_3]_2$ ...... In: Org.Comp.1-328/9

$B_4C_{12}FeH_{20}$ ..... $(C_6H_6)Fe[2,3\text{-}C_2B_4H_4(C_2H_5)_2\text{-}2,3]$...... Fe: Org.Comp.B18-54/5, 56/7, 67/8

– .... $(C_8H_{10})Fe[2,3\text{-}C_2B_4H_4(CH_3)_2\text{-}2,3]$...... Fe: Org.Comp.B18-54, 57, 68/9

$B_4C_{12}H_{14}N_2Sn$ ... $1\text{-}[2\text{-}(NC_5H_4\text{-}2)\text{-}NC_5H_4]\text{-}1,2,3\text{-}SnC_2B_4H_6$..... B: B Comp.SVol.4/4-195

$B_4C_{12}H_{15}^-$....... $[2,3\text{-}C_2B_4H_6\text{-}2\text{-}(CH_2\text{-}C_9H_7)]^-$ ...... B: B Comp.SVol.4/4-185, 190

$B_4C_{12}H_{16}$ .... $2,3\text{-}C_2B_4H_7\text{-}2\text{-}(CH_2\text{-}C_9H_7)$ ...... B: B Comp.SVol.4/4-185, 190/1

$B_4C_{12}H_{28}$ .... $2,3\text{-}C_2B_4H_6\text{-}2,3\text{-}(C_5H_{11}\text{-}i)_2$ ...... B: B Comp.SVol.4/4-185, 187

$B_4C_{12}H_{39}N_3O_3$ ... $[H_3B\text{-}N(CH_3)_2\text{-}CH_2CH_2\text{-}O\text{-}]_3B$ ...... B: B Comp.SVol.4/3b-15

$B_4C_{13}CoH_{27}Si_2$... $1\text{-}C_5H_5\text{-}1,2,3\text{-}CoC_2B_4H_4\text{-}2,3\text{-}[Si(CH_3)_3]_2$ ..... B: B Comp.SVol.4/4-196

$B_4C_{13}CrH_{21}$ ..... $(C_7H_7)Cr[(C_2H_5)_2C_2B_4H_4]$ ...... B: B Comp.SVol.3/4-162/3

$B_4C_{13}FeH_{22}$ ..... $(CH_3\text{-}C_6H_5)Fe[2,3\text{-}C_2B_4H_4(C_2H_5)_2\text{-}2,3]$...... Fe: Org.Comp.B18-55, 57, 68

$B_4C_{13}H_{22}$ .... $2,3\text{-}C_2B_4H_5\text{-}2,3\text{-}(C_2H_5)_2\text{-}4\text{-}(CH_2\text{-}C_6H_5)$ ...... B: B Comp.SVol.4/4-189/90

$B_4C_{14}CoH_{25}$ ..... $1,2,3\text{-}[C_5(CH_3)_5]Co(CH_3)_2C_2B_4H_4$ ...... B: B Comp.SVol.3/4-163

$B_4C_{14}FeH_{22}$ ..... $(C_8H_8)Fe[2,3\text{-}C_2B_4H_4(C_2H_5)_2\text{-}2,3]$...... Fe: Org.Comp.B18-69

$B_4C_{14}FeH_{24}$ ..... $(C_8H_{10})Fe[2,3\text{-}C_2B_4H_4(C_2H_5)_2\text{-}2,3]$ ...... Fe: Org.Comp.B18-54/6, 58, 69/70

$B_4C_{14}H_{15}^-$....... $[2,3\text{-}C_2B_4H_5\text{-}2,3\text{-}(C_6H_5)_2]^-$ ...... B: B Comp.SVol.4/4-185

$B_4C_{14}H_{16}$ .... $2,3\text{-}C_2B_4H_6\text{-}2,3\text{-}(C_6H_5)_2$ ...... B: B Comp.SVol.4/4-185, 188

$B_4C_{14}H_{20}I_2Ti$..... $(C_8H_8)Ti[(C_2H_5)_2C_2B_4H_2\text{-}4,5\text{-}(I)_2]$ ...... B: B Comp.SVol.3/4-163

$B_4C_{14}H_{21}ITi$ ..... $(C_8H_8)Ti[(C_2H_5)_2C_2B_4H_3\text{-}5\text{-}I]$...... B: B Comp.SVol.3/4-163

Formula	Compound	Reference
$B_4C_{14}H_{22}Ti$	$(C_8H_8)Ti[(C_2H_5)_2C_2B_4H_4]$	B: B Comp.SVol.3/4-162/3
		B: B Comp.SVol.4/4-196
$B_4C_{14}H_{22}V$	$(C_8H_8)V[(C_2H_5)_2C_2B_4H_4]$	B: B Comp.SVol.3/4-162
$B_4C_{14}H_{24}$	$1,6-C_2B_4H_3-2,3,4-(CH_3C{=}CHCH_3)_3$	B: B Comp.SVol.3/4-161
–	$2,3-C_2B_4H_5-2,3-(C_2H_5)_2-4-(CH_2C_6H_4CH_3-4)$	B: B Comp.SVol.4/4-189
$B_4C_{14}H_{32}$	$2,3-C_2B_4H_6-2,3-(C_6H_{13}-n)_2$	B: B Comp.SVol.4/4-185, 188
$B_4C_{15}FeH_{26}$	$[1,3,5-(CH_3)_3C_6H_3]Fe[2,3-C_2B_4H_4(C_2H_5)_2-2,3]$	Fe: Org.Comp.B18-55, 61, 70/1
$B_4C_{15}GeH_{22}N_2Si$	$1-[2-(NC_5H_4-2)-NC_5H_4]-2-(CH_3)_3Si-1,2,3-GeC_2B_4H_5$	B: B Comp.SVol.4/4-195
$B_4C_{15}H_{22}N_2SiSn$	$1-[2-(NC_5H_4-2)-NC_5H_4]-2-(CH_3)_3Si-1,2,3-SnC_2B_4H_5$	B: B Comp.SVol.4/4-195
$B_4C_{15}H_{26}$	$2,3-C_2B_4H_5-2,3-(C_2H_5)_2-4-(CH_2CH_2CH_2C_6H_5)$	B: B Comp.SVol.4/4-189/90
$B_4C_{16}Cl_4H_{36}N_4$	$[-BCl-N(C_4H_9-t)-BCl-N(C_4H_9-t)-BCl$ $-N(C_4H_9-t)-BCl-N(C_4H_9-t)-]$	B: B Comp.SVol.4/4-67
$B_4C_{16}Cl_6H_{40}N_8$	$[(-NCH_3-CH_2-CH_2-NCH_3-)_2B]_2 \cdot 2\ BCl_3$	B: B Comp.SVol.3/3-118
$B_4C_{16}DyH_{48}LiO_4$	$LiDy[BH_4]_4 \cdot 4\ C_4H_8O$	Sc: MVol.C11b-492
$B_4C_{16}ErH_{48}LiO_4$	$LiEr[BH_4]_4 \cdot 4\ C_4H_8O$	Sc: MVol.C11b-492
$B_4C_{16}FeH_{22}$	$(C_{10}H_8)Fe[2,3-C_2B_4H_4(C_2H_5)_2-2,3]$	Fe: Org.Comp.B18-54, 61, 71/2
$B_4C_{16}FeH_{24}$	$(C_{10}H_{10})Fe[2,3-C_2B_4H_4(C_2H_5)_2-2,3]$	Fe: Org.Comp.B18-55, 61
$B_4C_{16}FeH_{26}$	$(C_{10}H_{12})Fe[2,3-C_2B_4H_4(C_2H_5)_2-2,3]$	Fe: Org.Comp.B18-55, 61
$B_4C_{16}FeH_{26}O$	$(C_5H_5)FeH(CH_2{=}C{=}C(CH_3)C_2B_4H_5(C_2H_5)_2)CO$	Fe: Org.Comp.B16b-75, 78
$B_4C_{16}FeH_{28}$	$(C_6H_6)Fe[2,3-C_2B_4H_4(C_4H_9-n)_2-2,3]$	Fe: Org.Comp.B18-54, 57
–	$[(CH_3)_6C_6]Fe[2,3-C_2B_4H_4(CH_3)_2-2,3]$	Fe: Org.Comp.B18-62
$B_4C_{16}FeH_{32}S_2$	$[C_2B_2S(C_2H_5)_2(CH_3)_2]_2Fe$	B: B Comp.SVol.4/4-199
$B_4C_{16}GdH_{48}LiO_4$	$LiGd[BH_4]_4 \cdot 4\ C_4H_8O$	Sc: MVol.C11b-492
$B_4C_{16}GeH_{24}N_2Si$	$1-[2-(NC_5H_4-2)-NC_5H_4]-2-(CH_3)_3$ $Si-3-CH_3-1,2,3-GeC_2B_4H_4$	B: B Comp.SVol.4/4-195
$B_4C_{16}H_{17}^-$	$[2,3-C_2B_4H_6-2-(CH_2-C_{13}H_9)]^-$	B: B Comp.SVol.4/4-185, 190
$B_4C_{16}H_{18}$	$2,3-C_2B_4H_7-2-(CH_2-C_{13}H_9)$	B: B Comp.SVol.4/4-185, 190/1
$B_4C_{16}H_{19}^-$	$[2,3-C_2B_4H_5-2,3-(CH_2-C_6H_5)_2]^-$	B: B Comp.SVol.4/4-185, 189, 190
$B_4C_{16}H_{20}$	$2,3-C_2B_4H_6-2,3-(CH_2-C_6H_5)_2$	B: B Comp.SVol.4/4-185, 190
$B_4C_{16}H_{24}N_2SiSn$	$1-[2-(NC_5H_4-2)-NC_5H_4]-2-(CH_3)_3Si-3-CH_3-$ $1,2,3-SnC_2B_4H_4$	B: B Comp.SVol.4/4-195
$B_4C_{16}H_{39}N_5$	$1,3,5-(CH_3)_3-2-[2,2,4,4-(CH_3)_4-NC_5H_6-1-BH$ $-N(C_4H_9-t)]-1,3,5,2,4,6-N_3B_3H_2$	B: B Comp.SVol.4/3a-190
$B_4C_{16}H_{40}I_6N_8$	$[(-NCH_3-CH_2-CH_2-NCH_3-)_2B]_2 \cdot 2\ BI_3$	B: B Comp.SVol.3/3-118
$B_4C_{16}H_{48}LaNaO_4$	$La[BH_4]_3 \cdot NaBH_4 \cdot 4\ C_4H_8O$	Sc: MVol.C11b-478
$B_4C_{16}H_{48}LiLuO_4$	$LiLu[BH_4]_4 \cdot 4\ C_4H_8O$	Sc: MVol.C11b-491/2
$B_4C_{16}H_{48}LiO_4Y$	$LiY[BH_4]_4 \cdot 4\ C_4H_8O$	Sc: MVol.C11b-491/2
$B_4C_{16}H_{48}N_4Sn_4$	$[-B(CH_3)-N(Sn(CH_3)_3)-]_4$	B: B Comp.SVol.4/3a-199
$B_4C_{17}FeH_{17}$	$(C_5H_5)Fe(2,4-C_2B_4H_5(C_{10}H_7)-5)$	Fe: Org.Comp.B16b-55
$B_4C_{17}H_{22}$	$2,3-C_2B_4H_5-2,3-(CH_2-C_6H_5)_2-4-CH_3$	B: B Comp.SVol.4/4-189/90

$B_4C_{18}F_4H_{32}Ni$ .... $[(1,4,6-(CH_3)_3-2,3-(C_2H_5)_2-2,3,5-C_3B_3H)Ni(C_8H_{12})][BF_4]$ B: B Comp.SVol.4/4-182

$B_4C_{18}F_{12}H_{22}N_{12}P_2$ $[H_2B(-N_2C_3H_3-)_2B(-N_2C_3H_3-)_2B(-N_2C_3H_3-)_2BH_2][PF_6]_2$ B: B Comp.SVol.4/3b-39

$B_4C_{18}FeH_{24}$ ..... $(C_6H_5-C_6H_5)Fe[2,3-C_2B_4H_4(C_2H_5)_2-2,3]$ ...... Fe: Org.Comp.B18-54/5, 62, 72, 73

– .... $(C_8H_{10})Fe[2,3-C_2B_4H_5(CH_2CH_2-C_6H_5)-2]$ ..... Fe: Org.Comp.B18-54, 58, 70

$B_4C_{18}FeH_{32}$ ..... $(C_8H_{10})Fe[2,3-C_2B_4H_4(C_4H_9-n)_2-2,3]$ .... Fe: Org.Comp.B18-54/6, 59

– .... $[(CH_3)_6C_6]Fe[2,3-C_2B_4H_4(C_2H_5)_2-2,3]$ .... B: B Comp.SVol.4/4-198
Fe: Org.Comp.B18-55, 62, 72, 73

$B_4C_{18}GeH_{30}N_2Si_2$ 1-[2-$(NC_5H_4$-2)-NC_5H_4]-1,2,3-$GeC_2B_4H_4$-2,3-$[Si(CH_3)_3]_2$ .... B: B Comp.SVol.4/4-195/6

$B_4C_{18}H_{27}N_3O_3$ ... $[H_3B \cdot NC_5H_4-2-CH_2-O]_3B$ .... B: B Comp.SVol.4/3b-13

$B_4C_{18}H_{30}$ .... 1,6-$C_2B_4H_2$-2,3,4,5-$(CH_3C{=}CHCH_3)_4$ .... B: B Comp.SVol.3/4-161

$B_4C_{18}H_{30}N_2Si_2Sn$ 1-[2-$(NC_5H_4$-2)-NC_5H_4]-2,3-$[(CH_3)_3Si]_2$-1,2,3-$SnC_2B_4H_4$ .... B: B Comp.SVol.4/4-195

$B_4C_{18}H_{32}MoO_2S_2$ [3,4-$(C_2H_5)_2$-2,5-$(CH_3)_2$-1,2,5-SB_2C_2-1$]_2Mo(CO)_2$ Mo: Org.Comp.5-192

$B_4C_{19}CrH_{20}O_3$ ... 2,3-$C_2B_4H_6$-2-$[CH_2-C_6H_5(Cr(CO)_3)]$-3-$(CH_2-C_6H_5)$ B: B Comp.SVol.4/4-190, 196

$B_4C_{19}FeH_{24}$ ..... $(C_{13}H_{10})Fe[2,3-C_2B_4H_4(C_2H_5)_2-2,3]$ .... B: B Comp.SVol.4/4-197/8
Fe: Org.Comp.B18-54/5, 63, 73

$B_4C_{19}FeH_{26}$ ..... $(C_8H_{10})Fe[2,3-C_2B_4H_5(CH_2CH_2CH_2-C_6H_5)-2]$ Fe: Org.Comp.B18-54, 59, 70

$B_4C_{20}Cl_2H_{46}N_4$ ... $[-BCl-N(C_4H_9-t)-B(C_2H_5)-N(C_4H_9-t)-B(C_2H_5)-N(C_4H_9-t)-BCl-N(C_4H_9-t)-]$ .... B: B Comp.SVol.4/4-67

$B_4C_{20}FeH_{24}$ ..... $(C_{14}H_{10})Fe[2,3-C_2B_4H_4(C_2H_5)_2-2,3]$ .... Fe: Org.Comp.B18-54/5, 63, 64, 73

$B_4C_{20}FeH_{26}$ ..... $(C_{14}H_{12})Fe[2,3-C_2B_4H_4(C_2H_5)_2-2,3]$ .... Fe: Org.Comp.B18-55, 64

$B_4C_{20}H_{30}MnO_{22}$ .. $Mn[B_2O(CH_3COO)_5]_2$ .... Mn: MVol.D8-4

$B_4C_{20}H_{36}Ni_3$ ..... $[(C_3H_5)Ni((CH_3)_4HC_3B_2)]_2Ni$ .... B: B Comp.SVol.4/4-199

$B_4C_{20}H_{48}N_4$ ..... $[-B(CH_3)-N(C_4H_9-t)-]_4$ .... B: B Comp.SVol.3/3-136
B: B Comp.SVol.4/3a-198

– .... $[-B(C_4H_9-t)-N(CH_3)-]_4$ .... B: B Comp.SVol.4/3a-198

$B_4C_{20}H_{48}N_4O_4$ ... $[-B(OCH_3)-N(C_4H_9-t)-]_4$ .... B: B Comp.SVol.4/3a-199

$B_4C_{20}H_{60}N_8Sn_4$ .. $[-B(N(CH_3)_2)-N(Sn(CH_3)_3)-]_4$ .... B: B Comp.SVol.4/3a-199

$B_4C_{21}H_{33}N_3O_3$ ... $[BH_3 \cdot NC_5H_4-2-CH_2CH_2-O]_3B$ .... B: B Comp.SVol.4/3b-72

– .... $[H_3B \cdot NC_5H_3-6-CH_3-2-CH_2O-]_3B$ .... B: B Comp.SVol.4/3b-13

– .... $[H_3B \cdot NC_5H_4-2-CH(CH_3)-O]_3B$ .... B: B Comp.SVol.4/3b-13

$B_4C_{22}Cr_2H_{19}O_6^-$ $[2,3-C_2B_4H_5-2,3-(CH_2-C_6H_5(Cr(CO)_3))_2]^-$ .... B: B Comp.SVol.4/4-185, 190

$B_4C_{22}Cr_2H_{20}O_6$ .. 2,3-$C_2B_4H_6$-2,3-$[CH_2-C_6H_5(Cr(CO)_3)]_2$ ...... B: B Comp.SVol.4/4-185, 190, 196

$B_4C_{22}FeH_{30}$ ..... $(C_{16}H_{16})Fe[2,3-C_2B_4H_4(C_2H_5)_2-2,3]$ .... Fe: Org.Comp.B18-54/5, 65

$B_4C_{22}FeH_{32}$ ..... $(C_{16}H_{18})Fe[2,3-C_2B_4H_4(C_2H_5)_2-2,3]$ .... Fe: Org.Comp.B18-54, 65

$B_4C_{22}H_{23}^-$ .... $[2,3-C_2B_4H_5-2,3-(CH_2-C_9H_7)_2]^-$ .... B: B Comp.SVol.4/4-185, 190

$B_4C_{22}H_{24}$ .... $2,3-C_2B_4H_6-2,3-(CH_2-C_9H_7)_2$ .... B: B Comp.SVol.4/4-185, 190/1

$B_4C_{23}H_{26}$ .... $2,3-C_2B_4H_5-2,3,4-(CH_2-C_6H_5)_3$ .... B: B Comp.SVol.4/4-189/90

$B_4C_{24}DyH_{64}LiO_6$ .. $LiDy[BH_4]_4 \cdot 6\ C_4H_8O$ .... Sc: MVol.C11b-492

$B_4C_{24}ErH_{64}LiO_6$ .. $LiEr[BH_4]_4 \cdot 6\ C_4H_8O$ .... Sc: MVol.C11b-492

$B_4C_{24}FeH_{28}$ ..... $(C_8H_{10})Fe[2,3-C_2B_4H_4(CH_2-C_6H_5)_2-2,3]$ ...... Fe: Org.Comp.B18-54/6, 59/60

$B_4C_{24}Fe_3H_{34}O_6$ .. $(CO)_3Fe[C_3B_2H(C_2H_5)_2(CH_3)_2]Fe[C_3B_2H(C_2H_5)_2(CH_3)_2]Fe(CO)_3$ .... B: B Comp.SVol.4/4-199

$B_4C_{24}GdH_{64}LiO_6$ $LiGd[BH_4]_4 \cdot 6\ C_4H_8O$ .... Sc: MVol.C11b-491/2

$B_4C_{24}H_{24}N_{16}$ .... $(N_2C_3H_3)_2B_2[-N_2C_3H_3-]_4B_2(N_2C_3H_3)_2$ .... B: B Comp.SVol.4/3b-36

$B_4C_{24}H_{47}Rh$ ..... $[C_3B_2(C_2H_5)_4(CH_3)]Rh[C_3B_2H(C_2H_5)_4(CH_3)]$ ... B: B Comp.SVol.4/4-199

$B_4C_{24}H_{56}N_4$ ..... $[-B(C_3H_7-i)-N(C_3H_7-i)-]_4$ .... B: B Comp.SVol.4/3a-198

$B_4C_{24}H_{64}HoLiO_6$. $LiHo[BH_4]_4 \cdot 6\ C_4H_8O$ .... Sc: MVol.C11b-491/2

$B_4C_{26}CoFe_2H_{42}S_2$ $[(C_5H_5Fe)((C_2H_5)_2C_2B_2(CH_3)_2S)]_2Co$ .... B: B Comp.SVol.3/4-163

$B_4C_{26}F_{12}H_{38}N_{12}P_2$ $[(C_2H_5)_2B(-N_2C_3H_3-)_2B(-N_2C_3H_3-)_2B(-N_2C_3H_3-)_2B(C_2H_5)_2][PF_6]_2$ .... B: B Comp.SVol.4/3b-40

$B_4C_{26}Fe_3H_{42}S_2$... $[(C_5H_5Fe)((C_2H_5)_2C_2B_2(CH_3)_2S)]_2Fe$ .... B: B Comp.SVol.3/4-163

$B_4C_{26}H_{58}N_4$ ..... $t-C_4H_9-NH-B(C_2H_5)-N(C_4H_9-t)-B(C_2H_5)-CC-B(C_2H_5)-N(C_4H_9-t)-B(C_2H_5)-NH-C_4H_9-t$ ... B: B Comp.SVol.4/3a-168

$B_4C_{26}H_{60}N_4$ ..... $1,2,4,5-[(i-C_3H_7)_2N]_4-1,2,4,5-B_4C_2H_4$ .... B: B Comp.SVol.4/3a-246/7

$B_4C_{27}CrFeH_{28}O_3$ $(C_8H_{10})Fe[3-(C_6H_5CH_2)-2,3-C_2B_4H_4-2-CH_2C_6H_5Cr(CO)_3]$ .... Fe: Org.Comp.B18-59/60

$B_4C_{28}ClH_{31}NiP_2$ .. $[(C_6H_5)_2PCH_2CH_2P(C_6H_5)_2]NiCl-2,3-C_2B_4H_7$ .. B: B Comp.SVol.4/4-196

$B_4C_{28}Co_2FeH_{44}$ .. $[(C_5H_5Co)(H(C_2H_5)_2C_3B_2(CH_3)_2)]_2Fe$......... B: B Comp.SVol.3/4-163

$B_4C_{28}Co_2H_{44}Ni$... $[(C_5H_5Co)(H(C_2H_5)_2C_3B_2(CH_3)_2)]_2Ni$ .... B: B Comp.SVol.3/4-163

$B_4C_{28}Co_2H_{44}Zn$ .. $[(C_5H_5Co)(H(C_2H_5)_2C_3B_2(CH_3)_2)]_2Zn$......... B: B Comp.SVol.3/4-163

$B_4C_{28}Co_3H_{44}$ .... $[(C_5H_5Co)(H(C_2H_5)_2C_3B_2(CH_3)_2)]_2Co$ .... B: B Comp.SVol.3/4-163

$B_4C_{28}CuH_{31}P_2$ ... $1-[(C_6H_5)_2PCH_2CH_2P(C_6H_5)_2]-1,2,3-CuC_2B_4H_7$ B: B Comp.SVol.4/4-199

$B_4C_{28}DyH_{72}LiO_7$.. $LiDy[BH_4]_4 \cdot 7\ C_4H_8O$ .... Sc: MVol.C11b-491/2

$B_4C_{28}ErH_{72}LiO_7$.. $LiEr[BH_4]_4 \cdot 7\ C_4H_8O$ .... Sc: MVol.C11b-491/2

$B_4C_{28}GdH_{72}LiO_7$ $LiGd[BH_4]_4 \cdot 7\ C_4H_8O$.................... Sc: MVol.C11b-492

$B_4C_{28}H_{30}NiP_2$.... $1-[(C_6H_5)_2PCH_2CH_2P(C_6H_5)_2]-1,2,3-NiC_2B_4H_6$ B: B Comp.SVol.4/4-196

$B_4C_{28}H_{60}N_4$ ..... $1,2,5,6-[(i-C_3H_7)_2N]_4-1,2,5,6-B_4C_4H_4$ .... B: B Comp.SVol.4/3a-247

$B_4C_{28}H_{72}LiO_7Tm$ $LiTm[BH_4]_4 \cdot 7\ C_4H_8O$ .... Sc: MVol.C11b-491/2

$B_4C_{29}FeH_{28}$ ..... $(C_{13}H_{10})Fe[2,3-C_2B_4H_4(CH_2-C_6H_5)_2-2,3]$ ..... B: B Comp.SVol.4/4-198/9; Fe: Org.Comp.B18-54/5, 63

$B_4C_{29}H_{51}NiPt$ .... $(C_5H_5)Ni[(C_2H_5)_4(CH_3)C_3B_2]Pt[(C_2H_5)_4(CH_3)C_3B_2]$ B: B Comp.SVol.4/4-199

$B_4C_{30}Cr_2FeH_{28}O_6$ $(C_8H_{10})Fe[2,3-C_2B_4H_4(CH_2-C_6H_5Cr(CO)_3)_2-2,3]$ Fe: Org.Comp.B18-59/60

$B_4C_{30}FeH_{30}$ ..... $(C_{14}H_{12})Fe[2,3-C_2B_4H_4(CH_2C_6H_5)_2-2,3]$ ...... Fe: Org.Comp.B18-55, 64

$B_4C_{30}H_{27}^-$....... $[2,3-C_2B_4H_5-2,3-(CH_2-C_{13}H_9)_2]^-$ .... B: B Comp.SVol.4/4-190

$B_4C_{30}H_{28}$ .... $2,3-C_2B_4H_6-2,3-(CH_2-C_{13}H_9)_2$ .... B: B Comp.SVol.4/4-190/2

$B_4C_{32}ClCoH_{38}P_2$. $1-[(C_6H_5)_2P-CH_2CH_2-P(C_6H_5)_2]-1-Cl-2,3-(C_2H_5)_2-1,2,3-CoC_2B_4H_4$.............. B: B Comp.SVol.4/4-196

$B_4C_{32}ClFeH_{38}P_2$.. $1-[(C_6H_5)_2P-CH_2CH_2-P(C_6H_5)_2]-1-Cl-2,3-(C_2H_5)_2-1,2,3-FeC_2B_4H_4$.............. B: B Comp.SVol.4/4-196

$B_4C_{32}CoH_{38}IP_2$ ... 1-[$(C_6H_5)_2P-CH_2CH_2-P(C_6H_5)_2$]-1-I-2,3-$(C_2H_5)_2$-1,2,3-$CoC_2B_4H_4$ ... B: B Comp.SVol.4/4-196
$B_4C_{32}FeH_{34}$ ... $(C_{16}H_{16})Fe[2,3-C_2B_4H_4(CH_2-C_6H_5)_2$-2,3] ... Fe: Org.Comp.B18-54/5, 65
$B_4C_{32}FeH_{36}$ ... $(C_{16}H_{18})Fe[2,3-C_2B_4H_4(CH_2C_6H_5)_2$-2,3] ... Fe: Org.Comp.B18-54, 66
$B_4C_{33}CoH_{38}NP_2$ ... 1-[$(C_6H_5)_2P-CH_2CH_2-P(C_6H_5)_2$]-1-CN-2,3-$(C_2H_5)_2$-1,2,3-$CoC_2B_4H_4$ ... B: B Comp.SVol.4/4-196
$B_4C_{34}H_{56}Ni_2Pt$ ... $(C_5H_5)Ni[(C_2H_5)_4(CH_3)C_3B_2]Pt[(C_2H_5)_4(CH_3)C_3B_2]Ni(C_5H_5)$ ... B: B Comp.SVol.4/4-199
$B_4C_{35}FeH_{64}N_4O_3$ $(CO)_3Fe[C_4B_2H_4(N(C_3H_7\text{-}i)_2)_2]_2$ ... B: B Comp.SVol.4/4-199
$B_4C_{36}Co_2H_{76}N_4Si_4$
[4,5-$(C_2H_5)_2$-2,2,3-$(CH_3)_3$-1,2,5-$NSiBC_2$-1$]_4Co_2$ ... B: B Comp.SVol.4/3a-253
$B_4C_{36}Fe_2H_{76}N_4Si_4$ [4,5-$(C_2H_5)_2$-2,2,3-$(CH_3)_3$-1,2,5-$NSiBC_2$-1$]_4Fe_2$ ... B: B Comp.SVol.4/3a-253
$B_4C_{36}H_{39}N_3O_3$ ... $[H_3B \cdot NC_5H_4\text{-}2\text{-}CH(C_6H_5)\text{-}O]_3B$ ... B: B Comp.SVol.4/3b-13
$B_4C_{37}H_{38}OOsP_2$ ... $(CO)Os(B_4H_8)[P(C_6H_5)_3]_2$ ... Os: Org.Comp.A1-214
$B_4C_{38}CuH_{37}P_2$ ... 1,1-$[(C_6H_5)_3P]_2$-1,2,3-$CuC_2B_4H_7$ ... B: B Comp.SVol.4/4-199
$B_4C_{38}H_{43}Mn_2N_{24}O_4$
$[((1,2\text{-}N_2C_3H_3\text{-}1)_3BH)_2Mn(O)_2(CH_3COO)Mn((1,2\text{-}N_2C_3H_3\text{-}1)_3BH)_2]$ ... Mn: MVol.D8-13/4
$B_4C_{40}ClCoH_{38}P_2$. 1-[$(C_6H_5)_2P-CH_2CH_2-P(C_6H_5)_2$]-1-Cl-2,3-$(C_6H_5)_2$-1,2,3-$CoC_2B_4H_4$ ... B: B Comp.SVol.4/4-185
$B_4C_{40}CuH_{41}P_2$ ... 1,1-$[(C_6H_5)_3P]_2$-2,3-$(CH_3)_2$-1,2,3-$CuC_2B_4H_5$ ... B: B Comp.SVol.4/4-199
$B_4C_{40}F_{16}H_{60}N_4Sn$ [4-$(CH_3)_2(C_2H_5)N-C_6H_4]_4Sn[BF_4]_4$ ... B: B Comp.SVol.3/3-335
$B_4C_{42}H_{38}N_{12}O$ ... $[C_6H_5\text{-}B(N_2C_3H_3)(\text{-}N_2C_3H_3\text{-})_2B(C_6H_5)\text{-}]_2O$ ... B: B Comp.SVol.4/3b-59/60
$B_4C_{54}H_{69}N_3$ ... $B[NH\text{-}B(C_6H_2\text{-}2,4,6\text{-}(CH_3)_3)_2]_3$ ... B: B Comp.SVol.4/3a-153
$B_4C_{55}F_4H_{48}O_2OsP_3$
$[(H)(CO)Os(H_2O)(P(C_6H_5)_3)_3][BF]_4 \cdot C_2H_5OH$. Os: Org.Comp.A1-135/6
$B_4C_{57}F_4H_{54}O_3OsP_3$
$[(H)(CO)Os(H_2O)(P(C_6H_5)_3)_3][BF]_4 \cdot C_2H_5OH$. Os: Org.Comp.A1-135/6
$B_4C_{141}H_{128}Mn_2N_{12}$
$[Mn_2(NC_5H_4\text{-}2\text{-}CH{=}N\text{-}(CH_2)_3\text{-}N{=}CH\text{-}2\text{-}C_5H_4N)_3][B(C_6H_5)_4]_4$ ... Mn: MVol.D6-189/90
B_4Ce ... CeB_4 ... Sc: MVol.C11a-6/15, 18
Sc: MVol.C11b-1/4
– ... CeB_4 solid solutions
$CeB_4\text{-}ThB_4$ ... Sc: MVol.C11b-3
$CeB_4\text{-}UB_4$ ... Sc: MVol.C11b-3
B_4CeO_8 ... $CeO_2 \cdot 2\ B_2O_3$
doped with Sm ... Sc: MVol.C11b-411
$B_4CeO_{12}Sc_3$ ... $CeSc_3[BO_3]_4$ ... Sc: MVol.C11b-419
B_4Cl_4 ... B_4Cl_4 ... B: B Comp.SVol.3/4-34
B: B Comp.SVol.4/4-35
B_4Dy ... DyB_4 ... Sc: MVol.C11a-6/21
Sc: MVol.C11b-316, 317/9
$B_4DyH_{16}Li$ ... $LiDy[BH_4]_4 \cdot n\ C_4H_8O$ ... Sc: MVol.C11b-491/2
B_4Er ... ErB_4 ... Sc: MVol.C11a-6/21
Sc: MVol.C11b-330/6
$B_4ErH_{16}Li$ ... $LiEr[BH_4]_4 \cdot n\ C_4H_8O$ ... Sc: MVol.C11b-491/2

B_4EuO_7 . . . . EuB_4O_7 . . . . Sc: MVol.C11b-447/9
$B_4EuO_{12}Sc_3$ . . . . . $EuSc_3[BO_3]_4$ . . . . Sc: MVol.C11b-419
B_4EuPd_6 . . . . $EuPd_6B_4$ . . . . Pd: SVol.B2-259
$B_4Eu_5O_{12}$ . . . . $Eu_5(BO_3)_4$ . . . . Sc: MVol.C11b-440
$B_4F_{12}H_8N_2O$ . . . . . $[NH_4]_2[OB_4F_{12}]$ . . . . B: B Comp.SVol.4/3b-237
B_4Gd . . . . GdB_4 . . . . Sc: MVol.C11a-6/21
Sc: MVol.C11b-264/5, 267/72
$B_4GdH_{16}Li$ . . . . . . $LiGd[BH_4]_4 \cdot n\ OC_4H_8$ . . . . Sc: MVol.C11b-491/2
$B_4H_4S_6$ . . . . $(\text{-HB-S-BH-S-S-})_2$ . . . . B: B Comp.SVol.3/4-111
$B_4H_{10}S$ . . . . $[\text{-BH}_2\text{-H-BH}_2\text{-}]S[\text{-BH}_2\text{-H-BH}_2\text{-}]$ . . . . B: B Comp.SVol.4/4-128
$B_4H_{12}N_2O_9$ . . . . . . $[NH_4]_2[B_4O_5(OH)_4] \cdot 2\ H_2O$ . . . . B: B Comp.SVol.3/3-225
$B_4H_{16}HoLi$ . . . . . . $LiHo[BH_4]_4 \cdot n\ OC_4H_8$ . . . . Sc: MVol.C11b-491/2
$B_4H_{16}LaNa$ . . . . . . $La[BH_4]_3 \cdot NaBH_4 \cdot 4\ OC_4H_8$ . . . . Sc: MVol.C11b-478
– . . . . $NaLa[BH_4]_4$ . . . . Sc: MVol.C11b-492
$B_4H_{16}LiLu$ . . . . $LiLu[BH_4]_4 \cdot n\ OC_4H_8$ . . . . Sc: MVol.C11b-491/2
$B_4H_{16}LiTm$ . . . . . . $LiTm[BH_4]_4 \cdot n\ OC_4H_8$ . . . . Sc: MVol.C11b-491/2
$B_4H_{16}LiY$ . . . . $LiY[BH_4]_4 \cdot n\ OC_4H_8$ . . . . Sc: MVol.C11b-491/2
$B_4H_{16}NaSc$ . . . . . . $NaSc[BH_4]_4$ . . . . Sc: MVol.C11b-492/3
B_4Ho . . . . HoB_4 . . . . Sc: MVol.C11a-6/21
Sc: MVol.C11b-323/4, 325/8
– . . . . HoB_4 solid solutions
HoB_4-YB_4 . . . . Sc: MVol.C11b-329
B_4La . . . . LaB_4
Chemical reactions . . . . Sc: MVol.C11a-21, 167
Crystallographic properties . . . . Sc: MVol.C11a-8/12, 166
Electrical properties . . . . Sc: MVol.C11a-18, 167
Electrochemistry . . . . Sc: MVol.C11a-21
Formation . . . . Sc: MVol.C11a-6/8, 162/5
Magnetic properties . . . . Sc: MVol.C11a-15, 167
Mechanical properties . . . . Sc: MVol.C11a-12, 166
Preparation . . . . Sc: MVol.C11a-6/8
Spectra . . . . Sc: MVol.C11a-166, 167
Thermal properties . . . . Sc: MVol.C11a-13/5, 167
Uses . . . . Sc: MVol.C11a-167
Valency . . . . Sc: MVol.C11a-12
$B_4Li_2O_7$ . . . . $Li_2B_4O_7$ systems
$Li_2B_4O_7$-$(YBO_3)_2$ . . . . Sc: MVol.C11b-453/5
B_4Lu . . . . LuB_4 . . . . Sc: MVol.C11a-8/17, 21
Sc: MVol.C11b-377, 378
$B_4Na_2O_7$ . . . . $Na_2B_4O_7$ systems
$Na_2B_4O_7$-CeO_2 . . . . Sc: MVol.C11b-465/6
$Na_2B_4O_7$-Nd_2O_3 . . . . Sc: MVol.C11b-465/6
B_4Nd . . . . NdB_4 . . . . Sc: MVol.C11a-6/21
Sc: MVol.C11b-115/7
$B_4NdO_{12}Sc_3$ . . . . . $NdSc_3[BO_3]_4$ . . . . Sc: MVol.C11b-419
$B_4O_{12}PrSc_3$ . . . . . $PrSc_3[BO_3]_4$ . . . . Sc: MVol.C11b-419
$B_4O_{12}Sc_3Sm$ . . . . . $SmSc_3[BO_3]_4$ . . . . Sc: MVol.C11b-419
B_4Pm . . . . PmB_4 . . . . Sc: MVol.C11b-136
B_4Pr . . . . PrB_4 . . . . Sc: MVol.C11a-6/21

B_4Pr .... PrB_4 .... Sc: MVol.C11b-92/6
$B_4S_{10}^{8-}$ .... $[B_4S_{10}]^{8-}$ .... B: B Comp.SVol.4/4-125
B_4Sc .... ScB_4 .... Sc: MVol.C11a-6, 125/6
B_4Sm .... SmB_4 .... Sc: MVol.C11a-6/21
Sc: MVol.C11b-137, 138/41
B_4Tb .... TbB_4 .... Sc: MVol.C11a-6/21
Sc: MVol.C11b-307/8, 309/14
B_4Th .... ThB_4
Mößbauer spectra .... Th: SVol.A4-142
– .... ThB_4 solid solutions
ThB_4-CeB_4 .... Sc: MVol.C11b-3
B_4Tm .... TmB_4 .... Sc: MVol.C11a-6/21
Sc: MVol.C11b-340/1, 342/3
B_4U .... UB_4 solid solutions
UB_4-CeB_4 .... Sc: MVol.C11b-3
B_4Y .... YB_4
Chemical reactions .... Sc: MVol.C11a-21, 138/9
Crystallographic properties .... Sc: MVol.C11a-8/12, 134
Electrical properties .... Sc: MVol.C11a-17/21, 136/8
Formation .... Sc: MVol.C11a-6/8, 129/31, 133/4
Magnetic properties .... Sc: MVol.C11a-15, 135/6
Mechanical properties .... Sc: MVol.C11a-12, 135
Preparation .... Sc: MVol.C11a-6/8
Thermal properties .... Sc: MVol.C11a-13/5, 135
Uses .... Sc: MVol.C11a-139
Valency .... Sc: MVol.C11a-12
– .... YB_4 solid solutions
YB_4-HoB_4 .... Sc: MVol.C11b-329
B_4Yb .... YbB_4 .... Sc: MVol.C11a-6/17
Sc: MVol.C11b-346/9
$B_5BrC_2ClH_5$ ..... 5-Cl-6-Br-2,4-$C_2B_5H_5$ .... B: B Comp.SVol.4/4-214
$B_5BrC_2H_6$ .... 1-Br-2,4-$C_2B_5H_6$ .... B: B Comp.SVol.4/4-204/5, 207, 212
– .... 3-Br-2,4-$C_2B_5H_6$ .... B: B Comp.SVol.4/4-202/5, 207, 214
– .... 5-Br-2,4-$C_2B_5H_6$ .... B: B Comp.SVol.3/4-165/6
B: B Comp.SVol.4/4-202/5, 207, 211/2, 214
– .... 6-Br-2,4-$C_2B_5H_6$ .... B: B Comp.SVol.4/4-204
– .... 7-Br-2,4-$C_2B_5H_6$ .... B: B Comp.SVol.4/4-204
– .... Br-2,4-$C_2B_5H_6$ .... B: B Comp.SVol.4/4-207, 208
$B_5BrC_5H_{15}N$ ..... 5-Br-2,4-$C_2B_5H_6$ · $N(CH_3)_3$ .... B: B Comp.SVol.3/4-166
$B_5BrEu_2O_9$ ...... $Eu_2B_5O_9Br$ .... Sc: MVol.C11b-450/3
$B_5Br_2C_2H_5$ ...... 1,3-Br_2-2,4-$C_2B_5H_5$ .... B: B Comp.SVol.4/4-205, 207, 211/2

$B_5Br_2C_2H_5$ ...... 1,5-Br_2-2,4-$C_2B_5H_5$ ...... B: B Comp.SVol.4/4-205, 207, 211/2
– ...... 1,7-Br_2-2,4-$C_2B_5H_5$ ...... B: B Comp.SVol.4/4-207, 211/2
– ...... 2-Br_2B-1,6-$C_2B_4H_5$ ...... B: B Comp.SVol.4/4-184
– ...... 3,5-Br_2-2,4-$C_2B_5H_5$ ...... B: B Comp.SVol.4/4-205, 207, 211/2
– ...... 5,6-Br_2-2,4-$C_2B_5H_5$ ...... B: B Comp.SVol.4/4-203, 205, 207, 211/2
$B_5Br_2C_5H_{14}N$ ...... 5,6-Br_2-2,4-$C_2B_5H_5$ · $N(CH_3)_3$ ...... B: B Comp.SVol.4/4-214
$B_5CH_6^-$ ...... $[CB_5H_6]^-$ ...... B: B Comp.SVol.4/4-200
$B_5CH_6^{3-}$ ...... $[CB_5H_6]^{3-}$ ...... B: B Comp.SVol.3/4-164
B_5CH_7 ...... 1-CB_5H_7 ...... B: B Comp.SVol.3/4-164
B_5CH_9 ...... 2-CB_5H_9 ...... B: B Comp.SVol.3/4-164
$B_5CH_{12}NaS_4$ ...... $Na[B_5CH_{12}S_4]$ · 3 (1,4-$O_2C_4H_8$) ...... B: B Comp.SVol.4/4-133
$B_5CH_{12}S_4^-$ ...... $[B_5CH_{12}S_4]^-$ ...... B: B Comp.SVol.4/4-133
$B_5C_2ClH_6$ ...... 1-Cl-2,4-$C_2B_5H_6$ ...... B: B Comp.SVol.3/4-165/6
B: B Comp.SVol.4/4-204, 207/9, 211/4
– ...... 3-Cl-2,4-$C_2B_5H_6$ ...... B: B Comp.SVol.3/4-165/7
B: B Comp.SVol.4/4-202/4, 207/9, 211/4
– ...... 5-Cl-2,4-$C_2B_5H_6$ ...... B: B Comp.SVol.3/4-165/7
B: B Comp.SVol.4/4-202/4, 207/9, 211/4
– ...... 6-Cl-2,4-$C_2B_5H_6$ ...... B: B Comp.SVol.4/4-204
– ...... 7-Cl-2,4-$C_2B_5H_6$ ...... B: B Comp.SVol.4/4-204
– ...... Cl-2,4-$C_2B_5H_6$ ...... B: B Comp.SVol.4/4-203, 207, 208
$B_5C_2Cl_2H_5$ ...... 1,3-Cl_2-2,4-$C_2B_5H_5$ ...... B: B Comp.SVol.3/4-167
B: B Comp.SVol.4/4-207/8, 211/4
– ...... 1,5-Cl_2-2,4-$C_2B_5H_5$ ...... B: B Comp.SVol.3/4-165
B: B Comp.SVol.4/4-207/8, 211/4
– ...... 1,7-Cl_2-2,4-$C_2B_5H_5$ ...... B: B Comp.SVol.4/4-207/8, 211/4
– ...... 3,5-Cl_2-2,4-$C_2B_5H_5$ ...... B: B Comp.SVol.3/4-165/7
B: B Comp.SVol.4/4-207/8, 211/4
– ...... 5,6-Cl_2-2,4-$C_2B_5H_5$ ...... B: B Comp.SVol.3/4-167
B: B Comp.SVol.4/4-203, 207/8, 211/4
$B_5C_2FH_6$ ...... 1-F-2,4-$C_2B_5H_6$ ...... B: B Comp.SVol.4/4-204
– ...... 3-F-2,4-$C_2B_5H_6$ ...... B: B Comp.SVol.4/4-204, 214
– ...... 5-F-2,4-$C_2B_5H_6$ ...... B: B Comp.SVol.4/4-204, 214
– ...... 6-F-2,4-$C_2B_5H_6$ ...... B: B Comp.SVol.4/4-204
– ...... 7-F-2,4-$C_2B_5H_6$ ...... B: B Comp.SVol.4/4-204

$B_5C_2F_2H_5$ .... $3,5-F_2-2,4-C_2B_5H_5$ .... B: B Comp.SVol.4/4-214
$B_5C_2H_5I_2$ .... $1,3-I_2-2,4-C_2B_5H_5$ .... B: B Comp.SVol.4/4-205, 207, 211/2
– .... $1,5-I_2-2,4-C_2B_5H_5$ .... B: B Comp.SVol.4/4-205, 207, 211/2
– .... $1,7-I_2-2,4-C_2B_5H_5$ .... B: B Comp.SVol.4/4-207, 211/2
– .... $3,5-I_2-2,4-C_2B_5H_5$ .... B: B Comp.SVol.4/4-203, 205, 207, 211/2, 214
– .... $5,6-I_2-2,4-C_2B_5H_5$ .... B: B Comp.SVol.4/4-205, 207, 211/2
$B_5C_2H_5Li_2$ .... $Li_2[2,4-C_2B_5H_5]$ .... B: B Comp.SVol.3/4-168
$B_5C_2H_6I$ .... $1-I-2,4-C_2B_5H_6$ .... B: B Comp.SVol.4/4-204/5, 207, 212
– .... $3-I-2,4-C_2B_5H_6$ .... B: B Comp.SVol.4/4-202/5, 212, 214
– .... $5-I-2,4-C_2B_5H_6$ .... B: B Comp.SVol.4/4-202/5, 207, 211/2, 214
– .... $6-I-2,4-C_2B_5H_6$ .... B: B Comp.SVol.4/4-204
– .... $7-I-2,4-C_2B_5H_6$ .... B: B Comp.SVol.4/4-204
– .... $I-2,4-C_2B_5H_6$ .... B: B Comp.SVol.4/4-207, 208
$B_5C_2H_7$ .... $1,2-C_2B_5H_7$ .... B: B Comp.SVol.4/4-200/2
– .... $1,7-C_2B_5H_7$ .... B: B Comp.SVol.4/4-201/2
– .... $2,3-C_2B_5H_7$ .... B: B Comp.SVol.4/4-200/2
– .... $2,4-C_2B_5H_7$ .... B: B Comp.SVol.3/4-165/8
B: B Comp.SVol.4/4-202, 203, 211
– .... $2,4-C_2B_5H_5-2,4-D_2$ .... B: B Comp.SVol.3/4-168
– .... $C_2B_5H_7$ .... B: B Comp.SVol.3/4-165
$B_5C_2H_{11}$ .... $2-CB_5H_8-2-CH_3$ .... B: B Comp.SVol.3/4-164
– .... $2-CB_5H_8-3-CH_3$ .... B: B Comp.SVol.3/4-164
– .... $2-CB_5H_8-4-CH_3$ .... B: B Comp.SVol.3/4-164
$B_5C_3ClH_8$ .... $1-CH_3-3-Cl-2,4-C_2B_5H_5$ .... B: B Comp.SVol.4/4-206/8, 209/10, 212/4
– .... $1-CH_3-5-Cl-2,4-C_2B_5H_5$ .... B: B Comp.SVol.3/4-166
B: B Comp.SVol.4/4-206/8, 209/14
– .... $1-CH_3-6-Cl-2,4-C_2B_5H_5$ .... B: B Comp.SVol.4/4-207, 211
– .... $1-CH_3-7-Cl-2,4-C_2B_5H_5$ .... B: B Comp.SVol.4/4-206/8, 209/10, 212/4
– .... $3-CH_3-1-Cl-2,4-C_2B_5H_5$ .... B: B Comp.SVol.4/4-206/10, 212/4
– .... $3-CH_3-5-Cl-2,4-C_2B_5H_5$ .... B: B Comp.SVol.4/4-203, 206/10, 212/4
– .... $5-CH_3-1-Cl-2,4-C_2B_5H_5$ .... B: B Comp.SVol.4/4-206/14
– .... $5-CH_3-3-Cl-2,4-C_2B_5H_5$ .... B: B Comp.SVol.4/4-206/10, 212/4
– .... $5-CH_3-6-Cl-2,4-C_2B_5H_5$ .... B: B Comp.SVol.3/4-165

Formula	Compound	Reference
$B_5C_3ClH_8$	$5\text{-}CH_3\text{-}6\text{-}Cl\text{-}2,4\text{-}C_2B_5H_5$	B: B Comp.SVol.4/4-206/14
–	$6\text{-}CH_3\text{-}1\text{-}Cl\text{-}2,4\text{-}C_2B_5H_5$	B: B Comp.SVol.4/4-211
–	$6\text{-}CH_3\text{-}5\text{-}Cl\text{-}2,4\text{-}C_2B_5H_5$	B: B Comp.SVol.3/4-166/7
		B: B Comp.SVol.4/4-203, 211
–	$7\text{-}CH_3\text{-}3\text{-}Cl\text{-}2,4\text{-}C_2B_5H_5$	B: B Comp.SVol.4/4-207
–	$7\text{-}CH_3\text{-}5\text{-}Cl\text{-}2,4\text{-}C_2B_5H_5$	B: B Comp.SVol.4/4-207
–	$7\text{-}CH_3\text{-}6\text{-}Cl\text{-}2,4\text{-}C_2B_5H_5$	B: B Comp.SVol.4/4-207
$B_5C_3H_8I$	$6\text{-}CH_3\text{-}5\text{-}I\text{-}2,4\text{-}C_2B_5H_5$	B: B Comp.SVol.4/4-203
$B_5C_3H_9$	$1\text{-}CH_3\text{-}2,4\text{-}C_2B_5H_6$	B: B Comp.SVol.3/4-167
		B: B Comp.SVol.4/4-203, 206, 209, 212/4
–	$3\text{-}CH_3\text{-}2,4\text{-}C_2B_5H_6$	B: B Comp.SVol.3/4-167
		B: B Comp.SVol.4/4-203, 206, 209, 212/4
–	$5\text{-}CH_3\text{-}2,4\text{-}C_2B_5H_6$	B: B Comp.SVol.3/4-166/7
		B: B Comp.SVol.4/4-203, 206, 209, 211/4
$B_5C_3H_{10}O_3Re$	$(CO)_3ReB_5H_{10}\text{-}2,2,2$	Re: Org.Comp.1-115
$B_5C_3H_{13}$	$2\text{-}CB_5H_7\text{-}2,3\text{-}(CH_3)_2$	B: B Comp.SVol.3/4-164
–	$2\text{-}CB_5H_7\text{-}2,4\text{-}(CH_3)_2$	B: B Comp.SVol.3/4-164
–	$2\text{-}CB_5H_8\text{-}3\text{-}C_2H_5$	B: B Comp.SVol.3/4-164
–	$2\text{-}CB_5H_8\text{-}4\text{-}C_2H_5$	B: B Comp.SVol.3/4-164
$B_5C_4ClH_{10}$	$1,5\text{-}(CH_3)_2\text{-}6\text{-}Cl\text{-}2,4\text{-}C_2B_5H_4$	B: B Comp.SVol.4/4-206
–	$3,5\text{-}(CH_3)_2\text{-}6\text{-}Cl\text{-}2,4\text{-}C_2B_5H_4$	B: B Comp.SVol.4/4-206
$B_5C_4H_{11}$	$1,3\text{-}(CH_3)_2\text{-}2,4\text{-}C_2B_5H_5$	B: B Comp.SVol.4/4-209, 212/4
–	$1,5\text{-}(CH_3)_2\text{-}2,4\text{-}C_2B_5H_5$	B: B Comp.SVol.4/4-209, 211/4
–	$1,7\text{-}(CH_3)_2\text{-}2,4\text{-}C_2B_5H_5$	B: B Comp.SVol.4/4-209, 212/4
–	$2,4\text{-}(CH_3)_2\text{-}2,4\text{-}C_2B_5H_5$	B: B Comp.SVol.3/4-169
–	$3,5\text{-}(CH_3)_2\text{-}2,4\text{-}C_2B_5H_5$	B: B Comp.SVol.4/4-209, 212/4
–	$5,6\text{-}(CH_3)_2\text{-}2,4\text{-}C_2B_5H_5$	B: B Comp.SVol.3/4-166/7
		B: B Comp.SVol.4/4-203, 209, 211/4
–	$5\text{-}C_2H_5\text{-}2,4\text{-}C_2B_5H_6$	B: B Comp.SVol.3/4-166/7
		B: B Comp.SVol.4/4-203
$B_5C_4H_{13}N^-$	$[(CH_3)_2NBH\text{-}2,4\text{-}C_2B_4H_6]^-$	B: B Comp.SVol.4/4-202
$B_5C_4H_{14}N_3O_2$	$2,4,6,8\text{-}(CH_3)_4\text{-}1,7,3,5,9,2,4,6,8,10\text{-}O_2N_3B_5H_2$	B: B Comp.SVol.4/3a-203/5
$B_5C_4H_{15}$	$2\text{-}CB_5H_7\text{-}2\text{-}CH_3\text{-}3\text{-}C_2H_5$	B: B Comp.SVol.3/4-164
–	$2\text{-}CB_5H_7\text{-}2\text{-}CH_3\text{-}4\text{-}C_2H_5$	B: B Comp.SVol.3/4-164
–	$2\text{-}CB_5H_7\text{-}2\text{-}C_2H_5\text{-}3\text{-}CH_3$	B: B Comp.SVol.3/4-164
$B_5C_4H_{20}N$	$(CH_3)_3N\text{-}CB_5H_{11}$	B: B Comp.SVol.3/4-164
$B_5C_5ClH_{12}$	$1,3,5\text{-}(CH_3)_3\text{-}6\text{-}Cl\text{-}2,4\text{-}C_2B_5H_3$	B: B Comp.SVol.4/4-206
–	$1,5,7\text{-}(CH_3)_3\text{-}6\text{-}Cl\text{-}2,4\text{-}C_2B_5H_3$	B: B Comp.SVol.4/4-206
$B_5C_5ClH_{15}N$	$1\text{-}Cl\text{-}2,4\text{-}C_2B_5H_6 \cdot N(CH_3)_3$	B: B Comp.SVol.3/4-165/6
–	$3\text{-}Cl\text{-}2,4\text{-}C_2B_5H_6 \cdot N(CH_3)_3$	B: B Comp.SVol.3/4-165
–	$5\text{-}Cl\text{-}2,4\text{-}C_2B_5H_6 \cdot N(CH_3)_3$	B: B Comp.SVol.3/4-165

Formula	Compound	Reference
$B_5C_5ClH_{15}N$	$5-Cl-2,4-C_2B_5H_6 \cdot N(CH_3)_3$	B: B Comp.SVol.4/4-214
$B_5C_5ClH_{15}P$	$5-Cl-2,4-C_2B_5H_6 \cdot P(CH_3)_3$	B: B Comp.SVol.3/4-166
$B_5C_5F_6H_7O$	$5-[HO(CF_3)_2C]-2,4-C_2B_5H_6$	B: B Comp.SVol.3/4-166
$B_5C_5H_8O_5Re$	$2-(CO)_5ReB_5H_8$	Re: Org.Comp.2-1
$B_5C_5H_{13}$	$1-CH_3-5-C_2H_5-2,4-C_2B_5H_5$	B: B Comp.SVol.4/4-203
–	$5-CH_3-6-C_2H_5-2,4-C_2B_5H_5$	B: B Comp.SVol.4/4-203
$B_5C_5H_{15}IN$	$5-I-2,4-C_2B_5H_6 \cdot N(CH_3)_3$	B: B Comp.SVol.4/4-214
$B_5C_5H_{15}N^+$	$[5-(CH_3)_3N-2,4-C_2B_5H_6]^+$	B: B Comp.SVol.3/4-166
$B_5C_6ClH_{14}$	$1,3,5,7-(CH_3)_4-6-Cl-2,4-C_2B_5H_2$	B: B Comp.SVol.4/4-206
$B_5C_6H_{13}$	$1-[CH_3-CH=C(CH_3)]-2,4-C_2B_5H_6$	B: B Comp.SVol.3/4-167/8
–	$3-[CH_3-CH=C(CH_3)]-2,4-C_2B_5H_6$	B: B Comp.SVol.3/4-167/8
–	$5-[CH_3-CH=C(CH_3)]-2,4-C_2B_5H_6$	B: B Comp.SVol.3/4-167/8
$B_5C_6H_{15}$	$2,3-C_2B_5H_5-2,3-(C_2H_5)_2$	B: B Comp.SVol.4/4-201
–	$2,4-C_2B_5H_5-2,4-(C_2H_5)_2$	B: B Comp.SVol.4/4-201
–	$2,4-C_2B_5H_5-5,6-(C_2H_5)_2$	B: B Comp.SVol.3/4-166/7
		B: B Comp.SVol.4/4-203
$B_5C_6H_{16}^-$	$[3,4-C_2B_5H_6-3,4-(C_2H_5)_2]^-$	B: B Comp.SVol.4/4-201
$B_5C_6H_{17}N^+$	$[5-(CH_3)_3N-6-CH_3-2,4-C_2B_5H_5]^+$	B: B Comp.SVol.3/4-165
$B_5C_6H_{17}N^-$	$[(C_2H_5)_2NBH-2,4-C_2B_4H_6]^-$	B: B Comp.SVol.4/4-202
$B_5C_6H_{19}Si_2$	$2,4-C_2B_5H_5-2,4-[Si(CH_3)_2H]_2$	B: B Comp.SVol.3/4-168
$B_5C_7H_{17}$	$1-CH_3-5,6-(C_2H_5)_2-2,4-C_2B_5H_4$	B: B Comp.SVol.4/4-203
$B_5C_8F_{12}H_7O_2$	$5-[HOC(CF_3)_2C(CF_3)_2O]-2,4-C_2B_5H_6$	B: B Comp.SVol.3/4-166
$B_5C_8H_{11}$	$3-C_6H_5-2,4-C_2B_5H_6$	B: B Comp.SVol.4/4-203
–	$5-C_6H_5-2,4-C_2B_5H_6$	B: B Comp.SVol.4/4-203
$B_5C_8H_{21}N^-$	$[(i-C_3H_7-i)_2NBH-2,4-C_2B_4H_6]^-$	B: B Comp.SVol.4/4-202
$B_5C_8H_{23}Si_2$	$2,4-C_2B_5H_5-2,4-[Si(CH_3)_3]_2$	B: B Comp.SVol.3/4-168
$B_5C_9CoH_{16}$	$1,2-(CH_3)_2-3,1,2-(C_5H_5)CoC_2B_5H_5$	B: B Comp.SVol.3/4-169
$B_5C_{10}H_{19}$	$1,3-[CH_3-CH=C(CH_3)]_2-2,4-C_2B_5H_5$	B: B Comp.SVol.3/4-167/8
–	$1,5-[CH_3-CH=C(CH_3)]_2-2,4-C_2B_5H_5$	B: B Comp.SVol.3/4-167/8
–	$1,7-[CH_3-CH=C(CH_3)]_2-2,4-C_2B_5H_5$	B: B Comp.SVol.3/4-167/8
–	$3,5-[CH_3-CH=C(CH_3)]_2-2,4-C_2B_5H_5$	B: B Comp.SVol.3/4-167/8
–	$5,6-[CH_3-CH=C(CH_3)]_2-2,4-C_2B_5H_5$	B: B Comp.SVol.3/4-167/8
$B_5C_{10}H_{28}N$	$[(CH_3)_4N][3,4-C_2B_5H_6-3,4-(C_2H_5)_2]$	B: B Comp.SVol.4/4-201
$B_5C_{13}CoH_{28}Si_2$	$5-C_5H_5-5,1,8-CoC_2B_5H_5-1,8-[Si(CH_3)_3]_2$	B: B Comp.SVol.4/4-216
$B_5C_{13}H_{36}NaO_6S_4$	$Na[B_5CH_{12}S_4] \cdot 3\ (1,4-O_2C_4H_8)$	B: B Comp.SVol.4/4-133
$B_5C_{14}H_{25}$	$1,3,5-[CH_3-CH=C(CH_3)]_3-2,4-C_2B_5H_4$	B: B Comp.SVol.3/4-167/8
–	$1,5,6-[CH_3-CH=C(CH_3)]_3-2,4-C_2B_5H_4$	B: B Comp.SVol.3/4-167/8
–	$1,5,7-[CH_3-CH=C(CH_3)]_3-2,4-C_2B_5H_4$	B: B Comp.SVol.3/4-167/8
–	$3,5,6-[CH_3-CH=C(CH_3)]_3-2,4-C_2B_5H_4$	B: B Comp.SVol.3/4-167/8
$B_5C_{15}FeH_{25}$	$(CH_3-C_6H_5)Fe[(CH_3)_4-C_4B_5H_5]$	Fe: Org.Comp.B18-81, 88, 94
$B_5C_{16}H_{41}P_2Pt$	$4,4-[(C_2H_5)_3P]_2-1,7-(CH_3)_2-1,4,7-CPtCB_5H_5$	B: B Comp.SVol.3/4-169
$B_5C_{18}H_{31}$	$1,3,5,6-[CH_3-CH=C(CH_3)]_4-2,4-C_2B_5H_3$	B: B Comp.SVol.3/4-167/8
–	$1,3,5,7-[CH_3-CH=C(CH_3)]_4-2,4-C_2B_5H_3$	B: B Comp.SVol.3/4-167/8
–	$1,5,6,7-[CH_3-CH=C(CH_3)]_4-2,4-C_2B_5H_3$	B: B Comp.SVol.3/4-167/8
$B_5C_{19}FeH_{38}N_3$	$4,4,4-[(CH_3)_3CNC]_3-1,7-(CH_3)_2-1,4,7-CFeCB_5H_5$	B: B Comp.SVol.3/4-169
$B_5C_{19}H_{37}Ni$	$[1,4,6-(CH_3)_3-2,3-(C_2H_5)_2-2,3,5-C_3B_3H]Ni$ $[1,3-B_2C_3H-1,3-(CH_3)_2-4,5-(C_2H_5)_2]$	B: B Comp.SVol.4/4-216
$B_5C_{22}H_{37}$	$1,3,5,6,7-(CH_3C=CHCH_3)_5-2,4-C_2B_5H_2$	B: B Comp.SVol.3/4-167/8

$B_5C_{24}FeH_{42}Ni$ .... $C_5H_5Fe(1,3-(CH_3)_2-4,5-(C_2H_5)_2C_3HB_2-1,3)$
$NiC_3HB_3(C_2H_5)_2(CH_3)_3$ .... Fe: Org.Comp.B17-177

$B_5C_{24}H_{42}Ni_2$ ..... $[1,4,6-(CH_3)_3-2,3-(C_2H_5)_2-2,3,5-C_3B_3H]Ni$
$[1,3-B_2C_3H-1,3-(CH_3)_2-4,5-(C_2H_5)_2]Ni(C_5H_5)$ B: B Comp.SVol.4/4-216

$B_5C_{25}H_{32}PS_4$ .... $[(C_6H_5)_4P][B_5CH_{12}S_4]$ .... B: B Comp.SVol.4/4-133

$B_5C_{26}Co_2H_{67}P_4$ .. $4-(C_2H_5)_3P-1,7-(CH_3)_2-[Co(H)_2(P(C_2H_5)_3)_2$
$P(C_2H_5)_2]-1,4,7-CCoCB_5H_4$ .... B: B Comp.SVol.3/4-169

$B_5C_{28}H_{71}P_4Pt_2$ ... $1,1,6,6-[(C_2H_5)_3P]_4-4,5-(CH_3)_2-1,4,5,6-PtC_2PtB_5H_5$
B: B Comp.SVol.3/4-169

$B_5C_{37}H_{39}OOsP_2$ .. $(CO)Os(B_5H_9)[P(C_6H_5)_3]_2$ .... Os: Org.Comp.A1-214

$B_5C_{45}ClH_{50}OOsP_3Pt$
$(H)(CO)Os(B_5H_8-PtCl-P(CH_3)_2C_6H_5)[P(C_6H_5)_3]_2$
Os: Org.Comp.A1-214/5, 216/7

$B_5C_{55}H_{52}OOsP_3Pt$ $[(C_6H_5)_3P](CO)Os(B_5H_7-C_6H_5)[P(C_6H_5)_2][Pt-P(C_6H_5)_3]$
Os: Org.Comp.A1-215, 217/8

$B_5ClEu_2O_9$ ...... $Eu_2B_5O_9Cl$ .... Sc: MVol.C11b-450/3

B_5CoNdO_{10} ...... $NdCoB_5O_{10}$ .... Sc: MVol.C11b-439

$B_5CoO_{10}Sm$ ..... $SmCoB_5O_{10}$ .... Sc: MVol.C11b-440

$B_5F_{20}MoN_5O_5$ .... $[Mo(NO)_5][BF_4]_5$ .... B: B Comp.SVol.4/3b-217

B_5Gd_2 .... Gd_2B_5 .... Sc: MVol.C11a-6
Sc: MVol.C11b-264/5, 266/7

$B_5H_8NO_{10}$ .... $[NH_4][B_5O_6(OH)_4] \cdot 2\ H_2O$ .... B: B Comp.SVol.4/3b-77

$B_5H_8N_5$ .... $1,3,5,7,9,2,4,6,8,10-N_5B_5H_8$ .... B: B Comp.SVol.4/3a-190

B_5H_9Na .... $Na[B_5H_9]$ .... B: B Comp.SVol.3/4-243

$B_5Li_2O_{10}Y$ .... $Li_2YB_5O_{10}$ .... Sc: MVol.C11b-453/5, 463

– .... $Li_2YB_5O_{10}$ systems
$Li_2YB_5O_{10}-Li_6Y(BO_3)_3$ .... Sc: MVol.C11b-453/5

B_5Nd_2 .... Nd_2B_5 .... Sc: MVol.C11a-6
Sc: MVol.C11b-115

B_5Pm_2 .... Pm_2B_5 .... Sc: MVol.C11b-136

B_5Pr_2 .... Pr_2B_5 .... Sc: MVol.C11a-6
Sc: MVol.C11b-92

B_5Sm_2 .... Sm_2B_5 .... Sc: MVol.C11a-6
Sc: MVol.C11b-137/8